THE EARTH

THE EARTH

ITS ORIGIN
HISTORY AND PHYSICAL
CONSTITUTION

SIR HAROLD JEFFREYS

D.Sc., F.R.S.

*Emeritus Plumian Professor of Astronomy and Experimental Philosophy
and Fellow of St John's College, Cambridge*

SIXTH EDITION

CAMBRIDGE UNIVERSITY PRESS

CAMBRIDGE

LONDON · NEW YORK · MELBOURNE

CAMBRIDGE UNIVERSITY PRESS
Cambridge, New York, Melbourne, Madrid, Cape Town, Singapore, São Paulo, Delhi

Cambridge University Press
The Edinburgh Building, Cambridge CB2 8RU, UK

Published in the United States of America by Cambridge University Press, New York

www.cambridge.org
Information on this title: www.cambridge.org/9780521206488

First Edition 1924
Second Edition 1929
Third Edition 1952
Fourth Edition 1959
Reprinted with additions 1962
Fifth Edition 1970
Sixth Edition 1976
This digitally printed version 2008

A catalogue record for this publication is available from the British Library

Library of Congress Catalogue Card Number: 74–19527

ISBN 978-0-521-20648-8 hardback
ISBN 978-0-521-08518-2 paperback

TO THE MEMORY OF
SIR GEORGE HOWARD DARWIN

PREFACE TO THE SIXTH EDITION

Amplifications are made throughout the book. The major changes are the following.

(1) The dynamical ellipticities of the Moon are redetermined and it appears that they are consistent with a nearly uniform density.

(2) The damping of the 14-monthly nutation is redetermined. Wako's explanation of the Kimura term is given.

(3) New data relevant to the secular acceleration of the Moon's motion and the rotation of the Earth are discussed, and it appears that the three determinations least likely to be affected by disturbances not yet considered are reasonably consistent; they are also consistent with evidence derived from fossils.

(4) The arguments against convection in the Earth in its present state, continental drift, and now plate tectonics are greatly expanded.

(5) The statistical methods used in developing results in earlier chapters are described more fully.

My wife has given me great help with the editing, and has revised and extended the bibliography and subject index.

I should like to thank the staff of the Cambridge University Press for their courtesy and patience in the production of the successive editions, now extending over fifty years.

HAROLD JEFFREYS

ST JOHN'S COLLEGE
1974 December 4

PREFACE TO THE FIFTH EDITION

In the fifth edition considerable additions have been made. Study of the orbital motions of artificial satellites has now determined at least the zonal and tesseral harmonics up to degree 8 in the gravitational potential, and has shown definite discrepancies with the hydrostatic theory of the figure of the Earth; previous estimates of departures from isostasy and strength at considerable depths are confirmed in general magnitude.

Much attention is given to Lomnitz's law of imperfection of elasticity and a slight modification of it. It accounts for many facts that were previously hard to understand, notably how to reconcile the Moon's rotation with the persistence of its excess ellipticities. It forbids convection and continental drift in the Earth's present state, though the former, in a formerly fluid state, provides the nearest approach to a satisfactory theory of the origin of continents. Many features, notably the upward concentration of radioactivity, seem to demand former fluidity.

Recent work has determined small but definite regional differences in travel times of seismic waves. Study of free vibrations of the Earth as a whole provides valuable checks on the mechanical structure. Their damping, that of the free nutation, and that of surface waves give information on the distribution of imperfection of elasticity with depth.

The latest revision of the constants of the Solar System is explained, and the results are included.

When the first and second editions of this book were written it seemed that Jeans's theory of the origin of the Solar System, with some modifications, was sufficiently established to be taken as a datum for geophysics. When the third was written I was sufficiently doubtful to omit most of the cosmogonical matter. But much recent work has adopted alternatives that seem to me no better, and I have added a section to explain why I consider no theory satisfactory. However it seems, with the increased estimates of the age of the Earth, that radioactivity of ^{40}K and ^{235}U in the early history may well have produced fusion for a time even if the Earth was originally cold.

Other substantial alterations are made throughout the book. Explicit forms for the kinetic and potential energies of an elastic sphere are given in a way that should permit small corrections easily by Rayleigh's principle.

Most of the present revision was done while I was holding a post jointly at the Southern Methodist University and the Southwest Center for Advanced Studies, Dallas, Texas. I am grateful for much help to the staffs, especially Professors J. E. Brooks, A. L. Hales, and M. Landisman. Dr Stoneley has helped me with the proofs as on previous occasions. I have had useful advice from Professors M. Ewing, F. Hoyle, and others.

Preface

I take this opportunity of expressing thanks to various agencies for financial support to my collaborators and myself during the last few years: Carnegie Institution of Washington, Harry Oscar Wood Award in seismology; U.S. Air Force Office of Aerospace Research, under AFOSR Grant 62.303; Air Force Cambridge Research Laboratories under Contract AF 19(604)-7376, Project Vela-Uniform; N.A.S.A. Grants NGR 445.63 and NGR 33.008.037; The National Science Foundation Fellowship for distinguished Foreign Scientists; the John Simon Guggenheim Foundation grant for the study of the Earth's core.

The collaborators that have helped me are Dr M. Shimshoni, Dr E. P. Arnold, and Dr M. L. Gogna.

AUTHOR'S NOTE

Section numbering. The sections are numbered according to the decimal system; of any two sections, that with the smaller number comes earlier in the book. The integral part of a section number is the number of the chapter. Equations have as a rule been numbered consecutively through each section; but in some cases they have been numbered consecutively through several closely related sections. In cross-references, where reference is made to another equation in the same section, only the number of the equation is given; but where the reference is to a different section, the numbers of that section and of the equation are given; e.g. 14·61 (3).

References. As there are many instances in this book where several references are made to the same book or paper, they are made to a bibliography at the end to reduce repetition. The system is approximately that used by the Royal Society, which itself differs slightly from the 'Harvard system'. The outstanding recommendation of these systems is that, by making the date second in importance only to the author's name in identifying a paper, they help the reader to see whether the results of one paper can be discussed in another—a paper of 1927 is unlikely to make use of results published in 1938. On the other hand it is often useful to know the length of a paper, and these systems make no explicit provision for giving it; I give both initial and final pages. For instance, the reader may wish to know whether a paper is an extended discussion or a brief note, possibly an abstract. When a reference is to a particular page in a paper, the page is indicated in the text; it will not be necessary to scan a 40-page paper for something discussed in one paragraph in the middle. In the absence of such an indication of page it may be taken that the reference is to the paper as a whole.

I do not enclose the date in brackets. In many years' experience the British Association Mathematical Tables Committee and its successsor, the corresponding committee of the Royal Society, have found that legibility is improved by having white spaces around the entry that is meant to catch the eye. Brackets about the date delay the eye in finding it.

Where there are several references to the same author, the author's name is not repeated. To identify a paper it is necessary to find first the name and then the date; and the second step is easier if the reader looks down a series of entries, in each of which the date is the first item, than if he has to look down the interior of a block of type and make sure on the way that the name has not changed. Nobody would think of repeating the name of a substance in every determination of its properties quoted in a table of physical constants.

As both these departures from the practice recommended tend to reduce the cost of printing there may be other reasons for recommending them for wider adoption.

The method of numbered references is unsatisfactory, because insertion of a new one entails renumbering, and therefore unnecessary labour and risk of mistakes.

Symbols for approximations. $\sim$ is used for asymptotic equality in Poincaré's sense.

O and o are used in their usual mathematical senses. Note that $y = O(x)$ with x small means only that as $x \to 0$, y/x is bounded—it is not necessary that y and x should be of the same dimensions. $a \doteqdot b$ (read 'a is approximately equal to b') means that $|\, a{-}b \,|$ is small compared with a or b. How small it needs to be for the statement to be interesting depends on the particular problem.

$a \eqsim b$ (read 'a and b are of the same order of magnitude') is a useful physical expression but has no precise mathematical definition. It says that a and b are of the same dimensions and that from our general experience in computation (notably the rarity of large numerical coefficients in the solution of a differential equation) neither is likely to be more than 10 times the other. In application it usually means that some numerical factors, which are unlikely to be outside the range 1/3 to 3, have been taken equal to 1. It is used either to identify the most important terms in an equation or to test whether some hypothesis affords any reasonable hope of an explanation of a given fact.

The sign $\sim$ is used in all these senses in physical papers, and the above definitions (none of which is due to me) are proposed as a way of reducing confusion.

$\eqsim$ is in fact seldom used; where rough values have been assumed for calculation, $=$ refers to their consequences, but for application to the actual Earth $=$ should be interpreted as $\eqsim$.

CONTENTS

PLATES

NOTE

CHAPTER I

The Mechanical Properties of Rocks

'So far', admitted Lao Ting, 'it is more in the nature of a vision. There are, of necessity, many trials, and few can reach the ultimate end. Yet even the Yangtze-Kiang has a source.' ERNEST BRAMAH, *Kai Lung's Golden Hours*

1·01. In studying the constitution and history of the Earth we are largely concerned with the phenomena of change of dimensions, especially of shape. Hence the physical properties that we most need to know are the elastic properties. Perfectly elastic solids are the subject of an extensive theory, which is well confirmed by experiment when the strains are small. When the strains are large enough different substances behave very differently. The information available is mainly experimental, but progress is being made towards its coordination into a general theory under the name of rheology.

1·02. Stress; equations of motion. For all types of material it is possible to define the stress at a point in terms of nine components, six of which are equal in pairs. It is convenient to use rectangular coordinates x_1, x_2, x_3, or, more shortly, x_i, where it is understood that i can take any of the values 1, 2, 3.* If we take a plane of x_i constant through the point, the force per unit area between parts of the material on opposite sides of it, tending to displace the matter with smaller x_i in the direction of increasing x_k, is the stress-component denoted by p_{ik}. It is shown in works on elasticity and hydrodynamics that $p_{ik} = p_{ki}$, and that the equations of motion are

$$\rho f_i = \rho X_i + \sum_{k=1, 2, 3} \frac{\partial p_{ik}}{\partial x_k}. \tag{1}$$

Here ρ is the density, f_i a component of acceleration, and X_i the external force per unit mass (usually gravity). These relations assume nothing about the properties of the material except that it is continuous and that there are no infinite accelerations near the point considered. Also we have the equation of continuity, expressing the fact that the rate of increase of the mass within an element of volume is equal to the net rate of inflow of mass into it. This equation is

$$\frac{\partial \rho}{\partial t} = -\sum_i \frac{\partial}{\partial x_i}(\rho v_i), \tag{2}$$

where v_i is the component of velocity in the direction of x_i.

Full analysis of the hypothesis that the material is continuous would require consideration of atomic theory. We may, however, remark briefly

---

* For a fuller discussion see H. and B. S. Jeffreys (1972, Chapters 2, 3).

that it requires the quantities in the equations to be determinate within the accuracy appropriate to the thermal agitation. The temperature is determined to an accuracy of order $0\cdot1\%$ if the volume contains about 10^6 atoms, and as the spacing between atoms in a solid or liquid is of order 10^{-8} cm., we should expect our results to be right as long as we do not try to apply them to regions of dimensions less than about 10^{-6} cm. Essentially the notion of stress supposes the atomic structure smoothed out in such a way as to preserve the actual mass, momentum and force when regions more than 10^{-6} cm. in linear extent are considered.

If a plane whose normal has direction cosines l_i is considered, the stress across it is obtained from the p_{ik} by the rule that the component in the direction x_i is $\sum\limits_{k} l_k p_{ik}$. In particular, if a surface is free, that is, in contact with no matter on one side, there can be no stress across it, and these three sums are zero at every point of the surface.

Another property of importance is as follows. If we take an inclined set of axes x'_j, such that the cosine of the angle between x_i and x'_j is l_{ij}, the stress-components with regard to the new axes are p'_{jl}, where j and l can take the values 1, 2, 3 and

$$p'_{jl} = \sum_{i,k} l_{ij} l_{kl} p_{ik}. \tag{3}$$

(l as a suffix specifies an axis, l on the line indicates a direction cosine. There need be no confusion if this is borne in mind.) To abbreviate writing it is customary to omit the sign of summation in such equations as (1), (2) and (3). It is then understood that in an expression containing a repeated suffix, that suffix is to be given all possible values and the results are to be added. Thus the occurrence of a repeated suffix implies the summation.

A set of nine quantities w_{ik} that transform to new axes according to the rule (3) is called a tensor of the second order. If it also has the property $w_{ik} = w_{ki}$ it is called *symmetrical*. Such a tensor has the remarkable property that it is possible to choose a set of axes such that all components with j, l unequal are zero; then the components with $j = l$ are the roots of the cubic equation in λ (taking p_{ik} as the typical case):

$$\begin{vmatrix} p_{11}-\lambda & p_{12} & p_{13} \\ p_{21} & p_{22}-\lambda & p_{23} \\ p_{31} & p_{32} & p_{33}-\lambda \end{vmatrix} = 0. \tag{4}$$

These roots are all real and are called the *principal stresses*. Since p_{ik} in general vary from place to place, the principal stresses do so too. We denote them by P_1, P_2, P_3. Thus any system of stresses at a point can be represented as the resultant of three simple tensions or pressures at right angles. The mean of the roots P_1, P_2, P_3 is also the mean of p_{11}, p_{22}, p_{33}. We denote it by $-p$ (because it is nearly always negative in our problems). If $P_1 = P_2 = P_3$ at a point the stress there is called *hydrostatic*, because this condition is satisfied in a fluid at rest. Accordingly, it is often useful to regard the stress as composed of a hydrostatic stress equal to the mean of the principal stresses together with a part known as the *deviatoric stress*, which measures the departure of the stress from symmetry. The reason is that any substance, even a gas, behaves almost as perfectly elastic under hydrostatic stress alone; the density depends only on the pressure and the temperature at the actual time, not on their rates of change or on the past history, and the relation is the same whatever the deviatoric stress may be. For gases it is expressed by the laws of Boyle and Charles. There is actually a slight dependence on the rate of change (Liebermann, 1949). This gives extra damping of sound, and is important in shock waves. It would lead to a modification of the relation between dilatation and mean pressure analogous to that given for shear in (30)

$$k\frac{d\Delta}{dt} = \left(\frac{d}{dt}+\frac{1}{\tau_1}\right)(\tfrac{1}{3}p_{ii}),$$

and $k\tau_1$ would behave as a second coefficient of viscosity. For some liquids it is about 100 times the ordinary viscosity, and in both liquids and gases leads to a much greater damping of sound waves as they travel than is given by the ordinary viscosity (Rosenhead *et al.* 1954). A full theoretical discussion of its effects, especially on the formation of shock waves, is given by M. J. Lighthill (in Batchelor and Davies, eds., 1956). On the other hand, changes of shape are related to the deviatoric stress and only to a subordinate extent to the mean stress. Hence it is convenient to introduce the set of quantities δ_{ik}, known as the *substitution tensor*, and defined by the equations
$$\delta_{ik} = 1 \quad (i = k); \qquad \delta_{ik} = 0 \quad (i \neq k). \tag{5}$$

Then the deviatoric stress-components are given by

$$p'_{ik} = p_{ik} + p\,\delta_{ik}. \tag{6}$$

1·03. Rotation and strain; stress-strain relations for liquid and perfectly elastic solids. The motion in the neighbourhood of a point is specified if we have the velocity at the point and all the nine derivatives of the velocity with respect to the coordinates. That is, we need the velocity components v_i and the derivatives $\partial v_i/\partial x_k$. It is best to use these in the combinations
$$e_{ik} = \frac{1}{2}\left(\frac{\partial v_k}{\partial x_i}+\frac{\partial v_i}{\partial x_k}\right), \quad \xi_{ik} = \frac{1}{2}\left(\frac{\partial v_k}{\partial x_i}-\frac{\partial v_i}{\partial x_k}\right). \tag{7}$$

This is because if the e_{ik} are zero at all points it can be shown that distances between given particles are not changing and therefore the whole is moving as a rigid body; and then the rate of rotation is completely determined by the ξ_{ik}, which are constant through the body. If we consider two neighbouring particles, the rate of change of the distance between them depends on the velocities only through the e_{ik}, which are therefore called the *rates of strain*. They form a symmetrical tensor of the second order, like the p_{ik}.

The sum e_{ii} is called the rate of *dilatation*, because it gives the relative rate of increase of volume $-d\rho/\rho\,dt$ by the equation of continuity.

Both for solids and fluids there is a relation between p_{ii} and e_{ii}, because for a material of given composition ρ is determined by the pressure and temperature. If the variations are small, as they are in our problems, the relation can be written

$$e_{ii} = \frac{1}{3k}\frac{\partial p_{ii}}{\partial t} + 3\alpha\frac{\partial \vartheta}{\partial t}, \tag{8}$$

where k is a constant of the material called the *bulk-modulus* or *incompressibility* ($1/k$ being the compressibility), α is the coefficient of linear thermal expansion, and ϑ the temperature. For the present we can neglect changes of temperature.

It is convenient to separate e_{ik} into dilatational and deviatoric parts, representing changes of volume and shape respectively;

$$e_{ik} = \tfrac{1}{3}e_{mm}\delta_{ik} + e'_{ik}. \tag{9}$$

In a normal fluid (that is, not a 'liquid crystal') there is a simple relation between p'_{ik} and e'_{ik}, namely,

$$p'_{ik} = 2\eta e'_{ik}, \tag{10}$$

where η is the coefficient of viscosity. Then we have relations connecting all the components of stress with those of rate of strain and therefore with the velocity; and substituting these in the equations of motion we get differential equations for the velocities.

An isotropic, perfectly elastic solid is a substance such that the rates of change of the p'_{ik} are proportional to the e'_{ik}; that is, there is a constant μ, called the *rigidity*, such that

$$\frac{\partial p'_{ik}}{\partial t} = 2\mu e'_{ik}. \tag{11}$$

This is not true for crystals. It is true with considerable accuracy for most of the Earth. This is probably because rocks are usually aggregates of crystals oriented in all directions, and the differences of properties in different directions average out when we consider specimens of the sizes that usually concern us in geophysics. It may be expected to fail in stratified rocks, but little is known about their elastic properties, and what is known has so far defied summary.

On account of the similarity between (10) and (11) it is convenient in the theory of elasticity to regard the particle at x_i as having originally been at a point with coordinates $x_i - u_i$, where the u_i are components of *displacement* and are regarded as small, so that we can neglect their squares and products. Then to this accuracy $v_i = \dfrac{\partial u_i}{\partial t}$. We can then define a set of components of strain

$$e_{ik} = \frac{1}{2}\left(\frac{\partial u_k}{\partial x_i} + \frac{\partial u_i}{\partial x_k}\right), \tag{12}$$

and
$$e_{ik} = \frac{\partial \epsilon_{ik}}{\partial t}. \tag{13}$$

At present we take the material to have been originally in a state of zero stress and under no external forces. Then

$$p_{ii} = 3k\epsilon_{ii}, \quad p'_{ik} = 2\mu\epsilon'_{ik}, \tag{14}$$

$$p_{ik} = \tfrac{1}{3}p_{mm}\delta_{ik} + p'_{ik}$$
$$= k\epsilon_{mm}\delta_{ik} + 2\mu(\epsilon_{ik} - \tfrac{1}{3}e_{mm}\delta_{ik}) \tag{15}$$
$$= (k - \tfrac{2}{3}\mu)\,\epsilon_{mm}\delta_{ik} + 2\mu\epsilon_{ik}. \tag{16}$$

We write
$$k - \tfrac{2}{3}\mu = \lambda. \tag{17}$$

λ has no particular name, but its introduction simplifies the writing. λ and μ together specify the elastic properties of the material and are called Lamé's constants.

Another description of the properties is obtained by considering the case where p_{11} is given and all other stress-components are zero; this corresponds to a bar under longitudinal stress, the sides being free. The bar is stretched along x_1, but contracts in the same ratio along x_2 and x_3. Then if $\epsilon_{11} = e$, $\epsilon_{22} = \epsilon_{33} = -\sigma e$, $\epsilon_{12} = \epsilon_{23} = \epsilon_{31} = 0$, we have

$$\epsilon_{mm} = e(1 - 2\sigma), \tag{18}$$

$$p_{11} = \lambda e(1 - 2\sigma) + 2\mu e, \tag{19}$$

$$p_{22} = p_{33} = \lambda e(1 - 2\sigma) - 2\mu\sigma e = 0, \tag{20}$$

whence
$$\sigma = \frac{\lambda}{2(\lambda + \mu)}, \quad p_{11} = \frac{3\lambda + 2\mu}{\lambda + \mu}\mu e. \tag{21}$$

The coefficient of e in (21) is *Young's modulus*, denoted by E:

$$E = \frac{3\lambda + 2\mu}{\lambda + \mu}\mu. \tag{22}$$

σ is *Poisson's ratio*. Alternatively, k, λ and μ can be expressed in terms of E and σ. But σ is difficult to measure accurately. The easiest quantities to measure are E and μ. E is found by measuring the extension of a rod under given tensions, and μ by measuring the torsion under given couples applied at the ends. k can also be measured directly as follows. A cylinder of the material is placed in a slightly larger container, which is then filled with a liquid. Pressure is applied to the container by a piston, and the liquid transmits hydrostatic pressure to the solid. The pressure is measured by an instrument immersed in the liquid, and the change of volume is also measured. It is necessary to allow for the compression of the liquid. This method was introduced by Bridgman at Harvard, and has been followed by the Geophysical Laboratory at Washington.

If we substitute (16) in the equations of motion we find, again neglecting squares of displacements,

$$\rho \frac{\partial^2 u_i}{\partial t^2} = (\lambda + \mu) \frac{\partial \Delta}{\partial x_i} + \mu \nabla^2 u_i, \tag{23}$$

where
$$\Delta = \frac{\partial u_m}{\partial x_m}, \quad \nabla^2 = \frac{\partial^2}{\partial x_1^2} + \frac{\partial^2}{\partial x_2^2} + \frac{\partial^2}{\partial x_3^2}. \tag{24}$$

Extensions when squares of the displacements are not neglected have been given by several authors, especially F. D. Murnaghan. But according to these the higher terms do not become important till the stresses are of the order of λ and μ. In actual materials they are noticeable at much smaller stresses. In geophysics this concerns rocks that are on the verge of fracture anyhow, when imperfections of elasticity are more important. At great depths in the Earth the symmetrical part of the stress is of the order of k; the deviatoric part is much smaller. An adequate approximation is to assume a linear relation between departures of stress and strain from an initial state, often hydrostatic.

1·04. Imperfections of elasticity. Imperfections of elasticity under hydrostatic stress are usually negligible, and in what follows we shall be concerned mainly with the deviatoric parts of the stress and strain. Most materials ordinarily regarded as solid are not quite perfectly elastic, even at small stresses. If a tension is applied to a bar and then kept constant a considerable variety of behaviour may ensue. Vibrations are set up, but these usually lose their energy by transmission to the surroundings in a much shorter time than concerns us here, and we may ignore them. If the tension is great enough there may be immediate fracture. If not, there is an initial extension, which appears to be accurately proportional to the tension. This has been verified for rocks by D. W. Phillips (1931, 1932, 1934). But in most cases the extension continues to increase slowly. The duration of experiments by Phillips and D. T. Griggs (1936) ranged from hours to months. Usually the rate of strain became slower as time went on; but, sometimes, especially under large stresses, the rate began to increase again and ultimately the specimen broke. If it did not break and the stress was removed, there was an immediate recovery by the amount of the original elastic displacement, but further changes followed. There is thus a considerable variety of behaviour.

In *perfect elasticity* the strain depends only on the stress, however long this is maintained. When the stress is removed the original form is recovered at once.

In order of increasing departure from perfect elasticity the following occur:

Elastic afterworking or viscoelasticity: The strain under constant stress increases at a decreasing rate; it may or may not tend to a finite limit. When the stress is removed, there is immediate recovery by the amount of the initial strain, but the strain decreases further and ultimately tends to zero. It can be seen in certain biscuits containing treacle if slightly stale. Such a biscuit can be slightly bent, but when released it can be seen slowly creeping back towards its original form, and ultimately becomes as nearly flat as we can see. The term *creep* is often used.

Elastic hysteresis: Part of the strain persists when the external stress is removed, even after a long time, but does not increase with duration of stress. An example is a screw driven into wood; it is not driven out again by the stresses in the wood, these being balanced by the friction of the screw.

Flow and plasticity: The strain under given stress may or may not increase beyond any limit if the stress is maintained long enough; and only a small fraction of it disappears after removal of the stress. This is typical of such materials as clay and dough. The distinction between this and elastic hysteresis is really one of degree. In a highly ductile material like copper the strain may increase beyond any limit if a sufficient stress is maintained long enough.

It usually arises only when the stresses have reached some value characteristic of the material. A moulded body, when the applied stress is removed, is still under stress due to gravity, but does not in general flatten out.

In *fracture* the displacement components within the body cease to be continuous functions of position. The body may separate into two or more parts, moving with different velocities. When the distribution of stress is not uniform an internal crack may form without reaching the boundary.

All these imperfections of elasticity follow complicated laws, usually non-linear, and the study of their consequences has been severely limited by mathematical difficulties. They can, however, be related to fairly simple models, which suggest qualitative explanations. The earliest suggestion of this kind was due to Maxwell and expressed as a quantitative stress-strain relation by J. G. Butcher (1876). The solid is regarded as composed of two types of molecules (or aggregates of molecules), one of which is regarded as perfectly elastic, while the other is capable of sustaining only a symmetrical pressure or tension. The former type, when the body is kept under a constant deformation, are supposed to break down at a rate proportional to their number and to be converted to the other type, so that the total shearing stress, measured as an average over all, diminishes exponentially. Then for a constant deformation component ϵ the corresponding component P of the shear stress has the form

$$P = 2\mu\epsilon\, e^{-t/\tau}, \tag{25}$$

where μ, τ are constants of the material.

A variable ϵ can be built up by considering successive deformations applied at times $u = t_1, t_2, \ldots$, and then

$$P = 2\mu \int_{u=0}^{t} e^{-(t-u)/\tau} d\epsilon, \qquad (26)$$

which is found to satisfy the differential equation

$$\frac{dP}{dt} + \frac{P}{\tau} = 2\mu \frac{d\epsilon}{dt}. \qquad (27)$$

An alternative method is to regard the strain as composed of two parts ϵ_1 and ϵ_2, the first related to the stress according to the elastic law, the second according to the viscous law, so that

$$\frac{dP}{dt} = 2\mu \frac{d\epsilon_1}{dt}; \quad P = 2\eta \frac{d\epsilon_2}{dt}, \qquad (28)$$

whence

$$\frac{d\epsilon}{dt} = \frac{d\epsilon_1}{dt} + \frac{d\epsilon_2}{dt} = \frac{1}{2\mu} \frac{dP}{dt} + \frac{1}{2\eta} P, \qquad (29)$$

$$2\mu \frac{d\epsilon}{dt} = \frac{dP}{dt} + \frac{P}{\tau}, \qquad (30)$$

where

$$\tau = \eta/\mu. \qquad (31)$$

This rule, applied to all the deviatoric components of stress and strain, is called the *elasticoviscous* law. Evidently, if a problem has been solved for the case of perfect elasticity, the solution of the corresponding problem for elasticoviscosity can be found by replacing μ by the operator

$$\frac{\mu}{1 + Q/\tau}, \qquad (32)$$

where Q is the operator of definite integration defined by

$$Qf(t) = \int_{-\infty}^{t} f(u) \, du. \qquad (33)$$

If P is zero up to time 0 and thereafter is constant, (27) leads to

$$\epsilon = 0, \quad t < 0, \qquad (34)$$

$$2\mu\epsilon = (1 + t/\tau) P \quad (t > 0). \qquad (35)$$

The deformation therefore increases discontinuously at time 0 as for a perfectly elastic substance, but thereafter increases uniformly with time at a rate proportional to the stress. The rule therefore represents an extreme case of plasticity. It is, however, too simple to represent actual materials. The actual rate of strain is not proportional to the stress. It is usually unnoticeable below a certain stress, and at higher stresses increases more rapidly than the stress. This could be represented by taking τ in (27) to depend on the stress, decreasing with increasing stress, and becoming infinite below a certain critical stress. In that case the material will behave as perfectly elastic if this critical value is not exceeded.

The behaviour can be represented better if we regard the solid as composed of pieces, all perfectly elastic when the stress is not too great, but flowing when it is great enough, the transitions being at different stresses for different portions. A sufficiently small stress will then give only an elastic displacement, but above a certain value of the stress more and more of the portions will begin to flow. Thus the rate of deformation will increase more rapidly than in proportion to the stress. Such a structure may well correspond to that of most rocks and even of metals as they usually occur. The stronger portions in rocks appear to be individual crystals and the weaker the interfaces, but the latter would not all be equally strong. In many metals a perfect crystal has hardly any strength, but acquires it on deformation. Under sufficient stress even the stronger parts flow and deformation can increase indefinitely. But if the stress is in an intermediate range the weaker parts may adjust themselves to a hydrostatic state, leaving the whole deforming stress to be borne by the stronger parts. If this happens the displacement will tend to a finite limit, the rate of approach being limited by the viscosity of the weak parts. If the stress is removed the strong parts will tend to spring back and, in doing so, apply deforming stresses to the weak ones, which will again flow, but in reversed direction. Thus we shall have slow recovery, which will continue until the stresses in all the weak parts have been brought below the critical stresses. There may be a residual displacement if the weak parts are not absolutely devoid of strength, so that elastic hysteresis is explained, and elastic afterworking is explained by the viscosity of the weak parts.

Such a model also makes delayed fracture intelligible. The strong parts may be such that under sufficient stress they break at once and do not flow. Suppose that a stress less than this has been applied. At first the strengths of all parts are available to resist it. But as the weaker parts adjust themselves towards a hydrostatic state more stress is thrown on the stronger ones, which will break if the increase is sufficient.

Unfortunately, little has been done towards stating the qualitative argument in quantitative terms and comparing with experimental data. Two points, however, may be mentioned at once. In the first place, the elastico-viscous law becomes that of viscosity if μ is put infinite, $\mu\tau$ remaining finite. It seems queer that a normal liquid should be regarded as a solid of infinite rigidity, even though the viscosity is finite. But if we consider water, for instance, η is about 0·01 c.g.s. and the bulk modulus is about 10^{10} c.g.s. In solids the rigidity is usually of the same order of magnitude as the bulk modulus. But if water has a rigidity of order 10^{10}, τ in (27) will be of order 10^{-12} sec. Thus in any experiment that tests the behaviour over intervals large compared with 10^{-12} sec., the flow will be much greater than the elastic deformation, and the insertion of an elasticity term in the stress-strain relation will affect nothing observable. There is some positive evidence for a rigidity in liquids from the facts that the conduction of heat appears to depend on the scattering of transverse waves and that study by X-rays shows traces of crystalline structure. Raman and Venkateswaran (1939) sent ultrasonic waves with a period of $0 \cdot 65^8 \times 10^{-10}$ through glycerine. The velocity was definitely higher than $\sqrt{(k/\rho)}$. Interpreted as $\sqrt{\{(\lambda + 2\mu)/\rho\}}$ it gives an estimate of μ of order 3×10^6 dynes/cm.^{2}, which, combined with the viscosity 20 c.g.s., leads to a relaxation time of order 10^{-5} sec.

The idea of a rigidity in fluids goes back to Maxwell's dynamical theory of gases (*Scientific Papers* 2, 26). He found the rigidity of air appreciable. With a coefficient of rigidity of the order of the pressure (like the bulk modulus) he estimated that in air the time of relaxation is of the order of 10^{-10} sec.

The other point is that the distinction between plasticity and elastic afterworking is very largely a matter of the spatial distribution of the weak places. In an example (1932e) I considered two differently constituted cylinders, partly of perfectly elastic material and partly of elasticoviscous material. In one case the weak material formed a core enclosed by a case of the stronger. A torsional pair of couples was applied. The result was found to be an immediate elastic yield, followed by a further deformation in the same direction, tending exponentially to a finite limit. In the other case a cylinder was supposed to be built up of successive bands of the two materials. In this case a constant pair of torsional couples led to torsion increasing without limit and still representable by the elasticoviscous law with different constants. The former case illustrates elastic afterworking and suggests that in some cases it can be taken into account by replacing μ in the elastic solution by an operator of the form

$$\mu \frac{1 + Q/\tau'}{1 + Q/\tau} \tag{36}$$

with $\tau' > \tau > 0$. The response to a stress suddenly applied and afterwards maintained would be

$$\frac{P}{\mu} \left\{ \frac{\tau'}{\tau} - \left(\frac{\tau'}{\tau} - 1 \right) e^{-t/\tau'} \right\}. \tag{37}$$

I call this *exponential afterworking*.

Another way of representing imperfections of elasticity was suggested to me by Sir J. Larmor. We may regard the stress as balancing an elastic stress proportional to the strain, and a viscous one proportional to the rate of strain. Then we should have

$$P = \mu \left(\epsilon + \tau \frac{d\epsilon}{dt} \right). \tag{38}$$

I gave the name *firmoviscosity* to this type because it supposes a resistance additional to the elastic one. Prof. M. Reiner has called my attention to the fact that the response of a sponge under water to stress is very like that for firmoviscosity. This can, I think, mean only that the initial elastic response is very small compared with the ultimate one; that is, in (36) τ' may be a few seconds and τ a small fraction of a second.

The firmoviscous law was stated by Kelvin and is usually known by his name. It should be emphasized that he was considering torsion of wires, and that he found the experimental results in flagrant disagreement with it and therefore rejected it. The form (36) does not fit his results perfectly, but is less violently discordant (Jeffreys, 1932e).

Within a limited range of time-scale, (38) is a good approximation to the effects of irregularities of structure on wave propagation. We shall return to this later (2·04).

In rocks the elastic law breaks down at deviatoric strains of the order of 10^{-3}. This is a small strain and quite difficult to measure. Geologists are mostly concerned with much larger inelastic strains. However, there is abundant geophysical evidence that there are imperfections of elasticity at much smaller strains. Those in the bodily tide in the Earth are of order 10^{-7}, those in the Moon of order 10^{-8}. The free nutation of the Earth's axis involves strains of order 10^{-8}. A displacement of 1 cm. on a seismogram with a magnification of 200 is large; it changes sign in a distance of order 10 km.; so the strain is of order $1/(200 \times 10^6)$ or $0·5 \times 10^{-8}$. All these phenomena show damping, but the strains are far too small to be detected by laboratory experiment. If the latter is to be used, the most promising procedure is to adopt a linear law, since departures from linearity in general increase with the stresses. There has, however, been much difficulty in finding a linear law that did not give results in flagrant disagreement with the available facts.

C. Lomnitz (1956, 1957) found for rocks in the laboratory that at strains in the range 10^{-5} to 10^{-4} the stress-strain relation is of the form

$$\epsilon = \frac{P}{\mu}\{1 + \phi(t)\}, \tag{39}$$

with
$$\phi(t) = q \log(1 + at), \tag{40}$$

where q and a are constants. The essential point is that the relation between ϵ and P is linear and can therefore presumably be applied however small ϵ may be. For variable stress the deformation can be built up by superposition analogously to (26). In igneous rocks a appears to be of order 6000/sec. (later workers have found values up to 10^6/sec.), and q of order 10^{-3} to 10^{-2}. Over a wide range of periods the dissipation per period is close to $\frac{1}{2}\pi^2 q$ of the total energy. Logarithmic laws had been proposed previously, but had the defect that as $t \to 0+$, $d\epsilon/dt$ would tend to infinity. Lomnitz's introduction of a avoids this. For small t this law tends to that for elasticoviscosity. But a is so large that for ordinary intervals of time the logarithmic rule is a good approximation. Pandit and Savage (1973) get linear response for strains from 10^{-8} to 10^{-5} in sandstone. Q, as defined on p. 48, is close to 50 for frequencies from 900 to 50,000/sec., a is $> 10^6/\text{sec}$.

P. Phillips (1905) had given an equivalent rule for glass, some metals and rubber. Lomnitz's experiments referred to both vibrations and maintained stresses, Phillips's to long-continued stresses.

Phillips verified the recovery after the stress was removed, and mentioned that the singularity at $t = 0$ is avoided by measuring it from a time somewhat earlier than that of application of the stress. He thus had the complete Lomnitz law, but the paper seems to have been overlooked. My attention was called to it by Professor E. Orowan.

I found that the Lomnitz law as it stood led to some unacceptable results when applied to the Earth as a whole, but the modification

$$\epsilon = \frac{P}{\mu}\left[1 + \frac{q}{\alpha}\{(1 + at)^\alpha - 1\}\right], \tag{41}$$

with α about 0·25, fitted many data very well (Jeffreys, 1958 a, b; Jeffreys and Crampin, 1960). If $\alpha = 1$ it reduces to elasticoviscosity; if $\alpha \to 0$ it tends to the logarithmic form. The dynamical equations are linear but will not be differential equations with coefficients independent of the time, and must be replaced by integral equations. Further work suggests that q, a, and α are not constant within the Earth. But the result found for the twisted cylinder showed that according to the situation of one elastico-viscous region we might get either elasticoviscosity or exponential after-working; and presumably with many such regions of different sizes any intermediate law would be possible. It seems possible that the whole of the dissipation at these small stresses could be attributed to a suitable distribution of elasticoviscous regions.

If a stress is applied for time T and removed, then under any linear law of the form (40) the strain for $t > T$ is

$$\epsilon = \frac{P}{\mu}\{1 + \phi(t)\} - \frac{P}{\mu}\{1 + \phi(t - T)\}$$

$$= \frac{P}{\mu}\{\phi(t) - \phi(t - T)\}. \tag{42}$$

Thus for any law such that $\phi'(t) \to 0$ as $t \to \infty$, the residual strain when the stress is removed will tend to zero. Thus it will do so even if $\phi(t)$ for large t behaves like t^α with $\alpha < 1$.

Such behaviour is included in the comprehensive term *viscoelasticity*. I shall not use this term, on account of the possible confusion with elasticoviscosity, which would be the case $\alpha = 1$ and would leave a residual strain proportional to T. The latter is often known as Newtonian viscosity, but Newton lived long before Poiseuille found the ordinary law for liquid viscosity. If personal names are to be used, it should be called Maxwell–Butcher–Darwin viscosity.

With the Lomnitz law or the modification, the times of any ordinary experiment are long enough for $1 + at$ to be replaced by at, and then the correcting term in $\mu\varepsilon/P$ is approximately $q(at)^\alpha/\alpha$, the Heaviside operator corresponding to which is $qa^\alpha(\alpha-1)!\,p^{-\alpha}$. Thus for small departures from perfect elasticity μ can be replaced by the operator

$$\mu\{1 - qa^\alpha(\alpha-1)!\,p^{-\alpha}\}.$$

Consider a vibrating instrument that for perfect elasticity would satisfy $\ddot{x} + n^2x = 0$, where n^2 is proportional to μ. With small damping we can replace p in the correcting term by in, and the equation becomes

$$\ddot{x} + n^2\{1 - qa^\alpha(\alpha-1)!\,n^{-\alpha}\cos\tfrac{1}{2}\pi\alpha\}x + n\{qa^\alpha(\alpha-1)!\,n^{-\alpha}\sin\tfrac{1}{2}\pi\alpha\}\dot{x} = 0.$$

The term in x gives a small increase of period. That in $\dot{x}$ introduces a damping factor e^{-kt} with

$$2k = qa^\alpha(\alpha-1)!\,n^{1-\alpha}\sin\tfrac{1}{2}\pi\alpha.$$

The index in a period $2\pi/n$ is

$$-\frac{2\pi k}{n} = -\pi qa^\alpha(\alpha-1)!\,n^{-\alpha}\sin\left(\tfrac{1}{2}\pi\alpha\right)$$

which, when $\alpha \to 0$, tends to $-\tfrac{1}{2}\pi^2q$, and is thus independent of the period. In the other extreme $\alpha = 1$ the approximations will break down if the period is too short; but for long periods the index for a period is proportional to the period.

1·05. Critical stress. So far we have indicated that for any solid there is a critical stress such that at larger stresses elasticity ceases to be perfect, but the stress contains six different components, and an adequate theory must state the criterion more precisely. Three criteria need consideration: First, greatest principal stress, secondly difference between the greatest and least principal stresses, and thirdly distortional energy per unit volume.

There are appreciable differences in behaviour according as the greatest principal stress is a pressure or a tension. Under a sufficient tension a bar of brittle material usually breaks nearly straight across; under longitudinal thrust it breaks at an angle, usually about 45°. The crushing strength is usually much greater than the tensile strength. In D. W. Phillips's experiments on the bending of bars the upper side of the bar was under tension and the lower under thrust. When the bar broke, the fracture surface was perpendicular to the length at the top, but changed direction abruptly when part of the way across and proceeded at about 45°. Some such difference might be expected. In a bar broken by tension the pieces separate at once and will move each in the direction of the force acting on it. But under thrust, if the pieces moved in the direction of the forces they would be driven into each other. Fracture under thrust must be inclined to the thrust if this is to be relieved.

The criterion stated by Coulomb (1776) was that fracture takes place by sliding on a plane of greatest tangential stress, which is also a plane of greatest shear. W. Hopkins (1849) showed that this plane passes through the direction of the intermediate principal stress and bisects the angle between the extreme stresses, and that the shear stress on it is half the difference between the extreme stresses. For a proof of this see Appendix A.

The positive difference between the greatest and least principal stresses is called the *stress-difference*.

A modification of the Coulomb–Hopkins theory has been made by E. M. Anderson (1905). In the original form it is supposed that fracture occurs when the stress-difference attains a value S characteristic of the material, and the slip is in the direction of greatest shear stress. Anderson supposes that slip is also resisted by solid friction proportional to the normal stress (aided by this when it is a tension).

When bars are broken by longitudinal thrust, fracture usually takes place at about 45° to the axis of thrust, but there is geological evidence that when the fracture has taken place under heavy load from above this angle is substantially reduced, often to 30° or less. This is explained by Anderson's theory. There is also evidence that fractures under tension are at angles greater than 45° to the tension. But the theory does not explain tension cracks at right angles to the greatest tension. If the coefficient of friction, again, is taken constant the theory would also, apparently, imply an increase of critical shear stress with depth that is not otherwise supported. There is experimental evidence that the critical shear stress increases under a general pressure, and Anderson's theory is in accordance with this, but it appears to make the effect too great, at least if we use ordinary values of the coefficient of friction, say 0·2–0·6.

The nature of the effect can be seen as follows. Take the greatest thrust to be in the direction of x_3 and let all the principal stresses be thrusts. Then the normal stress across any plane is a thrust, and is greater in magnitude the more steeply the plane is inclined to x_3. But the shear stress is the same for planes at θ and $\frac{1}{2}\pi - \theta$ to x_3 ($\theta < \frac{1}{4}\pi$). The friction is greater for $\frac{1}{2}\pi - \theta$, and therefore slip will take place at a smaller stress-difference on the θ plane than at $\frac{1}{2}\pi - \theta$; thus friction favours slip at a smaller angle than $\frac{1}{4}\pi$ to the greatest thrust.

A possible interpretation of Anderson's argument rests on the fact that fracture, however it begins, involves discontinuous motion. If, as under tension, the parts separate, friction does not arise. But if they do not, friction will be the principal resistance to their relative motion as long as slip continues. Then the Coulomb–Hopkins theory may be correct for the initiation of slip, but Anderson's may still be right for determining in what direction slip will proceed when once started.

The Mises criterion. This rests on the assumption that a material cannot store more than a given amount of distortional strain energy per unit volume. The total elastic energy per unit volume is given by

$$2W = p_{ik}e_{ik}. \tag{1}$$

Henceforward we denote strain components by e instead of ϵ.

We write $\qquad p_{ik} = \tfrac{1}{3}p_{mm}\delta_{ik} + p'_{ik}, \quad e_{ik} = \tfrac{1}{3}e_{mm}\delta_{ik} + e'_{ik},$ (2)

and $\qquad\qquad p_{mm} = ke_{mm}, \quad p'_{ik} = 2\mu e'_{ik}.$ (3)

Hence $\quad 2W = (\tfrac{1}{3}p_{mm}\delta_{ik} + p'_{ik})\left(\dfrac{1}{3k}p_{nn}\delta_{ik} + \dfrac{1}{2\mu}p'_{ik}\right)$

$$= \frac{1}{9k}p_{mm}p_{nn}\delta_{ik}\delta_{ik} + \frac{1}{6\mu}p_{mm}\delta_{ik}p'_{ik} + \frac{1}{3k}\delta_{ik}p_{nn}p'_{ik} + \frac{1}{2\mu}p'_{ik}p'_{ik}. \tag{4}$$

But $\qquad\qquad \delta_{ik}p'_{ik} = p'_{ii} = 0, \quad \delta_{ik}\delta_{ik} = \delta_{ii} = 3.$ (5)

Hence $\qquad\qquad 2W = \dfrac{1}{3k}p_{mm}p_{nn} + \dfrac{1}{2\mu}p'_{ik}p'_{ik}.$ (6)

The first part depends wholly on the symmetrical part of the stress, which does not produce any sort of failure of elasticity. The second depends wholly on the departures from symmetry. The Mises function (Mises, 1928) is

$$F = p'_{ik}p'_{ik} \tag{7}$$

$$= p'^2_{11} + p'^2_{22} + p'^2_{33} + 2p'^2_{12} + 2p'^2_{23} + 2p'^2_{31}. \tag{8}$$

But $\qquad\qquad p'_{11} = \tfrac{1}{3}(2p_{11} - p_{22} - p_{33}), \quad p'_{12} = p_{12},$ (9)

with symmetrical relations; hence

$$F = \tfrac{2}{3}(p^2_{11} + p^2_{22} + p^2_{33} - p_{11}p_{22} - p_{11}p_{33} - p_{22}p_{33}) + 3p^2_{12} + 3p^2_{13} + 3p^2_{23}$$

$$= \tfrac{1}{3}\{(p_{11} - p_{22})^2 + (p_{22} - p_{33})^2 + (p_{33} - p_{11})^2 + 6p^2_{12} + 6p^2_{13} + 6p^2_{23}\}. \tag{10}$$

The suggestion of Mises is that failure of perfect elasticity occurs when F reaches a value characteristic of the material. It should be noticed that F is nearly, but not quite, proportional to the square of the stress-difference. For if we refer to principal axes and take

$$p_{33} = p_{11} + S, \quad p_{22} = p_{11} + \alpha S \quad (0 < \alpha < 1), \tag{11}$$

the stress-difference is S, and

$$F = \tfrac{1}{3}S^2\{1 + \alpha^2 + (1 - \alpha)^2\}, \tag{12}$$

which varies from $\tfrac{1}{2}S^2$ for $\alpha = \tfrac{1}{2}$ to $\tfrac{2}{3}S^2$ for $\alpha = 0$ or 1. It should be possible to decide by experiment whether S or F provides the better criterion for yield, that is, which is the more nearly constant for a given material on the verge of fracture or flow, but the test is difficult to apply. If, for instance, the material ceased to be perfectly elastic at a given value of S, the corresponding values of F for different values of the intermediate principal stress would vary only by a factor $\frac{4}{3}$, which is within the range of variation for different specimens of the same material tested under similar conditions. A test has nevertheless been made by G. I. Taylor and H. Quinney (1931), using copper cylinders. It was possible with these to apply tension or thrust and torsion simultaneously. If we take the axis x_3 along the cylinder the non-vanishing components of the stress where the x_1 axis meets the surface are $p_{33} = P$, $p_{23} = p_{32} = Q$, say, and

$$F = \tfrac{2}{3}P^2 + 2Q^2. \tag{13}$$

The equation for the principal stresses is

$$\begin{vmatrix} -\lambda & 0 & 0 \\ 0 & -\lambda & Q \\ 0 & Q & P-\lambda \end{vmatrix} = 0, \tag{14}$$

whence $\lambda = 0$, $\tfrac{1}{2}P \pm \tfrac{1}{2}\sqrt{(P^2 + 4Q^2)}$. Hence

$$S = \sqrt{(P^2 + 4Q^2)}. \tag{15}$$

The point of the method is that it was possible to carry out successive tests on the same specimen, thus eliminating the usual differences between different specimens. The rod could be stretched until flow started, and the value of P recorded. It could then be released, and twisted until flow began again; or tension and torsion could be applied in any intermediate ratio. The result was that when flow began $P^2 + 3Q^2$ was nearly constant, while $P^2 + 4Q^2$ varied appreciably. Consequently it appears that at least for the initiation of flow in copper the Mises function affords a better criterion than the stress-difference.

However, this result does not solve the whole of the problem, for three reasons. In the first place, the imperfection of elasticity was flow, not fracture; consequently, it remains an open question whether the stress-difference is not a better criterion for fracture. Clearly repeated tests on the same specimen are impossible if the specimen is broken. Secondly, the material used was unusually uniform. Most materials have flaws in all directions. Thirdly, the Mises function is a scalar function of position and can by itself give no indication of the direction of fracture if it occurs. Taylor (1934) therefore considered the problem of a small elliptical crack in the material, at any angle to the greatest principal stress, and supposed that the crack would extend when the Mises function reached a definite value near it. He considered two hypotheses, according to which the crack was supposed to be empty or filled with fluid of the same bulk-modulus as the rest of the material. In the former case he found that the Mises function would be largest if the crack was at right angles to the greatest principal tension, so that it would be expected that if tension produced holes it would select for most rapid extension those in this direction. In the latter case he found that it would be largest for cracks at 45° to the greatest principal tension and would then depend chiefly on the average stress-difference in the whole region. Thus the results indicate that (1) the Mises function gives the better criterion for flow in a uniform substance, (2) in a heterogeneous substance, tension will tend to produce holes, developing into cracks at right angles to the greatest principal tension, (3) if there are cracks filled with weak material, such as the interfaces between crystals, and general pressure prevents these from opening, the first to extend will be those at 45° to the extreme principal stresses. Thus even on the hypothesis of Mises it may be expected that in natural materials the greatest tension theory will be the better when the material is under general tension, and that the stress-difference theory will be the better when all the principal stresses are pressures. The agreement with general observation is satisfactory. For geological applications see E. M. Anderson (1936) and Jeffreys (1936 d).

Many materials become stronger when deformed (as in work-hardening) and some rocks undergo changes of crystal form and even chemical changes (metamorphism), apparently producing stronger constituents. I, therefore (1959 d), adapted Taylor's argument to a solid containing rigid elliptical inclusions. The result was that the greatest value of the Mises function would occur when the greatest principal stress (without regard to sign) was along the flaw and would be near the ends, so that long flaws would presumably become longer in the direction of the greatest principal stress. Numerical examination, however, suggested that the energy likely to be available in this way would be much less than that of the chemical changes, and that metamorphism is likely to be due to direct effects of pressure and temperature.

Anderson's theory is probably correct at small depths. But the mean stress at a depth of about 3 km. in the Earth is already of the order of the strength, and ordinary laboratory measures of the coefficient of friction are of order 1. If this persists to great depths, we should expect the stress-difference needed to produce fracture at a depth of 300 km. to be 100 times that of surface rocks. Earthquakes occur at such depths, but are not as strong as the strongest shallow ones. So to have consistency with the facts about deep earthquakes we should require the coefficient of friction to decrease greatly with mean pressure. This has been verified for granite up to a normal pressure of 2×10^{10} dynes/cm.^{2} by J. D. Byerlee (1967). The apparent coefficient for unbroken material drops from about 1 to 0·5.

Some writers have supposed that the coefficient of friction *must* be of order 1 at all depths, and hence that deep earthquakes cannot be fractures.

This ignores the facts that they send out transverse waves, that the beginnings of these waves are as clear as or clearer than in shallow earthquakes, and that many of them are strong enough to give recognizable movement at all distances, even to the anticentre.

Riecker and Seifert (1964) report that the strength of olivine was 1.77×10^9 dynes/cm.^{2} at 5.5×10^9 dynes/cm.^{2} pressure, rising to 15.2×10^9 at 50×10^9 dynes/cm.^{2}. This could be interpreted as due to a coefficient of friction about 0.3. The yield was by fracture of grains; there was no evidence for recrystallization or gliding. Apparently the lower strengths indicated by isostatic adjustment are due to temperature.

Coulomb (1776) mentions both cohesion and sand, and discusses cases of no cohesion and no friction separately. It looks as if he had both the Anderson terms, but the paper is not very readable.

In our problems the choice between the Mises function and the stress-difference is not usually important, since one is proportional to the square of the other within the variation shown by crushing tests on different specimens of the same material; the crushing strengths may easily vary by a factor of 2. Hence our choice will usually be determined by convenience of calculation. The solution of the cubic equation for the principal stresses is sometimes easy, but often tiresome. The calculation of the Mises function, however, is always straightforward.

1·06. Stress-strain relation for plastic flow.

The elasticoviscous law has been successively modified by E. C. Bingham (1922), M. Reiner (1943) and J. G. Oldroyd (1947) to take account of the existence of a non-zero strength. It is supposed that the behaviour is perfectly elastic so long as the Mises function does not exceed a fixed value F_0; but if $F > F_0$ flow takes place. The deviatoric stress in flow is regarded as composed of two parts:

$$p'_{ik} = \vartheta_{ik} + 2\eta_1 e'_{ik}, \tag{1}$$

where e'_{ik} is the rate of strain, η_1 is a constant and ϑ_{ik} is a symmetrical tensor satisfying

$$\vartheta_{ik}\vartheta_{ik} = F_0. \tag{2}$$

The principles can be supposed to be that in flow part of the stress is maintaining elastic strain, limited in amount by the critical value of F for the elastic state, and the rest maintains viscous flow. The constancy of η_1 ensures that under sufficiently high values of F the rate of flow will become approximately proportional to the deforming stress. Oldroyd introduces a further hypothesis, which is equivalent to saying that the ratio p'_{ik}/e'_{ik} is the same for all components, which is reasonable. Then ϑ_{ik}/e'_{ik} is also the same for all components, and there is an A independent of i, k such that

$$\vartheta_{ik} = Ae'_{ik}. \tag{3}$$

Then
$$F_0 = \vartheta_{ik}\vartheta_{ik} = A^2 e'_{ik}e'_{ik} \tag{4}$$

and $\qquad p'_{ik} = \{2\eta_1 + (F_0/e'_{ik}e'_{ik})^{\frac{1}{2}}\}\, e'_{ik} = 2\eta_1\{1 - (F_0/p'_{ik}p'_{ik})^{\frac{1}{2}}\}^{-1}\, e'_{ik}.$ $\qquad$ (5)

The hypotheses therefore lead to a variable viscosity depending on the stress or on the rate of flow, and containing two constants of the material, η_1 and F_0. As the relation between stress and rate of strain is non-linear the solution is always difficult, and has so far been carried out only for a few specially simple cases (which would be quite difficult enough for most people). The effective viscosity tends to infinity when the Mises function tends to F_0 through larger values, so that the rule explains why the rate of flow increases more rapidly than in proportion to the stress. The improvement over the elasticoviscous rule in representing flow is so great that we may expect considerable advances to result. See also Griggs (1940), Goranson (1940).

One curious result is for the flow through a pipe under a pressure gradient. The velocity is longitudinal and must be a maximum in the centre line; hence the e_{ik} are zero there. Thus the relations for a solid hold there, and also for some distance from the centre line, and a central plug is pushed through without distortion. There is of course great distortion in the surrounding material. Such properties may explain why a contorted mass of rock sometimes contains an uncontorted inclusion, and at the same time how a fossil may have its dimensions altered in very different ratios while remaining recognizable.

The classical theory of plasticity has much in common with that just described. Many problems have been solved by its methods. A detailed account is given by R. Hill (1950). It can be regarded as a limiting case of (5) when e'_{ik} is small enough for η_1 to be neglected.

1·07. Effects of pressure and temperature on imperfections of elasticity.

Most of the experiments that have been made on imperfections of elasticity naturally concern small pressures and temperatures. The effect of high pressure has been studied by Bridgman. In general, it increases both k and μ, and also the strength. There is some conflict of opinion as to whether it favours failure by fracture or flow. In some classical experiments on rocks F. D. Adams and E. G. Coker obtained flow at high pressures in materials that were brittle at low pressures, but Griggs points out that the arrangement of the experiment forbade fracture; the broken material would have had nowhere to go. Griggs found that his rock specimens broke before they flowed at all pressures that he considered. Heat certainly favours flow, especially in impure materials, which usually have no sharp melting-point but are pasty over a considerable range of temperature.

The experimental evidence obviously cannot cover the whole ranges of temperature and pressure that are found in the Earth. What it can do is to distinguish characteristic types of behaviour, so that we can recognize them when the Earth itself calls attention to them. It should be noticed that 1·06 (5) approximates to the viscous rule when F is large compared with F_0.

In other words, a liquid can be regarded as a plastic solid whose strength is less than any shear stress likely to be applied in an actual experiment. We have already seen that it can be regarded as an elasticoviscous solid with a rigidity of the order of magnitude to be found in ordinary solids but a very low viscosity. Detailed explanations of the crystalline state and the gaseous state exist, and are well confirmed by experiment, but hitherto there has been no satisfactory explanation of the fact that for most substances there is a range of temperature such that both these states are impossible or unstable and are replaced by the compromise that we call the liquid state, in which the density and bulk-modulus are comparable with those of a solid, but the low strength and viscosity show more affinity with a gas. The most promising attempt seems to be one by M. Born and H. S. Green (1946, 1947 *a, b, c*).*

The best simple explanation of the change from gas to liquid is still van der Waals's model. The solid–liquid transition has been discussed by Alder and Wainwright (1957) and by Longuet-Higgins and Widom (1964). I am indebted to Dr T. E. Faber for the references.

For our purposes there is a great need for such a theory on account of the geological importance of glasses, which resemble crystals in having high rigidity and in their apparently considerable strength, but resemble liquids in having no obvious crystalline structure, though traces of one are revealed both for liquids and glasses by sufficiently sensitive methods. A liquid cooled sufficiently rapidly through the melting-point often fails to crystallize and passes into a highly viscous amorphous state called the *vitreous* state. Plastic sulphur affords, perhaps, the most accessible example for experiment. Its apparent consistency is similar to that of rubber, but if left to itself, its surface becomes flat, and in time it proceeds to crystallize. Shoemaker's wax and pitch are other well-known examples. Even water has been obtained in this state (L. Hawkes, 1929).† Silica and silicates are specially liable to exist in the vitreous state. It is not experimentally proved that they have any finite viscosity when cold, though presumably it might be so high as not to be revealed by any experiment in the times actually available. In a sense silica glass is stronger than any known crystal. A. A. Griffith (1920) maintained a specimen of it at room temperature for a week at an elastic extension of 5 % without detecting any flow. If there had been an extension of 0·005 % he could have detected it. This corresponds to a far greater stress-difference than could be withstood for so long by any substance in the crystalline state. Rayleigh (the younger, 1940) applied some very sensitive tests to ordinary glass and found no sign of

---

* References to other theories are given. Some features not yet discussed by them are mentioned by Jeffreys (1928*e*).

† But B. M. Cwilong (1947, p. 53) found for water down to − 60° a viscosity of the same order as at room temperature. This does not agree with Hawkes, whose observations down to − 20° seem quite definite, as did Beilby's (1921, p. 195), to − 12°.

imperfection of elasticity. In practical intervals of time, therefore, some vitreous substances may be treated as elastic solids. In general, they are less dense than the corresponding crystals. They are liable to crystallize or 'devitrify' when kept a long time, especially at temperatures not far below the melting-point. The time needed in at least some cases, however, seems to be of the order of several geological periods.

1·08. Principal rock types. The rocks exposed at the surface of the Earth fall into two main divisions, the igneous and sedimentary rocks. The latter are derived from the former, which will therefore be considered first. Igneous rocks are composed mainly of silicates and free silica. The silicates themselves are of three main types: orthosilicates, containing the quadrivalent radical SiO_4; metasilicates, containing the divalent radical SiO_3; and trisilicates, containing the quadrivalent radical Si_3O_8. Comparing them, we see that four metallic valencies correspond in an orthosilicate to one silicon atom, in a metasilicate to two, and in a trisilicate to three. The order is therefore one of increasing acidity, and can be continued by the addition of silica, the free oxide, at the end. Some of the principal rock-forming minerals and their properties are given in the table below, which is mostly extracted from F. W. Clarke's *Data of Geochemistry* (1924); G. W. Tyrrell's *Principles of Petrology* (1926); Landolt-Börnstein's *Physikalische Tabelle* (1923), and H. E. Boeke and W. Eitel's *Physikalische-chemischen Petrographie* (1923).* Only a brief summary of the vast literature of petrology can of course be attempted here, but the solution of the problems treated in the present work has not gone so far as to make greater detail necessary.

With reference to this table, quartz is the commonest natural form of silica, but at low pressures it is transformed into tridymite at about 830° and to cristobalite about 1430°, melting about 1700°. Quartz melts directly at pressures over about $1\cdot3 \times 10^9$ dynes/cm.^{2} and temperatures about 1150°. The phase diagram is given by Clark (1966, p. 362). At very high pressures it changes structure to the dense forms stichovite and coesite. Albite and anorthite form the isomorphous mixtures known as the plagioclase felspars. The common mineral augite is diopside, with part of the magnesium replaced by ferrous iron, and with some alumina (Al_2O_3) in solution. Enstatite, hypersthene and augite are known as pyroxenes; all are somewhat variable in composition and properties on account of their tendency to form isomorphous mixtures. Forsterite and fayalite mixed form the important mineral olivine. Natural specimens of this usually melt near 1400°. Muscovite and biotite are the two commonest micas. Hornblende is a mixture of the form $m\text{-}Ca(Mg, Fe)_3(SiO_3)_4 + n\text{-}CaMg_2Al_2(SiO_4)_3$. Several of the above minerals decompose at high temperatures, notably orthoclase, enstatite and the

* A very full compendium of the properties of rocks is the *Handbook of Physical Constants* of Birch, Schairer and Spicer (1942) and in the revision by S. P. Clark (1966).

	Name	Composition	Density	Melting-point (° C.)
Silica	Quartz	SiO_2	2·65	—
	Tridymite	SiO_2	2·3	—
	Cristobalite	SiO_2	2·348	1710
Trisilicates	Orthoclase	$KAlSi_3O_8$	2·56	1200(?)
	Albite	$NaAlSi_3O_8$	2·605	1100
Metasilicates	Diopside	$CaMg(SiO_3)_2$	3·275	1391
	Enstatite	$MgSiO_3$	3·1	1557(?)
	Hypersthene	$(Fe, Mg)SiO_3$	3·4–3·5	1500–1550(?)
Orthosilicates	Anorthite	$CaAl_2(SiO_4)_2$	2·765	1550
	Forsterite	Mg_2SiO_4	3·2	1890
	Fayalite	Fe_2SiO_4	4–4·14	1055–1075
	Muscovite	$KH_2Al_3(SiO_4)_3$	2·85	—
	Biotite	$KHMg_2Al_2(SiO_4)_3$	2·7	—
	Garnets	$(Ca, Mg, Fe'')_3(Al, Fe''')_2(SiO_4)_3$	3·4–4·2	—

micas and garnets, and the melting-points given are really temperatures of decomposition. The value 1557° given for enstatite corresponds to its isomer clinoenstatite; this partly decomposes on melting.

The principal rocks we shall need to consider are granite, granodiorite, diorite, syenite, basalt, eclogite, peridotite and dunite. A typical granite consists mainly of orthoclase and free quartz, with small amounts of micas. Syenite is roughly similar but does not contain free quartz. Diorite is essentially plagioclase felspar with some biotite, hornblende, or augite. In a granodiorite comparable amounts of quartz, orthoclase and plagioclase are associated. A vitreous rock similar in composition to granite is known as obsidian, and one largely glassy but containing small crystals is rhyolite.

Tachylyte, basalt, dolerite and gabbro are rocks of similar composition, tachylyte being glassy, basalt largely glassy, dolerite fine crystalline, and gabbro coarsely crystalline, the differences in texture presumably corresponding to increasing time available for crystallization. Dolerite is called diabase in American works. Tyrrell has suggested restricting the latter term to somewhat metamorphosed dolerites, but in geophysical literature it is synonymous with dolerite. The principal minerals in them are plagioclase, the pyroxenes, and often quartz or olivine (not both together) and magnetite (Fe_3O_4). Peridotites are largely composed of olivine, dunite almost wholly so.

Eclogite consists largely of garnets and jadeite. It is chemically equivalent to an olivine gabbro, according to equations of the type

$$CaAl_2(SiO_4)_2 + Mg_2SiO_4 = CaMg_2Al_2(SiO_4)_3,$$
$$\text{anorthite} \qquad \text{forsterite} \qquad \text{garnet}$$

$$CaAl_2(SiO_4)_2 + CaMg(SiO_3)_2 = Ca_2MgAl_2(SiO_4)_3 + SiO_2,$$
$$\text{anorthite} \qquad \text{diopside} \qquad \text{garnet}$$

$$NaAlSi_3O_8 = NaAl(SiO_3)_2 + SiO_2.$$
$$\text{albite} \qquad \text{jadeite} \qquad \text{silica}$$

The silica may be regarded as absorbed by the excess olivine. The interest of eclogite arises from its high density: the garnets have densities of about

3·7 and jadeite 3·34. Consequently eclogite is widely believed to be the normal form of gabbro at high pressures (L. L. Fermor, 1913).

Rocks are not constant in chemical composition. Every intermediate stage exists between a granite and a dunite. The names given represent a classification that geologists have found convenient, but rocks in the same class may differ appreciably in chemical composition.

The densities of typical rocks in the above types are: granite, 2·64; granodiorite, 2·73; diorite, 2·85; syenite, 2·78; gabbro, 2·94; dunite, 3·3; eclogite, 3·4. The general correlation of density with basicity is to be specially noticed. With regard to the glassy forms, the density of obsidian is 2·33–2·41, and that of tachylyte 2·85. More basic rocks are not known in the glassy state.

The commonest igneous rocks exposed at the surface are, in order of abundance, granite (with granodiorite), basalt (with rocks of similar composition) and andesite (a fine-grained, often largely glassy rock consisting mainly of plagioclase). Now all these rocks have at some time or other come up from some depth within the crust, unless indeed some of the granite was originally at the top, and may be expected to be representative of the material at various depths. This hypothesis will be seen to need some modification, but we shall expect the material to be more or less stratified according to density, and therefore, comparing these facts with the densities, we infer provisionally that the uppermost igneous layer in the Earth is probably granitic, with a basalt layer below it, possibly separated by a layer of diorite. However, this does not turn out to represent some of the facts most definitely revealed by seismology, and especially some denser rock such as dunite is needed at great depths.

The idea of a general granitic layer, forming the primitive upper crust of the continents, was given by E. Suess, in *Das Antlitz der Erde* (1885–1909). He based it on the widespread granitic areas that form the great continental shields, largely of pre-Cambrian ages, and on the abundance of sandstone, which implies a source from a rock containing much free quartz. The denser rocks have been intruded through the granite layer from greater depths. In addition, large granite intrusions of later age are known, formed presumably within the granitic layer itself and then driven up to the surface. The existence of sandstone is the better evidence. It is widely believed now that the old shields are mainly metamorphosed sediments. But whether that is so or not, sandstone implies a silica-rich source rock at the surface or near it.

Suess introduced the names *sal* (modified to *sial* by Wegener) and *sima* for the upper and deep-seated materials on account of the supposed preponderance of aluminium and magnesium silicates respectively. I shall not use these terms, because, although geophysical evidence leaves some latitude of interpretation, it does exclude some interpretations of *sial* and *sima* that are still to be found in geological literature. In particular *sima* is sometimes interpreted as crystalline gabbro and sometimes as tachylyte, neither of which has the mechanical properties of the matter at depths over 50 km.

None of the rocks considered likely to be prevalent near the surface has a density over 3·4 or so. The mean density of the Earth, on the other hand, is about 5·5. It has been shown that compression is unable to increase the mean density of the materials to anything like this extent, and a still denser material is indicated near the centre. Reasons will be given later for supposing this to be a core of a liquid heavy metal, probably mainly iron.

Experimental information is available about the elastic properties and strengths of the principal rock types, but is most conveniently presented in relation to seismology and to the strength of the Earth.

1·09. Structure of minerals. Much has been found out about the structure of silicates by the use of X-rays (W. L. Bragg, 1937). The universal rule among them is that each silicon atom is surrounded by four oxygen atoms, in directions towards the vertices of a regular tetrahedron. This leaves four negative valencies, which can be satisfied in two extreme ways: by sharing with another silicon atom, so that the average composition of the substance is SiO_2; or by sharing with a metallic atom, whose other valencies, if any, are also connected with oxygen atoms, so that the substance is an orthosilicate. Metasilicates and trisilicates represent mixtures of the two types of structure. The silicon-oxygen distances are always about 1·6 A. and the oxygen-oxygen distances 2·6 A. (A. = Ångström unit = 10^{-8} cm.). It appears that the SiO_4 group is practically unaltered in form and dimensions by the influence of neighbouring groups, and consequently determines the whole scale of any structure that contains it. Now a given metallic ion has its own symmetry properties, and these must be possessed by the neighbouring set of oxygen atoms; and it also has a characteristic size, though this may be more disturbed by its neighbours than is the case for the SiO_4 group. These considerations impose severe limitations on the structures possible for a given compound, and are found to explain why, for instance, there are no trisilicates containing magnesium. Size is important in a curious way. Evidently the metallic ion must not be too large to fit into the pattern. But also it must not be too small, for it carries an electric charge and would be attracted to one side or another of the cavity, thus upsetting the symmetry of the oxygen atoms about it, and thus the structure would be radically altered. The difference in form between albite and orthoclase appears to be due to the fact that the sodium ion is smaller than the potassium ion. The existence of isomorphous mixtures of albite and anorthite, which appear to have quite dissimilar chemical formulae, is explained by the fact that the calcium replaces sodium in the structure, while an extra aluminium replaces one silicon in three.

1·10. Sedimentary rocks. Igneous rocks exposed at the surface are slowly acted upon chemically by water, aided by dissolved carbon dioxide. In a granite, for instance, the orthoclase is converted into kaolin and muscovite:

$$2KAlSi_3O_8 + 2H_2O + CO_2 = 2HAlSiO_4 . H_2O + 4SiO_2 + K_2CO_3,$$
$$\text{orthoclase} \qquad\qquad\qquad \text{kaolin}$$

$$3KAlSi_3O_8 + H_2O + CO_2 = KH_2Al_2(SiO_4)_3 + 6SiO_2 + K_2CO_3.$$
$$\text{muscovite}$$

The potassium carbonate and silica are removed in solution, and the kaolin and muscovite in suspension. When redeposited they form clay. The disintegration of the quartz and micas left is mainly mechanical, and gives sand with micaceous flakes. Plagioclase is decomposed similarly. Iron may be dissolved, but if so it is usually oxidized and reprecipitated as ferric hydroxide or basic carbonate, with the texture of a clay. Typical sandstones and shales are formed from the sands and clays produced in these ways. The dissolved calcium and magnesium in the ocean are continually removed to form limestone and dolomite. There is a curious difference between the behaviours of sodium and potassium. Though they are about equally common in surface rocks, sodium is far more abundant in rivers. The reason, according to Clarke, is that potassium is strongly absorbed by clays and silts, so that much of it never gets far from its source. What reaches the sea is largely precipitated as glauconite, a substance of variable composition of the general type $KFe'''(SiO_3)_2$.

The densities of sedimentary rocks are naturally very variable. When freshly deposited they are loose and the pore space is filled with water; consequently they are much less dense than the individual particles. When they are raised to the surface they dry without much contraction and the density becomes still less. If a rock is metamorphosed by heat and pressure, pores close up. The density then increases, and in extreme cases approaches that of a crystalline material of the same composition. The following data are taken mainly from Kempe's *Engineer's Year-Book* and the *American Mechanical Engineer's Pocket-Book*:

Clay, 1·8–1·93; slate, 2·77. Shales may have any intermediate density.
Sand, 1·39–1·9; sandstone, 2·2; quartzite, 2·6 (cf. quartz, 2·65).
Portland stone (oolitic limestone), 2·37; limestone, 2·48–2·53; marble, 2·70 (cf. calcite, 2·72).

1·11. Strengths of rocks. The crushing strengths of igneous rocks at ordinary pressures and temperatures are mostly between 10^9 and 2×10^9 dynes/cm.^{2}. Those of sandstone and limestone are about half these (Birch, Schairer and Spicer, 1942, p. 116). The tensile strengths are much smaller, about 10^8 dynes/cm.^{2}, in accordance with the difference in behaviour on fracture discussed on pp. 16–17.

Rigidities at small depths are of the order of 3×10^{12} dynes/cm.^{2}; they increase greatly with depth. In geological papers the word *rigidity* usually means *strength*.

CHAPTER II

The Theory of Elastic Waves

'One, two! One, two! And through and through
The vorpal blade went snicker-snack!'
LEWIS CARROLL, *Through the Looking-Glass*

2·01. The Earth is continually undergoing deformation under the influence of stresses developed within it. Elastic and plastic deformations and also fractures occur. The first two give no sudden changes. Fracture involves a sudden change of stress at the place where it takes place, and this gives rise to elastic waves travelling through the substance. The resulting motion may be of destructive intensity at short distances, and may be large enough to be recorded by suitably designed instruments at any distance. The disturbance as a whole is an earthquake. The explanation in terms of fracture was given by Fielding Reid (1910). It seems to have been supposed previously that an earthquake was the result of a sudden blow or explosion. Reid's theory is simply that under long-continued changes the stresses grow gradually and fracture occurs when the stress-difference reaches the strength of the material. The place of origin is called the *focus*.

2·02. The elastic equations. We have had (1·03 (23)) the equations of motion of a uniform, initially unstressed isotropic elastic solid (u_i being the displacement)

$$\rho \frac{\partial^2 u_i}{\partial t^2} = (\lambda + \mu) \frac{\partial}{\partial x_i} \frac{\partial u_m}{\partial x_m} + \mu \nabla^2 u_i. \tag{1}$$

Let us look for solutions in the form of plane waves, that is, disturbances such that each component displacement is a function only of the time and the distance from some fixed plane. As the form of the equations is independent of the choice of axes we may take the fixed plane to be $x_1 = 0$. Then the displacements are independent of x_2, x_3, and

$$\frac{\partial u_m}{\partial x_m} = \frac{\partial u_1}{\partial x_1}, \quad \nabla^2 u_i = \frac{\partial^2 u_i}{\partial x_1^2}, \tag{2}$$

which also are functions of x_1 and t only. Then the equations become

$$\rho \frac{\partial^2}{\partial t^2}(u_1, u_2, u_3) = (\lambda + \mu)\left(\frac{\partial}{\partial x_1}, 0, 0\right)\frac{\partial u_1}{\partial x_1} + \mu \frac{\partial^2}{\partial x_1^2}(u_1, u_2, u_3), \tag{3}$$

that is, $\qquad \rho \dfrac{\partial^2 u_1}{\partial t^2} = (\lambda + 2\mu)\dfrac{\partial^2 u_1}{\partial x_1^2}, \quad \rho \dfrac{\partial^2}{\partial t^2}(u_2, u_3) = \mu \dfrac{\partial^2}{\partial x_1^2}(u_2, u_3).$ \hfill (4)

These have the form of the equation of wave propagation, and if we write

$$\rho \alpha^2 = \lambda + 2\mu, \quad \rho \beta^2 = \mu, \tag{5}$$

the solutions are

$$u_1 = f_1(x_1 - \alpha t) + g_1(x_1 + \alpha t),$$
$$u_2 = f_2(x_1 - \beta t) + g_2(x_1 + \beta t),$$
$$u_3 = f_3(x_1 - \beta t) + g_3(x_1 + \beta t),$$

(6)

where the f's and g's may be any twice differentiable functions. If we imagine g_1 to be zero, and we increase t by τ and x_1 by $\alpha\tau$, we do not alter u_1, so that after an interval τ every value of u_1 reappears at a place where x_1 is increased by $\alpha\tau$; hence f_1 represents a wave travelling with velocity α in the direction of x_1 increasing. Similarly, g_1 represents a wave travelling with velocity α in the direction of x_1 decreasing. Similar considerations apply to u_2 and u_3, the characteristic velocity being β instead of α. The variations of the displacements in the directions of the three axes are independent. The displacement corresponding to f_1 is in the direction of travel of the wave, which is therefore called *longitudinal*. That corresponding to f_2 or f_3 is perpendicular to the direction of travel, and the wave is therefore called *transverse*. Since $\alpha > \beta$ for actual materials, the longitudinal waves are also called *primary* and the transverse ones *secondary*, usually abbreviated to P and S, following G. v. dem Borne (1904) and E. Wiechert. H. H. Turner very appropriately called them the *push* and the *shake*—though P is as often as not a *pull*!

As the u_2 and u_3 motions are independent but travel with the same velocity, we may include them both under the description S, but S shows polarization. It is often convenient to distinguish whether the displacement in an S wave is horizontal or in a vertical plane through the direction of propagation; in the former case it is called an SH wave, in the latter an SV wave. The two types are reflected differently at an interface between different materials.

The existence of the longitudinal and transverse elastic waves was first predicted theoretically by Poisson (1829, 1831). He took $\lambda = \mu$, making $\alpha/\beta = \sqrt{3}$. This case is of theoretical interest, and is near the truth for much of the Earth.

The first longitudinal motion is often called a compression when it is outwards, a dilatation when it is inwards. This is presumably intended to refer to the places being reached; if the whole region between the focus and the station was considered, the meanings would be reversed. The International Seismological Summary used a for outwards and upwards, k for downwards and inwards (a for anaseism, k for kataseism). These are unambiguous. Unfortunately the *Bulletin of the International Seismological Centre* has reverted to C and D, thus restoring the ambiguity; a and k were, I think, suggested by F. J. W. Whipple.

The independence of the P and S movements can be seen also without the restriction to plane waves. If we put

$$\Delta = \frac{\partial u_m}{\partial x_m}, \quad 2\xi_1 = \frac{\partial u_3}{\partial x_2} - \frac{\partial u_2}{\partial x_3}, \quad 2\xi_2 = \frac{\partial u_1}{\partial x_3} - \frac{\partial u_3}{\partial x_1}, \quad 2\xi_3 = \frac{\partial u_2}{\partial x_1} - \frac{\partial u_1}{\partial x_2}, \quad (7)$$

we easily derive from (1) the equations

$$\rho \frac{\partial^2 \Delta}{\partial t^2} = (\lambda + 2\mu) \nabla^2 \Delta, \quad \rho \frac{\partial^2 \xi_i}{\partial t^2} = \mu \nabla^2 \xi_i,$$

so that the dilatation Δ and the three components of rotation ξ_i separately satisfy the wave equation, with the characteristic velocities α and β. An immediate consequence is that a local disturbance gives rise to disturbances travelling radially outwards with velocities α and β.

The ratio α/β is of direct interest. It is customary to state Poisson's ratio σ instead. The relation is

$$\sigma = \frac{\alpha^2 - 2\beta^2}{2(\alpha^2 - \beta^2)}.$$

For an incompressible substance or one of zero rigidity $\lambda/\mu = \infty$, $\alpha/\beta = \infty$, $\sigma = \frac{1}{2}$. For a Poisson substance $\lambda = \mu$, $\alpha/\beta = \sqrt{3}$, $\sigma = \frac{1}{4}$. σ has no direct application in seismology and is an inconvenience. However, a principle of much application in geophysics seems to be: if there are two usages, choose the less convenient.

2·021. The equations (1) have been drastically simplified by the assumptions of uniformity and of absence of initial stress. These departures from the actual conditions are not usually important, though they rise into importance in special circumstances. So long as we consider only regions in which the variations of properties are small we can use the relations for a homogeneous medium. Then the mass as a whole can be treated by imagining it divided into such elements and treating them separately. A sufficient condition that this procedure may lead to a good approximation turns out to be that the quantities ρ, λ, μ change only by small fractions of their mean values within a wave-length.

The inner parts of the Earth are under considerable stress from the weight of the matter resting on them. The displacements from equilibrium are therefore not from an unstressed state, and some attention should be paid to the effect of this on the equations of motion. The stresses far exceed any stress-difference that is suggested by any information that we have about the strength, and consequently we may suppose that the equilibrium stress is approximately hydrostatic. We may also suppose that *changes* of displacement and stress are related as for those of an isotropic solid from equilibrium. With this hypothesis the elastic equations can be found without much difficulty.

If we use index o to indicate the equilibrium state, the equations 1·02 (1) reduce to

$$0 = \rho^o X_i^o + \frac{\partial p_{ik}^o}{\partial x_i}. \quad (1)$$

Also the X_i are gravitational accelerations derived from a potential U. Hence in the equilibrium state we can take

$$0 = \rho^o \frac{\partial U^o}{\partial x_i} + \frac{\partial p}{\partial x_i}. \qquad (2)$$

It follows that p, and therefore ρ^o, are constant over surfaces of constant U^o. Hence if there is an equation of state such that the density is a definite function of the pressure, equilibrium is possible. Then if we subtract (2) from $1 \cdot 02(1)$ we are left with three equations connecting the acceleration with changes of ρ, X_i and p_{ik} from the equilibrium state, and therefore linear in the changes when these are small. Also if the original state is hydrostatic, and the stress-strain relation is isotropic for departures from it, the relation between changes of stress and strain keeps the same form as for initial absence of stress. However, ρ, λ and μ are no longer constant, partly because the composition of the Earth is not uniform, partly because a large hydrostatic pressure leads to considerable variations of properties in large regions. But in certain conditions that does not matter much. This can be seen physically as follows. We can imagine the region divided up into small parts such that the variables ρ, λ, μ do not vary within any part by more than given small fractions of their mean values, and then replace the actual value at each point by the mean value in the chosen neighbourhood. Then within each part the waves, and in particular the front of any sudden disturbance, travel with the velocities α, β characteristic of that part. Consequently, if a sudden disturbance takes place at P, its time of travel to another point Q is $\Sigma \delta s/c$, where δs is the distance travelled within any region, c is the velocity of travel in that region, and the summation is for all regions traversed. Hence if we are given that the wave is of P or S type throughout the path, and we take the region divided into indefinitely small parts, we see that P cannot arrive at Q in time less than $\int ds/\alpha$ taken from P to Q along the path that makes this integral least, and similarly S cannot arrive in time less than $\int ds/\beta$ along an analogous path. Further, the waves actually will arrive at these times because we could consider a set of parts of the region including the paths of minimum time. Thus even in a nonuniform body there is a characteristic time of travel of a sudden disturbance, determined by the local values of α or β according as the disturbance was originally of P or S type. The rule satisfied by the size of the disturbance is more difficult and must be postponed until we have attended to problems of refraction.

In mathematical terms the rule that the disturbance is propagated with the local velocity α or β follows at once from the fact that the rate of travel depends wholly on the coefficients of the second derivatives of u_i in the equations of motion. In other words, the velocity of a sudden disturbance is the limiting value of that of a disturbance periodic in time when the period tends to zero.

We shall generally speak of a periodic or approximately periodic disturbance as a *wave*, and of a disturbance starting at a definite instant as a *pulse* or *phase*.

There is a general analogy between the propagation of elastic waves and that of light waves, and the notions of rays and wave fronts arise in both in the same way. The wave fronts are the loci of points reached simultaneously by a pulse of given type, and cut the rays at right angles. Refraction, reflexion, diffraction and dispersion all occur and are of considerable importance.

2·022. Reflexion at a free surface. We take the coordinates to be (x, y, z), the free surface being the plane $z = 0$ and the medium on the side of positive z. We denote the corresponding components of displacement by u, v, w, and suppose all independent of y. The equations of motion are then

$$\rho \frac{\partial^2}{\partial t^2}(u, v, w) = (\lambda + \mu)\left(\frac{\partial}{\partial x}, \frac{\partial}{\partial y}, \frac{\partial}{\partial z}\right)\Delta + \mu \nabla^2(u, v, w). \tag{1}$$

The boundary condition is that there is no stress over the surface $z = 0$; hence at $z = 0$

$$p_{zx} = \mu\left(\frac{\partial w}{\partial x} + \frac{\partial u}{\partial z}\right) = 0, \tag{2}$$

$$p_{zy} = \mu\left(\frac{\partial w}{\partial y} + \frac{\partial v}{\partial z}\right) = 0, \tag{3}$$

$$p_{zz} = \lambda\Delta + 2\mu\frac{\partial w}{\partial z} = 0. \tag{4}$$

If we take ϕ, ψ independent of y, and

$$u = \frac{\partial \phi}{\partial x} + \frac{\partial \psi}{\partial z}, \quad w = \frac{\partial \phi}{\partial z} - \frac{\partial \psi}{\partial x}, \quad v = 0, \tag{5}$$

the first and third equations of motion reduce to

$$\rho \frac{\partial^2}{\partial t^2}\left(\frac{\partial \phi}{\partial x} + \frac{\partial \psi}{\partial z}\right) = (\lambda + \mu)\frac{\partial}{\partial x}\nabla^2\phi + \mu \nabla^2\left(\frac{\partial \phi}{\partial x} + \frac{\partial \psi}{\partial z}\right), \tag{6}$$

$$\rho \frac{\partial^2}{\partial t^2}\left(\frac{\partial \phi}{\partial z} - \frac{\partial \psi}{\partial x}\right) = (\lambda + \mu)\frac{\partial}{\partial z}\nabla^2\phi + \mu \nabla^2\left(\frac{\partial \phi}{\partial z} - \frac{\partial \psi}{\partial x}\right). \tag{7}$$

Hence these equations are satisfied if

$$\rho \frac{\partial^2 \phi}{\partial t^2} = (\lambda + 2\mu)\nabla^2\phi, \quad \rho \frac{\partial^2 \psi}{\partial t^2} = \mu \nabla^2\psi. \tag{8}$$

v does not affect these equations if it is independent of y; and if Δ and v are independent of y

$$\rho \frac{\partial^2 v}{\partial t^2} = \mu \nabla^2 v. \tag{9}$$

The boundary conditions reduce to

$$\frac{\partial u}{\partial z} + \frac{\partial w}{\partial x} = 2\frac{\partial^2 \phi}{\partial z \partial x} + \frac{\partial^2 \psi}{\partial z^2} - \frac{\partial^2 \psi}{\partial x^2} = 0, \tag{10}$$

$$\mu \frac{\partial v}{\partial z} = 0, \tag{11}$$

$$\lambda \Delta + 2\mu \frac{\partial w}{\partial z} = \lambda \nabla^2 \phi + 2\mu \left(\frac{\partial^2 \phi}{\partial z^2} - \frac{\partial^2 \psi}{\partial z \partial x} \right) = 0. \tag{12}$$

We propose to consider incident waves of various types and see whether we can satisfy the equations (10), (11) and (12). We see that (10) and (12) contain only ϕ and ψ and give motions parallel to the plane $y = 0$; (11) contains only v and gives motion perpendicular to $y = 0$. Thus the (ϕ, ψ) and v motions are independent. We take free surfaces and interfaces as horizontal unless the contrary is stated.

Incident P wave. We first take the incident wave to be given by

$$\phi = A \exp[i\kappa(x - z\tan e - ct)]$$

(where we can take the real or the imaginary part after solution). Then there may be reflected waves of both P and S types, and we assume

$$\phi = A \exp[i\kappa(x - z\tan e - ct)] + A_1 \exp[i\kappa(x + z\tan e - ct)], \tag{13}$$

$$\psi = B_1 \exp[i\kappa(x + z\tan f - ct)]. \tag{14}$$

If we did not take the coefficients of x and t the same in all waves, it would not be possible to satisfy the conditions at $z = 0$ for all values of x, t. Then from (8) we find

$$c^2 = \alpha^2 \sec^2 e = \beta^2 \sec^2 f, \tag{15}$$

whence

$$\frac{\cos e}{\alpha} = \frac{\cos f}{\beta}. \tag{16}$$

e is known as the *angle of emergence*, the complement of the angle of incidence, and $\frac{1}{2}\pi - f$ is the *angle of reflexion* for S waves. Then the boundary conditions give

$$\lambda(A + A_1)\sec^2 e + 2\mu \tan^2 e(A + A_1) - 2\mu B_1 \tan f = 0, \tag{17}$$

$$-2(A - A_1)\tan e + B_1(\tan^2 f - 1) = 0. \tag{18}$$

Using (15) we have

$$A + A_1 = \frac{2\tan f}{\tan^2 f - 1} B_1, \quad A - A_1 = \frac{\tan^2 f - 1}{2\tan e} B_1, \tag{19}$$

and

$$\frac{A - A_1}{A + A_1} = \frac{(\tan^2 f - 1)^2}{4\tan e \tan f}. \tag{20}$$

If e is a real angle, $\sec^2 f > \sec^2 e$ and f also is a real angle. Hence there is always a solution. An incident P wave gives both a reflected P and a reflected S. This complication does not occur in light or sound, but the relation (16) between e and f is an adaptation of the law of refraction to this kind of composite reflexion.

Note that $A_1 = A$ only if $\tan^2 f = 1$, which would imply $\sec^2 e = \frac{2}{3}$ if $\alpha = \beta \sqrt{3}$. Thus with ordinary values of α/β we cannot have $A_1 = A$. We can have $A_1 = -A$, namely, if $\tan e = 0$, $\tan f = 0$, or $\tan f = \infty$. $\tan e = 0$ corresponds to grazing incidence, $\tan f = \infty$ to normal incidence. $\tan f = 0$ would give $\sec^2 e = \frac{1}{3}$, which is impossible.

We can also have $A_1 = 0$, so that there is no reflected P at all. In the case $\alpha/\beta = \sqrt{3}$ this leads to the equation, with $\tan e = \xi$,

$$81\xi^8 + 108\xi^6 + 6\xi^4 - 20\xi^2 + 1 = 0. \tag{21}$$

This has two positive roots (in ξ^2), namely, $\frac{1}{3}$ and $0\cdot051$. This condition cannot arise with all values of α/β; even with $\alpha/\beta = 1\cdot8$ the reflected P never disappears for real ξ.

The surface movement is given by

$$\left. \begin{aligned} u &= (A + A_1 + B_1 \tan f) \exp[i\kappa(x - ct)], \\ w &= -\{(A - A_1)\tan e + B_1\} \exp[i\kappa(x - ct)]. \end{aligned} \right\} \tag{22}$$

We speak of an *apparent angle of emergence* $\bar{e}$ such that

$$\tan \bar{e} = -\frac{w}{u}, \tag{23}$$

and we find

$$\tan \bar{e} = -\frac{1 - \tan^2 f}{2 \tan f} = -\cot 2f, \tag{24}$$

$$\bar{e} = 2f - \tfrac{1}{2}\pi. \tag{25}$$

This relation makes it theoretically possible, given the ratio of the observed horizontal and vertical displacements at the surface, to calculate e. In practice, however, the amplitudes cannot be read with great accuracy, and the structure near the surface introduces other complications.

If $e = 0$, $B_1 = 0$, $A + A_1 = 0$ and $u = w = 0$. Hence a P wave at grazing incidence would produce no motion at the boundary. The conditions cannot be realized physically, but the crucial point is seen if we consider small non-zero values of e, the incident wave keeping the same amplitude. The smaller e is, the more closely the displacements given by the reflected waves at the boundary cancel those given by the incident wave.

Incident S V wave. We take $z = 0$ horizontal, and

$$\psi = B \exp[i\kappa(x - z\tan f - ct)] + B_1 \exp[i\kappa(x + z\tan f - ct)], \tag{26}$$

$$\phi = A_1 \exp[i\kappa(x + z\tan e - ct)], \tag{27}$$

and proceed similarly. We find

$$B + B_1 = -\frac{2\tan e}{\tan^2 f - 1} A_1, \quad B - B_1 = -\frac{\tan^2 f - 1}{2\tan f} A_1, \tag{28}$$

$$\frac{B - B_1}{B + B_1} = \frac{(\tan^2 f - 1)^2}{4\tan e \tan f}. \tag{29}$$

This ratio is identical with (20) for the same e. We get complete reflexion as SV if $\tan e = 0$ or ∞, corresponding to $\sec f = \alpha/\beta$ or ∞. It can likewise happen that there is no reflected SV wave, corresponding to the same two values of e as are given by (21).

If, however, $\cos f > \beta/\alpha$, (16) gives $\cos e > 1$ and e is imaginary. This is a new feature. The coefficient of z in the exponent in (27) is now real, and the reflected P is no longer a harmonic wave travelling downwards. If the coefficient were positive ϕ would increase exponentially with depth. Hence we must take

$$\tan e = i \tanh h, \tag{30}$$

where h is real and positive, and $\cosh h = (\alpha/\beta) \cos f$. In this case $(B+B_1)/A_1$ is imaginary, $(B-B_1)/A_1$ is real, and therefore iB/A_1 and iB_1/A_1 are conjugate complexes. Hence if f is small enough to make $\cos f > \beta/\alpha$ the reflected SV is equal in amplitude to the incident one, but differs in phase; and the reflected P is not a harmonic wave but a movement confined to the neighbourhood of the surface. This condition may be called *total reflexion in the original type*.

The ratio w/u defines, for e real,

$$\tan \bar{f} = -\frac{w}{u} = \frac{2 \tan e}{1 - \tan^2 f}. \tag{31}$$

For e imaginary, u and w are out of phase, and surface particles describe elliptic paths.

Incident SH wave. We take

$$v = C \exp\left[i\kappa(x - z \tan f - ct)\right] + C_1 \exp\left[i\kappa(x + z \tan f - ct)\right]. \tag{32}$$

Then the boundary condition is simply

$$\frac{\partial v}{\partial z} = 0, \tag{33}$$

whence $C_1 = C$. The reflected wave is equal in amplitude to the incident wave, and the surface amplitude is $2C$.

Energy transmission. Consider a P wave approaching the surface at angle of emergence e; take the real part

$$\phi = A \cos \kappa(x - z \tan e - ct) = A \cos \kappa\theta \quad \text{say}. \tag{34}$$

Then
$$u = -\kappa A \sin \kappa\theta, \quad w = \kappa A \tan e \sin \kappa\theta, \tag{35}$$

$$\dot{u} = \kappa^2 cA \cos \kappa\theta, \quad \dot{w} = -\kappa^2 cA \tan e \cos \kappa\theta. \tag{36}$$

The kinetic energy per unit volume is $\tfrac{1}{2}\rho(\dot{u}^2 + \dot{w}^2)$, and its mean value is $\tfrac{1}{4}\rho\kappa^4 c^2 A^2 \sec^2 e$. In a harmonic motion the mean potential and kinetic energies are equal: hence the mean energy per unit volume is $\tfrac{1}{2}\rho\kappa^4 c^2 A^2 \sec^2 e$. The waves advance at rate α, and the energy on unit area of the wave front is spread over an area $\operatorname{cosec} e$ of the surface. Hence we may regard the wave as bringing energy to the surface at a rate $\tfrac{1}{2}\rho\kappa^4 c^2 \alpha A^2 \sec^2 e \sin e$ per unit time per unit area. The reflected P removes energy at a rate $\frac{1}{2}\rho\kappa^4 c^2\alpha A_1^2\sec^2 e\sin e$, the reflected S at $\frac{1}{2}\rho\kappa^4 c^2\beta B_1^2\sec^2 f\sin f$. Since there is no stress across the free surface, no work is being done at the surface, and we should expect that the supply of energy to the surface by the incident P would just balance that removed by the reflected P and S; that is, that

$$\alpha(A^2-A_1^2)\sec^2 e\sin e = \beta B_1^2\sec^2 f\sin f. \tag{37}$$

From (19) by multiplication

$$A^2-A_1^2 = \frac{\tan f}{\tan e}B_1^2, \tag{38}$$

and from (16)
$$\alpha\sec e = \beta\sec f. \tag{39}$$

Hence (37) follows.

Similar considerations apply to an incident SV wave when $\tan e$ is real. If $\tan e$ is imaginary the existence of the surface P movement introduces a modification. In a developed harmonic movement this gives only a constant amount of energy near the surface and needs no supply of new energy; and the energy balance can be examined by considering the supply to a surface where z is large enough for the P movement to be negligible. But at such a surface the movement consists of the incident and reflected SV waves, which are equal in amplitude though not in phase. Hence the energy balance still holds, the superficial P movement not appearing in it.

2·023. Reflexion and refraction at a general interface.

Incident P and SV waves. We suppose the waves approaching from positive z and take the interface to be $z=0$. We use A to denote coefficients in ϕ, B for coefficients in ψ, suffix 1 for reflected waves, accents for refracted waves. Then we have, if all e, f are real,

$$\phi = A\exp[i\kappa(x-z\tan e-ct)]+A_1\exp[i\kappa(x+z\tan e-ct)], \tag{1}$$

$$\psi = B\exp[i\kappa(x-z\tan f-ct)]+B_1\exp[i\kappa(x+z\tan f-ct)], \tag{2}$$

$$\phi' = A'\exp[i\kappa(x-z\tan e'-ct)], \tag{3}$$

$$\psi' = B'\exp[i\kappa(x-z\tan f'-ct)]. \tag{4}$$

The conditions that u, w, p_{zx}, p_{zz} must be continuous at $z=0$ give

$$A+A_1-\tan f.(B-B_1) = A'-\tan f'.B', \tag{5}$$

$$\tan e.(A-A_1)+(B+B_1) = \tan e'.A'+B', \tag{6}$$

$$\rho\beta^2\{(\tan^2 f-1)(A+A_1)+2\tan f.(B-B_1)\}$$
$$= \rho'\beta'^2\{(\tan^2 f'-1)A'+2\tan f'.B'\}, \tag{7}$$

$$\rho\beta^2\{-2\tan e.(A-A_1)+(\tan^2 f-1)(B+B_1)\}$$
$$= \rho'\beta'^2\{-2\tan e'.A'+(\tan^2 f'-1)B'\}. \tag{8}$$

These have to be solved numerically. The energy equation

$$\rho\{\tan e.(A^2-A_1^2)+\tan f.(B^2-B_1^2)\} = \rho'\{\tan e'.A'^2+\tan f'.B'^2\} \tag{9}$$

is available as a check. We are usually taking only one incident wave, so that A and B do not both arise.

If some of e, f, e', f' are imaginary we must replace the corresponding tangents by i times positive hyperbolic tangents, the sign being taken so that the disturbance considered tends to zero away from the interface, and therefore $+i$ for both reflected and transmitted disturbances. The corresponding terms disappear from the energy equation and the others must be replaced by their moduli.

Numerous cases have been worked out in detail (Knott, 1899; Wiechert, 1907; Jeffreys, 1926a; Gutenberg, 1944). The results are in some cases very remarkable. Thus for the reflexion of P waves at a free surface, with $\alpha/\beta = \sqrt{3}$, at least half the energy goes into the reflected S wave for angles of emergence between 12° and 63°. For normal and grazing incidence there is complete reflexion in the original types. When an SV wave is reflected, more than half the energy goes into the reflected P if the angle of emergence is between 55° and 75° and the reflected S disappears altogether for angles of emergence of 60° and 55° 44'; but for any angle of emergence less than 54° 44' there is complete reflexion in the original type. SH waves at a free surface are of course always completely reflected in the original type.

The behaviour of waves for incidence on an interface is so varied as to defy general description. It is convenient to describe a P derived from an S, or an S from a P, as a transformed wave. The nearest approach to a general rule is that the transmitted transformed wave is usually small so long as the two ratios $\alpha/\beta, \alpha'/\beta'$ do not differ greatly.

If one of the media is a liquid, it can transmit P only. A large P wave can in this case be derived from an SV incident on a surface between a solid and a liquid.

Incident SH wave. We take

$$v = C \exp\left[i\kappa(x - z\tan f - ct)\right] + C_1 \exp\left[i\kappa(x + z\tan f - ct)\right] \quad (z > 0), \quad (10)$$

$$v = C' \exp\left[i\kappa(x - z\tan f' - ct)\right] \qquad\qquad (z < 0). \quad (11)$$

The continuity conditions are

$$C + C_1 = C', \tag{12}$$

$$\mu \tan f . (C - C_1) = \mu' \tan f' . C'. \tag{13}$$

If f' is real, C' is real; if f' is imaginary there is total reflexion with a change of phase.

It is possible in this case to have complete transmission, $C_1 = 0$. Then we must have

$$\mu \tan f = \mu' \tan f'. \tag{14}$$

But

$$\beta \sec f = \beta' \sec f'. \tag{15}$$

Suppose $\beta' > \beta$. Then $f' = 0$ is possible and, for small f', $\mu \tan f > \mu' \tan f'$. If $\tan f'$ is large, on the other hand,

$$\tan f \doteq \frac{\beta'}{\beta} \tan f', \tag{16}$$

$$\mu \tan f \doteq \frac{\mu \beta'}{\beta} \tan f' < \mu' \tan f', \tag{17}$$

if $\mu \beta' < \mu' \beta$, i.e. if $\rho \beta < \rho' \beta'$. This condition is usually satisfied if $\beta < \beta'$. Hence as a rule $\mu \tan f = \mu' \tan f'$ for some intermediate value of f' and complete transmission is possible.

Refraction into a liquid involves a different boundary condition; the displacement can be discontinuous and the shear stress is zero.

2·024. Interface between slightly different media. We suppose that the differences in density and elastic properties are small quantities of the first order. We also suppose that the amplitude of the transmitted wave differs from that of the incident wave of the same type by a small quantity of the first order, and that the amplitudes of the other derived waves are all small quantities of the first order. On these hypotheses we can convert 2·023 (5), (6), (7) and (8) into equations for $A' - A$, $B' - B$, A_1, B_1, the coefficients of A and B on the right being small of the first order. The determinant of the coefficients on the left is found to be $4\mu^2 \tan e \tan f \sec^4 f$,* and is therefore not small unless either e or f is small. Hence the hypothesis made about the amplitudes of the derived waves is correct, except possibly in these two cases. It then follows that if the original wave is a P, the reflected P and S and the transmitted S take away energy at a rate that is of the second order of small quantities, and therefore the energy of the transmitted P differs from that of the incident P by a second-order quantity. Similarly if the incident wave is an S, all the energy except a second-order small quantity goes into the transmitted S. Similar considerations apply to SH.

If e or f is small these conclusions do not follow. On examination it is found that for an incident P, A_1 need not be small, and for an incident S, B_1 may not be small. These results were to be expected because in these conditions the transmitted wave is on the verge of being replaced by one that dies away exponentially at a distance from the boundary, and its energy is therefore largely or wholly transferred to the reflected wave of the original type. We may therefore infer that if e' or f' becomes imaginary the energy of the incident wave is nearly all transferred to the reflected wave of the original type.

2·025. Transmission through a continuously varying medium. If the medium has properties varying continuously from place to place we can regard it as replaced by a succession of n uniform layers, where n is

---

* Cf. Jeffreys (1926a, p. 333), where $\alpha/\beta = \sqrt{3}$ is assumed.

large enough for the differences between consecutive layers with respect to any property to be of the order of $1/n$ of the whole variation. If a wave enters the region at an angle such that neither e nor f becomes zero in the region, two cases may arise. If the total thickness is small compared with the wave-length, the derived waves arriving at any place will be in nearly the same phase and their amplitudes will add up. Hence the total amplitude of each may not be a small fraction of that of the incident wave, and the loss of energy to the derived waves may be considerable. Thus we infer that if a considerable change of properties occurs within a wave-length an incident wave may lose a considerable fraction of its energy to the derived waves, the extreme case being that of a sharp change at a single interface. On the other hand, if the total change is spread through a thickness large compared with the wave-length, in such a way that there is never a great change within a wave-length, the reflected waves will reach a given point in phases spread over many periods, and the total amplitude of each will be of the order of that from only one of the interfaces. Then the loss of energy to the reflected waves will be of the order of $1/n^2$ of the whole. For an incident P, the transmitted SV waves might apparently arrive at a given point in nearly the same phase and add up. Special investigation is desirable to determine how important this transformation and that from SV to P are likely to be. Probably it will not be great because even for a sharp interface it seems to be small in most cases.

If the original angle of entry is such that e or f vanishes within the region the conditions are altered. In that case at the interface where this takes place the wave will be completely reflected in the original type and will emerge on the side where it entered, carrying nearly all its original energy. In the optical case this corresponds to the formation of mirage. These effects have now been examined (Jeffreys, 1957 e). So long as the change of properties in a wave-length is small, the separation of P, SV and SH types remains very accurate. Even in the case of total internal reflexion there is no appreciable transformation of P to SV or SV to P. There is, however, a phase shift of $\tfrac{1}{2}\pi$ in comparison with the time indicated by the elementary theory. This has been found for electromagnetic waves by D. R. Hartree (1931).

For a pulse internally reflected in a layer of low velocity the theory applies only to the shorter wave components; the longer ones may be largely transmitted, and thus a type of dispersion may arise, converting a simple pulse into an oscillating train. It would, however, be premature to accept this as an explanation of oscillatory movement in seismograms.

An equivalent problem of waves in a string of varying thickness was solved by Rayleigh (1894, 1, 235–239). If for normal incidence the velocity varies linearly from c_1 to c_2 in a distance l, and the period is $2\pi/\gamma$, we write

$$p = \frac{\gamma l}{c_1 - c_2}, \quad m = \sqrt{(p^2 - \tfrac{1}{4})}, \quad m' = \sqrt{(\tfrac{1}{4} - p^2)}, \quad \alpha = c_2/c_1.$$

Then the ratio of the reflected and incident amplitudes is

$$\left|\frac{A_1}{A}\right| = \frac{|\sinh(m'\log\alpha)|}{\{\sinh^2(m'\log\alpha)+4m'^2\}^{\frac{1}{2}}} \quad m' \text{ real,}$$

$$= \frac{|\sin(m\log\alpha)|}{\{4m^2+\sin^2(m\log\alpha)\}^{\frac{1}{2}}} \quad m \text{ real.}$$

The ratio reduces to $\left|\dfrac{c_2-c_1}{c_2+c_1}\right|$ for l small; as l increases, it decreases until $m\log\alpha = \pi$, when it vanishes, and it is small for larger m. If α is near 1 the corresponding value of l approximates to half the wave-length. Hence the presence or absence of reflexions, if αA would be large enough to be observed, may be taken as an indication that the thickness of the layer of transition is less or more than half the wave-length.

The conclusion is that so long as the properties do not vary greatly within a wave-length the energy can be regarded as transmitted along the rays, in the sense that if we take a cone of rays the rate of transmission of energy across any cross-section of the cone is the same.

The argument does not assume all interfaces to be parallel, and therefore can be applied to waves in a sphere, travelling along rays inclined to the radii.

2·03. Surface waves. In addition to the P and S waves, a solid with a free surface can transmit two kinds of surface waves, which are confined to the neighbourhood of the surface and give little movement at great depths, like gravity waves on deep water. These are specially important at great distances, for the following reason. A pulse travelling symmetrically outwards from a point in three dimensions gives a displacement proportional to $1/r$, r being the distance. But if the disturbance is confined to a given depth h, its energy over a circle at distance r is distributed over an area $2\pi rh$, and therefore the amplitude will vary as $r^{-\frac{1}{2}}$. Hence even though the body waves may be the more prominent at short distances the surface waves may be the larger at great distances, and this is actually found to happen.

2·031. Rayleigh waves. We return to 2·022 (5) and assume

$$\phi = A(z)\exp[i\kappa(x-ct)], \quad \psi = B(z)\exp[i\kappa(x-ct)]. \tag{1}$$

Then from 2·022 (8)

$$-\kappa^2c^2A = \alpha^2\left(\frac{d^2A}{dz^2}-\kappa^2A\right), \quad -\kappa^2c^2B = \beta^2\left(\frac{d^2B}{dz^2}-\kappa^2B\right), \tag{2}$$

whence $\qquad A = Ce^{-rz}, \quad B = De^{-sz}, \tag{3}$

where C, D are constant and r, s are the positive values satisfying

$$\frac{r^2}{\kappa^2} = 1-\frac{c^2}{\alpha^2}, \quad \frac{s^2}{\kappa^2} = 1-\frac{c^2}{\beta^2}. \tag{4}$$

A condition for the existence of a surface wave is therefore that c^2 shall be less than β^2 and therefore less than α^2. If these conditions are not satisfied ϕ or ψ or both will not represent a surface wave but a wave travelling upwards or downwards.

The boundary conditions 2·022 (10), (12) now become

$$-2i\kappa r C + (s^2 + \kappa^2) D = 0, \tag{5}$$

$$\{-\lambda \kappa^2 + (\lambda + 2\mu) r^2\} C + 2\mu i \kappa s D = 0. \tag{6}$$

Since

$$\lambda + 2\mu = \rho \alpha^2, \quad \mu = \rho \beta^2,$$

the condition of consistency is

$$\left(2 - \frac{c^2}{\beta^2}\right)^2 = 4 \left(1 - \frac{c^2}{\alpha^2}\right)^{\frac{1}{2}} \left(1 - \frac{c^2}{\beta^2}\right)^{\frac{1}{2}}, \tag{7}$$

the positive signs being taken for the roots. Squaring, we have

$$\frac{c^2}{\beta^2} \left\{ \frac{c^6}{\beta^6} - 8 \frac{c^4}{\beta^4} + \frac{c^2}{\beta^2} \left(24 - 16 \frac{\beta^2}{\alpha^2}\right) - 16 \left(1 - \frac{\beta^2}{\alpha^2}\right) \right\} = 0. \tag{8}$$

If $c = 0$, the motion is not periodic in t. Hence we have $r = s = \kappa$, $C = -iD$; and then substituting in the expressions u and w we find $u = w = 0$. Hence this solution does not give even a standing wave. This is otherwise obvious, for $c = 0$ would mean a static distortion of an elastic solid under no forces and thus would violate the conservation of energy.

The other factor is negative for $c = 0$. For $c^2 = \beta^2$ it is $+1$. Hence there is always a root between 0 and β. It makes both sides of (7) positive and is therefore not a root of the equation obtained by reversing the sign of the right side of (7). Hence so long as $\beta < \alpha$ there can always be surface waves; the velocity is less than that of S. There are not three zeros of the second factor for c between 0 and β. For if there were, the second derivative with regard to c^2/β^2 would vanish somewhere in this interval; but we find easily that it vanishes at $c^2/\beta^2 = \frac{8}{3}$ and nowhere else. Hence precisely one value of c gives a surface wave.

When the solid is incompressible $\alpha = \infty$, and the root is

$$c^2/\beta^2 = 0\cdot 91275, \tag{9}$$

whence, if we take the real part of the solution,

$$c = 0\cdot 9554\beta, \quad r = \kappa, \quad s = 0\cdot 2954\kappa, \tag{10}$$

$$u = E(\mathrm{e}^{-\kappa z} - 0\cdot 5433\, \mathrm{e}^{-sz}) \sin \kappa(x - ct), \tag{11}$$

$$w = E(\mathrm{e}^{-\kappa z} - 1\cdot 840\, \mathrm{e}^{-sz}) \cos \kappa(x - ct), \tag{12}$$

E being a constant. The motion at the surface is given by

$$u = 0\cdot 4567 E \sin \kappa(x - ct), \quad w = -0\cdot 840 E \cos \kappa(x - ct). \tag{13}$$

The particles move in ellipses, the amplitude of the vertical movement being about 1·9 times that of the horizontal. $\partial u/\partial t$ has the same sign as w, and w is measured downwards. Hence in a wave travelling from left to right the particles move counter-clockwise. (In a surface wave on water they move clockwise.)

The other two roots are complex:

$$c^2/\beta^2 = 3 \cdot 5436 \pm 2 \cdot 2301i \tag{14}$$

and $\qquad r = \kappa, \quad 4s/\kappa = (2 - c^2/\beta^2)^2 = -2 \cdot 7431 \pm 0 \cdot 8846i. \tag{15}$

Thus e^{-sz} tends to infinity with depth, and these solutions are physically inadmissible. We could reverse s, but only at the cost of reversing r, and then e^{-rz} would tend to infinity with depth. Thus for incompressible material the surface wave is the only one that satisfies the conditions.

When $\alpha/\beta = \sqrt{3}$ all the roots are real; they are

$$\frac{c^2}{\beta^2} = 4, 2\left(1 \pm \frac{1}{\sqrt{3}}\right). \tag{16}$$

The smallest of these gives

$$c = 0 \cdot 9194\beta, \quad r = 0 \cdot 8475\kappa, \quad s = 0 \cdot 3933\kappa, \tag{17}$$

$$u = E(e^{-rz} - 0 \cdot 5773\, e^{-sz}) \sin \kappa(x - ct), \tag{18}$$

$$w = E(0 \cdot 8475\, e^{-rz} - 1 \cdot 4679\, e^{-sz}) \cos \kappa(x - ct). \tag{19}$$

The ratio of the axes of the orbit described by a surface particle is reduced to about 1·5.

The other roots make both r and s purely imaginary and therefore do not represent surface waves. If for either we take the positive imaginary values for both r and s we find that they satisfy (7) with the sign changed, and hence it has been supposed that they do not represent solutions at all. But if r and s are imaginary there is no reason why we should not take the positive imaginary for one and the negative imaginary for the other; consequently these solutions represent a P wave approaching the surface and wholly reflected as S, or an S wave wholly reflected as P, and the two solutions correspond to the two angles of emergence that make either possible. It is easy to verify that if $c^2 = \alpha^2(1 + \xi^2)$, (8) reduces to 2·022 (21).

The surface waves considered here were discovered by Lord Rayleigh (1887) and are usually named after him. The motion is entirely in vertical planes through the direction of propagation.

2·032. Love waves. After it had been verified that waves corresponding to P, S and Rayleigh waves were traceable in the records of actual earthquakes, it was noticed (Zöppritz, 1907; Knott, 1908) that the Rayleigh waves were associated with horizontal displacements at right angles to the plane of propagation, for which the theory provides no place. (Nor does

any theory of a uniform crust.) An explanation of these was provided by
A. E. H. Love (1911). He supposed a surface layer of thickness T overlying
a deep layer, with no slipping over the interface. We take as before the
axis of z downwards and x in the direction of travel; the interface is taken
as $z = 0$, so that the outer surface is $z = -T$. We take the displacement v
to be wholly in the direction of the y axis. If unaccented letters refer to the
upper layer, accented ones to the lower, and

$$v = V(z) \exp\left[i\kappa(x - ct)\right], \tag{1}$$

the equations of motion not satisfied identically are

$$-\rho\kappa^2 c^2 V = \mu\left(-\kappa^2 V + \frac{\partial^2 V}{\partial z^2}\right) \quad (-T < z < 0), \tag{2}$$

$$-\rho'\kappa^2 c^2 V = \mu'\left(-\kappa^2 V + \frac{\partial^2 V}{\partial z^2}\right) \quad (0 < z), \tag{3}$$

and the admissible solutions are

$$V = A\cos\kappa\sigma z + B\sin\kappa\sigma z \quad (-T < z < 0), \tag{4}$$

$$V = C\exp\left[-\kappa\sigma' z\right] \quad (0 < z), \tag{5}$$

where
$$\sigma^2 = c^2/\beta^2 - 1, \quad \sigma'^2 = 1 - c^2/\beta'^2. \tag{6}$$

It follows that for any surface wave of this type $c < \beta'$. The surface con-
ditions are that V and $\mu\,\partial V/\partial z$ are continuous at $z = 0$, and $\mu\,\partial V/\partial z = 0$ at
$z = -T$. Hence

$$C = A, \quad -\mu'\sigma'C = \mu\sigma B, \quad A\sin\kappa\sigma T + B\cos\kappa\sigma T = 0, \tag{7}$$

whence
$$\tan\kappa\sigma T = -\frac{B}{A} = -\frac{B}{C} = \frac{\mu'\sigma'}{\mu\sigma}. \tag{8}$$

This is an equation for c.

If $c < \beta$, σ is imaginary, $\sigma\tan\kappa\sigma T$ is negative. But $\mu'\sigma'/\mu$ is positive.
Hence $c < \beta$ is impossible and all values of c satisfy $\beta < c < \beta'$. Incidentally
this kind of wave exists *only* if the velocity of S is greater in the lower
layer.

For any value of c in this interval, (8) determines a set of possible values
of $\kappa\sigma T$, one in 0 to $\frac{1}{2}\pi$, one in π to $\frac{3}{2}\pi$, and so on. These solutions correspond
to differences in the number of nodal planes, that is, planes where $v = 0$ for
all time. Evidently A and C are equal and B has the opposite sign. Hence
V never changes sign in the lower region; and in the upper region it never
changes sign if $0 < \kappa\sigma T < \frac{1}{2}\pi$. If $\pi < \kappa\sigma T < \frac{3}{2}\pi$ it changes sign once, since V is
reversed if z is changed by $\pi/\kappa\sigma < T$, and so on. Thus the solutions are
classified into families according to the number of nodal planes.

Love waves can be interpreted in terms of SH waves; for on account of
the harmonic variation with z in the upper layer they can be regarded as
SH waves continually reflected between the outer surface and the interface.
The condition that σ' is real then becomes the condition that the waves

are totally reflected at the interface, and therefore that they can travel horizontally without continual loss of energy downwards.

For waves with no nodal plane, for given T, if c increases steadily from β to β', κ decreases steadily from ∞ to 0. Hence different wave-lengths correspond to different wave-velocities. This is a new feature, since in all our previous problems the wave-velocity has been the same for all wave-lengths.

If the number of nodal planes is given and not zero, there is a difference because $\kappa \sigma T$ no longer tends to zero with σ'. For any fixed number ($\geqslant 1$) of nodal planes, therefore, there is a least possible value of κ, and therefore a greatest possible wave-length.

2·033. Stoneley waves. An extension of the theory of surface waves is due to Stoneley (1924 b). When solids in welded contact have the same α and β but different densities, a wave of Rayleigh type in both media can travel along the interface. These solutions may not exist if α and β are greatly different in the two media. Love waves in an internal stratum are also considered. (This is the first paper where the convenient notation α and β for the velocities of P and S was used.)

These waves, at first considered a theoretical curiosity, have acquired much attention recently. Incidentally, in some of the slower free oscillations of the Earth as a whole, an unexpectedly large fraction of the energy is near the boundary of the core.

2·034. Dispersion. The velocity c in 2·032 is the rate of advance of an unlimited train of harmonic waves all of the same length. An arbitrary initial disturbance, and in particular a sudden local disturbance, requires all possible wave-lengths to represent it, and therefore the velocity c can be identified with the rate of travel of a disturbance only if it is independent of the period. This is true for light *in vacuo*, sound in a homogeneous medium, and for P and S waves in a uniform solid. It is not true for light in matter, nor for gravity or capillary waves on water, nor for Love waves. It is true for Rayleigh waves on a uniform solid, but not if the properties of the solid vary with depth. Even when the velocity depends on the wave-length, however, it is possible to give useful approximations to the solution by methods due to Stokes, Kelvin and Debye. The problem reduces to the evaluation of integrals of the form (H. and B. S. Jeffreys, 1972, §§ 17·08, 17·09; H. Jeffreys, 1968 b)

$$u = \frac{1}{2\pi} \int_0^\infty \sin\left(\gamma t \pm \kappa x + \alpha\right) \phi(\kappa)\, d\kappa, \tag{1}$$

where $\phi(\kappa)$ has to be chosen to satisfy the initial conditions, which will in the seismological case require that at time 0 the displacement u and the velocity $\dot{u}$ are zero except in a small interval about $x = 0$. If t is large such integrals are in general of the order of $1/t$; but there is an important class of cases where they are of order $t^{-\frac{1}{2}}$ or larger, and identification of such cases determines the places and times when the disturbance is greatest. The condition for this to occur is that there shall be a value of κ within the range of integration where the argument of the sine is stationary with regard to κ. If x (which we may identify with the horizontal distance) is positive, then since γ always increases with κ in the seismological case there can be no such value of κ if the upper sign is taken before κx. If we take the negative sign and there is such a value of κ, an approximation to

$$u = \frac{1}{2\pi} \int_0^\infty \phi(\kappa) \sin(\gamma t - \kappa x + \alpha) \, d\kappa \qquad (2)$$

is

$$u \sim \left\{ \frac{1}{2\pi t \left| d^2\gamma/d\kappa^2 \right|_0} \right\}^{\frac{1}{2}} \phi(\kappa_0) \sin(\gamma_0 t - \kappa_0 x + \alpha \pm \tfrac{1}{4}\pi). \qquad (3)$$

Here the suffix 0 indicates that the quantity is evaluated for the value of κ, if any, that satisfies

$$\frac{\partial}{\partial \kappa}(\gamma t - \kappa x) = 0, \qquad (4)$$

that is,

$$\frac{d\gamma}{d\kappa} = \frac{x}{t}. \qquad (5)$$

The upper or the lower sign before $\tfrac{1}{4}\pi$ is to be taken according as $d^2\gamma/d\kappa^2$ is positive or negative at $\kappa = \kappa_0$. Then $2\pi/\kappa_0$ and $2\pi/\gamma_0$ are the wave-length and period of the waves near the point x at time t. The ratio γ_0/κ_0 is c_0, the wave-velocity of the individual waves in this neighbourhood, and is also the velocity of an infinite train of waves of length $2\pi/\kappa_0$. $(d\gamma/d\kappa)_0$ is the ratio of the distance travelled by a given wave-length to the time taken and can therefore be regarded as the velocity of the period. It is not in general equal to c_0; a given wave usually changes its length and period as it travels. We call $d\gamma/d\kappa$ the *group-velocity* and denote it by C. It is connected with the *wave-velocity* c by the equation

$$C = \frac{d\gamma}{d\kappa} = \frac{d}{d\kappa}(\kappa c) = c + \kappa \frac{dc}{d\kappa}. \qquad (6)$$

In a dispersive system the train of waves from a localized disturbance lengthens as it advances, in the sense that if we attend to places where the waves have two different given periods, the distance between them increases in proportion to the time, the number of crests and troughs between them also increasing. The energy associated with the waves over a given range of period, however, varies little with time.

The approximation (3) is generally valid, but attention is needed to some critical cases. The group-velocity may have upper or lower limiting values, not ∞ or 0. If x/t is near one of these a different type of approximation is needed and the amplitude decreases less rapidly than as $t^{-\frac{1}{2}}$, usually as $t^{-\frac{1}{3}}$. Consequently waves near these periods will become more prominent in comparison with the rest of the train as it proceeds. This is well shown in water waves (H. and B. S. Jeffreys, 1972, §17·09).

For Love waves with no nodal plane we can derive from 2·032 (7) approximate solutions for c and κT in terms of σ' for σ' small, and of σ for σ small, and hence for c in terms of κT. We find for κ small

$$c = \beta'(1 - a\kappa^2 T^2 + \dots),$$

where a is positive; and therefore

$$C = \beta'(1 - 3a\kappa^2 T^2 + \dots).$$

For κ large,

$$c = \beta(1 + b\kappa^{-2} T^{-2} + \dots), \quad C = \beta(1 - b\kappa^{-2} T^{-2} - \dots),$$

where b is positive. Hence as the wave-length increases from 0, C decreases from β to a minimum and then rises again to β'. We should therefore expect the front of the train of Love waves with no nodal plane to consist of long waves advancing with velocity β'; later a new entry should take place of short waves advancing with velocity β. But after the second set have arrived the short waves should be followed by longer ones, the long ones of the first set by shorter ones, and the two periods will coalesce and give rise to a maximum amplitude at a time corresponding to the arrival of waves with the minimum group-velocity. At later times there is no solution of $x/t = C$, so that the train should stop abruptly. There are signs of such a phenomenon in some records of earthquakes at short distances, but these have not yet been studied sufficiently for us to say whether an explanation can be found on these lines. (See Zürich record, Plate I.)

For Love waves with n (> 0) nodal planes the group-velocity still has limits β and β' as $c \to \beta$ or β', with a minimum for an intermediate period. There are a greatest possible wave-length and a greatest possible period. Similar results for Rayleigh waves have been found by Miss M. Newlands (1950). One peculiarity is that some distributions give a secondary minimum group-velocity, previously suspected by A. W. Lee and myself, but determined accurately by Newlands.

In a layered crust Rayleigh waves also will show dispersion, and again there is theoretically a minimum group-velocity.

The hypothesis of two uniform layers is not necessary to the existence of Love waves; it is sufficient that the velocity of S waves shall increase with depth. It seems, however, that a slow increase in the lower layer does not affect the group-velocity much, though it has a substantial effect on the wave-velocity (Jeffreys, 1928c).

Numerical calculation for the case where the lower layer is infinitely deep and the velocity of S increases uniformly with depth suggested that for very long waves c might tend to ∞ but C to a finite value. J. C. P. Miller (private communication) got an infinite limit for C, but his working has unfortunately been lost. Jeffreys and Hudson (1965) showed that c and C both tend to infinity but C very much more slowly. The period tends to a finite value. However, at great wave-lengths the curvature of the Earth is no longer negligible and the limiting case would be the slowest torsional free oscillation of a sphere.

2·04. Scattering. It is possible that wave propagation may be affected by small irregularities of structure which we may call grains. If the properties, especially the wave-velocity, vary, a plane wave will be irregularly refracted, and when it has crossed several grains it may be broken up into waves travelling in widely different directions. This can occur for waves of any length, but is most noticeable for very short waves. The derived waves from different grains interfere so as to give an irregular movement similar to thermal agitation and are ultimately converted into heat. A rough treatment can be given by methods derived from the theory of gases. A gas molecule from a region where the mean velocity in a given direction is high, if it enters a region where the mean velocity is lower, is greatly deflected by its first collision, and the excess momentum is shared among the molecules in the second region. This gives rise to the phenomenon of viscosity. Similarly, molecules from a region where the irregular differences of velocity from molecule to molecule are great will pass on their variation of velocity to other regions, and the result is heat conduction. Both the kinematic viscosity and the thermometric conductivity are of the order of magnitude of $c\delta$, where c is the velocity of agitation and δ is the mean free path. For scattering the momentum is carried with the velocity of an elastic wave. The important thing about δ in a gas is that it is the distance a molecule, on an average, travels before undergoing a great deflexion in direction. If in a solid l is the average grain diameter and the wave-velocities vary from grain to grain by a factor ϵ, the analogue of δ will be of order l/ϵ, so that there will be an effective kinematic viscosity of order $\alpha l/\epsilon$. The numerical factor is likely to be about $\frac{1}{3}$. Now if we consider a region containing many grains the differences of mean stress over its boundary may be regarded as partly accelerating the whole region, and partly as generating scattered waves; and hence the relation between mean stress and mean strain in such a region may be represented approximately by the firmoviscous law 1·04 (38), with

$$\tau \simeq l/3\alpha\epsilon. \tag{1}$$

The firmoviscous law does not hold accurately in such a case, for if the maximum velocity of a longitudinal wave is $\alpha(1+\epsilon)$ no motion at all can be transmitted with a velocity greater than this, whereas according to the strict form of the law, however high a velocity we choose, some motion is transmitted with that velocity. But so long as we are considering velocities not too far from the mean wave-velocities the approximation should be tolerable.

For a transverse harmonic disturbance satisfying the firmoviscous law

$$\frac{\partial^2 u}{\partial t^2} = \beta^2 \left(1 + \tau \frac{\partial}{\partial t}\right) \frac{\partial^2 u}{\partial x^2};$$ (2)

with a time factor $e^{i\gamma t}$, if γt is small, the distance factor is $e^{-i\kappa x}$, where

$$\kappa = (\gamma/\beta)(1 + \tau i\gamma)^{-\frac{1}{2}} \doteqdot (\gamma/\beta)(1 - \tfrac{1}{2}\tau i\gamma),$$ (3)

whence
$$u \propto \exp\left\{i\gamma\left(t - \frac{x}{\beta}\right) - \frac{1}{2}\frac{\tau\gamma^2}{\beta}x\right\}.$$ (4)

The shorter waves are therefore the most damped in travelling a given distance.

The effect on a sudden impulse is peculiar. If the disturbance for $\tau = 0$ is a sudden change of u from 0 to 1 at time x/β, the solution with τ not zero is, when $x - \beta t$ is small, approximately (Jeffreys, 1931b)

$$u = \frac{1}{2}\left(1 + \mathrm{erf}\frac{t - x/\beta}{(2t\tau)^{\frac{1}{2}}}\right).$$ (5)

The displacement is therefore not sharp; it begins a little before time x/β and increases continuously, tending to 1, the greater part of the change being spread over an interval of order $(2t\tau)^{\frac{1}{2}}$. This corresponds to the fact that in a heterogeneous medium some of the disturbance will travel a little faster, some a little slower than the mean. The most rapid increase of displacement at a given place will be at time x/β.

Longitudinal disturbances do involve distortion and the theory still applies to them, with α replacing β. It is not clear whether τ will have the same value, but it will be of the same order of magnitude.

2·041. Other types of dissipation.
Any sort of linear damping can be represented by replacing μ by $\mu\{1 + f(p)\}$, where p is the Heaviside operator, and if f is small, β by $\beta\{1 + \tfrac{1}{2}f(p)\}$. Then for an S wave, if $p = i\gamma$, $i\gamma\left(t - \dfrac{x}{\beta}\right)$ is replaced by

$$i\gamma\left[t - \frac{x}{\beta}\{1 - \tfrac{1}{2}f(p)\}\right].$$ (1)

Thus for small damping an S wave proportional to $e^{i\gamma t}$ as it travels will acquire a factor
$$\exp\left\{-\frac{\gamma}{2\beta}x\Im f(i\gamma)\right\},$$ (2)

$\Im$ denoting the imaginary part. The real part will give a small change in the apparent velocity. It is not necessarily negligible.

In a longitudinal wave we assume the bulk-modulus to be unaffected; then $\lambda + 2\mu = k + \tfrac{4}{3}\mu$ is replaced by, for $\lambda = \mu$,

$$(\lambda + 2\mu)\left\{1 + \frac{\mu}{\lambda + 2\mu} \cdot \tfrac{4}{3}f(p)\right\} \doteqdot (\lambda + 2\mu)\{1 + \tfrac{4}{9}f(p)\}.$$ (3)

Then $i\gamma(t-x/\alpha)$ is replaced by

$$i\gamma\left[t-\frac{x}{\alpha}\{1-\tfrac{2}{9}f(i\gamma)\}\right] \tag{4}$$

and the damping factor for P is

$$\exp\left\{-\frac{2}{9}\frac{\gamma}{\alpha}x\Im f(i\gamma)\right\}. \tag{5}$$

P will in general be damped less than S in travelling a given distance. This neglects a possible analogue of the 'second coefficient of viscosity' but its influence has not yet been detected.

For the modified Lomnitz law (§1·04), except for very short periods, the extra factor for S is

$$\exp\left\{-\frac{qx}{2\beta}a^{\alpha}(\alpha-1)!\,\gamma^{1-\alpha}(\sin\tfrac{1}{2}\pi\alpha+i\cos\tfrac{1}{2}\pi\alpha)\right\}. \tag{6}$$

The greatest damping in travelling a given distance is for short waves, except for $\alpha = 1$, when the damping in a given distance is independent of the period.

The usual method is to define a dimensionless number Q such that $1/Q$ is the relative loss of elastic energy (potential + kinetic) when the wave advances by a radian. The increase of distance is then β/γ and the real part of the exponent increases by $-\tfrac{1}{2}qa^{\alpha}(\alpha-1)!\,\gamma^{-\alpha}\sin\tfrac{1}{2}\pi\alpha$, which is $-1/2Q$. Q is generally assumed independent of period ($\alpha \to 0$), but if $\alpha = \tfrac{1}{5}$ it changes only by a factor 10 if γ changes by one of 10^5. But also the imaginary part makes a phase shift equal to $-(1/2Q)\cot\tfrac{1}{2}\pi\alpha$, equivalent to a change of velocity, which is far from negligible—indeed it tends to ∞ if $\alpha \to 0$. The approximation ceases to be uniform. (See also pp. 147, 312, 356–361, and Jeffreys, 1975a.)

For P, if the dilatation is not subject to damping, $\lambda + 2\mu$ must be written as

$$(\lambda + \tfrac{2}{3}\mu) + \tfrac{4}{3}\mu, \tag{7}$$

and the ratio of the μ term to the whole is approximately $\tfrac{4}{9}$. Thus the required factors for P are $\tfrac{4}{9}$ of those for S.

For exponential afterworking μ is replaced by

$$\mu\left\{\frac{p+1/\tau'}{p+1/\tau}\right\} \doteq \mu\left\{1-\left(\frac{1}{\tau}-\frac{1}{\tau'}\right)p\right\} \tag{8}$$

for small damping. The exponent for distance x is

$$i\gamma\left(t-\frac{x}{\beta}\right)-\frac{x}{2\beta}\left(\frac{1}{\tau}-\frac{1}{\tau'}\right). \tag{9}$$

The damping factor is nearly independent of period. This is true also of the elasticoviscous law, and is strong evidence against both. It is observed that at short distances periods of a small fraction of a second are observed in P waves, but at greater distances these short periods disappear; the same holds more strongly for S. A law that gives extra damping on short periods was badly needed.

2·05. Bodily waves in a sphere. We suppose that the velocity c of a pulse, or of a wave of short period, depends only on the distance r from the centre of the sphere. The pulse is supposed at present to originate at the surface, at a point A, and the radius is R. Take polar coordinates r, θ, the initial line being the radius OA. The wave front will at every instant be symmetrical about OA. The time taken to reach a given point P (Fig. 1) is

$$t = \int \frac{ds}{c} = \int \frac{1}{c} \left\{ \left(\frac{dr}{d\theta} \right)^2 + r^2 \right\}^{\frac{1}{2}} d\theta, \tag{1}$$

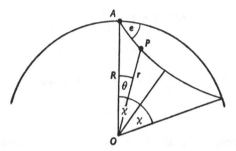

Fig. 1.

where ds is an element of length along the path. The actual path is such that this integral is stationary for small variations of the path. Putting for a moment V for the integrand in (1) and ρ for $dr/d\theta$, we know from the calculus of variations that this is true if r satisfies the differential equation

$$\frac{\partial V}{\partial r} - \frac{d}{d\theta} \left(\frac{\partial V}{\partial \rho} \right) = 0, \tag{2}$$

a first integral of which is known to be

$$V = \rho \frac{\partial V}{\partial \rho} + p, \tag{3}$$

where p is constant for a given ray. Substituting for V and simplifying we find

$$\frac{r^2}{c} = p \left\{ \left(\frac{dr}{d\theta} \right)^2 + r^2 \right\}^{\frac{1}{2}}, \tag{4}$$

whence

$$\frac{dr}{d\theta} = \pm r \left(\frac{r^2}{p^2 c^2} - 1 \right)^{\frac{1}{2}}, \tag{5}$$

$$\theta = \pm \int_R^r \frac{p \, dr}{r (r^2/c^2 - p^2)^{\frac{1}{2}}}. \tag{6}$$

If θ begins by increasing, then as r begins by decreasing we must take the negative sign. But $dr/d\theta$ vanishes when the ray reaches its deepest point, and if χ is the corresponding value of θ

$$\chi = \int_{r/c=p}^{r=R} \frac{p \, dr}{r (r^2/c^2 - p^2)^{\frac{1}{2}}}. \tag{7}$$

After passing this point the ray bends upwards, remaining symmetrical about the line joining the deepest point to the centre, and reaches the surface again at the point $(R, 2\chi)$. We also write $2\chi = \Delta$ and call Δ the epicentral distance of a point on the surface. Then if we know c as a function of r we can calculate the time of travel of the pulse to any distance Δ that it reaches; and conversely, if the time of travel is known from observation as a function of Δ one of our problems is to find what distribution of c is consistent with it.

If the ray makes an angle i with the radius we have for small displacements along the ray

$$r\,d\theta = \sin i\,ds = \sin i\left\{\left(\frac{dr}{d\theta}\right)^2 + r^2\right\}^{\frac{1}{2}}d\theta, \tag{8}$$

whence from (4)

$$p = \frac{r}{c}\sin i; \tag{9}$$

and if e is the angle made by the ray with the outer surface, and c_0 is the value of c there,

$$p = \frac{R}{c_0}\cos e. \tag{10}$$

Now if the ray emerges at P at time T, e will also be the angle of emergence at P. Let P' (Fig. 2) be a neighbouring point $(R, \Delta + d\Delta)$ in the same plane,

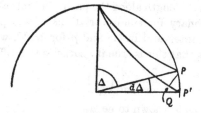

Fig. 2.

$T + dT$ the time needed by the wave to reach P'. Draw PQ perpendicular to the ray that reaches P'. Then PQ is nearly a part of a wave front and the times to P and Q differ by a quantity of order PQ^2. Hence, to the first order

$$QP' = c_0\,dT, \quad PP' = R\,d\Delta, \quad QP' = PP'\cos e, \tag{11}$$

whence

$$\frac{R}{c_0}\cos e = \frac{dT}{d\Delta}. \tag{12}$$

Thus the ray parameter p is identified as $dT/d\Delta$.

More than this is true; for we can apply the same argument to the rays to a pair of neighbouring points at the same distance r from the centre and use (9); then if t is given as a function of r, θ

$$\frac{\partial t}{\partial \theta} = p. \tag{13}$$

If we also compare a pair of neighbouring points on the same radius we get

$$\frac{\partial t}{\partial r} = \frac{\cos i}{c} = \pm \left(\frac{1}{c^2} - \frac{p^2}{r^2}\right)^{\frac{1}{2}}, \tag{14}$$

whence

$$r^2 \left(\frac{\partial t}{\partial r}\right)^2 + \left(\frac{\partial t}{\partial \theta}\right)^2 = \frac{r^2}{c^2}, \tag{15}$$

which is a differential equation satisfied by the time to a general point (r, θ).

If $T = t(R, \Delta)$ is a known function of Δ, p is also found in terms of Δ by differentiation, by (10) and (12). Then we put

$$r/c = \eta, \tag{16}$$

and (7) becomes

$$\tfrac{1}{2}\Delta = \int_p^{R/c_0} \frac{p}{\sqrt{(\eta^2 - p^2)}} \frac{d}{d\eta} (\log r) \, d\eta. \tag{17}$$

The left side is a known function of the parameter p; the right contains $\log r$, which is so far an unknown function of η. Hence (17) can be regarded as an integral equation to determine $\log r$ as a function of η. It has been solved by G. Herglotz (1907) and H. Bateman (1910). An elementary solution has been given by G. Rasch (during a joint work with I. Lehmann) and was communicated to me privately. Take a quantity μ such that (17) holds for $\mu \leqslant p \leqslant R/c_0$. Multiply (17) by $dp/\sqrt{(p^2 - \mu^2)}$ and integrate from $p = \mu$ to R/c_0. Then

$$\frac{1}{2} \int_\mu^{R/c_0} \frac{\Delta \, dp}{\sqrt{(p^2 - \mu^2)}} = \int_\mu^{R/c_0} \frac{p \, dp}{\sqrt{(p^2 - \mu^2)}} \int_p^{R/c_0} \frac{d}{d\eta} (\log r) \frac{d\eta}{\sqrt{(\eta^2 - p^2)}}. \tag{18}$$

Change the order of integration; the limits for p become μ to η, and those for η from μ to R/c_0. But if $\eta > \mu$

$$\int_\mu^\eta \frac{p \, dp}{\sqrt{(p^2 - \mu^2)} \sqrt{(\eta^2 - p^2)}} = \tfrac{1}{2}\pi. \tag{19}$$

Hence the integral is equal to

$$\tfrac{1}{2}\pi \int_\mu^{R/c_0} \frac{d}{d\eta} (\log r) \, d\eta = \tfrac{1}{2}\pi \log \frac{R}{r(\mu)}, \tag{20}$$

and

$$\log \frac{R}{r} = \frac{1}{\pi} \int_\mu^{R/c_0} \frac{\Delta \, dp}{\sqrt{(p^2 - \mu^2)}}, \tag{21}$$

where r has the value that corresponds to $r/c = \mu$. This gives r as a function of r/c and hence c as a function of r.

The solution can be simplified by a transformation due to Wiechert and L. Geiger (1910). In (13) η and r vary along the ray considered, p remaining constant. But in (21) we are considering a set of rays, $p = \mu$ corresponding to that which makes $\eta = \mu$ at the deepest point, and $p = R/c_0$ to one that grazes the surface. p itself is a function of the angle of emergence of a ray that descends at an intermediate angle. Then with a change of notation we write Δ for the value of Δ corresponding to $p = \mu$, and Δ_1 for a general p between μ and R/c_0. Put

$$p = \mu \cosh q. \qquad (22)$$

Then

$$\cosh q = \frac{(dT/d\Delta)_1}{(dT/d\Delta)} = \frac{\cos e_1}{\cos e}, \qquad (23)$$

and q ranges from 0 to the value q_0 found by taking the limiting value for short distances, R/c_0, for the numerator in the second expression. Then

$$\log \frac{R}{r} = \frac{1}{\pi} \int_0^{q_0} \Delta_1 \, dq$$

$$= \left[\frac{1}{\pi} \Delta_1 q \right]_{\Delta_1=0}^{\Delta} - \frac{1}{\pi} \int_\Delta^0 q \, d\Delta_1. \qquad (24)$$

But q vanishes at one limit and Δ_1 at the other; hence the integrated part is 0 and

$$\log \frac{R}{r} = \frac{1}{\pi} \int_0^{\Delta} q \, d\Delta_1. \qquad (25)$$

The advantage of this form over (21) is that if T is given in terms of Δ by a table, it will usually be at equal intervals of Δ over large parts of the table; but it will not be at equal intervals of p. Numerical formulae of integration are much more manageable for equal intervals.

C. G. Knott (1919) used (21); but (25) had already been applied to an empirical table by S. Mohorovičić (1914, 1916).

As q behaves like $(\Delta - \Delta_1)^{\frac{1}{2}}$ at one terminus the usual Gregory and central-difference formulae of integration are unsatisfactory; but formulae suited to this case are known and can be applied to a few intervals at the end, while the usual ones are used for the rest of the range (Jeffreys, 1939e, p. 597; H. and B. S. Jeffreys, 1972, § 9·092). Other formulae of the same type are available for $\int_0^x f(x) \, dx$, where $f(x)$ behaves like $x^{-\frac{1}{2}}$ for x small.

If the velocity distribution is known we may still want to calculate times for parts of rays, especially for foci not on the surface. From (5), (6) and (1) we deduce

$$t = \int \frac{r \, |dr|}{c^2 (r^2/c^2 - p^2)^{\frac{1}{2}}}. \qquad (26)$$

The procedure is to adopt a set of values of p and to work out θ, t for each by integration. In (6) and (26) the integrand behaves like $(\eta - p)^{-\frac{1}{2}}$ if the ray is nearly horizontal at some point, and therefore needs considerable accuracy in the computation. But

$$t - p\theta = \int \left(\frac{r^2}{c^2} - p^2 \right)^{\frac{1}{2}} \frac{|dr|}{r}, \qquad (27)$$

and its calculation is more manageable. In some cases, especially for rays that have not penetrated deeply, the right side of (27) varies slowly with p, and then even if there is a small error in the calculated θ we can use (27) to calculate t, which will be nearly correct for the calculated θ. This device often avoids the need for specially close intervals.

Hitherto, when the adopted travel times have been altered, the velocities have been completely recalculated. This can be avoided as follows. Suppose that a trial solution gives $dt/d\Delta = p$ and that the modification gives $dt/d\Delta = p + \delta p$; then to the first order in the changes

$$q + \delta q = \cosh^{-1}\frac{p_1 + \delta p_1}{p + \delta p}.$$

We find
$$\delta q = \frac{p\,\delta p_1 - p_1\,\delta p}{p(p_1^2 - p^2)^{\frac{1}{2}}}.$$

In most intervals this will vary smoothly, and as $\Delta_1 \to \Delta$, δq behaves like $(\Delta - \Delta_1)^{\frac{1}{2}}$ so long as δp is differentiable. Thus the integration presents no difficulty. It is of course necessary that the calculation for the trial solution shall have been done sufficiently accurately for accumulation of rounding off errors not to matter. This is not true for the J.B. solution of 1940; when the times of PcP were calculated, the time of P at grazing incidence on the core (where it coalesces with PcP) differed by $0\cdot6^{s}$ from the original P time. This exceeds the standard errors now attained in some studies. It was probably due to insufficient accuracy in integration formulae used in the early part of the work. No similar discrepancy was found for S.

Bullen (1960a) has suggested the use of his power law (2·082) as a first approximation, but as it makes d^2c/dr^2 positive and the actual values are mostly negative the above method is probably easier.

2·051. Exceptional cases. To apply (25) it is necessary that q and therefore $(dT/d\Delta)_1$ shall be known for all values of Δ_1 from 0 to Δ, both inclusive. This condition may break down in several ways. There may be a shadow zone within the interval, in which no pulses of the type arrive. There may be refraction so strong that rays penetrating to different depths emerge at the same distance. In the latter case, if these are all observed, there is no theoretical difficulty. (21) requires Δ to be a single-valued function of p. If on account of strong refraction several values of p correspond to the same Δ, nothing in the argument is affected, but for given μ we must include in the integral all values of p from μ to R/c_0, and therefore in (25) q is not a single-valued function of Δ_1; we must integrate over increasing q and therefore proceed around all the loops in the time curve. There is, however, a practical difficulty. When this happens the time intervals between the arrivals at a given distance are usually short and they cannot be satisfactorily separated. Consequently the observations usually give no information about the later branches and direct evaluation of the integral becomes impossible.

The method will not apply to reflected waves, because if there is a reflecting surface of radius R_1 the lower limit in (7) will be replaced by $r = R_1$, where $r/c \neq p$, and the solution fails.

As a rule times are shown as tables or graphs in order of increasing distance, which is usually the order of decreasing $p = dT/d\Delta$. In the exceptional cases where Δ decreases as p decreases we speak of the corresponding part of the table or graph as a *receding branch*.

2·052. In general η is a continuous increasing function of r. In any case it will vanish for $r = 0$. So long as these conditions are satisfied any p less than R/c_0 will be matched by η for precisely one value of r, and the theory is straightforward. But suppose that over an interval η increases with depth, that is, that the velocity decreases downwards faster than r does. Let R_1 correspond to the top of such an interval, and suppose that, for $R_2 < r < R_1$, $\eta > R_1/c_1 = \eta_1$. For $p = \eta_1$ the ray will touch the sphere $r = R_1$. Since in $R_2 < r < R_1$ we have $\eta > \eta_1$, no η in this interval will be equal to any $p < \eta_1$. In such a case no ray will have its deepest point in this interval of r, and therefore the method leads to no determination of the velocity distribution in this interval. Actually, since θ always increases along the ray, any ray entering this shell will traverse a certain angular distance before it comes out again, and Δ will have a discontinuity at the corresponding value of p. Shadow zones can arise in this way.

The extreme case of this phenomenon is where c has a discontinuous decrease with increasing depth (Fig. 3). It is of the first importance because the greatest discontinuity within the Earth is of this type. A ray grazing the discontinuity is sharply refracted downwards, but rays striking it more steeply may be refracted less and actually emerge at a shorter distance; there will in this case be a minimum distance for emergence of rays that have entered the core. The analogous case in light is the focusing by a sphere, and the minimum distance corresponds to the caustic surface. One would therefore expect it to be associated with specially large amplitudes, and it is.

There is some movement within the shadow zone because a little energy is diffracted around the boundary. A discussion for the cylindrical case is given by F. G. Friedlander (1954). The spherical case has been done by L. Knopoff and F. Gilbert (1961). For a simplified theory of caustics and shadows see Jeffreys (1968b). (But see also p. 109.)

2·053. It is an immediate consequence of the principle of least time that if the velocity increases with depth the rays must be curved upwards. For suppose a ray to be straight between two points A, B. If we take a path of small curvature connecting A to B, entirely on the side where the velocity is greater than on AB, its length exceeds that of AB by a small quantity of

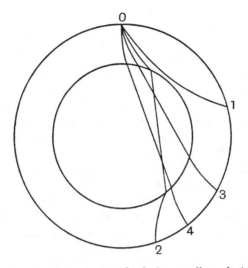

Fig. 3. Rays at a discontinuity of velocity, smaller velocity below.

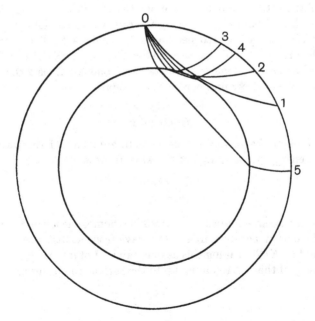

Fig. 4. Rays at a discontinuity of velocity, greater velocity below.

the second order; but the velocity at any point of the curve exceeds that at the corresponding point of the line by a small quantity of the first order. Hence the time of travel along the curve is less than along the straight line.

If c increases rapidly with depth the upward curvature may be very strong. Suppose, in fact, that c increases much more rapidly between R_3 and R_4 than above or below that range. Then a ray penetrating slightly below R_3 may be so strongly refracted that it emerges at a shorter distance than a ray that just grazes R_3. But if we consider steeper rays we shall arrive at one that penetrates below R_4, and the distance reached will increase again. Thus there will be a range of Δ where three rays meet the surface. The graph of T against Δ will in this case have a loop. Since η varies continuously with r there is no discontinuity in $dt/d\Delta$; hence the loop ends in cusps.

The extreme case is again that of a discontinuity, the larger values of the velocity being now on the lower side. In this case the receding part of the (T, Δ) curve is seen to become the wave reflected at the discontinuity, the angles of incidence on the discontinuity being such that the wave cannot penetrate (Fig. 4).

In both cases of discontinuity there will be further complications from the partial conversion of P to S or S to P. All these phenomena have been discussed by I. Lehmann (1934) and K. E. Bullen (1945).

2·06. Amplitude relations. We saw that, in general, energy in a continuous medium is transmitted along the rays with little loss. Let us suppose that the energy emitted per unit solid angle is K. Then the energy sent out in a range of inclinations de to the surface is $2\pi K \cos e\, de$. The area of the surface between epicentral distances Δ and $\Delta + d\Delta$ is $2\pi R^2 \sin \Delta\, d\Delta$. The energy received per unit surface is therefore

$$\frac{K}{R^2} \frac{\cos e}{\sin \Delta} \frac{de}{d\Delta}. \tag{1}$$

But unit area of the surface corresponds to area $\sin e$ of the wave front. Hence the energy per unit area of the wave front is

$$\frac{K}{R^2} \frac{\cot e}{\sin \Delta} \frac{de}{d\Delta}. \tag{2}$$

If a shock of the same intensity occurred in a homogeneous body of infinite extent, the energy per unit area of the wave front at distance R from the focus would be K/R^2. Taking this as our standard of intensity, we have for the intensity of the wave coming up to the surface the formula

$$E = \frac{\cot e}{\sin \Delta} \frac{de}{d\Delta}. \tag{3}$$

Also e satisfies the relation $\quad \cos e = \dfrac{c_0}{R} \dfrac{dT}{d\Delta}. \tag{4}$

The amplitude of the wave coming up to the surface is proportional to $\sqrt{E}$. In Fig. 5, the great variation of Δ over a small range of e will correspond to a range of small amplitudes. To obtain the motion of the ground we must apply a factor to allow for the ratio of the movement of the ground to that in the emergent wave. It may be necessary to allow for losses of energy at discontinuities and by absorption on the way.

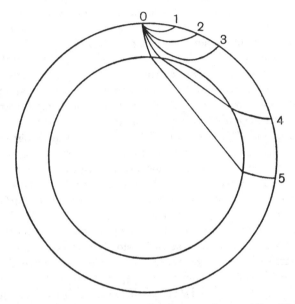

Fig. 5. Rays when velocity is continuous, but increases less rapidly with depth below a certain sphere than just above it. The region between rays 3 and 5 is one of small amplitudes.

As a ray may leave the focus at any angle to the surface there are advantages in regarding e as the fundamental variable. In general Δ increases with e and so does T. But in cases of strong refraction or reflexion Δ may decrease with e. By (4), $d^2T/d\Delta^2$ has the sign of $-de/d\Delta$. Hence in normal cases $d^2T/d\Delta^2$ is negative. But if Δ decreases with increasing e, $d^2T/d\Delta^2$ is positive and the time curve is curved upwards. Thus if strong refraction leads to a maximum or minimum of Δ within a range of e, the time curve has a cusp at the corresponding value of Δ. In such cases we must take the absolute value in (3), and as $de/d\Delta$ tends to infinity at the cusp the amplitudes will be large there. The optical analogue is a caustic zone. There will be some diffracted movement beyond the geometrical caustic, but this appears in seismological conditions to be confined to too narrow a zone to be distinguished.

If the amplitudes were accurately known it would theoretically be possible to use (3) as a differential equation to give e as a function of Δ, and then (4) to give T; thus amplitudes would give an independent way of finding T as a function of Δ. In practice amplitudes are not sufficiently well known for this purpose, but they sometimes provide a useful check. A theory may predict large amplitudes at a certain distance, and if so direct examination of records at that distance gives a crucial test.

The amplitudes at short distances present a difficulty. We have seen that in an infinite medium the amplitude should be nearly inversely proportional to the distance. This also follows for transmission over a sphere from (3). For at short distances for a surface focus e is nearly proportional to Δ, and therefore E to $1/\Delta^2$. But this relation is considerably modified in the conditions of near earthquakes (p. 88). Pg (see p. 86), if the layer transmitting it came right up to the surface, would be near grazing incidence, and the ratio of the ground displacement to the displacement in the incident wave would be of order e. Thus the amplitude would tend to a finite limit for small Δ. The same would apply to SVg but not to SHg, for which the amplitude would vary like $1/\Delta$. When there is depth of focus, there will be an increase of amplitude at distances less than the focal depth, and this has been used to give estimates of focal depth, especially from macroseismic data.

The above argument takes account of the curvature of the surface. If it is treated as plane, e is proportional to $1/\Delta$ for finite focal depth, and the amplitude of Pg would behave like $1/\Delta^2$ except at distances of order h.

For indirect waves e is finite and varies slowly with Δ, so that the effect of the outer surface is nearly constant. If Δ_0, e_0 are the values of Δ and e at, for instance, the shortest distance where P (p. 84) can exist, Δ' the distance travelled in the lower layer, and e' the angle of emergence of a ray just after entering or before leaving it, e' is nearly proportional to Δ', and $e - e_0$ to e'^2 or to Δ'^2. Hence the distance travelled in the upper layers for small Δ is nearly Δ_0, and $de/d\Delta$ is proportional to e' or Δ'. Thus $E \propto \Delta'/\Delta$. Thus E will vanish where P first exists, but when Δ' has reached, say, Δ_0, E will only increase slowly.

This does not allow for loss of energy on the way. When e' is small, only a fraction of order e' of the energy in the incident wave will enter the lower layer, and only a similar fraction can leave it; this is for a sharp interface. The rest is reflected. Thus E contains an extra factor e'^2, and the amplitude should vary like $(\Delta'^3/\Delta)^{\frac{1}{2}}$. For $\Delta \gg \Delta_0$ the amplitude will be nearly proportional to Δ.

If the irregularities in the interfaces, including that between the sedimentary and granitic layers, make variations of inclination large compared with the estimated e, the transmission will be nearly independent of e and the amplitudes will be nearly as $1/\Delta$ but with an irregular variation superposed. Some confirmation of this is provided by a remark of Byerly and Wilson (1935), that in some Californian earthquakes P fails to reach certain stations on account of the shadow cast by the root of an intervening mountain range; this would be an extreme instance of the phenomenon. But it is quite normal for the amplitudes to differ by factors of 2 or 3 from any smooth function of Δ.

Actual data on amplitudes seem to be seldom published. The following are Gutenberg's on the 1911 South German earthquake and mine on the Jersey and Hereford earthquakes (Jeffreys, 1926b, pp. 400–401; 1927b, pp. 490–492). They all refer to the first swing in each phase on mechanically or optically recording instruments; for this the nominal magnification should be correct, even for differences in period and damping, within a moderate factor. Distances are given in kilometres.

South German earthquake

Δ	$u(Pg)$	$u(P)$	$(\Delta/100)\,u(Pg)$
43	700	—	300
56	37	—	21
94	> 400	—	> 376
104	> 300	—	> 312
164	22	—	36
204	5	8	10
225	120	11	270
262	220	23	576
303	9	1	27
342	20	3	68
350	20	14	70
351	15	18	52
365	20	4	73
384	4	11	15

Jersey earthquake

Δ	$u(Pg)$	$u(P)$	$u(Sg)$	$u(S)$	$(\Delta/100)\,u(Pg)$	$(\Delta/100)\,u(Sg)$	$(\Delta/100)\,u(P)$	$(\Delta/100)\,u(S)$
297	5	—	18	7	15	53	—	21
469	6	1	20	10	28	74	4·7	47
474	2	—	12	3	10	57	—	14
689	3	0·1?	15	4	21	103	0·7	28
776	0·5	0·06	6	1·2	4	47	0·5	9
784	0·4	—	5	1	3	39	—	8
1300	—	—	9	<0·5	—	117	—	<6

Hereford earthquake

Δ	$u(Pg)$	$u(P)$	$u(Sg)$	$u(S)$	$(\Delta/100)\,u(Pg)$	$(\Delta/100)\,u(Sg)$	$(\Delta/100)\,u(P)$	$(\Delta/100)\,u(S)$
63	7	—	20	—	4·4	13	—	—
135	2	—	8	—	2·7	11	—	—
181	1	—	8	—	1·8	14	—	—
414	—	—	1·4	—	—	6	—	—
504	0·6	0·1	5	1	3·2	25	0·5	5
827	0·3	0·05	1·3	0·7	2·8	11	0·5	6
945	—	0·1	1·1	0·06	—	9	0·9	6
1380	—	—	0·05	0·1	—	7	—	1·4

There is great irregularity in all cases; the values depart from any simple law by factors up to 3 and sometimes by 10. For Pg the variation agrees better with $1/\Delta$ than with any other simple power, at any rate up to about 700 km., beyond which Pg becomes indistinct. For Sg the agreement with $1/\Delta$ persists to large distances, but at the largest it must be merging with the surface waves.

In the 1911 earthquake P shows no significant variation with distance. In the Jersey one there are only three observations, and there is an apparent drop between 469 and 689 km. In the Hereford one there is fair agreement with $1/\Delta$ between 504 and 945 km. In the Jersey earthquake the amplitude of S decreases somewhat like Δ^{-2}, but Δ^{-1} agrees better with the Hereford earthquake. The larger distances are within the range of small amplitudes for P and S. There is however nowhere agreement with Δ^{-3}.

Richter (1935) proposed an instrumental scale of magnitude based originally on the amplitude of P at small distances, as recorded on a standard instrument. The observations indicated that the amplitude varies as Δ^{-3}, which appears to be confirmed by the study of the Gnome explosion (Carder *et al.* 1962). If there is damping, introducing a factor varying exponentially with Δ, it might mimic Δ^{-3} (Wright, Carpenter and Savill, 1962). They use a formula approximating to $\Delta^{-2} \exp(-\Delta/\lambda)$ and compare with the Logan and Blanca explosions, with $\lambda = 300$ km. Perhaps there is another regional difference!

The method has been extended to observations at large distances and to surface waves. The last are unsatisfactory, of course, for deep foci. The scales have been made comparable, and compared with the energy (Gutenberg and Richter, 1956 *a*, *b*).

There are, of course, many complications, such as differences of instrumental type, especially sensitivity to different periods, sheltering of a station by the root of a mountain range, and focal depth. Variation of direction and amplitude of movement with azimuth is extensively used in the location of fault planes, especially by Dr J. H. Hodgson.

It would ordinarily be expected that when there is triplication of a pulse the curve of t against Δ will have cusps at both ends of the receding branch, and hence that $d^2t/d\Delta^2$ passes through infinity and large amplitudes will occur. Bullen (1960 *b*, 1963 *a*) shows that this will not always happen. In the extreme case of a reflexion at a discontinuity of velocity there may not be large amplitudes near either cusp. If the velocity changes continuously with depth there will be large amplitudes near the nearer cusp, but not necessarily near the further one; it can happen that at the latter $d^2t/d\Delta^2$ changes discontinuously without ever becoming large. A valuable account of methods of assigning magnitudes to earthquakes is by M. Båth (1967).

2·07. Focal depth, compound phases and discontinuities. In spite of the difficulties mentioned in the last two sections, it is possible to make use

of observations of pulses that have not started at the surface and that may also have crossed one or more discontinuities. Far the best method is one that F. J. W. Whipple suggested in a discussion at the Royal Astronomical Society on 1933 February 24. His remarks were not included in the report in *The Observatory*. Bullen and I used it in our 1935 paper, but the first publication was by K. Wadati and K. Masuda (1933, 1934). It rests directly on the principle of stationary time, and may be illustrated by considering the pulse *PS*; that is, a pulse that starts as *P* but is partly converted into *S* by reflexion at the outer surface. Suppose that we have the times of *P* and *S* given as functions of Δ; we want those of *PS* as a function of Δ. The crucial point is the use of the relation

$$\frac{\partial t}{\partial \theta} = p, \tag{1}$$

which rests directly on the principle of stationary time and is a way of stating the laws of reflexion and refraction. Suppose the angular distance travelled as *P* is Δ_1, that as *S* is Δ_2, and the corresponding times are t_1, t_2. Then for given Δ_1 we look for a value Δ_2 such that

$$\frac{\partial t_2}{\partial \Delta_2} = \frac{\partial t_1}{\partial \Delta_1}. \tag{2}$$

The distance travelled as *S* will then be Δ_2 and the time t_2. Hence the distance from the origin where *PS* emerges will be $\Delta_1 + \Delta_2$, and the time $t_1 + t_2$. Thus by matching gradients and simple addition we obtain times of *PS* for as many distances as we like.

To illustrate the errors that may arise from inaccurate matching, suppose that we make an inaccurate match of values of p, so that we take

$$t_1(\Delta_1 + \delta) + t_2(\Delta_2 - \delta).$$

Then the estimate of $t(\Delta)$ to order δ^2 is

$$t_1(\Delta_1) + \delta\, dt_1/d\Delta_1 + \tfrac{1}{2}\delta^2 d^2 t_1/d\Delta_1^2$$
$$+ t_2(\Delta_2) - \delta\, dt_2/d\Delta_2 + \tfrac{1}{2}\delta^2 d^2 t_2/d\Delta_2^2.$$

The terms in δ cancel and the error is

$$\tfrac{1}{2}\delta^2 \left(\frac{d^2 t_1}{d\Delta_1^2} + \frac{d^2 t_2}{d\Delta_2^2} \right).$$

If the tabular intervals are h, δ can reach $\tfrac{1}{2}h$. But $h^2 d^2 t_1/d\Delta_1^2$ approximates to the central second difference of the table of t_1. Hence we are safe if $\tfrac{1}{8}$ of each central difference is numerically less than half the error to be tolerated.

If this condition is not satisfied the best procedure is to interpolate more closely by Bessel's formula and match again. The alternative would be to take four values of δ, form the required sums, and interpolate for a stationary value, but this is usually more troublesome.

If we increase Δ_1 by δ_1, Δ_2 by δ_2, the time of *PS* will be increased by

$p(\delta_1 + \delta_2)$. Hence (1) also holds for PS, with the same p. If therefore we had the times of P and PS, we could reverse the process; we could match gradients for P and PS, and then by subtraction derive times and distances for S.

The argument does not depend on the supposition that reflexion is at the surface, but merely that it is on a concentric sphere. It also covers refraction at discontinuities. Take, for instance, the case of a central core traversed by the pulse denoted by PKP in Fig. 6. This pulse is of P type all the way, the part within the core being denoted by K. The part CD differs only from the ray BE, reflected at the core boundary, by being displaced through an angular distance equal to that traversed as K. Hence the distance and time traversed by PKP are the sums of those traversed by K (path BC) and by PcP (path ABE), the pulse reflected at the core boundary, the partition of the distance again being determined by matching gradients. If we have found the velocities of P down to the

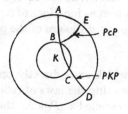

Fig. 6.

core, we can calculate the times of PcP; then if we know the times of PKP from observation we can find those of K. This process is known as 'stripping'.* It does not usually give times at very small distances in the inner sphere and therefore does not permit a direct application of the method for calculating the velocities, but at any rate the time must vanish for zero distance and satisfactory methods of interpolation to small distances are usually available.

Earthquakes normally occur at depths of a few kilometres, but have been known to occur at depths down to about a tenth of the radius. The place where an earthquake actually happens is called the *focus*, and the point of the surface vertically above it the *epicentre*. The opposite point of the surface is the *anticentre*. If the focus is at H, the epicentre E, and the velocities of P are known for depths less than EH, the times can be calculated for any ray that leaves H in a direction with an upward component. Then if the times of P are known between points on the surface, those for rays leaving H in directions with a downward component are obtained by subtraction.

The applications of this principle in the calculation of travel times and velocities are very numerous.

If HA, HB are two rays leaving the focus in opposite directions, AHB is a possible ray, and the angles of emergence at A and B are equal. Thus the (T, Δ) curve for the focus H has parallel tangents at the distances of A and B. Hence it has an inflexion at the distance corresponding to a ray that leaves H horizontally. Attempts have been made to use this principle to estimate focal depth, but accurate location of the point of inflexion from observational data is impossible and much better methods are available.

* The term is due to J. B. Macelwane.

In the 1940 tables times of P and S are given in full for focal depths to $0\cdot12R$. For other phases 'depth allowances' are given for each Δ and h, to be subtracted from or added to the times for a surface focus. These vary more slowly with Δ than the total times do. Another method is given by Shimshoni (1967). He tabulates the depth allowance for PKP as a function of h and $p = \partial t/\partial \Delta$. This has the advantage that for any phase starting as P and going downwards the allowance is the same.

Another method is used by Arnold and Gogna. For any phase a value of p specifies a Δ_1 and t_1 for a ray ascending to the surface. Then if Δ_2 and t_2 for a descending ray with the same p are found, $\Delta_1 + \Delta_2$ and $t_1 + t_2$ are the corresponding distance and time for a surface focus.

2·08. Special velocity distributions. There are a few cases where the times can be found in finite terms.

2·081. The circular ray solution. Assume that a ray is a circle of radius a about a point at distance b from the centre of the planet. Then its polar equation is

$$r^2 - 2br\cos\theta + b^2 = a^2, \tag{1}$$

and

$$\frac{dr}{d\theta} = -\frac{br\sin\theta}{r - b\cos\theta}, \tag{2}$$

$$\left(\frac{dr}{d\theta}\right)^2 + r^2 = \frac{a^2 r^2}{(r - b\cos\theta)^2}. \tag{3}$$

But

$$\cos\theta = \frac{r^2 + b^2 - a^2}{2br}, \quad r - b\cos\theta = \frac{a^2 + r^2 - b^2}{2r}. \tag{4}$$

Hence

$$\left(\frac{dr}{d\theta}\right)^2 + r^2 = \frac{4a^2 r^4}{(a^2 + r^2 - b^2)^2}, \tag{5}$$

and also by 2·05 (4),

$$= \frac{r^4}{p^2 c^2}. \tag{6}$$

Hence

$$c = \frac{1}{2ap}(b^2 - a^2 - r^2), \tag{7}$$

the sign being taken so that the velocity will increase with depth. But c is the same function of r for all rays; hence $b^2 - a^2$ and ap are the same for all rays. The former shows that the tangent from the centre of the Earth to the ray has the same length for all rays. If

$$c = c_0\{1 + \alpha(R^2 - r^2)\}, \tag{8}$$

$$b^2 - a^2 = R^2 + 1/\alpha, \tag{9}$$

$$a = 1/2\alpha R\cos e. \tag{10}$$

Hence a and b are found in terms of e. But by trigonometry

$$\frac{\sin \tfrac{1}{2}\Delta}{\sin e} = \frac{a}{b};$$ (11)

whence on simplifying

$$\tan \tfrac{1}{2}\Delta = \frac{a \sin e}{\sqrt{(b^2 - a^2 \sin^2 e)}} = \frac{1}{\lambda}\tan e,$$ (12)

where

$$\lambda = 1 + 2\alpha R^2.$$ (13)

For comparison of different rays

$$c_0 \, dT = R \cos e \, d\Delta$$

$$= \frac{R \, d\Delta}{\sqrt{(1 + \lambda^2 \tan^2 \tfrac{1}{2}\Delta)}},$$ (14)

whence

$$c_0 T = \frac{2R}{\sqrt{(\lambda^2 - 1)}} \sinh^{-1}\{\sqrt{(\lambda^2 - 1)} \sin \tfrac{1}{2}\Delta\}.$$ (15)

Thus T is found as an explicit function of Δ. This result is due to Wiechert (1907). It certainly does not apply to the whole of the Earth, but it has been useful as a trial solution in the past and in some parts is as near the truth as we can tell.

2·082. Velocity proportional to a power of r. It was noticed by K. E. Bullen (1945) that the equations are also soluble in finite terms if

$$c = c_0(r/R)^{-k},$$ (1)

where $k > -1$. The solution is found to be

$$T = \frac{2}{k+1}\frac{R_0}{c_0} \sin \tfrac{1}{2}(k+1)\Delta.$$ (2)

This is remarkably simple, but has the peculiarity that T is not necessarily an increasing function of Δ, for $0 < \Delta < \pi$, and in general has not a simple maximum at $\Delta = \pi$. This is due of course to the fact that with positive k (1) makes c tend to infinity at the centre. But over ranges of depth that exclude $r = 0$ this solution can be useful.

The condition $k > -1$ is not necessary to the integrability of the equations, but if it is not satisfied r/c does not decrease with depth.

Times for a focus at some depth can also be calculated fairly easily. Put

$$r^{k+1} = pc_0 R^k \sec u.$$

The angular distance travelled between two values of r on an ascending or descending ray is

$$[\theta] = \int \frac{p \, dr}{r(r^2/c^2 - p^2)^{\frac{1}{2}}} = \frac{[u]}{k+1}$$

and, from 2·05 (27), $\qquad [t - p\theta] = \dfrac{p}{k+1} [\tan u - u].$

At the deepest point of the ray $u = 0$.

2·083. The transitional case $k = -1$ is of some interest. If

$$c = c_0 r/R,$$

the solution is

$$\theta = \frac{p}{(R^2/c_0^2 - p^2)^{\frac{1}{2}}} [\log r], \quad t = \frac{R^2/c_0^2}{(R^2/c_0^2 - p^2)^{\frac{1}{2}}} [\log r].$$

The rays are equiangular spirals.

2·084. Times at short distances. If we take Δ small in either 2·081 (15) or 2·082 (2), we find that t is of the form

$$t = a\Delta - b\Delta^3 + \dots. \tag{1}$$

There is no term in Δ^2. This is a general result and requires only that the velocity c shall have a continuous derivative with regard to r in the neighbourhood of the surface. Its importance is enough to make a direct examination of the effect of a term in Δ^2 worth while. Suppose that there is a Δ^2 term and that Δ is small enough to make higher terms negligible, that is

$$t = a\Delta - b\Delta^2 + \dots. \tag{2}$$

Apply 2·05 (25); the first two terms correspond exactly to

$$\log \frac{R}{r} = \frac{1}{2\pi b} \left\{ a \cosh^{-1} \frac{a}{\mu} - \sqrt{(a^2 - \mu^2)} \right\}. \tag{3}$$

If $\mu = a(1-u)$, with u small, this is approximately $\dfrac{a}{12\pi b} \sqrt{(2u^3)}$ and

$$\mu \doteq a \left\{ 1 - \left(\frac{12\pi b}{\sqrt{2}\,a} \frac{R-r}{R} \right)^{2/3} \right\}. \tag{4}$$

It follows that c is continuous, but dc/dr tends to infinity like $\{R/(R-r)\}^{1/3}$ as $r \to R$. Thus the presence of a Δ^2 term in T would imply a most remarkable velocity distribution.

Consequently if we have times at fairly small distances that fit a formula of the form (1), we can use it to interpolate to smaller distances.

A distribution similar to (2) does however appear to have been found in some explosion work (Bullen, 1965b).

2·09. Generation of simple P and S in a uniform medium. If the focus consists of a small sphere of radius a, and at time 0 a symmetrical pressure A is applied to the sphere from the inside, the subsequent motion outside the sphere is easily found. The displacement is wholly radial and is (Jeffreys, 1931e, pp. 410–412)

$$u = \frac{a^3 A}{\mu r^2} \left[\frac{1}{4} + \frac{1}{4a} \exp\left[-\tfrac{2}{3}(\alpha t - r + a)/a\right] \left\{ (r - \tfrac{1}{2}a)\sqrt{2} \sin\left(\frac{2}{3}\sqrt{2}\frac{\alpha t - r + a}{a}\right) \right. \right.$$
$$\left. \left. - a\cos\left(\frac{2}{3}\sqrt{2}\cdot\frac{\alpha t - r + a}{a}\right)\right\} \right],$$

when $\alpha t > r - a$; before this time the displacement is zero. The motion at a given distance is a heavily damped harmonic oscillation, with a time scale proportional to a. For large r the maximum displacement is nearly proportional to $1/r$.

It is possible to generate a simple S pulse by application of a couple over the interior of a sphere. In this case the transverse displacement for large r is of the form (Jeffreys, 1931e, pp. 414–416)

$$\frac{2}{\sqrt{3}} \frac{a^3 A}{\mu r} \sin\theta \exp\left[-\frac{3}{2a}(\beta t - r + a)/a \right] \sin\frac{\sqrt{3}}{2a}(\beta t - r + a),$$

where θ is the angle of the ray with the vertical. The motion consists practically of a single swing and return. This is an artificial type of disturbance because it supposes a couple generated internally; any motion produced by relief of internal stress would need at least two such motions superposed to represent it.

2·10. Generation of Love waves.

The region is here taken to be a superficial layer resting on a semi-infinite lower layer, the greater value of β being that for the lower layer. A disturbance of the type just considered for S is applied about a vertical axis over a small sphere in the upper layer. A solution in the form of a complex integral is obtainable by standard methods. One method of evaluation is to expand the integrand in a series of exponentials. It is then found that each term represents a disturbance arriving at the theoretical time for one of the reflexions and refractions—an S pulse travelling horizontally in the lower layer can be imagined to have been reflected up and down between the free surface and the interface any number of times. But each such pulse when evaluated turns out to be the head of an infinite train of waves, and it is the superposition of these trains that gives the Love waves. These may therefore be regarded as a residual phenomenon, arising from the slowness of recovery after a sudden disturbance when the greater part of the energy is in any case confined to the neighbourhood of a given plane. On the other hand, if instead of expanding the integrand we approximate to the integral directly by the method of steepest descents, we are led to the period equation for Love waves and to the dispersion formula in terms of the group-velocity; but this solution is not valid near the beginning of the movement. The two methods are therefore complementary; the wave expansion method gives the beginnings of pulses predictable by the methods of geometrical optics, while the method of steepest descents gives the later parts of the movement (Jeffreys, 1931 a).

The generation of P in a fluid layer resting on a different fluid can be studied by substantially the same analysis (Jeffreys, 1926 e; Muskat, 1933). I think that some of the integrals in my treatment diverge but that the results are substantially right.

For a plane interface the methods of geometrical optics predict zero amplitude for the refracted waves. This has presented a difficulty both in the study of near earthquakes and in seismic survey. Muskat's theory and mine give a motion starting at the same time, but it is diffracted—which is only a way of saying that it is not predicted by the geometrical theory. The diffraction introduces considerable distortion; for instance, if the original movement starts with a sudden change of velocity, the diffracted movement will start with a finite acceleration. This appears consistent with many observations in seismic survey, but not with those of near earthquakes, where P shows short periods even at the shortest distances where it is observed. It seems likely that the movement is truly refracted, but at an irregular interface, so that a good deal gets through, as for light grazing ground glass (Jeffreys, 1962 c).

2·11. Generation of Rayleigh waves.
The theory of this is very difficult. It has been treated for a uniform semi-infinite solid by H. Lamb (1904), H. Nakano (1925) and E. R. Lapwood (1949). The fundamental difficulty in both this problem and the last is that for a focus at finite depth the curvature of the wave front where it meets the surface must be taken into account. No combination of an original P or SV wave with reflected P or SV will give zero stress over the boundary when the wave front is curved. But it is found that the conditions can all be satisfied if a system of Rayleigh waves is included.

The solution was put into a more compact form by Cagniard, using the Laplace transform and turning the data into an integral equation. The operator has square roots in the denominator, but Pekeris (1955, 1956) makes a substitution that rationalizes these, and when this is done the Bromwich integral can be evaluated by the partial fraction rule. Ben-Menahem and Vered (1973) deal with more complicated cases by a transformation of Pekeris's method.

When the disturbance at the focus is periodic, if the depth of focus is h, it is found that the amplitude of the Rayleigh waves contains a factor e^{-rh} if the original disturbance is radial, and e^{-sh} if it is transverse, where r and s have the meanings given them in 2·031. The deeper the focus, then, the smaller are the Rayleigh waves produced. The same applies to Love waves for foci in the lower layer.

Besides the P, S and Rayleigh waves, the analysis shows that several subsidiary phases should exist. These have been most fully studied by Lapwood. One noteworthy feature is that in some cases an S movement should begin gradually a little before the time calculated for the geometrical path. This gives a hint of an explanation of some of the anomalies found in the travel times of S at short distances, though even Lapwood's very complicated problem is still much simpler than the real one. The extension of Lapwood's analysis to a two-layer crust has been studied in great detail by M. Newlands (1953).

All these phenomena are cases of *diffraction*. The classical wave theory, as in geometrical optics, deals only with waves that travel at right angles to their fronts, so that energy within a given cone or cylinder of rays remains within that cone or cylinder. It is seldom exact, practically only for plane and spherical waves; and even for spherical waves it is impossible to satisfy the boundary conditions over a plane interface between two media in the optical case. Refraction of light from a point source does not give a point image. The nearest optical analogue of reflexion at a plane free surface would be reflexion at a plane mirror; in this case there is a point image for light, but not for elastic waves. The complications arise from the need to superpose additional movements to satisfy the boundary conditions. Another reason why they are more serious for elastic waves than for optics is that the wave-lengths are greater. In both cases there are often extensive regions where the elementary theory is approximately true, but the transition regions (e.g. at the edge of a shadow or a caustic) are broader for elastic waves. A source within a wave-length of a free surface is rare in optics, but can easily occur in seismology.

2·12. The ellipticity correction. So far we have treated the Earth as spherically symmetrical. This is not quite true, and even in a homogeneous Earth the effects of ellipticity would not be negligible with modern accuracy of observation. This was noticed independently by Gutenberg and Richter (1933) and by L. J. Comrie. It may be seen in two ways. We must first distinguish between geographic and geocentric latitudes. The geocentric latitude ϕ' is the angle between the radius to a place and the plane of the equator. The geographic latitude ϕ is the angle between the normal at the place and the plane of the equator. In general, ϕ is numerically greater than ϕ'. If λ is the longitude and r the radius to a place, consider two neighbouring places separated by $\delta\phi'$ in geocentric latitude and $\delta\lambda$ in longitude respectively. Then to the first order in $\delta\phi'$, $\delta\lambda$ the distances from the original place are $r\delta\phi'$, $r\cos\phi'\,\delta\lambda$. The ratio of these is the same as for a sphere, but r varies from equator to pole by ea, where e is the ellipticity and a the equatorial radius. But for two places separated by $\delta\phi$ in geographic latitude the distance is $\rho\delta\phi$, where ρ is the radius of curvature of the meridian. At the equator and the pole respectively $\rho = b^2/a$, a^2/b, where $b = a(1-e)$ is the polar radius. The ratio is nearly $1-3e$. Hence for displacements through an arc ds of meridian $ds/d\phi$ varies three times as much as $ds/d\phi'$ does. The calculation of distances on the Earth as if it was a sphere is inaccurate in any event, but the inaccuracy is smaller with geocentric than with geographic coordinates. It is also more uniformly distributed in azimuth when the distance is short. On this ground Gutenberg and Richter proposed that geocentric latitudes should be used in seismology for the calculation of distances instead of geographic ones, which had previously been used.

At large distances the effect of ellipticity can be seen by considering the arrival at the anticentre of a wave from the surface, according as it starts at a pole or on the equator. In either case the geocentric and geographic latitudes are equal; but the lengths of the paths differ by about 1 part in 300. The actual time of travel of PKP to the anticentre is about 20 min., so that differences of about 4 sec. could arise in this way.

Similar ambiguities could arise according as geocentric or geographic distances are used in calculating distances, as we can see by taking distances about 90°. Consider an epicentre at 45° N., 0° E., and two stations at 45° S., 0° E. and 45° N., 180° E., in geocentric latitudes. The computed distances with geocentric latitudes are both 90°. But the geographic latitudes are 0·2° greater, and the distances found would be 90·4° and 89·6°. $dT/d\Delta$ for P at 90° is about $5^{s}/1°$, so that the calculated differences in travel times to the two stations would differ by 4^{s}.

Until about 1933 such differences were tolerable because transmission times were not known to such accuracy. But the standard error of one observation is between 1 and 2 sec., and it became clear that if full use was to be made of the observations a systematic error that might reach 4 sec. must be removed.

The actual method of allowing for ellipticity given by Gutenberg and Richter was found unsatisfactory (Jeffreys, 1935b), and the construction of a suitable one is due to K. E. Bullen. The same causes that maintain the ellipticity of the outer surface also make internal layers of equal density elliptical, and presumably also the layers of equal wave-velocity, since both depend on the pressure. The internal ellipticities are treated in the theory of the figure of the Earth. The solutions that Bullen (1937a, b) found most convenient for computation were derived from mine, but can be obtained more directly as follows.

We take as standard of comparison a sphere such that the surfaces of equal c have the same volume as in the actual Earth. We call this the *mean sphere*. If r is the radius vector in the mean sphere, for given c, that in the actual Earth is $r + \delta r$, where

$$\delta r = r\epsilon S_2, \quad S_2 = \tfrac{1}{3} - \sin^2\phi', \tag{1}$$

and ϵ is the ellipticity of the layer of equal c. We take θ to be the angle

J E I

subtended at the centre and therefore calculated with geocentric co-ordinates. For the mean sphere consider a surface focus A and let B be the point of emergence and P a point on the ray. The points with the same geocentric latitude and longitude for the actual Earth are A', B', P'. Now since by 2·05 (13)

$$\left|\frac{\partial t}{\partial r}\right| = \left(\frac{1}{c^2} - \frac{p^2}{r^2}\right)^{\frac{1}{2}}, \tag{2}$$

the fact that the ray is from A' to B' instead of from A to B lengthens T by the sum

$$\left(r\epsilon S_2 \left|\frac{\partial t}{\partial r}\right|\right)_A + \left(r\epsilon S_2 \left|\frac{\partial t}{\partial r}\right|\right)_B = [\epsilon S_2 (\eta_0^2 - p^2)^{\frac{1}{2}}]_A + [\epsilon S_2 (\eta_0^2 - p^2)^{\frac{1}{2}}]_B. \tag{3}$$

This takes account only of the changes in the termini, not of the deformation of the path. We have therefore still to compare the times along AB for the Earth and the mean sphere. Denote them by t_1 and t_m. For the actual Earth t_1 is stationary for small variations of the path, and differs only by a quantity of order ϵ^2 from the time for the actual Earth along APB. But at P' the velocity in the Earth is that at radius $r - \delta r$ in the mean sphere. Hence to the first order in δr

$$t_1 = \int_A^B \frac{ds}{c - (dc/dr)\,\delta r} = t_m + \int_A^B \frac{1}{c^2}\frac{dc}{dr}\delta r\,ds. \tag{4}$$

But
$$ds = \sec i\,|\,dr\,| = (1 - p^2 c^2/r^2)^{-\frac{1}{2}}\,|\,dr\,| \tag{5}$$

$$t_1 - t_m = \int_A^B \frac{r^2}{pc^3}\frac{dc}{dr}\left(\frac{r^2}{p^2 c^2} - 1\right)^{-\frac{1}{2}} \epsilon S_2\,|\,dr\,|. \tag{6}$$

The effect of ellipticity is the sum of (3) and (6), and must be subtracted from observed times to give times over the mean sphere.

For short distances S_2 will have the same sign over most of the path, and (6) will partly cancel (3) on account of the negative sign of dc/dr.

For a given ray the azimuth Az is constant. In seismology the azimuth is the angle made by the plane OAB with the meridian plane through A, and is measured from north through east. Then if ϕ' is the geocentric latitude of A, ψ that of P,

$$\sin \psi = \sin \phi' \cos \theta + \cos \phi' \sin \theta \cos \text{Az}$$

and $S_2 = \frac{1}{3} - \sin^2 \psi$.

The calculation is naturally extremely laborious, since the correction was needed for all possible values of ϕ', θ and Az. Fortunately, it could be done with wide intervals.

Bullen actually found that to an accuracy of $0 \cdot 2^s$ for P and $0 \cdot 3^s$ for S the correction could be put in the simple form

$$\delta t = f(\Delta)\,(h_A + h_B),$$

where h_A, h_B are the heights of A, B above the mean sphere and $f(\Delta)$ is tabulated for each pulse. As this accuracy is within that of the present tables this form is adequate.

$f(\Delta)$ and h were tabulated as critical tables in the 1951 tables of geocentric direction cosines, but are not now readily available. $f(\Delta)$ was computed by Bullen, h by Comrie. These were originally given as 'critical tables', h being given to 1 km. and $f(\Delta)$ to 0·01, printed on intermediate lines. The same result can however be obtained by printing in the usual way and understanding that, for instance, if $36° 48' < \phi \leqslant 39° 33'$, h is to be taken as -1 km.

			P		S		SKS	
ϕ (deg.)	(min.)	h (km.)	Δ (deg.)	$f(\Delta)$ (sec./km.)	Δ (deg.)	$f(\Delta)$ (sec./km.)	Δ (deg.)	$f(\Delta)$ (sec./km.)
0	0	+7	9	0·005	9	0·005	79	0·105
10	11	+6·5	13	0·015	10	0·015	83	0·115
16	12	+5·5	22	0·025	13	0·025	106	0·125
20	42	+4·5	41	0·035	16	0·035	116	0·135
24	29	+3·5	63	0·045	22	0·045	124	0·145
27	53	+2·5	72	0·055	32	0·055	140	0·155
31	00	+1·5	83	0·065	44	0·065		
33	58	+0·5	124	0·075	54	0·075		
36	48	−0·5	140	0·085	66	0·085		
39	33	−1·5	PKP	0·09	71	0·095		
42	15	−2·5			76	0·105		
44	35	−3·5			82	0·115		
47	37	−4.5			103	0·125		
50	17	−5·5			123	0·135		
53	01	−6·5						
55	50	−7·5						
58	46	−8·5						
61	51	−9·5						
65	11	−10·5						
68	52	−11·5						
73	11	−12·5						
78	50	−13·5						
90	00	−14·5						

Still greater accuracy can be obtained with a similar form if instead of the geocentric latitude a special 'seismological latitude' is used, namely (Bullen, 1937b),
$$1·1\phi' - 0·1\phi.$$

This has not been adopted, however, as ϕ' is already satisfactory.

Bullen (1938d) has also provided a table to convert geographic distances to geocentric ones. With this, data for any earthquake that has already been studied with geographic distances are readily converted to geocentric distances.

For waves derived at discontinuities it is necessary to take account of terms similar to (3) that arise at the discontinuities. The allowance for deep-focus earthquakes has been examined and does not differ appreciably from that for surface ones (Bullen, 1938c).

Bullen and Haddon (1973b) recalculate ellipticities on the hydrostatic theory for recent models. There is no important change in travel times. The core ellipticity remains about $1/400$.

2·13. Calculation of epicentral distances. If ϕ_0', λ_0 are the geocentric latitude and longitude of the epicentre, ϕ', λ those of a station, the distance Δ is given by

$$\cos \Delta = \sin \phi_0' \sin \phi' + \cos \phi_0' \cos \phi' \cos (\lambda - \lambda_0). \tag{1}$$

Calculation from this formula is rather troublesome in large quantities (over 100 stations may observe one earthquake), and a more compact method was given by H. H. Turner (1915a). If we take rectangular axes at the centre of the Earth, that of x being to $0°$ N., $0°$ E., y to $0°$ N., $90°$ E. and z to $90°$ N., the direction cosines of the epicentre are

$$(A, B, C) = (\cos \phi_0' \cos \lambda_0, \cos \phi_0' \sin \lambda_0, \sin \phi_0'), \tag{2}$$

and those of the station are

$$(a, b, c) = (\cos \phi' \cos \lambda, \cos \phi' \sin \lambda, \sin \phi'). \tag{3}$$

Then (1) is equivalent to

$$\cos \Delta = aA + bB + cC, \tag{4}$$

the ordinary formula for an inclination in terms of direction cosines. Equivalent formulae are

$$2(1 - \cos \Delta) = (A - a)^2 + (B - b)^2 + (C - c)^2, \tag{5}$$

$$2(1 + \cos \Delta) = (A + a)^2 + (B + b)^2 + (C + c)^2, \tag{6}$$

which also follow from simple geometrical considerations. The values of (a, b, c) for the stations can be tabulated once for all.* Then if A, B, C are calculated for the suggested epicentre any one of equations (4), (5) or (6) will give Δ. (4) is calculated by direct accumulation of products on a multiplying machine, (5) or (6) by using a table of squares and tables of $2(1 - \cos \Delta)$ given by Turner (1915a) and Bullen (1934). Comrie recommends converting (5) or (6) into a formula for $\cos \Delta$ and using an ordinary table of cosines, but this is not worth while if Turner's and Bullen's tables are available.

As the values of a, b, c in the tables are rounded off to three or four figures they are not exact direction cosines. Bullen, in the paper just mentioned, has studied in detail the consequences of errors of computation, of which this is the most important. If the maximum error of an entry produced by rounding off is q, the mean square error is σ, where

$$\sigma^2 = \tfrac{1}{3} q^2. \tag{7}$$

* Originally with geographic latitudes by Turner (Brit. Ass. Gray-Milne Trust, 1921–30) (3 figures) and by Bullen (1933) (4 figures); with geocentric latitudes by Comrie (1938) (4 figures). Some stations given in the earlier lists are not in Comrie's, and geocentric (a, b, c) for them must be calculated if old data are being used. Later lists have been issued by the International Seismological Summary.

Then from (4) the error in $\cos \Delta$, if δA, ..., δc are the separate errors, is $A\delta a + a\,\delta A + ...$, and the expectation of its square is

$$(A^2 + B^2 + C^2 + a^2 + b^2 + c^2)\,\sigma^2 = 2\sigma^2. \tag{8}$$

From (5) the error in $\cos \Delta$ is

$$(A - a)(\delta A - \delta a) + ..., \tag{9}$$

and the expectation of its square is

$$2\{(A-a)^2 + (B-b)^2 + (C-c)^2\}\,\sigma^2 = 4\sigma^2(1 - \cos \Delta). \tag{10}$$

Similarly from (6) the expectation of the square of the error in $\cos \Delta$ is $4\sigma^2(1 + \cos \Delta)$.

It follows that (5) is the most accurate formula for $0 < \Delta < 60°$, (4) for $60° < \Delta < 120°$, and (6) for $120° < \Delta < 180°$. With direction cosines of four-figure accuracy, $q = 0.00005$, and

$$\sigma_\Delta = \sigma_{\cos \Delta}/\sin \Delta. \tag{11}$$

Then with the best formula in each case the standard error of Δ ranges from $0.0023°$ to $0.0026°$. In practice distances are usually needed to $0.1°$, but in special cases to $0.01°$. The formulae cover these in all cases. Even if (4) is used down to $\Delta = 20°$ the standard error of Δ only reaches $0.007°$, and this method is quicker if a multiplying machine is available.

Comrie points out that (4), (5) and (6) can be used to check one another over the whole range $0 < \Delta < 180°$. For, if

$$D = \tfrac{1}{2}(A^2 + B^2 + C^2 - 1), \quad d = \tfrac{1}{2}(a^2 + b^2 + c^2 - 1), \tag{12}$$

where by A, a, etc., we now mean the values as given to three or four figures,

$$Aa + Bb + Cc = D + d + 1 - \tfrac{1}{2}\{(A-a)^2 + (B-b)^2 + (C-c)^2\} \tag{13}$$
$$= -D - d - 1 + \tfrac{1}{2}\{(A+a)^2 + (B+b)^2 + (C+c)^2\}. \tag{14}$$

d is tabulated with a, b, c in Comrie's table; D can be found and is, in fact, found in checking the computation of A, B, C; and the differences of the three calculated values are therefore predicted. The values actually calculated should agree with them.

Turner also associated with an epicentre the two orthogonal directions $(0, \lambda_0 - \tfrac{1}{2}\pi)$, $(\phi_0' - \tfrac{1}{2}\pi, \lambda_0)$ and denoted their direction cosines by $(D, E, 0)$, (G, H, K). These are given in the International Seismological Summary. Then we find

$$Da + Eb = -\sin \Delta \sin \text{Az}, \tag{15}$$
$$Ga + Hb + Kc = -\sin \Delta \cos \text{Az}. \tag{16}$$

Hence the azimuth also can be calculated. It is useful in forming equations of condition when a small correction to the epicentre is sought, but is not generally needed to great accuracy. The usual practice in the work of the International Seismological Summary was to calculate the azimuths of the nearer stations and read those of the more distant ones on a large globe. This can be done to 1°.

With modern high-speed calculating machines a different procedure is often found better. The original latitude and longitude are two data; the direction cosines are three. When many stations are used, the storage capacity needed with direction cosines may exceed that of the machine. It is then found more practicable to go back to the original formula (1) and calculate all the sines and cosines directly by power series! Even with this complication the modern machine may be able to calculate a solution in a few seconds, whereas an ordinary machine may take a day. Nevertheless plenty of problems remain where an old-fashioned machine can make a solution in minutes, and it takes hours to write a programme and sometimes days to get access to a fast one.

2·14. Determination of epicentres. This is done by the method of least squares (Jeffreys, 1967d, pp. 147–161). For any earthquake there are three or four unknowns, the time of occurrence t_0, the latitude and longitude of the epicentre, and possibly the focal depth. The data are the times of arrival of the identifiable phases at the stations. It is usual to rely entirely on P, with PKP when available, but other phases can be used when well observed.

A trial time of occurrence and epicentre may be known from observations in the region where the earthquake was felt. When such data are not available two methods can be used if a globe is accessible. (1) The interval between P and S is a known function of distance. Hence the observed intervals give trial distances. We draw a circle about a station with the trial distance as radius, and the intersections of several such arcs give an approximate epicentre. (2) With a rough t_0 we can get a trial Δ for each station from $t(P) - t_0$, and proceed as just described, if necessary adjusting t_0 to give a better agreement. If t_0 is taken too early or too late it will be shown by the small circles overlapping or failing to meet. If none of these methods is available it is necessary to resort to calculation from the start.

Given a trial t_0, epicentre, and focal depth, let Y be the correction needed by t_0, x, y the displacements of the epicentre to the south and east, and h' the correction to the focal depth. The tables give the transmission time of P as a function of Δ and the focal depth. Let t_1 be the calculated time of arrival of P according to the trial solution; then for the change of epicentre, if small, the change of distance is $x \cos \mathrm{Az} - y \sin \mathrm{Az}$, and a typical equation of condition is

$$Y + \frac{\partial t}{\partial \Delta}(x \cos \mathrm{Az} - y \sin \mathrm{Az}) + \frac{\partial t}{\partial h} h' = t(\mathrm{obs.}) - t_1. \qquad (1)$$

These equations are then solved for Y, x, y, h' by the method of least squares. I find myself that it is usually sufficiently accurate to read the azimuths by means of a thread stretched over a small globe or a map. When the trial epicentre is taken from the I.S.S. the azimuths also can be taken from it.

Comrie (1941) proceeds as follows. His method avoids the introduction of the azimuth altogether. We have

$$\cos \Delta = Aa + Bb + Cc, \tag{2}$$

and for variations of ϕ_0', λ_0

$$- \sin \Delta \delta \Delta = (-a \sin \phi_0' \cos \lambda_0 - b \sin \phi_0' \sin \lambda_0 + c \cos \phi_0') \delta \phi_0'$$
$$+ (-a \cos \phi_0' \sin \lambda_0 + b \cos \phi_0' \cos \lambda_0) \delta \lambda_0, \tag{3}$$

whence from 2·13 (2), (16)

$$\sin \Delta \, \delta \Delta = (aG + bH + cK) \delta \phi_0' + (aB - bA) \delta \lambda_0, \tag{4}$$

and the correction of the time of P for the displacement of the epicentre is

$$\frac{\partial t_P}{\partial \Delta} \delta \Delta = \frac{1}{\sin \Delta} \frac{\partial t_P}{\partial \Delta} \{(aG + bH + cK) \delta \phi_0' + (aB - bA) \delta \lambda_0\}. \tag{5}$$

Comrie gives tables of

$$Q = \frac{1}{10 \sin \Delta} \frac{\partial t_P}{\partial \Delta}, \quad \frac{1}{10 \sin \Delta} \frac{\partial t_S}{\partial \Delta}, \tag{6}$$

and writes (5) as

$$Q(M \delta \phi_0' + N \delta \lambda_0). \tag{7}$$

Note that $x = -\delta \phi_0'$, $y = \cos \phi_0' \delta \lambda_0$.

Ordinarily seismological observations show a considerable departure from the normal law of error, and it is usually desirable to restrict the solution to the best stations and even for them to reject P residuals over about 5^s. This can be done at an early stage if the stations are taken in groups, but usually a second or even a third approximation is needed if the best possible determination is being attempted.

A fuller solution does not reject any observation but uses a continuous system of weighting according to the size of the residuals (Jeffreys, 1967 d, p. 214). It is best (as E. P. Arnold points out) to treat all observations at equal weight in the first approximation and use the variable weights in later approximations. If the latter are used from the start they can give high weights to a few irrelevant observations and negligible ones to all the good ones, and a totally erroneous solution may be found.

For normal earthquakes the terms in h' can be omitted.

Standard errors of about $0.05°$ in ϕ_0' and λ_0 are attainable for well-observed earthquakes, but even if only three or four stations are used the standard errors may be under $0.2°$ if they are in widely different azimuths.

2·15. Estimation of small changes in a function.

In such problems as the use of surface waves or free vibrations to improve distributions of density and velocity, the functions required are treated as initially unknown within fairly wide limits, but in general continuous. We require alterations that will give a better fit to the data and will have intelligible uncertainties.

Suppose the range of argument is $(0, 1)$. Divide this into

$$(0, x_1), (x_1, x_2) \ldots (x_{n-1}, x_n = 1)$$

and take $f_r = 1$ in (x_{r-1}, x_r) and otherwise zero. The x_r may be chosen for convenience. Replace the function to be estimated by the step function

$$f = \sum_1^n a_r f_r.$$

For a given datum (the calculated value of which may depend on f for all values of x) the residual against the case $f = 0$ can be written $g_s = \Sigma b_{rs} a_r f_s$, where the b_{rs} are found by making small changes in the separate a_r. These are in the ordinary form of equations of condition and the a_r can be found by least squares. In the seismological cases they will refer to specified ranges of depth.

The estimates so found will not all be equal, and the resulting f will have simple discontinuities at the arbitrary points x_r. These can be removed by a smoothing process, such as adding $\frac{1}{4}$ of the second difference or subtracting $\frac{1}{12}$ of the fourth, and then interpolating. We can then repeat the process, using intervals of half the length. The approximation will stop when the changes are not significantly more than the standard errors. It is (Jeffreys, 1966) undesirable to use too many intervals at the start, since the inevitable uncertainties in short adjacent ranges may mask a real change in average properties over several ranges.

2·16. Rayleigh's principle.

This principle, stated originally by Rayleigh as a property of the free periods of oscillation of a dynamical system about equilibrium, has much wider application. Wherever a quantity estimated for a system is stationary for small variations in the solutions, the change in it due to small changes of the system itself can be calculated.

The principle in its original form is that if Lagrangian coordinates in the system are q_r $(r = 1$ to $n)$,

the kinetic energy $\qquad T = \frac{1}{2} a_{rs} \dot{q}_r \dot{q}_s,$ $\qquad\qquad\qquad$ (1)

the potential energy $\qquad V = \frac{1}{2} b_{rs} q_r q_s,$ $\qquad\qquad\qquad\qquad$ (2)

to the second order, then the ratio

$$\lambda = \frac{b_{rs} q_r q_s}{a_{rs} q_r q_s} \qquad\qquad\qquad (3)$$

is stationary for small variations of the q_r whenever they are in the ratios corresponding to a normal mode of the system; and λ is the square of the corresponding speed of vibration. That is, if the q_r are changed to $q_r + \delta q_r$,

$$\lambda a_{rs}(q_r + \delta q_r)(q_s + \delta q_s) - b_{rs}(q_r + \delta q_r)(q_s + \delta q_s) = O(\delta q_r)^2. \qquad (4)$$

All first order terms in the δq_r cancel.

Much more is true. Suppose that we have found a solution for a particular set of values of a_{rs} and b_{rs}, and we make small alterations in them to

$a_{rs} + \delta a_{rs}, b_{rs} + \delta b_{rs}.$ λ and the ratios of the q_r will in general be altered by small quantities of the first order. (3) still holds with λ replaced by $\lambda + \delta\lambda$. The change in (4) to the second order will be

$$\delta\lambda \, . \, a_{rs} q_r q_s + q_r q_s (\lambda \, \delta a_{rs} - \delta b_{rs}) + O(\delta q_r)^2. \tag{5}$$

Hence the change in λ to the first order can be found from the changes in the system and the original solution for the q_r. Complete recalculation is unnecessary.

The modifications to the system may contain small gyroscopic or frictional terms. These need not even be proportional to the velocities. Suppose that

$$\delta b_{rs} = f_{rs}(p), \tag{6}$$

where $f_{rs}(p)$ is a small Heaviside operator. Then we shall have

$$\delta\lambda \, . \, a_{rs} q_r q_s = -f_{rs}(p) \, q_r q_s \tag{7}$$

and, on putting $p = i\gamma$, $\lambda = \gamma^2$, we have for what was a harmonic oscillation

$$2\gamma \, \delta\gamma \, . \, a_{rs} q_r q_s = f_{rs}(i\gamma) \, q_r q_s. \tag{8}$$

A real part of $\delta\gamma$ will give a change of period, an imaginary one a damping factor.

Even if there is a large gyroscopic term a generalization of Rayleigh's principle exists (H. and B. S. Jeffreys, 1972, §4·093), and can be adapted to small alterations of the properties.

2·161. We consider the case of Love waves, where ρ and μ may vary with z. Suppose that for some value of κ the displacement has been found to be $V(z)\cos(\gamma t - \kappa x)$. The mean kinetic and potential energies per unit surface are

$$\tfrac{1}{4}\gamma^2 I_0 \quad \text{and} \quad \tfrac{1}{4}\kappa^2 I_1 + \tfrac{1}{4} I_2, \tag{1}$$

where

$$I_0 = \int_0^\infty \rho V^2 \, dz, \quad I_1 = \int_0^\infty \mu V^2 \, dz, \quad I_2 = \int_0^\infty \mu \left(\frac{dV}{dz}\right)^2 dz. \tag{2}$$

Make small changes in κ, ρ and μ, but keep the same form for V. Then we shall have by Rayleigh's principle

$$(I_0 + \delta I_0)(\gamma + \delta\gamma)^2 = (I_1 + \delta I_1)(\kappa + \delta\kappa)^2 + I_2 + \delta I_2 + \{(\delta\kappa)^2, (\delta\rho)^2, (\delta\mu)^2\}. \tag{3}$$

If $\delta\rho = 0$, $\delta\mu = 0$ the first-order terms in $\delta\kappa$, $\delta\gamma$ give

$$I_0 \gamma \, \delta\gamma = I_1 \kappa \, \delta\kappa, \tag{4}$$

whence the group velocity C is given by

$$cC = \frac{\gamma}{\kappa} \frac{d\gamma}{d\kappa} = \frac{I_1}{I_0}. \tag{5}$$

This exact result was obtained by E. Meissner (1926, p. 7). If we find $V(z)$ for a given value of c, the determination of C is reduced to that of a pair of integrals. It does not require numerical differentiation, which loses accuracy.

But further, if we do not vary κ, but vary ρ and μ, we are led to

$$(I_0 + \delta I_0)(\gamma + \delta\gamma)^2 = (I_1 + \delta I_1)\kappa^2 + I_2 + \delta I_2 \tag{6}$$

determining

$$c + \delta c = (\gamma + \delta\gamma)/\kappa \tag{7}$$

with a second-order error; and now

$$(c + \delta c)(C + \delta C) = (I_1 + \delta I_1)/(I_0 + \delta I_0) \tag{8}$$

giving C also with a second-order error.

For Rayleigh waves, similarly, we have

$$(u_1, u_2, u_3) = U(z)\sin(\gamma t - \kappa x), \quad 0, \quad W(z)\cos(\gamma t - \kappa x). \tag{9}$$

We find, apart from equal factors,

$$2T = \int \rho\gamma^2 (U^2 + W^2)\,dz, \tag{10}$$

$$2V = \int \{\lambda(W' - \kappa U)^2 + 2\mu(\kappa^2 U^2 + W'^2) + \mu(U' + \kappa W)^2\}\,dz \tag{11}$$

and

$$\gamma^2 I_0 = \kappa^2 I_1 + 2\kappa I_2 + I_3, \tag{12}$$

where

$$I_0 = \int \rho(U^2 + W^2)\,dz, \tag{13}$$

$$I_1 = \int \{(\lambda + 2\mu)\,U^2 + \mu W^2\}\,dz, \tag{14}$$

$$I_2 = \int \{-\lambda W'U + \mu U'W\}\,dz, \tag{15}$$

$$I_3 = \int \{(\lambda + 2\mu)\,W'^2 + \mu U'^2\}\,dz. \tag{16}$$

Varying κ as for Love waves leads to

$$I_0 cC = I_1 + I_2/\kappa \tag{17}$$

and varying ρ, λ and μ gives

$$(I_0 + \delta I_0)(\gamma + \delta\gamma)^2 = (I_1 + \delta I_1)\kappa^2 + 2(I_2 + \delta I_2)\kappa + I_3 + \delta I_3, \tag{18}$$

whence

$$c + \delta c = (\gamma + \delta\gamma)/\kappa, \tag{19}$$

$$(I_0 + \delta I_0)(c + \delta c)(C + \delta C) = I_1 + \delta I_1 + 2(I_2 + \delta I_2)/\kappa + (I_3 + \delta I_3)/\kappa^2 \tag{20}$$

(H. Jeffreys, 1961 a; Takeuchi, Saito and Kobayashi, 1962). The latter take observed values from Brune, Benioff and Ewing (1961).

In practice the order of decreasing sensitivity to small changes in the elastic properties is β, ρ for Love waves, β, α, ρ for Rayleigh waves. It is therefore convenient to substitute for λ and μ in terms of them, and find the effects of their changes separately in this order.

The method can be adapted to investigation of the effects of small changes in the properties (including damping and gyroscopic effects) on periods of free vibrations of the Earth as a whole.

Explicit forms for the kinetic and potential energies of the Earth as a whole are given by Jeffreys and Vicente (1967); see also Appendix B·6.

Data for the use of Rayleigh's principle for the lowest 4 modes of Love waves are calculated in detail by Anderson and Harkrider (1968).

2·2. Interpolation for small depths in the lower layer. If the crust was uniform the travel time for a focus at small depth h would be $\sqrt{(h^2 + \Delta^2)}/c$. The singularity at $h = 0$, $\Delta = 0$ makes it impossible to use any of the ordinary formulae for interpolation when h and Δ are both small, and it is difficult when a thin crust is present and the focus is at a small depth in the lower layer. However, if the velocities in the crust and the lower layer are c and $c' > c$, consider a ray from depth h in the lower layer, meeting the interface at distance Δ_1. The angle of emergence at the epicentre is $\frac{1}{2}\pi$ and is never less than $\cos^{-1} c/c'$; it and the distance travelled in the upper layer vary smoothly with Δ_1. Then if the thickness of the upper layer is T, the time to the interface is $(h^2 + \Delta_1^2)/c'$, and the correction for the upper layer varies smoothly. Then $(t - h/c')^2$ will be $(T/c)^2$ at $\Delta = 0$ and will vary smoothly with regard to both h and Δ_1, and therefore with regard to Δ. Allowance for variability of velocities will not alter this result except that h/c' must be replaced by the time of vertical travel to the interface. This avoids the need for separate calculations at very close intervals.

2·3. Spline interpolation. If $f(x)$ and its first derivative are given at the ends of an interval $(0, 1)$ a formula of interpolation, more accurate than Bessel's formula with third differences, is (Jeffreys and Jeffreys, 1972, p. 705; Jeffreys, 1968d; Curtis and Shimshoni, 1970)

$$f(x) = f(0)(1 - 3x^2 + 2x^3) + f(1)(3x^2 - 2x^3) + f'(0)x(1-x)^2 + f'(1)x^2(1-x).$$

If $f(0), f(1), f'(1)$ are given, a quadratic formula is

$$f(x) = f(0)(1-x)^2 + f(1)(2x - x^2) - f'(1)x(1-x),$$

and if $f(0), f(1), f'(0)$ are given, a similar formula is

$$f(x) = f(0)(1 - x^2) + f(1)x^2 + f'(0)(x - x^2).$$

These formulae do not require equal intervals; but if the tabular intervals are unequal they can be used to interpolate to equal intervals.

Most of the usual formulae for interpolation, with equal intervals, make the derivative of the interpolate discontinuous at the tabular values. This can be avoided by using in these formulae

$$f'(0) = \tfrac{1}{2}\{f(1) - f(-1)\}.$$

Substitution of this at the tabular values gives an interpolate

$$f(x) = f(0) + x\Delta f_0 - \tfrac{1}{4}x(1-x)(\Delta^2 f_{-1} + \Delta^2 f_0) + \tfrac{1}{2}x(x-1)(x - \tfrac{1}{2})\Delta^3 f_{-1}.$$

80   *Spline interpolation*

It differs from Bessel's formula in having $\frac{1}{2}$ instead of $\frac{1}{6}$ in the coefficient of the third derivative; consequently it is not exact for $f(x) = x^3$. But it has the advantage that the derivative is continuous at the tabular values. It is known as spline interpolation. There is a further adaptation to make both f' and f'' continuous.

CHAPTER III

Observational Seismology

As if this earth in fast thick pants were breathing.
S. T. COLERIDGE, *Kubla Khan*

3·01. Seismographs. Instruments for recording ground motion in earthquakes are known as seismographs. The principle of their design, in its simplest form, is that if the surface of the Earth acquires a displacement x and another body is not displaced, we can measure x as the difference of the two displacements. The main difficulty, of course, is to find a body that will not move when the ground does, but it is possible to make one whose displacement is small compared with that of the ground. One method is based on the Euler pendulum, an ordinary pendulum hanging from a movable support. If the support vibrates horizontally in a period short compared with the natural period of the pendulum, the bob will hardly move, and if a record is made of the slope of the shaft it will provide a way of determining the displacement of the support. If the support vibrates in a period long compared with the natural period, on the other hand, the bob will move with it, and the slope of the shaft will hardly vary. The P and S oscillations in earthquakes have periods up to about 7 sec. Hence if we want to record them with an Euler pendulum it would have to be about 12 m. long. Consequently a device for lengthening the period has to be sought. This is done by making the pendulum support nearly, but not quite, vertical, so that the pendulum lies nearly horizontal but can swing slowly like a farmyard gate about a position of equilibrium. If in the undisturbed position the pendulum rod, or boom, points to the north, it will be twisted about the support by a displacement of the ground to the east or west. This principle is the basis of the Milne-Shaw, Mainka and Galitzin seismographs.

If a small cylinder with its axis vertical is attached along a generator to a fine vertical wire, it will have a natural position of equilibrium maintained by the torsional rigidity of the wire. Conversely, a motion of the wire, if sufficiently sudden, gives a twist. The wire is mounted in a frame attached to the ground. This principle is used in the Wood-Anderson and Nikiforov instruments and, in a different form, in that of Benioff.

In the Wiechert instrument a heavy mass is supported on a stiff spring, clamped at its lower end, and is kept stable by springs on four sides. In this case a single instrument is sensitive to displacements in both horizontal components.

Records of the vertical displacement are also needed. It is still more difficult to obtain a suitably long natural period for vertical displacements, but it has been done by Galitzin, Wiechert and Benioff.

There are three ways of recording the movement in a permanent form. In all a roll of sensitive paper covers a steadily revolving drum. In the absence of ground movement the record is a straight line, and the arrangement is such that any displacement of the ground in the component to be recorded displaces the line at right angles, so that the result is a distorted graph of the displacement of the ground against time. Magnification is needed to make the displacement visible. In the Wiechert and Mainka instruments the paper is smoked, and the magnification is done by a system of levers, ending in a pointer, which scratches the soot off. The Milne-Shaw, Wood-Anderson and Nikiforov instruments carry small mirrors, which reflect a narrow beam of light on to photographic paper. In the Galitzin and Benioff instruments, and in most of those used for seismic survey, the moving part carries a coil of wire, which vibrates in a magnetic field, or *vice versa*, and the resulting induced current passes through a mirror galvanometer, the displacement of which is recorded optically.

All seismographs must be damped if they are to be satisfactory. Otherwise the early movements in an earthquake would set up free vibrations, which would make later movements indistinguishable. It is best for the damping to be such that after a sharp displacement of the ground the second swing is small compared with the first. It may be achieved by including a vane immersed in oil, by a piston moving with small clearance in a cylinder filled with air, or electromagnetically. Ineffective damping was the chief defect of the old Milne instrument; the introduction of electromagnetic damping and improved magnification by J. J. Shaw made it reasonably serviceable.

Solid friction must be minimized because it prevents a linear relation between the ground movement and the displacement of the record. Instruments with optical and electric recording are reasonably free from it, but when the record is made on smoked paper through a system of levers the friction is serious. If the magnification is λ, a frictional force F on the paper is equivalent to one of λF on the bob, and very massive bobs are needed if it is to be negligible in comparison with the product of the mass and the acceleration of the ground. The optically and electrically recording instruments have bob masses of a few grams to a few kilograms, whereas the Wiechert, with a magnification of about 250 for short periods, has a mass of about a ton, and instruments on the same principle with magnifications of about 2000 have masses of about 20 tons.

On the other hand, a smoked paper record gives a very fine line with sharp edges, whereas a photographic record is always somewhat fuzzy at the edges. Consequently, a displacement is often clearly readable under a magnifying glass on a smoked paper record when it is quite unidentifiable on a photographic one of the same magnification. Recent machines using photographic recording give much clearer traces than the older ones did.

The clock that drives the drum is made by various methods to mark exact minutes on the record. The rates of movement differ; specimen values are: about 8 mm./min. (Milne-Shaw), 15 mm./min. (Wiechert), 30 mm./min. (Galitzin) and 60 mm./min. (20-ton machines, Benioff, and short period Wood-Anderson). With the greater speeds the cost of upkeep is considerable, especially when recording is photographic, but they are necessary if high accuracy is to be attained. Uniformity of the drum rate is also of the first importance.

For further information on recording methods see P. Byerly, 1942, Chapter 8; H. P. Berlage in B. Gutenberg, *Handb. d. Geophys.* 4, 1932, 299–544; F. W. Sohon in Macelwane and Sohon (1936).

In small earthquakes recorded at short distances (technically known as near earthquakes), the movements include vibrations with periods of a second or less; this is true of P at all distances. Consequently there are advantages in using instruments with natural periods of this order and high magnification for the recording of near earthquakes. The difficulty of making a vertical instrument with long enough period is also much less. It was found afterwards that the same combination of properties favours the accurate recording of P at large distances and P through the core.

The original Galitzin instruments were made with natural periods of the order of 20 sec. This is too long for most purposes, though highly skilled observers get excellent results with them. So long a period gives an undue magnification to the surface waves and makes the smaller body waves much more difficult to identify.

3·02. The International Seismological Summary.

There are approximately 1000 seismological observatories in the world. Many of these are purely seismological, but others carry on seismology in association with astronomy or meteorology. The observer at each station reads the most prominent movements on his own records. Little can be done, however, with the records of a single station, in comparison with what can be got from comparison of a large number. Many stations or groups of stations produce their own bulletins, but the systematic collection of the world's data is done in the International Seismological Summary (henceforward referred to as the I.S.S.). The observations used in this are compiled from the bulletins and from readings sent in directly by the stations. The work was started by John Milne and continued by Prof. H. H. Turner at Oxford University Observatory, under the auspices of the British Association, and the publication was called the *B.A. Seismological Bulletin*. In 1922 the scheme was approved by the International Seismological Association, which contributed a large part of the cost and the name was changed accordingly. Up to Turner's death in 1930 he was assisted by J. S. Hughes and Miss E. F. Bellamy, who continued the work under the supervision of Prof. H. H. Plaskett. It was later transferred to Kew Observatory. I was Director from 1945 to 1957 and was succeeded by Dr Stoneley from 1957 to 1963; and by

Dr P. L. Willmore in 1963. Mr Hughes was chief assistant until his death in 1965.

The cost has been met from various sources, especially U.N.E.S.C.O., H.M. Treasury, the National Science Foundation of U.S.A., and the Government of Canada. When I was Director about 600 earthquakes were analysed every year; since then the number has greatly increased, and so has the average number of stations reporting an earthquake. Calculations had long been done on desk machines, but increase in the number of observations to be handled has necessitated the use of electronic machines. From the readings received the epicentres and times of occurrence of the earthquakes are worked out, and also the focal depth if there is clear sign that it is appreciable. An iterative method of solution is now used, such that the weight attached to an observation depends on the residual (see p. 114).

The increase in the number of stations and the demand for rapid solution have made a new system necessary; and an International Seismological Centre was created at Edinburgh under Dr Willmore. The Bulletins began in January, 1964. The I.S.S. work was also carried out at the Centre until 1963. The present director is Dr E. P. Arnold.

The general accuracy of the data is so important in their interpretation that it is best stated at once. Observations of P in favourable conditions have a standard error of about $1 \cdot 5^s$; those of S about $2 \cdot 5^s$ or so. It was not realized until about 1930 that they were so accurate, because much larger residuals were normal; but these turned out to be mainly due to the inaccuracy of the tables used for comparison.

Many of the great advances in seismology have been made by 'special studies', in which the research worker collects the original records from the stations, reads them himself, and does the solution for the elements of the earthquake before attending to the main subject-matter of his investigation. Others have been made by relying as far as possible on the routine observations submitted to the I.S.S. Both methods have their advantages. Direct comparison of the readings has shown no appreciable difference in accuracy of individual observations; if two competent observers agree on what to read they hardly ever differ by more than 1 sec. This needs emphasis, because it was long believed that observations in special studies were much the more accurate, and a combination of underestimates of uncertainty with inadequate statistical treatment led to numerous inconsistencies. The chief advantage of using the I.S.S. is that the most laborious parts of the work, the reading of the records and the computation of approximate elements of the earthquake, have already been done by the observatory and the I.S.S. staff; and consequently it is possible to discuss several earthquakes and test the results for consistency in a time that would hardly suffice for one in a special study. Another is that the I.S.S. observations of P are more numerous and permit better determinations of the epicentres; and an error in the epicentre is transmitted to the whole of the data and therefore should be

PLATE I

West Bromwich (0·6°)

Stonyhurst (1·6°)

Records of Hereford earthquake. The indications of phases on the Stonyhurst record, shown above the trace, are mine. Those given below the trace were made at the Observatory.

Record of Jersey earthquake at Zürich (7·0°). Uccle, Strasbourg and Zürich all show an unidentified displacement between Pg and S. This may be a compressional wave in the sedimentary layer. The building-up of the movement to a maximum about 30s after Sg, followed by a sudden drop in amplitude, suggests a minimum group-velocity.

PLATE II

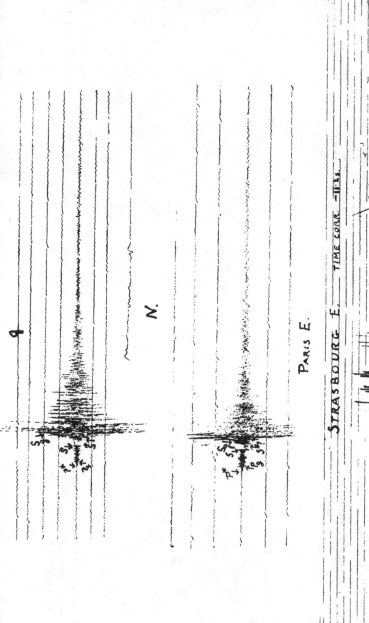
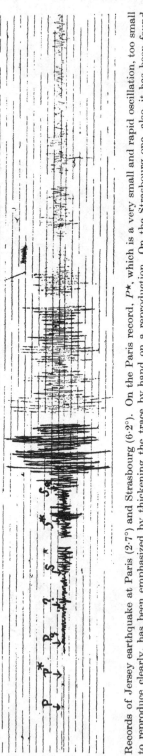

Records of Jersey earthquake at Paris (2·7°) and Strasbourg (6·2°). On the Paris record, P^*, which is a very small and rapid oscillation, too small to reproduce clearly, has been emphasized by thickening the trace by hand on a reproduction. On the Strasbourg one, also, it has been found desirable to ink in the trace for two minutes.

kept as small as possible. The chief advantage of the special study is that the investigator knows what he is looking for, and if it is presen; he will find it—and if he is wise he will equally be able to recognize when it is not. The routine observers are not looking for it and consequently may not report it; often very few do, and the analysis of a large number of earthquakes becomes necessary to provide as much directly relevant material. But this is partly compensated by the possibility of personal error, which, if present, is systematic in a special study but is randomized in routine observations because every record is read by a different observer, and by the fact that, if several independent observers find something when not expecting it, it carries more conviction than if one finds it when he is expecting it. In either case there is a possibility of systematic differences between regions and even between different earthquakes in the same region, and before results are combined some check is needed to test the hypothesis that the residuals can be treated as random error. If the hypothesis is not satisfied it will be necessary to look for explanations and possibly to revise the estimates of uncertainty accordingly.

A method that combines the advantages of both methods is to select from the I.S.S. earthquakes that are likely to be able to provide information on the special point, revise the I.S.S. epicentres, and collect from the stations the records directly relevant. This has been done on several occasions, but not so often as it might, partly because the I.S.S. observations frequently turn out to settle the point after all, and partly because attention is first drawn to the point by a particular earthquake, and the observer proceeds to investigate that earthquake. The I.S.S. will not then be available because of the interval between the occurrence of the earthquake and the appearance of the data for it in the I.S.S., which amounted to several years. The remedy is for stations to send in their readings earlier. A detailed history of the I.S.S. is given by Stoneley (1951).

The work has always been limited by considerations of cost. When there were fewer observations all were published, and the 'additional readings' contributed valuable information on the less studied phases. However, the increasing emphasis on speed of preparation has gone with increasing economy on publication of information on phases that do not contribute directly to the determination of epicentres and focal depths. Speed may be desirable for its own sake, but in the long run contributes little to the advancement of the subject. The reduction of reports on the less studied phases is a serious loss; these are precisely those that may lead to more knowledge of the structure of the Earth.

The tendency of the last twenty years has been to concentrate on the recording of P, because it is the most useful phase for determining epicentres. Short-period instruments of high magnification have become much more plentiful. The times of P are now better known than they were, and it is possible to compare those for epicentres in different regions more satisfactorily. However, there is no corresponding improvement in the recording of other fundamental phases, especially S and SKS, for which the first swing takes 6 sec. or so, and is usually unreadable on an instrument with a period of 1 sec. The standard error of one observation has in fact increased since 1940.

3·03. General structure revealed by seismology. Although the P and S waves had been predicted theoretically by Poisson in 1829, and the Rayleigh waves in 1887, it was not until 1900 that they were clearly distinguished in actual records. Many seismologists had identified the surface waves with S and missed the true S. This was found by R. D. Oldham (1899, 1900) from the records of the Assam earthquake of 1897 June 12. The delay was probably due to the low damping of the early instruments. P and S show sudden commencements as the theory predicts, but are followed by long trains of oscillations with periods of 4–7 sec., the nature of which is not yet understood. With low damping these can build up a large amplitude on the instrument, and anything after the first movement loses its distinctness. The two types of surface waves were not distinguished until 1907. They give a long train of large waves of longer periods, the period diminishing from about 40 to 10 sec. as time goes on, and are collectively known as L; when the Rayleigh and Love types are to be distinguished they are denoted by LR and LQ (from German *Querwellen*). Oldham constructed the first useful tables of times of transit of P and S. He noticed that the times increase with distance less rapidly than they would in a uniform sphere, and inferred that the velocities increase considerably with depth. The beginning of the surface waves travelled with uniform velocity over the surface, and this was his main reason for rejecting their identification with S.

The next great advance was made also by Oldham (1906), who studied the records made near the anticentre and found that P arrives there at a time too late to be reconciled with the time near $\Delta = 90°$ unless there is a central region where the velocity is much less than that just outside it.

K. Zöppritz (1907) collected the available information for three well-observed earthquakes (India, 1905 April 4; Calabria, 1905 September 8; and California, 1906 April 18) covering distances from under 1° to 100°, and put it together to give a new set of transit times. He extrapolated the results to 117°, and they were later reduced to tables by H. H. Turner and further extrapolated to 150°.

A. Mohorovičić (1909), while studying the records of an earthquake in the Kulpa valley, Croatia, on 1909 October 8, found that two distinct P and two S pulses were present. The faster pair were identified with the P and S traceable to large distances. The new pair were denoted by $\bar{P}$ and $\bar{S}$, a notation still extensively used, but for printing and typing the notation Pg and Sg is more convenient. Near the epicentre only Pg and Sg could be observed. At greater distances they were overtaken by P and S, which were smaller but faster, so that four distinct arrivals could be read on one record. At larger distances Pg and Sg died out, but P and S continued. The interpretation placed on these facts was that the focus of the earthquake was in an upper layer of the crust, and that Pg and Sg were pulses that had travelled directly to the observing station, while P and S had been refracted down into a lower layer where the velocities of propagation were greater, and afterwards refracted up again. Similar features were detected in the records of two earthquakes near Stuttgart, on 1911 November 16 and 1913 July 20, by S. Mohorovičić (1914, 1916) and B. Gutenberg (1915).

Estimates of the velocity distribution were made by E. Wiechert, S. Mohorovičić and B. Gutenberg, using somewhat different data and methods.

By 1914, therefore, a general picture of the interior of the Earth had been derived from seismology, and the velocities of travel of P and S were approximately known. The main features were a gradual increase of the velocity of P from about 8 km./sec. near the surface to about 13 km./sec. at about half-way to the centre, with a sudden drop to about 9 km./sec. in a central core; the velocity of S increased from about 4·4 to about 7·5 km./sec., but S through the core was not identified. Nearer to the surface the velocity of P must be about 5·6 km./sec. and that of S a little over 3 km./sec. to explain Pg and Sg. The existence of this outer layer was confirmed by the existence of Love waves and Love's proof that both Love and Rayleigh waves should show dispersion, thus explaining why the surface waves should take the form of a long train. Curiously, the probable connexion of the surface structure with that inferred by Suess in *The Face of the Earth* from geological considerations does not appear to have received mention at that time.

The latter (which I first heard of from Arthur Holmes) first appeared in seismology, apparently, in the study of the Oppau explosion by Wrinch and me (1923). So far as I know none of the continental work mentions Suess, but we were in turn unaware of continental work later than that of Zöppritz.

3·04. Compound phases.

In addition to P and S, several reflexions were identified. These have been denoted by different symbols at different times. The modern practice is thoroughly systematic. The pulse is considered in terms of its type in each part of its path between successive discontinuities, the chief discontinuities being the outer surface and the core boundary. Thus PP means a pulse that has started as P and been reflected as P at the outer surface. PS means one that has started as P and been reflected once, but has been of S type from the place of reflexion to the observing station. SP on a spherical Earth would have the same travel time as PS for a surface focus. Pulses twice reflected on the way are PPP, PPS, ..., SSS. Reflexion can also take place at the core boundary and is indicated by the letter c; thus PcP has travelled down to the core and been reflected up again as P. PcS, ScP and ScS also exist.

To distinguish pulses that have travelled through the core as P the letter K was introduced by Sohon (Macelwane and Sohon, 1936). Thus the distant P studied by Oldham is now denoted by PKP, indicating that it has been of P type through its whole path and has been twice refracted at the core boundary.* The next most important core wave is SKS, which has travelled twice through the outer part as S but has been of P type in the core. PKS (or SKP, which has the same travel time for a surface focus) and $SKKS$ are also often observed. $SKKS$ has been once reflected internally in the core, and was the first core reflexion to be identified. Pulses that have been of S type in the core have not been definitely recognized; Z in place of K for them, if they exist, has been suggested by Bastings. A similar notation is used for the inner core (p. 110) with i for c and I for K.

3·05. Near earthquakes. Technically a near earthquake is one well observed at a considerable number of stations near enough to record Pg and Sg, that is, within about 6°. It is usually a small earthquake, because in a large one at small distances the movement is too violent for anything but the first displacement to be accurately read; though this has been partially remedied, especially in California, by the use of 'strong motion seismographs' of low magnification.

After the pioneer work of A. and S. Mohorovičić and Gutenberg the next great advance was made by V. Conrad (1925) in a study of the Tauern (Austria) earthquake of 1923 November 28. In addition to P, S, Pg and Sg, he found a fifth phase, which he called $P^\star$, with a velocity between those of P and Pg. This appeared to have travelled in an intermediate layer. On 1926 July 30 an earthquake occurred in Jersey, and on 1926 August 14 one near Hereford, and these were studied by the present writer (1927b). In the former $P^\star$ was again identified, but it was very indistinct in the latter. What appears to be the corresponding transverse wave $S^\star$ was identified in both. A further study by Conrad (1928) of the Schwadorf (Austria) earthquake of 1927 October 8 showed the same six pulses. Many later studies have been made of European near earthquakes, especially by W. Hiller, E. Wanner, H. Gräfe and P. Caloi.

The six pulses all travel with uniform velocity over the surface over the distances concerned, as closely as the observations can tell, indicating that the curvature of the Earth and the variation of velocity with depth are unimportant. In these conditions the solutions are simple. In the first place, a movement starting in any layer cannot give rise by refraction to a wave travelling horizontally in a layer where the velocity is lower. Hence the shocks showing Pg and Sg must have taken place in the layer that transmits Pg, or possibly in one where the velocities are smaller still. Again, P and S are found to be traceable to great distances in large earthquakes, so

* It has also been denoted by P', $[P]$ and $PcPcP$. The reflexion $PKPPKP$ is still denoted by $P'P'$. $[S]$ and $ScPcS$ have been used for SKS.

that the layer that transmits them must be the deepest. Accordingly, the structure suggested is an upper layer, also called granitic, transmitting Pg and Sg, an intermediate one transmitting $P^\star$ and $S^\star$, and a lower one of great depth transmitting P and S. The latter pair are often denoted by Pn and Sn, especially when $\bar{P}$ and S are being used for longitudinal and transverse waves in general. The surface that separates the region where P and S travel from the intermediate layer is often called the Mohorovičić discontinuity (now often abbreviated to *The Moho*); the interface between the upper and intermediate layers is sometimes called the Conrad discontinuity.

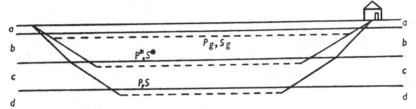

Fig. 7. Diagram of the probable paths of the six pulses observed in near earthquakes for a focus in the sedimentary layer. Broken lines indicate waves propagated along or near to boundaries. The horizontal scale is, of course, much smaller than the vertical; the angles are approximately correct. *aa*, Sedimentary layer; *bb*, upper layer; *cc*, intermediate layer; *dd*, lower layer.

The times of arrival would be expected to yield estimates of the velocities and of the thicknesses of the layers and the focal depths of the individual earthquakes. It is obvious that for a focus at depth h the time of transmission of Pg to a station at distance x (measured as a length, so that $x = R\Delta$) would be $(h^2 + x^2)^{\frac{1}{2}}/\alpha$, where α is the velocity of Pg. That of Sg would be $(x^2 + h^2)^{\frac{1}{2}}/\beta$. The time of occurrence, t_0, and the position of the epicentre have also to be found from the data, and thus there are four unknowns special to each earthquake. Many determinations have been made in this way, mostly leading to values of h in the neighbourhood of 35 km. The methods of solution, however, have often been extremely unsatisfactory. In the original paper of A. Mohorovičić several of the distances turned out to have been wrongly calculated, and the epicentre could not be determined within 25 km. (Jeffreys 1937e, p. 201.) In the two South German earthquakes the number of discrepant observations was much larger than would now be considered tolerable, and, in particular, there was a flagrant inconsistency between the two nearest stations, which would be the most sensitive to changes in h. The situation has considerably improved since the introduction of wireless time signals. It is easier now to check an observatory clock to 1 sec. than to 3 sec. before 1914 (unless the observatory was also astronomical). Several recent studies were rediscussed by G. Schmerwitz (1938), who found that though the general accuracy of the observations had improved it was still impossible to determine focal depths with anything like the precisions of the order of $\pm$ 5 km. that had been claimed. In some cases, however, estimates of focal depth were obtained that appeared genuinely different from zero when judged by ordinary statistical standards.

Formulae for the times in the intermediate and lower layers are easily obtained, or we can apply 2·05 (14). If the time to distance Δ in either is known, and each layer can be treated as uniform, then for a focus in the upper layer and a station at the surface this time will be increased by

$$\int \left(\frac{1}{c^2} - \frac{p^2}{r^2} \right)^{\frac{1}{2}} | dz |, \tag{1}$$

where dz is an element of the vertical distance travelled down and up in the higher layers. The contribution from the layer where the ray is nearly horizontal is the product of two small factors. As we can neglect the variations of p and r the effect is practically to increase the time of travel of each pulse by a constant. Also p/r is practically $1/c'$, where c' is the velocity in the layer traversed in the nearly horizontal part of the path. The coefficient $(c^{-2} - c'^{-2})^{\frac{1}{2}}$ was conveniently called the 'delay-depth coefficient' by A. W. Lee. The whole expression (1) may be called the 'apparent delay in starting'. The point is that to the times of arrival of any phase at the stations it should be possible to fit a formula of the form

$$t = t_r + p_r x \pm \sigma_r, \tag{2}$$

where t_r will differ for different phases and σ_r is the standard error of an observation, which need not be only the error of reading the record. For Pg and Sg in determining p_r it is safest to omit stations within $1°$ on account of the possible effect of focal depth; other phases are not read within $1°$. For waves in a surface layer t_r will be the actual time of the earthquake. As the determination of t_r is essentially an extrapolation to zero distance its uncertainty depends greatly on that of the velocity, and if this has to be found from the earthquake considered the uncertainty of t_r would in practice never be less than that of one observation. Consequently, it is necessary, if full use is to be made of the data, to get as accurate an estimate of the velocity as possible by combining values found from different earthquakes. Here, however, there was a serious difficulty because different earthquakes gave velocities differing by more than would be expected from their apparent accuracies obtained by the method of least squares. This was particularly noticeable for P, for which the earlier observations led to a velocity of about 7·8 km./sec.; but several recent studies led to about 8·1 km./sec. This was traced to the Tauern earthquake when the I.S.S. appeared. Only one station had itself confirmed Conrad's reading of P; hence it was clear that this was a very slight earthquake and a systematic error of reading was possible. When movement is small it may fail to be read at all; but what may happen is that the beginning is too small to be read, but the displacement increases with time and the first movement that the observer sees will be later than the true beginning of the movement. This effect will increase with distance because the displacement itself decreases with distance, and hence will lead to a systematic underestimate of the velocity. Special precautions have therefore to be taken when the movement is likely to be weak. But the readings made by Conrad were so consistent among themselves that this earthquake had received a specially high weight in the solution. In addition it agreed with observations of small Japanese earthquakes (which would be considered fairly large by European standards). But the later European studies agreed so well that they had to be accepted, and a new solution for Europe was based on them alone. The Kulpatal and South German earthquakes gave velocities with such large uncertainties, in comparison with the recent ones, that it makes little difference whether they are retained or not. Consequently, a new solution has been based on the later determinations alone. In this the stations were classified by quadrants because there was a strong suggestion that the velocities of the various phases vary from place to place, and this method gave such a variation a good chance of showing itself if it was there. The final result was that $dt/d\Delta$ for P at short distances in Europe is $(13 \cdot 73^s \pm 0 \cdot 13^s)/1°$, corresponding to a velocity of $8 \cdot 09 \pm 0 \cdot 08$ km./ sec. over the surface (Jeffreys, 1947 *b*).

On the other hand, several further studies have confirmed the original velocity in Japan, and there is clear evidence of a genuine regional difference, which had failed to be recognized earlier because it was balanced by a systematic error in the earthquake treated as most reliable for Europe.

For Pg, $dt/d\Delta$ is $(19 \cdot 85^s \pm 0 \cdot 11^s)/1°$, corresponding to a surface velocity of $(5 \cdot 598 \pm 0 \cdot 031)$ km./sec. There was no particular difficulty about combining the data, and therefore there is no evidence to show that the velocity of Pg is not uniform over an area extending from Lancashire to Yugoslavia and from Hamburg to central Italy.

The situation is quite different for S and Sg. At every point where comparison of different estimates was made, they differed by substantially more than would be expected if the errors were random, and there was no apparent explanation. The uncertainty is therefore based on the differences between the separate estimates. We have for Sg, $dt/d\Delta = (32 \cdot 66^s \pm 0 \cdot 19^s)/1°$, velocity $= (3 \cdot 402 \pm 0 \cdot 020)$ km./sec.; for S, $dt/d\Delta = (25 \cdot 67^s \pm 0 \cdot 24^s)/1°$, surface velocity $= (4 \cdot 328 \pm 0 \cdot 041)$ km./sec.

The constant terms in the expressions for the times of the pulses should be closely related to the depth of focus, the differences between those of P and Pg, and between those of S and Sg, being directly affected by it. Hence if the estimates of focal depth found from Pg and Sg directly are compared with these two differences there should be a strong correlation. None was found, and in some earthquakes the least squares solution from Pg and Sg gave negative h^2. The constant terms are the more reliable because they rest on many more observations. Hence it seems that the observations of Pg and Sg at short distances do not lead to reliable estimates of focal depth.

In my earlier studies I found that the constant term for Sg was up to about 2 sec. before that for Pg, and proposed the hypothesis that, in some cases at least, the movement sent out from the focus was entirely transverse and Pg was really sPg, generated by reflexion at the outer surface. This could not be a general rule, because primitive P movement is certainly sent out by deep-focus earthquakes. An alternative would be that P and S are both sent out from the focus, but S slightly earlier. Neither hypothesis was found satisfactory on further analysis (Jeffreys, 1937 e, pp. 207–209). Gutenberg and Richter (1943) have found the same phenomenon in California, and suggest that the fracture wave on a fault may travel faster than a transverse wave. This would explain the fact, but it is not clear that it is mechanically possible. The usual belief is that the readjustment after one fracture is due to a quasi-statical redistribution of stress and will travel with a much lower velocity than Sg. There is some evidence that it travels with the velocity of a Rayleigh wave. Any pulse loses amplitude by divergence as it travels. The stress-difference where the fracture starts is just over the strength, and in uniform material any body wave would not produce enough stress for fracture; though perhaps a Pg wave might locally add enough to a stress-difference, already near the strength, to bring it above the limit. L. Mansinha (1964) estimates that the rate of extension of a tensile fracture is $0{\cdot}631\beta$, of a shear fracture $0{\cdot}776\beta$. See also Bilby, Cottrell and Swinden (1963), Burridge (1969), Cottrell (1962, 1963).

What was even more surprising was that the differences between the constant terms in Pg and P showed no sign of anything but random variation when the uncertainties were taken into account. They were consistent with the hypothesis that the true difference was always the same, and the value indicated was consistent with that found for the explosion near Burton-on-Trent on 1944 November 27 (Jeffreys, 1947 a). It was $8{\cdot}7^s \pm 0{\cdot}7^s$. The corresponding comparison for Sg and S showed non-random variations, but these were not perceptibly correlated with either the P-Pg differences or the estimates of focal depth from near stations. Thus the S waves only lead to a further anomaly with no verified explanation. The main conclusion is that the apparent accuracy of results obtained by comparing P and Pg is genuine, but until the anomalies concerning the S waves are explained we cannot trust any times for S and Sg within 2 sec. and possibly more.

The P and Pg data, except for the occasional discrepancies at short distances, are consistent with the hypothesis that all the foci were near the surface. This appears to be tenable. The formation of mountains must have involved enormous stresses in the upper layers, and may have been associated with earthquakes more intense than any we have known. But in Europe at present mountain formation is not taking place, and denudation is. Stresses due to the carving out of ravines, for instance, would have their greatest values at small depths and may well be the principal cause of fracture; but fractures arising in this way would be in the sedimentary layer. This is probably comparatively thin, not more than a few kilometres. Velocities in it have been determined by seismic survey. Some of them appear to be more, some less, than that of Pg. Displacements that may possibly correspond to pulses in it have been noticed in natural earthquakes, but they are not found on many records, and it is likely that even for foci in the sedimentary layer transmission for hundreds of kilometres in it is made impossible by local irregularities. Thus there appears to be no serious difficulty in supposing that the foci of the near earthquakes considered were in the sedimentary layer.

The pulses $P^\star$ and $S^\star$ have velocities near 6·5 and 3·7 km./sec. Unfortunately, the data about them are scanty. $P^\star$ has been identified only in the Tauern, Jersey and Schwadorf earthquakes, $S^\star$ in the Jersey, Hereford and Schwadorf ones. In fact, the failure of later special studies to give series of observations of these pulses led me to have some doubt about whether the readings represented anything more than accidental alignments of points on a graph of t against Δ. Long series of routine observations in Japanese earthquakes, provisionally identified as $S^\star$, have appeared in the I.S.S. A conspicuous case is 1931 December 26, where there are 22 such observations, mostly with residuals of -3 to $+4$ sec. As the total travel time reaches 4 min. and the S–Sg interval about 1 min., the concentration looked like strong confirmation of the existence of $S^\star$. These series have been studied by E. R. Lapwood (1955). The stations report only P and S, and if the readings are later than the standard time the I.S.S. marks them as possibly $P^\star$ or Pg, $S^\star$ or Sg. Failure to read P can be only a consequence of weakness of movement or microseisms, but S is in any case a new impulse on a disturbed background, and if $S^\star$ or Sg were clearer than S a good reading of it might well be identified as S. Lapwood, after a preliminary adjustment of epicentres, plotted the whole of the recorded times against distance, but apart from P and S he found no convincing alignments showing other velocities. There was a possible significant clustering about 15·8^s $\pm$ 0·3^s after the time of S, which might represent a pulse once reflected between the surface and the Mohorovičić discontinuity. In any case the evidence for $P^\star$ and $S^\star$ in Japan has disappeared.

The fundamental difficulty is that if an error of $\pm 2^s$ is to be taken as normal, a pulse between P and Pg cannot be safely separated from both unless the interval between them is at least 8^s. According to the table this is reached at 2·6°. $P^\star$ and Pg are not observed beyond about 6°, so that only a short range of distance is available for fitting a linear function to the times of $P^\star$. Further, any random set of times between those of P and Pg would agree moderately well with a linear function.

The recognition of arrivals of new phases presents rather a difficult problem. From the arrival of the first P to a considerable time after the slowest predicted surface wave should have arrived the record is never at rest; and even P is usually superposed on a background of microseisms.

Three criteria are in use to test whether a movement represents a newly arrived phase. (1) The chief is a sharp increase in amplitude. (2) Sudden changes of period are also useful; for instance, P is often small, but can be distinguished from the microseisms because the time of the first swing is shorter than the prevailing period of the microseisms and the superposition of the shorter period is easily recognizable. The arrival of S phases can often be detected by an increase in period, and those of the surface waves are always so detected. (3) Sharp changes in the direction of movement on the horizontal components usually indicate the arrival of S phases (see p. 107). Even with such precautions movements may be read as new phases at one station that do not correspond satisfactorily with anything readable at other stations, even fairly near ones. For this reason it is fatal to read too many phases. Some stations, and even some makers of special studies, appear to have read every separate swing over a long interval, with the result that whatever hypothesis might be under discussion some reading would be found to agree with it.

The difficulties are greatest in near earthquakes. The most useful method is to read only the clearest phases and to pick out those that show a nearly linear variation of time with distance. This does not exclude the possibility that some alignments may be accidental; a satisfactory statistical test for this has not yet been devised (but see Willmore, 1949). It also does not exclude the possibility that something genuine may be ignored because it does not stand out clearly enough. Sometimes the records of two earthquakes from the same epicentre, recorded at the same station, are practically identical for every swing, so that apparently fine details of structure along the path are impressed on the record. But we do not know how to compare the records at different stations in such a way as to arrive at a physical interpretation of these details.

P. L. Willmore (1949) has made a detailed study of the Heligoland explosion of 1947 April 18 and of smaller explosions at Soltau, near Hamburg. For these the times of the explosions were decided in advance. This made it possible to establish many temporary stations, and the permanent stations greatly increased their paper speed. It was possible to make corrections for the local thickness of sediments, and the result was a great increase of general accuracy. His results for the velocities were, in km./sec.: P, $8 \cdot 18 \pm 0 \cdot 014$; $P^\star$, $6 \cdot 4 \pm 0 \cdot 16$; Pg, variable; S, $4 \cdot 36 \pm 0 \cdot 06$; Rayleigh (?), $3 \cdot 01 \pm 0 \cdot 04$. The S waves were unsatisfactory, but this is a usual feature of explosions; it looks as if SV in near earthquakes is always less satisfactorily recorded than SH. There was a series of readings, denoted by P_x, about $7 \cdot 0^{\mathrm{s}}$ after P; Willmore attributes these to an extra reflexion between the outer surface and the base of the upper layers. The remarkable feature is that this single study determined the velocity of P with greater accuracy than all previous ones taken together. The observations attributed to $P^\star$ were few, and Willmore does not insist strongly on the identification.

Great differences were found in some cases between the records at quite near stations. These may be due to irregularities of the various interfaces. In fact, it looks as if local irregularities forbid detailed application of the velocities except as averages over distances of some hundreds of kilometres.

3·051. Thicknesses from near earthquake studies. Comparison of the constant terms in the times of the near earthquake phases would be expected to give information about the thicknesses of the layers. It is, however, rendered difficult by the numerous anomalies that have been found when detailed comparison is made. At present the chief difficulty is probably the scarcity of good series of observations of $P^\star$. Petrologists have in fact remarked that on account of chemical action between minerals sharp transitions between layers are impossible, and that the properties must vary nearly uniformly with depth. This, however, is definitely inconsistent with the seismological evidence. For an assumed structure the times of travel are calculable. On the hypothesis of linear variation of velocity from 5·5 km./sec. at the surface to 7·8 km./sec. at a depth of 30 km., with a slowly increasing velocity at greater depths, the times are as shown in Fig. 8 (Jeffreys, 1937e, pp. 222–225). There is a single time curve, and there is no provision for the occurrence of Pg after P at the same distance. On an alternative hypothesis a uniform upper layer of thickness 15 km. was taken, but linear variation of velocity was assumed between 15 and 30 km., so that the intermediate layer is replaced by one of gradual transition. This gives a time-curve of the form shown in Fig. 9, with cusps at A and B. The branch OA gives a velocity corresponding to Pg, BDC to that of P. The two are connected by a receding branch with a cusp at B; this branch corresponds to rays whose deepest points are in the zone of transition. This is a little more satisfactory than the results shown in Fig. 8, but the nearer cusp would be associated with large amplitudes of P, which are not observed; in fact, one of the anomalies of near-earthquake studies is that P is often too small to be observed at short distances even when, according to the times inferred from greater distances, it should precede Pg. It seems, therefore, that irrespective of the observations of $P^\star$, those of P and Pg require the existence of fairly uniform layers with transitions between them not more than a few kilometres thick.

We denote the thickness of the upper layer by T, that of the intermediate layer by T'. The lower layer is of great depth.

The European near earthquakes where $P^\star$ had been observed led, after much discussion, to no useful information about T except an order of magnitude (Jeffreys, 1937e, pp. 196–212), so long as the focal depth had to be taken as one of the unknowns for each earthquake. Comparison of S with $S^\star$ and P with $P^\star$ suggested that $T' = 9\cdot1 \pm 3\cdot1$ km., but the uncertainty may be much too low on account of the various anomalies

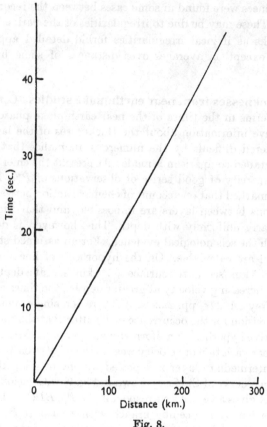

Fig. 8.

encountered on the way, and the estimate depends on the velocity of P in Europe, which has needed a change.

The comparison of near earthquakes with the Burton-on-Trent explosion has provided other information. The differences of the constant terms for Pg and P were the same in all, within the limits of random variation, and it seems probable that all the foci were within the sedimentary layer. On this hypothesis, the difference of the constant terms would be

$$0{\cdot}258T + 0{\cdot}184T',$$

the velocities being taken as 5·60, 6·50 and 8·09 km./sec. Then the observed difference of $8{\cdot}7^s \pm 0{\cdot}7^s$ leads to the equation

$$0{\cdot}258T + 0{\cdot}184T' = 8{\cdot}7 \pm 0{\cdot}7.$$

This difference should also be equal to the interval between Pn and its reflexion in the Heligoland explosion found by Willmore, which was $6{\cdot}95^s \pm 0{\cdot}45^s$. His estimate of the difference between the constant terms

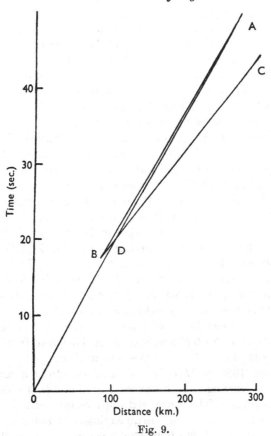

Fig. 9.

for Pg and Pn was $7{\cdot}37^{s} \pm 0{\cdot}13^{s}$. I have not used these data here, but they must be borne in mind.

This hypothesis also makes it possible to salvage some information from the comparison of the constant terms for Pg and $P^{\star}$. With velocities of $5{\cdot}57$ and $6{\cdot}50$ km./sec., which have needed little change, the differences of these terms were (Jeffreys, 1937 e, pp. 205, 206 foot)

Tauern, $+4{\cdot}6 \pm 0{\cdot}7$; Schwadorf, $+4{\cdot}5 \pm 0{\cdot}7$; Jersey, $+2{\cdot}7 \pm 1{\cdot}2$.

The weighted mean is $4{\cdot}3 \pm 0{\cdot}46$ and gives $\chi^2 = 2{\cdot}1$ on 2 degrees of freedom, so that the values are consistent. The theoretical value for a surface focus is $0{\cdot}182T$, whence

$$T = 23{\cdot}7 \pm 2{\cdot}5\,\text{km.}; \quad T' = 11 \pm 5\,\text{km.}$$

3·052. Near earthquakes in other regions. In California extensive studies have been made by Gutenberg, Byerly and their collaborators. Gutenberg's data (1932) for Pg correspond to a velocity of $5{\cdot}588 \pm 0{\cdot}023$

km./sec., for P, 7.92 ± 0.05 km./sec. (Jeffreys, 1937e, pp. 213–217). The latter gives $dt/d\Delta = 14.03^s \pm 0.10^s/1°$. His earthquakes, with few exceptions, show very little variation in the difference of the constant terms for Pn and Pg, but the average difference is only 5.8^s as against about 8.7^s for Europe. The velocities are close to the European ones. There appear to be two P's associated with intermediate layers, which Gutenberg denotes by Pm and Py, with velocities given as 6.83 and 6.05 km./sec. Neither of these agrees with the European $P^\star$ velocity. S waves with velocities of 3.66, 3.39 and 3.23 km./sec. are found, the first two of which agree with the European $S^\star$ and Sg. The interpretation is difficult; we might have expected agreement between Europe and California for both P and S waves in the intermediate layer, or disagreement for both, but not agreement for one and disagreement for the other. Byerly and Wilson (1935) do not confirm Pm and Py, but give waves with velocities of about 7.2 and 6.6 km./sec.

Extensive studies of near earthquakes in Japan have been made by Matuzawa *et al.* (1928, 1929a, b). In the last paper quoted (p. 257) velocities are given as follows in km./sec.: Pg, 5.0; $P^\star$, 6.2; P, 7.5; Sg, 3.15; $S^\star$, 3.7; Sn, 4.5. Uncertainties are not given. I am inclined to think that some of the differences from European values are due to graphical methods of computation. I found (1936b, p. 410) that routine observations of P in Japanese earthquakes gave at short distances $dt/d\Delta = (14.32 \pm 0.09)^s/1°$, corresponding to a velocity of 7.76 ± 0.05 km./sec. E. A. Hodgson, from the Tango earthquake of 1927 March 7, got 7.75 ± 0.02 km./sec.

E. A. Hodgson (1932, p. 275) gives what appears to be a very accurate determination by least squares of the velocity of a P in the Tango earthquake; it is 6.3 km./sec. This would naturally be interpreted as $P^\star$.

An unpublished collective work on an explosion at Isibuti, Japan, gave two P's with velocities (5.26 ± 0.007) km./sec. and (6.13 ± 0.017) km./sec. The constant terms, however, are so nearly equal that the writers attribute the former to a superficial layer not more than about 2 km. thick.

For New Zealand, studies have been made by R. C. Hayes (1935, 1936) and K. E. Bullen (1936a, 1938e, f, g, h, 1939a). In this case there is a special difficulty because only a few of the stations recorded absolute time. The method used was to select earthquakes whose epicentres as found from distant stations lay nearly on a great circle through Wellington and Christchurch, the two best stations; thus velocities at short distances could be found from the difference of time of arrival at these two stations, uncertainty of the epicentre making little effect on the result. For other stations it was necessary to use a standard velocity of P to give a standard of time for the other phases. In spite of these difficulties useful results were found:

P velocities: 5.28 ± 0.16; 5.93 ± 0.11; 6.31 ± 0.18; $8.10 \pm$? km./sec.

S velocities: 3.34 ± 0.20; 3.62 ± 0.14; 3.84 ± 0.12; 4.04 ± 0.10;
$\qquad\qquad\quad$ 4.40 ± 0.16 km./sec.

The slowest P is decidedly slower than the European and Californian Pg; the second value for P agrees with the Californian P_m, and the third with P^*. The first S value would agree with Sg, the second with S^*, and the fifth with S.

Thus near-earthquake data indicate substantial differences in structure for different parts of the world, even in land regions. Little progress has yet been made towards interpreting them.

In New England L. D. Leet (1938) finds an upper layer with a P velocity of 6·01 km./sec. and an S one of 3·45 km./sec., passing at a depth of about 15 km. into one where the velocities are 6·77 and 3·93 km./sec. F. Birch (1938) points out that the former pair agree (subject to some corrections for temperature and pressure) with laboratory values for ordinary granite.

For the Eureka earthquake (California) Neil R. Sparks (1936) found no Pg, indicating a focus in some intermediate layer; velocities of near-earthquake phases were 7·0 and 7·4 km./sec., which Stoneley (1938) re-determines as $7 \cdot 15 \pm 0 \cdot 065$ and $7 \cdot 53 \pm 0 \cdot 073$ km./sec. The former would agree with Gutenberg's Pm, and with a velocity found by Stoneley (1931a, p. 352) in some European near earthquakes reported in the I.S.S. He denoted the onset provisionally by P_Q. It has not been found in European special studies. The thicknesses derived by Stoneley from Sparks's readings are $12 \cdot 6 \pm 2 \cdot 5$ and $12 \cdot 2 \pm 1 \cdot 9$ km.

A disturbing feature appeared about 1950 or so in seismic survey in North America by means of explosions. The observations cover New England, Maryland, California, and parts of the Canadian shield. The superficial velocity of P is in each case found to be somewhat over 6 km./sec. Such a velocity was also found at short distances by Willmore in the North German explosions at Heligoland and Soltau. It would be characteristic of a true granite or granodiorite. The explosions show no sign of a layer where the velocity is 5·6 km./sec. or so, and some of the experimenters are disposed to think that there is no such layer. There would appear to be a fundamental difference between North America and Europe, where Pg is regularly found both in special studies and routine observations. In 500 km. a change of velocity from 5·6 to 6·0 km./sec. would make a difference of 7 sec. in travel time, and such a systematic error seems out of the question. On the other hand Willmore finds apparently genuine variations in its velocity in the Heligoland explosion.

If the velocity of Pg in Europe was 6·0 km./sec. instead of 5·5 km./sec., in the Jena record of the Schwadorf earthquake (Plate IV) there should not only be nothing genuine where P has been marked, but there should be something genuine about 7ˢ before it. Inspection of the record is a sufficient answer.

B. K. Balavadze and G. K. Tvaltvadze reported to the International Seismological Association at Toronto in 1957 that extensive work with explosions has been done in the U.S.S.R. and combined with gravity data. For the Kara depression in the Caucasus the following results are found:

Constitution of outer layers

Layer	Thickness (km.)	α (km./sec.)	β (km./sec.)
Sedimentary	0–8	3·5–4·4	2·2–2·6
Granite I	4–8	5·6	3·2
Granite II	8–10	6·0	3·4
Basalt	23–26	6·7	4·0
Substratum	—	7·9–8·1	4·6–4·7
Total	48–50		

The intermediate layer in this mountainous region varies in thickness at the expense of both the upper layers and the substratum. Similar work has been done in the Tian Shan and the Pamirs by P. S. Veystsman, I. P. Kosminska and Y. V. Reznichenko.

Kaminuma (1966) gives velocities of P from explosions in Japan, as follows:

Thickness (km.)	α (km./sec.)
1–3	2·3–2·7
5–10	5·5–5·8
10–20	6·0–6·2
?	6·5–6·9
?	7·5–8·0

Three peaks are shown for the deepest layer, the strongest being at about 7·7 km./sec. S velocities are derived by comparison with laboratory determinations, and the results are used to compute velocities of Rayleigh waves. The observed values agree well with a total crustal thickness of 30 km.

Aki and Kaminuma (1963) have found a curious anomaly, that the dispersions of Love and Rayleigh waves agree with different models. The observed velocities for Love waves are higher than the theoretical ones. Aki (private communication) finds that the data can be reconciled with a surface layer with $\beta = 1 \cdot 1$ km./sec. overlying one with $\beta = 4 \cdot 7$ km./sec.; or by a rigidity $7 \cdot 2 \times 10^{11}$ c.g.s. for shears across horizontal planes and $5 \cdot 4 \times 10^{11}$ c.g.s. for shears across vertical ones.

T. V. McEvilly (1964) studied phase velocities in the Central United States from dispersion of Love and Rayleigh waves. He found that both were rather insensitive to density and P velocity, but they indicated different velocities of S. His initial model is as follows.

Layer no.	Thickness (km.)	α (km./sec.)	β (km./sec.)	ρ
1	11	6·1	3·5	3·7
2	9	6·2	3·5	2·8
3	18	6·4	3·7	2·9
4	24	8·15	4·6	3·2
5	40	8·2	4·5	3·3
6	180	8·2	4·4	3·4
7	—	8·7	4·8	3·6

His results are as follows.

Records of Jersey earthquake at Uccle (4·2°; azimuth 66°). The amplitudes of P, $P\star$, and Pg are greater on the E.W. component than the N.S. one, but this relation is reversed for the S movements. This indicates that the former waves are longitudinal, and the latter mainly of type SH. P is most easily read on the vertical component, probably owing to its rather steep emergence.

PLATE IV

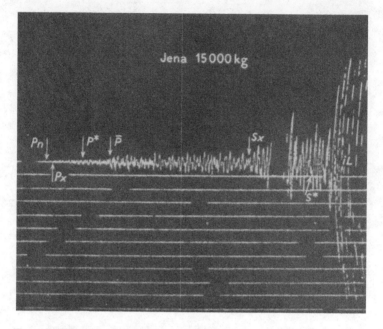

Record of the Schwadorf earthquake of 1927 October 8 at Jena (4·3°).
From V. Conrad (1928).

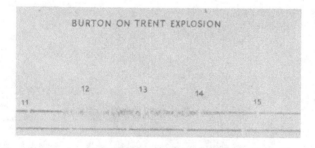

Record of the Burton-on-Trent explosion at Kew (1·4°). Short-period
vertical instrument.

PLATE V

(a)

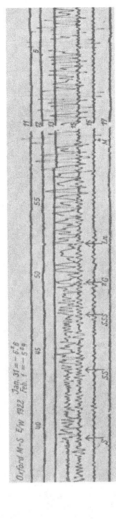

(b)

Normal earthquake 1922, Jan. 31 d. 13 h. ($\Delta = 75.7°$). (From Stoneley, 1931 b.)
Note the long train of large surface waves beginning about $13^h 50^m$ and contrast
with the smallness of the corresponding movement in the Stuttgart record of
a deep shock at about the same distance (Plate VI). The record has been cut in
half; the traces run from (a) to (b).

Layer no.	Thickness (km.)	Rayleigh β	Love β
3	18	3·67	3·94
4	24	4·67	4·75
5	40	4·47	4·83
6	180	4·45	4·80

According to this SH should travel a little faster than SV; that is, that down to depth 282 km. the shell is anisotropic.

On account of the fact that SH is usually larger than SV, and also more easily recognized, because of the change in direction of the movement, it is probable that most of the S observations used in determination of travel times refer to SH.

E. Nishimura and others (1956, 1958) get for the lower layer in Japan $\alpha = 8\cdot1 \pm 0\cdot1$ km./sec., $\beta = 4\cdot7 \pm 0\cdot2$ km./sec. They say that the amplitudes vary smoothly and that there is no notable increase about 15°. W. Stauder and G. Bollinger (1963) give the following velocities for South Missouri, in km./sec.

$$Pn \ 8\cdot24 \pm 0\cdot02 \qquad Sn \ 4\cdot72 \pm 0\cdot05$$
$$Pg \ 6\cdot41 \pm 0\cdot06 \qquad Sg \ 3\cdot65 \pm 0\cdot04$$

The difference of velocities in the east and west of North America, combined with the study of the Gnome explosion (Carder *et al.* 1962) seems to show that the difference comes in crossing the Rocky Mountains. See also Lehmann (1964). Hales and Sacks (1959) find evidence for an intermediate layer in the Transvaal. R. E. White (1971) finds regional differences in the travel times of P within Australia.

B. A. Bolt, H. A. Doyle and D. J. Sutton (1958), from studies of explosions in Australia, get for Pn, $8\cdot21 \pm 0\cdot005$ km./sec.; Sn, $4\cdot75 \pm 0\cdot01$ km./sec., and for Pg, $6\cdot03 \pm 0\cdot09$ km./sec., Sg, $3\cdot55 \pm 0\cdot04$ km./sec. With a single crustal layer the thickness inferred from P is 32 km., and from S, 39 km.

3·06. Constitution of the outer layers. The velocities of elastic waves found in near earthquakes could be compared with velocities in known rocks in the laboratory or in the field, and in this way the possible identifications of the materials would be severely limited. Alternatively, the velocities could be calculated from the density, compressibility and rigidity as measured in the laboratory. There are difficulties in the comparison, however we proceed. The elastic properties are appreciably affected by pressure, especially up to pressures of about 2×10^9 dynes/cm.², which would correspond to a depth of 7 or 8 km. L. H. Adams and E. D. Williamson (1923*a*) attributed this to the closing of flaws. From about 2×10^9 to 10^{10} dynes/cm.² the compressibilities varied smoothly with the pressure. Their method determined only the compressibility, not the rigidity. A possible method of comparison is to notice that if α, β are the velocities of P and S waves, k the bulk-modulus, and ρ the density,

$$\frac{k}{\rho} = \alpha^2 - \tfrac{4}{3}\beta^2.$$

J E I

Elastic properties of rocks

In this equation the left side can be found from the laboratory determinations for known materials, while the right side is given by the velocities of P and S in a given layer. In early comparisons it was necessary to assume a ratio of α to β, or, equivalently, a value of Poisson's ratio; but this is no longer necessary.* An advantage of this method of comparison is that it does not depend on the experimental measurement of the rigidity, partly because rigidity is difficult to measure at high pressures, and partly because, if there is any imperfection of elasticity, the measured rigidity will depend on the period of the applied forces. The bulk-modulus shows no such variation with period.

The seismological values adopted (Jeffreys and Bullen, 1940) were as follows:

	α (km./sec.)	β (km./sec.)	$\alpha^2 - \frac{4}{3}\beta^2$ ($\times 10^{-10}$)	α/β
Upper layer	5·57	3·36	15·9	1·69
Intermediate layer	6·50	3·74	23·6	1·74
Lower layer	7·76	4·36	34·9	1·78

In the work of Adams and Williamson it was found that the most reliable way of finding compressibilities at high pressures was to determine those of the component minerals separately and to form an average weighted according to the known composition by volume. The change of volume under pressure was compared with that of a standard cylinder of soft steel, the

Rock	Density	$1/(k \times 10^{-12})$	$k \times 10^{-12}$	$k/\rho \times 10^{-10}$
Granite	{2·61	2·10	0·476	18·3
	2·66}	1·86	0·538	19·5
Granodiorite	{2·69	1·81	0·552	20·5
	2·73}	1·64	0·610	22·3
Syenite	{2·61	1·85	0·540	20·7
	2·66}	1·66	0·602	22·7
Diorite	{2·74	1·60	0·625	22·8
	2·78}	1·47	0·680	24·3
Gabbro	{3·05	1·18	0·847	27·8
	3·08}	1·15	0·870	28·2
Pyroxenite	{3·40	1·01	0·99	29·0
	3·44}			
Peridotite	{3·40	0·95	1·05	30·7
	3·44}			
Dunite	{3·29	0·84	1·19	36·2
	3·32}	0·79	1·27	38·3
Pallasite	{5·65	0·75	1·33	23·5
	5·69}			
Siderite	7·9	0·58	1·72	25·0
Obsidian	2·333	2·86†	0·350	15·1
Tachylyte	2·851	1·45†	0·690	24·2
Eclogite	{3·7	0·73	1·37	37·0
	3·5}	0·83	1·20	34·2

† Bridgman (1925) gives about 2·5 (variable) for obsidian, 1·87 for tachylyte from Torvaig, Skye, and 1·33 for tachylyte from Kilauea.

* Provided, of course, that we are sure that the P and S considered travel in the same layer. This may be taken as certain for the lower layer and nearly certain for the upper layer. There is some room for doubt for the intermediate layer.

compressibility of which was taken as known. It was afterwards found that this had been taken too high by $0\cdot02 \times 10^{-12}$ per c.g.s. unit of pressure, and all their compressibilities need to be reduced by this amount in consequence. This has been done in the table above. In later papers Adams and R. E. Gibson (1926, 1929) gave values for dunite, tachylyte and eclogite. In the table, where two rows of results are given for a rock, the upper row corresponds to a pressure of 2×10^9 dynes/cm.^{2} and the lower to one of 10^{10} dynes/cm.^{2}. Where only one compressibility is given, it is a mean over the range between these pressures. The pyroxenite considered was half augite and half hypersthene, the pallasite half olivine and half metallic iron, and the siderite all metallic iron.

The value of k/ρ derived from experiment for granite is appreciably more than the seismological value for the upper layer in Europe; the latter would be in good agreement with obsidian. But the American values appear to agree with ordinary granite. Leet's values would give 20×10^{10}. The geological evidence for the existence of a granitic layer seems decisive,* and the difference may be due either to temperature (since the experiments were carried out at ordinary temperatures) or possibly to the presence of a considerable quantity of obsidian. The value for the lower layer would fit either dunite or an eclogite. The lower layer, however, is very deep, and eclogite is a mixed crystalline rock. It would be very remarkable if the constituents of such a rock had failed to separate under gravity. There is more direct verification of the identification with dunite from the density of the Moon (p. 197). More recent seismic data, however, give a somewhat larger value of $\alpha^2 - \frac{4}{3}\beta^2$, say 36×10^{10}; and this agrees with forsterite, according to Birch (1952), leaving no room for admixture with iron silicates, which have about the same elastic constants but higher densities. Birch gives 39×10^{10} for jadeite and $40-45 \times 10^{10}$ for various garnets. See also L. H. Adams (1931).

The value for the intermediate layer would be in reasonable agreement with either syenite, diorite or tachylyte. It seems definitely too low for gabbro. This is surprising because igneous rocks of the basalt-dolerite-gabbro group are the most abundant of all, but the fact is that (except possibly in California) there is no seismological evidence of the existence of any widespread crystalline layer of that composition. If we are to suppose that basalts are representative of the composition of a widespread layer, that layer must be tachylyte and not dolerite or gabbro. Syenite hardly arises here, because it contains too much potassium. Diorite, suggested by Holmes (1926a), would have the merit of accounting for eruptions of ande-

* See also, however, the discussion by E. C. Bullard, H. H. Read, R. Stoneley, G. M. Lees, C. E. Tilley and others reported in *The Observatory*, **71**, 1951, pp. 15–19. The geological speakers mostly regarded the large 'granitic' masses of the continental shields as metamorphosed sediments and not as exposures of the seismological upper layer. I have never myself regarded these as the principal geological evidence; the best is, I think, the prevalence of sandstone, which appears to imply a granitic parent rock.

site, but the basaltic rocks would have to be referred to a transition zone between the intermediate and lower layers.

The difficulty about the identification of the intermediate layer with tachylyte is to see why it did not crystallize, but this does not appear insuperable. Slow cooling certainly favours crystallization and rapid cooling the formation of glass. But E. M. Anderson and E. G. Radley (1915) have found glassy inclusions in an igneous rock in Mull, where crystallization seems to have been prevented because the water was unable to escape. Igneous rocks certainly contain a great deal of water when they reach the surface, and it seems at least possible that this water has prevented previous crystallization.

The velocities of transverse waves in rocks have been measured directly by F. Birch and D. Bancroft (1938, p. 126), who used a vibration method. Their results refer to a pressure of 4×10^9 dynes/cm.^{2}. β was found by a resonance method, α by combining β with measurements of the density and compressibility. As for the compressibility measures, these indicate velocities in granite rather greater than those found in European near earthquakes. (See p. 102 and Birch (1943 and 1961c for later work).)

Rock	α	β	α/β
Quartzite	6·08	4·00	1·52
Solenhofen limestone	5·54	3·08	1·80
Marble	6·51	3·49	1·87
Granite	{6·08	3·61	1·68
	{6·28	3·59	1·75
Norite*	6·49	3·65	1·78
Diabase*	6·96	3·85	1·81
Gabbro	{6·96	3·71	1·88
	{7·15	3·98	1·80
Pyroxenite	7·83	4·58	1·71
Dunite	8·05	4·57	1·76

* Norite is dolerite containing rhombic pyroxene (hypersthene); diabase is an alternative name for dolerite.

Note that many experimental studies calculate Poisson's ratio from α/β; this is an unnecessary complication since Poisson's ratio has little direct interest in seismology.

It has been argued that these results amount to a disproof that the upper layer is granite, and therefore that it must be a sedimentary rock, since all other igneous rocks (except obsidian) lead to higher velocities still. I find it inconceivable, however, that there can be a continuous sedimentary layer extending over most of western and southern Europe and showing no detectable variation of velocity of propagation of longitudinal waves. Both geological and seismological evidence suggest that the sedimentary layer is very heterogeneous. Gutenberg, Wood and Buwalda (1932) find, however, about 5·25 km./sec. for the velocity of P in granite in the Yosemite Valley. L. D. Leet and M. Ewing (1932a, b) have found $4·96 \pm 0·03$, $5·00 \pm 0·06$ and $5·08 \pm 0·02$ km./sec. (standard errors) in various surface granites from the study of explosions. Even igneous rocks, therefore, seem somewhat variable in this respect.

Gutenberg suggested that there may be a layer of low velocity within the upper layer, and that the Pg, Sg velocities really refer to this. Waves in such a layer are called channel waves. They are certainly mechanically possible (Jeffreys, 1957 e). For SV and SH they could travel to any distance; P would, however, suffer loss by conversion into SV by reflexion at the outer surface. The amplitudes at the surface would be small compared with those in the layer of low velocity unless the difference of velocity was fairly small.

On the whole it seems to me most likely that the upper layer is granite, probably mixed with obsidian in Europe, the intermediate one tachylyte, and the lower dunite. But other identifications would be possible. For many purposes we are concerned directly with the densities and not essentially with the petrological identification, and it is possible to give the densities more definitely because the experimental results show a very close correlation between density and velocity of propagation. Whatever the materials, if they give the right velocities, the density in the upper layer will be about 2·6, in the intermediate one about 2·85, and in the lower about 3·3. Where the actual composition is not immediately relevant to the problem under consideration I therefore prefer to continue to speak of the upper, intermediate and lower layers.

The laboratory measures of bulk-modulus are made at constant temperature, whereas seismic waves are practically adiabatic. It is therefore sometimes asked whether this distinction vitiates the comparison, as it does for sound in gases. The answer is in the negative (Jeffreys, 1930 c, 1931 g, Chapter VIII). If k, k' are the isothermal and adiabatic bulk-moduli,

$$k' = k + \frac{9k^2\alpha^2\theta}{\rho c},$$

where α is here the coefficient of linear expansion, θ the absolute temperature, ρ the density, and c the specific heat at zero strain. The latter is connected with c_p, the specific heat at zero stress, by

$$c_p = \frac{k'}{k} c,$$

whence
$$k' = \frac{k}{1 - 9k\alpha^2\theta/\rho c_p}.$$
With

$$k/\rho = 16 \times 10^{10} \text{ (cm./sec.)}^2, \; \theta = 300°, \; c_p = \tfrac{1}{3}J = 1·4 \times 10^7 \text{ erg./gm. } 1°,$$

$$\alpha = 10^{-5}/1°,$$

this makes $\qquad\qquad k' - k = 0·003k,$

so that the difference for rocks at ordinary temperatures is negligible.

The isothermal and adiabatic rigidities are equal.

The situation may be altered at great depths in the Earth, since θ may be multiplied by about 10 and k/ρ by about 3. The difference might then reach 10 % unless α decreases.

3·07. The core. Oldham's discovery of the late arrival of P near the anticentre (now called PKP) was followed up by Gutenberg. It is found that P and S are sharp movements up to about $\Delta = 105°$, though they become small at about 100°. Oldham had inferred from the drop in velocity of P that there would be a shadow zone. The phenomenon can be imitated by filling a spherical flask with water, illuminating it by a pinhole source of light, and catching the emergent beam on a photographic plate before it comes to a focus. There is a general illumination over a circle, but this is extremely intense over the circumference, where the plate cuts the caustic surface. Correspondingly in earthquakes the P at the far edge of the shadow zone, at about $\Delta = 143°$, has an enormous amplitude. At most distances P is a rather small movement, attracting attention chiefly because it comes first, but in this region it is often larger than the surface waves.

Gutenberg, having approximate velocities as data, proceeded to an extensive calculation of provisional times of other pulses that should exist if the core boundary is a sharp transition. Either a P or SV striking the core would give rise to two reflected waves whether the core is solid or liquid. The reflected waves are denoted by PcP, PcS, ScP, ScS respectively (the c being needed to distinguish them from the surface reflexions PP, PS, SP, SS). An SH wave would generate only ScS by reflexion. If the core is solid P and SV would each generate a transmitted P and SV, and an SH wave would give a transmitted SH. If the core is liquid there would be no transmitted S waves. Any transmitted wave would be broken up again when it strikes the core boundary from inside, giving a great variety of phases, emerging at the surface at different distances and times. The most important of these is SKS, which is of S type in both passages down to and up from the core, but of P type in the core. Gutenberg identified this phase and several other core phases, but all that he found had been of P type in the core. He also tried to find waves that had been of S type in the core, assuming the ratio of the velocities of P and S to be about the same as in the shell, about 1·8, but failed. Several later workers, notably Macelwane (1930) and Bastings (1935), have claimed to detect such waves, but the results are conflicting and capable of other interpretations. There is no satisfactory evidence that the core can transmit S waves, and if it cannot it is liquid. There is strong confirmatory evidence of this.

Gutenberg's predictions were completely verified by comparison with observation. Some of the pulses, indeed, were recognized by others who did not expect them. This was especially true of SKS, which arrives at the observing station at the same time as S when $\Delta = 83°$, and before S at greater distances. It is comparable in size with S, and stations reporting to the I.S.S. were liable to read as S the earliest wave arriving about the time when S would be expected; thus when Δ was over 83° they usually reported SKS as S, and there were hardly any genuine S readings. Abnormalities in the behaviour of S in this region were first noticed in 1915, and the symbol Y was given to the intruder (Turner, 1915*b*). It was identified as SKS in 1926 (Jeffreys, 1926*f*). $SKKS$ was identified in the I.S.S. in 1928 (issue for 1923 January–March; Jeffreys, 1928*a*, p. 521). A principle stated

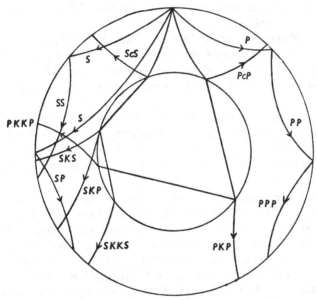

Fig. 10. Paths of the principal pulses observed in distant earthquakes.

by Lehmann and Plett has been of great service in identification of S in these and similar circumstances. As SKS has been of P type in part of its path, it must be wholly of SV type when it emerges. Consequently the horizontal displacement is in the same direction as that of P or opposite to it. But S contains SH, which is often the larger component, and the ground movement may be at any angle to that in P. Consequently S can be identified by the change of direction of the movement. Further, in normal earthquakes at the distances concerned the interval between P and SKS is nearly constant at about $10^{\mathrm{m}}\ 30^{\mathrm{s}}$. Lehmann and Plett (1932) therefore recommended that when this interval is found between P and S, it should be suspected that the apparent S is really SKS, and the true S should be sought after it.

The reflected waves PcP and ScS arrive after the surface waves at short distances and were not satisfactorily identified for a long time. They were finally found in deep-focus earthquakes by Scrase (1933) and Stechschulte (1932); in these the surface waves were small and the reflected waves were more easily seen. *ScS* is much the stronger, presumably because *SH* would be completely reflected at a liquid core. *PcP* has also been found in normal earthquakes from the records on short-period instruments, which give less magnification to the surface waves.

The first waves reflected at the core boundary to be recognized were *SKKS* and *PKKP*, each once reflected internally. The existence of all these reflexions is specially important because it shows that the core boundary is a discontinuity and not a gradual transition.

Several core phases, notably *PKKP*, *PKS* and *ScSPKP*, have caustics like that of *PKP*. It is possible that *ScSPKP* has provided some of the readings identified as *S* through the core. Others may be *SKSP* or *PPS*. Whipple tried to use the Lehmann-Plett criterion but did not publish any results.

At stations near the anticentre two waves of type *PKP* exist. The *PKP* described so far is the earlier. The later, denoted by PKP_2, has travelled by a path similar to *O2* in Fig. 3 (p. 55). It was first identified by Macelwane (1930) and Lehmann (1930), but it is less sharp than the earlier branch *O3*.

It is not strictly correct to say that there are no *P* waves between $\Delta = 105°$ and $\Delta = 143°$. It was originally supposed that there is a belt of shadow in the sense of geometrical optics, some energy being diffracted around the core boundary. There was a reading at about 150° at Abisko by Lehmann in the Buller earthquake of 1929 June 16. The beginning was not sharp. The times fitted a linear function of Δ, with $dt/d\Delta = (4·69 \pm 0·03)\text{s}/1°$. (Jeffreys 1938*a*, p. 293.) On account of the scarcity of *P* observations in $100° < \Delta < 105°$, $dt/d\Delta$ was uncertain, but was about 4·5s/1°. The extrapolated times joined satisfactorily at 105° and were combined. Theoretically the beginnings of diffracted *P* should be linear in Δ, the slope being that of the tangent to the *P* curve at its end. I interpreted the change of gradient as due to failure to read the beginning of a small movement.

Considerable doubt has now been thrown on this interpretation. Lehmann (1953) made a detailed study and found that though the amplitudes are small the beginnings show no difference of form before and beyond 105°. She therefore supposed that the pulse is refracted through a layer just outside the core where the velocity decreases a little with depth. Her inference was confirmed by Bolt (1970, 1972, pp. 302–6), who got

$$dt/d\Delta = (4·55 \pm 0·05)\text{s}/1°$$

for $105° < \Delta < 115°$.

These results do not explain the change of $dt/d\Delta$ that I found. Bolt attributes it to lower magnification and longer free periods of the earlier instruments. He quotes Scholte (1956) for diffraction in homogeneous shells, as giving the amplitude at distance $\Delta°$ in the shadow as

$$A = A_0 \exp\left(-0.20\Delta/T^{\frac{1}{3}}\right),$$

where T is the period in seconds. Periods as small as 1^s or 2^s are clear on sensitive seismographs deep in the shadow; some are shown on his Fig. 3 for $\Delta = 118.4°$.

Bolt prefers a velocity of 13·63 km./sec. at $\Delta = 92°$, with $r = 3621$ km., decreasing to 13·33 km./sec. at $r = 3475$ km. The layer of decreasing velocity would be about 150 km. thick. He proposed to call this movement Pc instead of first P and then diffracted P.

He reckons that if there is a decrease of 2 % in the seismic velocities an increase of 6 % in density is needed to account for it. It could be explained by an additional 10 % of iron with shell material.

R. D. Adams (1972) studies reflexions $PKKP$ up to $P4KP$ on the inner side of the core, especially their AB branches. The point A represents a pulse just able to enter the core, and it is inferred that $dt/d\Delta$ for it would be $(4.56 \pm 0.02)^s/1°$, implying a velocity 13.29 ± 0.06 km./sec. at the base of the shell, and in good agreement with Bolt's value.

Friedlander (1954) has investigated diffraction of a pulse by a cylinder theoretically. For a plane unit pulse the motion in the shadow zone, according to his Fig. 3, p. 722, rises from zero to a maximum and gradually falls off. The delay of the maximum appears to increase more rapidly than linearly with the distance travelled in the shadow. The rise when the distance travelled in the shadow is 30° is very sharp, and reaches a maximum when the time after the theoretical start is about 0·003 of the time needed to describe a distance equal to the radius of the core, which would be about 1^s in the seismological case. For a simple shadow the effects of diffraction are nothing like so striking as for a caustic. Friedlander's solution is for a rigid core, but no striking difference, except possibly a reversal in sign, is to be expected for a fluid core.

Similar results have been found by Knopoff and Gilbert (1961), who consider also the spherical case, both for rigid and fluid cores.

Other small waves, a few minutes later than those just mentioned, are found at distances between 110° and 143°; these appear to coalesce with PKP at 143°. These also were long believed to be diffracted, in this case from the caustic surface of PKP, but grave doubts were thrown on this explanation by Lehmann (1936) and Gutenberg and Richter (1938, 1939a). Their objection was that these movements show short periods, which are hard to understand if they are diffracted; and they maintained that the only explanation was refraction in very peculiar circumstances. I verified this quantitatively (1939d, p. 552) by applying Airy's theory of diffraction near a caustic, and found that periods of 1^s should be traceable back to 139°, and of 10^s to about 128°. Actually periods of 1^s are observed between

123° and 125°. The diffraction theory was therefore impossible. But to get
a branch of such length by refraction requires very exceptional conditions.
$dt/d\Delta$ along it varies little; it is about $2 \cdot 0^s/1°$, whereas it is about $3 \cdot 5^s/1°$
just beyond 143°. This is enough to suggest that the rays must pass much
nearer the centre than those that form the caustic. A sharp increase of
velocity, well within the core, would give the longest reflected branch that
any ordinary hypothesis will, but calculation showed that it could not be
long enough unless, in addition, the velocity decreased downwards for some
distance outside the discontinuity.

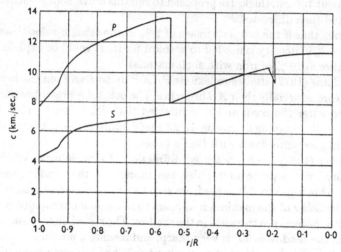

Fig. 11. Distribution of velocities of P and S from the base of the outer layers
to the centre. The outer layers are too thin to be shown on this scale.

Rays grazing what we must call the inner core would be reflected; those
striking it sufficiently steeply to penetrate would give a refracted branch
giving slightly smaller times of travel, for equal distances, than the
reflected branch. On this hypothesis the data could be fitted. Actual
records of this branch are reproduced by Nguyen Hai (1961). There is an
additional phase of short period, marked as X, about 10^s earlier. PKP_2 is
also very clear on some of the records.

With regard to the difference between Gutenberg's and my velocity
distributions around the inner core, he told me of an additional phase in
the relevant range of distance, and thought his results might refer to one
and mine to the other. His may be Hai's X.

My times for the receding branch of PKP rested on a single earthquake. Another (Jeffreys, 1942a) made them about 1^s longer; so does an
analysis of 25 earthquakes by B. A. Bolt (1959).

Considerable revision is however needed on account of recent work.

3·08. Deep-focus earthquakes. These were discovered by Turner (1922) using the I.S.S. data. His method of determining a preliminary epicentre was to draw an arc on a globe about the point that represented each station, with radius equal to the distance corresponding to the observed interval between P and S. These arcs usually intersected near one point and gave a rough epicentre. But he found that in some cases the arcs failed to meet by several degrees. The data could be reconciled fairly well by supposing P and S to have come from a focus at a considerable depth; for distant stations this would be nearer than the epicentre. When this happened, if PKP was also observed, it was found to be systematically early. A criticism of Turner's method was that the discrepancies in time were of the same order of magnitude as the errors known to exist, though not then well determined, in the Zöppritz-Turner tables, and that the possibility that they arose from errors in the tables could not be excluded.

K. Wadati (1928) studied the times of arrival of P and S at short distances in Japanese earthquakes, also relying principally on the interval between P and S. In a deep-focus earthquake it would be larger at the epicentre than in a shallow one, but would not increase so fast with distance from the epicentre. Wadati found substantial differences between earthquakes in this respect. Further, it happened several times that he and Turner found about the same focal depth for an earthquake by their different methods. His tables were much more accurate than those of Zöppritz, but it appeared possible that some of the upper layer pulses might have been mistaken for P and S. Consequently both methods, though suggestive, were not quite convincing.

Decisive evidence was obtained by Stoneley (1931*b*) and Scrase (1931). By a reciprocal theorem in the theory of small oscillations, a normal mode cannot be excited by an impulse at one of its nodes. Now a surface wave of given wave-length is a normal mode, and effectively the whole region more than a wave-length deep is composed of nodes. Hence a deep-seated earthquake should give rise to no appreciable surface waves of lengths less than the focal depth. The theory had already been worked out in detail by Lamb for the generation of Rayleigh waves. Now in the alleged deep-focus earthquakes the I.S.S. gave numerous readings in the columns devoted to surface waves, and this provided further reason for doubt. But Stoneley found that these readings were in fact much earlier than in shallow earthquakes, the difference reaching 10 min. They agreed well with the times expected for various reflected S movements (SS, SSS, etc.) which would exist with any depth of focus. He proceeded to examine a number of original records, and found that the true surface waves were untraceable at the normal time for them. It appeared therefore that deep foci did exist. Stoneley's work provided the first unambiguous evidence. It had previously been noticed by Byerly that some earthquakes showed disproportionately small surface waves, and Miss E. F. Bellamy paid special attention to records that showed 'violent commencement'. Her photographic copies of records of special interest were Stoneley's starting-point.

Scrase followed up a suggestion of G. W. Walker (1921). For a focus in the outer surface, PP is reflected midway between the focus and the observer. But if the focus is at some depth two phases of PP type are possible, one reflected about distance $\frac{1}{2}\Delta$ as in a shallow earthquake, but the other reflected close to the epicentre. The latter is actually the more familiar in the optical analogy, since it corresponds to the rays that form the image in a concave mirror. The phase reflected at half the distance is much less familiar in the optical case because the mirror seldom includes a sufficient part of the sphere, and the image is so distorted as not to deserve the name.

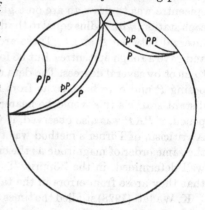

Fig. 12. Rays in a deep earthquake. S follows almost the same paths as P, SS as PP, sS as pP. sP is reflected somewhat nearer the epicentre than pP.

Scrase denoted the near reflexion by pP, keeping PP for the distant one, and similarly denoted other possible near reflexions by sP, pS and sS. He found, on examining the Kew records of some of Turner's deep shocks, that these reflexions were present and very clear. Many stations had read them for themselves and reported them as unidentified additional readings. The interval between P and pP, and that between S and sS, should vary slowly with distance by calculable amounts; this was verified and checked the identification. Thus the near reflexions provided another crucial test of the hypothesis of deep foci, and there was no further reason to doubt their reality. Stoneley's test and Scrase's, it must be noted, do not depend on having accurate times of transmission first, and are not subject to the possibility of errors of the types that cast doubt on the methods of Turner and Wadati. (See Plate VI.)

When the tables of transmission times had been improved I found it possible to make Turner's method much more accurate, and it is now possible to give a good determination of focal depth from P alone. The interval from P to pP, however, permits more rapid solution. Deep-focus earthquakes have also played an important part in solving problems that could not be settled from normal ones.

The geographical distribution of deep earthquakes appears to show systematic differences from that of normal ones. In particular, in Japan and South America the shallow earthquakes mostly fringe the coast on the oceanic side, whereas the deep ones lie in a belt on the continental side (A. Leith and J. A. Sharpe, 1936; B. Gutenberg and C. F. Richter, 1939b,

1941, 1945 d). There is evidence that they lie on steeply inclined faults, dipping toward the continental side. An extremely full discussion of geographical distribution is in *The Seismicity of the Earth* (Gutenberg and Richter, 1949).

3·09 The 1940 travel times. The early work on near earthquakes had shown the Zöppritz tables for P and S to be seriously wrong at short distances, since they correspond there to velocities of 7·2 km./sec. for P and 4·0 km./sec. for S. They were, however, used in the I.S.S. for determination of epicentres and times of origin until the issue for 1929, when they were superseded by the revised tables of K. E. Bullen and myself, published in 1935, and referred to later in this work as J.B. 1935. Already in 1915 Turner was classifying residuals in the hope of deriving corrections to the tables, and in 1926 he published an important paper, in which he classified the residuals for 1918–22 by distance and found that they varied systematically (Turner, 1926). The means varied with distance by about 20^s for P and 30^s for S. About the same time P. Byerly (1926) gave a set of times based on his special study of the Montana earthquake of 1925 June 28. I compared these (Jeffreys, 1928 a) with the corrections found by Turner and found that the agreement for P was practically perfect up to distance 55°. But at greater distances Byerly's times suggested little change in the Zöppritz times and Turner's a substantial reduction. Byerly had, however, only three observations beyond 80°. Some supplementary information was obtained from a paper by Macelwane (1923) on a Californian earthquake of 1922 January 31; in this a somewhat greater range of distance was available. Provisional corrections to S were also obtained. In a further paper (1931 d) I made a classification similar to Turner's of the I.S.S. residuals for 1923 January to 1927 March, except that I omitted a large number of earthquakes that for one reason or another seemed unlikely to yield much useful information. Actually eighty-five earthquakes were used. The general trend of the corrections was closely similar to that of the previous discussion. The corrected table was also fairly similar to those of A. Mohorovičić (1922), except that the times for P about 30° were about 10^s shorter. In all the comparisons a constant was added to the times so that the time extrapolated to distance 0° would be zero.

The corrected tables were interpolated and were published by the British Association in 1932. The matter was, however, still far from being at an end. The epicentres and times of occurrence in the I.S.S. had been derived by getting the best possible fit with the Z.-T. tables; hence if these tables were in error, it was possible that part of the error would be compensated by errors in the elements of the separate earthquakes. Consequently it was likely that the corrections, large as they were, were not large enough. It was therefore necessary to revise the solutions for the separate earthquakes, using the new tables, classify the residuals afresh, and form new corrections. Fortunately, at this stage I obtained the valuable assistance of K. E. Bullen. We found that further corrections reaching -6^s for P and -9^s for S were still needed, and proceeded by successive approximation till there was no further change. S was not found useful in determining epicentres, and the solutions for the individual earthquakes rest on P alone. Other phases were not treated till the solution for P was complete (Jeffreys and Bullen, 1935).

For S there were several troubles. Up to about $20°$ the S residuals were spread over about 20^s without any convincing concentration of frequency, and our final solution was based very largely on analogy with P in this interval, combined with a velocity of S at short distances derived from near-earthquake studies. From $25°$ to $83°$ it was found that the S residuals for a given earthquake varied fairly smoothly, but in comparison of different earthquakes there was a complication, which we called the Z phenomenon. S behaved as if its time at origin differed from that of P by a varying amount. The time of origin of P being t_0, the average S residual, taking the same t_0, might be anything from -12^s to $+8^s$. This was first noticed by Lehmann and Plett (1932) in their studies of the Peru and Marianne Islands earthquakes of 1928 July 18 and 1930 October 24. Bullen and I had to decide which of these should be regarded as a typical earthquake, and chose the Peruvian one, on the ground that the early S for the other might be due to focal depth. This, as it turned out, was wrong.

The complication introduced by SKS beyond $83°$ has already been described. We derived empirical times for it and for $SKKS$. Both showed the Z phenomenon.

For PKP the residuals showed a violently asymmetrical law of error. The normal law of error does not hold for any phase. The residuals usually show a concentration about a mode, this by itself suggesting a standard error of 2^s or 3^s, but there is a long range on both sides within which the numbers of residuals fall off very slowly up to about 10^s. A mean and a standard deviation found from the residuals in the usual way would give no idea of the real accuracy. The explanation lies chiefly in a difficulty of identification. All onsets are followed by irregular movements, and an observer may read part of the irregular movement and report it. When compared with the time of any genuine phase it will show a large residual. Even P does not usually start from rest, since there is a continual irregular motion of the ground, known as microseisms. These lead to uncertainties about the position of the true beginning of P and to large residuals of either sign. There is also occasional trouble from large clock errors. This was dealt with by detecting the offending stations and afterwards ignoring them. The result is that we have to deal with a modified law of error, with a sharp peak of correctly identified observations superposed on a nearly uniform background. We found that this could be treated without prohibitive trouble by the method of 'uniform reduction', in which the residuals are grouped by intervals and counted; then a constant is subtracted from the number in each interval, leaving the central group isolated, and a mean and standard deviation are found from the central group as reduced. Later work (Jeffreys, 1936a, 1967d, pp. 214–16) showed that this is a close approximation to the most accurate treatment, which supplies a weight to be attached to each observation as a function of the residual. This weight is approximately the probability that that particular observation is normal.

The usual methods, such as rejecting some observations while retaining others at full weight, are completely unsatisfactory. The decision about which observations to reject is arbitrary, and the choice about one observation may affect the mean by the full amount of its apparent standard error. If the study of the distribution of the residuals shows that the normal law is not followed, the correct procedure is to acknowledge the fact and devise a method of solution suitable to the actual law. This is not prohibitively difficult in practice—certainly not as difficult as making the observations.

For PKP there were a few early readings as for P, then a sharp rise to a maximum frequency within a range of about 2^s, but then the frequencies dropped off very slowly over about 20^s. We made a provisional solution, but without much confidence.

From the times of $SKKS$ and SKS it was possible to derive those of ScS, and from this PcP on the hypothesis that the velocities of P and S vary in proportion; and hence the times of other core waves. This method, however, was soon superseded because direct observations of ScS became available.

The resulting tables were adopted in place of the Z.-T. ones for the I.S.S. reductions from 1930 January.* The work since then has consisted mainly of tracing and removing systematic errors. The need to allow for the ellipticity of the Earth was already known, but we had postponed attention to it because it seemed unlikely to be one of the largest errors outstanding, and because its full calculation would take a long time. Bullen (1937a, 1938a, b, c), however, afterwards calculated it for all the relevant phases and allowance has been made for it. The correction of the P table was simple because the earthquakes fell into a few separate geographical groups: south-east Europe and western Asia, North and Central America, Japan, South America, and Pacific Ocean. For earthquakes of any one group the ellipticity correction would be nearly the same, and its values for a mean epicentre could be applied to the summaries already obtained. The corrected values were then combined as before, keeping the previous standard errors, since the correction was a known quantity. Thus the P table was reduced to a mean sphere.

A further correction was needed to P at distances over $90°$ and to PKP on account of the prevalence of late readings due to weakness. The difficult

* The allowance for depth of focus was not calculated for the J.B. 1935 tables, and until they were superseded the corrections based on the Z.-T. tables continued to be used.

ranges were treated by using only the best stations with vertical component instruments—the latter because P at large distances and PKP rise steeply and are most prominent on the vertical. I found that this selection gave nearly symmetrical distributions of residuals, with about the same standard error and background effect as at moderate distances, and the effect of skewness was satisfactorily eliminated. The greatest change from the J.B. 1935 times due to the correction of these two systematic errors was $-2\cdot7^s$ (Jeffreys, 1937*d*).*

For S and SKS the Z phenomenon was the outstanding difficulty, and there was something seriously wrong with the times at large distances. Scrase (1933) had already remarked that $dt/d\Delta$ for S beyond about 70° in his study was less than my 1932 times gave, but the 1935 values are greater than the 1932 ones. I thought that the difference might be due to some error in estimating the allowance for focal depth, but later work has fully confirmed it. The first work that convinced me (Jeffreys, 1935*c*) was on the Afghanistan deep-focus earthquake of 1929 February 1. A long series of European stations had read S, and many North American ones SKS. The former gave small residuals, the latter very consistent ones of about -12^s. The only possible explanation seemed to be that the J.B. times of SKS, and, therefore, of S at large distances, with which SKS had been accurately compared in the process of identification, were about 12^s too great.

The error appeared to be connected with the phenomenon of multiplicity. Many earthquakes do not consist of a single shock, but of several at intervals of a few seconds, and a later one may be stronger than the first. The interval may be even longer. In such a case most observers will read the P of the first shock; some of them may report that of some later shock as an additional reading. But in looking for the most conspicuous thing to identify as S they naturally take the largest, and thus get an excessive interval between P and S. Turner had found cases of excessive intervals and attributed them to foci higher than normal. This led to the conclusion that the normal earthquake had a focal depth of at least 100 km.; but this was contradicted by the Montana earthquake, which certainly had an upper-layer focus, since Pg and Sg were recorded, and gave transmission times as for normal earthquakes. The phenomenon was finally explained in terms of multiplicity by Stoneley (1937*b*, 1939*a*) after an extreme instance of it had been pointed out by Hughes (1936). Tillotson (1938) examined the additional readings in the I.S.S. in the cases where Turner had inferred 'high focus', and found that they also gave evidence of multiplicity. It seemed likely that all cases of positive Z could be explained on these lines; and the Peruvian earthquake, which we had taken as standard, showed signs of it. See also Pilani and Knopoff (1964).

The easiest procedure for S and SKS seemed to be to start again, using

---

* Bullen independently applied the ellipticity correction; his work was not published separately as the two papers were so similar.

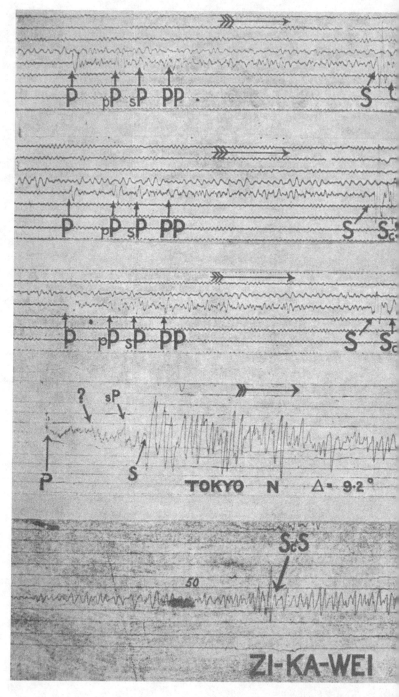

Records of the deep-focus earthqua[k]

PLATE VI

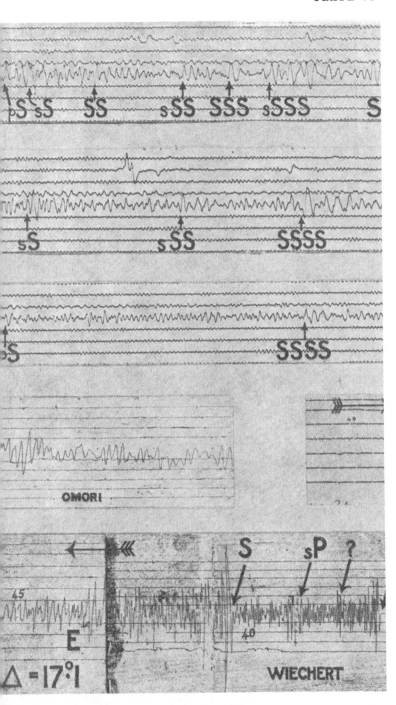

1931 February 20. From F. J. Scrase (1933).

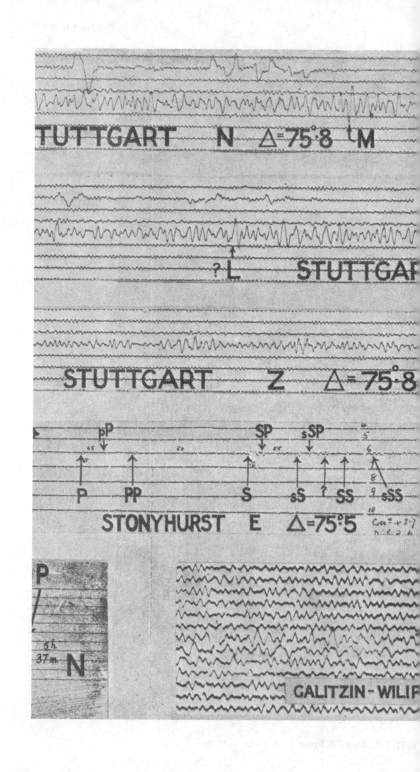

TUTTGART   N   △=75°·8 ᴸM

?L   STUTTGAᴿ

STUTTGART   Z   △=75°·8

pP   SP   sSP   ⁴⁄₅
+5   60   55   6
P   PP   S   sS ? SS 9 sSS   8
STONYHURST   E   △=75°·5   Cₙ=+2·7   10   n·l·a·h

P

5ʰ 37ᵐ   N

GALITZIN-WILIP

GALITZIN-WILIP

E   Δ=75°8   M⌐   GALITZIN-WILIP

M⌐   GALITZIN-WILIP

SSS   ?L

MILNE-SHAW   1931 Feb. 20

pP   P

BUFFALO   Z   Δ=87°6

geocentric distances throughout. The residuals were much more scattered in some earthquakes than in others, and in some cases seemed to fall into two or more groups separated by several seconds. A summary based on observations from two series would be representative of neither. At this stage, therefore, attention was confined to earthquakes with long series of *S* and *SKS* readings, consistent enough to ensure that nearly all stations over long ranges of distances had read the same thing (Jeffreys, 1939*a*, *b*). To get a sufficient number of observations of *SKS* at large distances it was necessary to use a number of earthquakes in the southern hemisphere (Jeffreys, 1938*a*). It has been recorded to 145°. Much of the information was supplied by a set of Japanese deep-focus earthquakes. The result was to confirm the need for the reduction at large distances. The mean residuals over 5° ranges of distance for the separate earthquakes were tested for consistency by means of χ^2, and found satisfactory, so that the correctness of the revised tables is adequately checked.

Reference has been made to the difficulty of identifying *S* up to 25° or so in normal earthquakes. It can be identified if attention is paid to the change of direction of the movement, but it is followed by several large movements, one or other of which is read by most stations. No satisfactory explanation has been given of these movements, but it is clear that they are not *S*. In deep shocks, however, the *S* residuals at these distances show a pronounced concentration, with only the usual amount of background effect. A study of ten deep shocks in Japan led to what appeared to be a highly satisfactory determination of the times of *S* up to 20°.

When this part of the work was complete it was possible to compute the velocities of *P* and *S* in the shell, and from them a set of calculated times for the core reflexions *PcP* and *ScS* for different trial depths of the core. Comparison with observed times gave an estimate of the radius of the core as a fraction of that of the outer radius of the lower layer. This had been obtained by Gutenberg in an early paper from the distance where *P* begins to be diffracted, but this was not very satisfactory because the shadow is not sharp, and the determination requires the slope of an empirical curve at its very end, always a matter of considerable uncertainty, but at that time no other method was possible. The data now used include determinations by Gutenberg and Richter (1939*a*, pp. 105–106) for *PcP* and for *ScS*, with focal depths from 100 to 200 km. and from 500 to 700 km., my determinations for *ScS* from the earthquakes studied by Scrase and Stechschulte, and Tillotson's (1939) for *PcP* for an earthquake in the African Rift Valley. The mean radius of the core was found to be 3473 ± 3 km.—actually in very good agreement with Gutenberg's value. With this datum the times of *PcP* and *ScS* can be computed and used to determine times in the core (Jeffreys, 1939*c*),

Times of *K* can be obtained by comparing *PcP* with *PKP*, or *ScS* with *SKS*. The ranges of distance overlap slightly, and the times for the range common to both were found to be consistent (an anxious moment). They

9

do not give directly values for very short distances, but they give values for distances short enough for the cubic formula to fit, and then those for shorter distances can be filled in. The velocities in the core can then be calculated. The inner core introduced a further complication; the solution obtained within it is not unique but at any rate reconciles the data. The resulting tables are referred to as J.B. 1940.

3·10. The 20° discontinuity. Byerly's work on the Montana earthquake and my earlier work showed that $d^2t/d\Delta^2$ is very large about distance 20°. This was again found in the early stages of my work with Bullen. F. Neumann (1933) also noted it in the Santiago (Cuba) earthquake of 1932 February 3. At short distances $dt/d\Delta$ in most regions is about 13·8^s/1°, decreases gradually to 12·3^s/1° between 18° and 19°, but drops to 10·4^s/1° between 20° and 21°. At larger distances it falls off gradually, reaching 4·4^s/1° beyond 100°. These facts indicate that between the depths reached by rays that emerge at 19° and 20° there is a great increase of the velocity of P. The suggestion of a new discontinuity was supported by some observations made by Lehmann (1934) on the Iceland earthquake of 1929 July 23 and the Azores one of 1931 May 20. She found what appeared to be a second P (Pd) in records made between 20° and 24°, just as Pg continues to be readable in near earthquakes at distances where it is preceded by P. Her readings, however, were a few seconds too late to lie on a smooth continuation of the P curve up to 19° (but see p. 124). What is certain is that if we take the first arrivals only, $dt/d\Delta$ decreases by about 20 % about 20°. This could be interpreted in two extreme ways: (1) In the 'lower layer' of near-earthquake studies the velocity may increase smoothly with depth, and then increase discontinuously at some deeper level, below which it again increases smoothly. Over a certain range of distance three P's would arrive. One would have travelled wholly above the discontinuity, one (Pr) would have been refracted into the deeper region and out again, and one would be reflected at it. (2) The velocity may increase continuously with depth and especially rapidly over a certain range, but not rapidly enough to give triplication. The time curve would have a continuous slope, but the curvature would be strong about 20°.

(1) would be confirmed if the later arrivals were definitely identified, but they are dubious. But even if (1) is correct the oscillations that follow P everywhere would make it impossible to identify with confidence a phase a few seconds after the first movement. In favour of (2) is the fact that amplitudes are large about 20°, which would be a natural consequence of (2) but not of (1).

It was clear that the question could not be decided by means of normal earthquakes alone. It was tested by means of the Japanese deep shocks. Separate calculations of P times were made for different focal depths, based on (1) and (2) respectively. The results differed by amounts up to 1·4^s for the depths of these earthquakes, and such differences were large enough to be tested with the quantity of information available. Comparison with observation gave agreement with (1) and definite disagreement with (2). (But see p. 124.) Nevertheless, (1) is not altogether satisfactory because it does not explain the large amplitudes. I finally adopted an intermediate hypothesis, using a continuous distribution of velocity, but with sufficiently rapid variation over a certain interval to give a triplication of P at the corresponding distances. In deep shocks the first arrivals of P and S at some distances correspond to parts of rays that do not give first arrivals in normal shocks, and consequently part of the range of triplication could be reconstructed from the data for the deep shocks. Some arbitrariness remained, but the solution reconciled the whole of the data.

Bolt (1970) made a study of I.S.S. data for $100° < \Delta < 115°$. Sixteen earthquakes were used. The times of travel, plotted against distance, showed strong concentrations near the calculated times of P (mostly diffracted), PP, and $PKiKP$, with scattered ones after P, attributable to pP and sP. From about 40^s to 2^m after P there were hardly any readings, but then the frequency became appreciable again. The edge of this band ran about 90^s before PP, not parallel to $PKiKP$. This suggests that it is associated with PP, and the interpretation suggested is that the readings are of a PP, called PdP, reflected on the underside of a first-order discontinuity at depth about 400 km. R. D. Adams (1968, 1969) had previously suggested this for early reflexions of $PKPPKP$. It is also about the depth 439 ± 21 km. that I estimated in 1936 (Jeffreys 1936b, p. 418) for the $20°$ discontinuity from the times of P. Several later studies cast doubt on this interpretation and indicated a rapid continuous transition instead. Bolt's evidence is more direct. Engdahl and Flinn (1969) in a study of $PKPPKP$ in the distance range $55°$ to $75°$ found a transition zone at a depth of 650 km.

The $20°$ discontinuity may be related to a peculiarity of electrical conductivity noticed by Chapman and further studied by Lahiri and Price (1939). The variations of the Earth's magnetic field are largely due to induced currents in the interior. The data for the diurnal variation and for magnetic storms appear to require a rather sharp increase of conductivity by a factor of about 10^4 at a depth of about 700 km. If the $20°$ discontinuity is really sudden, its depth is about 500 km. With the amount of smoothing that I have adopted, the specially rapid variation of the velocity of P is between depths $0·06R$ and $0·10R$, say 400 and 650 km., to which the thickness of the outer layers must be added. It seems possible that the sharp increase of conductivity is near the base of the region of rapid increase of velocity.

A possible explanation of the rapid increase of electrical conductivity found by Chapman and Price is suggested by some unpublished work of H. Hughes. For non-metallic solids conductivity depends mainly on the number of excited electrons, and if E is the lowest energy of excitation it follows the rule

$$\sigma = \sigma_0 \exp\left(-E/2kT\right),$$

where k is Boltzmann's constant, T the absolute temperature, and σ_0 another constant of the material. E can be measured by a study of the excitation by radiation or by actual measures of conductivity at high temperatures. The distribution of σ in the Earth therefore gives some idea of that of T. For olivines E ranges from 3·3 to 4 electron-volts per molecule according to the ratio of the amounts of Mg and Fe. The conductivity of the mantle would imply a temperature gradient of about 2° C./km. at a depth of 700 km.

H. P. Coster (1948) had measured electrical conductivities of seven different rocks at temperatures up to 1000° and inferred an increase by a factor of order 10^4 between the surface and 600 km. depth. Shimazu (1954) gives theoretical values.

From discussion with Hughes I gather that this is unlikely to have much to do with the changes of mechanical properties at depths of this order, because only a few per cent of the electrons are in the excited state even at the base of the shell and can hardly account for a drastic change of structure.

On the other hand these results require at least a considerable change in Ramsey's theory (p. 420). His estimate of the change of density at the core boundary assumes an excitation energy of 13 electron-volts per molecule, which is based on quartz and is much too high for olivine. If we accept the rest of the theory the discontinuity of density at the core boundary would be much too small. I do not think that the results altogether exclude the possibility that the core is due to pressure ionization, but they do require a complete revision of its quantitative basis.

The final stage is the computation of the times of various compound phases, especially those that have undergone one or more reflexions at the core boundary. This has been done for a surface focus (Jeffreys, 1939b, pp. 522–533; Bullen, 1939c, pp. 583–593) and for depths at intervals of $0·01R$ to $0·12R$. The greatest focal depth known so far is $(0·0969 \pm 0·0007) R$ for the earthquake of 1934 June 29 in the East Indies. This was originally reported as $0·11R$, but this turned out to be too great when more observations became available (Jeffreys, 1942a). Gutenberg and Richter (1949, p. 16) classify 9 shocks as of depth 700 km., and 25 as of 650 km., but I think that these need further check. A large number of compound phases are readable, but in general with less accuracy than the main phases P, S, PcP, ScS, PKP, SKS, and the deep-focus phases pP and sS. The latter pair provide a rapid means of estimating focal depth, since the intervals from P to pP and from S to sS are very sensitive to focal depth but not to distance. The interval from P to SKS is also useful over the range of distance about 80–100°, where both are observed. The direct evidence on focal depth given by these comparisons, even if they are possible for only a few stations, may give as good a determination as the whole of the P observations. Further, it facilitates the determination of the epicentre, because it effectively eliminates an unknown; and more drastic grouping of stations is possible when three unknowns instead of four have to be determined. The final tables were collected and republished by the British Association and are known as the J.B. 1940 tables. They were adopted for use in the I.S.S. for 1937 and later years.

In a long series of papers, over about the same period as was covered by mine and Bullen's, Gutenberg and Richter (1934, 1935, 1936 a, b, 1937) made independent determinations of the times of the principal phases and the velocity distributions. On the whole their results agree well with ours, and we were often able to settle doubtful points quickly by consultation with them.

3·101. Revised times. Gutenberg (1926) noticed that the P amplitudes in European earthquakes were small from about 5° to 15°, and later he and Richter (1935, 1939c) inferred from comparison of amplitudes at different distances that $d^2t/d\Delta^2$ is specially small from 12° to 14°. Lehmann, for the Azores earthquake, found small amplitudes about 15°. Gutenberg (1945, 1948; Gutenberg and Richter, 1939c) maintained that there is a layer of low velocity at a depth of about 80 km. An interesting consequence (pointed out privately by Gutenberg) would be an explanation of the clearness of S in a deep-focus earthquake. A pulse originating below the low-velocity layer could penetrate it easily; but one originating above it would be spread out in striking it. On the other hand, the amplitude data are not completely consistent, and may be capable of other interpretations. Besides $d^2t/d\Delta^2$ being small, small amplitudes could occur owing to peculiarities of magnification and orientation or absorption in a thin layer; slope of an interface near a station, for instance, could disturb the amplitude without appreciable change in the time, and it would be desirable to have a thorough analysis capable of separating out such effects, which would probably be strongly associated with azimuth of arrival.

In the construction of the P table adequate observations existed at distances up to 30° or so in European and Japanese earthquakes and appeared to be in satisfactory agreement. In a revision (Jeffreys, 1952d) I found that the data for the Mediterranean region supported a change of $dt/d\Delta$ at short distances (up to 8°) from about 14·3ˢ/1° to 13·8ˢ/1°, in agreement with the European near earthquakes; there is no evidence against the conclusion that this change is needed over the whole region from Spain to Turkey, though the evidence for it rests mainly on earthquakes near Italy. Changes beyond 8° are small, but the time curve is nearly straight up to 14°, and the accurate location of a sharp change in $dt/d\Delta$, if there is one, is more difficult than ever.

My results were:

	$dt/d\Delta$ (sec./1°)	Surface velocity (km./sec.)
Pg	$19·85 \pm 0·11$	$5·598 \pm 0·031$
Sg	$32·66 \pm 0·19$	$3·402 \pm 0·020$
S	$25·67 \pm 0·24$	$4·329 \pm 0·041$

Comparison of constant terms for P and Pg was satisfactory. The differences all agreed well with those from explosions, and indicate that the foci were very close to the surface, probably within the sedimentary layer. But comparison of Pg and Sg indicated a non-random variation of $\pm 1 \cdot 1^{\text{s}}$ common to all readings of each earthquake. Focal depth provides no explanation of this, since it introduces further difficulties in the comparison of P and Pg, which is satisfactory. The constant terms in $t(S) - t(Sg)$ led to a correction $-2 \cdot 2^{\text{s}} \pm 1 \cdot 1^{\text{s}}$, but the solution gave $\chi^2 = 23 \cdot 9$ on 5 degrees of freedom, and it could not be supposed that the times of S were known within 2^{s} and possibly more.

The appearance of records of near earthquakes has never been brought into proper relation to the theory of amplitudes given in § 2·06. If the interface is treated as plane and horizontal, a P pulse incident on it from above should not produce a pulse with a sharp beginning at all. It would give only a diffracted movement. If the original pulse starts with a finite velocity, the emergent P would start with a finite acceleration. If the curvature of the Earth is taken into account, a sharp movement could be refracted back, but the amplitudes would be in hopeless contradiction with the facts. The signs of consecutive reflexions would be reversed. For Sg, for which the theory is easier, there would be a sharp beginning, but successive reflexions would have the same sign and mount up, giving something like the G phase (the first swing in the Love waves) with a first swing of order 20^{s} at $20°$ (Jeffreys, 1962c). There would be nothing like the oscillations in 1^{s} or 2^{s} that actually appear.

It seems that the only way of reconciling theory with observation is to abandon the idea that the Mohorovičić discontinuity is smooth. If it is an irregular surface like ground glass some parts of the original wave front would be at angles that permit a pulse to penetrate.

The most direct evidence seems to come from explosions. E. Penttilä (1972) for Scandinavia (including Finland) gets values (Fig. 7) from 15 to 20 km. for the thickness of the 'granitic' layer, dropping to 10 km. just off the coast, and (Fig. 9) from 30 to 40 km. for the whole of the upper layers. He also shows a map (Fig. 21) of the thicknesses calculated from free-air anomalies of gravity, ranging from 30 to 45 km., and showing a general correspondence.

P. Mechler and his collaborators (esp. 1969) also get a variation of the thickness in three parts of France, partly from explosions and partly from earthquakes. Mechler gets variations of order 5 km. in the depth of the Moho, but mainly associated with mountains. One of his remarks (Mechler and Rocard, 1964) is relevant to variations with azimuth. The angle of emergence of P at the surface can range from about $45°$ to $90°$ for increasing Δ, and over this range the correction for the variation of the thickness of the upper layers can vary appreciably, especially if the interfaces are inclined. Consequently a variation with azimuth is possible, but it probably varies considerably with distance, unlike that given by Herrin and Taggart (1968). This is relevant also to Bolt's (Bolt and Nuttli, 1966) detection of such variation in California.

Mechler and his collaborators (1970, 1971) also find evidence for a discontinuity at a depth of about 600 km., which agrees with that inferred by Dorman and Lewis and by Engdahl & Flinn.

Landisman and Mueller (1966) comment on the fast Pg pulse noted by Willmore in the Heligoland explosion. They consider this the same as is called Pg in America. Their interpretation of the usual Pg, with a velocity of about 5·6 km./sec., is that the upper layer is underlain by a layer of lower velocity, about 5·6 km./sec., with one of high velocity below it. The European Pg is then interpreted as a series of branches produced by successive reflexions at the lower interface. They reproduce records from many parts of the world that show the early movement, which they consider to represent the true granitic layer. An alternative, however, is that it may be sedimentary. Some sediments (especially limestones) show higher P velocities than 5·6 km./sec., the lower elastic moduli being more than compensated by the lower density. A shock in a thin surface layer with an irregular interface could produce waves travelling horizontally. If what is usually called Pg (their Pc) is a series of reflexions we should expect gaps or overlapping in some ranges of distance; actually it looks like a single series.

In Japanese earthquakes the times up to 18° need practically no change and are about as well determined as the European ones; there is therefore clear evidence for regional differences. The times beyond 20°, relative to those up to 18°, need a reduction of 1^s or possibly 2^s. Unfortunately this vitiates the comparison used to test whether the 20° discontinuity is a genuine discontinuity (p. 119). At large distances the differences became small (Jeffreys, 1954a). In any case the strong curvature should be moved to 18° or 19°.

The evidence for North America is conflicting. I found (1940a) that the times of P were about 3^s longer near 20° than the tables indicate. Gutenberg found that the time is nearly a linear function of distance up to 15° and that a new branch, due to a pulse that has passed through the layer of low velocity, enters at 17°. Thus, though he claims that the 20° discontinuity does not exist, he effectively shifts it to 17° in an accentuated form. I found his evidence statistically unsatisfactory, and a fresh analysis of different sets of data left the need for a change of velocity at short distances in complete doubt. Comparison of times at greatly different distances is practically impossible for Californian earthquakes, because the stations concerned, when they differ much in distance, are usually in very different azimuths, so that an error in the adopted travel times would be compensated by an error in the epicentre.

Explosions in North America and South Africa, as in Europe, have usually given P velocities at short distances of 8·1 to 8·2 km./sec. (Tatel, Adams and Tuve, 1953). Lehmann (1955) made a study of five earthquakes in eastern North America, and I found (1961c) a suitable set, mainly near the west coast, in the I.S.S. For these a classification of stations into quadrants led to a useful solution. It appears that the North American times of P agree with the European ones within a second or so and definitely not with the Japanese ones. There is a considerable difference for S, according to Lehmann, who found $dt/d\Delta$ about 24·0^s/1° up to 14°, with a discontinuous increase by about 13^s.

The reduction of $dt/d\Delta$ at short distances, with little change at 20°, implies an increase of $dt/d\Delta$ near 20°, and it is now possible to connect the European times up to 15° with Lehmann's observations of Pd in the Azores earthquake while keeping $d^2t/d\Delta^2$ negative. These observations did not extend beyond 21°, though there were many stations between there and 25°, and it appeared that Pd actually terminates about 21°, either by a cusp or through being cut off by a discontinuity in velocity. On investigation it was found (Jeffreys 1958e) that the hypothesis of a cusp led to a contradiction, but that if an ordinary cubic formula holds up to 21° for Pd, Pr must enter between 17° and 18°. The Pd emerging at 21° would reach about 0·03 R below the Mohorovičić discontinuity. The times of Pr imply that at the transition there is a considerable discontinuity in velocity gradient but not in velocity.

Lehmann's Pd has not been confirmed in other earthquakes, but the published seismograms for this one look convincing. However the variation of velocity with depth is so likely to be sensitive to differences of temperature gradient that it would not be at all surprising if the phenomenon is a regional peculiarity. In any case, if the 20° discontinuity exists, it is at a distance appreciably less than 20° and the discontinuity is in dc/dr, not in c.

In various studies (Jeffreys, 1954a, 1961c, 1962b) I have since found the following values for different regions. They are averages between 2° and 10°.

	P		S	
	$dt/d\Delta$	Velocity	$dt/d\Delta$	Velocity
J.B. 1940	14·08	7·90	25·09	4·29
Europe	13·66 ± 0·07	8·140 ± 0·041	24·28 ± 0·15	4·576 ± 0·028
Central Asia	13·64 ± 0·10	8·146 ± 0·060	24·11 ± 0·10	4·608 ± 0·018
W. North America	13·95 ± 0·16	7·966 ± 0·091	—	—
E. North America	13·59 ± 0·10	8·176 ± 0·060	23·66 ± 0·17	4·969 ± 0·033
Pacific	13·654 ± 0·041	—	—	—
Japan	14·13 ± 0·04	7·870 ± 0·024	25·41 ± 0·20	4·373 ± 0·034

The data for eastern North America are from a paper by Lehmann (1955), the rest from the I.S.S. There was a total lack of S observations near the 1940 times about 8°; there were appreciable concentrations for Europe and Central Asia corresponding to the values given. The difference from my value of 1952, above, is striking. In a later paper with Arnold and Shimshoni

(1963) with much advice from Dr Lehmann, we traced S with this velocity to 17°. Regional differences are clear even for P. The Pacific data are from papers by Carder and Kogan (1960) on atomic explosions at Bikini and Eniwetok.

In the calculation of the velocities the radius of the mean sphere was taken as 6371 km. and the length of a degree on it 111·19 km.

Arnold (1966) made a mighty study on Japanese deep earthquakes. The main object was to improve the times of S at short distances. This seemed promising because the I.S.S. had reported many depths as small as 0·005R, with apparently good series of S observations, and less extrapolation would be needed to derive times for a surface focus. A slight correction was actually found necessary to the P times.

Further data on Pacific explosions have been published by D. S. Carder (1964) and reduced by M. L. Gogna (1967). The tables on p. 126 are extracted from the various results. The uncertainties given are standard errors of summary values over ranges; the rest are interpolates. Gogna's solution up to 18°·6 is
$$t = (4·27 \pm 0·20)^s + (13·628 \pm 0·019)\,(\Delta/1°)^s.$$

For all regions of epicentres except Japan the times of both P and S need slight reductions. The details are, however, very different. The times of P up to 15° (actually a little beyond it) indicate a larger $dt/d\Delta$ in Japan than in Europe but about the same constant term. Those for the Pacific give about the same $dt/d\Delta$ as in Europe but a smaller constant term. The smallness of Gogna's standard errors arises from two causes. The standard error of a P observation was 0·45^s instead of the usual 1·2^s or so; this appears to be due to exceptional clearness of the P arrivals from these explosions. Also since the epicentres and times of occurrence were accurately known, no allowance had to be made for their uncertainties.

Lapwood and Gogna (1970), Gogna (1973) give times of S, PcP and ScS from Pacific earthquakes.

Randall (1970) gives values for SKS, quoted from Hales and Roberts after a slight adjustment. He infers a velocity 8·26 km./sec. at the top of the core. He also (1971) gives times of S up to 80°. The source parameters were based on the 1968 seismological tables for P-phases (Herrin *et al.*).

When the observations of the first Bikini explosion were made at about 70°, they were found to be a little early, and this was immediately attributed to smaller thickness of the upper layers. Comparison with Europe, however, shows that the difference decreases with Δ and has practically disappeared by about 50°. The effect of change in a surface layer would increase with Δ, contrary to the results. There can be no doubt from the times up to 15°, where the differences are 3^s to 4^s, that the times through the upper layers are less in the Pacific; but the difference must be partly compensated by slightly lower velocities at greater depths. The large corrections about 60° found by Gogna are not shown by the other regions. They rest on many

New times of P and S

P

Δ	J.B. 1940	Europe (Jeffreys 1958c)–J.B.	Japan (Arnold)–J.B.	Pacific (Gogna 1967–J.B.
0	0 (6·8)	(+0·9)	(+0·8)	(−2·5)
5	1 18·1	−2·0	+0·9±0·2	−5·7±0·1
10	2 28·0	−3·6	+0·8±0·2	−7·4±0·1
15	3 35·0	−2·3	+0·1±0·2	−6·3±0·2
20	4 37·0	−3·8	−2·6	−2·8±0·1
25	5 26·8	−2·3±0·3	−1·7±0·4	−3·0±0·2
30	6 12·5	−1·6	−2·1	−2·5±0·3
35	56·1	−1·1±0·4	−2·0	−2·2±0·2
40	7 38·1	−1·0	−1·7	−1·6
45	8 18·9	−1·0	−1·8	−0·9±0·2
50	58·0	−1·0±0·5	−1·7±0·3	−1·2±0·2
55	9 35·4	−0·9	−1·5	−2·4±0·3
60	10 10·7	−0·7	−1·2	−3·4
65	44·0	−0·7±0·8	−0·9	−3·1±0·2
70	11 15·4	−1·1	−0·7	−1·6±0·1
75	45·0	−1·6±0·5	−0·6	−1·0
80	12 12·7	−1·7	−0·9	−1·2±0·1
85	38·5	−1·5	−0·9±0·3	−1·6±0·2
90	13 2·7	−1·2	−0·9	−1·9
95	25·7	−0·5±0·7	−0·9	−2·0±0·3
100	48·4	—	—	−2·4
105	14 10·6	—	—	−2·2

S

Δ	J.B. 1940	Japan (Arnold)–J.B.	Pacific (Gogna 1973)–J.B.	Randall–J.B.
0	(10·7)	(+0·8)	−3·3±1·0	—
5	2 17·5	+0·6±0·6	−6·6±0·9	—
10	4 22·2	+0·3±0·9	−7·9±1·0	—
15	6 22·9	−0·7±0·8	−5·2±0·8	—
20	8 17·1	−3·2±0·6	−2·1±0·6	—
21	8 37·4	—	—	+5·5
25	9 48·9	−0·5	−4·6±0·6	+3·4
30	11 10·2	+0·1±1·0	−2·9±0·6	+1·6
35	12 28·2	+0·9	−0·8±0·6	+0·5
40	13 44·5	−1·2	+0·1±0·7	−0·4
45	14 57·9	−1·3±0·8	+1·0±1·0	+0·1
50	16 8·6	−1·4	+1·7±0·9	+0·5
55	17 16·8	−1·2	+2·0±0·8	+1·7
60	18 22·6	−1·2	+1·9±0·5	+0·7
65	19 25·5	−1·1±0·9	+1·7±0·3	+1·0
70	20 25·6	−1·2	+1·5±0·3	+1·4
75	21 22·6	−1·3	+1·4±0·3	+1·1
80	22 16·5	−1·4	+1·6±0·3	+2·7
85	23 7·3	−1·5±0·9	+2·0±0·4	—
90	23 54·5	−1·5	+3·1±0·5	—
95	24 38·2	−1·4	+4·8±0·7	—
100	25 20·4	—	+5·2±0·8	—
105	26 2·1	—	+3·1±0·5	—

highly accordant observations at College (Alaska) and Watheroo (W. Australia). There was some doubt about the European times as there was an inconsistency between stations to the east and west, which might be due either to a variation of velocity within Eurasia or a systematic error. Arnold's Japanese results, however, should be thoroughly reliable.

The explosions have yielded little information on S. Shimshoni has started a study of European and Central Asian deep focus earthquakes on the same lines as Arnold's. The focal depths do not exceed about $0.025R$, and for Central Asia the near stations are not so numerous, but the data should be adequate.

A disturbing result on S at large distances is found by Cleary (1969) and partially confirmed by Bolt *et al.* (1970). Cleary estimated $dt/d\Delta$ for S at $98° < \Delta < 126°$ as about $8.9^s/1°$. Bolt for SH in $91° < \Delta < 114°$ got

$$(8.68 \pm 0.13)^s/1°.$$

The J.B. times in $100° < \Delta < 105°$ give $8.34^s/1°$. Bolt gives the time at $100°$ as $25^m (28 \pm 1)^s$; the J.B. time is $25^m 20.4^s$. Arnold had no data beyond $95°$ but from $90°$ to $95°$ would give $8.76^s/1°$. Extrapolating to $100°$ would give $25^m 20.6^s$. If $dt/d\Delta$ was constant beyond $95°$ it would agree with Bolt's value; but there is a difference of over 7^s in the travel time.

This recalls a difficulty found by Lehmann and Plett (1932) (see also p. 114). They had data for three earthquakes, in Mexico, Peru, and Marianne Islands. The times at large distances in the last two of these differed considerably, and I ultimately decided that the Marianne Islands values were right and that the Peru earthquake was multiple. The decision seems well confirmed by Arnold's study, which made little change from the J.B. times. But it does seem possible that $dt/d\Delta$ for both P and S is nearly constant beyond $95°$, as if the velocity decreased slightly with depth. The continuations beyond $105°$ would then not need to be due to diffraction. This has also been asserted by Lehmann (1953).

I am not sure whether any of these discrepancies is due to multiplicity, discussed on pp. 116–17.

Arnold's work had two main objects: to test Gutenberg's theory of a layer of low velocity and to improve the times of S at short distances. Gutenberg in his original paper proceeded by plotting the arrival times with regard to distance and looking for an inflexion. $dt/d\Delta$ at the inflexion would be equal to r/c at the focus, and if the depth of focus is known otherwise (as from $pP-P$ or from comparison of times of P at near and distant stations) this would determine c at a known depth. From inspection of Gutenberg's graphs I doubted whether $dt/d\Delta$ at the inflexion could be determined within 10%. Arnold proceeded by dividing the observations into three ranges of distance: an intermediate one about the inflexion, where the times could be represented by a cubic or quartic in Δ, a nearer one and a distant one. In the near and distant ranges my times of 1954 (adapted to deep foci) were used; it was in the intermediate range that any anomaly should show if there is a layer of low velocity. Actually for many earthquakes the coefficient of the cubic term had the wrong sign; and it was always too small for a good location of the inflexion.

The epicentres were recalculated for about 100 earthquakes, with the

1954 times as basis, but they depended too much on distant stations in Europe, so that an error in the tables might lead to a systematic error in the epicentres. Revised solutions were therefore made, with the depths given by the previous solution, but redetermining the times of occurrence and epicentres from stations within 15°. For them there was little doubt about the times. The result was that the epicentres were systematically shifted a little to the southeast. It therefore appeared that the P times at distances over 50° needed a slight reduction from the 1954 ones. A principle due to S. Mohorovičić was used. This depends on locating pairs of distances Δ_1, Δ_2 for ascending and decending rays, such that

$$\partial t/\partial\Delta_1 = \partial t/\partial\Delta_2;$$

then for a surface focus

$$t(\Delta_1 + \Delta_2) = t(\Delta_1) + t(\Delta_2) - 2\tau,$$

where τ is a correction to the origin time and constant for each earthquake. Forty-one earthquakes gave useful data and a least squares solution was made for the τ and 21 mean corrections by ranges of distance. This remained arbitrary by an additive constant, which might be added to all origin times and subtracted from all travel times and leave all the calculated arrival times unaltered. This was fixed by using the near reflexion pP, which has travelled up and down through the upper layers. The solution from P alone really gives the depth below the discontinuity and hence the time from the focus to the nearest point on it. The interval $pP-P$, extrapolated to vertical travel, is twice the time to the epicentre. Comparison permits elimination of the additive constant. Actually the earthquakes fell into two regions, for which this constant differs by about a second. The whole of the P times were then recalculated and revised solutions for the separate earthquakes were made.

In my 1939 work on Japanese deep foci, focal depths from $0.035R$ to $0.06R$ were used. Arnold had depths from $0.005R$ to $0.09R$. These would permit estimates for a surface focus with much less extrapolation. He used the 1940 times of S as a first approximation. Two methods were used. One was the matching technique as for P. The other concerned distances up to about 18°, where the effect of focal depth is small and its inaccuracy will produce an error of the second order. It was therefore supposed that the mean residuals could be applied directly as corrections to the surface times (again apart from a constant). The additive constant could be found from $sS-S$ as for $pP-P$, and also from the mean S residual, but the former turned out to have much the smaller uncertainty. The standard errors for S beyond 18°, rather surprisingly, came out about 0.9s as against 0.4s that I got in 1939 from fewer earthquakes. This is because the times are sums, and the number of observations for ascending rays is usually small.

Adapting to a surface focus gave for P up to 15°

$$7.58 \pm 0.18 + (14.34 \pm 0.29)\,\Delta - (0.0026 \pm 0.0010)\,\Delta^3,$$

and for S up to $17°$

$$t = 11{\cdot}46 \pm 0{\cdot}40 + (25{\cdot}39 \pm 0{\cdot}09)\,\Delta - (0{\cdot}00332 \pm 0{\cdot}00044)\,(\Delta - 0{\cdot}5°)^3.$$

The time of one vertical passage of P through the upper layers is $5{\cdot}44 \pm 0{\cdot}13^s$ in mid-Honshiu, about $1{\cdot}2^s$ longer in S.E. Hokkaido. In the expression for P the uncertainties of the coefficients are far from independent; the accuracy at intermediate distances is much better than they would suggest.

The Pacific explosions themselves being on islands, we should expect the contribution to travel times from one end of the path to be the same as at the other, and that for equal distance the travel times would be about 4^s shorter for observations on oceanic islands than on continents. Actually no systematic difference was found.

A linear formula was fitted up to $18°{\cdot}6$; beyond $25°{\cdot}5$ ten summary values with standard errors of about $0{\cdot}2^s$ were found. The gap between $18°{\cdot}6$ and $25°{\cdot}5$ was filled in by an interpolation based on data at $17°$, $20°$, $28°$ and $36°$. Results beyond $94°$ are extrapolated.

The linearity of the (Δ, t) relation at short distances, if exact, would imply the limiting case where the velocity is proportional to distance from the centre. But if we take the case of constant velocity, so that the rays are chords, we can work out how much the times would deviate from linearity. If they agree at distance Δ_1 we should have

$$ct = 2a \sin \tfrac{1}{2}\Delta, \quad ct_1 = b\Delta, \quad \frac{b}{a} = \frac{2 \sin \tfrac{1}{2}\Delta_1}{\Delta_1}.$$

$t - t_1$ is a maximum near $\Delta^2 = \tfrac{1}{3}\Delta_1^2$. If $ct = 140^s$ at $10° = 0{\cdot}17$ radian, the maximum departure of t from t_1 is $0{\cdot}12^s$ for $\Delta_1 = 10°$, $0{\cdot}4^s$ for $\Delta_1 = 15°$, $1{\cdot}0^s$ for $\Delta_1 = 20°$. The first two would be too small to be detected by observation; the third might just be if the range of Δ is well covered. The failure to detect a cube term at short distances in Europe and the Pacific therefore does not imply a velocity decreasing with depth. The detection of this term in Japan, however, implies one definitely increasing.

In all this work residuals in different quadrants of azimuth were compared whenever possible. No differences have been found. Shimshoni, however, points out that they must exist. If in the diagram the triangles are equilateral and there is no systematic difference between times along AC and AB, then none between $CA\,(=AC)$ and BC, CD and CE, it follows that there is none between AC and $EC = CE$; by repetition there can be no regional differences. Variations with

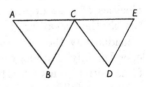

azimuth have in fact been detected in explosion work in North America; apparently there is a discontinuity in $dt/d\Delta$ for P in crossing the Rocky Mountains.

The J.B. times lead to a variation of the velocity of P by about 1% for

20 km. of depth. The revised times on the hypothesis that Pd is genuine would imply an increase of about 3% in the first 200 km., followed by an increase of about 7% per 100 km. at greater depths. The data for dunite on p. 102 would make a variation of 6% in k/ρ, or 3% in α, for depths between 6 and 30 km., or 2·5% in 20 km. The decline of compressibility between 2×10^9 and 10^{10} dynes/cm.^{2} may be greater than that at higher pressures, and it is possible that the variation due to pressure may be less at greater depths. Data for the effect of temperature on compressibility are rather scanty; data for the effect on rigidity are given by Birch, Schairer and Spicer (1942), and if the compressibility varies in proportion the reasonable temperature gradient of 7°/km. in the lower layer would indicate a reduction of velocity of about 2% in 20 km. (Jeffreys, 1952 d; an error in the allowance for pressure is corrected above). It appears that quite small errors in the estimates of the effects would make the difference between the J.B. velocity gradient and a decrease with depth.

Birch (1952), with more recent data, estimates that a temperature gradient of 6·6°/km. in the lower layer would produce a shadow zone.

D. S. Hughes and C. Mauretta (1957) estimate from direct measures of velocities that the velocities rise from the surface to a maximum and then decrease. Birch (private communication) estimates that a temperature gradient of 48–19°/km., according to depth, would give a velocity in the upper layer decreasing with depth. See also D. S. Hughes and J. H. Cross (1951).

Several authors have attempted to use data on dispersion of Rayleigh waves to test distributions of the velocity of S. The fullest treatment is by Dorman, Ewing and Oliver (1960). They take modified forms of distributions due to Bullen and me, Gutenberg and Lehmann, and find (see especially Fig. 1 of their paper) that Gutenberg's gives a very good fit for periods from 80^s to 220^s, and Lehmann's to about 180^s. Those based on the J.B. velocities depart systematically from observation over the whole range. Lehmann's values give a slight drop of β at depths from 120 to 220 km., Gutenberg's a decrease from about 4·7 km./sec. at 50 km. to 4·3 km./sec. at 150 km., rising gradually to 4·7 km./sec. at 350 km. For longer periods the observed group-velocities are less than the calculated ones for all models. This is difficult to understand, since for long periods they should depend on the velocities of S at depths over 400 km., for which the times of S agree. Possibly the curvature of the Earth is beginning to be important.

Incidentally the authors remark that the J.B. times of S for $\Delta \leqslant 20°$ depend mainly on analogy with P, and quote a passage from this book (p. 114 of this edition). This refers to the 1935 tables; in the 1940 tables, as stated on p. 117, much fuller material from Japanese deep-focus earthquakes was used.

3·11. Outstanding problems. It became clear at an early stage of the revision of the tables that the chief need if any substantial progress was to be made was for better methods of combining observations than had hitherto been used. Consequently the seismological work has been carried on simultaneously with investigations into the principles of scientific method and the development of probability theory (Jeffreys, 1973b, 1967d). Considerable advances had been made during the present century, especially by Karl Pearson and R. A. Fisher, mostly for biological purposes, but seismology was found to present a number of problems that had not been treated (not counting the treatment of serious departures from the normal law of error, the principles of which had been given by Fisher). The most frequent of these was perhaps that of smoothing an empirical table when the true values were not expected to agree with any pre-assigned mathematical function. The observed values would necessarily show some irregularity on account of observational error, and the question was how far this could be smoothed out without at the same time smoothing out some genuine irregularity. The solution of this problem (1967d, pp. 223-7) should have much more general application. The principles of significance tests have had to be considerably extended because, whereas biologists have usually been able to design their experiments so as to make a comparatively simple analysis of the results adequate, such design is impossible in seismology, and approximations have had to be developed. Much more, however, remains to be done.

It was hardly to be expected that the Earth's materials would have exactly the same properties from place to place; it has also been suggested that even within the same region there are differences between times of transmission that are not explicable simply by differences of epicentre and focal depth. On the whole it was found that such differences disappeared on statistical examination, being no more than random errors of observation would be likely to produce. However, there appear to be genuine differences in the velocities in near earthquakes in different regions, and the surface waves indicate a considerable difference between continents and ocean floors. The chief anomaly for comparison of earthquakes in the same region was the Z phenomenon, but most of this was explained by multiplicity and the ellipticity correction. A little of it was found to survive in deep-focus earthquakes. By a solution for the elements it was found possible to fit P at all distances, pP, sS, S up to 20° and SKS within their apparent uncertainties; but the mean residual of S from 20° to 90° fluctuated from earthquake to earthquake as if it had a standard error of about 1·5^s superposed on the random error. This was allowed for in estimating the uncertainty of the final tables. No explanation has been suggested of how such a variation affecting only part of one phase could arise. Several explanations have been suggested of the fluctuation of the differences of the constant terms in the times of the near-earthquake pulses, but none has appeared satisfactory.

If the Earth is not quite isotropic there may be small differences between the velocities of SV and SH. If S is identified by the Lehmann–Plett criterion the observations will nearly all refer to SH.

The greatest outstanding difficulty is one that possibly escapes attention because it is so familiar. According to Lamb's theory for a uniform crust, P, S and the Rayleigh waves from a sudden shock would all consist of single swings. The actual movement is a continual irregular oscillation. We know why the surface waves of both types show oscillation; it is due to dispersion. But no cause of dispersion has been suggested that would convert a sudden P or S into a long train. Gravitation, damping, and continuous variation of properties would all produce quite different effects (Jeffreys, 1931 e). The most hopeful explanation at one time appeared to be internal reflexion within the surface layers, but reflexions at their interfaces were sought, but not found, in the deep-focus earthquake of 1929 February 1. (But see p 147.) If the original disturbance at the focus was oscillatory it would explain oscillations in P and S; but the duration of the oscillations would be the same for all distances, whereas it increases with distance. This applies, in particular, to the suggestion sometimes made that the oscillation is due to non-linearity in the stress-strain relations at the focus.

A possibly related fact is that the oscillation of the ground in an earthquake continues long after the time when the slowest theoretical surface wave should arrive. This *coda* is still proceeding at a time when waves with velocities of 1–1·5 km./sec. should have reached the station, and this is true both for near and distant shocks.

Ordinary dispersion cannot explain the coda, because it necessarily makes the group-velocity a single-valued function of the period. The coda shows waves of apparently the same period with times of travel, and therefore group-velocities, differing by a factor of 2.

Ewing, Press and Jardetsky (1957) explain the coda as Rayleigh waves with group velocities about 1·2 km./sec. arising from oceans. Sediments have been appealed to similarly to explain it for continental paths. I do not see how this, as far as the oceans are concerned, can explain the ordinariness of the periods.

Another difficulty is the large movement that complicates the reading of S up to 20° in normal earthquakes. A discontinuous increase in the time of S at a certain distance, such as Lehmann finds in North American earthquakes, would imply some serious complication in the velocity distribution. It is very desirable that there should be more studies of S between 10° and 25° in normal earthquakes, and that the entry of S should be identified by the Lehmann-Plett criterion (p. 107) of change of direction of the movement. J. S. Hughes drew attention to the extreme scarcity of S readings at the normal time in the Italian earthquake of 1930 July 23. I collected the records in convenient azimuths (Jeffreys 1937 f, p. 33; 1937 g) and found that although the motion was small at the normal time of S, there was a definite change of direction. This suggests that the late S is an additional phase and not a continuation of the normal S. At many stations it showed a longer period. But this needs verification in other earthquakes.

The travel time of a compound pulse is stationary for small variations of the path, but not necessarily a minimum for all variations. It is easy to see that the time of *PP* in a homogeneous sphere would be a maximum for variations of the point of reflexion in the plane of the path. Attempts were made to derive empirical times for all predicted phases, but it was found that there was a qualitative difference between those whose times were true minima and those whose times were merely stationary. For the former, namely, *P*, *S*, *PKP*, *SKS*, *PcP*, *ScS*, *pP*, *sS* and *PKS*, the distributions of residuals showed pronounced humps corresponding to a standard error of 1ˢ to 3ˢ, with the usual background effect. The latter, typified by *PP*, *PS*, *SS*, *SKKS* and *SKSP*, gave distributions scattered over 20ˢ or more with no pronounced concentration, and were finally abandoned. Their published times are calculated, and the observed times are scattered on both sides of the calculated ones. Several attempts were made to construct a theory of a pulse of this class, but the mathematical difficulties were very great, and I finally adopted a suggestion of F. G. Friedlander, who pointed out that in the expansion of a cylindrical pulse the disturbance from the rear of the cylinder has the same property. An analytical solution of this problem was known, and there was no great difficulty in approximating to it (Jeffreys, 1942 e). The result was that the disturbance should be very large, but not sharp; it should begin a little before the theoretical time and subside gradually after it. E. R. Lapwood (1949) has carried out a thorough rediscussion of Lamb's problem and found that several subsidiary phases should exist in the case of a deep source in a semi-infinite solid. This clears up some points left in doubt by Lamb and Nakano; and, in particular, it is found that some phases whose times are not theoretically true minima have the same properties.

The mathematical point is as follows. In many diffraction problems a simple wave train suffers a phase change of $\frac{1}{2}\pi$; this happens in particular for a ray that touches a caustic. When the original pulse is represented by a Fourier or Bromwich integral, the continuation is approximately represented by the allied integral (H. and B. S. Jeffreys, 1972, §14·111), and it is related to the original function as the charge function is to the potential in two-dimensional electrostatics for a semi-infinite region. In particular if the original pulse gives a jump from 0 to 1 at the theoretical time, the corresponding modified pulse near its theoretical time τ will behave like $-(1/\pi) \log (t-\tau)^2$. It is large near $t = \tau$, but has not a sharp beginning and in fact grows gradually out of previous movements. This appears to apply to all reflexions at a spherical boundary whose travel times are not true minima, to rays that have touched a caustic, and to those that have been internally reflected within a medium of continuously varying properties. The same has been found for reflexion of a pulse within a sphere (Jeffreys and Lapwood, 1957).

Pekeris (1967) gives a transformation of co-ordinates that is directly related to the rays. For a once-reflected ray, from a focus at B, reflected at D, and arriving at P, in a homogeneous sphere, two co-ordinates are taken as $S = BD$ and $T = BD + DP$, and the other as azimuth. The line element squared has no product terms in these co-ordinates. Multiply reflected pulses can be treated similarly. He shows that 5·22 of Jeffreys and Lapwood (1957) needs a factor 2. The method could probably be extended to a non-uniform sphere.

The complete solution is in the form of an infinite series of oscillating functions, slowly converging. The summation has been carried out numerically for several models by Z. Alterman (1965, 1966). The results look astonishingly like actual seismograms. Lapwood and I used a transformation due to Watson, which greatly improved the convergence. These methods are useful, however, only for a uniform model. An approximate method based on Huyghens's principle has been given by R. Burridge (1962, 1963).

If the origin is in direction $(0, 0)$ from the centre and the station at $P(\Delta, \theta)$, then if the travel time t is a true minimum for reflexion or refraction at Q given by ray theory, and that for reflexion or refraction at points not given by ray theory is $t + \delta$, where δ is small, the curves of constant δ are closed, and integration through their interior presents no difficulty. With suitable modifications the same is true when t is a true maximum. If, however, it is a minimum for some changes of Q and a maximum for others, the curves of constant δ resemble hyperbolas and the integration becomes difficult. However, Burridge shows that if we still integrate over a circle about Q the parts outside this circle give only a slowly varying part, which would not affect the appearance of the onset on a record. Sharp variations still arise only for contributions from the interior of this circle. Since I found that all phases whose travel times were not true minima showed a wide scatter, and since this seemed to be explained fully by Friedlander's suggestion, it is remarkable that clear beginnings of the AB branch of PKP are shown on the records reproduced by Nguyen Hai and that clear reflexions on the inside of the core are found by other writers. A possible explanation may be in the increased use of instruments of short period. These phases should, as stated above, give displacements of the form $\log\{a|t - T|\}$, where T is the calculated time of arrival. If the instrument has a period short compared with $|t - T|$, when the expected displacement is varying slowly, it may be expected that little displacement would be recorded; but when $|t - T|$ is comparable with the period it might be appreciable. On the other hand, since the expected displacement is approximately symmetrical about T, it might become appreciable a little before T, and the recorded time may have a negative systematic error. I do not insist on this explanation, but one appears to be needed. Kennett and Bolt have done Hilbert transforms on pulses with periods 0·6s and get apparent beginnings about 0·3s early.

3·12. The use of amplitudes. We have seen that the square of the amplitude A contains $-d^2t/d\Delta^2$ as a factor, the remaining factors in general varying slowly. Thus we can assume over a considerable range of Δ

$$\frac{d^2t}{d\Delta^2} = -(c+d\Delta)A^2,\tag{1}$$

where c and d are constants, and integrating twice

$$\frac{dt}{d\Delta} = b - c\int_{\Delta_0}^{\Delta} A^2\,d\Delta - d\int_{\Delta_0}^{\Delta} A^2\Delta\,d\Delta$$

$$= b - cu_1 - du_2,\tag{2}$$

$$t = a + b\Delta - c\int u_1\,d\Delta - d\int u_2\,d\Delta.\tag{3}$$

Δ_0 can be taken at a convenient point in the interval used.

If t is known for 4 or more values of Δ, and the distribution of amplitude is known, (3) will determine a, b, c and d. The distribution of velocity depends on $dt/d\Delta$, and hence on an integration instead of a numerical differentiation as when only the times are known. Consequently much greater accuracy would be attainable. The amplitudes must, of course, be reduced first to those in a standard earthquake if several earthquakes are to be combined.

Suitable amplitude data were given by Vanek and Stelzner (1959, 1960 a, b) for the stations Prague, Jena, Collmberg and Potsdam and by Ruprechtova (1959). In the range $12°$ to $31°$ the amplitude varies by a factor of over 10. As the data cover the range of large amplitudes about $20°$ they should be valuable. Times at $14°$, $17°$, $20°$, $23°$ and $28°$ were taken from the European travel times of P (Jeffreys and Shimshoni, 1966). Actually the results differed very little from those obtained by fitting a cubic. This is rather surprising; it is because the range where $d^2t/d\Delta^2$ differs greatly in the two methods is rather short.

Shimshoni worked out a corresponding velocity distribution. Specimen values are as follows:

r/R	α
1·000	8·10
0·980	8·33
0·960	8·70
0·940	9·30
0·920	9·99
0·900	10·54

Amplitudes were given for S also, but no solution would fit the observed times within 4^s, and these remain rather unsatisfactory. See also p. 146.

The range of small amplitude need not be due to small $d^2t/d\Delta^2$. It could be due to absorption in a thin layer, and I think that this is the more likely explanation. More will be said about this in a later chapter.

3·13. Energy of earthquakes. If we had a satisfactory theory of the motion produced by an earthquake the computation of the energy from the ground movement would be straightforward. The kinetic energy per unit volume is given by the displacements and periods shown on the seismograms, and the potential energy will on an average be equal to the kinetic. Knowing the rate of variation of velocity with depth the determination of the energy would then be reduced to integration. Unfortunately, the coda, for which there is at present no theory, carries amplitudes comparable with the maximum, and it lasts a long time. The method usually adopted is simply to suppose that the energy per unit volume is uniform through the upper layer and zero below that. Then for any station we take the velocity associated with a point on the record (that is, the distance divided by the time of travel) and suppose the energy to be travelling out with this velocity. The whole transfer per unit width past a station is then found, and the whole energy of the train is obtained by multiplying by the circumference of the circle through the station centred on the epicentre. The whole energy of the earthquake is unlikely to be more than twice this.

Specimen values of the energy so obtained are as follows:

		ergs
Hereford	1926 August 14	5×10^{16}
Oppau explosion	1921 September 21	5×10^{16}
Jersey	1926 July 30	10^{19}
Montana	1925 June 28	10^{21}
Pamir	1911 February 18	10^{21}

According to Gutenberg and Richter the average annual release of energy in all earthquakes is about 10^{26} ergs, and more than 80 % of this comes from earthquakes whose energy is 10^{25} ergs or more. They estimate that nearly a third of the whole energy release between 1904 and 1939 came from the four largest earthquakes: Colombia, 1906 January 31; Tien-Shan, 1911 January 3; Kansu, 1920 December 16; and Chile, 1922 November 11.

The effects of earthquakes that can be detected without special instruments have long been used to give a rough scale of intensity; the extent of destruction, if any, the movement felt, the stopping of pendulum clocks, and so on, all give useful information. The chief such scales in use are those of Rossi-Forel and Mercalli (Byerly, 1942, p. 57; Bullen, 1947, p. 253). Richter (1935) proposed an instrumental magnitude scale. The standard of reference is the maximum amplitude in microns ($1\mu = 10^{-4}$ cm.) traced by a seismograph 100 km. from the epicentre, the instrument having statical magnification of 2000, free period 0·8 sec., damping coefficient 0·8. The magnitude is defined to be the logarithm to base 10 of the maximum amplitude so recorded. Empirical tables are given by Gutenberg and Richter (1942, 1945a, b) to reduce readings on other types of instrument and at other distances to the standard. The scale has also been adapted to deep-focus earthquakes (Gutenberg and Richter, 1945c). The correlation between energy and magnitude is fairly close. The greatest earthquakes have magnitudes of about 8·5, and the smallest reported of about 1·5, with energies of about 2×10^{11} ergs. The magnitude corresponding to the amplitude of the seismic waves from the Bikini atomic bomb was 5·5 (Gutenberg and Richter, 1946).

The Oppau explosion was produced by the detonation of 4500 tons of the double salt $2NH_4NO_3 . (NH_4)_2SO_4$. Assuming a heat of decomposition of 7500 cal./g.mol. of NH_4NO_3 (a value suggested to me by Sir Eric Rideal), the energy liberated would be about $1·5 \times 10^{12}$ cal. or 6×10^{19} ergs. Thus only a small fraction of the energy went into the seismic waves. The impulses upward on the air and downward on the ground must have been equal, so that the energies imparted to the ground and the air would be in the ratio of the vertical velocities, and probably hundreds or thousands of times as great for the air as for the ground. Thus most of the energy went into the sound wave in the air. In the Burton-on-Trent explosion the explosive was buried, but the cover blew out and the ratio of the energies was still about the same as in the Oppau explosion. The same applies to the Heligoland explosion, for which Willmore (1949) gives a detailed discussion.

In seismic survey it is usual to bury the charge to such a depth that the top does not blow out, and then a much larger fraction of the energy goes into the waves in the ground.

In deep-focus earthquakes the intensity of movement is less near the epicentre than in shallow ones, but falls off more slowly with distance. Thus the field observations alone often suffice to indicate focal depth.

It is possible that the distinction between P and S movements may require some adjustment of intensity scales. It has often been noticed that objects are overturned in an earthquake in two approximately perpendicular directions. This is what would be expected if some were upset by Pg, and some survived Pg but were upset by the SH component of Sg, which is usually larger than the SV component. On 1931 June 7 I was in bed lying approximately northeast to southwest, when the waves from the Dogger Bank earthquake arrived, and the motion began with rapid oscillation along the bed. After some time it changed to a larger but slower motion across the bed. The change was just what would be expected if the first part of the motion was Pg and the second SHg. This observation shows, at any rate, that the distinction can sometimes be noticed in non-instrumental observations. It does not appear to have been emphasized in macroseismic studies, however, and I think that some of the cases of 'twin earthquakes' that have been reported but not confirmed by instrumental observations may be explained in this way.

3·14. Surface waves. These consist of a long train of approximately harmonic waves of varying amplitude and period. The beginning of the train can usually be read fairly well and is denoted by L in reports. (But in near earthquakes the symbol iL is often applied to Sg.) The largest amplitude is denoted by M; the notations L, M were introduced by v. d. Borne in the same paper as P, S. M is not a definite phase, because in practice it means only the largest swing on the actual record, and as the magnification depends on the properties of the recording instrument M will be read at different times at the same place according to the period and damping.

The possibility of making use of the surface waves for geophysical purposes depends on the separation of the Rayleigh and Love components, which are propagated according to different laws. This is most easily done by selecting such a station that the waves reaching it are travelling toward one of the cardinal points. For instance, if the waves are coming from the east (that is, if the azimuth of the epicentre referred to the station is 90°) the Rayleigh waves will be recorded on the easterly and vertical components, the Love waves on the northerly component. Each begins with a long wave with a period of order 60^s, but the periods of successive waves shorten gradually to about 10^s. For an assumed crustal structure it is possible to work out the group-velocity for each period and to compare it with the observed velocity, which is the distance divided by the time of travel for that period. The velocity associated with the first movement in each type is nearly constant, irrespective of the nature of the path. Stoneley (1939b) gives, from an analysis of data from the I.S.S.,

$$LQ: \quad \frac{dt}{d\Delta} = 0 \cdot 4195 \pm 0 \cdot 0012 \text{ min.}/1°; \quad \text{velocity} = 4 \cdot 426 \pm 0 \cdot 013 \text{ km./sec.}$$

$$LR: \quad \frac{dt}{d\Delta} = 0 \cdot 4674 \pm 0 \cdot 0011 \text{ min.}/1°; \quad \text{velocity} = 3 \cdot 972 \pm 0 \cdot 009 \text{ km./sec.}$$

The first velocity is close to that of S near the top of the lower layer. The ratio of the two is 0·90, which is near the theoretical ratio of the velocity of long Rayleigh waves to the velocity of S.

Attempts have been made to use even the first swing in each type of surface wave to determine crustal structure. I should regard these with suspicion because the usual dispersion formula is an approximation and may go badly wrong near a maximum or minimum group-velocity. From analogous cases (notably the Stokes approximation to the Bessel functions) it seems likely to be a fairly good approximation for the second swing, but definitely not for the first. In particular, the first swing is likely to depend greatly on the form of $\phi(\kappa)$ of 2·033, when κ is small, about which we know hardly anything. For later swings $\phi(\kappa)$ affects several consecutive swings in nearly the same ratio as long as it is slowly varying, which is a reasonable supposition.

The easiest thing to observe is the group velocity, $C = d\gamma/d\kappa$ as a function of period $2\pi/\gamma$. The wave-length $2\pi/\kappa$ is not measured directly and would have to be calculated from

$$\kappa = \int d\gamma/C.$$

This contains a constant of integration. This could be fixed if we had information for either very short or very long periods. This is difficult. For short periods the velocities are much affected by the variable sedimentary layer; also there is in general a minimum group velocity, and waves of shorter periods are overlaid by those of longer periods and difficult to separate. For long periods the surface waves merge into free oscillations of the Earth as a whole, and the finite size of the Earth needs to be taken into account. The admissible values of γ are no longer a continuous set. An approximate way of doing this is as follows. If n is the degree of the harmonic involved, the terms in the differential equation that depend on latitude and longitude reduce to $n(n+1)/r^2$. This corresponds to κ^2. Then we may expect that for long periods κ can be taken from

$$a\kappa \doteqdot \sqrt{[n(n+1)]} \doteqdot n + \tfrac{1}{2}.$$

There is reason to suppose that $(n+\tfrac{1}{2})/a$ is actually the better approximation. Errors of order $1/n$ are to be expected in any case.

There have been many attempts to do a Fourier analysis on the surface waves. I doubt the usefulness of this. For one thing, the interval analysed must not contain the major bodily waves, and also must not contain the coda, which remains unexplained. Thus we are concerned with a record of finite length, which could be exactly represented by a Fourier series. If the length of the record is T, the periods of the terms are of the form T/m, where m is an integer. The usual method of harmonic analysis does not use these periods. An effective one was given by Turner, who also recommended attention to the changes of sign of the cosine and sine coefficients. The usual method uses only the amplitude, loses much information, and gives no estimate of uncertainty. I have given a better method (Jeffreys, 1964a, 1967d, Appendix C). A peculiarity of the problem is that, whereas if n is the number of observations, the standard error of an estimate by maximum likelihood usually decreases like $n^{-\frac{1}{2}}$, in this case that of a period decreases like $n^{-\frac{3}{2}}$.

There have been many attempts recently to determine phase velocities directly by Fourier analysis. The procedure is based on Fourier's integral theorem: if

$$h(\kappa) = \frac{1}{\sqrt{(2\pi)}} \int_{-\infty}^{\infty} f(y)\, e^{-i\kappa y}\, dy \tag{1}$$

then

$$f(y) = \frac{1}{\sqrt{(2\pi)}} \int_{-\infty}^{\infty} h(\kappa)\, e^{i\kappa y}\, d\kappa. \tag{2}$$

If the displacement at time t and distance Δ is $f(t, x)$

$$\kappa(\gamma, x) = \frac{1}{\sqrt{(2\pi)}} \int_{-\infty}^{\infty} f(\tau, x)\, e^{-i\gamma\tau}\, d\tau, \tag{3}$$

$$f(\tau, x) = \frac{1}{\sqrt{(2\pi)}} \int_{-\infty}^{\infty} \kappa(\gamma, x)\, \mathrm{e}^{i\gamma t}\, d\gamma. \tag{4}$$

Then (4) represents $f(t, x)$ as an integral over periodic components, and $\kappa(\gamma, x)$ can be determined for any γ, x by numerical integration. Now for an advancing wave κ will be the form $A(\gamma)\mathrm{e}^{-i\kappa x}$, and comparison for two different values of x will determine κ for each γ. Then c is given by γ/κ.

I see several difficulties in this procedure (and more complicated ones based on it). (3) requires numerical integration from $-\infty$ to ∞; thus the whole record, including all body waves, coda and microseisms contributes to κ. Thus it is primarily quite uncertain to what extent the solution represents the surface waves.

Also for the actual Earth surface waves may reach a station after travelling around the Earth several times. Each return will make its contribution, but the effective values of x differ. Attempts have been made to treat the returns separately, but again there are difficulties. Dr M. Båth showed me a record where surface waves were visible after 1, 2, 3, 6, 7 returns, so far as I remember, but nothing was visible at the times corresponding to 4 and 5 returns.

To eliminate the body waves and the coda, for which there is no adequate explanation, it is necessary (and usually done), to use only a finite length of the record. But then the ordinary Fourier series applies; if the length of the record is T, it is completely representable by a series of terms whose periods are T and its submultiples. Any results for intermediate periods are only interpolates and highly correlated. In comparison of results for different distances some way of allowing for this has to be found. Essentially the assumption made is that the displacement is zero outside the interval used.

In studies of surface waves, wave velocities are found for periods up to 500^s. The periods of actual swings on the records are up to about 50^s, and these refer only to the first swing, for which the usual approximations that lead to the group velocity approximation break down. What harmonic analysis does is to evaluate the sum of many terms of alternating sign, and the estimation of a small quantity as a difference of two large ones (though sometimes inevitable) can be very inaccurate. Dr Landisman has however shown me some examples where the wave velocity has been estimated from the group velocity, and the agreement is good.

The following more detailed treatment was given me by Dr Freeman Gilbert. Two stations are at Δ_1, Δ_2 in the same azimuth. The displacement at Δ is $f(\Delta, t)$ with the Fourier integral representation

$$f(\Delta, t) = \int F(\sigma)\, d\sigma \exp\{-i\sigma(t - \Delta/c)\}$$

and $c(\sigma)$ is the phase velocity. Let the covariance between the signals at Δ_1, Δ_2 be

$$\phi_{12}(\tau) = \int f(\Delta_1, t) f(\Delta_2, t-\tau)\, dt.$$

In terms of the Fourier representation

$$\phi_{12}(\tau) = \int dt \int d\sigma F(\sigma) \exp\left[-i\sigma\{t - \Delta_1/c(\sigma)\}\right]$$
$$\times \int d\omega\, F(\omega) \exp\left[-i\omega(t - \tau - \Delta_2/c(\omega))\right].$$

Integrating with respect to t, we get

$$\phi_{12}(\tau) = \int d\sigma\, F(\sigma) \exp\{i\sigma\Delta_1/c(\sigma)\}$$
$$\times \int d\omega\, F(\omega) \exp\{i\omega\tau + i\omega\Delta_2/c(\omega)\}\, \delta(\sigma + \omega).$$

Integrating with respect to ω we get

$$\phi_{12}(\sigma) = \int d\sigma\, F(\sigma) F(-\sigma) \exp\{-i\sigma\tau + i\sigma(\Delta_1 - \Delta_2)/c(\sigma)\}.$$

Let the Fourier transform of ϕ_{12} be

$$\Phi_{12}(\sigma) = \int \phi_{12}(\tau)\, e^{i\sigma\tau}\, d\tau.$$

Then apart from a factor 2π

$$\Phi_{12}(\sigma) = F(\sigma) F(-\sigma) \exp\{i\sigma(\Delta_1 - \Delta_2)/c(\sigma)\}$$
$$= A(\sigma) \exp\{i\psi(\sigma)\},$$

say. Therefore the phase velocity is determined from the phase of the spectrum of the covariance:

$$c(\sigma) = \sigma(\Delta_1 - \Delta_2)/\psi(\sigma).$$

The method is correct in principle; but it seems impossible to derive an estimate of the uncertainty of the results.

Rayleigh's principle has been used successfully for surface waves by Harkrider and Anderson (1966).

The relation of the group-velocity to crustal structure is complicated, and it is necessary to proceed numerically from an early stage. The velocities in the upper and lower layers are taken from near-earthquake studies, and appropriate densities corresponding to them are adopted. The thickness of the upper layer T is taken as a parameter, so that the results for c are found as functions of the numerical parameter κT. Then the group-velocity is given by

$$C = \frac{d\gamma}{d\kappa} = \frac{d(c\kappa T)}{d(\kappa T)} = \frac{d(c\kappa T)}{dc} \bigg/ \frac{d(\kappa T)}{dc}.$$

It is convenient to use as far as possible equal intervals of c, and then this formula is adapted to numerical differentiation. But to get C to three figures it is practically necessary to find κT to five, and the numerical work is consequently troublesome.

The method based on Rayleigh's principle avoids this difficulty and can also be used to study the effects of small changes in the model.

The apparent advantage of the use of surface waves is that observations at any distance can be used, and accurate timing is not needed. An error of a minute in the timing of a given period would produce less error in the estimated thickness than one of a second would produce in a near-earthquake study. The chief uncertainty is in the measurement of the period. It is not necessary to restrict the adopted structure to a single upper layer. If there is an intermediate layer of known mechanical properties and thickness T', it is still possible to adopt a definite ratio T'/T and extend the theory to this case, of course at the cost of extra complexity in the computation. As the group-velocities can be measured for many periods, the variation of group-velocity with period may even help to determine T'/T. Times of surface waves from Japanese earthquakes observed in western Europe should give information about average structure over the whole of Eurasia, whereas all other methods give estimates for special regions, which may not be representative. This advantage is qualified by the need to use the S velocities determined in those special regions, but at worst the velocities are not likely to vary relatively from place to place as much as the thicknesses.

Love waves have been used extensively in this way, especially by Stoneley. Rayleigh waves have been less used, chiefly on account of their more complicated theory. Rayleigh waves, however, have one advantage, that they alone are recorded on the vertical component, so that the relation between γ and C for them can be found for any station with a vertical instrument; it is not even necessary to attend to the direction of arrival. If the crust was uniform the precaution of using, say, the north-south component to record Love waves approaching from the east would be enough to separate them, but in practice there are changes of structure, especially, of course, at coasts, which produce considerable refraction, and this affects the direction of displacement more than the time of travel. The result is that even with this precaution the Love and Rayleigh waves may be superposed on the horizontal components. The anomalous association of vertical and transverse movement has been often noticed; G. W. Walker commented on it, and it has been emphasized by Macelwane (1933, p. 123), who argued that it proved the analysis into Love and Rayleigh waves inadequate. Stoneley's consideration of the effects of refraction (1935a, p. 268; 1935b), however, appears to provide an explanation.

Some representative values given by Stoneley (1935a, 1937a) from observations by W. Rohrbach (1932) of Love waves are as follows:

Period (sec.)	C (km./sec.)	c (km./sec.)	T (km.)
66·7	4·00	4·20	15
50	3·70	4·08	16
40	3·57	3·97	15
33·3	3·46	3·90	15
28·6	3·30	3·81	—
25	3·20	3·72	—
22·2	3·15	3·66	—
20	3·12	3·59	—
18·2	3·08	3·54	—
16·7	3·06	3·48	—

It was supposed that $T' = 2T$, and periods from 60^s to 33^s were used; the results were consistent with $T = 15$ km. for Eurasian paths, but for periods less than 30^s the group-velocity observed was well below the minimum group-velocity (3·27 km./sec.) calculated for this structure. Stoneley inferred that at such periods the dispersion was appreciably affected by the sedimentary layer, and proceeded to an analysis taking account of an extra layer. He found that the data could be reconciled fairly well by assuming a superficial layer about 3 km. thick with an S velocity probably below 2·7 km./sec. This would agree with the average thickness of the sediments indicated otherwise (9·02).

The records of the Jersey earthquake at Strasbourg and Zürich show a train of waves after Sg, building up to a maximum, and suddenly ceasing. This is just what would be expected if there is a minimum group velocity. But this velocity, according to the records, would have to be about 20 % less than that of Sg, which is a greater depression than has been found by calculation for Love waves. Possibly sediments or a low-velocity layer might explain this; but other records are needed, because only one component was available at each station and we cannot be sure whether the waves were of Love or Rayleigh type. A layer of low velocity always reduces the wave velocity, and must reduce the group-velocity over a range of periods, but it is not obvious that this range will include the period that gives the minimum group-velocity.

Detailed results for Rayleigh waves in many models are given by Stoneley and Hochstrasser (1956, 1961).

For Pacific paths the observed range of periods is from 14·7^s to 26·7^s, C ranging from 2·62 to 4·11 km./sec. and c from 4·01 to 4·27 km./sec. The data are in fairly good agreement with a single upper layer of granite, with a thickness of about 10 km.

The single crustal layer theory for Rayleigh waves has also been compared with observation (Jeffreys, 1935a; Bullen, 1939b; James T. Wilson and O. Baykal, 1948). Gutenberg gives group-velocities for Eurasia ranging from 3·0 to 3·8 km./sec. for periods from 20^s to 35^s. From these, with a granitic crust, I derived thicknesses in the region of 20 km. Bullen made a study of the Rayleigh waves from the Bering Sea earthquake of 1938 November 10, observed at Wellington. These were compared with the

hypotheses of a granitic outer layer and one corresponding roughly to the
European intermediate layer, calculations for which had been done by
K. Sezawa (1935). Results are as follows:

Period (sec.)	C (km./sec.)	T (if granitic) (km.)	T (if intermediate) (km.)
32	3·92	6	31
28	3·81	12	31
24	3·58	18	27
20	3·26	19	23
17	3·15	17·5	19

If the constitution adopted was right, all periods should lead to the same
value of T. The values found show variations with period in opposite
directions, and indicate an intermediate structure. This might mean that
the upper layer under the Pacific is mainly andesite, or that it is basaltic
and that the deep ocean sediments have an appreciable effect. The great
difficulty about the application of the method to the ocean floor is the lack
of near-earthquake studies, which would provide much more definite values
for the velocities in the upper layer. It can be stated from the comparisons
made so far that the granitic layer is thinner under the Pacific than in
Eurasia, and may be largely replaced by andesite or some similar material,
so that the evidence is qualitatively consistent with the indications of
geology and gravity. In the continents thicknesses of 15 km. for the upper
and 30 km. for the intermediate layer would be in good agreement with the
data, but some latitude of interpretation is possible because the estimates
depend on the adopted ratio of the thicknesses.

Wilson and Baykal compare the dispersion of Rayleigh waves under the
Atlantic with the theoretical values. They take the velocity of transverse
waves in the lower layer as 4·45 km./sec., and compare the observations
with three hypotheses according to which the velocity in the upper layer is
3·34, 3·70 or 4·02 km./sec. The last represents the variation of group-
velocity with period very well for periods from 18ˢ to 26ˢ, with a thickness
of 26 km. This would be consistent with a layer of gabbro. The correspondence for the others is definitely worse.

The greatest outstanding problem is the uncertainty about the velocity of
S at small depths below the oceans. Ordinary near-earthquake studies, as
in Europe and California, cannot solve it, on account of the sparsity of the
stations; but an expedition to install temporary stations on a group of
islands in a seismic region might well solve it for the Pacific in a year. The
Atlantic presents a greater difficulty, as natural earthquakes and islands
are both fewer.

The theoretical ratio of the vertical and horizontal amplitudes in
Rayleigh waves on a uniform crust is about 1·5. The observed values are
usually smaller, from 0·9 to 1·4. This appears to be partly due to inadequate
allowance for differences of instrumental constants for horizontal and
vertical instruments, and partly to the presence of Love waves in the

horizontal component, but, in addition, the ratio is fairly sensitive to crustal structure. In my solution I obtained values from 1·128 to 1·535. Lee (1934, p. 241), for various structures, gets theoretical values ranging from about 0·4 to nearly 2.

3·15. Layers of low velocity. Gutenberg's early study (1953) concerned P in Japanese deep focus earthquakes. I suspected, and Arnold showed (see p. 127), that the method was statistically unsatisfactory. The times of both P and S in shallow earthquakes fit a cubic formula well up to 18° or so, with a definitely negative coefficient of the cube term. This is consistent with velocities increasing steadily with depth. Gutenberg, in an earlier study of Californian earthquakes (1948) claimed that P actually dies out about 15° and that there is a gap before a new phase enters at 16°. In some papers he made the gap start at 6°, diffracted waves occurring at intermediate distances. This was interpreted as due to a shadow zone due to a layer of low velocity. Now in all my studies I have never found any sign of such a gap. Late readings of P sometimes occur, but they may be at any distance. In the 1953 paper Gutenberg always supposed that the decrease of velocity with depth is not so rapid as to prevent a ray leaving the focus horizontally from reaching the surface. Thus his gap in the P curve had apparently disappeared, and it should be possible to construct the table entirely from observations of shallow earthquakes.

It is in fact very difficult to find the sort of evidence that could test the existence of a layer of low velocity sufficiently intense to produce a shadow zone. We should in any case have to assume a region of higher velocity below it, so that rays refracted in it can penetrate to the surface. Suppose that rays for surface focus confined to the upper region can reach distance Δ_1, and that those refracted in the region of high velocity begin at Δ_2. If $\Delta_2 > \Delta_1$ there will be a gap, possibly occupied by weak diffracted waves. If $\Delta_2 < \Delta_1$ the branches will overlap and first arrivals will show a discontinuity in $dt/d\Delta$. In the first case it is most unlikely that a smooth continuation across the gap could be found. We should expect a range where the observed $dt/d\Delta$ increases with Δ, on account of weak readings. In the second, detection would need either (1) observations of the later branches, which is difficult since the difference of time is unlikely to exceed a few seconds, and separation from swings following the earlier movement would be difficult; or (2) location of the discontinuity in $dt/d\Delta$ and determining the values at the largest distance where the upper wave is first and the smallest where the deeper is first. Determination of a gradient at the end of an interval from observations has always a considerable uncertainty.

I have considered (1962 a) by numerical calculation what would happen in such cases if there is a focus within the layer of low velocity. The time of occurrence and the depth of focus occur as adjustable parameters. The result is that if only observations corresponding to ascending rays are used, nearly all the effects on arrival times would be absorbed in changes of these two parameters. If descending rays are also used there is some hope. Arnold's study for Japan showed no sign of an anomaly that could arise in this way. The focal depths were rather scarce about $0.02R$ to $0.03R$; but the possibility of representing t by single functions of Δ is strong evidence against one.

In Gutenberg's later work (1954, 1955, 1959) he starts with the small amplitudes from 6° to 14° being due to diffraction, with a possible focal point (caustic) just beyond 14°. But he finishes with r/c decreasing with depth all the way, and there would be no question of diffraction. Japanese, European, and Central Asian earthquakes are combined in his last paper; this is unfortunate as the Japanese region in any case differs from the other two.

For Europe, Shimshoni's calculation of the velocities, taking account of amplitudes, gives increase with depth everywhere for P. For S, however, while the amplitudes look good, the basic travel times seem to need corrections of about 4^s. I have hopes that a study of deep shocks may give an improvement.

Studies of free vibrations and surface waves have appeared to support layers of low velocity. However, the distributions of velocities and density attributed to me are not mine. All the velocities adopted for near earthquake phases are higher than mine, and are apparently based on North American data, and it is not clear that adjustments have been made in those at greater depths to make the travel times to larger distances correct. For P, it is probable that the Pacific explosions are representative of the greater part of the Earth; unfortunately S observations are scanty. For S there are many unsettled points; no travel times except Arnold's Japanese ones are trustworthy within 4^s, and Japan is certainly unrepresentative for P. Lehmann (1962) gives the following velocities of P. 6.58 km./sec. is used to depth 48 km.; below that her results are

Depth (km.)	
48 +	8·00
150 −	8·00
150 +	7·90
215 −	7·90
670	10·50

The dip is at depth $0.016R$.

Superposed on the main trains of Rayleigh and Love waves are movements of shorter periods. These have been attributed to surface waves of higher modes. Analyses have been made by several authors. Theoretically they are very sensitive to structure at small depths, and give evidence for layers of low velocity of S in some regions but not in others. They are superposed on the motion of long period, but can be detected by instruments with higher magnification for the relevant periods. (Kovach and Anderson, 1964; Dorman, Ewing and Oliver, 1960; Oliver and Ewing, 1957.)

It seems possible that weakness of S on arrival around 10°, the large movement later by something of the order of 20ˢ, and the evidence for a layer of low velocity, may all be parts of the same phenomenon. According to the modified Lomnitz law a pulse starts gradually at the theoretical time, but may take a long time to build up to the theoretical amplitude. This would accord with the low values of Q found at about depth 150 km.

3·16. Reflexions in the upper layer. If the main discontinuity is sharp, we should expect recognizable reflexions from it. I looked for reflexions of Pg and Sg in the records of the Oppau explosion. There were additional displacements, but any prominent one that could be interpreted as a reflected P was never associated with one at the time for the reflected S. In the deep earthquake of 1929 February 1, I sought for pP reflected at the under side of the discontinuity, and was sure that there were none with amplitudes more than a tenth of those of the main pP. But Tuve and Tatel got reflexions in explosion work in the United States, and Treskov reported at Helsinki that he had found the early pP. Båth and Stefansson (1966) find a movement about 6·3ˢ before S at 63° that appears to be due to conversion to P at the discontinuity. It is about $\frac{1}{2}$ as large as S. Byerly has long spoken of a 'curtsey' preceding S; this may be the same. I was in some doubt, with regard to the construction of the S table, as to whether observers had prevailingly read the main S or the curtsey. I made an appeal for this to be checked, but the only reply was from Bolt at Berkeley. On inspection of these records and some of Byerly's original ones I formed the opinion that the curtsey would seldom be read and that the observations do refer to the main S.

I may refer here to Lapwood's detection of what appears to be a pulse about 13ˢ after S in Japan. This might be a reflexion between the surface and the interface. See p. 93.

From the apparent existence of these reflexions in some cases and not in others it looks as if the transition is sharp in some places and continuous in others.

3·17. Damping. On the whole the agreement of amplitudes of body waves with those calculated from the travel times by the theory of 2·06 is not bad, but there are signs that damping is appreciable. It is more noticeable in surface waves. The shorter periods in these are found to disappear at increasing distances. If we are to attribute this to scattering according to the rule of 2·04, we may take the period as 10ˢ, $\beta = 3 \times 10^5$ cm./sec., and suppose that the damping factor for this period, in comparison with that for long periods, is e^{-1} for a distance of 5000 km. Then from 2·04 (4) we find
$$\tau = 0.003 \text{ sec.}$$

With this value of τ, and a period of 1ˢ, the damping factor would be e^{-1} in a distance of 50 km.; for a period of 3ˢ it would be e^{-1} at a distance of 450 km.

This corresponds fairly well with the prominence of short periods in the movement that follows Pg and Sg at short distances and their gradual disappearance as the distance increases. The fact that short periods are traceable by suitable instruments in PKP is understandable, since PKP travels nearly vertically and scattering may be appreciable only fairly near the surface.

The effect of scattering on a pulse also explains a peculiarity in the times of the near-earthquake phases. Here we must allow for the geometrical reduction of amplitude, since the formula 2·04 (4) was derived for propagation in one dimension. Accordingly, a factor $x^{-\frac{1}{2}}$ or x^{-1} should be inserted. Now a displacement must reach a certain threshold value before it is recognizable. Then at large enough distances the threshold, in a given earthquake, will not be reached at all, which accounts for the fact that the near-earthquake phases are not readable beyond a few hundred kilometres. But, in addition, if the form of the pulse is altered by scattering, the times of travel as recorded will be longer for small earthquakes than for large ones. If the true mean velocity is c, the velocity found from distant observations will be less than c. With the above value of τ and a travel time of 100^s, systematic errors of the order of $0\cdot5$–1^s are to be expected, which are comparable with the differences found between different studies. This indicates that, if the observations cease at a certain distance, observations for a degree or two short of that distance should be mistrusted, and explains why in some cases a smaller apparent velocity is found for greater distances than for smaller ones, contrary to the normal behaviour of travel times (Jeffreys, 1937 e, p. 220). But this explanation does not appear available for some anomalies at short distances found by Willmore in the North German explosions.

There are noticeable differences in quality between seismological stations. This is largely due to differences in the distinctness of the records given by different types of instrument. It is found that in strong earthquakes there is not much difference between the performances of the first- and second-class stations at intermediate distances; but in moderate ones at intermediate distances and in strong ones at great distances the second-class ones show a greater scatter of the residuals and the mean residual is systematically positive compared with that at the first-class ones. This is probably because the observer usually waits till he is sure that a displacement has occurred before reading it, and a higher threshold value, determined by the relative indistinctness of the record, will produce a larger scatter and a systematic error. In cases of doubt it is therefore best to rely wholly on the best stations.

Scattering explains these phenomena, but so does the Lomnitz law or the modified form of it.

It is not clear whether scattering in P and S pulses would give rise to the same value of τ; the values may differ in a ratio of order α/β, but we are only considering orders of magnitude.

As an indication of the sizes of the grains needed to give the requisite value of τ, we may take in 2·04(1)

$$\tau = 0\cdot003^s, \quad \alpha = 5 \text{ km./sec.}, \quad \epsilon = 0\cdot1.$$

Then $l \simeq 450$ cm.,

which is a fairly reasonable value for the sizes of crustal irregularities.

3·18. Thickness from deep-focus earthquakes. If the transitions in the outer crust are sharp enough to give definite reflexions it would be possible to derive estimates of thickness in two ways. In near earthquakes reflexions should be identifiable at short distances, and their travel times would lead to estimates of depth of the reflecting surfaces. It is, in fact, always possible to find displacements that could be interpreted as reflexions, but, for instance, if we interpret one as a reflexion of Pg, there should be one interpretable as a reflexion of Sg at the same depth, and this confirmation is lacking. For deep-focus earthquakes, again, besides the pP and sS reflected at the outer surface, there should be smaller and earlier reflexions at the interfaces, and the intervals between these and the main reflexions would give information about the thicknesses. These, however, have not been found. (I made a special search for them in copies of the records of the Afghanistan earthquake of 1929 February 1, and am sure that no early reflexion existed with an amplitude as much as a tenth of that of the main reflexion.) It appears therefore that the transitions are gradual, probably spread through a few kilometres of depth. But M. A. Tuve and H. Tatel have announced striking results found in seismic survey near Washington, D.C. They had seismographs at intervals of about 1 km. spread over about 200 km. A definite pulse could then be traced by the similarity of its beginnings on many successive records, and some were found that appeared to be reflected at the top of the lower layer. Willmore's reflected P in the Heligoland explosion is also relevant, as is Lapwood's delayed S in Japanese earthquakes (see p. 93). On the other hand A. A. Treskov at the Helsinki conference of 1960 showed very convincing records of such reflexions. There is a similar anomaly in explosion records. The reflexions are regularly found. But I looked specially for them in the records of the Oppau explosion published by Hecker and that at Heidelberg, which was sent to me privately. There were many readable movements besides Pg and Sg, but if any was identified as a reflexion of P there was nothing identifiable as that of S at the appropriate time, or conversely. It looks as if the nature of the boundary varies from place to place. An interesting feature is that the general irregular movement does not repeat itself in this way; this suggests that it is impressed in the immediate neighbourhood of the observing station. This is an interesting confirmation of the hypothesis of scattering.

Until more details are available on reflexions, any estimates of thickness will be subject to some ambiguity of interpretation, which we may hope to resolve some day but cannot now. We can only assume sharp transitions to make our problems definite, and do our best with the pulses that do exist. With this understanding we can do a lot with pP and sS. To put the matter roughly, we notice that for a focus in the upper layers any pulse in the lower layer has begun by descending or ascending steeply, and consequently differences of focal depth within the upper layers affect its time by nearly the same amount at all distances. Thus the study of P or S alone will determine only differences of times of travel for different distances; if a small constant is added to all times the changes in the residuals could be equally well interpreted as due to error in the time at origin or to focal depth within the upper layers. But if we have a consistent table for P for an upper-layer focus we can use it to calculate times for foci in the lower layer. For these the effect of focal depth varies greatly with distance, and consequently for deep foci it is possible to estimate focal depth from P alone. The depth so found is effectively the depth below the top of the lower layer. But for a given depth the times of pP also can be calculated for a given crustal structure; these are affected by the thickness of the surface layers because pP traverses them twice near the epicentre and P does not. Hence the interval from P to pP at a given distance provides an equation for the thicknesses. The effect of focal depth within the upper layers actually does vary a little with distance, and this effect must be taken into account, but it is only a small correction. This method was applied to the Japanese deep-focus earthquakes. For those such that the reflexion of pP would be under land, using trial values $T = 17$ km., $T' = 9$ km., I found (1939a, p. 451)

$$\delta T + 0 \cdot 85 \delta T' = + 5 \cdot 7 \pm 1 \cdot 3 \text{ km.}, \tag{1}$$

that is,
$$T + 0 \cdot 85 T' = 30 \cdot 4 \pm 1 \cdot 3 \text{ km.} \tag{2}$$

The estimates of thickness used in the J.B. tables of 1940 were obtained by combining this equation with one derived from early studies of surface waves by Stoneley and Tillotson (1928) and Stoneley (1928). They gave the following solutions:

$$T = 19 \text{ km.,} \quad T' = 0; \quad T = T' = 15 \cdot 7 \text{ km.;} \quad T = 13 \cdot 0 \text{ km.,} \quad T' = 2T.$$

These agree in giving
$$4T + T' = 78 \text{ km.}, \tag{3}$$

the standard error of which I estimated as 8 km. Combining this equation with (2) we have

$$T = 14 \cdot 9 \pm 2 \cdot 9 \text{ km.,} \quad T' = 18 \cdot 1 \pm 4 \cdot 0 \text{ km.} \tag{4}$$

As these values are incorporated in the 1940 tables they must be preserved as a standard of comparison, but they are subject to some corrections. In the two papers used for the surface waves the velocities of S adopted were not quite the same, so that the solutions are not strictly comparable. In later work, Stoneley (1948) solved the problem of Love waves in

Eurasia with later velocity data and different ratios of T' to T, and found instead of (3)

$$2T + T' = 66 \pm 6 \text{ km.} \tag{5}$$

Then we have four relevant equations:

		O − C	χ^2
Near earthquakes:	$T = 23{\cdot}7 \pm 2{\cdot}5$ km.,	− 1·5	0·4
	$0{\cdot}258T + 0{\cdot}184T' = 8{\cdot}7 \pm 0{\cdot}7$ km.,	+ 0·9	1·6
Deep earthquakes:	$T + 0{\cdot}85T' = 30{\cdot}4 \pm 1{\cdot}3$ km.,	− 1·0	0·6
Surface waves:	$2T + T' = 66 \pm 6$ km.	+ 8·3	1·9
			4·5

The least squares solution is

$$T = 25{\cdot}2 \pm 2{\cdot}3 \text{ km.}, \quad T' = 7{\cdot}3 \pm 3{\cdot}2 \text{ km.}$$

The residuals are satisfactory. Nevertheless, I should be disinclined to trust the apparent accuracy. In the first place, the standard error of the solution for T' is only just less than that derived from the near-earthquake data alone, and the anomalies found in deriving these equations suggested that the uncertainty was underestimated. In the second, the direct study of near earthquakes in western Europe and Japan has suggested that the upper layer is thinner and the intermediate one thicker in Japan, so that the legitimacy of combining the data is in doubt. The Love wave data are none too consistent. Stoneley's data are for two earthquakes studied by himself and two by Rohrbach. Each period yields a group-velocity and hence an estimate of thickness on the hypothesis of a particular ratio T'/T. If the value of T'/T was correct there should be no trend in the estimates of T for different group-velocities. Actually from Rohrbach's data the trends are in opposite directions for $T' = 0$ and $T' = T$, suggesting a ratio about $T'/T = 0{\cdot}7$. But for Stoneley's own data the trends are in the same direction and would suggest a negative value of T'/T, and the differences appear significant. I think that the correct conclusion is that T is near 30 km. and T' small.

Stoneley (1955) has solved the corresponding problem for Rayleigh waves from two Asiatic earthquakes observed at Göttingen. The results fit the formulae

$$T + 0{\cdot}453T' = 41{\cdot}2 \text{ km.}, \quad T + 0{\cdot}646T' = 36{\cdot}6 \text{ km.}$$

Stoneley thinks that this may correspond to a real difference, since the great circle track from the first epicentre to Göttingen passed through the Himalayas. The calculated values on the solution just given would be 28·5 and 29·9 km.

There is a decided suggestion that there is some further complication, possibly an effect of the sedimentary layer or non-homogeneity of the lower layer.

3·19. Travel times and velocity distribution. The times of the fundamental observed phases P, S, PKP and SKS are reduced to a surface focus and to the mean sphere, the surface structure being taken to consist of an upper layer 15 km. thick and an intermediate one 18 km. thick. The times of PcP and ScS were calculated (given the velocities of P and S), for a few trial values of the radius of the core. This radius was then chosen to fit the observed times as well as possible, and the times of PcS and ScS were interpolated to the adopted value. Times within the core are found by subtraction of PcP from PKP, and of ScS from SKS, and interpolated to shorter distances than the comparison gave directly. The inner core makes considerable complications. The times of K itself are of the type shown in Fig. 13, and so are those of many other core waves, notably SKS. For PKP and other core waves with prominent caustics the corresponding graphs are of forms similar to those in Figs. 14 and 15. In some (especially PKP) the extra cusp at B comes below the DF branch, in some above it. In any case the ECD portion of the curve arises from the complicated behaviour near the inner core, and the observed times refer to AEF in Fig. 13; from D to the distance of B, the range BE, and the range

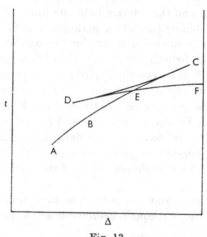

Fig. 13.

EF in Fig. 14; and DF in Fig. 15. Thus a certain amount of hypothetical reconstruction of the unobserved parts of the curves is necessary, and affects the estimation of the velocity of K at the corresponding depths. The hypothetical part, however, is designed to fit the times of the various core phases where they are observed, and any permissible changes in it must be such as will not affect these times.

The peculiar region where velocity decreases with depth was introduced because if c/r is constant the rays are equiangular spirals, and can proceed to any distance with constant $dt/d\Delta$. This appeared necessary to account for

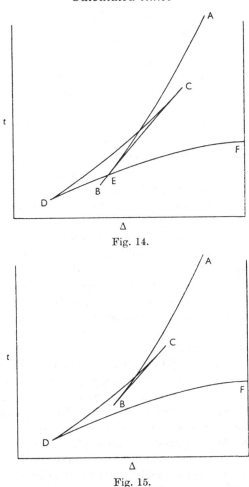

Fig. 14.

Fig. 15.

the length of the branch DE and its apparent straightness. The 1940 times
for this branch depended on a single earthquake. Another valuable one was
an Indonesian one of 1934 June 29 (Jeffreys, 1942a). This suggested that
the times needed an increase of $0.9^s \pm 0.3^s$.

Gutenberg took a single discontinuity, and I have never understood how
he reconciled this with the length of the receding branch. Landisman
(private communication) has found that no reconciliation is possible.

The tables of travel times are given in full in papers in vols. **4** and **5** of
the Geophysical Supplement, and are collected in a pamphlet issued by the
I.S.S. (Jeffreys and Bullen, 1940). Values for focal depths up to $0.12R$
are given. R for this purpose is the mean radius of the base of the crustal
layers in the continents; with a mean radius of 6371 km. for the outer

surface and a total thickness of the outer layers of 33 km., $R = 6338$ km. Here we give only specimen values.

Times for other phases are calculated from those of the fundamental ones by the principle of stationary time, with the exception of diffracted P beyond $105°$, for which they are derived from the observed times. For a surface focus they are shown in Fig. 16. It must be remarked again that only phases whose travel times are true minima are satisfactorily readable, and besides the fundamental phases the only one shown in the diagram that satisfies this condition is PKS. The rest are often large movements, but always diffuse. For deep-focus earthquakes pP and sS are sharp.

The standard errors of the times of the fundamental phases, as indicated by the consistency of the data used to form them, range from about $0 \cdot 3^s$ for most of the P table to about 1^s for some parts of the SKS table and ScS. It is not asserted that all observations must agree with the tables to these accuracies, but it is asserted that any greater change in the tables, while it might fit some sets of data better, would certainly fit others worse. There is much evidence, notably the redeterminations of the velocity of P in western Europe, Japan and the Pacific, that travel times are not quite the same everywhere, but the first requirement for any discussion of such variations is the use of a consistent standard of comparison. A discrepancy is usually capable of two or three interpretations, and a decision between them cannot be made without comparison with other material; but such a comparison will be invalid unless the standard itself is consistent.

It is an unsatisfactory feature that in a general table we cannot take account of the difference between continents and oceans, and a supplementary set of tables of the effects of this difference is needed. It is anomalous that the zero of level for focal depths of oceanic earthquakes should be taken as 33 km. below the mean sphere, which has no special importance below the oceans; but there can be no generally satisfactory standard until we know more about the structure below the oceans, and we have not the information needed for the supplementary tables suggested.

Standard times: surface focus

Near-earthquake phases:

Pg, $19 \cdot 95\Delta$; Sg, $33 \cdot 0\Delta$; P^*, $17 \cdot 10\Delta + 2 \cdot 8$; S^*, $29 \cdot 70\Delta + 3 \cdot 9$.

Δ	P	S	Δ	P	S
0	$(0^m\ 6 \cdot 8^s)^*$	$(0^m\ 10 \cdot 7^s)^*$	35	$6^m\ 56 \cdot 1^s$	$12^m\ 28 \cdot 2^s$
5	1 18·1	2 17·5	40	7 38·1	13 44·5
10	2 28·0	4 22·2	45	8 18·9	14 57·9
15	3 25·0	6 22·9	50	8 58·0	16 8·6
20	4 37·0	8 17·1	55	9 35·4	17 16·8
25	5 26·8	9 48·9	60	10 10·7	18 22·6
30	6 12·5	11 10·2			

* P and S do not exist at $\Delta = 0°$; the times given there are to facilitate interpolation.

Standard times: surface focus (cont.)

Δ	P	S	SKS	Δ	P	S	SKS
65	10m 44.0s	19m 25.5s	20m 36.2s	90	13m 2.7s	23m 54.5s	23m 32.8s
70	11 15.4	20 25.6	21 13.4	95	13 25.7	24 38.2	24 1.1
75	11 45.0	21 22.6	21 50.2	100	13 48.4	25 20.4	24 27.0
80	12 12.7	22 16.5	22 26.3	105	14 10.6	26 2.1	24 50.5
85	12 38.5	23 7.3	23 0.8				

Δ	SKS	PKP	Δ	SKS	PKP
110	25m 12.2s	18m 33.2s	145	26m 46.9s	19m 39.2s
115	25 32.0	18 43.0	150	26 53.8	19 47.4
120	25 50.0	18 52.7	155	26 59.7	19 54.5
125	26 5.9	19 2.4	160	27 4.7	20 0.8
130	26 19.9	19 12.0	170	27 11.5	20 9.2
135	26 30.8	19 21.4	180	27 13.5	20 12.2
140	26 39.1	19 30.5			

Δ	PcP	ScS	Δ	PcP	ScS
0	8m 34.3s	15m 35.7s	60	10m 56.6s	19m 58.8s
10	8 39.0	15 44.6	70	11 37.8	21 15.2
20	8 53.0	16 10.3	80	12 20.6	22 35.5
30	9 14.9	16 51.0	90	13 4.2	23 57.8
40	9 43.9	17 44.6	100	13 48.5	25 20.7
50	10 18.3	18 47.8			

Standard times: deep focus

$pP-P$

h/R

Δ	0.00	0.01	0.02	0.03	0.04	0.05	0.06	0.07	0.08	0.09	0.10	0.11	0.12
30	10	22	34	45	56	66	75	83	90	—	—	—	—
40	10	23	35	47	58	69	80	89	97	104	110	—	—
50	10	23	36	49	61	72	83	93	102	110	118	124	130
60	10	24	38	51	64	76	88	98	108	117	126	134	141
70	10	25	39	53	66	79	91	102	113	123	132	141	150
80	10	25	40	54	67	81	94	105	117	127	137	147	156
90	10	26	41	55	69	82	96	108	120	131	141	152	162
100	11	26	41	55	69	83	96	108	120	131	142	153	163

$SKS-P$

h/R

Δ	Surface	0.00	0.01	0.02	0.03
85	10m 22.3s	10m 18.5s	10m 12.1s	10m 5.8s	9m 59.6s
90	10 30.1	10 26.3	10 20.0	10 13.7	10 7.5
95	10 35.4	10 31.6	10 25.2	10 18.9	10 12.5
100	10 38.6	10 34.8	10 28.4	10 21.9	10 15.6

Δ	0.04	0.05	0.06	0.07	0.08
85	9m 53.7s	9m 47.8s	9m 42.1s	9m 36.7s	9m 31.3s
90	10 1.4	9 55.5	9 49.9	9 44.3	9 38.8
95	10 6.3	10 0.3	9 54.5	9 48.9	9 43.4
100	10 9.3	10 3.2	9 57.3	9 51.7	9 46.2

Δ	0.09	0.10	0.11	0.12
85	9m 26.0s	9m 20.9s	9m 16.2s	9m 11.7s
90	9 33.5	9 28.3	9 23.5	9 19.0
95	9 38.1	9 32.6	9 27.6	9 23.0
100	9 40.7	9 35.2	9 30.0	9 25.2

Velocity distribution

Standard times: deep focus (cont.)

$sS-S$

h/R

Δ	0·00	0·01	0·02	0·03	0·04	0·05	0·06	0·07	0·08	0·09	0·10	0·11	0·12
30	16	38	60	80	100	118	135	150	162	—	—	—	—
40	16	39	61	83	103	122	141	158	172	185	196	206	—
50	17	41	64	86	107	128	148	166	182	196	210	221	231
60	17	42	66	89	111	132	153	172	190	206	220	234	246
70	17	43	67	91	114	137	158	179	198	215	231	246	261
80	18	44	69	94	118	141	164	185	205	223	241	258	274
90	18	45	71	96	121	145	169	191	212	231	250	268	286
100	18	45	72	98	123	147	171	194	215	235	254	272	290

Velocities

Shell			Outer core			Inner core	
r/R	α (km./sec.)	β (km./sec.)	r/R	r/R_1	α (km./sec.)	r/R_2	α (km./sec.)
1·00	7·75	4·353	0·548	1·00	8·10	1·0	11·16
0·99	7·94	4·444	0·537	0·98	8·18	0·9	11·19
0·98	8·13	4·539	0·526	0·96	8·26	0·8	11·21
0·97	8·33	4·638	0·515	0·94	8·35	0·7	11·23
0·96	8·54	4·741	0·504	0·92	8·44	0·6	11·25
0·95	8·75	4·850	0·493	0·90	8·53	0·5	11·27
0·94	8·97	4·962	0·482	0·88	8·63	0·4	11·28
0·93	9·50	5·227	0·471	0·86	8·74	0·3	11·29
0·92	9·91	5·463	0·460	0·84	8·83	0·2	11·30
0·91	10·26	5·670	0·449	0·82	8·93	0·0	11·31
0·90	10·55	5·850	0·438	0·80	9·03		
0·88	10·99	6·125	0·427	0·78	9·11	($R_2 = 0.36R_1$	
0·86	11·29	6·295	0·416	0·76	9·20	$= 0.197R$	
0·84	11·50	6·395	0·406	0·74	9·28	$= 1250$ km.)	
0·82	11·67	6·483	0·395	0·72	9·37		
0·80	11·85	6·564	0·384	0·70	9·44		
0·78	12·03	6·637	0·373	0·68	9·52		
0·76	12·20	6·706	0·362	0·66	9·58		
0·74	12·37	6·770	0·351	0·64	9·65		
0·72	12·54	6·833	0·340	0·62	9·72		
0·70	12·71	6·893	0·329	0·60	9·78		
0·68	12·87	6·953	0·318	0·58	9·84		
0·66	13·02	7·012	0·307	0·56	9·90		
0·64	13·16	7·074	0·296	0·54	9·97		
0·62	13·32	7·137	0·285	0·52	10·03		
0·60	13·46	7·199	0·274	0·50	10·10		
0·58	13·60	7·258	0·263	0·48	10·17		
0·56	13·64	7·314	0·252	0·46	10·23		
0·55	13·64	7·304	0·241	0·44	10·30		
			0·230	0·42	10·37		
			0·219	0·40	10·44		
			0·208	0·38	9·92		
			0·197	0·36	9·40		

$(R_1 = 0.548R = 3473$ km.)

(See also Fig. 11, p. 110.)

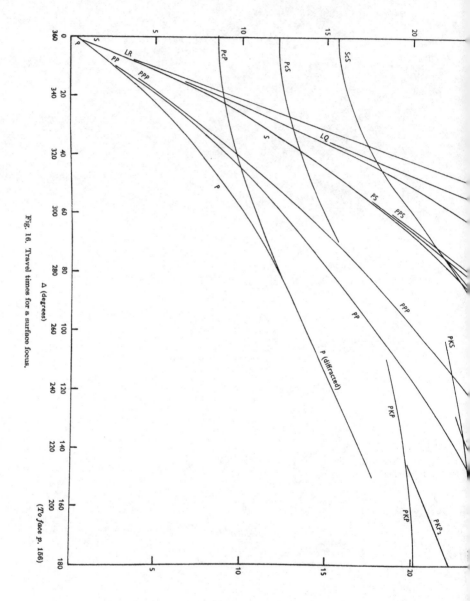

Fig. 16. Travel times for a surface focus.

(To face p. 156)

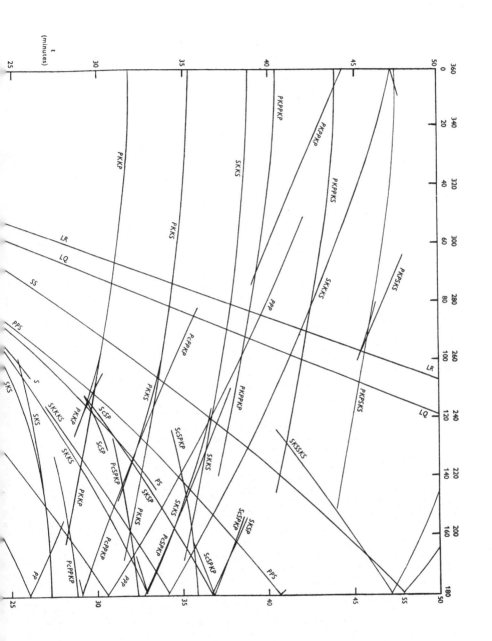

A general statement of accuracy would be difficult. As the times of P and S to $100°$ seemed known to about 1 part in 1000, some weighted average of the velocities would have a similar accuracy. The chief inaccuracy of the separate values comes from the numerical differentiation to get $dt/d\Delta$. Each table beyond $20°$ rests on determinations of time at distances so spaced as to give independent uncertainties. The average interval is about $10°$. Then the individual velocities may be taken to have standard errors of about 1 part in 100, but the uncertainties of the velocities at different depths are strongly negatively correlated. There are special difficulties about the $20°$ discontinuity, since part of the time curves here has been reconstructed from incomplete data. It can be shown that the effect of this is small compared with the general uncertainty of the table except for r/R between $0·95$ and $0·92$. All these estimates rest on the supposition that the variation is as smooth as possible while consistent with the observations. If this supposition is not maintained much larger changes would be possible. If, for instance, the times can be changed by $0·5^s$, we could always alter $1/c$ by $0·01$ sec./km. for 50 km. of the path without violating the conditions, and this would mean changes of c from 5 % to 14 %.

The time of transmission of K along a diameter is $11^m\ 37·9^s \pm 0·6^s$; thus the general accuracy for the core is comparable with that for the shell, but the inner core makes special complications. The assumption that just outside it there is a region where the velocity is proportional to r must apparently be nearly correct if we are to have any explanation of the existence of the receding branch of PKP and the remarkable constancy of $dt/d\Delta$ along it. It leads to a minimum distance of $110°$ for this branch, which is practically the observed value, though this was not used in the calculation of the thickness of the transition zone. The standard error of R_2 is not less than $0·002R_1 = 7$ km., and may possibly not be much more. Apparently the velocity in the inner core varies little and should be right to about 1 part in 200.

Buchbinder (1968) found that PcP reversed direction at about $32°$ and inferred a considerably smaller velocity of K, $7·5$ km./sec., and an inappreciable change of density at the core boundary. Berzon, Kogan and Passechnik (1972) find no such reversal. Randall (1970) however gets a higher velocity than mine, $8·26$ km./sec. I think any considerable change improbable.

In view of conflicting evidence on the velocity of K at small depths in the core, I have calculated the apparent uncertainties of the coefficients that I determined in 1940. The formula is

$$t = \Delta(7·479 - 0·0204(\Delta/10)^2),$$

with no term in Δ^5. Altering the times at $35°$, $47°$ and $60°$ by their standard errors and restoring this term makes the following changes in the coefficients:

	Δ	$\Delta(\Delta/10)^2$	$\Delta(\Delta/10)^4$
35°	0·035	0·0027	0·00005
47°	−0·017	0·0019	−0·00004
60°	0·005	−0·0007	+0·00002

On the face of it the standard error of the coefficient of Δ and therefore of the velocity at small depths is about 1 in 200.

D. K. Chowdhury and C. W. Frasier (1972, 1973) study P and PcP at LASA, Montana, for about 50 earthquakes at distances from 26° to 40°. They find them in the same phase at all distances, and make a comparison of ratios of amplitudes with various models.

3·20. Recent work on core structure. Considerable changes have been made by Nguyen-Hai (1961, 1963), B. Bolt (1959, 1964, 1965, 1968), Adams and Randall (1964) and Subiza and Båth (1964). Gutenberg had already commented on early readings of PKP in some ranges. Nguyen-Hai and Bolt found a definite series and explained it as due to a smaller discontinuity a little outside the inner core. Adams and Randall found two series, explicable by discontinuities 323 km. and 722 km. outside the inner core, the radius of the latter being 1257 km. Each gives rise to a receding and an advancing branch. The times on the main branches DE and ABF are not appreciably altered. There is a small decrease of velocity with depth between the discontinuities. Many seismograms are reproduced and look convincing.

The 1940 times in the inner core are based on the circular ray formula in the approximate form

$$t = (222 + a)\sin \tfrac{1}{2}\Delta - b\sin^3 \tfrac{1}{2}\Delta$$

with $a = +2\cdot1 \pm 1\cdot2, b = +1\cdot97 \pm 1\cdot12$. b is hardly significant, but is probably too low. The corresponding increase of velocity to the centre is 0·15 km./sec. The variation of pressure is roughly 0·15 of that in the outer core, and in the outer core the variation is about 12 %. Then we should expect the increase in the inner core to be about 1·8 % or, say, 0·2 km./sec. 0·3 km./sec. would certainly be permitted by the uncertainty of b.

Nguyen-Hai has determined epicentres by the unsatisfactory method of station pairs; Adams and Randall appear to have used too few distant stations. So long as there is a good comparison of PKP with P at 70° or so a small error in the epicentre does not greatly matter; but any correction to the P times will be transferred to PKP, and a definitive solution must wait for fuller data.

It is unfortunate that in special studies distances are often given in kilometres and focal depths nearly always so. The present tables are standard; they certainly need corrections, but results given in other forms tell us nothing about these corrections until they have been compared with the standard. I know of no place where a scale of depths in kilometres is compared with that in multiples of $0\cdot01R$. Probably the authors have done the computation from the present tables and converted to kilometres according to some unspecified scale. Any such conversion introduces loss of accuracy, owing to rounding-off errors. Kilometres for distance arise when distances are found from a local survey, and hence up to some tens of kilometres; but conversion of angular distances to kilometres is a useless complication.

It is possible that the radius of the core needs a slight increase. Kogan (1960) gave 7 residuals of *PcP* against the 1940 table. They range from $-3\cdot6^s$ at $42°$ to $-2\cdot8^s$ at $78°$. In this range the effect of changing the core radius by $0\cdot01R$ varies by a factor of 3, and it appears that they cannot be explained by a change in the size of the core. Gogna's correction to P at $95°$ for Pacific explosions is $-2\cdot0^s$ and the corresponding correction to *PcP* due to the shell alone would be about the same. It remains possible, however, that the core radius needs a slight increase, reducing the time of *PcP* at large distances by about 1^s; a change of the core radius by $0\cdot01R$ would change the time of *PcP* at $70°$ by about $2\cdot2^s$ (Jeffreys, 1939b, p. 517). The readings there could be fitted by an increase of the core radius by about $0\cdot005R$ or 30 km. This is not, however, very satisfactory as it does not account for the small variation between $42°$ and $78°$.

On the other hand Bullen and Haddon (1967a, 1969, 1970) in a study of free vibrations find evidence that the core radius needs an increase of 15 to 20 km. and possibly more. My standard error was $2\cdot5$ km. However, this is based only on the internal scatter of the data for *ScS* and might be increased if the various errors of comparison were taken into account.

A valuable series of observations of P and *PcP* from the nuclear explosion of 1963 September 13 is given by Buchbinder (1965). Residuals are given against the J.B. times from $19°$ to $88°$. *PcP–P* is the most directly related to the radius of the core. Using this for the permanent stations I get a correction to the radius of the core of $+3\cdot8\pm2\cdot1$ km.; if one large residual (Trinidad) is rejected this becomes $+4\cdot8\pm1\cdot6$ km. The standard errors of one difference are $0\cdot95^s$ and $0\cdot7^s$ according to the two solutions. A set of 8 temporary stations gave $+7\cdot8\pm3\cdot3$ km., standard error of one difference $1\cdot2^s$. The result is important because the 1939 radius depends entirely on *ScS*, *PcP* having shown inconsistencies. Bolt (1970) gets a core radius of 3479 ± 2 km. This compares with my 1939 estimate of $3473\pm2\cdot5$ km. and $3477\cdot8\pm1\cdot6$ km. based on Buchbinder's observations. I think that Bullen and Haddon's result is not due to error in the radius of the core but to the side effect of damping mentioned in §2·041.

The P residuals at the permanent stations (Trinidad omitted) give means: $25°$ to $30°$, $-1\cdot2^s\pm0\cdot2^s$; $30°$ to $40°$, $-1\cdot5^s\pm0\cdot3^s$; $40°$ to $50°$, $-1\cdot4^s\pm0\cdot3^s$; $50°$ to $60°$, $-1\cdot6^s\pm0\cdot6^s$; $70°$ to $90°$, $-1\cdot2^s\pm0\cdot4^s$. The variation in this range is not significant. The results are similar to those found for shocks in other regions.

Buchbinder remarks that *PcP* is larger relative to P than in the ratio calculated from the travel times.

A. F. Espinosa (1966) has studied ScS reflected up to five times at the core. Unfortunately the results are given only by graphs.

Anderson and Kovach (1964) studied an earthquake of 1962 December 8 in South America. The focal depth is given as 620 km. Consecutive reflexions of ScS and their surface reflexions are read. The dominant period of the swings is 20–30^s. Comparison of amplitudes indicates a Q for the upper mantle (really above the focus) of 151 to 185, and an average for greater depths of 1430. Two estimates of the travel time of ScS through the shell are given as 15^m 37^s and 15^m 40^s as against 15^m 35·7^s from the J.B. table.

Gogna (1968) has made a new study of PKP using I.S.S. data. He confirms the existence of a further branch as found by Nguyen-Hai and Bolt; the extra branch of Adams and Randall lies too near the others to be satisfactorily separated. He finds signs of a small discontinuity in the DE branch at about 135°; for smaller Δ the curvature is upwards. He quotes a result of Keilis-Borok and Yanovskaya (1967) that the theoretical amplitudes for the refracted branch should be too small to be observed, but the reflected ones should be larger; in that case PKP up to 135° would be reflected at the core boundary, not transmitted.

A much more complicated structure is derived by Ergin (1967).

S. S. Alexander (1967) finds diffracted P to show significant lateral variations, most pronounced in the deepest 100–200 km. of the shell.

There are some difficulties in the comparison of Gogna's results with those of Adams and Randall. At intervals of, usually, 5°, the differences for the various branches are as follows.

DEF			AB			GH and IJ		
Δ	$AR-G$		Δ	$AR-G$		Δ	$AR(GH)-G$	$AR(IJ)-G$
125	+1·4		145	+3·1		130	+9·8	−2·2
130	+1·2		150	+2·3		135	+7·1	−1·8
135	+0·9		155	+2·3		140	+4·2	−0·1
140	+0·4		160	+2·4		145	+1·5	+0·6
145	+0·1					150	−0·2	+1·7
150	+0·5					153	−0·3	+0·7
155	+0·7							
160	+0·9							

The times of origin depend mainly on observations of P for $\Delta > 70°$, and may differ on account of differences in the tables used for P. A systematic difference of about 1^s might be expected. This is true for the DEF branch.

Gogna's GH agrees better with their IJ than with their GH, but might be consistent with their GH beyond 145°.

From 110° to 135° Gogna's (1968) times give a slight positive value of $d^2t/d\Delta^2$. He quotes Keilis-Borok and Yanovskaya for the result that in this range the reflected phases would be much larger than the reflected ones. This is strongly confirmed by Buchbinder, Wright and Poupinet (1973) who trace the reflexion $PKiKP$ back to 44°.

It should be noted that all these reconstructions avoid the mysterious drop in the velocity of P just outside the inner core; refraction at their interfaces gives the required length of the receding branch.

Buchbinder (1971) gets branches for reflexions $PKKP$, $PKKKP$, etc. up to 5; depth to core 2892 km.; depth to transition zone 4550 $(6371 - 4550 = 1821)$; inner core boundary 5145 ± 10 $(6371 - 5145 = 1226)$. The radii are 3479, 1821, 1226 km.

Birch (1972) calls attention to the existence of three states of solid iron and to their thermodynamic relations with the liquid. His discussion is tentative, but indicates that the temperature at the core boundary might be not more than 4000 °K and that at the centre 5000 °K, and not less than 3000 °K and 4000 °K respectively.

Bolt has made a study that assembles many properties related to the core (1972, pp. 307-11) and gets the following results

Location	r (km.)	ρ (g./cm.^{3})	k (10^{12} dyne/cm.^{2})	μ (10^{12} dyne/cm.^{2})	α (km./sec.)	β (km./sec.)
Base of E	1782·0	12·00	12·10	0·00	10·03	0·00
Top of F	1782·0	12·00	12·10	0·00	10·05	0·00
Base of F	1216·0	12·30	12·90	0·00	10·20	0·00
Top of G	1216·0	12·70	13·31	1·05	10·80	2·85
Centre	0·0	13·00	15·05	1·25	11·35	3·10

Clear records have now been obtained for several PKP reflexions (up to 6) on the inner side of the boundary of the core (Bolt and Qamar, 1972). Reflexions are found that occur as a doublet about 140^s before the main phase, and are attributed to a reflexion at a depth of 650 ± 5 km. It is mysterious that a reflexion at about 400 km. depth, as indicated in his earlier paper, is not shown on the seismograms reproduced. Bolt thinks that there is a regional variation. Lehmann (1970) gives a detailed discussion, with many seismograms, indicating a discontinuity near 400 km. depth, but does not find similar behaviour on all records.

A. Qamar (1973) did a very detailed study of core waves, of both arrival times and amplitudes. He found that the amplitudes that would have resulted from Bolt's earlier model were not in accord with the observed amplitudes. He found that he could explain the travel times and amplitudes by having only a tiny velocity jump (about 0·6 km./sec.) at the boundary of the inner core, and one of 0·02 km./sec. about 500 km. outside it.

Müller (1973) studies amplitudes of core phases of long period and finds no evidence for discontinuities above the inner core boundary.

Cleary and Haddon (1972) doubt the additional phases GH and IJ; they prefer to suppose that they are due to scattering from irregularities in the core boundary near the caustic. I think that this does not agree with the clear separation between GH and DE at some distances in Gogna's study, and that it does not explain the extra readings found by several authors (especially Qamar) for $\Delta > 143°$. See also Bullen and Haddon (1973a), King, Haddon and Cleary (1974).

Stamou and Båth (1974) study SKP near the caustic, which they find at $\Delta = 128 \cdot 9°$. The times differ from those of Gutenberg and those of Jeffreys, extended by Shimshoni (1971), by from 5 sec. to -10 sec.

3·21. Local corrections.

Effect of elevation is given in the 1940 tables in the form $f(\Delta)\,(h_1 + h_2)$, where h_1 and h_2 are heights of the local sea level above the mean sphere at the epicentre and the station. h_1 and h_2 are functions of latitude only. This correction could reach about 2^s in an extreme case, that of an epicentre at a pole producing PKP at the opposite pole. However, most epicentres and stations are in middle latitudes and in fact the correction for P seldom reaches 1^s, and it has never been made in the I.S.S. It has of course been made in most more detailed studies. In addition the list of stations includes the height of the station above mean sea level and this may be expected to give a systematic effect. If the height is 1 km. there will be a direct effect of about $0 \cdot 18^s$ for vertical travel of P. But where isostasy holds, a height of 1 km. will imply an extra thickness of 5·5 km. of the upper layers and a depression of their base by 4·5 km., and the effect on vertical travel will be about

$$\left(\frac{5 \cdot 5}{5 \cdot 6} - \frac{4 \cdot 5}{8 \cdot 0}\right) \text{sec.} = 0 \cdot 42 \text{ sec.}$$

For an oceanic island we should expect the time of vertical travel from a depth of 30 km. to be about

$$\left(\frac{5}{7 \cdot 0} + \frac{25}{8 \cdot 0}\right) = 3 \cdot 8^s;$$

for a continent for the same depth about

$$\left(\frac{15}{5 \cdot 5} + \frac{15}{7 \cdot 0}\right) = 4 \cdot 8^s.$$

Thus we might expect a systematic difference of order 1^s between stations on continents and oceanic islands.

The curious fact is that these effects are not observed. I sought for systematic effects in my original comparison of seismological stations, and though some stations showed systematic anomalies they had no recognizable correlation with height. In my study of Pacific explosions no systematic difference between continents and islands was found. This is confirmed by Gogna (1967). Mintrop, in a discussion, reported absence of these effects within the Alps.

On the other hand detailed study within North America (Cleary and Hales, 1966; Doyle and Hales, 1967; Herrin and Taggart, 1962) for the range $32°$–$100°$ has shown systematic differences between stations, P being up to a second early in the central United States and up to a second late in the Basin and Range Province. They are correlated with region but not notably with height.

Doyle and Hales have extended the study to S from 28° to 82° and find effects similar to those for P, but larger. They find no significant departure from the 1940 times in this interval, apart from a possible additive constant.

As these papers concern earthquakes in widely separated regions there is some possibility that systematic errors in the epicentres may have contributed something. Gogna's data for Pacific explosions and Arnold's for earthquakes, however, each concern only one epicentral region, and the epicentres are first-rate. It is most desirable that Arnold's work should be published in full soon. The agreement as it stands is appreciable but not perfect (Jeffreys, 1968a).

The American results indicate an average difference of about 1^s between the coastal areas and the ancient shields. As there appears to be a slight difference in the velocity of P, this could be explained if this difference extends to a considerable depth; but the depth would need to be of the order of 300 km.

A curious feature is that, applying similar treatments to P and S, Doyle and Hales find

$$S \text{ residual} = (3 \cdot 72 \pm 0 \cdot 43) \ (P \text{ residual}) - (1 \cdot 50 \pm 0 \cdot 20).$$

It would naturally be expected that the S means would be about $1 \cdot 8$ times the P means; but they vary much more.

Another local peculiarity was shown by the Longshot (Aleutians) explosion (Carder, Tocher, Bufe, Stewart, Eisler and Berg, 1967). For $\Delta \leqslant 11°$ the times of P fitted $\Delta / 7 \cdot 85 + 4 \cdot 9 \pm 0 \cdot 5$. From 13° to 27° they average $1 \cdot 5$ to $4 \cdot 0^s$ less than the 1940 times. This would suggest that the Japanese velocity extends up to the Aleutians, but the constant term is much less. PcP is 2 to 4^s earlier than the J.B. times.

An extensive set of travel times for P, PP, pP, PcP and PKP is by Herrin and others (1968). It is not quite clear how they are meant to be used. I think it premature at present to attempt a new single set of tables. We already know that there are significant differences between Europe, Japan and the central Pacific, and between the east and west of North America. What is useful is a single standard of comparison such as is already provided by the 1940 tables. Small differences from the standard can be listed separately, but a multiplicity of standards is a continual source of troubles and mistakes, as astronomers know well. Comparison is in fact complicated further by focal depths in kilometres.

Passages in the statistical article by H. W. Freedman could lead to confusion between my method of uniform reduction and the rejection of large residuals, which my method explicitly avoids. It involves a continuous set of weights, low for large residuals, but does not absolutely reject any; the large shifts of the mean according to which observations are retained are avoided (Jeffreys, 1967d, pp. 214–216). A precaution not mentioned in the book just cited is that the weighting system should not be used in the first approximation to an epicentre. Cases have been found where a bad pre-liminary epicentre had led to attaching high weights to the wrong observa-tions, as Arnold has pointed out. This can be avoided if equal weights are given to all observations in the first approximation.

Kehar Singh and Jeffreys (Jeffreys and Singh 1973) have compared mean residuals of stations, taken from the studies of Arnold, Gogna and Herrin and Taggart ('968). Here the residuals are against the authors' standard travel times. The standard error of a mean, judged by comparison of residuals at the same station, is usually well under 0·5s, 0·2s being normal, but those judged from comparison of different stations are usually about 0·6s, some means reaching 2s. The stations were classified into regions. The correlations between the three series of observations, mostly independent, were usually about +0·7, indicating that there is some real effect but not that there would be much gain in accuracy by trying to allow for it.

Correlation with height was examined. The American data did not show it, but those for Europe and Asia were consistent with the suggested effect of 0·42 sec./km. In all these regions isostasy is a good approximation, and there is no theoretical reason for the direct effect alone, namely 0·18 sec./km. The failure to detect the effect for North America seems to be as follows. In both Arnold's and Gogna's data the coastal ranges, the great plains, and the Appalachians are in the order of increasing distance, so that the effect would be absorbed into the average travel times.

There was a positive correlation, doubtfully significant, with the depth to the Palaeozoic or older rocks; and there are signs of a correlation with the presence of thick recent sediments.

The mean S residuals are strongly correlated with those of P; in most of the variation the ratio is about 3, agreeing with Doyle and Hales. This also suggests that much of the variation is due to loose sediments, in which the ratio of the velocities will be much more than 1·8.

One peculiarity is that Arnold's P residuals for North America average about 0·4s, those for Europe and Asia about −0·2s. The explanation seems to be as follows. He started by calculating times for deep foci based on my paper (1954a) and determining focal depths and epicentres. It was noticed that the Japanese stations showed a significant variation with azimuth, and this was attributed to a systematic error in the table at large distances. The epicentres were therefore redetermined to fit the stations up to 15°. The result was a small systematic displacement to the south-east, which would increase the calculated times to Asia and Europe and hardly affect those to America. But the final tables combine these again. The result is the first verification of Shimshoni's prediction (p. 129) that if there are differences between epicentral regions there must also be variations with azimuth.

Niazi (1973) studies SH for two earthquakes in Turkey and Persia. The movements were resolved to give components along and across the direction of arrival, thus making *SH* much clearer. But there were curious differences between the times, even though the epicentres were not far apart.

3·22. Seismic survey. The methods of the study of near earthquakes have been extensively applied to the investigation of the sedimentary layers. The waves are generated by firing a charge of high explosive, which must be buried to prevent nearly all the energy going into the air wave. If it is deep enough the top does not blow out and there is hardly any change in the form of the surface afterwards. Broughton Edge and Laby (1931, pp. 221–222) give records showing that burial multiplies the amplitudes by a factor of order 10. Immersion in water produces similar effects, refraction into the bottom being considerable (Hill and Willmore, 1947). The position of the focus is known, and the time of the explosion can be recorded directly on the records, so that the time of transit of each pulse can be recorded directly; thus one of the chief difficulties of the study of natural earthquakes, the need to determine the epicentre and time of occurrence from the times of arrival themselves, does not arise. Further, the only way of making sure of getting a record of a natural earthquake is to keep the seismograph running all the time; to do this, and keep the apparatus manageable in size, limits the speed of the paper to about 1 mm./sec., and usually much less. But in recording an artificial earthquake we are free to start the drum whenever we like, and consequently can run it much faster. It is therefore as easy to read the records to 0·001ˢ as to do it to 1ˢ in a natural earthquake, and it becomes possible to determine not only mean depths of layers but variations about the mean. On the other hand, the energy available is usually small compared with that in a natural earthquake, and the phases recorded are usually all longitudinal and in the sedimentary layers, and readable only at distances of the order of 10 km. *S* is usually indistinct, possibly because an explosion can produce no *SH*; but *Sg* was well recorded in the Burton and Heligoland explosions. The latter was of the same order of magnitude as the Burton explosion, but as its time was decided in advance it was possible to adapt the seismic survey technique to it, with the result that by itself it gave determinations of the times of *P* of the same order of accuracy as had been obtained from all previous determinations together. Quarry blasts have been used extensively in the United States; in some of them *Pg* has been recorded, but not *P*.

The principle, for the simplest case, that of one uniform layer resting on another, is that if v is the velocity of the wave in the lower layer, the delay-depth coefficient is η, and the depths of the interface at the beginning and end of the path are H_1, H_2, the time of travel t is given (Jeffreys, 1935d), to the first order, by

$$t = x/v + \eta(H_1 + H_2). \tag{1}$$

For this purpose variations of *H* are small quantities of the first order. If several seismographs are arranged in a straight line, v and H_1 remain constant, but H_2 varies if the interface is not parallel to the outer surface. Suppose that two explosions are made at places A, B, distance l apart, where the depths are H_0, H_0', and that the seismographs are placed along the line joining them at distances x_1, x_2, ... from A, the depth at x_n being H_n. Then the time at x_n for the first explosion will be

$$t_n = \frac{x_n}{v} + \eta(H_0 + H_n), \tag{2}$$

and for the second $\qquad t_n' = \dfrac{l - x_n}{v} + \eta(H_0' + H_n). \tag{3}$

The difference of these is independent of H_n, so that if both times are observed for several values of x_n we can find v; then η is calculated, and the sums $t_n + t_n'$ vary only on account of the variation of H_n, which can therefore be found.

A direct wave in an upper layer of sediments is sometimes unobservable over a long distance if that layer is chalk or clay; damping is strong unless the rock is hard or under considerable pressure. It is usual to start from a place where the stratum to be mapped either appears at the surface or has its depth found from a boring (and without such a check the geological interpretation will be uncertain even if the form is known). Given a starting point, the method can be used to follow a given stratum for considerable distances. Extensions are obviously possible to find the depths of several superposed layers.

The method has been much used in oil prospecting, especially in the United States and Persia, since the oil collects in the anticlines in a limestone or a salt dome. The highest point in the oil-bearing layer is located and provides the most promising place for a boring.

The most thorough published description of the method is by E. C. Bullard, T. F. Gaskell, W. B. Harland and C. Kerr Grant (1940), who used it for following the top of the Palaeozoic rocks through eastern England; several Mesozoic strata were followed at the same time, and the results were checked by the existing borings. The agreement at these points was satisfactory, and the result is a substantial contribution to geology. Another application is to the structure of the continental shelf. This was begun in the United States by M. Ewing, A. P. Crary and H. M. Rutherford (1937), and applied to the mouth of the English Channel by Bullard and Gaskell (1941). A great advance was the use of depth charges exploded within the sea instead of resting on the bottom. It was found that the compressional pulse in the water was sufficiently refracted into the sea floor to give useful records. The work was unfortunately interrupted by the outbreak of war, but was continued by M. N. Hill and J. C. Swallow (1950) and by T. F. Gaskell and J. C. Swallow (1951). The method has been applied by Ewing, Worzel, Hersey, Press and Hamilton (1950) between the United States and Bermuda. R. W. Raitt and others have investigated the North Pacific

(1956). In general a succession of layers is found; two thin ones probably representing sediments in varying degrees of consolidation, and then an igneous one where the P velocity is about 6·8 km./sec., resting on a deep one with a velocity about 8·2 km./sec. Hill (1957) gave 8·09 km./sec. as a mean over all oceans. The last velocity is reached at a depth of 10·14 km. below sea level.

Reflexions have also been used, especially in California (Gutenberg, Wood and Buwalda, 1932). They can be used to locate fault planes, but as these are steep the analysis cannot be simplified as described above.

3·23. Oceanic structure. A great deal of work has been done on the structure of the ocean floor. The method is to make an explosion in deep water or on the bed, and to observe the times of arrival of pulses on a set of instruments placed in a line. Typical results are as follows:*

	Thickness (km.)	P Velocity (km./sec.)	Supposed rock type
Sea water	4–5	1·5	—
	0·45	2	Unconsolidated sediment
	1·75	4–6	Volcanics or consolidated sediment
	4–7	6·71	'Basaltic' rock
		8·09	Ultrabasic

There is appreciable variability, but there is confirmation of the idea, suggested originally by geology and the study of gravity, that the depth of the Mohorovičić discontinuity is considerably less than under the continents, say 7 to 10 km. of rock instead of about 30 km., and there is nothing corresponding to a granitic layer. Much of the North Atlantic, North Pacific and Indian Oceans has been surveyed, especially by Lamont Observatory, Cambridge Department of Geodesy and Geophysics, and the Scripps Institute, California.

Work has been done on the structure at and near Pacific Islands by R. W. Raitt (1957) and G. G. Shor (1964 a, b). There appear to be about six recognizable layers above the Moho; in the first two the velocities of P are about 2·44 and 3·0 km./sec. The two together are a little less than 2 km. thick. The depth of the Moho is about 16 km. at Eniwetok, 13 km. at Bikini. At Midway Atoll there seems to be about 1·5 km. of coral.

3·24. Microseisms. Seismographic records show that the ground is nearly always in motion. Part of this motion is definitely due to local disturbances, such as traffic and rocking of trees and buildings by wind; this part is usually characterized by periods of a second or so. Microseisms, in

---

* M. N. Hill (1957), *Phys. & Chem. of the Earth*, **2**, Pergamon Press.

the strict sense, are widespread and have periods of the order of 7ˢ. Their amplitude is certainly correlated with storms and with proximity to the sea-coast, but also there is some relation to the position of the recording station, the apparent period being somewhat different at different places even on the same day. They appear to be Rayleigh waves and to be of short enough period to be greatly affected by the sedimentary layer (A. W. Lee, 1932*a*, *b*, 1934, 1935; Whipple and Lee, 1935). The mechanism of their generation is doubtful, and several hypotheses have been proposed. The chief are: beating of the surf on the coast; pressure disturbances on the sea bottom as the swell enters shallow water; and pressure disturbance on the sea bottom in the storm area itself. There is some observational support for all, but there are two outstanding objections to the first two. N. F. Barber and F. Ursell (1948) did a detailed harmonic analysis of the waves reaching S.E. England; G. E. R. Deacon, by comparison with the microseisms at Kew, showed that the periods in the latter were about half those in the former, whereas we should expect the periods to be equal on any ordinary hypothesis. This had already been noticed by P. Bernard for the sea waves on the coast of Morocco and the microseisms at Algiers and European stations. Also they arrive from a storm before the earliest sea waves do, and the times agree with the hypothesis that they are really generated in the storm centre and have travelled along the sea bottom as elastic waves. This hypothesis has to face a serious theoretical difficulty, because the amplitude of the motion in a water wave decreases rapidly with depth and becomes inappreciable at a depth of a wave-length; and water waves of the periods concerned would not be appreciable at the bottom in mid-ocean. M. Miche (1944, p. 73), however, had already shown that in some circumstances there should theoretically be a second-order variation of the pressure (i.e. depending on the square of the amplitude of the surface motion) which has half the period of the waves themselves and is independent of the depth. M. S. Longuet-Higgins saw that this result was just what was needed to explain the facts. He also gave the following simple proof of Miche's main result. Consider a standing wave with surface elevation, to the first order,

$$\zeta = a \cos \kappa x \cos \gamma t,$$

and consider the mass of water between the bottom $z = -H$, the free surface $z = \zeta$, and the two vertical planes $x = 0$, $x = 2\pi/\kappa$. There is no flow across the vertical planes and no change of mass. The only vertical forces on the water in the region are gravity and the pressure over the bottom. If dm is an element of mass the total force due to gravity is $\Sigma g\,dm$ over the region; this is constant because the mass is not changing. The effective total vertical force is

$$F = \Sigma dm \frac{d^2z}{dt^2} = \frac{d^2}{dt^2} \Sigma z\,dm = \frac{d^2}{dt^2} \int_0^{2\pi/\kappa} \tfrac{1}{2}\rho\zeta^2\,dx,$$

since the mean elevation of the water in a column from $z = 0$ to $z = \zeta$ is $\frac{1}{2}\zeta$ and its mass per unit area $\rho\zeta$. The mean of $\frac{1}{2}\rho\zeta^2$ with regard to x is $\frac{1}{4}\rho a^2 \cos^2 \gamma t + O(a^3)$. The variation of this with time must be due to the variation of pressure over the bottom, which therefore contains a term independent of x, namely,

$$- \tfrac{1}{4}\rho\gamma^2 a^2 \cos 2\gamma t + O(a^3).$$

As $\gamma^2 = g\kappa \tanh \kappa H$, we can write this as

$$- \tfrac{1}{4}g\rho(\kappa a)\,(a \tanh \kappa H)\,\cos 2\gamma t.$$

For deep water ($\tanh \kappa H \doteqdot 1$) it is proportional to the product of the amplitude and the maximum slope.

For travelling waves with, say, $\zeta = a \cos(\gamma t - \kappa x)$, every variable term in the pressure will be harmonic in x and therefore will decrease rapidly with depth. But the Miche term will exist whenever two waves of equal length, not necessarily of equal amplitude, travel in opposite directions, and any approximation to this condition may occur in a storm centre.

In the elementary (first order) theory of water waves the pressure variation decreases exponentially with depth. The Miche term is of the second order and consequently is left out of this theory. The exact pressure equation contains a term $f(t)$, which has to be determined by the condition that the pressure over the free surface is atmospheric; this term is Miche's term. The effective transmission to the bottom requires the storm centre to be large compared with the depth. A full discussion taking account of various complications, including the compressibility of the water, is due to Longuet-Higgins (1950), who finds that the total transmission of energy to the sea floor is of the right order of magnitude to account for the microseisms.

Ocean waves can generate microseisms when they enter shallow water. The importance of the second-order theory is that it accounts for their generation when this explanation is not available.

The amplitude of microseisms often varies greatly along a given coast. This has been explained by J. and M. Darbyshire (1957) as due to horizontal refraction in crustal layers of varying thickness.

CHAPTER IV

The Theory of the Figures of the Earth and Moon

'My name means the shape I am—and a good handsome shape it is too.
With a name like yours, you might be any shape almost.'

LEWIS CARROLL, *Through the Looking-Glass*

4·01. Objects of the theory. The determination of the Earth's axes is the main object of geodesy, which is the largest branch of geophysics unless we include geology. Also, observable perturbations of the Moon are produced by the departure of the Earth's gravitational field from that of a sphere. Gravity at the surface of the Earth has been measured in considerable detail. The three sources of information are theoretically closely connected, because the standard of height is mean sea-level, which is defined by the ocean surface, and the ocean surface itself has the property that the direction of gravity is everywhere perpendicular to it. We first describe the problem as it was in 1959, but there have been so many striking developments since then that this description is followed by one of these also.

The theory depends largely on the fact that the surface is nearly a sphere, so that the departures from a sphere can be treated as small quantities of the first order. It is not, however, true that their squares can always be neglected. The ellipticity is about 1/298 and the denominator is uncertain by much less than 1. Thus e^2 is greater than the uncertainty of e. Terms arising from the rotation of the Earth are of the same order of magnitude as the first-order terms containing e. Consequently, if we are to make full use of the accuracy of the observations, it is necessary to retain in the theory quantities depending on squares and products of the ellipticity and the rotation terms. But a formula depending on latitude only would not represent accurately either the radius vector from the centre to the surface or the values of gravity. Irregularities in the former produce departures of the order of 10 km. from the mean over a parallel, and the corresponding departures for gravity reach about 400 mgal.* The products of these with the ellipticity are of the same order as the uncertainty of observation. Fortunately, the greatest departures are confined to areas very small compared with the whole surface of the Earth, and their effect on the general field is very much smaller than their local effect. It is therefore possible to separate the problem into two parts. One concerns the general field. In this we consider the ellipticity and the corresponding parts of gravity and the external field, keeping terms depending on the square of the ellipticity, and also other widespread disturbances; for the latter we can neglect their squares and their products by the ellipticity. The other part concerns local

* 1 gal = 1 cm./sec.²; 1 milligal = 0·001 gal.

[170]

disturbances; these are mostly treated separately, but need some standard general field for comparison.

The theory of the external field can be treated as a self-contained problem. Given the values of gravity over a complete level surface it is possible to infer the field at all external points, without making any special hypothesis about the state of the interior. Gravity is actually measured at the outer surface, which is far from level, but a simple adjustment reduces the actual problem to the theoretical one.

A further extensive theory concerns the form of surfaces of equal density inside the Earth. For a given law of density along a given radius and a given rate of rotation, if the stress at all internal points is hydrostatic, the forms of these surfaces are determinate, and they are approximately spheroids. The differential equation satisfied by their ellipticities turns out to have a remarkable property, which permits the external ellipticity to be determined from the precessional constant with apparently greater accuracy than is given by any of the equations of the external field theory. The weak point of the theory is in the hypothesis of hydrostatic stress. But all evidence available indicates at least that the departures from hydrostatic stress are extremely small compared with the pressure at the centre, and therefore the theory must be at least a good first approximation.

The departures of the external field from spherical symmetry, other than the main ellipticity term, are inconsistent with exact hydrostatic stress in the interior, and consequently provide a way of estimating how far this hypothesis is wrong.

The dimensions of the Earth are estimated in the first place from trigonometric survey. Essentially the method is to make a survey along a meridian or a parallel, determining the length and comparing it with the difference of latitude or longitude of the ends. Latitudes and longitudes are found by astronomical observations, and specify for each place the direction of the normal to the level surface there. If the level surfaces were exact spheres the normals would coincide with the radii. But any departure from spheres is associated with deflexions of the vertical from the radius. The most important consequence of this is that for arcs of meridian the distance travelled for given change of the direction of the vertical is greater in high latitudes than in low ones, on account of the smaller curvature, and comparison of estimates so obtained provided the first way of finding the ellipticity. Other methods now give the ellipticity with greater accuracy, and the measures of arcs are now best used to determine a general scale, the form being determined otherwise. All disturbances of the gravitational field, however, introduce further complications, and it is a difficult matter to decide how far each geodetic arc may be regarded as representative of the Earth as a whole. In particular such arcs are necessarily on land, and the question of a systematic difference between continents and oceans needs examination.

For a spherical Earth with no rotation, a knowledge of surface gravity, the radius, and the constant of gravitation (determined in the laboratory) would suffice to determine the mass and hence the mean density. Here again the effect of the ellipticity on the comparison is not negligible.

The Moon's distance provides a useful datum. The parallax has been measured directly by observing the Moon's position relative to the stars as seen from Greenwich and the Cape of Good Hope at the same time; if we knew the distance between these two observatories exactly the Moon's distance would follow. An alternative method is to calculate from surface gravity what the distance would have to be in order that the Moon should revolve in its actual period. The two results are usually stated in terms of the angle that the equatorial radius would subtend at the Moon's mean distance, and the values given by the two methods are called the visual parallax and the dynamical parallax. They should agree, and the theoretical agreement provides another equation connecting the equatorial radius and the ellipticity. Actually both are disturbed in different ways by departures of the field from spherical symmetry.

Again, on account of the ellipticity of figure, the attractions of the Sun and Moon on the Earth do not reduce to a force through the centre. They produce also a couple, the effect of which is to produce a motion of the axis of instantaneous rotation in a cone about the pole of the ecliptic. This motion is known as precession. The reaction on the Moon disturbs the Moon's orbit in three ways. It makes the perigee move forwards (i.e. in the direction of the Moon's general motion), the nodes backwards. It also produces a motion of the Moon in latitude with a period of a sidereal month. All these effects can be observed. The first two are rendered difficult to interpret by the fact that there are much larger effects due to the attraction of the Sun on the Moon, and small but not negligible ones due to the fact that the Moon also is not quite a sphere. Calculations by E. W. Brown have, however, determined the solar parts with sufficient accuracy to leave a good determination of the parts due to the figures of the Earth and Moon. There is independent evidence about the figure of the Moon, since the differences of its moments of inertia can be determined from the inclination of its axis and a certain disturbance of its rotation. Consequently, the analysis of the Moon's motion leads to a further set of equations relevant to the figures of both the Earth and the Moon.

The most striking recent development is due to the study of artificial satellites. If the external field was symmetrical about the centre, the motion of such a satellite would be in an elliptic orbit, apart from very small perturbations due to the Sun and Moon. Actually the main ellipticity term in the Earth's potential produces a large perturbation. The node of the orbit on the equator moves backwards, and the perigee moves (relatively to the node) forwards for inclinations less than about 60°. Both complete a revolution in a time of the order of a month. As far as I know, the first correct theory of these effects to the first order was given by R. E. Roberson (1957) and soon extended to the second order by D. G. King-Hele (1958). (A second-order theory had been given previously by D. Brouwer, but only for small inclinations.) The speeds for satellites of different inclinations to the equator give values also for the other even zonal harmonics. O'Keefe and others (1959 a, b) also noticed that the eccentricities vary systematically, being larger when the perigee was north of the equator than when it was south. This implies a difference between the northern and southern hemispheres. From this some odd zonal harmonics have been determined. These have been described as showing that the Earth is pear-shaped, but the difference from an oblate spheroid is small compared with the main ellipticity—a quince or tomato would be a better, but still exaggerated, analogy. Effects of terms depending on longitude seemed unlikely to be measurable, but they do give some long-period effects. The results were very inconsistent for a time but now seem to have settled down.

Improved measurements of the velocity of light have made it possible to measure the distances of the Moon, Venus and Mercury by observing the time of a reflected pulse, and also the velocities by means of the Doppler effect. The results are in good agreement. The curious fact is that the Sun's distance comes out about midway between Spencer Jones's trigonometrical determination and Rabe's original dynamical one. (See p. 210.)

According to de Sitter's estimate of the probable amounts of the terms neglected in E. W. Brown's theory of the Moon's motion, it seemed that the observed motions of the Moon's node and perigee could just be reconciled with the theoretical amounts. The theory has been extended by Eckert, and it now appears that theory and observation cannot be made to agree except with a most improbable constitution of the Moon.

Combining the radio distance of the Moon with the dynamical parallax gives the Earth's radius more accurately than before, and incidentally in agreement with the latest geodetic results.

A meeting in Paris in 1950 on the fundamental constants of astronomy decided to recommend no alterations to the standard values used for comparison. But in consequence of this recent work a further meeting in 1963* recommended a complete recalculation, which has been carried out.

The theory of the figures of the Earth and Moon must try to provide a consistent interpretation of these numerous types of evidence.

4·02. Theory of the external field. The Earth's gravitational potential is denoted by U; the acceleration produced by it in any direction is given by the gradient of U in that direction. For any body attached to the Earth the rotation produces an acceleration perpendicular to the axis; in fact, if (x, y, z) are coordinates with respect to non-rotating axes at the centre, z being along the polar axis, the component accelerations of such a body

* I.A.U. Symposium 21, *Bull. Astron.* **25**, 1965.

are $(-\omega^2 x, -\omega^2 y, 0)$, where ω is the rate of the Earth's rotation. But the acceleration components of a free body are $\partial U/\partial x$, $\partial U/\partial y$, $\partial U/\partial z$. Hence the difference between the accelerations of a free body and one initially coincident with it but attached to the Earth is

$$\left(\frac{\partial U}{\partial x} + \omega^2 x, \ \frac{\partial U}{\partial y} + \omega^2 y, \ \frac{\partial U}{\partial z}\right) = \left(\frac{\partial \Psi}{\partial x}, \ \frac{\partial \Psi}{\partial y}, \ \frac{\partial \Psi}{\partial z}\right), \tag{1}$$

where
$$\Psi = U + \tfrac{1}{2}\omega^2(x^2 + y^2). \tag{2}$$

Ψ is a function of position only and is called the *geopotential*. All measurements of gravity are measures of the acceleration of a free body with respect to a frame rotating with the Earth, and therefore of the gradient of Ψ; and the level surfaces are surfaces of constant Ψ. Resultant gravity is then the gradient of Ψ at right angles to the surface of constant Ψ through the place.

The position of the ocean surface is part of a level surface, apart from some minor disturbances due to small meteorological effects such as the drag of the trade winds and periodic tidal ones.

Then over the mean ocean surface

$$\Psi = C = \text{constant.} \tag{3}$$

In general write
$$\Psi = C + \Psi_1. \tag{4}$$

The solid surface is not a level surface, but there is an intimate relation between Ψ over it and the measured height. The standard of level is the spirit level, and therefore differences of height between consecutive points of observation are displacements normal to the level surfaces. Thus for neighbouring points on a survey route, ds apart, the difference of height is

$$dh = \sin \psi \, ds, \tag{5}$$

also
$$d\Psi = -g \sin \psi \, ds = -g \, dh, \tag{6}$$

where ψ is the slope of the line of sight. Hence for a path from the ocean

$$\Psi_1 = -\int g \, dh, \tag{7}$$

the integral being taken along the survey route. The integral will have the same value whatever the route, since Ψ is a function of position alone. But the elements of height dh are normal to the level surfaces, which are curved, and therefore they are not all in the same direction. The crude measure of height would be $\int dh$; but if we measured from sea-level to a mountain top this would have somewhat different values according to the route taken—h is not a single-valued function of position. The differences are actually small, since $\int g \, dh$ is the same for all routes and the differences can arise only from the variations of g, and therefore they are of the order of the crude height times the variations of g expressed as fractions of its mean value and are therefore of orders eh and h^2/a.

If g is known along the survey route Ψ_1 is completely determined along it, and can be used to define a unique height. $-\Psi_1/g_{45}$ is called the dynamic height, where g_{45} is the value of gravity at sea-level according to a standard formula in latitude $45°$. $-\Psi_1/g_l$ is the orthometric height, where g_l is gravity according to the same formula at the actual latitude. These devices eliminate the ambiguity of the crude height. They cannot be used, however, without fairly detailed knowledge of gravity along the route, and it often remains necessary to use the crude height and take simply $\Psi_1 = -gh$.

We do not know directly the form of the outer surface except along the trigonometric arcs, which cover an extremely small fraction of it. However, we do know gravity at many places in all continents and over many extensive parts of the ocean, and the problem can be stated as that of finding the form of the outer surface, given Ψ and g over it.

If for each point of the outer surface we take a normal to the local level surface and proceed along it downwards for a distance equal to the measured height, we get a locus called the co-geoid (Hunter, 1951). Slightly different co-geoids will be obtained according as we use the crude, dynamic, or orthometric height. Over the ocean the co-geoid coincides with mean sea-level. The choice of the most convenient standard is still a matter of discussion but the differences are too small to concern us here.

The analysis of observations of gravity requires a transformation known as the free-air reduction, which involves some approximations. The following theorem makes the nature of these clearer, I think, than they are in the usual treatments. It is a slight modification of one given previously (Jeffreys, 1931 i), which was intended only for a first-order theory, but with a slight change of notation it becomes exact.

If U is the external gravitation potential of a body bounded by a surface S, Green's theorem of the equivalent stratum is

$$4\pi U_P = \iint \left(-\frac{\partial U}{\partial n}\frac{1}{R} + U\frac{\partial}{\partial n}\frac{1}{R} \right) dS \qquad (8)$$

taken over the surface. U_P is the potential at an external point P, $\partial/\partial n$ indicates differentiation along the outward normal, and R is the distance from the element dS to P. In general substitute (4) in (8). Let g be the intensity of gravity at the surface and ψ the inclination of the surface to a surface of constant Ψ. Then

$$\partial\Psi/\partial n = -g\cos\psi \qquad (9)$$

and
$$4\pi U_P = \iint \left(g\cos\psi\,\frac{1}{R} + \Psi_1\frac{\partial}{\partial n}\frac{1}{R} \right) dS$$
$$+ \iint \left[\{C - \tfrac{1}{2}\omega^2(x^2+y^2)\}\frac{\partial}{\partial n}\frac{1}{R} + \frac{\partial}{\partial n}\{\tfrac{1}{2}\omega^2(x^2+y^2)\}\frac{1}{R} \right] dS. \qquad (10)$$

Since P is an external point, $\nabla^2(1/R) = 0$ for points in and on S; then by Green's theorem

$$\iint C \frac{\partial}{\partial n} \frac{1}{R} dS = \iiint C \nabla^2 \frac{1}{R} d\tau = 0 \tag{11}$$

through the interior of S; and

$$\iint \left[-\tfrac{1}{2}\omega^2(x^2+y^2) \frac{\partial}{\partial n} \frac{1}{R} + \frac{\partial}{\partial n} \{\tfrac{1}{2}\omega^2(x^2+y^2)\} \frac{1}{R} \right] dS$$

$$= \iiint \left[-\tfrac{1}{2}\omega^2(x^2+y^2) \nabla^2 \frac{1}{R} + \frac{1}{R} \nabla^2\{\tfrac{1}{2}\omega^2(x^2+y^2)\} \right] d\tau$$

$$= \iiint \frac{2\omega^2}{R} d\tau. \tag{12}$$

Then U_P is the potential of a surface density $g\cos\psi/4\pi f$ over S together with one of normally directed doublets and the potential due to a constant density $\omega^2/2\pi f$ through the interior. This result is exact.

To the first order in Ψ_1/R

$$g\cos\psi \frac{1}{R} + \Psi_1 \frac{\partial}{\partial n} \frac{1}{R} = g\cos\psi \frac{1}{R'}, \tag{13}$$

where R' is the distance from P of a point at depth $\Psi_1/g\cos\psi$ normally below dS. Call the locus of such points S'. Then with a possible error of order Ψ_1^2/R^2 the contribution of the first line of (10) to the potential is that of a surface density $g\cos\psi(dS/dS')/4\pi f$ over S'. The position of this density is independent of the direction of R. $\cos\psi\, dS$ would be the area of the projection of dS on a surface of constant Ψ through it.

In (9) g is gravity at the point of observation. The normal distance of dS' from dS is $\Psi_1/g\cos\psi$. Hence if we extrapolate Ψ along the normal till we reach dS', the result is simply C. To the first order, therefore, the whole external field is equivalent to that of a known surface density over a surface of constant Ψ, together with the small ω^2 term. The depth of this surface below the actual surface, now measured normally to the level surface at the point of observation, is $-\Psi_1/g$, that is, to the measured height if defined as $-\Psi_1/g$ in terms of actual gravity. Then S' is approximately the co-geoid.

Also if the mean radius of the Earth is a, and the slope ψ is small, dS'/dS is practically the ratio of a^2 and $(a+h)^2$. Thus to the first order the equivalent surface gravity is $g(1+2h/a)$, that is, gravity adapted to the new surface according to the inverse square law. This is the *free-air formula*. In a ridge with strong curvature, however, dS' is displaced systematically from dS towards the centre of the ridge. If ψ and h are both small this is another second-order effect. There are places where it is serious, but they are of very small extent in comparison with the surface as a whole.

For a known form of the surface, if the surface density or the potential is completely known, the external field is determinate. In our problem, however, the form of the surface is one of our unknowns, and we have to find a surface such that Ψ is constant and the equivalent surface density has the given value simultaneously.

In practice the free air formula is used to extrapolate gravity and the geopotential Ψ to a surface of constant Ψ, along the direction of g. The principle is that if $f(x)$ is given as a polynomial of degree n for $x \geqslant h > 0$, extrapolated values of $f(x)$ and its first n derivatives can be calculated for $x = 0$; and if the result is converted into a Taylor series in powers of x it will reproduce the correct values of $f(x)$ for $x \geqslant h$. No assumption is made at this stage about the physical interpretation of $f(x)$ for $x < h$; it is simply a mathematical device for making calculation easier.

It seems desirable that heights should be given in a way applicable to cases where they can be directly measured, as up a vertical pole or the interior of a tower, and also at sea level, where the surface $\Psi = C$ is accessible. The normal trajectories of the level surfaces are in general curved, but treating them as straight can produce only a third-order error.

It is worth while to retain some terms of the second order in the departure from a sphere. The chief such departure is the main ellipticity; a second-order solution for this alone is given later, and we need a reduction that will take account of departures from it.

The equations will be of the form

$$\Psi_1 = -g_c h + \tfrac{1}{2} g_1 h^2, \tag{14}$$

$$g_h = g_c - g_1 h + \tfrac{1}{2} g_2 h^2. \tag{15}$$

Then to order h^2 $\qquad\qquad \Psi_1 = -g_h h - \tfrac{1}{2} g_1 h^2 \tag{16}$

which can be taken as the definition of h (not agreeing with any of those mentioned above). g_c is then extrapolated gravity.

We are supposing h/a small. g_1 and g_2 are not constants, since they depend on the ellipticity of the Earth, e, and on its rate of rotation. But we shall neglect terms in $e^2 h$ and eh^2. Then we shall have

$$g_c = g_h + g_1 h - 3g_c h^2/a^2, \tag{17}$$

but g_1 will contain terms of the first order in the ellipticity. We can evaluate these when we have given the theory of the ellipticity. (See §4·023.)

The observed values of g appear only through the expression g_c. Thus, given this quantity over the Earth, gravity can yield no further information. It is called *free-air gravity* and the adjustment $+ 2gh/a$ is called the *free-air reduction*. It amounts, on an average, to 306 mgal. per kilometre of height. It was introduced by Stokes in a classical paper (1849), and later emphasized by Helmert. The reason for the name is that it amounts to using the observed g to calculate a value of g on the co-geoid, neglecting all intervening matter and treating the field as spherically symmetrical. (The actual value on the co-geoid is of course appreciably different; but the actual value is not observed, and if it were it would not lead directly to a determination of the external field because we should still have to allow for the gravitation of the matter above it, which on the whole gives an upward acceleration on the co-geoid and a downward one just outside the Earth.)

The errors of order h^2 are strictly local and can be safely neglected in discussing the widespread features of the field. Those of order eh need a little further discussion because they may keep the same sign over most of a continent. But the error in (12) due to them will be of order egh/a. Over most of the continents h is of order 1 km., and the error will be of order 2 mgal. This might apparently be more than the uncertainty of the mean of the observed values over a region where data are abundant, since a good observation has an uncertainty of 1 or 2 mgal., provided that the usual hypothesis of independence of the errors was correct. But when checks on this hypothesis are applied, correlations between neighbouring residuals are always found, and on present data the true uncertainty of a regional value is more like 5 or 10 mgal. at best.

4·021. Spherical harmonics.* The potential due to the Earth and outside it is a solution of Laplace's equation

$$\nabla^2 U = \frac{\partial^2 U}{\partial x^2} + \frac{\partial^2 U}{\partial y^2} + \frac{\partial^2 U}{\partial z^2} = 0.$$

Along any radius it can be expressed as a series of descending powers of r, the coefficients depending on the direction. Thus it can be written

$$U = \sum_{n=1}^{\infty} \frac{A_n}{r^{n+1}} S_n(\phi', \lambda), \tag{1}$$

where ϕ' is the geocentric latitude and λ the longitude, and the separate terms are also solutions of Laplace's equation.

If $K_n = r^n S_n(\phi', \lambda)$ is a solution of Laplace's equation, then $r^{-n-1} S_n(\phi', \lambda)$ is another, and conversely. K_n is called a spherical solid harmonic, S_n a spherical surface harmonic, of degree n. In the solutions that concern us K_n is a homogeneous polynomial of degree n in x, y, z; for given n there are $2n+1$ such solutions with the property that none of them can be expressed as a linear combination of the rest, and accordingly there are $2n+1$ linearly independent functions $S_n(\phi', \lambda)$. The most convenient way of choosing them in general is to put

$$\sin\phi' = \mu, \quad S_n = p_n^s(\mu)(\cos s\lambda, \sin s\lambda) \quad (s = 0 \text{ to } n), \tag{2}$$

$$p_n^s(\mu) = \frac{(n-s)!}{2^n(n!)^2} \cos^s\phi' \frac{d^{n+s}}{d\mu^{n+s}}(\mu^2 - 1)^n. \tag{3}$$

* Proofs of the statements in this section will be found in H. and B. S. Jeffreys (1972, Ch. 24).

In particular, we have the following explicit forms:

n	s	p_n^s	K_n
0	0	1	1
1	0	$\sin \phi'$	z
	1	$\cos \phi'$	x, y
2	0	$\frac{3}{2}\sin^2 \phi' - \frac{1}{2}$	$\frac{1}{2}(2z^2 - x^2 - y^2)$
	1	$\frac{1}{3}\sin \phi' \cos \phi'$	$\frac{1}{3}zx, \frac{1}{3}zy$
	2	$\frac{1}{3}\cos^2 \phi'$	$\frac{1}{3}(x^2 - y^2), 3xy$
3	0	$\frac{5}{2}\sin^3 \phi' - \frac{3}{2}\sin \phi'$	$\frac{1}{2}z(2z^2 - 3x^2 - 3y^2)$
	1	$\frac{1}{2}\cos \phi'(5\sin^2 \phi' - 1)$	$\frac{1}{2}x(4z^2 - x^2 - y^2), \frac{1}{2}y(4z^2 - x^2 - y^2)$
	2	$\frac{5}{2}\cos^2 \phi' \sin \phi'$	$\frac{5}{2}z(x^2 - y^2), 5xyz$
	3	$\frac{5}{2}\cos^3 \phi'$	$\frac{5}{2}(x^3 - 3xy^2), \frac{5}{2}(3x^2y - y^3)$
4	0	$\frac{1}{8}(35\sin^4 \phi' - 30\sin^2 \phi' + 3)$	$\frac{1}{8}\{8z^4 - 24z^2(x^2+y^2) + 3(x^2+y^2)^2\}.$
	1	$\frac{5}{8}\cos \phi' \sin \phi'(7\sin^2 \phi' - 3)$	$\frac{5}{8}z(x, y)(4z^2 - 3x^2 - 3y^2)$
	2	$\frac{5}{8}\cos^2 \phi'(7\sin^2 \phi' - 1)$	$\frac{5}{8}(x^2 - y^2, 2xy)(6z^2 - x^2 - y^2)$
	3	$\frac{35}{8}\cos^3 \phi' \sin \phi'$	$\frac{35}{8}z(x^3 - 3xy^2, 3x^2y - y^3)$
	4	$\frac{35}{8}\cos^4 \phi'$	$\frac{35}{8}\{x^4 - 6x^2y^2 + y^4, 4(x^3y - xy^3)\}$

With this specification the standard surface harmonics have the property that they are mutually orthogonal; that is, if S, T are any two different harmonics

$$\int_{-\frac{1}{2}\pi}^{\frac{1}{2}\pi} \int_0^{2\pi} ST \cos \phi' \, d\phi' \, d\lambda = 0. \tag{4}$$

If they are identical we have

$$\int_{-\frac{1}{2}\pi}^{\frac{1}{2}\pi} \int_0^{2\pi} (p_n^s)^2 (\cos^2 s\lambda, \sin^2 s\lambda) \cos \phi' \, d\phi' \, d\lambda = \frac{2\pi}{2n+1} \frac{(n-s)!\,(n+s)!}{(n!)^2}, \tag{5}$$

except for $s = 0$, when the right side is $4\pi/(2n+1)$. When $s = 0$ it is usual to write p_n^s as simply p_n.

The functions P_n^s considered in most accounts differ by constant factors:

$$p_n^s = \frac{(n-s)!}{n!} P_n^s. \tag{6}$$

In this form the mean square values of the functions over a sphere vary much more with s.

Bounded and integrable functions of position on a sphere can usually be expressed by infinite series in the standard spherical harmonics, the coefficients being found by integration as for Fourier's series. For some very peculiar functions this is not true, but the exceptions need not concern us.

The surface harmonics p_n^s change sign $n-s$ times as we pass from one pole to the other; while $\cos s\lambda$ and $\sin s\lambda$ change sign $2s$ times as we pass around a parallel. Hence increasing irregularity in a function will be represented by including higher values of n in its expansion. We may therefore speak of widespread disturbances and local disturbances as corresponding in general to low and high values of n. p_n is 1 at the north pole, $(-1)^n$ at the south pole.

In many recent treatments all harmonics are normalized by multiplying by factors to make the mean squares equal to 1. This makes it easier to see their relative importance, but of course complicates the algebra in theoretical work. Root mean square values of the complete functions (including the $\cos s\lambda$ or $\sin s\lambda$ factors) are as follows:

n	s	$P_n^s \cos s\lambda$	$p_n^s \cos s\lambda$
0	0	1	1
1	0	0·577	0·577
	1	0·577	0·577
2	0	0·447	0·447
	1	0·775	0·387
	2	1·549	0·774
3	0	0·377	0·377
	1	0·926	0·309
	2	2·928	0·488
	3	7·169	1·194
4	0	0·333	0·333
	1	1·054	0·264
	2	4·472	0·373
	3	16·73	0·697
	4	47·33	1·972

In terms of P_n^s the root mean square values of the functions vary much more with s. When P_n^s is used for $n = 4$ they differ by a factor of about 140, for p_n^s by about 6. There are advantages in retaining both p_n^s and the normalized functions, but since gravity contains an extra factor $n-1$ not in the potential, no complete normalization would be possible for the terms in both. In comparison it seems that P_n^s should disappear from mathematical literature, especially on account of the simple relation

$$K_n^s = r^{2n+1} \left(\frac{\partial}{\partial z}\right)^{n-s} \left(\frac{\partial}{\partial x} + i\frac{\partial}{\partial y}\right)^s \frac{1}{r}$$

$$= (-1)^n n! \, r^n p_n^s \, e^{is\lambda},$$

the simpler asymptotic relation to Bessel functions, and the simpler rule of transformation to different axes (B. Jeffreys, 1965).

It would be natural to represent the general magnitude of a function on a sphere by its mean square, say

$$\overline{f^2} = \frac{1}{4\pi} \iint f^2 \, d\omega.$$

An analogous representation of its general rate of variation over the sphere would be

$$\overline{g^2} = \frac{1}{4\pi} \iint \left\{ \left(\frac{\partial f}{\partial \phi}\right)^2 + \left(\frac{\partial f}{\cos\phi \, \partial\lambda}\right)^2 \right\} d\omega,$$

which is independent of the choice of axes. The irregularity of the function, irrespective of its general magnitude, would then be taken as the ratio of these two quantities. Denote it by Q. Then Q has the properties (Jeffreys, 1953, 1955):

(1) If f is a surface harmonic of degree n, $Q = n(n+1)$.

(2) Q is stationary for all small variations of f if and only if f is a surface harmonic.

(3) If f is a sum of harmonics up to degree n, $Q \leqslant n(n+1)$; and if f is a sum of harmonics of degrees n and more, $Q \geqslant n(n+1)$.

Thus spherical harmonics are closely related to the expression of the smoothness of a function of position on a sphere; omitting high harmonics is an effective way of smoothing the function, and has the great advantage that it is free from personal peculiarities of the smoother.

Instead of p_2 we shall usually work with the function $\tfrac{1}{3} - \sin^2 \phi'$ because it decreases by 1 from equator to pole.

4·022. Theory of the ellipticity. The values of r at the surface could be expressed by a series of spherical harmonics. A term in p_2 is prominent, but even if r contains no term in p_4, its powers will, and terms in p_4 will still appear in the external potential and in gravity. For comparison with observation we need consistent formulae with given coefficients for these quantities, and therefore must make some convention about the p_4 terms. The method adopted was suggested by geodesy, which takes an exact spheroid as standard of reference. The problem then becomes that of finding an external field, correct to quantities of order e^2, which will be consistent with the co-geoid being an exact spheroid. More detailed observation has disclosed departures from an exact spheroid requiring, among other spherical harmonics, a term in p_4 to represent them, but it will be possible to treat these by the first-order theory.

A formal solution of the problem in finite terms has been given by P. Pizzetti (1913) using solutions of Laplace's equation directly adapted to a spheroidal boundary (H. and B. S. Jeffreys, 1972, pp. 539–40), and the International Gravity Formula, due to G. Cassinis (1930), is based on Pizzetti's analysis. A summary is given by Lambert (1945). Pizzetti's theory has been extended to an ellipsoid with three unequal axes by C. Somigliana (1933 a, b). It leads to a simple exact formula for surface gravity, but for the astronomical comparisons we need an expansion of the potential in powers of r in any case, and I shall proceed by the method of expansion in spherical harmonics.

Consider first an ellipsoid of revolution of major semi-axis a and ellipticity e, so that the minor semi-axis is $a(1-e)$. The polar equation is

$$r^2 \left(\frac{\cos^2 \phi'}{a^2} + \frac{\sin^2 \phi'}{a^2(1-e)^2} \right) = 1, \tag{1}$$

whence $$r^2 = \frac{a^2(1-e)^2}{(1-e\cos^2\phi')^2 + e^2\cos^2\phi'\sin^2\phi'}, \tag{2}$$

and to order e^2

$$r = \frac{a(1-e)}{1 - e\cos^2\phi'}\{1 - \tfrac{1}{2}e^2\cos^2\phi'\sin^2\phi'\} + O(e^3), \tag{3}$$

$$\frac{a}{r} = 1 + e\sin^2\phi' + \tfrac{3}{2}e^2\sin^2\phi' - \tfrac{1}{2}e^2\sin^4\phi'. \tag{4}$$

Put
$$U = \frac{fM}{a}\left\{\frac{a}{r} + J\frac{a^3}{r^3}(\tfrac{1}{3} - \sin^2\phi') + \tfrac{8}{35}D\frac{a^5}{r^5}p_4\right\}, \tag{5}$$

$$\frac{\omega^2 a^3(1-e)}{fM} = m. \tag{6}$$

Then
$$\tfrac{1}{2}\omega^2 r^2\cos^2\phi' = \frac{1}{2}\frac{fMm}{a^3(1-e)}\cos^2\phi'\frac{a^2(1-e)^2}{(1-e\cos^2\phi')^2}$$

$$= \frac{1}{2}\frac{fMm}{a}\{1 + e - (1+3e)\sin^2\phi' + 2e\sin^4\phi'\}. \tag{7}$$

m is small of order e; so we do not need terms in me^2.

On the surface (1) Ψ of 4·02 (2) is constant; expanding to $O(e^2)$ we have

$$\Psi = \frac{fM}{a}[1 + (e + \tfrac{3}{2}e^2)\sin^2\phi' - \tfrac{1}{2}e^2\sin^4\phi' + J(1 + 3e\sin^2\phi')(\tfrac{1}{3} - \sin^2\phi')$$

$$+ D(\sin^4\phi' - \tfrac{6}{7}\sin^2\phi' + \tfrac{3}{35}) + \tfrac{1}{2}m\{1 + e - (1+3e)\sin^2\phi' + 2e\sin^4\phi'\}]. \tag{8}$$

Picking out the coefficients of $\sin^2\phi'$, $\sin^4\phi'$, we have

$$e + \tfrac{3}{2}e^2 - J(1-e) - \tfrac{6}{7}D - \tfrac{1}{2}m(1+3e) = 0, \tag{9}$$

$$-\tfrac{1}{2}e^2 - 3Je + D + me = 0. \tag{10}$$

To the first order
$$J = e - \tfrac{1}{2}m. \tag{11}$$

Substitute in (10); we have
$$D = \tfrac{7}{2}e^2 - \tfrac{5}{2}me, \tag{12}$$

and returning to (9)
$$J = e - \tfrac{1}{2}m + e(-\tfrac{1}{2}e + \tfrac{1}{7}m). \tag{13}$$

(12) and (13) determine U. To get g, we have first

$$-\frac{\partial U}{\partial r} = \frac{fM}{a^2}\left\{\frac{a^2}{r^2} + \frac{3Ja^4}{r^4}(\tfrac{1}{3} - \sin^2\phi') + \frac{8}{35}\frac{5Da^6}{r^6}p_4\right\}$$

$$= \frac{fM}{a^2}\{1 + 2e\sin^2\phi' + 3e^2\sin^2\phi' + 3J(\tfrac{1}{3} - \sin^2\phi')(1 + 4e\sin^2\phi')$$

$$+ 5D(\sin^4\phi' - \tfrac{6}{7}\sin^2\phi' + \tfrac{3}{35})\}, \tag{14}$$

$$\omega^2 r\cos^2\phi' = \frac{fMm}{a^2}\{1 + e - (1 + 2e)\sin^2\phi' + e\sin^4\phi'\}. \tag{15}$$

Subtracting and substituting for J and D we have

$$-\frac{\partial \Psi}{\partial r} = \frac{fM}{a^2}\left[1 + e - \tfrac{3}{2}m + e(e - \tfrac{27}{14}m) + \sin^2\phi'\{\tfrac{5}{2}m - e - e(\tfrac{13}{2}e - \tfrac{72}{7}m)\}\right.$$
$$\left. + e\sin^4\phi'(\tfrac{11}{2}e - \tfrac{15}{2}m)\right]$$
$$= \frac{fM}{a^2}\left[1 + e - \tfrac{3}{2}m + e(e - \tfrac{27}{14}m) + \sin^2\phi'\left(\tfrac{5}{2}m - e - e^2 + \tfrac{39}{14}em\right)\right.$$
$$\left. - \tfrac{1}{8}e\sin^2 2\phi'(11e - 15m)\right], \tag{16}$$

and to order e
$$-\frac{\partial \Psi'}{r\,\partial\phi'} = \frac{fM}{a^2}e\sin 2\phi'. \tag{17}$$

But
$$g^2 = \left(\frac{\partial\Psi'}{\partial r}\right)^2 + \left(\frac{\partial\Psi'}{r\,\partial\phi'}\right)^2, \tag{18}$$

whence $g = -\dfrac{\partial\Psi'}{\partial r}\left\{1 + \dfrac{1}{2}\left(\dfrac{a^2}{fM}\dfrac{\partial\Psi'}{r\,\partial\phi'}\right)^2\right\}$

$$= \frac{fM}{a^2}\left[1 + e - \tfrac{3}{2}m + e(e - \tfrac{27}{14}m) + \{\tfrac{5}{2}m - e - e(e - \tfrac{39}{14}m)\}\sin^2\phi'\right.$$
$$\left. - \tfrac{1}{8}e(7e - 15m)\sin^2 2\phi'\right]. \tag{19}$$

If g_0 is gravity on the equator

$$g = g_0[1 + (\tfrac{5}{2}m - e + \tfrac{15}{4}m^2 - \tfrac{17}{14}em)\sin^2\phi' - \tfrac{1}{8}e(7e - 15m)\sin^2 2\phi'] \tag{20}$$

and
$$\frac{fM}{a^2} = \frac{g_0}{1 + e - \tfrac{3}{2}m}\{1 + O(e^2)\}. \tag{21}$$

(20) expresses g as a function of geocentric latitude. The geographic latitude is given by

$$\tan(\phi - \phi') = -\frac{dr}{r\,d\phi'} = e\sin 2\phi' + O(e^2), \tag{22}$$

whence if B is of order e and C of order e^2

$$A + B\sin^2\phi' + C\sin^2 2\phi' = A + B\sin^2\phi + (C - Be)\sin^2 2\phi, \tag{23}$$

and in particular

$$g = g_0[1 + (\tfrac{5}{2}m - e + \tfrac{15}{4}m^2 - \tfrac{17}{14}em)\sin^2\phi + (\tfrac{1}{8}e^2 - \tfrac{5}{8}em)\sin^2 2\phi]. \tag{24}$$

The definition of m (6) appears to depend on e. But in the latitude where $\sin^2\phi = \tfrac{1}{3}$ let the values of r and g be r_m, g_m. Then

$$r_m^3 = a(1 - e) + O(e^2), \tag{25}$$

$$g_m = \frac{fM}{r_m^2} - \omega^2 r_m\cos^2\phi + O(e^2) = \frac{fM}{r_m^2} - \tfrac{2}{3}\omega^2 r_m + O(e^2). \tag{26}$$

Then
$$\omega^2 = \frac{fMm}{a^3(1-e)} = \frac{fMm}{r_m^3}\{1+O(e^2)\}, \tag{27}$$

$$g_m = \frac{fM}{r_m^2}\left(1 - \frac{2}{3}\frac{\omega^2 r_m^3}{fM}\right) = \frac{fM}{r_m^2}\left(1 - \tfrac{2}{3}m\right)(1+O(e^2)), \tag{28}$$

$$\frac{m}{1-\tfrac{2}{3}m} = \frac{\omega^2 r_m}{g_m}\{1+O(e^2)\}. \tag{29}$$

ω is extremely accurately known. Observations of g are based ultimately on a few absolute determinations, made in middle latitudes, and their adjustment to $\sin^2\phi = \tfrac{1}{3}$ needs a small correction in comparison with the whole variation. Determinations of r from geodesy cover a great range of latitude, and it is not obvious at first sight in what latitude the determination is best. But $r_m - a \doteq -\tfrac{1}{3}ea$, and if we use an approximate value for e the error in it and the contribution from e^2 will be multiplied by the small factor $\omega^2 r_m/g_m$ before m is calculated from (29); thus effectively any error in the calculation of m from (29) is of the third order in e, and m can be regarded as definitely determined in equations where its uncertainty is combined with that of e.

The above account (especially the definition of m in terms of r_m and g_m) is due substantially to de Sitter (1924a). A solution taking account of first-order terms was given by A. C. Clairaut, after whom, in particular, the formula for g to the first order is named.

The principle of Pizzetti's analysis is as follows. Oblate spheroidal coordinates are used, such that

$$z = c\sinh\xi\sin\eta, \qquad \varpi = c\cosh\xi\cos\eta.$$

$\xi = \text{constant} = \alpha$ over the surface. Solutions of Laplace's equation at external points are of the form

$$\Sigma A_n p_n(\sin\eta)\, q_n(i\sinh\xi).$$

q_n is a series of descending powers. The potential is taken to consist of terms in A_0 and A_2, for which q_n is real. Then $U + \tfrac{1}{2}\omega^2\varpi^2$ can be made constant over $\xi = \alpha$, and a complete solution exists. Gravity is $-\partial\Psi/\partial n$; $\partial n/\partial\xi$ is not constant but contains a square root. Curiously, the expression for g is simplest in terms of geographic latitude. The solution is compact, but for the effects on satellites it has to be transformed to the ordinary polar coordinates and Cook (1959a) has used it to develop the theory to $O(e^3)$; but terms of this order appear to be still negligible. Lambert (1960) gives a neat way of doing the transformation to polars. It appears, however, that to $O(e^2)$ it is still longer than the direct theory. It is not easily adapted to the internal theory.

A related point is as follows. We could imagine a *prolate* body, covered by a liquid of small density, and rotating at such a speed that the outer surface is an exact sphere. Then if U consists only of terms in $1/r$ and $p_2(\sin\phi)/r^3$, Ψ can be made exactly constant over the sphere; there need be no terms at all in p_4/r^5, and g also contains no term in p_4. Thus we have the theorem that all terms in p_4 and higher harmonics must contain the ellipticity as a factor. It will be noticed that this applies to all the second-order terms in 4·022 (12), (19), (24) above. It also applies to $e - \frac{1}{2}m$ in (13) (Jeffreys, 1964b).

The result for $e - \frac{1}{2}m$ and the second-order terms in $\sin^2\phi'$ in (19) depend on de Sitter's definition of m, (6). The usual definition in geodesy, adopted in the recent list of constants of the I.A.U., is m_1, defined by (in the present notation) $m_1 = a\omega^2/g_0$.

We find
$$\frac{m}{m_1} = 1 - \tfrac{3}{2}m - \tfrac{3}{7}em,$$

$$J = e - \tfrac{1}{2}m_1 - \tfrac{1}{2}e^2 + \tfrac{3}{4}m_1^2 + \tfrac{1}{7}m_1 e,$$

$$e - \tfrac{1}{2}m_1 = J + \tfrac{1}{2}J^2 + \tfrac{5}{14}Jm_1 - \tfrac{39}{56}m_1^2,$$

$$\frac{g_0 a^2}{fM} = 1 + J - m_1 + \tfrac{3}{2}J^2 - \tfrac{4}{7}Jm_1 + \tfrac{47}{56}m_1^2.$$

The change of m to m_1 introduces extra terms in m_1^2, and the terms of the second order no longer have e as a factor.

4·023. Corrected free air formula.

Differentiating Ψ twice with regard to r we get for the terms up to the first order in 4·02 (17) (Jeffreys, 1952c)

$$g_1 = \frac{fM}{a^3}\{2 + 4e - m - (6e - 5m)\sin^2\phi'\}$$

and general gravity is

$$g_c = \frac{fM}{a^2}\{1 + e - \tfrac{3}{2}m + (\tfrac{5}{2}m - e)\sin^2\phi'\}.$$

Then
$$\frac{g_1}{g_c} = \frac{1}{a}(2 + 2e + 2m - 4e\sin^2\phi').$$

Substituting in 4·02 (17) we have

$$g_c\left\{1 - \frac{2h}{a}(1 + e + m - 2e\sin^2\phi')\right\} = g_h\left(1 - \frac{3h^2}{a^2}\right)$$

or
$$g_c = g_h\left\{1 + \frac{2h}{a}(1 + e + m - 2e\sin^2\phi') + \frac{h^2}{a^2}\right\}.$$

For an observation at the top of Mount Everest the h term would be about 3000 mgal. The h^2 term would be about 3 mgal. It appears therefore that this term can be appreciable only in regions difficult of access and of very limited extent. The terms in eh and mh are appreciable over wider regions.

The definition of h is needed for consistency if differences of level are found both by triangulation and by means of levelling staves.

4·024. Departures from the spheroid. Suppose that U contains small terms not included in the foregoing theory, say

$$U'' = \Sigma A_{ns} \left(\frac{a}{r}\right)^{n+1} S_{ns}. \tag{1}$$

These must be compensated on $\Psi = C$ by a change in r; write the correction to r as ar'. Then (with neglect of factors e and m)

$$\frac{fM}{a} r' = U''; \quad r' = \frac{1}{ag} \Sigma A_{ns} S_{ns}. \tag{2}$$

g will contain terms

$$-\frac{2fM}{a^2} r' - \frac{\partial U''}{\partial r} = -\frac{\partial U''}{\partial r} - \frac{2U''}{r} = \frac{n-1}{a} \Sigma A_{ns} S_{ns}, \tag{3}$$

on substituting for r'. Then if gravity reduced to the co-geoid contains a term $g_{ns} S_{ns}$, the corresponding term in U'' is $ag_{ns} S_{ns}/(n-1)$.

It is supposed that U'' is small enough for its products by e and m to be neglected.

4·025. The international ellipsoid and related formulae. This solution rests on the determination of the major semi-axis and ellipticity made in the United States by J. F. Hayford in 1909, and on the value 978·049 gal. for g on the equator found in an analysis by Heiskanen. Hayford's values are

$$a = 6378\cdot388 \pm 0\cdot018 \,\text{km.}, \quad e^{-1} = 297\cdot0 \pm 0\cdot5 \text{ (probable error).} \tag{1}$$

The uncertainties given are definitely too low; Helmert (1911) increased them to 0·053 km. and 1·2 (standard errors). We take

$$r_m = 6371\cdot269(1+u) \,\text{km.}, \quad g_m = 979\cdot771(1+v) \,\text{gal.}, \quad e = 0\cdot00336700 + e', \tag{2}$$

where u, v, e' are small corrections, u, e' being equal to 0 for the adopted solution. The number of seconds in a sidereal day is 86164·1, whence

$$\omega = 7\cdot29211 \times 10^{-5}/\text{sec.}, \tag{3}$$

$$\frac{m}{1 - \frac{2}{3}m} = 0\cdot00345786(1+u-v), \tag{4}$$

$$m = 0\cdot00344991\{1 + 0\cdot9977(u-v)\}. \tag{5}$$

These lead to $\quad J = 0\cdot0016380 - 0\cdot001721(u-v) + e', \tag{6}$

$$D = 0.0000107, \tag{7}$$

$$g = g_0(1 + A\sin^2\phi - B\sin^2 2\phi), \tag{8}$$

where
$$A = 0.0052883 - e' + 0.008605(u - v), \tag{9}$$

$$B = 0.0000058, \tag{10}$$

$$g_1 = g_0\{1.0017576 - 0.33e' + 0.002868(u - v)\}. \tag{11}$$

If we take $g_0 = 978.049$, $e' = u = v = 0$, this gives $g_1 = 979.7680$, 3 mgal. less than the trial value. We adjust this by taking $v = -0.0000031$. With this value of v, and with $e' = u = 0$,

$$m = 0.0034499, \tag{12}$$

$$J = 0.0016375, \quad D = 0.0000107, \tag{13}$$

$$g = 978.049(1 + 0.0052883\sin^2\phi - 0.0000058\sin^2 2\phi). \tag{14}$$

The international formula was derived by Cassinis from Pizzetti's exact solution, and is

$$g = 978.049(1 + 0.0052884\sin^2\phi - 0.0000059\sin^2 2\phi). \tag{15}$$

The differences are less than 0·1 mgal. This gives the standard values used in the reduction of observational data, according to decisions of the International Geodetic Association. There is no claim that these values are exact, but they are close enough for departures to be treated as small. Observations are expressed as residuals against the formula, and furnish data for possible corrections. A table of gravity according to this formula is given by Lambert and Darling (1931). For the 1967 values of the constants see p. 261.

4·026. Mass and mean density. From 4·022(28)

$$\frac{fM}{r_m^2} = \frac{g_m}{1 - \frac{2}{3}m} = 982.032\{1 + v + 0.001537(u - v)\}. \tag{16}$$

From determinations by Boys and Heyl (Jeffreys, 1967d, p. 307),

$$f = (6.670 \pm 0.004) \times 10^{-8} \text{ c.g.s.}, \tag{17}$$

whence if $\bar{p}$ is the mean density $M/\frac{4}{3}\pi r_m^3$,

$$M = 5.977 \times 10^{27}(1 \pm 0.0006 + 2u + v) \text{ g.}, \tag{18}$$

$$\bar{p} = 5.517(1 \pm 0.0006 - u + v) \text{ g./cm.}^3. \tag{19}$$

As the estimates of r_m and g_m are unlikely to be in error by more than a few parts in 10^5, the uncertainties of M and $\bar{p}$ depend almost entirely on that of the constant of gravitation.

4·027. MacCullagh's formula. At a distant point P the gravitational potential due to any body is

$$U = f\left(\frac{M}{r} + \frac{A + B + C - 3I}{2r^3}\right) + O\left(\frac{1}{r^4}\right), \tag{1}$$

where r is the distance from the centre of mass O, A, B, C are the moments of inertia about the principal axes at the centre, and I is the moment of inertia about the line joining the centre of mass to the point considered. With this choice of origin there is no term in r^{-2}. If A, B, C are about the axes of x, y, z already taken, and these are principal axes, and l, m, n are the direction cosines of the line OP, we have

$$
\begin{aligned}
A + B + C - 3I &= A + B + C - 3(Al^2 + Bm^2 + Cn^2) \\
&= (C - \tfrac{1}{2}A - \tfrac{1}{2}B)(1 - 3n^2) - \tfrac{3}{2}(A - B)(l^2 - m^2) \\
&= 3(C - \tfrac{1}{2}A - \tfrac{1}{2}B)(\tfrac{1}{3} - \sin^2\phi') - \tfrac{3}{2}(A - B)\cos^2\phi'\cos 2\lambda. \quad (2)
\end{aligned}
$$

Both terms are surface harmonics. The first gives a term in the potential of the same form as the second term in $4\cdot022\,(5)$. The second depends on the longitude with respect to axes fixed in the Earth and therefore rotating in space. Its value at the Moon will on this account vary in a period of half a lunar day. Consequently it will give only a short-period disturbance of the Moon's motion. Further, $A - B$ is small in comparison with $C - \tfrac{1}{2}A - \tfrac{1}{2}B$. No influence of this term on the Moon's motion has in fact been detected, but it has a measurable effect on artificial satellites.

Comparing with $4\cdot022\,(5)$ we have

$$
\frac{C - \tfrac{1}{2}(A + B)}{Ma^2} = \tfrac{2}{3}J = 0\cdot0010917 - 0\cdot001147(u - v) + 0\cdot67e'. \quad (3)
$$

4·028. Gravity survey. From $4\cdot024\,(3)$ surface gravity contains no terms in $n = 1$. With the origin at the centre of mass, A_{1s} is zero by Mac-Cullagh's formula; but even without this choice of origin there would still be no first harmonics in g. This has a simple physical interpretation. The change of potential due to a rigid-body displacement, keeping a fixed origin, is of the form $A_1 S_1/r^2$; but such a displacement will not affect gravity at a given point of the surface.

Two of the five second harmonics must also be absent, for another reason. If terms in nl and nm were present in I, the axis of z would not be a principal axis. A rigid body can persist in a state of stable rotation about its axis of greatest moment of inertia, but if disturbed from this state it will oscillate so that the axis of rotation describes a cone in the body about the axis of greatest moment. The same is true for an elastic body, as we shall see in Chapter VII. A very small variation of this type exists, with a period of about $1\cdot2$ years, but on an average over a long time the axis of rotation coincides with that of greatest moment and therefore is a principal axis. The displacements reach about $0\cdot1''$ and therefore give contributions of the order of 2×10^{-6} of the terms that contain ω^2 as a factor, and they are periodic. They are therefore utterly negligible. Then the harmonics in $\sin\phi'\cos\phi'(\cos\lambda, \sin\lambda)$ will be absent.

Gravity is ordinarily measured from the period of a pendulum, but an absolute determination to 1 part in 10^6 requires the dimensions of the pendulum to similar accuracy. Actually comparatively few absolute determinations approach this accuracy, and until recently there was effectively only one, that made by Kühnen and Furtwängler at Potsdam in 1906. The practical method is usually to measure the period of a pendulum at a base station where there is an adopted absolute value of gravity, carry it to the place where gravity is wanted, measure the period again, and return to the base station and measure the period a third time. The two periods at the base station should agree. This is an important check because pendulums can alter in transit owing to rust on the knife edges, magnetism, and accidental damage. Many determinations have failed for this reason. The ratio of the periods determines the ratio of the values of g.

The Potsdam value being taken as standard, a base station for a survey in a new region must be compared with Potsdam by carrying a pendulum (or several pendulums) back and forth. Then other stations in the region are compared with the base station. Comparisons with base stations in other regions are also usually made so as to provide further checks.

Many modern gravimeters are based on the principle of the spring balance. The bending of an elastic body by the weight of a known mass depends on gravity, and if it can be measured a comparison of values of g can be made. The elastic displacements are very small, but methods have been devised for measuring them with extreme accuracy. A standard error of 1 mgal. for a pendulum comparison is normal, but 0·1 or even 0·001 mgal. can be obtained with a gravimeter, and takes less time and less cumbersome apparatus. The gravimeter, however, has some disadvantages. It is more likely than a pendulum to alter its properties through creep under stress. Also it needs calibration, which can best be done by comparison with a pendulum at places where gravity differs by the largest amount likely to be encountered.

The compactness of gravimeters adapts them to transport by aeroplane, and they have accordingly been used to give comparisons between base stations all over the world, especially by G. J. Woollard. They are also less sensitive than a pendulum to motion of the support, and measures have been made by the Cambridge and Lamont observers on a surface ship at sea. This independence of submarines should make determinations over the oceans much more abundant.

Determinations of the coefficients g_{ns} can be made only by analysis of gravity itself. Unlike geodetic survey, gravity observations can be made on the sea, by a technique invented by F. A. Vening Meinesz. The great difficulty arises from the waves; this had made measurements on board an ordinary ship almost impossible, and in any case much less accurate than land ones. Meinesz met this by working in a submarine and making use of the fact that wave motion in the sea falls off rapidly with depth. The disturbance is further reduced by having a pair of pendulums (actually two pairs) of the same period, swinging in opposite phases. By an optical method the difference of the displacements is recorded; from this most of the forced motion due to the waves cancels. Other corrections needed are for the rate of motion of the ship in longitude, which on a rotating Earth makes a contribution to the apparent g, and for a change of the pendulum period depending on the mean square of the acceleration produced by the waves, discovered by B. C. Browne (1937). Browne and Cooper (1950) and Worzel (1965) both estimated the standard error for a pendulum in a submarine as 3·5 mgal. On land 1 mgal. is normally achieved, and with the special care in modern comparisons of base stations something less. Loncarevič (1963) gives 2·1 mgal. Gravimeters have been used on surface ships and even on aeroplanes; on account of their short free periods they are much less sensitive to variations of motion of the vehicle. But they still do not approach the accuracy of submarine observations except when the sea is very smooth. Caputo *et al.* (1963), Caputo (in Orlin *et al.* 1966) estimate that 80 % of their results are within 10 mgal. of the true values. Similar results are given by Wall *et al.* (1966) and Bower (1966).

In consequence of the introduction of this method gravity has been measured over large parts of all the oceans as well as on land.* Extensive regions remain unsurveyed, notably most of the South Pacific and Siberia, while there are not many inland stations in Africa, except in East Africa, which was studied in detail by E. C. Bullard (1938). Nevertheless, there is enough material to make it worth while to attempt a statistical discussion to estimate some of the harmonics of low degree. As a solution made in this way yields only a finite number of terms the question of convergence of the series does not arise. The first attempt on these lines was made by Helmert, and much further work has been done by Heiskanen. Systematic variations of gravity with longitude were found and attributed to the terms in $p_2^2(\cos 2\lambda, \sin 2\lambda)$. These would correspond to an 'ellipticity of the equator', that is, the equator of the co-geoid would be elliptical instead of circular. Their physical importance is considerable, but the observations used were almost all in the northern hemisphere and use of data for the southern hemisphere led to some changes. Artificial satellites have yielded more data.

Stokes (1849) showed that 4·024 (3) could be used to derive the form of the co-geoid from observations of gravity by direct integration. In 4·024 (3) denote the harmonics of degree n together by G_n. If P is a point on the sphere and Q another point on it, at angular distance ϑ, then the value of G_n at P is, for $n > 0$,

$$G_n(P) = \frac{2n+1}{4\pi} \iint p_n(\cos\vartheta)\, g_2(Q)\, d\omega, \tag{1}$$

* Rapp (1970) gives details.

taken over the sphere. Then the part of U at an external point P' on the radius to P corresponding to g_2 is

$$U_2 = \frac{1}{4\pi} \sum_{2}^{\infty} \int\int \frac{a^{n+2}}{r^{n+1}} \frac{2n+1}{n-1} g_2(Q) p_n(\cos\vartheta) \, d\omega. \tag{2}$$

Now $\quad \sum_{n=2}^{\infty} \frac{a^{n+2}}{r^{n+1}} \frac{2n+1}{n-1} p_n(\cos\vartheta) = a^2 \Big\{ 2\sum \frac{a^n}{r^{n+1}} p_n + 3\sum \frac{a^n}{(n-1)r^{n+1}} p_n \Big\}. \tag{3}$

If $P'Q = R$, $\qquad\qquad R^2 = r^2 - 2ar\cos\vartheta + a^2,$ $\qquad\qquad$ (4)

$$\sum_{2}^{\infty} \frac{a^n}{r^{n+1}} p_n = \frac{1}{R} - \frac{1}{r} - \frac{a\cos\vartheta}{r^2}, \tag{5}$$

$$\begin{aligned}
\sum_{2}^{\infty} \frac{a^n}{(n-1)r^{n+1}} p_n &= \frac{1}{r^2} \int_{r}^{\infty} \sum \frac{a^n}{r^n} p_n \, dr = \frac{1}{r^2} \int_{r}^{\infty} r \sum \frac{a^n}{r^{n+1}} p_n \, dr \\
&= \frac{1}{r^2} \int_{r}^{\infty} \Big(\frac{r}{R} - 1 - \frac{a\cos\vartheta}{r} \Big) dr \\
&= \frac{1}{r^2} \Big[R + a\cos\vartheta \log(r - a\cos\vartheta + R) - r - a\cos\vartheta \log r \Big]_{r}^{\infty}.
\end{aligned} \tag{6}$$

For r large the expression in the bracket tends to $a\cos\vartheta(\log 2 - 1)$. Hence the sum is

$$\frac{1}{r^2} \Big(-R - a\cos\vartheta \log\frac{r - a\cos\vartheta + R}{2r} + r - a\cos\vartheta \Big), \tag{7}$$

and

$$U_2 = \frac{a^2}{4\pi r^2} \int\int \Big\{ \frac{2r^2}{R} + r - 3R - 5a\cos\vartheta - 3a\cos\vartheta \log\frac{r - a\cos\vartheta + R}{2r} \Big\} g_2(Q) \, d\omega. \tag{8}$$

There is no difficulty about convergence when $r > a$ because the p_n are always in the interval $-1 \leqslant p_n \leqslant 1$ and g_2 is bounded. Further, U_2 is continuous when $r \to a$, and so is the function under the integral sign, since $d\omega$ contains a factor $\sin\vartheta$. Hence we may put $r = a$, $R = 2a\sin\frac{1}{2}\vartheta$, and then

$$U_2(P) = \frac{a}{4\pi} \int\int f(\vartheta) g_2(Q) \, d\omega, \tag{9}$$

where

$$f(\vartheta) = \operatorname{cosec}\tfrac{1}{2}\vartheta + 1 - 6\sin\tfrac{1}{2}\vartheta - 5\cos\vartheta - 3\cos\vartheta \log(\sin\tfrac{1}{2}\vartheta + \sin^2\tfrac{1}{2}\vartheta). \tag{10}$$

The elevation of the co-geoid above the spheroid due to U_2 is U_2/g.

Another way of deriving the solution has been given by N. Idelson and N. Malkin (1931). This depends on the facts that, on $r = a$,

$$\frac{\partial U_2}{\partial r} + \frac{2U_2}{a} = \frac{\partial U_2}{\partial r} + \frac{2U_2}{r} = \frac{1}{r^2} \frac{\partial}{\partial r}(r^2 U_2), \tag{11}$$

and that if U_2 is a solution of Laplace's equation,

$$V_2 = \frac{1}{r} \frac{\partial}{\partial r}(r^2 U_2) \tag{12}$$

is another. V_2 can be considered given over $r = a$, and is then determined at external points by a known solution in potential theory. Then U_2 can be found from V_2 by a further integration with regard to r.

This yields

$$U_2 = \frac{a^2}{4\pi r^2} \iiint \frac{r^3 - ra^2}{R^3} \, dr \, g_2(Q) \, dS$$

$$= \frac{a^2}{4\pi r^2} \iint \left[\frac{2r^2}{R} - 3R - 3a\cos\vartheta \log \frac{r - a\cos\vartheta + R}{a} + \text{constant} \right] g_2(Q) \, dS.$$

(13)

Now U_2 is $O(1/r^3)$ and $r^2 U_2 = O(1/r)$. The bracket in the integrand for r large is

$$-r + 5a\cos\vartheta - 3a\cos\vartheta \log \frac{2r}{a} + O\left(\frac{1}{r}\right).$$

(14)

But $(r - \text{constant})$ times any S_n $(n > 0)$ gives 0 when integrated over a sphere; and $\cos\vartheta$ times any S_n $(n > 1)$ also gives 0. Hence if we subtract any multiple of (14) from the function in (13) we do not alter the integral. Then what remains is Stokes's solution. One interesting feature of this method is that it does not assume that g_2 can be expanded in spherical harmonics, only that it is integrable. Another is that it amounts to a direct proof that the determination of the external field depends only on the free-air anomaly.

Not much use has been made of Stokes's integral. The chief reason is that it requires a knowledge of g_2 everywhere, which we have not. Interpolations over the vast ranges where there are no observations of gravity are impossible. In fact the best way of doing it is to fit the smoothest possible function that shows no systematic departure from the data. But this function is the expansion in spherical harmonics up to the highest degree that give significant coefficients Thus the best way of interpolating presupposes spherical harmonic analysis. We do not, however, know any of the coefficients yet (apart from the main ellipticity term) with sufficient accuracy to be useful in providing a standard of reference, which would need to be used for a long time. It is possible that additional information on them could be obtained from comparison of surveys, and it is essential that the information given by the surveys should be preserved in such a way that the reductions do not introduce systematic error into these terms.

W. D. Lambert made two interesting suggestions. The first (1931) deals with the difficulty that $f(\vartheta)$ has an infinity at $\vartheta = 0$. If we take polar coordinates ϑ, χ at P,

$$d\omega = \sin\vartheta \, d\vartheta \, d\chi,$$

(15)

and we can tabulate

$$F(\vartheta) = \int_0^\vartheta f(\vartheta) \sin\vartheta \, d\vartheta.$$

(16)

Then

$$U_2 = \frac{a}{4\pi} \iint g_2(Q) \, dF(\vartheta) \, d\chi.$$

(17)

$F(\vartheta)$ is continuous and the interval of ϑ can be divided into intervals of $F(\vartheta)$. In each g_2 is replaced by its average with regard to χ, and the calculation is manageable. Formulae are also available for calculating the deflexion of the vertical in both directions (Sollins 1947). Full tables of F are given by Lambert and Darling (1936).

Lambert pointed out also that a survey is mainly concerned with the relative positions of places within the region and with the position of the co-geoid with reference to them; and that the differences of elevation of the co-geoid at different points of the region with regard to the standard spheroid will be mainly due to local irregularities. Thus Stokes's integral could be used to estimate the part of U_2 that is due to local irregularities; since the part due to distant parts of the Earth will at any rate vary smoothly, this would determine the local variation within the region. The disadvantage of this procedure is as follows. If we had a complete gravity survey, the integral would determine the complete U_2 and therefore the whole elevation of the co-geoid above the mean sphere; subtracting this elevation would therefore simply reproduce the mean sphere. If g_2 is taken to be the departure from a standard gravity formula, the modified form would be a spheroid with the ellipticity corresponding to that formula. This is a consequence of the fact that Stokes's function contains a term in p_2. In these idealized conditions we should not need to use surveys to determine anything but the radius, and no great harm would be done. But with incomplete gravity data any error in the distant zones would produce an error in the reduction, and it would be impossible to say what the results mean.

A possible alternative is as follows. All harmonics to degree 4 are now well determined from artificial satellites. Some higher ones, up to degree 8, and possibly a few special higher ones, are also known. Then the corresponding terms in gravity could be subtracted from observed gravity before Stokes's formula is applied. This has been done in a paper by W. M. Kaula (1966). But equally, if we omit terms up to p_4 from Stokes's formula, we shall still get estimates that are unaffected by harmonics up to degree 4. The effects of distant zones will be greatly reduced; and comparison of surveys would retain its full value as supplementary information towards estimating the low harmonics (Jeffreys, 1953, 1955). This suggestion was discussed at the Rome meeting of the International Geodetic Association (1954) and Dr Lambert approved the modification.

The values of F and F_4, where F_4 arises from the terms up to p_4 in f, are given on p. 194; they are rough and intended only for illustration. The suggestion made here is to use $F - F_4$ instead of F. It is much smaller, except for $\vartheta < 10°$, and changes sign more often, and the results would therefore depend much less on distant zones. The greatest absolute value beyond $90°$ is 0.106; the greatest for F is 0.786.

$\vartheta(°)$	F	F_4	$F - F_4$
0	0·000	0·000	0·000
10	0·414	0·159	+0·255
20	0·798	0·569	+0·229
30	1·048	1·025	+0·023
40	1·122	1·272	−0·150
50	1·016	1·202	−0·186
60	0·761	0·844	−0·083
70	0·411	0·348	+0·063
80	0·028	−0·114	+0·142
90	−0·325	−0·438	+0·113
100	−0·596	−0·606	+0·010
110	−0·743	−0·668	−0·075
120	−0·786	−0·680	−0·106
130	−0·707	−0·651	−0·056
140	−0·509	−0·573	+0·064
150	−0·356	−0·423	+0·067
160	−0·174	−0·225	+0·051
170	−0·046	−0·062	+0·016
180	0·000	0·000	0·000

4·03. Theory of the internal field. In this theory the principle that the stress is hydrostatic is applied to the whole of the interior. The equations of motion in the steady state reduce to

$$0 = \frac{\partial p}{\partial x} - \rho \frac{\partial \Psi}{\partial x}, \tag{1}$$

with similar equations; thus surfaces of constant Ψ are also surfaces of constant p and ρ. But the potential is itself determined by the distribution of ρ. Hence the condition leads to one satisfied by ρ at internal points. Ψ in any case departs from spherical symmetry, and therefore ρ also does. We use a first-order theory.

For a homogeneous nearly spherical body whose surface is

$$r = a\left(1 + \sum_{n=1}^{\infty} \epsilon_n S_n\right), \tag{2}$$

where S_n is a surface harmonic, the gravitational potential outside the body is

$$U_0 = \tfrac{4}{3}\pi f \rho a^3 \left(\frac{1}{r} + \sum_{n=1}^{\infty} \frac{3}{2n+1} \frac{a^n \epsilon_n}{r^{n+1}} S_n\right), \tag{3}$$

and that inside is

$$U_1 = \tfrac{4}{3}\pi f \rho a^3 \left(\frac{3a^2 - r^2}{2a^3} + \sum_{n=1}^{\infty} \frac{3}{2n+1} \frac{r^n \epsilon_n}{a^{n+1}} S_n\right), \tag{4}$$

a here being the mean radius, not the equatorial radius as in 4·022.

For a heterogeneous body we take the density to be uniform, equal to ρ', over a surface

$$r = a'(1 + \Sigma \epsilon_n S_n), \tag{5}$$

where ρ', ϵ_n will be functions of a'. Then the potential can be regarded as built up from that due to shells corresponding to small increases of a'. In the shell between a' and $a' + \delta a'$ the density varies from ρ' to $\rho' + \delta \rho'$, and the contribution to U lies between those found by taking the density uniform and equal to ρ' and $\rho' + \delta\rho'$ in turn. In either case the potential due to the shell is the difference of those due to two homogeneous bodies with surfaces determined by taking a' and $a' + \delta a'$ in (5). By adding and taking the intervals of a' small we have for the external potential

$$U_0 = \tfrac{4}{3}\pi f \int_0^a \rho' \frac{\partial}{\partial a'} \left(\frac{a'^3}{r} + \Sigma \frac{3}{2n+1} \frac{a'^{n+3}\epsilon_n}{r^{n+1}} S_n \right) da'. \qquad (6)$$

At internal points the potential must be separated into two parts. Let r_1 be the value of a' for the surface of constant density through the point considered. Then the contribution for $a' < r_1$ is built up from (3) and that for $a' > r_1$ from (4), and

$$U_1 = \tfrac{4}{3}\pi f \int_0^{r_1} \rho' \frac{\partial}{\partial a'} \left(\frac{a'^3}{r} + \Sigma \frac{3}{2n+1} \frac{a'^{n+3}}{r^{n+1}} \epsilon_n S_n \right) da'$$

$$+ \tfrac{4}{3}\pi f \int_{r_1}^a \rho' \frac{\partial}{\partial a'} \left(\tfrac{3}{2}a'^2 + \Sigma \frac{3}{2n+1} \frac{r^n}{a'^{n-2}} \epsilon_n S_n \right) da'. \qquad (7)$$

To get Ψ we must add

$$\tfrac{1}{2}\omega^2 r^2 \cos^2\phi' = \tfrac{1}{3}\omega^2 r^2 + \tfrac{1}{2}\omega^2 r^2 (\tfrac{1}{3} - \sin^2\phi'). \qquad (8)$$

As we are only attempting a first-order theory we can ignore the difference between ϕ and ϕ'. The condition that ρ and Ψ are constant over the same surfaces then reduces to saying that Ψ at internal points is a function of r_1 only.

The mass of the whole body is

$$M = 4\pi \int_0^a \rho' a'^2 da' = \tfrac{4}{3}\pi a^3 \bar{\rho}, \qquad (9)$$

and that within a given surface $r_1 = \text{constant}$ is given by replacing the upper limit by r_1. We write ρ_0 for the mean density within this surface; then

$$\rho_0 = \frac{3}{r_1^3} \int_0^{r_1} \rho' a'^2 da'. \qquad (10)$$

In the first integral in (7) we can substitute for $1/r$ in the first term from (5) and retain first-order small quantities in the expansion. All other terms are small except that arising from the first term in the second integral, which is already a function of r_1 only. Hence in other terms we can replace r by r_1. This gives

$$\tfrac{4}{3}\pi f \left[\frac{(1 - \Sigma\epsilon_n S_n)}{r_1} \int_0^{r_1} 3\rho' a'^2 da' + \Sigma \frac{3}{2n+1} S_n \left\{ \frac{1}{r_1^{n+1}} \int_0^{r_1} \rho' d(a'^{n+3}\epsilon_n) \right. \right.$$

$$\left. \left. + r_1^n \int_{r_1}^a \rho' d\left(\frac{\epsilon_n}{a'^{n-2}} \right) \right\} \right]$$

$$+ \tfrac{1}{3}\omega^2 r_1^2 + \tfrac{1}{2}\omega^2 r_1^2 (\tfrac{1}{3} - \sin^2\phi') = \text{function of } r_1 \text{ only}. \qquad (11)$$

In other words, the function on the left is constant for given r_1 and the coefficient of every S_n ($n \geqslant 1$) must vanish. This gives

$$-\frac{\epsilon_n}{r_1}\int_0^{r_1}\rho'a'^2da' + \frac{1}{2n+1}\left\{\frac{1}{r_1^{n+1}}\int_0^{r_1}\rho'd(a'^{n+3}\epsilon_n) + r_1^n\int_{r_1}^a\rho'd\left(\frac{\epsilon_n}{a'^{n-2}}\right)\right\} = 0, \quad (12)$$

except for $S_n = \frac{1}{3} - \sin^2\phi$, when there is an extra term $\frac{1}{3}\omega^2 r_1^2/\pi f$ on the left. We express this by writing the right side as $(0, -\frac{1}{6}\omega^2 r_1^2/\pi f)$. Multiply by r_1^{n+1} and differentiate. We can now write r for r_1 without confusion. Then on simplifying

$$-\left(r^n\frac{d\epsilon_n}{dr} + nr^{n-1}\epsilon_n\right)\int_0^r\rho'a'^2da' + r^{2n}\int_r^a\rho'\frac{d}{da'}\left(\frac{\epsilon_n}{a'^{n-2}}\right)da' = \left(0, -\frac{5\omega^2 r^4}{8\pi f}\right). \quad (13)$$

Divide by r^{2n} and differentiate again; then for any n

$$\left(\frac{d^2\epsilon_n}{dr^2} - \frac{n(n+1)}{r^2}\epsilon_n\right)\int_0^r\rho'a'^2da' + 2\rho r^2\left(\frac{d\epsilon_n}{dr} + \frac{\epsilon_n}{r}\right) = 0, \quad (14)$$

or, using (10), $\quad \rho_0\left(\frac{d^2\epsilon_n}{dr^2} - \frac{n(n+1)\epsilon_n}{r^2}\right) + \frac{6\rho}{r}\left(\frac{d\epsilon_n}{dr} + \frac{\epsilon_n}{r}\right) = 0. \quad (15)$

This is Clairaut's differential equation (1743). In early treatments, such as those of Legendre and Laplace, and later of Roche and Wiechert, special laws of density were assumed that made it soluble in terms of known functions; but a great deal can be inferred about the solutions by very general methods, which also lend themselves to rapid numerical approximation for any reasonable law of density.

We notice first that the density is a decreasing function of r, partly because the heaviest materials may be expected to have sunk to the centre, partly because the nearer the centre the greater the pressure, and therefore the more the matter is compressed. Hence $\rho_0 > \rho$ except for $r = 0$, where $\rho_0 - \rho \to 0$. Suppose that for r small ϵ_n behaves like r^p. Then the lowest powers of r give

$$p(p-1) + 6p - n(n+1) + 6 = 0, \quad (16)$$

whence $\qquad\qquad\qquad p = n-2 \quad\text{or}\quad -n-3. \quad (17)$

Now a surface of constant density is displaced outwards from a sphere by $r\epsilon_n S_n$. All the solutions $p = -n-3$ would make this tend to infinity at the centre, and are therefore impossible. For the others, take first $n = 1$. We find that $\epsilon_n \propto r^{-1}$ is an exact solution of (15). In this case the radial displacement is proportional to S_1, and is the same at all distances from the centre. Hence $n = 1$ gives only a rigid-body displacement and need not be considered further.

If $n = 2$, ϵ_n is bounded and not zero near the centre. Suppose that for small r

$$1 - \rho/\rho_0 = Hr^k + \dots \quad (H, k > 0), \qquad \epsilon_2 = A + Br^s + \dots. \quad (18)$$

Substitute in (15) and again pick out the terms of lowest degree in r:

$$s(s+5)Br^{s-2} - 6Hr^{k-2}A + \dots = 0. \quad (19)$$

Absence of all harmonics but P_2 197

This can be true only if $s = k$, and then B has the sign of AH, and therefore of A.

If $n > 2$, ϵ_n behaves like r^{n-2} for small r. Hence in all non-trivial cases ϵ_n increases numerically with r for small r.

We can now show that ϵ_n increases numerically all the way to the surface. For if it ceased to increase there would be an r such that $d\epsilon_n/dr$ vanishes. But then

$$\frac{d^2\epsilon_n}{dr^2} = \left\{n(n+1) - 6\frac{\rho}{\rho_0}\right\}\frac{\epsilon_n}{r^2}, \tag{20}$$

This has the sign of ϵ_n, and therefore $|\epsilon_n|$ would immediately proceed to increase again.

We show next that for all harmonics with $n > 1$ except $\frac{1}{3} - \sin^2\phi'$ we must have $\epsilon_n = 0$. This does not follow from the differential equation, but we still have the two relations (12) and (13), which must hold for $r = a$. Then (12) reduces to the condition that the outer surface is a level surface. It becomes (suffix a being used for surface values)

$$-\tfrac{1}{3}\epsilon_{na}a^2\bar{\rho} + \frac{1}{(2n+1)a^{n+1}}\int_{a'=0}^{a}\rho'd(a'^{n+3}\epsilon_n) = \left(0, -\frac{\omega^2 a^2}{8\pi f}\right). \tag{21}$$

Denote the integral by I; assume ϵ_n positive (we know that ϵ_n cannot change sign). Then

$$I = \rho_a\epsilon_{na}a^{n+3} - \int_{a'=0}^{a}a'^{n+3}\epsilon_n d\rho'. \tag{22}$$

Since ρ' is a decreasing function of a' the integral is negative and

$$I > \rho_a\epsilon_{na}a^{n+3}. \tag{23}$$

But also, since $\epsilon_n \leqslant \epsilon_{na}$,

$$-\int_{a'=0}^{a}a'^{n+3}\epsilon_n d\rho' < -\epsilon_{na}\int_{a'=0}^{a}a'^{n+3}d\rho',$$

$$I < \epsilon_{na}\left(\rho_a a^{n+3} - \int_{a'=0}^{a}a'^{n+3}d\rho'\right)$$

$$= \epsilon_{na}\int_{a'=0}^{a}\rho'da'^{n+3}$$

$$= \epsilon_{na}\left(\bar{\rho}a^{n+3} + \int_{a'=0}^{a}(\rho'-\bar{\rho})da'^{n+3}\right)$$

$$= \epsilon_{na}\left(\bar{\rho}a^{n+3} + \frac{n+3}{3}\int_{a'=0}^{a}(\rho'-\bar{\rho})a'^n da'^3\right). \tag{24}$$

For $n = 0$ the last integral is 0. Again, since ρ' is a decreasing function, there is an a_0 such that $\rho' - \bar{\rho} > 0$ for $a' < a_0$, and < 0 for $a' > a_0$; then for $n > 0$

$$\int_{a'=0}^{a}(\rho'-\bar{\rho})a'^n da'^3 = \int_{a'=0}^{a}(\rho'-\bar{\rho})(a'^n - a_0^n)da'^3 < 0.$$

Hence $\qquad\qquad \epsilon_{na}\rho_a a^{n+3} < I < \epsilon_{na}\bar{\rho}a^{n+3},$

Hence $\qquad\qquad \epsilon_{na}\rho_a a^{n+3} < I < \epsilon_{na}\bar{\rho}a^{n+3},$

J E I

and whether ϵ is positive or negative

$$I = \theta \epsilon_{na} \bar{\rho} a^{n+3}$$

for a θ satisfying $0 < \theta < 1$ (and in fact $\rho_a/\bar{\rho} < \theta < 1$).

Then
$$\left(-\tfrac{1}{3} + \frac{\theta}{2n+1}\right) \epsilon_{na} a^2 \bar{\rho} = \left(0, -\frac{\omega^2 a^2}{8\pi f}\right). \tag{25}$$

This cannot be satisfied for any $n > 1$ unless $\epsilon_{na} = 0$, except for the harmonic $\tfrac{1}{3} - \sin^2 \phi'$; and for this ϵ_{na} must be positive. Hence we can have equality provided that ϵ_2 is positive at the surface and therefore everywhere. We thus have our first main result: on the hydrostatic theory the radius of a surface of constant density contains no harmonics other than that representing the ellipticity; the ellipticities increase all the way from the centre to the surface, and the surface is oblate.

If in (15) for $n = 2$ we put
$$\epsilon_2 = \epsilon = r^3 \lambda, \tag{26}$$

we find
$$\frac{d^2 \lambda}{dr^2} + 6 \left(\frac{\rho}{\rho_0} + 1\right) \frac{1}{r} \frac{d\lambda}{dr} + \frac{24\rho}{\rho_0} \frac{\lambda}{r^2} = 0. \tag{27}$$

λ behaves like r^{-3} when r is small and therefore begins by decreasing outwards. It cannot increase again because at the minimum $d^2\lambda/dr^2$ would have the opposite sign to λ. Hence λ decreases all the way from the centre to the surface.

Now in (13), for $S_n = S_2 = \tfrac{1}{3} - \sin^2 \phi$, put $r = a$. We have

$$-\tfrac{1}{3} \bar{\rho} a^3 \left(a^2 \frac{d\epsilon}{dr} + 2a\epsilon_a\right) = -\frac{5\omega^2 a^4}{8\pi f}. \tag{28}$$

To the first order,
$$m = \frac{\omega^2 a^3}{fM} = \frac{\omega^2}{\tfrac{4}{3}\pi f \bar{\rho}}. \tag{29}$$

Hence
$$a \left(\frac{d\epsilon}{dr}\right)_a + 2\epsilon_a = \tfrac{5}{2} m. \tag{30}$$

Now introduce a new dependent variable η, defined by

$$\eta = \frac{d \log \epsilon}{d \log r} = \frac{r}{\epsilon} \frac{d\epsilon}{dr}. \tag{31}$$

Then
$$\frac{d\epsilon}{dr} = \frac{\eta \epsilon}{r}, \quad \frac{d^2 \epsilon}{dr^2} = \left(\frac{1}{r} \frac{d\eta}{dr} + \frac{\eta^2 - \eta}{r^2}\right) \epsilon. \tag{32}$$

Substitute in (15); then

$$r \frac{d\eta}{dr} + \eta^2 - \eta - 6 + \frac{6\rho}{\rho_0}(\eta + 1) = 0. \tag{33}$$

Now
$$\rho r^2 = \frac{1}{3} \frac{d}{dr}(\rho_0 r^3), \quad \frac{\rho}{\rho_0} = 1 + \frac{1}{3} \frac{r}{\rho_0} \frac{d\rho_0}{dr}, \tag{34}$$

and then
$$\frac{r \, d\eta}{dr} + \eta^2 + 5\eta + 2 \frac{r}{\rho_0} \frac{d\rho_0}{dr}(1 + \eta) = 0. \tag{35}$$

We have the identity, by logarithmic differentiation,

$$\frac{\frac{d}{dr}\{\rho_0 r^5 \sqrt{(1+\eta)}\}}{\rho_0 r^5 \sqrt{(1+\eta)}} = \frac{1}{\rho_0}\frac{d\rho_0}{dr} + \frac{5}{r} + \frac{1}{2(1+\eta)}\frac{d\eta}{dr}. \tag{36}$$

This may be used to eliminate $d\eta/dr$ from (35). Then

$$\frac{2\sqrt{(1+\eta)}}{\rho_0 r^4}\frac{d}{dr}\{\rho_0 r^5 \sqrt{(1+\eta)}\} = 10(1 + \tfrac{1}{2}\eta - \tfrac{1}{10}\eta^2), \tag{37}$$

that is
$$\frac{d}{dr}\{\rho_0 r^5 \sqrt{(1+\eta)}\} = 5\rho_0 r^4 \psi(\eta), \tag{38}$$

where
$$\psi(\eta) = \frac{1 + \tfrac{1}{2}\eta - \tfrac{1}{10}\eta^2}{\sqrt{(1+\eta)}}. \tag{39}$$

This equation is due to Radau (1885; Tisserand, 1891, **2**, 225). On the face of it this complicated first-order equation looks no easier to solve than the linear second-order equation that we started with; but $\psi(\eta)$ has a remarkable property. We have

$$\frac{1}{\psi}\frac{d\psi}{d\eta} = \frac{1}{20}\frac{\eta(1-3\eta)}{(1+\eta)(1+\tfrac{1}{2}\eta - \tfrac{1}{10}\eta^2)}, \tag{40}$$

so that ψ has a minimum for $\eta = 0$ and a maximum for $\eta = \tfrac{1}{3}$. Now as ϵ/r^3 is a decreasing function

$$\frac{1}{\epsilon}\frac{d\epsilon}{dr} - \frac{3}{r} < 0, \text{ that is, } \eta < 3. \tag{41}$$

With actual approximate values $m = \tfrac{1}{288}$, $\epsilon_a = \tfrac{1}{297}$, and from (30)

$$\eta_a = 0\cdot 57.$$

We then have the following representative values of $\psi(\eta)$:

η =	0	$\tfrac{1}{3}$	0·57	3
$\psi(\eta)$ =	1·00000	1·00074	0·99961	0·8

When $r = 0$, $\eta = 0$. Hence $\psi(\eta) = 1$ at $r = 0$, rises to 1·00074 at the layer where $\eta = \tfrac{1}{3}$, and sinks to 0·99961 at the surface. Except in the very improbable event that η makes a wide excursion beyond the values it attains at the ends of the range, it follows that ψ never differs from 1 by more than 8 parts in 10,000. Then to an accuracy of this order

$$\frac{d}{dr}\{\rho_0 r^5 \sqrt{(1+\eta)}\} = 5\rho_0 r^4. \tag{42}$$

Now the moment of inertia C is, apart from small quantities,

$$C = \tfrac{8}{3}\pi \int_0^a \rho r^4 dr \tag{43}$$

$$= \tfrac{8}{9}\pi \int_0^a \left(3r^4\rho_0 + r^5 \frac{d\rho_0}{dr}\right) dr$$

$$= \tfrac{8}{9}\pi \left\{\bar{\rho}a^5 - 2\int_0^a \rho_0 r^4 dr\right\}, \tag{44}$$

where we have integrated the second term by parts. But by integration of (42)

$$\int_0^a \rho_0 r^4 dr = \tfrac{1}{5}\bar{\rho}a^5 \sqrt{(1+\eta_a)}. \tag{45}$$

Thus

$$C = \tfrac{8}{9}\pi\bar{\rho}a^5\{1 - \tfrac{2}{5}\sqrt{(1+\eta_a)}\}, \tag{46}$$

$$\frac{C}{Ma^2} = \tfrac{2}{3}\{1 - \tfrac{2}{5}\sqrt{(1+\eta_a)}\}. \tag{47}$$

Note that a similar relation will hold for the mass and moment of inertia of matter within any surface of constant density. But from the external theory, to the first order,

$$\frac{C - \tfrac{1}{2}(A+B)}{Ma^2} = \tfrac{2}{3}(\epsilon_a - \tfrac{1}{2}m), \tag{48}$$

and therefore

$$H = \frac{C - \tfrac{1}{2}(A+B)}{C} = \frac{\epsilon_a - \tfrac{1}{2}m}{1 - \tfrac{2}{5}\sqrt{(1+\eta_a)}}, \tag{49}$$

while η_a is given in terms of ϵ_a and m by (30), which may be written

$$\eta_a = \frac{5}{2}\frac{m}{\epsilon_a} - 2. \tag{50}$$

The ratio H is known as the precessional constant or the dynamical ellipticity. It is a factor in the theoretical value of the rate of the precession of the equinoxes, which is well known from observation. (The other factors are also well known.) For a homogeneous spheroid it is equal to the ellipticity. Then we have a very accurate relation connecting m, H and ϵ_a, expressed in terms of the parameter η_a. Of the three, ϵ_a is the least accurately known from other sources, so that, given m and H, ϵ_a can be found with high accuracy independently of the distribution of density. The easiest way of solution appears to be to write (49) as

$$\frac{H}{m} = \frac{\epsilon_a/m - \tfrac{1}{2}}{1 - \tfrac{2}{5}\sqrt{(1+\eta_a)}}, \tag{51}$$

assume a number of values of ϵ_a/m, work out H/m for each, and then determine ϵ_a/m to fit the actual H/m by interpolation.

In the theory as originally developed by Clairaut, the variation of density with r was not assigned, and it was hoped that knowledge of ϵ_a, H and m

would lead to useful information about the distribution of density. It was found, however, that with several widely different laws of density the known H and m led to nearly the same value of ϵ_a, and the method led to no discrimination between them. Radau's approximation led to the explanation, but at the same time it led to a contradiction, emphasized by Poincaré and Tisserand. Taking $m = 1/288\cdot4$, $H = 1/305\cdot6$, they derived $\epsilon_a \leqslant 1/297\cdot3$. The current value at the time was $1/293\cdot5$, found by A. R. Clarke in 1880. The redetermination by Hayford and Helmert gave $1/297\cdot0 \pm 1\cdot2$, which is consistent with the hydrostatic theory. But meanwhile it had become clear that the hydrostatic theory (always supposing it true) would give a more accurate value than was likely to be obtained by survey; and first Callandreau (1889, 1897) and then Sir G. H. Darwin (1900) extended it by making allowance for terms of order e^2 and for the variation of ψ within the Earth. They used as independent variable the equatorial radius of a surface of equal density. De Sitter (1924a, b, 1927b; De Sitter and Brouwer, 1938) showed that there were decided advantages in replacing this by the mean radius. He also showed that on the theory of general relativity there would be a small precession not due to the Earth's figure, and this would also be present in the secular motions of the Moon's node and perigee. Unfortunately, in his paper of 1927 he applied a correction for this with the wrong sign and got

$$H = 0\cdot0032770 \pm 0\cdot00000100, \quad 1/\epsilon_a = 296\cdot96 \pm 0\cdot10, \qquad (52)$$

which are replaced in the 1938 paper by

$$H = 0\cdot00327943 \pm 0\cdot00000100, \quad 1/\epsilon_a = 296\cdot753 \pm 0\cdot086 \qquad (53)$$

(probable errors). The greater part of the uncertainty arises from that of the mass of the Moon, which enters into the determination of H from the observed rate of precession. The adopted value was $M(1 \pm 0\cdot0005)/81\cdot53$ (probable error), following Hinks's determination from the observations of Eros during the opposition of 1900–1.*

Recent work has necessitated further changes. Spencer Jones (1941, 1942), from the Eros opposition of 1931, got the ratio of the masses $81\cdot271 \pm 0\cdot021$ (probable error), which I modified in a rediscussion (1942c) to $81\cdot291 \pm 0\cdot027$ (standard error). There seems to be some error in Hinks's determination, which Brouwer is disposed to attribute to the difficulty (mentioned already by Hinks) of separating the uncertainty of the predicted path from systematic errors in the positions of the comparison stars. In

---

* The main object of these observations was the determination of the solar parallax. But the attraction of the Moon on the Earth, in the elementary way of looking at it, makes the centre of the Earth move monthly in an ellipse about the centre of mass of the Earth and Moon together, and therefore through a distance depending on the mass of the Moon. This displacement of the Earth produces an apparent displacement of a planet with respect to the stars; consequently analysis of the observations gives the mass of the Moon as a by-product.

the 1931 data the material was adequate to separate these and the uncertainty is probably genuine. Meanwhile seismology and the theory of the figure of the Earth together have led to fairly definite knowledge of the density distribution (see p. 213) and the function ψ can be calculated. Inserting the new data in de Sitter's theory, Bullard (1948) gets, with Spencer Jones's mass of the Moon,

$$H = 0{\cdot}00327237 \pm 0{\cdot}00000059,$$

$$1/\epsilon_a = 297{\cdot}338 \pm 0{\cdot}050 \quad \text{(probable errors).} \tag{54}$$

My value gives
$$H = 0{\cdot}00327293 \pm 0{\cdot}00000075,$$

$$1/\epsilon_a = 297{\cdot}299 \pm 0{\cdot}071 \quad \text{(standard errors).} \tag{55}$$

The change from de Sitter's value is considerable in relation to the apparent uncertainties of both. But I think that we are now entitled to claim that (55) is the conclusion given by the hydrostatic theory. This theory is plainly false, because if it was true the solid surface would be a level surface, the ocean would cover it, and we should have no interest in the matter. There is, however, other evidence that indicates how far it is wrong. The solution is still useful, however, because the discrepancy gives important information on the strength of the shell.

In the second-order theory it is necessary to allow also for the variation of ψ. Two integral equations are found, one for the ellipticities of internal layers of equal density and the other for their departure from ellipsoids. In the usual treatment these are converted into differential equations, and a pair of boundary conditions for each, which can be solved. The problem arises also for Jupiter and Saturn; here the second-order terms in the figures are large enough to have an appreciable effect on the motions of the inner satellites, and the departure from uniformity of density is large enough for the variation of ψ to be serious. In this case it is found to be just as easy to solve the integral equations directly by successive approximation (Jeffreys, 1954b; Message, 1955).

The theory was given by de Sitter, but his equations contain some errors in the coefficients. I have given a modified theory (1963b). In this I define b as the mean radius, I as the mean moment of inertia. It is assumed that for changes of rotation the volume of a surface enclosing the same mass would remain the same. J is defined as before; $H = (C - A)/C$. Then it is a definite problem to say what J and H would be for a given rate of rotation on the hydrostatic hypothesis. I divide the problem into two parts. In the first, values are calculated according to the theory of 4·03 when second-order terms and $\psi - 1$ are neglected. In the table $e_0(1)$ is the surface ellipticity, and

$$m = 0{\cdot}00\,34498, \quad J = e(1) - \tfrac{1}{2}m, \quad H = \frac{3J}{3C/2Ma^2}.$$

$3I/2Mb^2$	$\eta_0(1)$	$100e_0(1)$	$100\{e_0(1)-\frac{1}{2}m\}$	$100H_0$
0·48	0·6900	0·32061	0·14912	0·30858
0·49	0·6256	0·32848	0·15599	0·31835
0·50	0·5626	0·33657	0·16408	0·32816
0·51	0·5006	0·34490	0·17241	0·33806
0·52	0·4400	0·35346	0·18047	0·34802

Apart from the correction for $\psi - 1$ these are really the terms of order m in the complete expression in powers of m.

The corrections for second-order terms and $\psi - 1$ are in any case small, and any model with some resemblance to the actual distribution of density will give them sufficiently accurately. I took quadratics in r for the core and shell, matching in each the mean density and moment of inertia. These made $\bar{\rho} = 5\cdot515$, $3I/2Mb^2 = 0\cdot5006$. It was found that $\psi - 1$ reduces e and J by 6×10^{-7}, H by 18×10^{-7}.

The effect of second-order terms was found from the integral equations. The crude solution $e_0(r)$ was taken as a first approximation, and an approximation

$$e = e_0 + \delta e, \quad \delta e = A - 10^4 B e_0^2,$$

with constant A, B, was chosen so that the residuals at $r = 1$, $0\cdot55$ and 0 would have equal moduli: a rough further approximation was applied. The total correction to e at the surface is 16×10^{-7}; that to H is -120×10^{-7}.

The best values of $J_2 = \frac{2}{3}J$ and H are probably about $10^{-6}(1082\cdot78 \pm 0\cdot005)$ and $0\cdot0032730(1 \pm 0\cdot00018)$, and adapting to equal volume of the comparison sphere we have

$$\frac{3}{2}\frac{I}{Mb^2} = 0\cdot496262(1 \pm 0\cdot00018).$$

Interpolating from the table and applying the corrections we find that on the hydrostatic theory we should have

$$100e = 0\cdot33370 \pm 0\cdot00006, \quad e^{-1} = 299\cdot67 \pm 0\cdot05,$$

$$100H = 0\cdot32379 \pm 0\cdot00006$$

and above all $\qquad J_2 = 10^{-6}(1072\cdot1 \pm 0\cdot4).$

The difference between this value and the observed one is clear, about 20 times the standard error. A comparison on these lines was first made by Henriksen (1960) and J. A. O'Keefe (1959).

4·04. Moment of inertia and density distribution. According to the above solution

$$\frac{I}{Mb^2} = 0\cdot330841(1 \pm 0\cdot00018),$$

C/Ma^2 would be about $0\cdot0001$ larger. It is clear that the ratio is substantially less than $0\cdot400$, which would be the value for a uniform sphere, and we have definite evidence that there is a strong increase of density towards the centre.

The simplest hypothesis to make at this point is that of Wiechert (1897), also considered by Kelvin and Tait (1883, §831) but not worked out by them to quantitative results. In this the Earth is supposed composed of a shell of uniform density, resting on a core also with uniform density. The shell may be supposed rocky and the core metallic, by analogy with the two commonest types of meteorites. Both investigations gave formulae for the outer ellipticity and the precessional constant to be associated on a hydrostatic theory with a given rate of rotation, and it appeared that these might lead to estimates of both the density of the shell and the radius of the core. Wiechert found, however, that the data were equally well fitted by a wide range of shell densities, and this is what we should expect from Radau's approximation. Wiechert's theory was a second order one. The simplest approach is to use 4·027 (3) directly, and it has the advantage that it does not assume hydrostatic conditions.

Wiechert's word for the rocky part was *Mantel*, which I have always translated by *shell*. Most recent writers use *mantle*. I consider this unfortunate. *Mantle* in English usually means a floppy outer garment, and is quite inappropriate to a material with a rigidity comparable with that of steel.

Let a and α be the mean radii of the outside and the core, the densities ρ_0 and ρ_1, where

$$\rho_1 = \rho_0(1+\mu). \tag{1}$$

Then the mass and mean density are

$$M = \tfrac{4}{3}\pi\rho_0 a^3(1+\mu\alpha^3), \quad \bar{\rho} = \rho_0(1+\mu\alpha^3), \tag{2}$$

and the mean moment of inertia

$$C = \tfrac{8}{15}\pi\rho_0 a^5(1+\mu\alpha^5). \tag{3}$$

Then, with $C/Ma^2 = \tfrac{1}{3}$,
$$\frac{2}{5}\frac{1+\mu\alpha^5}{1+\mu\alpha^3} = \frac{1}{3}, \tag{4}$$

whence
$$\frac{1}{\mu} = 5\alpha^3 - 6\alpha^5. \tag{5}$$

Wiechert adopted a mean density of 5·58 in one of his solutions, and a density of 3·20 for the shell. The latter was near that of the rocks that seemed most likely. Then (2) and (4) gave a pair of equations for α and μ. The solution was $\alpha = 0\cdot779$, $\mu = 1\cdot63$; the densities are then 3·20 and 8·206. The latter is near that of iron and seemed plausible.

This solution, however, met with difficulties. A considerable change of density at $r = 0\cdot78a$ would be expected to be associated with a sharp change of the velocities of seismic waves, and in any case should give strong reflexions. As seismic data increased it became clear that there was no sharp change in properties at this depth. It did become clear that there is one at about $r = 0.545a$, and accordingly it was worth while to make an alternative solution taking α as known and calculating both densities. If $\alpha = 0.545$, $\bar{\rho} = 5.53$, we get

$$\mu = 1.923, \quad \rho_0 = 4.22, \quad \rho_1 = 12.33. \tag{6}$$

The densities are greater than those found by Wiechert, but there is a simple explanation. The pressure half-way to the centre is about 1.2×10^{12} dynes/cm.². The bulk-modulus of the lower layer of seismology is also about 1.2×10^{12} dynes/cm.². Thus it would be expected that pressure alone would raise the mean density of the shell from 3·2 or so to over 4 g./cm.³. Similarly, from the estimated pressure at the centre, about 3×10^{12} g./cm.³, and the velocity of 9 km./sec. for longitudinal waves in the core we can show that pressure would increase the density of iron from about 8 to about 12. Accordingly, the Wiechert core and the Oldham core are identical; the difference in the sizes found is attributable to Wiechert's neglect of compressibility, which was almost inevitable at the date.

4.041. Density allowance for pressure. A much more accurate determination of the density distribution should be obtained by attending to compressibility in greater detail, and this has been done by Bullen in a remarkable series of papers. The method was originally due to L. H. Adams and E. D. Williamson (1923 b). Bullen's aim in the first place was simply to find the ellipticities of internal strata of uniform density as a preliminary to a calculation of the ellipticity correction in seismology. As this was small it was quite enough to aim at an accuracy of 10 % or so in the ellipticities, and therefore it was sufficient to use the distribution of velocities given by Gutenberg in 1929, which were at any rate right within 1 or 2 %, and the densities accordingly should have a similar accuracy. In calculating the density distribution for mean latitude the distribution could be taken as spherical; then if M', C' denote the mass and moment of inertia of the matter within distance r of the centre, we have

$$\frac{dp}{dr} = -\rho \frac{dU}{dr} = -\rho \frac{fM'}{r^2}. \tag{1}$$

Also if k is the bulk-modulus and α, β are the velocities of P and S,

$$\rho \frac{dp}{d\rho} = k, \quad \alpha^2 - \tfrac{4}{3}\beta^2 = \frac{k}{\rho}, \tag{2}$$

whence

$$\frac{d\rho}{dr} = -\frac{f\rho M'}{r^2(\alpha^2 - \tfrac{4}{3}\beta^2)}. \tag{3}$$

Also

$$\frac{dM'}{dr} = 4\pi\rho r^2. \tag{4}$$

(3) and (4) are a pair of differential equations for M' and ρ. These have known values M, ρ_a at the outer surface, and it is possible to integrate inwards. Also

$$\frac{dC'}{dr} = \tfrac{8}{3}\pi\rho r^4,\tag{5}$$

so that C' also can be found, given its surface value. Alternatively, if the surface value C is not given it can be found by integration from the centre after ρ is found. This condition arises in applications to other planets. For these the bulk-modulus is taken to be the same function of pressure as in the Earth.

Bullen (1936b) took values for the crustal layers indicated by near-earthquake studies and adopted 3·32 g./cm.³ for the density at depth 35 km. Proceeding inwards he found a density of 5·00 just outside the core, the depth of which was taken to be 2900 km. = 0·455 a. But then the mass and moment of inertia of the core were found to satisfy

$$C' = 0\cdot57\,M'r^2;$$

thus the core would have to be much denser near the boundary than near the centre. The coefficient in fact approaches that for a spherical shell, which would be $\tfrac{2}{3}$. The result was wholly unacceptable, but could be avoided only by adopting higher densities in the shell. One possibility might be to take a higher density at 35 km. depth; but this had no support. The other was suggested by the 20° discontinuity already found in seismology. The first estimates of the depth of this suggested that it was between 300 and 400 km.; Bullen adopted 350 km., and supposed that the jump in velocity of pulses there was associated with one in density, the amount of which would have to be determined. Accordingly, his problem became, given M and C, to find the densities just below 350 km. and just inside the core. Given these the rest of the problem would be determinate. He found that at 350 km. depth the density must jump from about 3·6 to 4·0. That in the core would range from 9·9 to 12·3.

The interpretation of this jump in density was difficult. I later (1937b, p. 54) redetermined the depth of the discontinuity as 480 km., and found the jump to be from 3·69 to 4·23. Extrapolating the density of the lower material to zero pressure gave 3·8. There seemed to be no plausible material with a density in this region, MgO, with a density of 3·6, being the nearest (1937b). At this point I consulted Prof. J. D. Bernal with respect to the possibility of a high-pressure modification of olivine. I had thought only of a general compression of the lattice without change of arrangement, but Bernal (1936) called attention to another possibility with experimental support. At low pressures olivine is a hexagonal lattice of oxygen atoms, with the metallic ones and silicon included in a somewhat asymmetrical pattern. The crystal as a whole is rhombic. The next element in the silicon series, germanium, has similar properties to silicon, and there is a magnesium germanate, Mg_2GeO_4, isomorphous and chemically analogous with for-sterite, Mg_2SiO_4. Now if ordinary olivine can undergo a change of crystal structure at high pressures, magnesium germanate should undergo a corresponding change at lower pressures, on account of the larger size of the germanium atom. It was made by V. M. Goldschmidt (1931), who also made a cubic form with a structure analogous to spinel, $MgAl_2O_4$. The crystals of this were too small to permit a direct determination of the density, but from the lattice dimensions given Bernal calculated that the density would be about 9 % higher than for the rhombic form. High pressure would favour the denser form, and it appeared probable that the jump in density at the 20° discontinuity might represent a change of olivine from rhombic to cubic form under high pressure. The suggested difference would be about 0.3 g./cm.^{3}, while the geophysical value is 0.54 g./cm.^{3}, but possibly agreement within a factor 2 is as much as could be expected. There is evidence also that such a change would explain the increase of wave-velocities, because magnetite, Fe_3O_4, which has a very high bulk-modulus, has the spinel structure.

Doubts were cast on this hypothesis, since other workers failed to reproduce Goldschmidt's preparation of cubic Mg_2GeO_4 and his original specimens were not available. However, A. E. Ringwood (1956) studied solutions of different concentrations of Mg_2SiO_4 in Ni_2GeO_4, the latter having the spinel structure. By extrapolation of the densities he inferred that the density of Mg_2SiO_4 in the spinel state would exceed that in the rhombic state by 11 ± 3 %. He later went into much more detail ($1958a, b, c, d$, 1959, $1962a, b$, $1966a, b$).

In subsequent work he and Major (Ringwood, 1969, 1970; Ringwood and Major, 1970) have gone further. Sufficient pressures were available to produce the transformation in olivine directly. The transformation is first to cubic spinel, each silicon atom being surrounded by 4 oxygen atoms at the corners of a tetrahedron, and then to a modified β form with somewhat lower symmetry. They estimate the transitions to occur in an interval of about 27 km. of depth, centred on 397 km., at a temperature of 1600 °C. In addition enstatite ($MgSiO_3$) changes at about 400 km. to forsterite and stishovite (the latter being a dense form of silica, stable above 120 kilobars (1.2×10^8 dynes/cm.^{2}) at 1200 °C). Silica appears to change from tetrahedral to octahedral symmetry at high pressures. At 900 to 1050 km. spinel changes to stishovite $+$ MgO. The total increase of density is 0.6 to 0.7 g./cm.^{3}. The metasilicates under somewhat smaller pressure change to a kind of garnet structure. This was first observed in $CaGeO_3$, which transforms to a garnet with the formula $Ca_3(CaGe)Ge_3O_{12}$; the CaGe replaces Al_2. All the increases of density are about 10 %.

At still greater pressures further changes are predicted from analogy with related compounds. Mg_2SiO_4 would become $2MgO + SiO_2$, with further increase of density. This is likely to occur at a depth of about 1000 km.

One difficulty in interpretation arises from the fact that the high values of $d^2t/d\Delta^2$ for P now appear to begin at 15° to 17° instead of at 20°; the corresponding depth reached would be about 200 km. and decidedly less than that indicated experimentally for the above transitions. Ringwood and Major (1966) have however crystallized glasses of composition

$$(Mg, Ca)SiO_3 + x(Al_2O_3)$$

at pressures $(1–2) \times 10^{11}$ dynes/cm.², corresponding to depths 300–600 km.; they give solutions of pyroxenes in garnets. It seems possible that these are related to the 20° discontinuity.

Ringwood considers the possible mode of formation of basaltic rock. He finds that a melted mixture of olivine and basalt does not separate on solidification unless the basalt amounts to at least a quarter of the whole. On this basis the two types of transition would both occur. The undifferentiated form is called pyrolite.

Bridgman, who discovered many polymorphic transitions under high pressure, has found (1951) a discontinuous change in the shearing strength of olivine at a pressure of 85,000 kg./cm.² (0.84×10^{11} dynes/cm.²). This pressure is rather less than is indicated for the 20° discontinuity, but of the right order of magnitude.

More detailed discussion, mainly from the standpoint of crystallography, is by Liebau and his collaborators (Liebau, 1971; Fuchs, Mayer-Rosa and Liebau, 1971; Liebau and Hesse 1971).

Bullen (1963b) has given a way of allowing for possible departures from the Adams–Williamson condition. He shows that for a chemically homogeneous region

$$\frac{dk}{dp} = 1 + g^{-1}\frac{d\phi}{dz},$$

where $\phi = \alpha^2 - \frac{4}{3}\beta^2$. But more generally, in any region where $d\rho/dz$ exists,

$$\frac{d\rho}{dz} = \frac{g\rho}{\phi}\frac{dk}{dp} - \frac{\rho}{\phi}\frac{d\phi}{dz}.$$

This can be written

$$\frac{d\rho}{dz} = \eta\frac{g\rho}{\phi}$$

where

$$\eta = \frac{dk}{dp} - g^{-1}\frac{d\phi}{dz} = \phi\frac{d\rho}{dp}.$$

The Adams–Williamson condition is $\eta = 1$. η thus takes account of chemical variation and departure of temperature from adiabatic.

Applications are given in this and later papers.

4·05. Constitutions of the Moon and inner planets. Comparison with the Moon leads to a test, which appears decisive. There are three possibilities:

(1) The density at 35 km. depth may be 3·8 or so instead of 3·3.

(2) The matter just below the discontinuity may be a new material.

(3) It may be a high pressure form of the old material.

Now it is reasonable to suppose that other inner planets, including the Moon, are made of similar materials, though not necessarily in the same proportions. If (1) or (2) is right we should expect the Moon to consist mostly of material of density 3·8 or so; allowance for a core, if any, and for compressibility would increase this. On the other hand, the pressure at a depth of 480 km. in the Earth is not reached anywhere in the Moon; hence if (3) is correct there would be no material of this density in the Moon, and the mean density of the Moon, if there is no core, would be near that of ordinary olivine, 3·3. The latter result agrees with observation; the density of the Moon is about 3·33. Incidentally it also appears to be too low for eclogite.

The actual density of the Moon is as low as would be possible on any hypothesis and implies that the Moon has no core or an extremely small one. Further, it may be regarded as confirming the rather indirect evidence from seismology and laboratory measurements that the lower layer of near-earthquake studies is either olivine or something mechanically indistinguishable from it.[*]

The argument has also been applied to the other inner planets. None of them leads to evidence directly testing the three hypotheses; but on Bernal's hypothesis it leads to information on the density distribution. The Moon, on account of its small size, is little compressed. If entirely olivine the density would range from 3·28 near the surface to 3·41 at the centre. There are probably outer layers similar to those in the Earth. On reasonable hypotheses about the range of possible thicknesses for these we get for the Moon (Jeffreys, 1937 a)

$$\tfrac{3}{2}C/Ma^2 = 0\cdot5956 \pm 0\cdot0010.$$

This is an important equation in connecting the data concerning the Moon's rotation and gravitational field.

If Mercury is similar to the Moon, the mean density would be $3\cdot39 \pm 0\cdot01$ (Jeffreys, 1937 a, c), whereas direct determination from the measured diameter and mass left it uncertain by a factor of $\pm 0\cdot22$.

Mars must have a small core and Venus one relatively nearly as large as the Earth's.[†]

[*] For more detail see Jeffreys (1937 c).

[†] Much revision has been done. E. Rabe (1950) applied certain corrections to the rotation of the Earth and derived masses of the inner planets and the solar parallax from perturbations of Eros. The accuracy appeared very high, but the parallax

It was considered likely that the Earth's core would be nearly all iron. A test was difficult; the density seemed to provide the only clue. Those of iron and nickel, which occur in meteorites, are 7·86 and 8·9 g./cm.^{3}. That of liquid iron at the melting point is 6·92 ± 0·07 g./cm.^{3}. A simple linear extrapolation from an early model of Bullen indicated that the density of core material at zero pressure would be about 8·16 g./cm.^{3} (1937c). In a later attempt I fitted cubics in the pressure to Bullen's solution A'' (1975, p. 173) and Birch's (p. 215). These gave ρ at $p = 0$ equal to 8·45 and 7·64 g./cm.^{3}. Birch (1952) extrapolated from Bullen's table (1942) on the supposition that k/ρ (= α^2) is a linear function of p and got ρ at $p = 0$ about 6·4 g./cm.^{3}. The differences are very large, though the densities in the core itself used in the extrapolations differed only by about 0·1 g./cm.^{3}, and it was clear that extrapolation was precarious.

Later work used experimental evidence in shock waves, which reached pressures up to about 4×10^{12} dynes/cm.^{2}, as large as those in the core. It was thus possible to compare densities at the same pressure. Birch (1963, pp. 144–145) found that the densities at equal pressures were about 1 g./cm.^{3} less than for pure iron. Later (1968, 1972, 1973) he has given much more detail and infers from the data of Balchan and Cowan (1966) that an alloy of iron with about 15 % by weight of silicon would satisfy the requirements of the core. Stewart (1973) gives 15 % to 19 %.

The greatest revisions for other planets, on various hypotheses on their compositions, are discussed by Bullen (1975, pp. 380–390).

disagreed seriously with Spencer Jones's trigonometrical one. However the radar determination gave a result about midway between the two. E. Rabe and Francis (1967) now use data for 1926–65 and get a value very close to the radar one. The error in the previous value was due to making a correction for the same thing twice. Their new value for $S/(E + M)$ is 328890 ± 16; the I.A.U. adopted value, based on the radar determination, is 328912.

Perhaps it is worth while to arrange treatments of small corrections in increasing order of carefulness. (1) Ignore them. (2) Make them once. (3) Apply them with the wrong sign. (4) Apply them twice. There are many instances of all cases.

Radar determinations give for the radius of Mercury 2434 ± 2 km. or 2440 ± 2 km. according as relativity or Newtonian theory is used (Smith, Shapiro and Ash 1966). Dr Carl Sagan derives from this a density 5·45 g./cm.^{3}. Lyttleton has shown that this density would imply that Mercury is over 60 per cent iron. Many people think that there must be something wrong somewhere.

There seems little doubt that there is a systematic error in Spencer Jones's determination of the solar parallax; no explanation has been published, but R. d'E. Atkinson suggests that it is due to flexure of an equatorial telescope when used far out of the meridian.

Radar methods have also redetermined (Colombo, 1965; Liu, 1966; Dyce, Pettengill and Shapiro, 1967) the rotations of Mercury and Venus. The period of rotation of Mercury is close to two-thirds of the orbital period. This may be due to a libration related to the high orbital eccentricity (Colombo and Shapiro, 1966). The rotation of Venus seems to be retrograde in about 247 days. The odd thing is that it appears to present the same face to the Earth at each inferior conjunction. The explanation seems more difficult than ever.

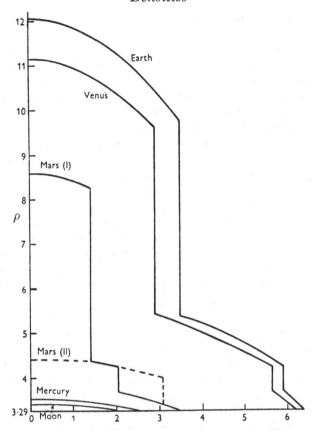

Fig. 17. Densities in the terrestrial planets according to 1937 c models.

Equation 4·041 (1) allows only for variation of density with pressure.* As temperature presumably increases downwards the variation of density within any one layer may be less than the solutions indicate. If we kept the mean density in each region the same, this would amount to displacing matter outwards and therefore increasing C. But C is given; also the density at the outside is given. Thus to keep M and C the same we must also reduce the mean density between the discontinuities and increase that of the core. The effects will be (1) a reduction of the density just above the 20° discontinuity, (2) an increase just below it, partly compensated by the general reduction in mean density between the discontinuities, (3) a general increase in the core. The amount of the discontinuity of density needed will be greater than when variations of temperature are ignored.

* Since seismic waves are adiabatic, the equation assumes adiabatic compression. This paragraph really concerns the excess of the actual variation of temperature with depth over the adiabatic variation.

4·06. Density distribution. In a further series of papers Bullen (1937c, 1939a, d, 1940a, b, 1941, 1942) has revised the density distribution in accordance with the latest data on the velocity distributions of seismic waves. The chief complication is the inner core, with the curious transition region outside it where the velocity of P diminishes towards the centre. This introduces a formal indeterminacy into the solution, but the volume of the region concerned is so small compared with that of the Earth that reasonable changes of the densities adopted within it make comparatively little difference to those in the shell and the main part of the core. As extreme values he takes densities from 12·3 to 21·6 at the centre, and finds that the corresponding densities just inside the core are 9·76 and 9·15.

The ellipticities of layers of equal density increase from about 0·00256 at the core boundary to 0·00337 at the outside. The value at the centre depends a good deal on the density of the inner core. In Bullard's solution (1948) it is about 0·0021. It is verified that the quantity η of Radau's approximation remains between 0 and 0·56 at all depths and therefore that the approximation is valid. This was under some doubt, because, though it is true with Wiechert's original values, it is not true if the α of 4·04 is less than $1/\sqrt{2}$, and it was doubtful whether it would hold for the actual Earth (Jeffreys, 1924). However, it is now clear that the variation of density in each layer restores the validity of the approximation.[*]

From analogy with meteorites Goldschmidt had inferred that the Earth would contain a good deal of ferrous sulphide. The density of this would imply that it would form a liquid layer on the outside of the core. This should be detectable by seismological methods. PcP and ScS would be reflected at the outside of the sulphide layer, and there would be associated reflexions from the bottom. These are not found. Also the time of K, inferred from ScS and SKS, would tend to a positive value at distance 0, but it does not. It appears therefore that if there is a sulphide layer it is not much more than a tarnish.

For the densities I gave in the last edition means between two solutions from Bullen (1963a, p. 231). In these the densities adopted for the inner core differed by 10 g./cm.³, and the upper value is certainly far too high. In a revised solution A″ Bullen and Haddon (1967b, c) and Bullen (1975, p. 173) used the modern value $C/Ma^2 = 0·3308$ instead of 0·3340. This increases the densities in the core a little and reduces those in the shell. The radius of the core is unaltered. Haddon has kindly given me his solution in detail. See p. 213. In the core k and μ are calculated from the interpolated densities and the velocities on p. 156 on the supposition that β in the inner core is 3 km./sec.

---

[*] In papers on the constitution of Jupiter and Saturn I applied the approximation to them also. In a discussion of bodies of Wiechert type, however, I found that Radau's approximation is seriously wrong for the degree of central condensation that they appear to possess (Jeffreys, 1951).

Shell

	r/R	ρ	k $(10^{12}\times)$	μ $(10^{12}\times)$	σ	g	p $(10^{12}\times)$
	1·00	3·320	1·16	0·629	0·270	985	0·009
	0·99	3·379	1·24	0·667	0·272	987	0·030
	0·98	3·435	1·33	0·708	0·275	988	0·051
B	0·97	3·490	1·42	0·751	0·276	990	0·073
	0·96	3·543	1·52	0·796	0·277	992	0·095
	0·95	3·594	1·62	0·845	0·278	994	0·118
	0·94	3·643	1·74	0·897	0·280	996	0·141
	0·93	3·800	2·04	1·038	0·283	998	0·164
	0·92	3·942	2·30	1·177	0·282	998	0·189
	0·91	4·071	2·54	1·309	0·280	999	0·214
C	0·90	4·186	2·75	1·432	0·278	999	0·240
	0·88	4·373	3·09	1·640	0·275	999	0·294
	0·86	4·504	3·36	1·785	0·274	997	0·350
	0·84	4·586	3·56	1·875	0·276	996	0·408
	0·82	4·660	3·74	1·958	0·277	995	0·466
	0·80	4·732	3·93	2·039	0·279	994	0·525
	0·78	4·803	4·13	2·116	0·281	994	0·585
	0·76	4·873	4·33	2·191	0·284	994	0·646
	0·74	4·942	4·54	2·265	0·286	995	0·708
	0·72	5·009	4·76	2·339	0·289	996	0·771
	0·70	5·075	4·98	2·411	0·292	999	0·835
D	0·68	5·140	5·20	2·485	0·294	1003	0·900
	0·66	5·204	5·41	2·559	0·296	1008	0·965
	0·64	5·268	5·61	2·636	0·297	1014	1·033
	0·62	5·331	5·84	2·716	0·299	1023	1·101
	0·60	5·394	6·04	2·796	0·300	1033	1·171
	0·58	5·457	6·26	2·875	0·301	1046	1·242
	0·56	5·516	6·33	2·951	0·298	1060	1·310
	0·548	5·560	6·39	2·966	0·299	1073	1·361

Core

	r/R_1	ρ	k $(10^{12}\times)$	g	p $(10^{12}\times)$	μ
	1·00	9·98	6·55	1073	1·36	0
	0·90	10·51	7·65	992	1·73	0
	0·80	10·97	8·95	902	2·08	0
E	0·70	11·36	10·12	804	2·41	0
	0·60	11·69	11·18	700	2·71	0
	0·50	11·96	12·20	591	2·98	0
	0·40	12·17	13·26	478	3·20	0
F	0·36	12·25	{10·82 / 13·8}	431	3·28	{0 / 1·1}
G	0·20	12·43	14·3	242	3·52	1·1
	0·00	12·51	14·5	0	3·62	1·1

In A'' the upper layers are replaced by a single layer 33 km. thick with $\rho = 2\cdot84$ g./cm.^{3}, $\alpha = 6\cdot30$ km./sec., $\beta = 3\cdot35$ km./sec. These may be fairly representative as averages. In Bullen's book further complications, such as effects of temperature and variation of composition, are discussed in great detail.*

* He had allowed (1956) for possible effects of variation of temperature from the adiabatic distribution and of composition. The changes in the shell were all less than

It will be noticed that in the core the ratio k/p lies between 4 and 5. This disposes of the possibility that the core is a perfect gas, for which this ratio would be 1. At such pressures it is not clear what criterion could be used to distinguish between an imperfect gas and a liquid, since the bulk-modulus is 5·5 to 10 times that of olivine at atmospheric pressure, and it is quite possible that there is no sharp distinction.

The approximate constancy of gravity in the shell is a curious feature, first noticed by A. E. Benfield.

Some revision is needed for various reasons. (1) The 1940 velocities at small distances were mainly based on Japan. It now appears that Japan is a rather exceptional region, and somewhat higher velocities have been found in all other regions examined. (2) To account for the length of the receding branch DE of PKP I introduced the peculiar region F, where α decreases with depth. Other explanations are given by Bolt (1964), Nguyen-Hai (1961, 1963) and Adams and Randall (1964); these replace my hypothesis by having one or two discontinuities in the outer core. Model A″ retains the region F. I cannot decide at present between the various alternatives, and have therefore retained it. Other solutions are due to Birch. He avoids as many complications as possible. In the first place, for $1·00 > R > 0·84$ he assumes an empirical relation $\rho = a + b\alpha$. From $R = 0·84$ to the core boundary he uses the Adams–Williamson relation. No discontinuity is used within the core. Two solutions are given. In the first, an empirical value of b is used, 0·328 g./cm.^{3} per km./sec.; in the second a and b are chosen so that the density at $r = 1·00$ is 3·32 and b is adjustable. The results are on p. 215 (Birch, 1964): ρ is in g./cm.^{3}, α and β in km./sec., P = pressure in dynes/cm.^{2}.

No allowance is made for a change of material for the inner core; and such allowance must be very arbitrary. But Birch (1964) has given strong reasons for supposing that the density at the centre cannot exceed 13·6. Bullen (1965a) using Bolt's velocities in the core infers that if the inner core has rigidity its density can be $\leqslant 12·6$. Without rigidity he gets $\geqslant 14·7$.

It has been pointed out that according to my solution (Jeffreys, 1939e) the variation of α with pressure is about half as rapid in the inner core as in the outer. This is not very plausible. But it may be remarked that the variation depends on the coefficient of r^2, which is about 1·8 times its standard error. It would not be very surprising if it should be doubled.

O. L. Anderson and J. E. Nafe (1965) gave an empirical law for oxides connecting the bulk modulus with other parameters, but later found that

0·1 g./cm.^{3}. Bullard (1957) considered other departures but still got a maximum departure of $\pm 0·3$ g./cm.^{3} in the shell and $\pm 0·5$ g./cm.^{3} in the core, except in the inner core.

Shell

r/R	α	β	Solution 1			Solution 2		
			ρ	g	$P(10^{12}\times)$	ρ	g	$P(10^{12}\times)$
1·00	8·10	—	3·425	984	0·009	3·320	984	0·009
0·98	8·13	—	3·435	988	0·052	3·332	989	0·051
0·96	8·38	—	3·517	992	0·095	3·426	994	0·093
0·94	8·97	—	3·710	995	0·141	3·650	998	0·138
0·92	9·91	—	4·018	997	0·190	4·006	1001	0·186
0·90	10·55	—	4·228	997	0·242	4·248	1001	0·239
0·88	10·99	—	4·372	996	0·296	4·415	1000	0·294
0·86	11·29	—	4·471	995	0·352	4·529	998	0·350
0·84	11·50	6·40	4·560	994	0·409	4·608	997	0·408
0·82	11·67	6·48	4·613	993	0·466	4·681	995	0·467
0·80	11·85	6·56	4·684	993	0·525	4·752	994	0·526
0·78	12·03	6·64	4·755	993	0·584	4·823	994	0·586
0·76	12·20	6·71	4·824	994	0·645	4·892	994	0·648
0·74	12·38	6·77	4·892	995	0·705	4·960	995	0·710
0·72	12·54	6·83	4·958	997	0·768	5·026	996	0·773
0·70	12·71	6·89	5·023	1001	0·831	5·091	999	0·837
0·68	12·88	6·95	5·088	1005	0·896	5·156	1002	0·902
0·66	13·01	7·01	5·152	1011	0·961	5·220	1007	0·968
0·64	13·16	7·07	5·216	1018	1·028	5·284	1013	1·035
0·62	13·32	7·14	5·278	1027	1·096	5·346	1021	1·103
0·60	13·46	7·20	5·341	1039	1·165	5·409	1032	1·173
0·58	13·60	7·26	5·405	1053	1·236	5·473	1044	1·245
0·56	13·64	7·31	5·468	1069	1·310	5·536	1059	1·318
0·548	13·64	7·30	5·508	1081	1·354	5·576	1070	1·363

Core

r/R_c	α	Solution 1			Solution 2		
		ρ	g	P	ρ	g	P
1·0	8·10	10·05	1081	1·35	9·96	1070	1·36
0·9	8·53	10·59	999	1·73	10·49	989	1·73
0·8	9·03	11·05	909	2·08	10·94	899	2·08
0·7	9·44	11·45	811	2·42	11·33	802	2·41
0·6	9·78	11·78	706	2·73	11·66	698	2·71
0·5	10·10	12·05	596	3·00	11·92	589	2·97
0·4	10·44	12·27	482	3·22	12·13	476	3·20
0·3	11·20	12·43	364	3·40	12·29	360	3·37
0·2	11·24	12·53	244	3·54	12·39	241	3·50
0·1	11·28	12·60	122	3·62	12·46	121	3·58
0·0	11·31	12·62	0	3·64	12·48	0	3·61

two of the data used were wrong, and Anderson and Soga (1967) find that Birch's law (Birch, 1961*b*) which is generally quoted in the form

$$\alpha = A + B\rho,$$

where A depends on the mean atomic weight but B is largely independent of it, should be a valuable guide in studying the physics of the Earth's interior. This is also discussed by Liebermann and Ringwood (1973) and O. L. Anderson (1973). D. L. Anderson (1967*a*) proposes

$$\rho = A\bar{M}\Phi^n,$$

where ρ is the density, $\bar{M}$ the mean atomic weight, n a constant between $\frac{1}{4}$ and $\frac{1}{3}$, $\Phi = \alpha^2 - \frac{4}{3}\beta^2$. The paper is rather too detailed to summarize here.

Further evidence has been given by Dziewonski and Gilbert (1971) and by Julian, Davies and Sheppard (1972). Dziewonski and Gilbert find that some higher normal modes are fairly sensitive to rigidity in the inner core, and infer from their periods that the average velocity of S in it is 3·517 $(1 \pm 0\cdot0012)$ km./sec. Julian, Davies and Sheppard use arrays, which could detect $PKJKP$ at distances 200° to 290°. They infer from five earthquakes that β is $(2\cdot95 \pm 0\cdot1)$ km./sec. The results are not very consistent, but at least the evidence for a solid inner core is stronger. They think that there is damping corresponding to a Q between 500 and 1000.

The statement of damping in terms of a constant Q may be nearly right for a solid inner core. But for the liquid outer core the damping must be due to ordinary viscosity. Estimates of constant Q have been given for it, and must be misleading.

Bullen in his later solutions has used the correct value 0·3308 for C/Ma^2, instead of 0·334 as in his early solutions, but at the same time he has increased the radius of the core by 15 km. by use of periods of free oscillations, which I think have a systematic error as a side effect of damping. Birch uses 0·3308, and the core radius 3473 km. The latter is the same as mine, but he increases the velocities of seismic waves at small depths and reduces them at depth $0\cdot04R$ with no change from depth $0\cdot06R$ to the core boundary. This may be more representative of average structures, but it is not stated what changes it would make in travel times.

This is highly relevant to the need for a standard Earth model, which is at present being discussed by the I.A.S.P.E.I. Different authors give their results in comparison with different tables of travel times and densities; some give depths in kilometres, some as fractions of the radius of the base of the upper layers. The result is that no reader can see whether two results are consistent without doing a lot of unnecessary arithmetic. Similar conditions have existed in astronomy and geodesy, and have been found intolerable. Some of the most important properties of a standard, as it seems to me, are as follows.

1. Where present values do not differ greatly, the standard should be close to them.

2. A table should be as smooth as is convenient. Besides the boundaries of the main and inner cores, and possibly the Mohorovičić discontinuity, there should be no discontinuities of the function or its first derivative. They may exist in the actual Earth, but in computation it is much easier to introduce a new irregularity than to displace one that is already there.

3. It should be made perfectly clear that regional differences exist, and that statements that any table is universally applicable are false; it was clear enough in 1940 that no such table can exist. Definite differences are established between Eurasia, the Pacific and Japan. It may be worth while

to construct separate models to fit them. The properties of the upper layers differ even within the same region, and in my later work I have used a single upper layer in each case.

4. Travel-times and velocities should be consistent within 0.1^s in the former. Densities should be consistent with the ratio $C/Ma^2 = 0.3308$.

5. No use should be made of periods of free vibrations; these are subject to an effect of damping, which not only gives a decay of amplitude but can make an appreciable alteration of period. Much further work is needed to give more detail.

6. Most results would be conveniently stated as differences from the standard model; in many cases a table could consist of 1 or 2 figures instead of 4, and the behaviour of the differences would catch the eye.

I think that either Bullen's A'' or Birch's model would satisfy these conditions.

4·07. Figure of the Moon: external theory. Like the theory of the figure of the Earth, that of the figure of the Moon separates into an external and an internal theory. But the Moon's rotation is much slower, and as the Moon always keeps nearly the same face towards the Earth, the Earth's attraction maintains a sort of permanent tide in it, and actually contributes more to the function Ψ than the Moon's rotation does. The axis that points nearly towards the Earth is the axis of least moment of inertia. The principal moments of inertia at the centre are all different, and their differences correspond to observable dynamical consequences. On the other hand, the differences are relatively smaller than for the Earth, and the ellipticity of the visible disk is too small to be separated by observation from the irregularities of topography; the ellipticities of sections through the line of sight are of course still smaller in relation to the uncertainty of measurement.

We take the principal axes at the centre of mass of the Moon as axes of x, y, z; the axis of z is the one that nearly coincides with the axis of rotation and that of x points nearly to the Earth. We use unaccented letters for quantities relating to the Earth and accented ones for the Moon. The Moon's mean angular velocity n is the same for rotation and revolution. Then the work function for the Earth and Moon together is

$$W = f\left\{\frac{MM'}{R} + M'\frac{A+B+C-3I}{2R^3} + M\frac{A'+B'+C'-3I'}{2R^3}\right\} + O\left(\frac{1}{R^4}\right), \quad (1)$$

I, I' being moments of inertia about the line of centres and R the distance between the centres. The second and third terms depend on the orientations of the Earth's and Moon's axes respectively in space, and when the equations of motion are formed by Lagrange's method they make contributions to the rates of change of the angular velocities. As they also depend on R and on the direction of the line of centres they also affect the relative orbit. There are also large terms depending on the attraction of the Sun on both bodies. The terms $O(R^{-4})$ are totally negligible. a/R is about $\frac{1}{60}$, and the low harmonics in gravity, other than the main ellipticity term, are of the order of $\frac{1}{100}$ of that term even at the Earth's surface. Hence the terms $O(R^{-4})$ produce effects on the relative motion of the order of $\frac{1}{6000}$ of the effects of the terms in R^{-3}.

The consequences of the presence of terms depending on the orientations of the bodies are considered in full in works on celestial mechanics. The chief effects on the Earth are the precession and the 19-yearly nutation of the axis, on which a number of smaller nutations are superposed. The reaction on the Moon gives three observable perturbations of the orbit. The effects of the figure of the Moon on its rotation may be described roughly by saying that the axis of least moment of inertia performs small oscillations in both directions about the line of centres. As the Moon's orbit is itself steadily changing owing to the action of the Sun, the consequences are rather complicated. The plane of the orbit maintains a fixed angle to the ecliptic, but rotates backwards about the pole of the ecliptic, completing the revolution in 18·6 years, which is also the period of the largest term in the Earth's nutation. The Moon's axis of greatest moment keeps near the plane containing the poles of the ecliptic and of the Moon's orbit, oscillating about a position in this plane whose inclination to either depends on the ratio $\beta = (C' - A')/C'$. As the direction of this axis is well observed, this ratio is very accurately determined. The ellipticity of the Moon's equator would lead to a free oscillation in longitude, the attraction of the Earth on the protuberances playing the same sort of part as gravity plays in the pendulum. If these free oscillations were observed the determination of $\gamma = (B' - A')/C'$ from the period would be straightforward; but they are too small to have been detected. The periodic parts of the Moon's motion in longitude, however, produce some forced oscillations, and one (possibly two) of these is large enough to be measured. It is small, and there is some conflict of evidence about its amount.

The dynamical theory is mostly due to F. Hayn; I have made some small corrections and taken some additional small terms into account (Jeffreys, 1961 b). The observations consist of measures of the displacements in latitude and longitude of a small crater near the centre of the disk. The former depend mostly on β, the latter on γ. The most elaborate treatment so far is by Koziel (1966, 1967), based on four series of observations. I was not altogether satisfied with this, for reasons stated previously (Jeffreys, 1957 b, p. 477; 1961 b, pp. 430–1), and carried out a rediscussion (Jeffreys, 1971). Koziel got (1967)

$$\beta = 0 \cdot 000\,6294 \pm 0 \cdot 000\,0006$$

which I revise to $\beta = 0 \cdot 000\,6271 \pm 0 \cdot 000\,0010.$

My previous solution of 1961, based on a comparison of ten determinations by various authors, was

$$\beta = 0{\cdot}000\,6279 \pm 0{\cdot}000\,0015.$$

The treatment of γ is complicated. The important terms in the longitude of the crater are an annual one and one with argument 2ω, twice the angular distance between the Moon's node and perigee. The latter has a period about 3 years. As its amplitude contains the squares of the eccentricity and inclination as factors it is necessarily very small except near resonance, which would occur at $\gamma = 0{\cdot}000\,2131$. Till recently all analyses referred to the annual term and the results fell into two groups, with γ about $0{\cdot}5\beta$ or $0{\cdot}2\beta$. The reason was pointed out by Banachiewicz and Koziel (1948). If a set of trial values of γ is taken and the 2ω term is computed, it tends to infinity at the critical value $0{\cdot}000\,2131$. Consequently with any observed values the sum of squared residuals has two minima, one on each side of the critical value. The two results depend on which of these is chosen. A possible treatment is to assume a form $a_1 \sin \odot + a_2 \sin 2\omega$ and estimate a_1 and a_2 by least squares; then we should have two equations for γ. Koziel's method is equivalent to this. His solution is equivalent to

$$\gamma = 0{\cdot}000\,2310 \pm 0{\cdot}000\,0032.$$

My rediscussion of his data gives

$$\gamma = 0{\cdot}000\,2362 \pm 0{\cdot}000\,0082.$$

Yakovkin (1952) analysed a series of 35 years' observations, which he thought referred to a free vibration. I interpreted his data as referring to the 2ω term and derived (Jeffreys, 1957b)

$$\gamma = 0{\cdot}000\,2049 \pm 0{\cdot}000\,0009.$$

However, a mistake in interpretation was pointed out by Habibullin and Schruttka-Rechtenstamm (Jeffreys, 1971), and the corrected result is

$$\gamma = 0{\cdot}000\,2308 \pm 0{\cdot}000\,0092.$$

By comparison of 20 published estimates of the annual term, which were reasonably consistent, I got

$$\gamma = 0{\cdot}000\,2274 \pm 0{\cdot}000\,0088.$$

The results are now consistent, if we have regard to the uncertainties.

The effects of $C' - A'$ and $B' - A'$ on the orbital motion depend on their ratios to $M'a'^2$; their effects on the rotations on their ratios to C'. The chief of the former are secular motions of the node and perigee. Far the greatest parts of these are due to the Sun; much smaller parts to the ellipticity of the Earth; but when these are allowed for a balance remains, which leads to an estimate of $C'/M'a'^2$. This was done by de Sitter (1927b, p. 61).

His value of $3C'/2M'a^2$ (g' in his notation) was $0{\cdot}65 \pm 0{\cdot}13$ (probable error), which cannot be considered a good determination; the values for a homogeneous sphere (0·6) and for the Earth (0·50) would both be within the standard error of the estimate. Several corrections were made later. I found (Jeffreys, 1961*b*, p. 432) that the motions of the node and perigee differed from the calculated values for a homogeneous Moon by just about the uncertainties. However, the greater part of the uncertainties was not observational but based on Brown's estimate (1904) of the possible errors arising from the incompleteness of his calculation; W. J. Eckert (1965) has now extended Brown's calculation. The motion of the perigee is in good agreement with theory. The node however leads to $g' = 0·96 \pm 0·07$. This is serious. Eckert seriously considers the possibility that the density decreases inwards. He does not emphasize a further consequence; given the mean density and g' it is possible to find a lower bound to the density at the outside, which comes out at about 13 g./cm.[3], rather more than that of lead (Jeffreys, 1967*a*). The result is difficult to understand anyhow; of four possibilities I think that the least unlikely is a systematic error in the observed motion of the node or in the reduction of the observations.

Other information has been provided by the artificial satellite Luna 10, which made a close approach to the Moon. Goudas (1967) refers to Akim (1966) for determinations for the second harmonics in the Moon's potential. Comparing with the values of β and γ Goudas derives

$$\frac{3C'}{2M'a^2} = 0·56 \pm 0·18,$$

the uncertainty being a 'maximum error'. This is in agreement with the value for a homogeneous Moon and definitely inconsistent with Eckert's.

A further discussion by Michael (1970) led to

$$C'/M'a'^2 = 0·4015 \pm 0·0030.$$

$C' - A'$ as determined from a satellite and from librations should differ slightly because the former contains a part due to elasticity and the latter does not. Allowing for this decreases $C'/M'a'^2$ by about 0·0014. The result would be consistent with either uniform density or with 0·397, which I estimated in 1936 (Jeffreys, 1937*a*).

Cook (1970) gives the method of calculation of $C'/M'a'^2$ for the Moon but the data need slight alterations.

The motion of the node has been redetermined by F. M. Sadler and L. V. Morrison (1969) from occultations 1960–66. They infer no significant change in the notion of the perigee; the discrepancy for the node is reduced but remains large. D. H. Sadler (personal communication) thinks that an error may survive in Brown's computation of the planetary contribution to the motion of the node.

4·08. Figure of the Moon: hydrostatic theory. We take the same axes at the centre of the Moon, x being measured away from the Earth; if c is the mean distance between the centres

$$R^2 = (c+x)^2 + y^2 + z^2, \tag{1}$$

and the gravitational potential due to the Earth is

$$\frac{M}{R} = \frac{fM}{c} - \frac{fMx}{c^2} + \frac{fM}{2c^3}(2x^2 - y^2 - z^2) + O\left(\frac{1}{c^4}\right). \tag{2}$$

We take the motion to be one of steady revolution, the Moon always keeping the same face to the Earth, so that the motion of each part of the Moon is one of revolution with angular velocity n about an axis perpendicular to the plane of the orbit and through the centre of mass of the Earth and Moon together. The distance of the latter point from the centre of the Moon is $Mc/(M+M')$ or $c/(1+\mu)$, where

$$\mu = M'/M. \tag{3}$$

The effect of the steady motion on acceleration relative to the Moon is then equivalent to that of a potential

$$\tfrac{1}{2}n^2\left\{\left(\frac{c}{1+\mu}+x\right)^2 + y^2\right\}. \tag{4}$$

Then the total potential due to the Earth and the orbital motion is

$$\frac{fM}{c} - \frac{fMx}{c^2} + \frac{fM}{2c^3}(2x^2 - y^2 - z^2) + \frac{1}{2}\frac{n^2c^2}{(1+\mu)^2} + \frac{n^2cx}{1+\mu} + \tfrac{1}{2}n^2(x^2+y^2). \tag{5}$$

But

$$n^2c^3 = f(M+M') = fM(1+\mu). \tag{6}$$

Hence the terms in x cancel, and we have a disturbing potential

$$\frac{n^2}{2(1+\mu)}(2x^2 - y^2 - z^2) + \tfrac{1}{2}n^2(x^2+y^2). \tag{7}$$

We require to know what effect this will have on the figure of the Moon. It is enough to treat the Moon as of uniform density and to neglect μ. If we add a term $\lambda(x^2+y^2+z^2)$, it will at most give a small uniform expansion and not affect the ellipticities to the first order. We determine λ to make the whole potential a solid harmonic; the condition is

$$n^2 + 3\lambda = 0. \tag{8}$$

Then the potential reduces to

$$\tfrac{1}{6}n^2(7x^2 - 2y^2 - 5z^2). \tag{9}$$

If the surface of the Moon is deformed until its equation is

$$r = a'(1+\epsilon S_2), \tag{10}$$

the external potential due to it is, by 4·03 (3),

$$fM'\left(\frac{1}{r} + \frac{3}{5}\frac{a'^2\epsilon S_2}{r^3}\right). \tag{11}$$

For the outer surface to be a level surface the sum of this and (9) must reduce to a constant when (10) is satisfied. This leads to the equation of the surface

$$r = a\left\{1 + \frac{5}{12}\frac{M}{M'}\frac{a'^3}{c^3}\frac{7x^2 - 2y^2 - 5z^2}{a'^2}\right\}. \tag{12}$$

The semi-axes, x, y, z of the Moon are respectively

$$a'\left(1 + \frac{35}{12}\frac{M}{M'}\frac{a'^3}{c^3}\right), \quad a'\left(1 - \frac{10}{12}\frac{M}{M'}\frac{a'^3}{c^3}\right), \quad a'\left(1 - \frac{25}{12}\frac{M}{M'}\frac{a'^3}{c^3}\right). \tag{13}$$

Then the principal moments of inertia of the Moon satisfy, to the first order,

$$\beta = \frac{C' - A'}{C'} = 5\frac{M}{M'}\frac{a'^3}{c^3} = 0.0000375, \tag{14}$$

$$\gamma = \frac{B' - A'}{C'} = \frac{15}{4}\frac{M}{M'}\frac{a'^3}{c^3} = 0.0000281, \tag{15}$$

taking $M/M' = 82$, $c/a' = 221$. These values are in total disagreement with observation. The observed value of $(C' - A')/C'$ is near 0.000629, about 17 times as large. The observed value of γ/β is about 0.33; the hydrostatic theory makes it $\frac{3}{4}$.*

If the Moon solidified when considerably nearer the Earth than it is now, and if it was rotating freely at the time (i.e. not keeping a fixed face to the Earth) the ratios given in (14), (15) would be larger. The justification for such a hypothesis would lie in the approximate circularity of the equator. Then we have to consider only the Moon's rotation, and it is found that the rotation period would have to be about 3.5 days (Jeffreys, 1937a, p. 11). This, however, is purely an *ad hoc* hypothesis with no independent confirmation.

If, in fact, $1 - \gamma/\beta$ was $\frac{1}{4}$ or near it we could explain the facts by supposing the Moon to have solidified when considerably nearer the Earth than it is now, for γ/β is independent of distance. The distance at the time was estimated at 140,000 km., and the period as 6.3 of our present days.

The outstanding result is that the Moon at present is in far from a hydrostatic state. As the inequality in question corresponds to a second harmonic in the figure, it serves as a warning against the acceptance of a hydrostatic theory for low harmonics in the Earth; direct evidence for the Earth has been much longer in coming.

---

* In my 1915a paper and in earlier editions of this book I supposed the ratio of $C' - A'$ to $C' - B'$ to be in agreement with theory. I seem to have been misled by the early values quoted from Bouvard and Nicollet in Routh's *Advanced Rigid Dynamics*, namely

$$\frac{B - A}{C} = 0.000564, \quad \frac{C - B}{C} = 0.000035, \quad \frac{C - A}{C} = 0.000599,$$

whence $\gamma/\beta = 0.94$. This appears to explain why I thought the observed values consistent with this ratio being 0.75.

CHAPTER V

The Figures of the Earth and Moon.
Discussion of Observations

In the last chapter we discussed the theories of the external fields of the
Earth and Moon and the hypothesis of internal hydrostatic stress. Observational data were used only to indicate the kind of information that exists
and the degree of accuracy that is worth attempting. We must now go into
more detail, and in the first place we must pay more attention to local
irregularities, the study of which has led to important results, though their
range of applicability has sometimes been exaggerated.

**5·01. Effect of uncompensated surface inequalities on the plumb-
line.** The older geologists and geodesists regarded mountains as composed
of matter of much the same density as the rest of the crust, and it was not

recognized that their weight would be expected to produce any deformation of the
matter below them, nor that the density of
the matter below a mountain range might
differ systematically from that of the matter
at an equal depth below a plain or even an
ocean. Now if a mountain is considered
merely as an extra mass superposed on a
previously uniform crust, and its deforming
effect on the interior is ignored, it is possible
to compute all components of its contribution to gravity on bodies in its neighbourhood. The attraction can also be found
experimentally, and the result compared

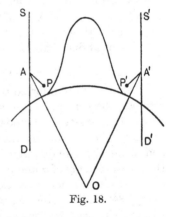

Fig. 18.

with that calculated. The experiment was carried out on several mountains
during the eighteenth century with unexpected results.

The principle of the measurement of the attraction of a mountain may
be illustrated by means of the above figure. Let A and A' be two pivots on
opposite sides of the mountain, carrying two plumb-lines, AP, $A'P'$. Let O
be the centre of the Earth, supposed spherical. Let AS, $A'S'$ be lines joining
A and A' to a fixed star in the plane AOA', and let AD, $A'D'$ be their pro-
longations past A, A'. Then the angle AOA' is nearly AA'/a, where a is the
radius of the Earth. Now AA' and a are found from surveying operations, so

that AOA' is determinable. Let its value be α. But we have, since AS and $A'S'$ are parallel,

$$OAD + OA'D' = AOA' = \alpha.$$

The zenith at A is defined to be on the prolongation of the plumb-line past its point of support. Hence the zenith distance of the star is the angle PAD. Thus we can find PAD and $P'A'D'$ from observation. Let their sum be β. Then we have by subtraction

$$OAP + OA'P' = \beta - \alpha.$$

Then $\beta - \alpha$ is the sum of the deflexions of the plumb-line at the two stations due to the attraction of the mountain.

Now if g is the intensity of gravity, and the bob of the pendulum AP is exposed to a small horizontal acceleration γ due to the mountain OAP, the deflexion of the pendulum is γ/g. If γ' (again measured towards the mountain) is the acceleration at P' due to the mountain,

$$\gamma + \gamma' = g(\beta - \alpha).$$

γ and γ' are calculable from the law of gravitation, if the region has been surveyed, and therefore this relation affords a test of the theory.

The first attempt to use this method was apparently made by Bouguer (1749). The sum $\gamma + \gamma'$ of the deflexions due to the mountain Chimborazo (Andes) was found to be very much less than that calculated from the law of gravitation. At that time the constant of gravitation was known only vaguely, but the deflexion of the plumb-line was much less than could be reconciled with any reasonable value. Maskelyne, in 1774, repeated the experiment at Schiehallion,* in Perthshire. His results gave

$$\alpha = 41'', \quad \beta = 53''.$$

The value of $\gamma + \gamma'$ inferred from this was 0.06 cm./sec.^{2}, and was used to estimate the constant of gravitation and hence the mean density of the Earth. The density was found to be 4.7. This is decidedly lower than modern estimates, but the discrepancy is much less than that found by Bouguer, and in the opposite direction. It seems doubtful whether the accuracy of the measurements of α and β or of the density within the mountain would have permitted a determination within 10% in any case. A repetition by Petit (1849) in the Pyrenees showed that their attraction was not only small, but actually negative; the plumb-line appeared to be deflected away from the mountains. Indeed, the attraction of mountains was generally found to be nearer to zero than to the values calculated on the supposition that the underlying matter was of normal density. Schiehallion was an exception. The reason was probably that it was of smaller horizontal extent, so that the attraction due to abnormal density below was mainly vertical and did not deflect the pendulum much.

---

* *Sch* pronounced as in German, not as in Gaelic. Professor K. H. Jackson kindly informs me that the Gaelic spelling is *Sidh Chaillean*, meaning the Fairy Hill of the Caledonians.

The modern development began with a discussion by Archdeacon Pratt (1855) of the deflexions of gravity observed by Everest in India. Everest had found discrepancies between the measured inclinations of the verticals at different places and the inclinations calculated from the known distances between the places, and attributed them to deflexions of the plumb-line due to the Himalayas; but Pratt found that they were only about a third of what would be produced by the attraction of the mountains, treated as additional masses.

Similar results were found in the United States by Hayford (1909, 1910). These covered a much larger area.

5·02. Effect on the intensity of gravity.

This also was discussed by Bouguer. For our immediate purpose it will be enough to replace the mountain by a low plateau, since the width of a mountain or range of mountains is always much greater than its height. More generally, the approximation is good when the slopes are small. If the mass of the Earth is M, its mean density $\bar\rho$, and its radius a, the intensity of gravity at the surface in the absence of disturbance is g_0, which is nearly

$$fM/a^2 = \tfrac{4}{3}\pi f\bar\rho a. \tag{1}$$

At a height h above the surface, in the free air, the intensity is

$$g_0 a^2/(a+h)^2 = g_0(1 - 2h/a), \tag{2}$$

nearly. If instead of the interval between sea-level and height h being empty it is filled by a plateau of density ρ', it is known by the theory of attractions that it will add an amount $2\pi f\rho'h$ to the intensity of gravity above it. Hence the total intensity of gravity at the top of the mountain, on the hypothesis again that the matter below it is of normal density, is

$$g_0\left(1 - \frac{2h}{a}\right) + 2\pi f\rho'h = g_0\left(1 - \frac{2h}{a} + \frac{3}{2}\frac{\rho'}{\bar\rho}\frac{h}{a}\right) \tag{3}$$

by (1). Hence the excess of gravity at height h over its value at sea-level should be $-\dfrac{2gh}{a}\left(1 - \dfrac{3}{4}\dfrac{\rho'}{\bar\rho}\right)$. As in the case of the deflexion of the plumb-line, this formula when tested was found to give considerable errors; the actual anomaly of gravity (against a formula taking account of the ellipticity) on the top of a mountain was found to be nearer to $-2gh/a$ than to the Bouguer formula.

It is seen from an examination of the formula that only the last term arises from the attraction of the mountain itself. Thus the statement that the gravity anomaly on the top of a mountain is equal to $-2gh/a$ implies that it is the same as if the mass was zero, the height remaining the same. The hypothesis that the mountain is an egg-shell gives better accordance with gravity than the hypothesis that it is simply extra matter. A hypothesis virtually identical with the former was offered by Boscovich (1755, cf. Todhunter (1873), **1**, 313), when he suggested that mountains were swellings caused by the Earth's internal heat, no extra matter being added. Cavendish also suggested (1772–4) that the matter below the stations in the Andes was of abnormally low density. The hypothesis that surface gravity is given by simply neglecting the attraction of all surface irregularities, but taking account of varying distance from the centre of the Earth, was introduced by Faye (1880) and may be called the free-air hypothesis. It must be distinguished from the free-air reduction, in spite of the formal similarity; the latter is simply a mathematical device used to simplify the calculation of the external field from observed gravity, and depends on no hypothesis about the *explanation* of irregularities of gravity. The elevations certainly have mass, but the free-air hypothesis is a useful standard of comparison because it represents an extreme case such that the internal anomalies of density are just enough to cancel the extra mass of the visible inequalities and are as near the surface as they could possibly be. The Bouguer formula can be regarded as the opposite extreme, where the internal anomalies are supposed to be at the centre of the Earth and therefore cancel one another. It is found that in many regions the two formulae differ from the truth in opposite ways, and together provide an indication that a mountain system is associated with a defect of density below it, but at an appreciable depth. The free-air hypothesis has the further utility that the residual against it is the same as the residual of free-air gravity against the gravity formula used as a standard, and consequently these residuals form the raw material for further investigation of the figure of the Earth.

5·03. Isostasy. A qualitative explanation of the facts discovered by Pratt was given by Airy (1855). He considered the Earth as having a thin solid crust, supported on a weak, but not necessarily fluid, substratum, and showed that the weight of the mountains, if they were simply added loads, would bend the crust and introduce stresses that would certainly break it if the strength was anywhere near that of rocks in the laboratory. He inferred that the crust *has* been broken; the extra load would then have pushed the crust down till it was balanced almost wholly by the upward pressure of the material below. The smallness of the disturbance of gravity associated with the surface topography follows as a natural consequence; to a first approximation a mountain system is floating on the substratum like an iceberg on water, the extra mass being equal to the mass displaced. Airy remarks indeed that Pratt's results ought to have been expected, for the notion of a crustal layer with a magmatic substratum was a familiar one at the time (cf. Sir John Herschel, 1837). It was inferred from the fact that the vertical gradient of temperature in the crust, if it continued to a depth of some tens of kilometres, would lead to fusion there. The argument is much too simple a statement of the problem, but does at any rate suggest a thickness of the right order of magnitude. Airy considered thicknesses up to 160 km.

Pratt (1859) did not accept Airy's explanation, on three grounds, which are now of historic interest only (Jeffreys, 1928*f*). If the lower layer was of the same density as the upper one, equilibrium would not be attained until the surface inequalities were quite effaced. If the lower layer was the less dense, the broken part of the crust would simply sink to the bottom of it. Pratt, considering the lower layer to be of the same material as the upper, but liquid instead of solid, inferred that it would be the less dense and that the mechanism would fail. But now that seismology has indicated considerable increase of density with depth in the top 50 km. or so the objection has lost its force. It is not obvious that the weak layer extends to the top of the lower layer of seismology; it may be at a considerable depth below it, and its depth can be determined, if at all, only by examining its effects on gravity.

Pratt proposed the alternative hypothesis that the development of surface features is due to the vertical expansion of columns of rock down to some uniform depth; the expansion is the same at all points of the same column, but differs from one column to another. There is no change of mass within any column, and the smallness of the disturbance of gravity agrees with this hypothesis as well as with Airy's. The fatal objection to it is that it does not account for the chief facts in the history of mountains. Their formation is not a matter of simple uplift, considerable horizontal movement having taken place too. Denudation has altered the surface by amounts comparable with the present heights, and the matter removed has been deposited at places hundreds of miles away. Pratt's hypothesis in fact assumed that there are no horizontal movements, and fails to explain the smallness of disturbances of gravity in places where great horizontal displacements of mass have occurred. Airy's theory does so without trouble.

The terms *compensation* and *isostasy* are used to denote the hypothesis of approximate uniformity of mass per unit area over the Earth's surface, measured from some standard level surface in the interior. *Isostasy* was coined by Major C. E. Dutton (1889). Some writers restrict its meaning to the physical process that tends to the establishment of this state.

The history of further work on the subject was curious. Geologists seem on the whole to have accepted Airy's mechanism, some even with exaggerations. Geodesists accepted Pratt's, and it was not until the work of Heiskanen (1924) that a serious attempt was made to test Airy's theory, and even then it tested a point that is not essential to Airy's mechanism. Airy, in fact, was supposed to have assumed the compensation to be concentrated in a single depth, Pratt that it was uniform at all levels down to a given depth. In fact, either would be consistent with Airy's mechanism. Suppose that the density of additional matter on top is ρ', its thickness k. Then the additional mass per unit area is $\rho'k$. Let the density at the level where outflow takes place be ρ_0. Then compensation would be attained if the

depression of the crust was $\rho' k/\rho_0$ (leaving a surface elevation $(1 - \rho'/\rho_0)k$).
Suppose first that the density down to the level where outflow takes place
is everywhere ρ'. Then if we neglect compressibility there is no change of
density anywhere except at the base, where the change is $-(\rho_0 - \rho')$
through a thickness $\rho' k/\rho_0$. The compensation is concentrated near one level.
But, on the other hand, suppose that the density increases continuously
downwards. Then the density after deformation at a given height above
a fixed equipotential surface is equal to the original density at a height
$x + \rho' k/\rho_0$, and is therefore $\rho + k\,d\rho/dx$, where ρ is the original density at
height x. The effect of the deformation is therefore to increase the density
at a given level by $k\,d\rho/dx$, which is in general negative, and will be constant
in a given column if ρ is a linear function of x. Thus uniformity of compensation down to a given depth is to be expected on Airy's theory if the
variation of density down to that depth is linear, but if the density changes
suddenly at any level the deficiency of mass will appear to be concentrated
near that level. The decision between uniform compensation and compensation at a given depth does not depend on the mechanism of isostasy, but
only on the normal distribution of density with depth. Now the evidence
of near earthquakes makes it clear that the variation is largely concentrated
near particular levels. These transitions may be gradual, but the changes in
properties within each of the upper layers are small compared with those
from one layer to the next. Compensation would therefore, on Airy's
theory brought up to date, be concentrated near the bases of the granitic
and intermediate layers. If the mountain systems are bounded by inclined
faults the compensation need not be strictly vertically below the inequality
compensated; it is likely to be somewhat spread out horizontally (regional
compensation), and its gravitative effect will imitate that of local compensation at a somewhat greater depth.

The values of gravity to be expected for no compensation, or for local
compensation, either uniformly distributed to a depth called the depth of
compensation, or concentrated at a single depth, have been computed for
many regions and compared with observation. The full calculation requires
a numerical integration, even for the Bouguer theory. The work was mainly
done under the supervision of Prof. W. Heiskanen at the Isostatic Institute
of the International Geodetic Association, Helsinki, later at the University
of Ohio. Values on various types of regional compensation have also been
worked out for some regions. I think that the chief interest of these hypotheses is that they provide definite standards of comparison; enough information has been gathered to show that none of them is right, but it
remains important to know how far they are wrong.

5·04. Comparison with observation. This is done most easily from
the intensity of gravity. Either of the deflexions of the plumb-line and the
disturbances of gravity could theoretically be calculated from the other if it

was completely known, so that in a full account they would give the same information. But the intensity of gravity has the advantage that disturbances in it are more closely concentrated about the masses producing them. It must be remembered, however, that the first big analyses indicating compensation were those of Pratt and Hayford on deflexions.

W. Bowie (1917) gives the following summaries of the residuals of gravity against a trial formula for the United States, on the basis of uniform compensation down to depth H. The unit is 1 mgal. $H = \infty$ corresponds to the Bouguer hypothesis, $H = 0$ to the free-air one. The residual for $H = 114$ km. is usually called the Hayford anomaly because this was Hayford's estimate of H.

	$H = \infty$	$H = 0$	$H = 114$ km.	$H = 60$ km.
Mean residuals without regard to sign				
Coast stations	21	22	18	12
Stations within 325 km. of coast	25	23	21	20
Inland stations, not in mountainous regions	33	20	19	19
Stations in mountainous regions, below general level	108	24	20	18
Stations in mountainous regions, above general level	111	59	17	22
Mean residuals with regard to sign				
Coast stations	+17	+17	−9	− 3
Stations within 325 km. of coast	+ 4	+17	−1	+ 2
Inland stations, not in mountainous regions	−28	+ 9	−1	− 1
Stations in mountainous regions, below general level	−107	− 8	−3	0
Stations in mountainous regions, above general level	−110	+58	+1	+16

In the first place, we see by inspection of these tables that no hypothesis makes the mean residual without regard to sign less than 18 mgal. But repeated observations at the same place generally gave agreement to about 3 mgal., which must be regarded as the ordinary range of observational error. The mean residuals are far too great to be regarded as observational error and represent real variations. Bowie seems to have paid little attention to this fact, at least in his published work, and his under-estimation of its importance has led to a widespread belief that isostasy is verified to a degree that the observations do not in the least warrant.

Next, we notice that the mean Bouguer anomaly with regard to sign is systematically negative at inland stations, increasing in magnitude with height. Comparison with the values taken without regard to sign shows that in the mountainous regions it must have been negative at nearly every station. This is sufficient to make the Bouguer hypothesis untenable.

JEI

The free-air hypothesis shows nothing unusual till the last lines of the tables, where it gives positive residuals at nearly all stations. But the mean with regard to sign can be annulled by taking compensation at an inter-

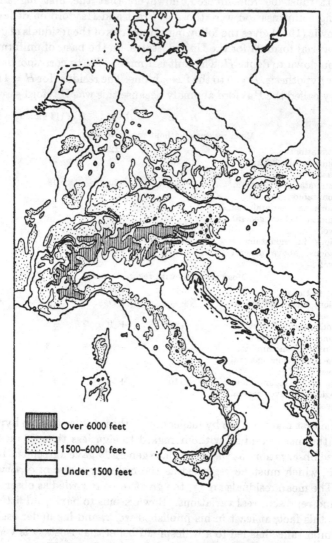

Fig. 19. Contours of the land in Central Europe.

mediate depth; Bowie finds that the data are best represented by taking the depth of compensation to be 96 km.

Heiskanen compared the same observations with those calculated on various hypotheses of compensation concentrated at the base of a single layer. He found that the average depth that fits the facts best is about 50 km., and that the fit is slightly better than is given by the hypothesis

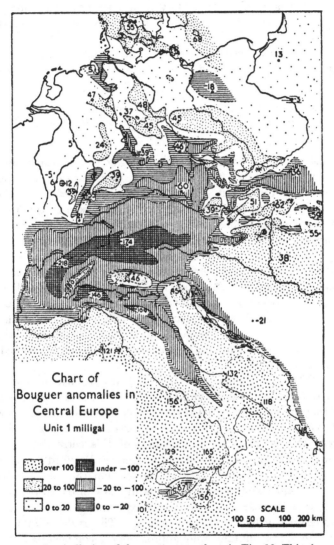

Fig. 20. Note the similarity of the contours to those in Fig. 19. This shows directly the presence of an extra thickness of light matter in rough proportion to the surface elevation. (*After F. Kossmat.*)

of uniform compensation at its best. He extended the comparison to the Swiss Alps, the Harz Mountains and the Caucasus, and in each case found that compensation concentrated at a single level fitted the facts as well as or slightly better than uniform compensation. The depth, according to him, is 50 km. below sea-level in the United States, 41 km. in the Alps, and 77 km. in the Caucasus. Theoretically, uniform compensation to depth H should have nearly the same effect on gravity as compensation concentrated at depth $\frac{1}{2}H$. This is why Heiskanen's values are much less than Hayford's and Bowie's.

Data for India have been collected by J. de Graaff Hunter (1932, p. 48 and Chart F). The summaries of residuals for the Himalayan region are as follows:

Without regard to sign			With regard to sign		
$H = \infty$	$H = 0$	Hayford	$H = \infty$	$H = 0$	Hayford
243	85	36	-243	$+5$	$+29$

Hunter does not give summaries corresponding to the rest of Bowie's, but from his Chart F it is seen that Hayford anomalies away from the Himalayas range from about $+40$ to -80 mgal. The general conclusions to be drawn from the Indian results are similar to those from the American ones; the Bouguer rule is consistently wrong in elevated regions, the free-air rule gives large residuals of both signs, the Hayford rule smaller ones of both signs, but still several times the uncertainty of observation. The actual magnitudes of the mean Hayford residuals are not very different. This needs emphasis because the American workers have tended (with the great exception of J. Barrell) to assert that observation supports isostasy completely, those in India to deny this, and consequently there is a current belief that this corresponds to a great difference in the observational data. The actual differences between the observational data are not great.

Barrell (1914–1915) made a careful study of the distribution of the deflexions of the vertical in Hayford's papers and called attention to the fact that they are not distributed at random, as observational errors would be; on the contrary, they keep the same sign over distances of hundreds of kilometres. It was, in fact, the existence of this correlation between neighbouring residuals that led Helmert to increase Hayford's estimates of uncertainty (p. 186). Barrell drew the conclusion that though isostasy may be a first approximation to the facts in the United States, it is not exact, and that there are regions of great extent where the crust is systematically heavier or lighter than the average. The evidence indicated, in fact, that the high mountains were approximately compensated, but that the compensation was not exact and that there were appreciable residual loads that could be supported only by appreciable strength in what had been supposed to be a region of complete weakness. Barrell coined for this the name *asthenosphere*; but the whole purport of his papers is that it is not a region of no strength but merely one of less strength than the outer rocks. He inferred a strength of the order of a sixth of that of surface rocks. The measures of gravity show similar systematic tendencies and point to the same conclusion.*

Meinesz's measures at sea (1932, Meinesz *et al.* 1934) showed greater departures from the hypothesis of isostasy. Along a strip on the seaward side of Sumatra and Java, passing Timor and curling north between Celebes and Halmaheira, he found systematic negative Hayford anomalies reaching about −100 mgal. In the northern part about −200 mgal was reached. Heiskanen's theory gave little difference. Meinesz also considered a theory of his own, which considered the problem as one of elastic bending of a floating crust. The distribution of residuals still followed a similar pattern. The negative residuals on the whole followed a belt of deep water, but, surprisingly, did not take their extreme values at the deepest places; on the contrary, they were usually somewhat to one side and often over a submarine ridge or even on land. It would be natural to regard these great departures from isostasy as evidence of great strength in this part of the crust, so that the surface could be much altered without fracture; but if that was the explanation we should expect a Bouguer formula, modified to allow for the attraction of the water, to be right, and it is no better than the others.

Similar results have been obtained in the West Indies (Meinesz and F. E. Wright, 1930).

5·05. Disturbances of density. The tendency of observations has therefore been to show increasing complexity of the relation of gravity to topography; no simple relation such as is given by the Bouguer, free-air, or any of the isostatic hypotheses is true. The proper line of progress now is to start from what we know, namely, the heights and the values of gravity, and try to find out what distributions of density underground are needed to explain them. The first step is to calculate and remove the contribution from matter above sea-level, or, in the case of sea observations, to allow for the presence of water in place of rock. We can also allow for the extra distance from the centre; but when we have done this we have simply subtracted the Bouguer correction, and are left with the Bouguer anomaly. (Strictly we should also allow for the alteration of the geoid, but this hardly matters if we are considering regions small compared with the whole size of the Earth.) The question then becomes that of finding what distribution of density will explain the Bouguer anomaly. A good deal has been done on

---

* We have so far defined the Bouguer and free-air anomalies only for land stations. For submarine observations, observed gravity is reduced to the surface by allowing for the reversed attraction of the water between the submarine and the surface. Various instrumental corrections and one for the motion of the ship are also applied. Then gravity as found is essentially free-air gravity for height zero. The Bouguer method, as on land, modifies the international formula to allow for the altered attraction of visible inequalities and therefore for the fact that the ocean is water and not rock. The isostatic hypothesis at sea allows also for increased quantity of the underlying dense rocks to keep the whole mass per unit area constant.

these lines (C. Tsuboi, 1937; Tsuboi and Fuchida, 1937; E. C. Bullard and R. I. B. Cooper, 1948). Meinesz explained some of his observations in this way, and E. A. Glennie has done the same for India. The method is extensively used on a small scale in geophysical prospecting to locate buried masses of specially dense or light material. It must be noticed, however, that the solution can never be unique unless some further hypothesis is made. By a theorem due to Green, if a surface S surrounds a distribution of mass, then there is at least one distribution of masses and doublets over S that gives exactly the same field at points outside S. In particular, if a sphere of radius a encloses a point mass m distant b from the centre, the potential outside the sphere due to the point mass is exactly the same as that due to a surface density $m(a^2 - b^2)/4\pi a r^3$ on the sphere, where r is the distance from m to the point on the surface considered. This surface density has a maximum on the radius through m and decreases continuously as we proceed from this radius. If, however, $a < b$, the field due to the particle cannot be represented by any distribution over the sphere, because it tends to infinity at the particle, and no distribution over the sphere could give a field with that property. This theorem has a number of important consequences. First, if a field is explained by a variation of mass over a surface at some depth, exactly the same external field would be produced by replacing each element of mass on the surface by a surface density over a sphere at a smaller depth. In particular, local compensation is exactly equivalent to a regional compensation at a smaller depth. Meinesz has in fact proposed a regional compensation, not quite according to the rule just given, and much computation has been devoted to it. So far as inequalities of density may be due to inclined faults there is a good deal to be said for it. But it hardly needs separate treatment. Secondly, in general a distribution of mass over a surface does not give a field that can be represented by any distribution at a greater depth; and for a given external field there will usually be an extreme possible depth for the masses capable of producing it. This can be seen in another way. Let a be the radius of a sphere and suppose gravity over it expressed in a series of spherical harmonics. If it arises from a surface density $\Sigma \sigma_n S_n$ over a surface at depth h, we have

$$U = \Sigma \frac{4\pi f a}{2n+1} \sigma_n S_n \left(\frac{a-h}{r} \right)^{n+1}, \qquad (1)$$

$$g = \Sigma 4\pi f \frac{n+1}{2n+1} \sigma_n S_n \left(\frac{a-h}{a} \right)^{n+1} = \Sigma g_n S_n. \qquad (2)$$

Now if $\Sigma \sigma_n S_n$ converges there is a quantity M independent of n and of the direction such that $| \sigma_n S_n | < M$. Then

$$| g_n S_n | < 4\pi f \frac{n+1}{2n+1} M \left(\frac{a-h}{a} \right)^{n+1} < N \left(\frac{a-h}{a} \right)^{n+1}, \qquad (3)$$

where N is independent of n. Speaking roughly, the expansion of g in spherical harmonics converges at least as fast as a geometric series, in which the ratio of consecutive terms is $1 - h/a$. If an arbitrary function of position on a sphere is expanded in spherical harmonics the maxima of the terms usually decrease like n^{-s}, where s is a small positive number; but if g had this property it could not arise from any distribution of mass except a surface one. If g does satisfy (3) for some h and N, let h_0 be the upper limit of the values of h such that N can be found for each of them. Then we can say that some of the inequalities of density producing variations in g are at depths not greater than h_0. Complete knowledge of g would therefore determine a maximum depth for the density inequalities. If we assume a depth less than h_0, g will determine a unique distribution of density over it.

In practice, we have g only at selected points, but the general argument still has its consequences. If we try to account for the observed values of g by a density variation at too great a depth, this will become wildly irregular; we shall be led to the major horror of negative densities or to the minor horror of a density in places decreasing with depth.

L. M. Dorman and B. T. R. Lewis (1970, 1972) attack the problem more generally. For a disturbance of density around each depth they work out the disturbances of gravity; then each observation gives an equation connecting the possible amounts of disturbances of density at different depths. Combining these equations they get a solution. They estimate that there is on the whole about 20 % overcompensation at about 50 km. depth, and that this is taken up by a change at about 400 km. The standard residual of a gravity observation is reduced to about 13 mgal. This is a rough summary of their results; they give also a continuous distribution and compare it with seismic data. The correspondence does not seem close; but in any case I think the seismic velocities have a serious systematic error due to the neglect of the side effect in the Lomnitz and the modified Lomnitz laws of damping.

5·06. Compensation of oceans. All these results really concern regions where there are great differences of height with respect to sea-level within distances of the order of 100 km. The mean depth of the oceans, however, is some kilometres, so that we have differences of height that persist over horizontal distances of the order of several thousand kilometres. We therefore are led to ask whether these also are compensated. For the greatest of them we can give an immediate answer in the affirmative. We have seen that when the Earth's centre of mass is taken as origin the gravitational potential contains no first harmonics and the centre of the co-geoid coincides with the centre of mass. But the Pacific Ocean occupies practically half the surface and has a mean depth of about 5 km. If the rocks below it had the same density as continental ones the extra height on the other side would correspond to a shift of the centre of mass relative to the centre of the co-geoid, which is impossible. The only solution is that rocks below the Pacific must be denser than continental ones. This does not depend on detailed observations of gravity; it is a direct consequence of the law of gravitation itself, the fact that the Pacific is an ocean, and the fact that the co-geoid in the west of Europe and the east of America is determined by the Atlantic Ocean.

The absence of granite from Pacific islands is in accordance with this result, and it is worth while to examine whether the difference of height is consistent with reasonable assumptions about the difference of structure between continents and oceans. We take a typical continental structure to be represented by 12 km. of granite of density 2·6, 24 km. of intermediate rocks of density 2·9, resting on dunite of density 3·3.* The upper layers would have the same mass per unit area as a column of dunite of thickness 30·6 km., giving a difference of level of 5·4 km., or by one of crystalline basalt of density 3·0 and thickness 33·7 km., giving a difference of level of 2·3 km. These estimates need some increase to allow for the sedimentary layer, which is mainly confined to the continents and has an average thickness of a few kilometres. Also the mass of the water must be taken into account. The average height of the continents above sea-level being taken as 0·5 km., our first hypothesis would put the ocean bottom at a depth of 4·9 km., and it would be further depressed by $4·9/2·3 = 2·1$ km. by the weight of the water, giving 7·0 km. in all. For a basaltic ocean floor the extra depression would be $1·8/2·0 = 0·9$ km., giving a depth of 2·7 km. The average depth of the Pacific lies between these two values, and would be consistent with an intermediate structure. Hence there is no difficulty in accounting for the depth of the Pacific by a permissible hypothesis about the structure of its floor, itself suggested by petrology and seismology.

The greatest depths in the Pacific, however, exceed 7 km.; 8 km. is reached in the Japan and Tonga deeps, 9–10 km. in the Guam and Philippine deeps. These cannot be compensated locally on the scheme just given. But it appears from Meinesz's work that they are associated with negative gravity anomalies and that the equality of mass per unit area does not hold for them anyhow. They are very restricted in width.

Inequalities of height on an intermediate scale, especially the difference between the Atlantic and its neighbouring continents, need more detailed treatment. In dealing with strips 100 km. or so in width the curvature of the Earth makes little difference; the comparison is really between the specially elevated or depressed region and the more or less ordinary ones on each side of it. Then for traverses across the strip the radii to different points are nearly parallel. But for widespread inequalities this is not true. If A, B are on the surface and far apart and A', B' are vertically below them, B' may actually be nearer to A than B is, and the general disturbance of gravity by topography and compensation together is seriously altered by this fact.

---

* These thicknesses differ somewhat from those suggested by the latest seismological work. The latter, however, still has a considerable uncertainty, and for a rough test of plausibility the values taken are near enough.

We take a surface density $\sigma_n S_n$ over $r = a$, and another $k_n \sigma_n S_n$ over $r = a - h$. The effect on the gravitational potential is

$$U' = \frac{4\pi f}{2n+1} \frac{\sigma_n S_n}{r^{n+1}} \{a^{n+2} + k_n(a-h)^{n+2}\}, \tag{1}$$

and that on free-air gravity is

$$-\left(\frac{\partial U'}{\partial r} + \frac{2U'}{r}\right)_{r=a} = 4\pi f \frac{n-1}{2n+1}\left\{1 + k_n\left(\frac{a-h}{a}\right)^{n+2}\right\} \sigma_n S_n. \tag{2}$$

If there is no compensation $k_n = 0$. According to the usual form of compensation each element of extra mass on the surface is associated with an equal defect of mass at $r = a - h$. Allowing for the difference of areas we have

$$k_n = -a^2/(a-h)^2. \tag{3}$$

Then for no compensation the disturbance of gravity (reduced to the co-geoid by the free-air formula) is

$$g'_n = 4\pi f \frac{n-1}{2n+1} \sigma_n S_n, \tag{4}$$

and for compensation it is

$$g'_n = 4\pi f \frac{n-1}{2n+1} \sigma_n S_n \left\{1 - \left(\frac{a-h}{a}\right)^n\right\}. \tag{5}$$

h is of the order of 50 km. Then if n is much more than 120 the term in $(a-h)/a$ is small and the two expressions approach equality. If n is more than about 10 the factor $(n-1)/(2n+1)$ approaches $\frac{1}{2}$. Hence for very local irregularities of topography the effect on gravity is nearly the same whether they are compensated or not, and approximates to the simple Bouguer term $2\pi f\sigma$. To indicate the magnitude of this, take a layer of rock, 1 km. thick, of density 2·5 g./cm.³. We have

$$2\pi f\sigma = 2 \times 3\cdot14 \times 6\cdot67 \times 10^{-8} \times 10^5 \times 2\cdot5$$
$$= 0\cdot105 \text{ cm./sec.}^2. \tag{6}$$

A kilometre of uncompensated rock under 4000 km. in horizontal extent will correspond to a disturbance of gravity over it of about 100 mgal.; if it covers a region much less than 50 km. across it will produce about the same disturbance whether it is compensated or not.

For smaller values of n we give the values of $g'_n/2\pi f\sigma_n S_n$:

n ...	1	2	3	4	5	6
Uncompensated	0	0·40	0·57	0·67	0·73	0·77
Compensated at 50 km.	0	0·006	0·013	0·021	0·029	0·036

n ...	10	20	30	50	100
Uncompensated	0·86	0·93	0·95	0·97	0·99
Compensated at 50 km.	0·068	0·15	0·20	0·31	0·54

On either hypothesis extra height will be associated with increase of g', but for the smaller values of n the effect of g' is much smaller if compensation exists, as we should naturally expect. Now the main features of the distribution of land and sea can be fairly well represented by low harmonics, up to degree 10 or so. The difference in surface density between the Atlantic and Europe consists in a difference of level of about 4·5 km., 4 km. of which is filled with water, so that the whole would be equivalent to the mass of about 2·9 km. of rock of density 2·5. Then for no compensation we should expect gravity to be systematically low over the Atlantic in comparison with Europe and America by something of the order of 300 mgal. With compensation it would still be low but only by about 10 mgal. Both conclusions disagree with observation. One of the remarkable features disclosed by early observations of gravity at sea was that positive anomalies were prevalent. Certainly the hypothesis of no compensation is wildly wrong. The early observations might possibly have been confined to exceptional regions, but a sufficient fraction of the oceans has now been surveyed to give a test. Mean free-air residuals against the international formula have been derived for regions at intervals of 10° of latitude and longitude (Jeffreys, 1941a, pp. 18–22). From this table we extract those whose standard errors are 25 mgal. or less, in three sections: (1) land regions in longitudes 0–60° E., (2) the Atlantic, (3) North and South America. The means are respectively + 8·4 ± 3·6 mgal., + 6·4 ± 4·7 mgal., − 1·6 ± 5·7 mgal. The statement that the Atlantic mean is less than the mean for the continent on each side by 10–20 mgal. is contradicted. The situation again is that the hypothesis of no compensation is hopeless, and the usual type of compensation is much better but still not in accordance with the facts. We can infer that the light upper layers under the Atlantic, as in the Pacific, are much thinner than in the continents, but the total mass per unit area is not quite constant, being slightly *more* under the Atlantic than in the continents.

Such a result is not altogether surprising. Suppose that compensation was complete at the end of the last period of mountain formation. At this time the systematic difference in gravity would exist. Since then much material has been removed from the continents and deposited in the Atlantic. If this has not yet proceeded so far as to break the crust again, it will have reduced the mass in the continents and increased that in the Atlantic, thus reducing and possibly even reversing the systematic difference of gravity.

An alternative way of looking at the matter is to consider an elevated strip on an infinite plane. We take the extra mass per unit area to be σ for $-l < x < l$, otherwise 0; the axis of z is upwards. The disturbance of gravity due to a surface density $\cos \kappa x$ is $2\pi f\, e^{-\kappa z} \cos \kappa x$. With a mass deficiency at depth h everywhere equal to the surface excess, since

$$\frac{1}{\pi} \int_0^\infty \{\sin \kappa(x+l) - \sin \kappa(x-l)\} \frac{d\kappa}{\kappa} = 1 \quad (-l < x < l) \\ = 0 \quad (x < -l,\ x > l) \Bigg\} \tag{7}$$

and also
$$= \frac{2}{\pi} \int_0^\infty \cos \kappa x \sin \kappa l \frac{d\kappa}{\kappa}, \tag{8}$$

the disturbance of g is

$$g' = 4f\sigma \int_0^\infty \cos \kappa x \sin \kappa l (1 - e^{-\kappa h}) \frac{d\kappa}{\kappa}$$

$$= 2\pi f \sigma (0, 1, 0) - 2f\sigma \left(\tan^{-1} \frac{x+l}{h} - \tan^{-1} \frac{x-l}{h} \right), \tag{9}$$

the values 0, 1, 0 in the first portion being taken according as $x < -l$, $-l < x < l$ or $x > l$. We find for various values of x the following values of g', h/l being supposed small:

x	g'
0	$4f\sigma h/l$
$\frac{1}{2}l$	$\frac{16}{3}f\sigma h/l$
$l-h$	$\frac{1}{2}\pi f\sigma + f\sigma h/l$
$l+h$	$-\frac{1}{2}\pi f\sigma + f\sigma h/l$
large	$-2f\sigma \left(\dfrac{h}{x-l} - \dfrac{h}{x+l} \right).$

The difference in the Bouguer term is $2\pi f\sigma$, so that over most of the strip the average value of g' would be of the order of h/l times the Bouguer term. For the Atlantic, in comparison with the neighbouring continents, the Bouguer terms differ by about 300 mgal. With $h = 50$ km., $l = 3000$ km., we should therefore expect the free-air anomalies in the middle of the Atlantic to be less than in the continents by, on an average, about 5 mgal., the differences becoming larger in the outer quarters of the width.

The tendency of observations has therefore been to show increasing complexity of the relation of gravity to topography. None of the comparatively simple relations given by the Bouguer, free-air and isostatic hypotheses is true. This has a twofold importance. We must revise our notions of the mechanical processes involved in compensation, which remains a first approximation to the facts as far as the greater inequalities of height are concerned. Also we can no longer calculate the disturbances of gravity and of the plumb-line from the observed inequalities of height, and this entails a revision of the method of analysis of survey data. In any case direct study of the distribution of free-air gravity is needed, not simply to correct the mean value and the main ellipticity term, but to determine the widespread anomalies.

5·07. Distribution of free-air gravity. The first attempt to analyse measures of gravity for harmonics other than the ellipticity term seems to have been by G. W. Hill; others were made by A. Berroth and F. Helmert, and further solutions taking account of more data have been given by Heiskanen (1938). They inferred a pair of second harmonics of the form $\cos^2\phi(\cos 2\lambda, \sin 2\lambda)$, which make the equator of the co-geoid an ellipse. The chief difficulty about the analysis is a statistical one. We cannot proceed by integration, but we can fit a finite set of spherical harmonics by the method of least squares. But gravity is a continuous function of position. The problem may be stated more simply in terms of that of fitting a set of cosines and sines to a continuous function $f(x)$ of one variable. The obvious way is to choose a set of ordinates and fit by least squares, using the residuals to provide an estimate of uncertainty. But the usual rule for estimating uncertainty supposes the errors independent. For a continuous function this is impossible, because the difference between $f(x)$ and the approximation is itself a continuous function and the correlation between its values for x and $x + h$ tends to 1 when h tends to 0. The error of the approximation is not a random error but is more like the sort of variation we should get if we assigned a set of values at random and interpolated between them by one of the standard interpolation formulae. Thus we really have to introduce a new unknown, the interval between consecutive values such that the errors may be taken as independent.

The method that I have adopted is as follows. The surface was first subdivided into regions bounded by meridians and parallels at 1° intervals. These are called 1° squares because one must call them something. To reduce the risk of overweighting special regions only the most northerly station available was used in each 1° square. As the free-air anomaly in limited regions is correlated with height, I fitted a linear formula in the height to all the values for 1° squares in each 10° square and used it to estimate a value for the mean height of the 10° square, with an estimate of uncertainty from the residuals.* It was found that the 10° means fluctuated far more than their apparent uncertainties would lead one to expect; also there was a decided positive correlation between the means and those for adjacent 10° squares. Consequently the 10° squares were combined to give 30° squares. The variation of the free-air anomaly was thus divided into three nearly independent parts: a variation between 1° squares, one between 10° squares, and one between 30° squares. The first differed greatly in different regions, naturally being greatest in mountainous regions. The second and third were called the τ_1 and τ_2 variations, and their mean-square values were about 420 and 150 mgal.^{2}. The whole variation could be regarded as due to

---

* The need for this adjustment requires emphasis. Prospecting companies are finding values of g over special regions in vast numbers, and many of them have kindly agreed to make summaries of their data available. It is most important that such summaries should be stated as for the mean height of the regions surveyed, or at least that the mean height of the stations should be given; otherwise there will be a systematic error arising from the fact that low-lying places are more accessible than high ones, and the mean height of the stations consequently less than that of the region.

the superposition of these three parts. It is known that heavy grouping leads to little loss of accuracy (Jeffreys, 1967d, p. 217). If values of a function of θ from 0 to 2π are given, we can get a good estimate of a term in $\cos\theta$ by comparing means over ranges $-\frac{1}{4}\pi$ to $\frac{1}{4}\pi$ and $\frac{3}{4}\pi$ to $\frac{5}{4}\pi$, and of one in $\sin\theta$ by comparing $\frac{1}{4}\pi$ to $\frac{3}{4}\pi$ with $\frac{5}{4}\pi$ to $\frac{7}{4}\pi$. If we are interested in a term in $\sin 3\theta$, which keeps its sign over ranges of 60°, we lose little information by taking ranges of 30°. The main aim was to get values over ranges of 30° with known accuracy. It was possible to examine whether any harmonics up to degree 4 were worth taking into account. A term similar to Heiskanen's second harmonic was found, but with a smaller coefficient, three third harmonics, and no fourth harmonics. The reason for the change in Heiskanen's term was that he had hardly any data for the southern hemisphere, whereas I had many from a cruise of Meinesz in 1939. The second and third harmonics $p_2^2\cos 2\lambda$, $p_3^2\cos 2\lambda$ add up in the northern hemisphere but largely cancel in the southern, and Heiskanen's material could not separate them. My final corrections to the international formula were, in milligals. (Jeffreys, 1943a) (transformed to the p_n^s notation)

$$+\ 2\cdot5 \pm 1\cdot9 - (6\cdot0 \pm 5\cdot0)\,p_2 + (8\cdot0 \pm 2\cdot8)\,p_2^2\cos 2\lambda$$
$$+\ (7\cdot8 \pm 4\cdot1)\,p_3^2\cos 2\lambda + (12\cdot6 \pm 7\cdot2)\,p_3^1\cos\lambda + (2\cdot7 \pm 1\cdot6)\,p_3^3\sin 3\lambda.$$

The corresponding terms in the potential would require division by $n-1$. If these are normalized by multiplying by the mean square values of the harmonics we get, multiplied by 10^{-6},

$$p_2^2\cos 2\lambda,\ 6\cdot2 \pm 2\cdot2;\quad p_3^2\cos 2\lambda,\ 1\cdot9 \pm 1\cdot0;$$
$$p_3^1\cos\lambda,\ 1\cdot9 \pm 1\cdot1;\quad p_3^3\sin 3\lambda,\ 1\cdot7 \pm 1\cdot0.$$

The uncertainties of the longitude terms are considerable, but are quite likely to be too low, because they are based on the supposition that all other harmonics would contribute at random to the estimates of those retained. There is strong evidence for the existence of some low harmonics other than the constant and p_2 terms, but without observations in some new regions we cannot be sure how much of it is to be attributed to terms of degrees 2 and 3 and how much to those of degrees 4, 5 and 6. Another criticism is that the comparison of some of the base stations with Potsdam was still uncertain, notably those of India and Argentina. Accordingly, the values of the longitude terms may differ considerably from those just given. This does not affect the reality of the τ_1 and τ_2 variations, which are established for variations within single countries, so that possible errors at the base stations do not enter. The comparison of Meinesz's observations with Potsdam is satisfactory.

If the longitude terms are omitted the solution becomes

$$+\ 1\cdot5 \pm 1\cdot8 - (1\cdot4 \pm 4\cdot6)p_2.$$

If they are genuine they imply departures of the geoid from the spheroid of the order of 100 m. They also imply mean square departures of gravity,

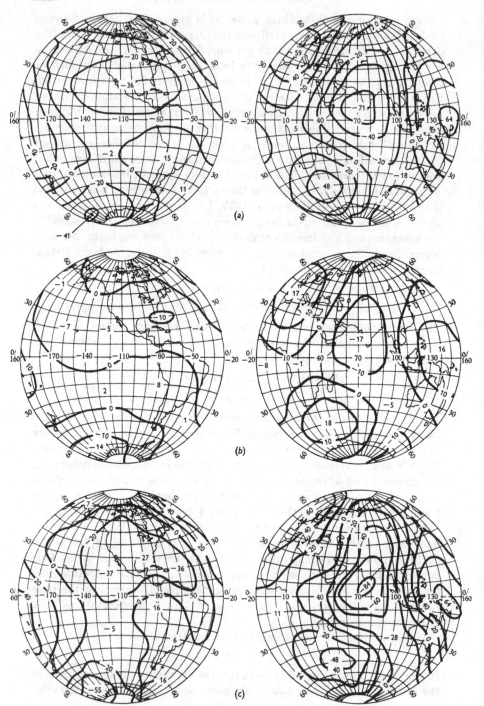

Fig. 21. (For legend see p. 244.)

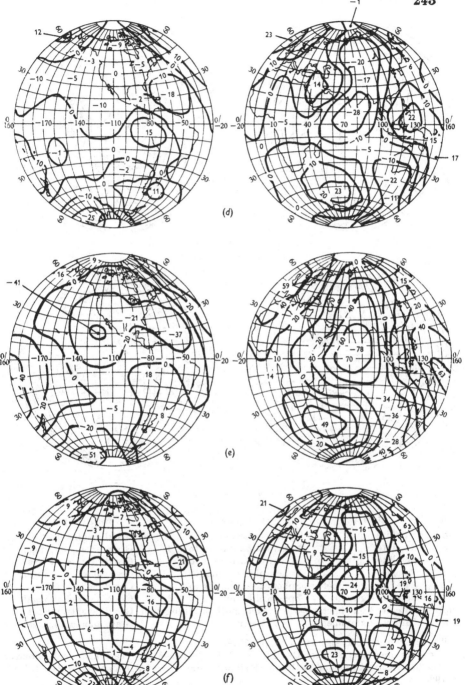

(d)

(e)

(f)

Fig. 21 (cont.).

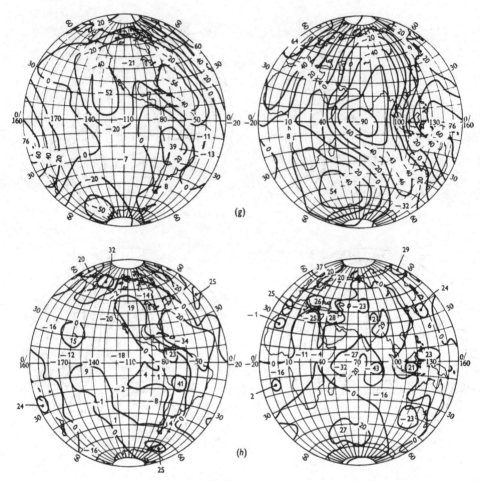

Fig. 21 (cont.).

Map no.	Data	Computation	Represented
(a)	Satellite	Spherical harmonics to 4, 4	Geoid heights, metres
(b)	Satellite	Spherical harmonics to 4, 4	Grav. anomalies, mgal.
(c)	Satellite	Spherical harmonics to 7, 2	Geoid heights, metres
(d)	Satellite	Spherical harmonics to 7, 2	Grav. anomalies, mgal.
(e)	Satellite + gravimetry	Spherical harmonics to 7, 2	Geoid heights, metres
(f)	Satellite + gravimetry	Spherical harmonics to 7, 2	Grav. anomalies, mgal.
(g)	Satellite + gravimetry	5° × 5° Means + Stokes's function	Geoid heights, metres
(h)	Satellite + gravimetry	5° × 5° Means	Grav. anomalies, mgal.

Satellite = Solution C, Tables 4 and 5, pp. 5305–6.
Satellite + Gravimetry = Solution CA, Tables 4, 5, 7, pp. 5305–6, 9 in 'Tests and Combination of Satellite Determinations of the Gravity Field with Gravimetry', *J. Geophys. Res.* **71**, 5303–14, 1966.
(By courtesy of W. M. Kaula, 1967 January 3.)

expressible by low harmonics, of about 46 mgal.[2]. If the Bouguer or isostatic rule held for these harmonics (which are certainly present in the Earth's external form) they would have the same sign as the corresponding terms in the external form, which have been given by A. Prey (1922, Table VII). Actually three out of four have opposite signs. This is a more searching analysis than the comparison of the Atlantic with the adjacent continents given on p. 238. The results may be appreciably in error, but they are good evidence against the coefficients having appreciable values of opposite signs to those found.

If I was doing the work again I should proceed somewhat differently. In the first approximation τ_2 included the variation that I was trying to estimate, and this was removed in the second. It would have been better to analyse from the 10° squares directly and look for correlations between the residuals at different separations, introducing additional uncertainties if any showed up. This would approach the ordinary treatment of internal correlation (Jeffreys, 1967d, pp. 295, 311).

Also the analysis was carried out piecemeal. In the first place estimates were made as if the normal equations had no non-diagonal terms, and terms clearly not significant were dropped. A least squares solution was then made from the survivors. Now that electronic machines are available it would be easy to do the direct analysis for n up to degrees 3 or 4.

I must comment on several later attempts. The number of observations and the comparisons of base stations have greatly improved, and a far better solution than mine is possible. However some results are ruined by bad statistical methods.

First, there is the use of isostatic reductions. A depth of compensation is chosen to make the residuals as small as possible. This is an extra unknown. In fact it usually makes little change in the residuals except in mountainous regions where the free air anomalies are irregular. But it overweights the mountain regions by making their mean values of calculated g agree with the plains, and begs the question whether mountains are a little lighter or heavier than the plains. Further, the reduction makes systematic changes and the external field cannot be reproduced until these are removed. This, as far as I know, is never done.

In my method regions of great irregularity automatically received little weight, but were not totally neglected.

Zhongolovitch (1952, 1957) used a method something like mine and derived 30 solutions. In half of these the simple mean of the observed anomalies in a square is used. Thus the adjustment to the mean height of the square is ignored. In the other half it is made. But of the fifteen solutions only two (one up to third harmonics, the other to fourth) are least squares solutions. Even in these, all squares have received equal weight, regardless of the number and consistency of observations within them. This of course overweights the scanty observations south of latitude 30° S. There are of course many more now in Argentine, S. Africa and Australia. For a detailed criticism see Jeffreys (1961 d, 1963 a).

Hirvonen (1956) used a different system of weighting. He noticed that the correlation of gravity residuals falls off rapidly with distance. 'If there is only one observed station inside a square with side 2°·6 it is better to put the anomaly of the square equal to zero than to use the observed anomaly as representative of the square. However, the best value obviously is the weighted mean of them both (observed value and zero).' But if there is, for instance, a term in $p_2^2 \cos 2\lambda$ it gives positive correlations between observations up to 90° separation. What the argument says is that it is best to assume no correlation at distances over 2°·6, in other words that the widespread harmonics that we want to estimate are absent. In fact, Heiskanen (1957) using this principle gets runs of constant sign in the 10° means for many consecutive squares; the hypothesis is contradicted by the conclusion. If the low harmonics are not ruled out as *a priori* impossible the 'anomaly' must be reinterpreted; it is simply the difference between observed gravity and the trial solution. Hirvonen's weights would give a weighted mean of the trial solution and that given by the observed values alone. Call this solution 1. If we are prepared to make any change from the trial solution, we must take solution 1 as a new trial solution and repeat the process, giving a new set of corrections. Call this solution 2. So long as the observed values alone give larger coefficients than the trial solution, the next approximation will increase them. The solution can repeat itself only when it has become identical with that given by the observed values alone (Jeffreys, 1959 c).

Kaula proposes to use covariances for different, including large, ranges of distance. This is less objectionable than the Hirvonen–Heiskanen method, but it is still true that the covariances would be completely calculable from the coefficients of spherical harmonics if they were known, and any revision of their values would alter the covariances and require a revision. My approach is far more direct.

Some writers have claimed to use many more stations than I did. This is often because they use many stations in a 1° square, whereas I used only one except in a few cases where another was needed to get a good range of height in a 10° square. This is nearly pointless. If we have two sets of measures both of standard error σ, that of the difference is $\sigma\sqrt{2}$. If we reduce one standard error by multiplying observations, we still cannot reduce that of the difference to less than σ. A wide distribution of stations is far more important than their number. But the argument has been accompanied by a bad method of reduction. My method was to get an estimate for a 10° square that would represent its mean free air anomaly as accurately as possible. What has now been done is to fit a form $a + bh$ to each 1° square, reduce to mean height, and take the mean of the results as representative of a 10° or 5° square. Now in a normal region, within a 1° square, the compensation varies little and differences of the free air value arise mainly from the Bouguer term. Many values of b found actually approach this. But over a 10° square compensation is largely effective and the variation with height is much less. The result will be that if, as is usual, the mean height of the 10° square is much more than that of the occupied 1° squares the mean anomaly is grossly underestimated. Several analyses have totally neglected the adjustment to mean height; others have over-estimated it for the same reason. I have not seen a single analysis of gravity since my 1941 papers that has not a systematic error for one or other of these reasons. See E. M. Gaposchkin (ed.) 1974, *Smithsonian Astrophysical Observatory Special Report 353*.

If we had four observations in every 10° square we could ordinarily get a mean for gravity with a standard error in the neighbourhood of 15 mgal. There would be about 600 such squares, and a normalized spherical harmonic could, apart from further complications, be estimated with a standard error of $15/\sqrt{1600} = 0.6$ mgal. The standard errors of the coefficients found from artificial satellites are already less than this, and comparable accuracy from gravity is unlikely to be attained for a long time. Nevertheless, the attempt is worth making, because any additional check on a complicated theoretical investigation is worth while.

5·08. Correction of survey solutions. The geographic latitude and longitude of a place specify the direction there of the normal to the level surface through the place. This surface lies above the co-geoid (except for such places as the Dead Sea), and therefore the apparent horizontal distance between two consecutive survey points differs a little from the distance between corresponding points of the co-geoid. In fact, the survey measures the linear distance, and this differs from the co-geoidal distance in two ways. First, it is inclined at an angle ψ to the level surface; this is corrected by multiplying by $\cos \psi$. Secondly, it is at a distance h above the co-geoid, the ratio of the two horizontal distances being, to the accuracy required, equal to $1 + h/a$. Hence the distance along the co-geoid is $(1 - h/a) \cos \psi \, ds$. The correcting factors are found from the survey measurements themselves and their application involves no difficulty, even in regions where slopes are considerable. They constitute the free-air reduction for this purpose.

The determination of the fundamental constants a, e from surveys involves a further difficulty because they refer to the spheroid. According as the survey is made along a meridian or a parallel, it determines the distance along the co-geoid for a given change of ϕ or λ. If the co-geoid and the spheroid were identical there would be no trouble in principle about con-verting the ratio into an equation connecting a and e; $ds/d\phi$ along a meridian measures the radius of curvature of the meridian, and $ds/d\lambda$ along a parallel in latitude gives the radius of that parallel. Comparison of results in different latitudes will then determine a and e. The original determinations were in fact made in this way. It might appear that the survey of a single long arc of meridian would suffice to give both a and e from comparison of curvatures in different latitudes, but in practice the differences are small and no single arc estimates $1/e$ with a standard error less than about 10, whereas one in the neighbourhood of 1 can be obtained by several other ways. Hayford's success in the United States was essentially due to the fact that the survey included long arcs in both directions, so that he was in a position to get e from comparison of the curvatures of a meridian and a parallel. At any rate if the co-geoid and spheroid were identical, comparison of surveys would lead directly to estimates of both a and e.

Actually the co-geoid and spheroid differ appreciably. Any departure of the gravitational field from that given by the theory of 4·022, which takes them identical, disturbs the direction of the plumb-line and also the intensity of gravity. The former alters the relation between the measured ds and $d\phi$ or $d\lambda$, and the results will not be correct for the spheroid. In India the actual form of the co-geoid has been measured, and it has been found to depart vertically by quantities of the order of 10 m. from the spheroid that agrees best with it (J. de Graaff Hunter, 1932). As this is in a range of latitude of only about 25° a considerable error in the inferred radius might result if the curvatures of the co-geoid were our only data. Fortunately, the analysis of gravity gives valuable supplementary information, which is continually being increased. From gravity the disturbances of external potential U' can be inferred as in 4·026; then the elevation of the co-geoid above the nearest spheroid is $U'/g = \zeta$, say. To the second order in e, for a meridian survey (Jeffreys, 1948b),

$$\frac{ds}{d\phi} = a\{1 - \tfrac{1}{2}e(1 + 3\cos 2\phi) + e^2 - \tfrac{15}{8}e^2\sin^2 2\phi\} + \zeta + \frac{\partial^2 \zeta}{\partial \phi^2}, \quad (1)$$

and for a parallel

$$\frac{ds}{\cos\phi\, d\lambda} = a(1 + e\sin^2\phi + e^2\sin^4\phi - \tfrac{1}{8}e^2\sin^2 2\phi) + \zeta - \tan\phi\frac{\partial\zeta}{\partial\phi} + \sec^2\phi\frac{\partial^2\zeta}{\partial\lambda^2}. \quad (2)$$

Denote the coefficient of ae by w. In practice the values of s, for given ϕ, are given as residuals against values computed with trial values of a and e, which are sufficiently accurate for changes in the terms in e^2 to be neglected. Then the equation given by a survey effectively becomes one for the change of $a(1 + we)$, which can be corrected by allowing for the terms in ζ. If e is kept unchanged it gives an estimate of a, which is said to be reduced to the trial value of e, say e_0. It is an estimate of a for the spheroid of ellipticity e_0 that best fits the data, and knowledge of w for the mean latitude of the arc permits adaptation to other ellipticities. Allowance for the terms in ζ is straightforward if they are known, which is hardly true at present. The low harmonics are now well determined from satellites, but the high ones would require means of gravity at 10° intervals or so all round each survey arc, and at present this is not true for any of the arcs. Accordingly, the best we can do is to continue to treat the 10° and 30° variations of gravity as random variations, whose average amount we know, but not their distribution in detail. This requires an estimate of the average deflexions of the vertical corresponding to them. It is found that such deflexions will introduce into an estimate of the radius from a survey an additional uncertainty represented by standard errors as follows:

Length (°)	10	20	30	40	50	100
σ (metres)	800	420	230	200	160	70

It was actually found that when the solutions for the principal survey arcs were combined, using the apparent uncertainties, they gave an excessive value of χ^2 against the least squares solution, indicating that some disturbance of the observations had not been allowed for. When allowance was made for the 10° and 30° variations of the deflexions of the vertical, however, it was found that they agreed among themselves as well as there was any reason to expect. The allowance also cleared up a discrepancy between the visual and dynamical parallaxes of the Moon. This had given trouble previously (not very serious as it was only 1·4 times the apparent standard error). Other information about the radius and the ellipticity is provided by the analysis of gravity and by the relation connecting e with the precessional constant. The uncertainty of the latter determination has to be corrected to take account of the extent of possible departures from the hydrostatic state, but this can be estimated from the τ_1 and τ_2 variations and the longitude terms of low degree in gravity. The result of combining all the methods, which are found to be satisfactorily consistent (Jeffreys 1948*b*), is

$$a = 6378 \cdot 097 \pm 0 \cdot 116 \text{ km.}, \quad e = 0 \cdot 0033638 \pm 0 \cdot 0000029,$$

$$1/e = 297 \cdot 28 \pm 0 \cdot 25,$$

if we reject the longitude terms in gravity, but becomes

$$a = 6378 \cdot 103 \pm 0 \cdot 116 \text{ km.}, \quad e = 0 \cdot 0033680 \pm 0 \cdot 0000046,$$

$$1/e = 296 \cdot 91 \pm 0 \cdot 40,$$

if we keep these terms. In either case Hayford's value 297·0 for e^{-1} is permissible, but his value of a, 6378·388 km., is definitely too large. As its stated uncertainty was only 0·018 km., increased to 0·053 km. by Helmert (p. 186), an explanation is needed. The chief difference is that I have used the reduction of the measured distance to the co-geoid by the free-air method. Hayford used isostatic reductions and has, I think, thereby introduced a systematic error. These reductions depend on calculating the deflexions of the vertical on the supposition that all the surface inequalities are compensated, and subtracting them from the observed deflexions before using the balance to evaluate a and e. The argument in favour of this method would be that surveys are made on land, and it is not certain how far results so obtained are representative of the sea. We get in fact an estimate of the radius of curvature of the co-geoid along the route. If the surface inequalities are excess loads, and to a smaller extent if they are isostatically compensated, the co-geoid will be raised with respect to the spheroid in land areas and depressed below it in oceanic ones. Consequently the radius of curvature of the co-geoid in a land area will be less than over the ocean, and correction of the deflexions on the Bouguer hypothesis or that of isostasy, as the case may be, will remove the systematic difference and make the surveys representative of the spheroid. The trouble is that both hypotheses are wrong, since they lead to wrong conclusions about the average difference of gravity between land and sea. This is, in fact, negligible and shows that the supposed difference of curvature does not exist, and correction for it has introduced systematic error instead of removing one.

It must be emphasized that the difference is not in the observational material. Hayford gave several other solutions besides that based on the isostatic hypothesis, and his final values on the free-air hypothesis were (1910, p. 62)

$$a = 6378 \cdot 062 \pm 0 \cdot 036 \text{ km.}, \quad e^{-1} = 298 \cdot 2 \pm 1 \cdot 0.$$

The uncertainties of course need an increase. But Hayford's data provided three important equations in my solution.

The use of isostatic reductions of gravity before using the observations to estimate the ellipticity introduces other systematic errors. In the first place, the older reductions took no account of the term $2U'/a$ in the reduction of gravity to the co-geoid and consequently introduce the impossible first harmonics.* (See 5·06.) This term has been included in more recent reductions. But even when this is done, the reductions remove the effects of topography and compensation from gravity and therefore give the values of gravity over a body whose density distribution differs from that of the Earth. Hence the use of the reduced values will introduce systematic error into calculations of the external field. Theoretically it would be possible to calculate the corrections over the whole Earth and by applying them to the results in the opposite sense to correct the error; but this has not been done and in any case would only reproduce the free-air values at places where data exist.

Many geodesists maintain that isostatic reductions are desirable because the residuals vary more smoothly than the free-air ones, and consequently average values over a region are more likely to be representative. This, if true, would be really a reason why they should not be used; for the inevitable systematic error will be combined with an underestimate of uncertainty. In fact, the free-air residuals within a 10° square show a systematic variation with height; but when this is found and used to reduce the mean at the stations to the average height of the square, the outstanding residuals are on an average nearly the same as on the isostatic hypothesis.

* J. Goguel (1951) has shown that on an isostatic theory this would happen unless the density of surface rocks is exactly $\frac{2}{3}$ of the mean density. It would be interesting to know the corresponding result for the second harmonics corresponding to zx, zy.

Hunter (1935) has discussed in detail the use of gravity observations and Stokes's formula to make the measurement of geodetic arcs more nearly comparable among themselves and hence to reduce the uncertainty of a. He finds, among other results, that with one gravity station per 5° square over the surface it would be possible to determine a to 10·4 m. (probable error). With one per 10° square I estimate that the coefficient of p_2 in gravity could be found to 2 mgal. This would be a practical possibility.

One common misconception needs correction at this point. In the standard methods of calculation of isostatic reductions a constant density of the surface rocks is assumed. It may be inferred from this that the residuals after the reductions have been applied may arise from local variations of density. This is not so for inequalities whose horizontal extent is more than a few times the depth of compensation. If the mass per unit area is kept constant by compensation, the effect of specially great density for a surface projection of given height will be a specially great depression of the layer of compensation, and the two disturbances of gravity will still nearly cancel. For widespread inequalities of height the remaining disturbance will vary only in the ratio of the densities assumed, and these can hardly lie outside the range 2–3 g./cm.³. The possibility of errors in the adopted densities of the outer rocks does not affect the main conclusion that if isostasy was true mean free-air gravity over wide regions would be noticeably positively correlated with height; and it is not.

We can use the actual variation of gravity to estimate the distribution of mass per unit area. So long as the horizontal extent is large compared with the depth of compensation we can again use the Bouguer formula to estimate the variation of total mass per unit area; for 100 mgal. of gravity over a region we can infer that the extra mass per unit area is at least equivalent to that of 1 km. of rock of density 2·5 g./cm.³; if placed at a great depth more would be needed.

This has an immediate application to the state of stress, for the weight of any extra load must be supported by stresses in the interior. The pressure at sea-level where the height is 1 km. above normal will be g times the extra mass per unit area, that is, about $2·5 \times 10^8$ dynes/cm.², which is comparable with the strengths of most rocks. But if we find a free-air gravity anomaly of 100 mgal., whatever the local height may be, we can still infer that an excess vertical stress in this region would have to be of this amount to support the associated load. We shall discuss the support of surface inequalities more fully later, but it is worth while to call attention at once to the relation between mass per unit area, gravity anomalies, and variation of stress across level surfaces.

5·09. The geoid. In this chapter and the last I have used Hunter's term 'co-geoid' for the locus of points at depths below the outer surface equal to the measured height. This corresponds to the usual practice in geodesy.

Geodesists usually define a geoid as a level surface defined within the land by the condition that the geopotential has the same value as on the ocean surface. Consequently its distance from the centre depends on the density of the matter above it, and gravity in a small cavity on the geoid would not be equal to gravity reduced according to the free-air formula; for the matter above the geoid is attracting downward at the outer surface but upward on the geoid. It appears to me that this definition leads only to many needless complications. The only purpose of any geoid is to serve as a standard of reference in describing the topography and the external field; observations on it are impossible, and it is in practice only a mathematical intermediary. It is plainly undesirable that the definition of such an intermediary should depend on assumptions about the density distribution; for if the work is done correctly (it has not always been) any effect due to them must cancel when the results for the geoid are used to calculate anything for the outer surface.

The difference between the geoid and co-geoid is in fact of the second order. For if Ψ_c, g_c are the values of Ψ, g at the outer surface we have to the second order in variations of r

$$\Psi = \Psi_c - g_c\,\delta r + \frac{1}{2}\frac{\partial g}{\partial r}\,(\delta r)^2,$$

whether δr is positive or negative (g being continuous in crossing a finite discontinuity of density); with the proviso that we must use a value of $\partial g/\partial r$ at points just outside or just inside according as δr is positive or negative. These two values will differ by about $4\pi f\rho$. If we take δr for different points of the surface in such a way as to make Ψ constant, the effect of the last term according to the sign of δr will differ only by a quantity of the order of $(\delta r)^2$ and is consequently of the order of quantities we have neglected.

In the reduction of survey data, I think that it is usually assumed that the measured heights are heights above the geoid; but if this was so a correction for deformation of the geoid by the attraction of neighbouring matter (depending on the assumed density) would be applied in the process. This does not appear to be done; that is why I say that the co-geoid corresponds to the practical geoid and the geoid does not. If it was done the correction would have to be taken out again if the theory of the external field was taken to the second order in the elevation.

5·10. General adjustment of the figures of the Earth and Moon. Various results suggesting slight corrections to the adopted values of the constants in the theory have been mentioned above. In addition, the absolute value of gravity has recently been redetermined at Teddington and Washington, and it appears that the standard value for Potsdam is about 13 mgal. too high. In a general adjustment of the data (1948b) I got the following results:

$a = 6378 \cdot 099 \pm 0 \cdot 116 \, \text{km.}$,

$e = 0 \cdot 0033659 \pm 0 \cdot 0000041$,

$e^{-1} = 297 \cdot 10 \pm 0 \cdot 36$,

$m = 0 \cdot 003449787 \pm 0 \cdot 000000063$,

$g_1 = 979 \cdot 7565 \pm 0 \cdot 0021 \, \text{gal.}$,

$g = g_0(1 + \beta \sin^2 \phi + \gamma \sin^2 2\phi)$,

$g_0 = 978 \cdot 0373(1 \pm 0 \cdot 0000024)$; $\quad \beta = 0 \cdot 0052891 \pm 0 \cdot 0000041$;

$\quad \gamma = -0 \cdot 0000059$,

$M/M' = 81 \cdot 278 \pm 0 \cdot 025$; $\quad H = 0 \cdot 00327260 \pm 0 \cdot 00000069$;

sine lunar equatorial parallax $= 3422 \cdot 419'' \pm 0 \cdot 024''$,

$J = 0 \cdot 0016370 \pm 0 \cdot 0000041$,

$L' = 0 \cdot 0003734 \pm 0 \cdot 0000016$,

$K' = 0 \cdot 000070 \pm 0 \cdot 000034$.

I consider these values better than those then in use, though some of the uncertainties given are larger; the uncertainties given for some of the current adopted values were too low. Where observational data are to be given as residuals against a standard formula, however, the present standard should be retained until international agreement is obtained to adopt some new one. It is much easier to combine sets of observations that have been reduced according to a uniform standard than sets reduced according to several different standards; the former can be done in one stage but the latter cannot. It should be noticed, however, that the dimensions taken in astronomy and geodesy are different, as D. Brouwer has emphasized, and so long as the current standards are maintained there will always be difficulties when need arises to compare astronomical and geodetic data.

As already stated, improvements have been made recently in nearly all the data on the constants related to the Earth and Moon. A complete revision has now been carried out. Some of the principles may be described as follows. Neglecting many small corrections, we have, if n, n' are the mean motions (angular velocities) of the Moon and Sun, and a, a' the mean distances,

$$n^2 a^3 = f(E + M), \quad n'^2 a'^3 = f(S + E + M).$$

Hence
$$\left(\frac{a'}{a} \right)^3 = \frac{S + E + M}{E + M} \left(\frac{n}{n'} \right)^2.$$

n and n' being extremely accurately known, a knowledge of $(E + M)/S$ is equivalent to one of a/a'. Study of the perturbations of the other planets by the Earth and Moon determines $(E + M)/S$. The most relevant for this purpose is the asteroid Eros, which comes closest to the Earth of those often observed.

The acceleration of the Moon due to the Earth is $an^2 E/(E + M)$. If R

denotes the Earth's radius this should be equal to gR^2/a^2. Thus we have an equation connecting R, a, and M/E.

The Moon's monthly motion is associated with one of the centre of the Earth in an ellipse with major semiaxis $Ma/(E+M)$, about two-thirds of the Earth's radius. This displacement gives a parallactic effect of monthly period on the apparent position of a near planet, especially Eros near opposition. This can be measured in the same programme as that for the determination of the solar parallax from the same planet. Ratios of distances between planets and the Sun being well known, this determines

$$Ma/(E+M)a'.$$

Thus we have three equations, which effectively determine R/a, a/a', and M/E without direct appeal to the observational estimate of R/a'. The latter is made in two ways. Comparison of apparent declinations of Eros at observatories in the northern and southern hemispheres, whose distance apart is known, gives one. The rotation of the Earth produces a diurnal effect for observations at the same station. The latter appeared more satisfactory. However the dynamical value of R/a' does not agree with the observational one. Spencer Jones's observational value of the solar parallax was $8\cdot7888'' \pm 0\cdot0011''$. Rabe's dynamical determination (1950) was $8\cdot79835'' \pm 0\cdot00039''$. (See pp. 209–10.) Other checks are available. Brouwer (1950) determined from the parallactic inequality of the Moon

$$8\cdot7981'' \pm 0\cdot0026'' - 0\cdot180''z$$

(probable error), where

$$E/M = 81\cdot53(1+z).$$

z is about -2×10^{-3}, so that it hardly matters. Brouwer also quotes spectroscopic determinations (i.e. from the Doppler effect on stars) of $8\cdot803'' \pm 0\cdot0004''$ (Spencer Jones, 1928) and $8\cdot805'' \pm 0\cdot007''$ (W. S. Adams, 1941) and $8\cdot796''$ and $8\cdot809''$ from the constant of aberration. On the other hand Kulikov, in the same publication as Brouwer's, gave $8\cdot788'' \pm 0\cdot001''$ from the constant of aberration.

The Doppler effect and the constant of aberration both measure the ratio of the Earth's velocity relative to the Sun to the velocity of light. Both are therefore annual effects. But aberration is subject to systematic errors of meteorological origin, and determinations of the parallax from it are open to some suspicion. The spectroscopic determinations, however, should be reliable. The Earth's velocity relative to the Sun being about 30 km./sec., an error of 1 in 1000 would imply an error in the velocity of 30 metres/sec. Meteorological effects would be far below this!

It appeared therefore that all the most reliable checks confirmed Rabe's parallax rather than Spencer Jones's. Adopting it made (Jeffreys and Vicente, 1957a, b)

$$E/M = 81\cdot356, \quad H = 0\cdot00327468.$$

A different approach is now made by using recent determinations of the velocity of light, which is now known to a few parts in a million. The mean distance of the Moon can be determined from the time taken by a reflected radio wave, and is $384402 \cdot 0 \pm 2 \cdot 1$ km. This depends somewhat on the diameter of the Moon and especially that in the semiaxis towards the Earth, which is likely to be about 1 km. longer than the polar one.

This was first done by Hey and Hughes (1959) and repeated by Bruton, Craig and Yaplee (1959). A compromise between the visual and dynamical parallaxes, $3422 \cdot 571'' \pm 0 \cdot 047''$, given by Yaplee and others, led to the value $6378 \cdot 125 \pm 0 \cdot 008$ km. for the equatorial radius of the Earth.

The last, however, needs further modification, since observed gravity enters into the dynamical parallax. The more direct way is to use the Moon's mean distance and mean motion to estimate fE. Then for an artificial satellite we have the mean motion and the range of height above the surface of reference. But knowing fE and the mean motion we can calculate the major semiaxis; subtracting the mean height above the surface we are left with the radius of the Earth. (Corrections have, of course, to be made for the ellipticity.)

Calculated mean gravity differs from observed mean gravity by about 4 mgal. This is not serious. The standard of reference for the absolute value of gravity is Potsdam, but several recent determinations of absolute gravity have been made. One particularly interesting method, initiated by C. Volet in Paris, is to time a freely falling body. These can be compared with Potsdam by means of transported pendulums. An account is given by A. Thulin (1960). The results are expressed as corrections to the Potsdam standard. The odd thing is that determinations in the Eastern hemisphere mostly indicate corrections of about -12 mgal., those in the western about -15 mgal. (A. H. Cook 1965).

There is no known explanation of the difference. This is emphasized particularly by A. H. Cook. Cook himself (1967) has made a new determination at Teddington, $981181 \cdot 75 \pm 0 \cdot 13$ mgal., by means of a freely falling body. This looks the most accurate yet made, and lies about midway between the two sets.

The error in the Potsdam value is not experimental; it is known to be one of reduction, being due to the correction of a systematic error that did not exist. A revision (Jeffreys, 1949c) gave a correction $-10 \cdot 7 \pm 2 \cdot 2$ mgal. to the adopted value.

In addition there is the uncertainty of the comparisons in each survey of field stations with the base station and of the base station with Potsdam. It would therefore be quite conceivable that average gravity over the Earth as yielded by analysis of observations should be in error by 4 mgal.

The geodetic data have been improved, since there is now a connexion all the way from Scandinavia to South Africa. Other new arcs are available. The results have been analysed by I. Fischer (1959, 1960) and W. M. Kaula, whose final conclusion is $a = 6378 \cdot 165 \pm 0 \cdot 025$ km. It did not appear that allowance had been made for departures of the co-geoid from a spheroid, discussed in § 5.08. Fischer informs me in a letter that she uses astro-geodetic deflexions of the vertical, defined as the angle between the vertical to the geoid and the normal to the reference spheroid. This is used to calculate the geoidal slope by numerical integration. This process is equivalent to the one I suggest in § 5·08. Accounts of some of the surveys used have not yet been published. There is no doubt, however, that Hayford's a is too large.

Timing of radar reflexions from Venus and Mercury has given a new estimate of the mean distance of the Sun, and the Doppler effect has determined their velocities. These agree in indicating a solar parallax about 8·794″, agreeing with the revised determination of Rabe and Francis (p. 210).

5.11. Artificial satellites. The orbit of an artificial satellite is primarily a Keplerian one about the Earth's centre, but is modified in several ways. Far the most important is due to the J term in the Earth's field. The period of revolution of a satellite close to the surface is about 90 minutes. But the motion is not in a fixed plane. The node of the orbit on the equator moves backwards (i.e. opposite to the satellite's motion in longitude), completing a revolution in something of the order of three months. The perigee moves forwards (relative to the node) for inclinations I such that $\sin^2 I < \frac{4}{5}$, backwards for greater inclinations. These changes accumulate so long as the satellite remains in orbit and are accurately measurable. It is necessary to retain J^2 in the calculation; this was first done by D. G. King-Hele for the node. All even zonal harmonics affect these speeds, but their effects depend on I, and hence it is possible to determine them separately by comparing satellites with different inclinations. In general the node is more useful because it can be more accurately measured than the perigee.

It was also noticed by O'Keefe that as the perigee moves the eccentricity alters, being greatest when the perigee is in the northern hemisphere. This is explained by the presence of odd zonal harmonics and has led to good estimates of terms in p_3 and p_5. There are now some satellites at such elevations that their periods approximate to a sidereal day, or an integral fraction $(1/s)$ of it. The perturbations by terms containing $\cos s\lambda$ can then become large.

At large heights perturbations by the Sun and Moon cease to be quite negligible; and since most satellites are illuminated for half a day effects of solar radiation pressure are appreciable.

Harmonics with $s = 0$ produce some effects that accumulate so long as the body remains in orbit and others for half or a quarter of the period of revolution of the perigee. It appeared at first that any effect of a harmonic with $s \neq 0$ would change sign at least twice during every revolution and always remain small. However the inclination of the orbit to the equator introduces complications. If a harmonic $p_n^s (\sin \phi') e^{is\lambda}$ is transformed to axes such that one is normal to the orbital plane, and the inclination is I, it is found that the result is of the form $\Sigma C_{r,m,s} p_n^m e^{im\lambda}$ referred to the new axes, where m ranges from $-n$ to n (B. Jeffreys, 1965). Thus, when referred to the mean orbital plane, the potential for $I \neq 0$ contains terms with $m = 0$ and ± 1. The periods of revolution of most artificial satellites are about $1\frac{1}{2}$ hours, and thus short compared with a day.

If $n - m$ is odd, $p_n^m (\sin \phi')$ contains a factor $\sin \phi'$, and thus vanishes on the plane of reference, but its derivative normal to the plane does not. If $n - m$ is even p_n^m does not contain this factor and is stationary for displacements normal to this plane. Hence for even n the term $m = 0$ gives a perturbation of long period in the longitude in orbit (a day being a long period for this purpose) and those with $m = \pm 1$ terms of long period in the node and inclination. For odd n the term $m = 0$ gives long period terms in the node and inclination, those with $m = \pm 1$ in the eccentricity and longitude of perigee. All of them acquire large amplitudes since they have divisors of order ω/n (sometimes $(\omega/n)^2$), where $2\pi/\omega = 1$ sidereal day and $2\pi/n$ is the orbital period.

All the perturbations depend greatly on the inclination, in different ways, and hence the solution requires comparison of satellites with different inclinations.

The essential point is that for a particle in a circular orbit under a potential fE/r the fundamental speeds for free oscillations about the steady motion are two each of $0, n, -n$. The two of speed 0 represent constant changes in the mean distance and the longitude of the node. If a disturbing potential introduces forces near these speeds the corresponding displacements will acquire small divisors by resonance.

It is for this reason that, though most of the tesseral coefficients in the potential are of the order of 10^{-6} of the main term, some of the displacements due to them are of the order of 1 km., equivalent to $1\cdot5 \times 10^{-4}$ in circular measure. King-Hele and his collaborators express the potential in the form

$$U = \frac{fE}{r} \left\{ 1 - \Sigma J_n \left(\frac{a}{r}\right)^n p_n \right\}.$$

Determinations of the even zonal harmonics gave inconsistencies for $n \geqslant 8$. King-Hele and Cook (1965) point out that although the solutions differ appreciably, the corresponding potentials differ little for latitudes more than about $20°$, and suggest the need for a few satellites with inclinations less than this. I think the point is that the higher even zonal harmonics, near the equator, behave very similarly; and the zeros nearest the equator are at not very different latitudes. Consequently they are difficult to separate. Their recommended solution is

$$10^6 J_2 = 1082\cdot64 \pm 0\cdot02,$$
$$10^6 J_4 = -1\cdot52 \pm 0\cdot03,$$
$$10^6 J_6 = +0\cdot57 \pm 0\cdot07,$$
$$10^6 J_8 = +0\cdot44 \pm 0\cdot11.$$

The uncertainties are based on the assumption that higher harmonics are negligible.

King-Hele, Cook and Scott (1967) give a detailed discussion of the odd zonal harmonics. For these the difficulty about separation of higher harmonics is less; they give two solutions as follows. They are based on 'seventeen satellites chosen to give the widest and most uniform distribution in inclination and major semiaxis':

$$10^6 J_3 = -2\cdot53 \pm 0\cdot02, \quad -2\cdot50 \pm 0\cdot01,$$
$$10^6 J_5 = -0\cdot22 \pm 0\cdot04, \quad -0\cdot26 \pm 0\cdot07,$$
$$10^6 J_7 = -0\cdot41 \pm 0\cdot06, \quad -0\cdot40 \pm 0\cdot02,$$
$$10^6 J_9 = +0\cdot09 \pm 0\cdot06, \quad 0 \pm 0\cdot06,$$
$$10^6 J_{11} = -0\cdot14 \pm 0\cdot05, \quad -0\cdot27 \pm 0\cdot06,$$
$$10^6 J_{13} = +0\cdot29 \pm 0\cdot06, \quad +0\cdot36 \pm 0\cdot08,$$
$$10^6 J_{15} = -0\cdot40 \pm 0\cdot06, \quad -0\cdot65 \pm 0\cdot10,$$
$$10^6 J_{17} = \qquad\qquad +0\cdot30 \pm 0\cdot08,$$
$$10^6 J_{19} = \qquad\qquad 0 \pm 0\cdot11,$$
$$10^6 J_{21} = \qquad\qquad +0\cdot58 \pm 0\cdot11.$$

The solution for 7 coefficients gave $\chi^2 = 7\cdot7$ on 8 degrees of freedom; that for 10 gave $\chi^2 = 4\cdot8$ on 4 degrees of freedom. (Two satellites were omitted because they would give little additional information on these harmonics, though they may for still higher ones.) As for the even ones the standard errors depend on the assumption that higher harmonics are negligible.

The greatest departure from pure gravitational theory is due to air resistance. Its chief effect is near perigee and the reduction of velocity there reduces the distance at apogee till the orbit is nearly circular. Then it acts nearly equally all round the orbit and brings the satellite rapidly to the ground. The effect is complicated by the fact that the air is in motion, with a prevailing westerly wind at great heights. This produces a secular change of inclination. Fortunately the changes of inclination and node affect observations in nearly independent ways, and the node retains its usefulness for determining the zonal harmonics and the inclination for studying winds at great heights.

There has been so much work on the subject that it is impossible to mention more than a small fraction of the relevant papers. Important contributors have been D. Brouwer (esp 1959), Y. Kozai, I. Izsak, D. G. King-Hele, R. H. Merson, R. J. Anderle, E. M. Gaposchkin, G. Hori (1960), and many others. I quote from a list given by W. M. Kaula (1966) based on comparison of several solutions. The harmonics are normalized so that the mean square is 1.

Zonal coefficients $C_n^0 \times 10^6$:

n	
2	$-584\cdot17 \pm 0\cdot01$
3	$0\cdot97 \pm 0\cdot02$
4	$0\cdot54 \pm 0\cdot02$
5	$0\cdot04 \pm 0\cdot02$
6	$-0\cdot18 \pm 0\cdot02$
7	$+0\cdot01 \pm 0\cdot02$

Tesseral coefficients C_n^s, $S_n^s \times 10^6$ (CA of Kaula's table):

C_{22}	$+2\cdot42$		S_{43}	$-0\cdot19$
S_{22}	$-1\cdot36$		C_{44}	$-0\cdot06$
C_{31}	$1\cdot79$		S_{44}	$+0\cdot32$
S_{31}	$0\cdot18$		C_{51}	$0\cdot00$
C_{32}	$0\cdot78$		S_{51}	$-0\cdot02$
S_{32}	$-0\cdot75$		C_{52}	$0\cdot44$
C_{33}	$0\cdot57$		S_{52}	$-0\cdot28$
S_{33}	$1\cdot42$		C_{53}	$-0\cdot31$
C_{41}	$-0\cdot56$		S_{53}	$0\cdot03$
S_{41}	$-0\cdot46$		C_{54}	$0\cdot02$
C_{42}	$0\cdot30$		S_{54}	$0\cdot11$
S_{42}	$0\cdot60$		C_{55}	$0\cdot10$
C_{43}	$0\cdot92$		S_{55}	$-0\cdot49$

Results from satellites are given for all C_n^s, S_n^s to $n = 6$. A complete list up to $n = 12$, $s = 12$, is given by combination with gravimetry. In comparison with my 1943 determinations from gravity, C_{22} and C_{32} are smaller, C_{31} and S_{33} about the same; but the standard errors, mostly, apparently, about $0\cdot02$, are far smaller. All agree in sign. The most recent determination is by Gaposchkin and Lambeck (1970, 1971).

5·12. Adopted values of constants. The following list of fundamental constants, many of which are relevant to geophysics, has been adopted by the International Union of Astronomy.* See also Kovalevsky (1965), Wilkins (1965). The main object of the system is consistency; values of the constants used in dynamical and positional astronomy have not always been consistent. Adoption of different values as standard in different portions of the theory has led to difficulties in the interpretation of observations. A secondary object is to make the constants agree satisfactorily with those that appear to be the most accurate. Besides the suggested standard values, estimates of the true values and their uncertainties are given.

* *Trans. I.A.U.* **12** B (Hamburg, 1964), 591–628.

Defining constants
Ephemeris seconds in 1 tropical year: 31 556 925·9747.
Gaussian gravitational constant k, defining A.U. (astronomical unit of length) by (AU) (ephemeris day)$^{-1}$ (Sun's mass)$^{-\frac{1}{2}}$.
$k = 0·017 202 098 95$.

Primary constants
1 A.U. $= A = 149 600 \times 10^6$ m.
Velocity of light: $299 792·5 \times 10^3$ m./sec.
Equatorial radius of Earth: $a = 6 378 160$ m.
$J_2 = \frac{2}{3} J = 0·001 082 7$.
$fE = 398 603 \times 10^9$ m.^{3} sec.$^{-2}$.
$\mu = M/E = 1/81·30 = 0·012 3001$.
Sidereal mean motion of Moon (1900):

$$n = 2·661 699 489 \times 10^{-6} \text{radians/sec.}$$

Precession in Longitude (1900):

$$p = 5025·64''/\text{tropical century.}$$

Obliquity of ecliptic (1900): $23° 27' 8·26''$.
Nutation constant: $\qquad N = 9·210''$.

Auxiliary constants
$k/86 400 = k' = 1·900 983 675 \times 10^{-7}$.
Radian/$1'' = 206 264·806$.
Factor for constant of aberration: $F_1 = 1·000 142$.
Factor for mean distance of Moon: $F_2 = 0·999 093 142$.
Factor for parallactic inequality: $F_3 = 49 853·2''$.

Derived constants
Solar parallax: $\pi_\odot = 8·79 405''$.
Light time for unit distance: $\tau_A = A/c = 499·012^s$.
Constant of aberration: $F_1 k' \tau_A = \kappa = 20·4958''$.
Ellipticity of Earth: $0·003 352 9 = 1/298·25$.
$fS = A^3 k'^2: 132 718 \times 10^{15}$.
S/E: $\qquad 332 958$.
$S/(E + M)$: $328 912$.
Perturbed mean distance of Moon:

$$F_2 \{ fE(H\mu)/n_{\mathbb{C}}^{*2} \}^{\frac{1}{3}} = a_{\mathbb{C}} = 384 400 \text{ km.}$$

Constant of sine parallax of Moon: $a/a_{\mathbb{C}} = \sin \pi_{\mathbb{C}}$; $\pi_{\mathbb{C}} = 3422·608''$; (1 radian) $\times \sin \pi_{\mathbb{C}} = 3422·451''$.
Constant of lunar inequality:

$$\frac{\mu}{1+\mu} \frac{a_{\mathbb{C}}}{A} = L = 6·43 987''.$$

Constant of parallactic inequality:

$$F_3 \cdot \frac{1-\mu}{1+\mu} \frac{a_{\mathbb{C}}}{A} = P_{\mathbb{C}} = 124 \cdot 986''.$$

Earth's angular velocity: $\omega = 0 \cdot 000\,072\,921$ radians/sec.

Mass of atmosphere: $\mu_a E$, $\mu_a = 10^{-6}$.

$m = a_e \omega^2 / g_e$.

$e = \frac{3}{2}J_2 + \frac{1}{2}m + \frac{9}{8}J_2^2 + \frac{15}{28}J_2 m - \frac{39}{56}m^2$.

$g_e = \dfrac{fE}{a_e^2}(1 - \mu_a + \frac{3}{2}J_2 - m + \frac{27}{8}J_2^2 - \frac{6}{7}J_2 m + \frac{47}{56}m^2)$.

The latest international gravity formula is (AIG, 1967):

$g = 978 \cdot 0318(1 + 0 \cdot 005\,302\,4\sin^2\phi - 0 \cdot 000\,005\,9\sin^2 2\phi)$ gal. and by comparison with § 4·025 (15).

$g_{1967} - g_{1930} = (-17 \cdot 2 + 13 \cdot 6\sin^2\phi)$ mgal.

The true values (apart from some powers of 10) are 'believed to lie' in ranges as follows:

$A = 149\,599 \pm 2,$	$\kappa = 20 \cdot 4957 \pm 0 \cdot 0003,$
$c = 299\,792 \cdot 5 \pm 0 \cdot 5,$	$fS = 132\,716 \pm 5,$
$a_e = 6\,378\,160 \pm 80,$	$S/E = 332\,951 \pm 16,$
$J_2 = 0 \cdot 0010\,826 \pm 3,$	$S/E(1+\mu) = 328\,906 \pm 16,$
$\mu^{-1} = 81 \cdot 30 \pm 0 \cdot 01,$	$e^{-1} = 298 \cdot 26 \pm 0 \cdot 06,$
$p = 5026 \cdot 65 \pm 0 \cdot 25,$	$a_{\mathbb{C}} = 384\,400 \pm 1,$
$\epsilon = 23° 27' 8 \cdot 26'' \pm 0 \cdot 10'',$	1 radian $\times \sin \pi_{\mathbb{C}} = 3422 \cdot 450'' \pm 0 \cdot 053,$
$N = 9 \cdot 205'' \pm 0 \cdot 005'',$	$L = 6 \cdot 4399 \pm 0 \cdot 0099,$
$\pi_{\odot} = 8 \cdot 79\,411 \pm 23,$	$P_{\mathbb{C}} = 124 \cdot 986 \pm 0 \cdot 003.$
$\tau_A = 499 \cdot 008 \pm 0 \cdot 008,$	

Planetary masses (reciprocals, Sun $= 1$)

	Adopted	Best value, suggested by Clemence
Mercury	6,000,000	6,110,000 ± 40,000
Venus	408,000	408,539 ± 12
Earth + Moon	329,390	328,906 ± 6
Mars	3,093,500	3,050,000 ± 50,000
Jupiter	1,047·355	1,047·41 ± 0·02
Saturn	3,501·6	3,499·6 ± 0·4
Uranus	22,869	22,930 ± 6
Neptune	19,314	19,070 ± 21
Pluto	360,000	400,000 ± 40,000

The present values of planetary masses used in calculating ephemerides are proposed to be retained, since some great calculations are partly done. In particular $E + M$ differs from the value proposed for general adoption.

The nutation constant proposed (for the 18·6 year nutation in obliquity) continues to be that of Newcomb, and the associated theory takes the Earth to be a rigid body. On this hypothesis it could be calculated from the rate of precession, but the calculated value is higher and the observed value lower. The difference is fairly well explained by allowance for fluidity of the core and elasticity of the shell.

The definition of m is different from that used in this book (p. 182). e and g_e are expressed directly in terms of J_2 and m. This seems to make unnecessary computation; with the definition of m used in chapter 4, and if e is used as an auxiliary constant all second-order terms are multiples of e (Jeffreys, 1964 b). This recommendation disagrees with Cook (1959 b).

The present value of the rate of precession is to be retained; altering it would entail complete revision of the proper motions of stars. But adoption of a corrected value would entail a slight change in the constant $(C - A)/C$ for the Earth.

The second of time is defined in terms of the tropical year rather than of the rotation of the Earth because the year fluctuates less.

Values of g according to the Pizzetti formula, with the new standard constants, are given by M. Caputo and L. Pieri (1968) at 1' interval. See also Caputo (1965, 1967).

CHAPTER VI

Stress-Differences in the Earth

'As regards Chou-hong, who is stated to be denying that the
Earth is upheld by a celestial tortoise, ...'
ERNEST BRAMAH, *The Moon of Much Gladness*

6·01. Existence of stress-differences. The easiest way to see that
there must be stress-differences in the Earth is to consider the consequences
of supposing them absent, that is, of supposing the Earth to be a perfect
liquid. As a matter of ordinary experience, irregularities of the surface of
a liquid do not persist if left to themselves, but give rise to waves travelling
out in both directions. The theory of these waves is discussed in works on
hydrodynamics. We need not enter into it fully here; the main result is
that on a uniform deep liquid an elevated region 100 km. in width would be
drastically altered in appearance in a few minutes. Even the basin of the
Atlantic would disappear in about half an hour. If we allow for heterogeneity the results are somewhat altered (Lamb, 1932, pp. 370–372). If
a layer of liquid rests on another of higher density, two types of harmonic
waves can travel without change of form. In one the displacements of the
outer surface and the interface are nearly equal; in the other they are of
opposite signs, and for waves long compared with the depth of the upper
liquid the isostatic condition of uniformity of mass per unit area is nearly
maintained. The latter type travel more slowly; but in any case the conclusion is that the Earth's main surface features would be destroyed in
a day at most. How long it would take the waves to be damped out by viscosity is irrelevant for our purposes; no mountains and mountains travelling
round the Earth in a few days are equally contradicted by the facts.

Consequently there must be stress-differences. A simplification is possible
because in an elastic solid the free periods are even shorter than in a liquid.
Their order of magnitude for the Earth is indicated by the times taken by
PKP to travel to the anticentre and back, about 40 min., and for *SSSS* to
travel round the Earth, about 100 min. For disturbances lasting longer
than this we can neglect the acceleration terms in the equations of motion
in comparison with the stress terms, and thus use a statical theory. The
equations in the interior are then derived from

$$\frac{\partial p_{ik}}{\partial x_k} + \rho X_i = 0, \tag{1}$$

where p_{ik} are the stress-components, ρ is the density, and X_i the bodily
force per unit mass. This is usually gravitational; if the gravitation potential
is U we have
$$X_i = \partial U / \partial x_i. \tag{2}$$

The outer boundary is free, so that the stress-components across it are all zero.

6·02. Principles of estimation of stress-differences. As six stress-components are independent and (1) are only three equations connecting them, there is a threefold infinity of solutions that will satisfy the conditions. We can make the problem definite in three ways. The first, initiated by Darwin, is to suppose that the interior was originally in a state of hydrostatic stress and has been deformed elastically by loads applied to the surface. Then we have also in the interior the elastic relations between the additional stress and strain. This is essentially a Bouguer theory corrected for elasticity. It can be modified to take account of possible weakness below a certain depth by supposing hydrostatic stress below this depth, and then gives something resembling isostasy. An objection to this method is that mountain systems have not been formed in the way assumed; the large faults and thrusts in them are inelastic displacements. But denudation and redeposition will give elastic displacements until they become too great. The second method is to assume some special method of formation and work out its consequences, and this procedure may become of great importance in the future. At present, however, the methods of formation are hardly known in sufficient detail to make the method repay its great mathematical difficulty. The third method is to restate the problem. For any distribution of stress consistent with the surface load, there is some place where the stress-difference is greatest, say S_1. Then for this distribution we can say that the stress-difference reaches S_1 somewhere. For some other distribution consistent with the surface load, the greatest stress-difference may be less than S_1. But there can be no distribution that makes the greatest stress-difference zero, for if so the stress everywhere would be hydrostatic, which is impossible. If we find that for *every* stress-system consistent with the surface load a stress-difference at least S is attained somewhere, we know that somewhere in the Earth the stress-difference S is reached and therefore that the material can stand this stress-difference without fracture —it can be regarded as a minimum estimate of the strength of the Earth. This method of analysis also can be modified in the direction of a theory of isostasy by assuming that stress below a certain depth is hydrostatic. As this involves a restriction on the distributions of stress that we are willing to consider, the value of S will ordinarily be increased. For the distribution that gives the least maximum stress-difference if we are free to assume stress-differences everywhere will in general imply non-zero stress-differences in the region where we are now assuming that they are zero. Physically this point can be interpreted as follows. The surface load is supported by stress-differences in the interior. If the whole Earth is strong, every part can make its contribution to the support. But if it is completely devoid of strength below a certain depth, the matter there is making no such contribution, and a given surface load must be supported by the outer parts alone, which therefore need extra strength. This point needs emphasis because it is habitually assumed (explicitly or implicitly) by advocates of the theory of continental drift that the ocean floor is weak right up to the sea bottom: if this was so the bed would be perfectly flat, and it is not.*

6·03. Elastic theory. This has the advantage that on its hypothesis a formal solution exists, due originally to Kelvin. The other methods require the adoption of a set of trial hypotheses, and numerical calculation is needed almost from the start. It is often found that the least maximum stress-difference given by them for the same distribution of surface load is not much less than that given by the elastic theory; there is a theoretical reason for this, which we shall discuss later. The elastic theory is also quite often true; denudation and redeposition are actually removal and application of loads, which must produce elastic deformation so long as the changes are not too great. Further, the analysis for it is very similar to that needed for the study of the bodily tide, the latter being simply a consequence of a distribution of forces through the interior instead of at the surface.

It is not necessary to attempt high accuracy, because the data on the surface loading are not very detailed in the regions where the excess loads, positive or negative, are greatest. It will be enough to take the case of incompressibility. This in itself does not simplify the analysis very much† so long as we are dealing with uniform bodies and neglecting gravity; but in a compressible body the changes of density introduce disturbances of gravity and additional complications. The general conclusion is that for a load expressed by a surface harmonic $a_n S_n$, the greatest stress-difference is about $\lambda_n a_n \sqrt{\bar{S}_n^2}$, where $\bar{S}_n^2$ is the mean of S_n^2 over a sphere, and λ_n is not far from 1.‡ For $n = 2$ the greatest stress-difference is at the centre; for the harmonics with $n = 3$ it is at depths near 0·4 of the radius, and therefore still perilously near the core. For higher harmonics it is at a depth near a/n. Many problems are treated by Niskanen (1942).

6·04. Elastic theory neglecting curvature of Earth. Special solutions can be made if we neglect the curvature of the Earth, thus restricting ourselves to inequalities with a horizontal scale of some hundreds of kilometres at most. We are then dealing with normal stress applied to the surface of a semi-infinite solid. We take a two-dimensional case, so that, in Cartesian coordinates,

$$v = 0, \quad \partial/\partial y = 0, \tag{1}$$

* Fuller accounts of the subject of this chapter are in Jeffreys (1932b, d, 1943b). See also *Collected Papers*, **3**, 440 for a note on corrections in 1932b, d.

† If λ is infinite, then $\Delta = 0$ for any finite stress; but the stresses contain the product $\lambda\Delta$, which is in general comparable with the rest of the stresses and has to be introduced as an extra unknown, corresponding to the pressure in a viscous liquid.

‡ Jeffreys (1943b); the solution does not allow for the attraction of the load itself and therefore, for a given load, should be multiplied by $2(n-1)/(2n+1)$.

and we take z downwards. Then

$$\frac{\partial p_{13}}{\partial x} + \frac{\partial p_{33}}{\partial z} = 0, \quad \frac{\partial p_{11}}{\partial x} + \frac{\partial p_{31}}{\partial z} = 0, \tag{2}$$

whence there is a function χ such that

$$p_{11} = \frac{\partial^2 \chi}{\partial z^2}, \quad p_{13} = -\frac{\partial^2 \chi}{\partial x \, \partial z}, \quad p_{33} = \frac{\partial^2 \chi}{\partial x^2}. \tag{3}$$

But

$$p_{11} = (\lambda + 2\mu)\frac{\partial u}{\partial x} + \lambda \frac{\partial w}{\partial z}, \tag{4}$$

$$p_{13} = \mu \left(\frac{\partial u}{\partial z} + \frac{\partial w}{\partial x} \right), \tag{5}$$

$$p_{33} = \lambda \frac{\partial u}{\partial x} + (\lambda + 2\mu)\frac{\partial w}{\partial z}, \tag{6}$$

whence from (4) and (6)

$$\frac{\partial u/\partial x}{(\lambda + 2\mu)\,p_{11} - \lambda p_{33}} = \frac{\partial w/\partial z}{(\lambda + 2\mu)\,p_{33} - \lambda p_{11}} = \frac{1}{4\mu(\lambda + \mu)}. \tag{7}$$

Also

$$\frac{\partial^2 p_{13}}{\partial x \, \partial z} = \mu \left(\frac{\partial^3 u}{\partial x \, \partial z^2} + \frac{\partial^3 w}{\partial x^2 \, \partial z} \right)$$

$$= \frac{1}{4(\lambda + \mu)} \left[\frac{\partial^2}{\partial z^2} \{ (\lambda + 2\mu)\,p_{11} - \lambda p_{33} \} + \frac{\partial^2}{\partial x^2} \{ (\lambda + 2\mu)\,p_{33} - \lambda p_{11} \} \right]. \tag{8}$$

Substituting from (3) we have

$$\left(\frac{\partial^2}{\partial x^2} + \frac{\partial^2}{\partial z^2} \right)^2 \chi = 0. \tag{9}$$

At the free surface $z = 0$, p_{33} is given and $p_{13} = 0$. Hence χ is determined except for a linear function of x, which contributes nothing to the stresses, and $\partial\chi/\partial z$ is independent of x, and can be taken zero since a constant addition to it does not affect the stresses. At great depths the displacements must remain finite.

6·041. Harmonic loading. At $z = 0$ take

$$p_{33} = g\sigma \cos \kappa x; \tag{1}$$

then

$$\chi = -\frac{g\sigma}{\kappa^2} \cos \kappa x, \quad \frac{\partial \chi}{\partial z} = 0 \quad (z = 0) \tag{2}$$

and

$$\left(\frac{\partial^2}{\partial z^2} - \kappa^2 \right)^2 \chi = 0. \tag{3}$$

The solution that satisfies these conditions and has bounded second derivatives as $z \to \infty$ is

$$\chi = -\frac{g\sigma}{\kappa^2} (1 + \kappa z)\, e^{-\kappa z} \cos \kappa x. \tag{4}$$

Then
$$p_{11} = g\sigma(1 - \kappa z)\,\mathrm{e}^{-\kappa z}\cos\kappa x, \tag{5}$$

$$p_{13} = -g\sigma\kappa z\,\mathrm{e}^{-\kappa z}\sin\kappa x, \tag{6}$$

$$p_{33} = g\sigma(1 + \kappa z)\,\mathrm{e}^{-\kappa z}\cos\kappa x, \tag{7}$$

$$p_{22} = \lambda\Delta = \frac{\lambda}{2(\lambda+\mu)}(p_{11}+p_{33})$$

$$= \frac{\lambda}{\lambda+\mu}g\sigma\,\mathrm{e}^{-\kappa z}\cos\kappa x, \tag{8}$$

$$p_{12} = p_{23} = 0. \tag{9}$$

p_{22} is a principal stress. The others are in planes of constant y and are

$$g\sigma\cos\kappa x\,\mathrm{e}^{-\kappa z} \pm g\sigma\kappa z\,\mathrm{e}^{-\kappa z}. \tag{10}$$

At small depths p_{22} is an extreme stress, and the stress-difference is

$$g\sigma\,\mathrm{e}^{-\kappa z}\left|\frac{\mu}{\lambda+\mu}\cos\kappa x \pm \kappa z\right|, \tag{11}$$

whichever is the greater. When $\kappa z = 1$, p_{22} is the intermediate stress, and the stress-difference in this neighbourhood is

$$2g\sigma\kappa z\,\mathrm{e}^{-\kappa z}, \tag{12}$$

the maximum of which is $2g\sigma/\mathrm{e}$ when $\kappa z = 1$. This is greater than (11) when $\kappa z > \frac{1}{2}$, for $\lambda = \mu$; hence if the strength is uniform failure will begin at a depth $1/2\pi$ times the wave-length, and will tend to flatten out the inequalities. If, however, the strength is smaller near the surface, λ and μ being still uniform, failure may take place by faulting or thrusting at the crests and troughs, in directions through the axis of x or z. This may be relevant to faults where the displacement is mainly horizontal.

We shall need the vertical displacement at the surface later. It is enough to take λ/μ large. Using 6·04 (7) we have

$$w = -\int_z^\infty \frac{p_{33}-p_{11}}{4\mu}\,dz = -\frac{g\sigma}{2\mu\kappa}(1+\kappa z)\,\mathrm{e}^{-\kappa z}\cos\kappa x. \tag{13}$$

The surface displacement is therefore

$$-(g\sigma/2\mu\kappa)\cos\kappa x. \tag{14}$$

In the same conditions $\partial u/\partial x = 0$ at $z = 0$, and therefore there is no horizontal displacement at the free surface.

6·042. Raised strip of uniform height. If the load is $g\sigma$ from $x = -l$ to $+l$, the solution is

$$\chi = -\frac{g\sigma}{2\pi}\left(r_1^2\tan^{-1}\frac{z}{x-l} - r_2^2\tan^{-1}\frac{z}{x+l}\right), \tag{1}$$

r_1, r_2 being the distances from $(\pm l, 0)$ to (x, z). Then

$$p_{11} = -\frac{g\sigma}{\pi}\left(\tan^{-1}\frac{z}{x-l} - \tan^{-1}\frac{z}{x+l} + \frac{(x-l)z}{r_1^2} - \frac{(x+l)z}{r_2^2}\right), \qquad (2)$$

$$p_{33} = -\frac{g\sigma}{\pi}\left(\tan^{-1}\frac{z}{x-l} - \tan^{-1}\frac{z}{x+l} - \frac{(x-l)z}{r_1^2} + \frac{(x+l)z}{r_2^2}\right), \qquad (3)$$

$$p_{13} = -\frac{g\sigma}{\pi}\left(\frac{z^2}{r_1^2} - \frac{z^2}{r_2^2}\right), \qquad (4)$$

as may be verified directly. The stress-difference is

$$S = \frac{4g\sigma l z}{\pi r_1 r_2}. \qquad (5)$$

This takes its maximum value at all points of the semicircle $x^2 + z^2 = l^2$ and is then $2g\sigma/\pi$.

6·043. Raised strip of triangular section. We take the maximum load to be $g\sigma$ at $x^{\cdot} = 0$, falling off uniformly to zero at $x = \pm l$. The solution can be constructed by superposing uniform loads, and gives

$$p_{33} - p_{11} = -\frac{g\sigma z}{\pi l}\log\frac{(l^2 + z^2 - x^2)^2 + 4z^2 x^2}{(z^2 + x^2)^2}, \qquad (1)$$

$$p_{13} = -\frac{g\sigma z}{\pi l}\tan^{-1}\frac{2l^2 zx}{l^2(z^2 - x^2) + (z^2 + x^2)^2}. \qquad (2)$$

The greatest stress-difference is $0\cdot512g\sigma$ at a depth $0\cdot513l$ below the crest.

Comparing these solutions, which refer to very different cases, we see that the greatest stress-difference on the elastic theory is from $\frac{1}{2}$ to $\frac{2}{3}$ of the range of load. The third, which makes the ratio nearly $\frac{1}{2}$, most closely resembles an actual mountain chain. It requires, however, that the requisite strength shall be available at a depth of about $\frac{1}{4}$ of the width of the chain.

6·05. Non-elastic solutions. We retain the equations of equilibrium but no longer assume that the stresses are related to a set of displacements by the rules of elasticity; we continue to take the density uniform. The problem is to find a distribution of stress, consistent with the surface load, that makes the maximum stress-difference as small as possible. A formal solution has not been obtained except in one special case. The method adopted is to start with the elastic solution, and modify the stresses in such a way that the stress-difference is reduced at the place or places where the elastic solution makes it greatest. In most cases the reduction possible is not great, because the modified solution increases the stress-differences in other places, and before much reduction has been made where the stress-difference is a maximum the stress-difference at some other place has risen to meet it. Further adjustment may make a little reduction at both places, but usually very little. There is a general reason for this. We have seen that the stress-difference is closely related to the Mises function M, and the latter to the distortional strain energy per unit volume. In an incompressible body the latter is the whole elastic energy, and in most of our problems is nearly the whole.

By Castigliano's principle (Appendix C), the whole strain energy of the body, for given surface conditions, is least when the stress is derived by elastic displacement from an unstressed state. Thus what we are doing is to make such changes in the stresses as will reduce the maximum value of the Mises function; but they inevitably increase the volume integral of the same function. We can do it only by spreading values of M comparable with the maximum through a greater volume.

The solutions should in some cases be very like the actual stress distribution. If stresses grow gradually, failure will first take place where the stress-difference first reaches the strength, and this stage is determined by the elastic theory. With further growth the region of failure extends, and the stress-difference will have the same value over it.

For loads expressed by harmonics of degrees 2 and 3, the reduction of the requisite strength from the elastic case is from 5 to 30 %. In all but one of them a new maximum appears at the surface, the position of the old maximum remaining nearly where it was.

For the two-dimensional case of harmonic loading over a plane boundary the best solution found made the stress-difference greatest at somewhat greater depths than in the elastic case, an equal value appearing at the surface at the crests and troughs. The reduction of the maximum from the elastic solution was only about 7 %.

For the ridge of uniform height there is a simple solution. If for all values of z we take
$$p_{11} = \tfrac{1}{2}g\sigma, \quad p_{13} = 0 \quad \text{(all values of } x\text{)}, \tag{1}$$
$$p_{33} = 0 \quad (x < -l, \, x > l), \tag{2}$$
$$p_{33} = g\sigma \quad (-l < x < l), \tag{3}$$
the equations of equilibrium are satisfied, and the stress-difference everywhere is $\tfrac{1}{2}g\sigma$, which is $\tfrac{1}{4}\pi = 0.78$ of the elastic solution. This is not, however, the smallest value attainable. Prandtl (1920) discussed the problem on the hypothesis that the maximum stress-difference is attained over a region and found that the stress-difference needed is $\dfrac{g\sigma}{1 + \tfrac{1}{2}\pi} = 0.39 g\sigma$. (See also R. Hill 1950, p. 254, 1954 and § 6·11.)

For the ridge of triangular section the reduction compared with the elastic theory is only about 2 %.

It has been found possible (Jeffreys, 1932d, pp. 63–64) to construct a ridge with a special form of section such that for a given maximum height support can be achieved with arbitrarily small strength. But it is the sort of exception that proves the rule, because the surface elevation is specially chosen, and the smaller the strength permitted the greater must be the distance to any other elevation; effectively we can do it with one ridge, but not with two.

As a general guide it is safe to say that a variation of surface load with range $g\sigma$ implies a stress-difference of about $\frac{1}{3}g\sigma$ to $\frac{1}{2}g\sigma$ somewhere. This may be compared with the fact that differences of height of about 5 km. occur between peaks in the Himalayas and neighbouring valleys; taking the density as 2·7 we infer that the stress-differences below them reach

$$\tfrac{1}{3}(5 \times 10^5) \times 2\cdot7 \times 980 = 4\cdot5 \times 10^8 \text{ dynes/cm.}^2.$$

This is about half the crushing strength of granite in the laboratory. The horizontal distances are of the order of 100 km., and if this strength is sufficient for the support it must be available to a depth of 20 or 30 km.

For the Moon, which appears to be a very homogeneous body, the elastic theory (Appendix B), indicates a stress-difference at the centre of 2×10^7 dynes/cm.^{2}. With the modified stress distribution this can be reduced to about $1\cdot3 \times 10^7$ dynes/cm.^{2}, say 13 atm. or the pressure of 130 m. of water, which would still need good masonry to hold it.

6·06. Loading of a floating crust: elastic solution. We consider a surface layer of density ρ, thickness H, resting on a substratum of negligible strength and density ρ'. For disturbances of surface load with a horizontal scale of order H or less the greatest stress-difference on the elastic theory will be well within the surface layer and those at depth H or more small. Hence for such horizontal scales the theory of 6·04 needs little alteration.

Detailed treatment for harmonic loading of any horizontal scale was given in Chapter VIII of the first and Chapter X of the second edition of this book. But at present it is enough to consider the other extreme, where the horizontal scale is large compared with the thickness. In this case the problem reduces to that of the bending of a thin plate, for which simple approximations exist.

The plane $z = 0$ is taken half-way down the undisturbed position of the layer. The upward displacement of the median plane is w, and

$$D = \frac{1}{12}\frac{EH^3}{1-\sigma^2} = \frac{1}{3}\frac{\mu(\lambda+\mu)}{\lambda+2\mu}H^3. \tag{1}$$

If Z is the upward force on the layer, per unit area, w satisfies*

$$D\left(\frac{\partial^2}{\partial x^2}+\frac{\partial^2}{\partial y^2}\right)^2 w = Z. \tag{2}$$

We take the layer to be of great horizontal extent and therefore are not concerned with its edges, if any. If the surface stress is P, we have

$$Z = P - g\rho'w. \tag{3}$$

---

* The simplest derivation is from the principle of virtual work. Cf. Rayleigh (1894, **1**, Chapter 10), Jeffreys (1941b), J. G. Oldroyd (1946). The last paper gives an improved treatment of the boundary conditions.

For harmonic loading $P = g\sigma \cos \kappa x$; then

$$w = \frac{g\sigma \cos \kappa x}{D\kappa^4 + g\rho'}. \tag{4}$$

Taking $\lambda = \mu = 3\cdot3 \times 10^{11}$ dynes/cm.^{2}, $H = 50$ km., $g = 980$ cm.^{2}/sec., $\rho' = 3\cdot3$ g./cm.^{3}, we have

$$D = 9 \times 10^{30} \text{ c.g.s.} \tag{5}$$

Then $D\kappa^4 = g\rho'$ when $2\pi/\kappa = 460$ km. For this value the maximum displacement of the crust is $\frac{1}{2}\sigma/\rho'$. If the load arises from material of density $2\cdot5$ and maximum thickness 1 km., the amplitude of $w = 0\cdot38$ km., and the visible change of level of the surface is $0\cdot62$ km. For κ very small the corresponding values are $0\cdot75$ and $0\cdot25$ km. The latter corresponds to the isostatic condition. Hence even for disturbances of rather long wave-length the elastic theory leads to considerable departures from isostasy; the ratio of the depression of the crust to the height of the visible inequalities is changed from 3 to $0\cdot6$.

The stress component p_{11} is $- Ez\,\partial^2 w/\partial x^2$, which, with the data adopted, reaches

$$4 \times 10^{18} \frac{g\sigma}{D\kappa^2 + g\rho'/\kappa^2}.$$

This is greatest if $D\kappa^4 = g\rho'$, and for a load due to 1 km. of material of density $2\cdot5$ reaches 3×10^9 dynes/cm.^{2}. $g\sigma$ itself is $2\cdot5 \times 10^8$ dynes/cm.^{2}, and p_{33} is of this order. Hence in a floating crust the stress-differences may be of the order of 10 times the surface load. This is the principle embodied in Airy's theory.

Consider instead a long elevated region of uniform height, with a straight edge at $x = 0$. We have to satisfy

$$\left.\begin{aligned} D\frac{d^4 w}{dx^4} &= g\sigma - g\rho' w \quad (x > 0), \\ &= 0 \qquad\qquad (x < 0). \end{aligned}\right\} \tag{6}$$

Put

$$g\rho'/D = 4\alpha^4. \tag{7}$$

w must be bounded for $|x|$ large, and w and its first three derivatives must be continuous at $x = 0$. We find

$$w = \frac{\sigma}{2\rho'} e^{\alpha x} \cos \alpha x \quad (x < 0), \tag{8}$$

$$w = \frac{\sigma}{2\rho'} (2 - e^{-\alpha x} \cos \alpha x) \quad (x > 0). \tag{9}$$

With the values adopted $1/\alpha$ is nearly 100 km., $w = 0$ at $x = -157$ km., and is σ/ρ' at $x = 157$ km. Outside this range the exponential factors are small and w varies little. The greatest stress-differences are at $x = \pm 240$ km., and are $1\cdot4 \times 10^{10}$ dynes/cm.^{2} when $\sigma = 2\cdot5 \times 10^5$ g./cm.^{2}.

The main conclusions from this comparison are that if the strength is uniform with regard to depth, at any rate to a depth of about half the horizontal scale of the surface inequalities, the laboratory strengths of igneous rocks would be enough to support inequalities 5 km. in height; but in a floating crust inequalities only 1 km. in height would lead to fracture if their horizontal extent is a few times the thickness.

It therefore becomes important, in stating a mechanics of isostasy, not only to specify that the stress below a certain level is hydrostatic, but to consider also the state of stress in the crust. If a harmonic surface load is supposed to give only elastic deformation in the crust, it would give very imperfect compensation, except for wave-lengths of 1000 km. or more. For abrupt changes of level the compensation will be spread horizontally over a distance large enough to imitate compensation at a depth of order 100–200 km., unless we very much reduce the adopted thickness of the crust and assume the existence of matter of density 3·3 far nearer the surface than other evidence suggests.

The meaning of 'crust' has changed from time to time. It was formerly believed that the Earth was fluid at all depths greater than about 50 km., and the solid overlying layer was then called the crust. Kelvin (1878) showed that such a structure would make the oceanic tides inappreciable. In modern usage the word is sometimes used to denote the seismological outer layers, sometimes for the strong outer part as opposed to the asthenosphere of Barrell. For a time these could be regarded as synonymous but it now appears that the material of the seismological lower layer, at least at its top, must be stronger than the more acidic material in contact with it (see p. 417). This chapter is largely devoted to the consequences of a strong outer region resting on a weak region, and the former is called the crust, without any special identification being assumed at the outset.

6·07. Non-elastic solutions for floating crust. We continue for purpose of illustration to neglect the curvature of the Earth. If the surface load is $g\sigma$, where σ is a function of x and y, the equations of equilibrium are satisfied if

$$p_{33} = g\sigma, \tag{1}$$

and all other stress components are zero. This condition is satisfied if the base of the crust is depressed by σ/ρ', so that the pressure at the top is balanced by the extra hydrostatic pressure at the bottom. Thus in the isostatic condition the stress-difference need not exceed the magnitude of the load. This condition would be realized in a set of vertical piles of different lengths, floating in a denser fluid, so that any change of load is transmitted to the bottom without support at the sides.

It is therefore possible to avoid the difficulties of the elastic solution by adopting an alternative distribution of stress, which not only makes the greatest stress-difference smaller but makes the compensation more complete and local, and therefore is in better agreement with the distribution of gravity. We can now specify the condition of isostasy more closely. The form just given is not immediately applicable when the curvature of the Earth is not neglected. We shall say that isostasy means (1) below a certain depth the stress is hydrostatic, (2) above that depth the maximum stress-difference is as small as possible, consistently with the surface load. This is a mechanical definition. We have to examine its consequences and see how closely it corresponds to the facts.

The above solution (1) clearly does not always satisfy these conditions for wave-lengths small compared with the depth H, because we can then arrange for harmonic loading with the elastic solution, which gives maximum stress-difference $2g\sigma/e$. The next question is whether we can always make the greatest stress-difference $< g\sigma$. In any case $p_{13} = 0$ at $z = 0$ and $z = H$. Suppose then that

$$p_{33} = f(z)\cos\kappa x, \quad p_{13} = Az(H-z)\sin\kappa x, \tag{2}$$

where $\kappa H < 1$, and $p_{33} = g\sigma\cos\kappa x$ at $z = 0$. The form for p_{13} is the simplest that vanishes for $z = 0$ and $z = H$. Then from

$$\frac{\partial p_{33}}{\partial z} + \frac{\partial p_{13}}{\partial x} = 0, \quad \frac{\partial p_{13}}{\partial z} + \frac{\partial p_{11}}{\partial x} = 0, \tag{3}$$

we get
$$p_{11} = \frac{A}{\kappa}(H - 2z)\cos\kappa x, \tag{4}$$

$$p_{33} = \{g\sigma - \kappa A(\tfrac{1}{2}Hz^2 - \tfrac{1}{3}z^3)\}\cos\kappa x. \tag{5}$$

If S is the stress-difference

$$S^2 = \left\{g\sigma - \kappa A(\tfrac{1}{2}Hz^2 - \tfrac{1}{3}z^3) - \frac{A}{\kappa}(H-2z)\right\}^2\cos^2\kappa x + 4A^2z^2(H-z)^2\sin^2\kappa x. \tag{6}$$

This is stationary with regard to x when $\kappa x = 0, \tfrac{1}{2}\pi$. Take $x = 0$. Then for $x = 0, z = 0,$
$$S^2 = (g\sigma - AH/\kappa)^2, \tag{7}$$

and for $x = 0, z = H$
$$S^2 = (g\sigma - \tfrac{1}{6}\kappa AH^3 + AH/\kappa)^2. \tag{8}$$

For these both to be less than $(g\sigma)^2$ we must have for positive σ

$$A > 0, \quad A(H/\kappa - \tfrac{1}{6}\kappa H^3) < 0. \tag{9}$$

These lead to
$$\tfrac{1}{6}\kappa^2 H^2 > 1, \quad \frac{2\pi}{\kappa} < \frac{2\pi}{\sqrt{6}}H = 2{\cdot}6H. \tag{10}$$

Hence unless the wave-length is less than about $2{\cdot}6H$ the assumption that $p_{13} \neq 0$ will always increase the greatest stress-difference. Similar results are found by taking $p_{13} = A\sin(\pi z/H)\sin\kappa x$.

The intermediate range, where the wave-length is between, say, H and $2{\cdot}5H$, may need separate investigation.

To take more rapid variation of p_{13}, for instance,

$$p_{13} = B \sin(n\pi z/H) \sin \kappa x,$$

makes the permissible values of $2\pi/\kappa$ still smaller and gives no help.

6·08. Floating crust, curvature of the Earth taken into account.

The surface is taken to be $r = a$ and the base of the crust $r = \alpha a$. The extra surface densities are taken to be σS_n and $\sigma_1 S_n$. For given σ, σ_1 is at our disposal. The stress-components are now denoted by p_{rr}, $p_{r\theta}$, $p_{r\lambda}$, and so on. The solution in the plane case for a thin crust corresponds to taking all components zero except p_{rr}. This does not satisfy the equations of equilibrium in the spherical case, but there is an exact solution (Jeffreys, 1932d, p. 65)

$$p_{\theta\theta} = p_{\lambda\lambda} = -\rho U + C, \tag{1}$$

$$p_{rr} = -\rho U + C + \psi, \tag{2}$$

where U is the gravitation potential, C is a constant, the other stress-components are zero, and ψ satisfies

$$\frac{\partial \psi}{\partial r} + \frac{2\psi}{r} = 0. \tag{3}$$

With this type of solution the stress-difference is therefore proportional to r^{-2}. It would approximate to that of 6·07 when the crust is thin, but if α is not near 1 it would give larger stress-differences than the elastic solution.

Another type of solution makes the stress-difference independent of r; in this $p_{\theta\theta}$ and $p_{\lambda\lambda}$ are no longer equal, and $p_{\theta\lambda} \neq 0$, but $p_{r\theta} = p_{r\lambda} = 0$. For similar reasons to those given in the plane case it appears that the stress-differences are always increased by taking $p_{r\theta}$, $p_{r\lambda}$ different from zero in a thin crust. Effectively, therefore, the problem reduces to finding, for given surface load, the combination of the two exact solutions that keeps the maximum stress-difference least. Even with this simplification the problem remains troublesome (Jeffreys, 1943b, pp. 80–89). One peculiar result is that σ_1 is usually numerically less than σ, and $\alpha^2 \sigma_1$ is always less than σ. The usual statement of isostasy, that mass per unit area is uniform, does not quite correspond to the distribution of stress that gives the smallest maximum stress-difference; and if we specify the meaning of isostasy in accordance with the latter criterion the disturbance of gravity for given disturbance of surface load will be more than we found in 5·07. The factor required is not the same for all harmonics but is usually about 2.

In applying this result to the support of compensated mountain systems we must recall that inequalities of level at distances of the order of 50 km. or so can be supported according to the theory of 6·03 or 6·06. We are now concerned only with features 100 km. and more in width, and must therefore use the average height, not the maximum. For the Himalayas we may take the average height to be about 5 km. Then the load is about $1·2 \times 10^9$

dynes/cm.², and the stress-differences reach about the same value. Those arising from inequalities of smaller width are superposed irregularly on them, and there will be places where the two add up. Hence there are places within the Himalayas where the stress-difference is about 1·6 or 1·7 × 10⁹ dynes/cm.². For the great Pacific deeps comparable strengths are needed. For the Alps, Andes and Rocky Mountains the stress-differences need be only about half as large.

The stress-differences indicated are somewhat greater than the crushing strengths of the strongest igneous rocks as measured, but it is probable that the strength in the crust is somewhat increased by pressure. The effect of pressure on strength was investigated by Bridgman, who found considerable differences between different substances. In an early paper he found that paraffin wax under sufficient pressure became stronger than steel. For most of the materials he has tested (1935), however, the increase of strength over a range of 50,000 atm. ranges from 2 to 20 %. We should expect such stress-differences if the greater surface inequalities were originally larger than they are now, and were enough to produce fracture or flow below them; for if so they would sink until the stress-differences were everywhere reduced below the strength, and thus the actual heights of the great ranges are explained. Since fracture does not take long to occur when the stress-difference reaches its critical value, it is safe to suppose that the excess was never large, and that the formation of mountains and their compensation proceeded simultaneously. Gravity accordingly played a dominant part.

6·09. Gravity anomalies of small horizontal extent. Analysis of gravity shows variations that are not consistent with either a Bouguer theory or with exact compensation. One possible way for them to arise is as follows. Suppose a deep depression to exist in the ocean floor; the missing matter may be taken to be representative of the ocean floor, and accordingly to have a density of about 3·0. Now let the depression be filled up with sediments of density 2·5 without fracture. Then no surface inequality of shape remains, and computed gravity is the same as for uniform structure; but actually there is a deep region where the density is 0·5 g./cm.³ below normal, and observed gravity will be low. Such anomalies may occur in other ways. The common feature is that regions of specially light or dense matter may be detected from the gravity anomalies even though there are no noteworthy differences in height. But from the variation of free-air gravity we can still infer the variation of mass per unit surface, and then the distribution of stress over a sphere can be inferred from the surface loads; the only difference is that the sphere now considered is not the external surface.

The use of gravity anomalies to infer variations of density may be illustrated as follows. The attraction of a hemisphere of density ρ and radius a at the centre of the plane surface is easily shown to be $\pi f \rho a$. Thus a hemisphere of radius 1 km. and density 2·5 would produce an acceleration of 50 mgal. at its centre. Near Seattle there are local anomalies reaching − 90 mgal., which could be explained by a hemispherical vacuum of radius 1·8 km. As this explanation is not available, we try the supposition that the normal density of 2·7 or so is replaced through a hemisphere by 2·2, corresponding to light sediments. The radius would have to be about 9 km., comparable with the greatest depths in the ocean, but it is not impossible.

Now consider a narrow strip of negative anomalies. We compare them with the greatest attraction of a semicircular cylinder, which is $4f\rho a$. The normal density of the region may be taken as 2·9; if a semicircular groove of depth a has been filled with sediments of density 2·2 anomalies of − 100 mgal. imply a radius of about 5 km.

Structures of widths 18 or 10 km. respectively could be supported by stress-differences above the level of compensation. Consequently the mere occurrence of isostatic gravity anomalies of the order of 100 mgal. is not evidence against the mechanical theory of isostasy; if they are of sufficiently small width they can be attributed to anomalies of density near the surface, which are not taken into account in the calculation of gravity. It is only when the anomalies persist in sign over distances substantially greater than 50 km. that the matter becomes serious.

6·10. Gravity anomalies of large horizontal extent. We now consider again an inequality represented by a surface harmonic elevation hS_n. For no compensation, if the density near the surface is ρ, the additional potential is

$$\frac{3}{2n+1}\frac{g\rho h}{\bar{\rho}}\left(\frac{a}{r}\right)^{n+1}S_n, \tag{1}$$

the measured height

$$h\left(1-\frac{3}{2n+1}\frac{\rho}{\bar{\rho}}\right)S_n, \tag{2}$$

and the disturbance of free-air gravity

$$\frac{3(n-1)}{2n+1}\frac{g\rho}{\bar{\rho}a}hS_n. \tag{3}$$

For given measured height $h'S_n$ the additional potential is

$$U_n = \frac{3g\rho h'}{(2n+1)\bar{\rho}-3\rho}\left(\frac{a}{r}\right)^{n+1}S_n \tag{4}$$

and the disturbance of gravity

$$g_n = \frac{3(n-1)g\rho h'}{\{(2n+1)\bar{\rho}-3\rho\}a}S_n. \tag{5}$$

For compensation at depth H, on the supposition (not far wrong if H is of the order of 50 km.) that the stress-difference is proportional to r^{-2}, it is found that for the same measured height, to order H, (5) is replaced by

$$g'_n = \frac{3(n-1)gHh'}{a^2} \frac{\rho}{\bar{\rho}} \frac{(n+2)\bar{\rho}-3\rho_0}{(2n+1)\bar{\rho}-3\rho_0} S_n \qquad (6)$$

and the disturbance of potential

$$U'_0 = \frac{3gHh'}{a} \frac{\rho}{\bar{\rho}} \frac{(n+2)\bar{\rho}-3\rho_0}{(2n+1)\bar{\rho}-3\rho_0} \left(\frac{a}{r}\right)^{n+1} S_n. \qquad (7)$$

ρ_0 being the density in the lower layer.

The surface load in both cases can be taken as $g\rho h S_n$. (This is not the true load on the sphere; but in calculating the internal stresses the difference is cancelled by the field due to the surface inequality.) ρ appears as a factor in all these expressions; hence to order H the load on either hypothesis can be inferred directly from the inequalities of gravity, irrespective of the densities of any inequalities near the surface, as follows, for $\rho = 2\cdot7\,\mathrm{g./cm.^3}$, $h' = 10^5$ cm. For the coefficient in the potential these must be divided by $(n-1)/a$. g_n, g'_n are in gals.

n	1	2	3	4	5	10	30	100
g_n	0	0·064	0·082	0·090	0·095	0·104	0·109	0·112
g'_n	0	0·0012	0·0021	0·0031	0·0040	0·0094	0·027	0·088

Kaula's list makes it possible to check my estimate of τ_2 from gravity. This removed harmonics up to degree 3. It represented otherwise variations over 30° squares uncorrelated with their neighbours. Now a harmonic containing a factor $\cos 3\lambda$ would be uncorrelated with neighbours differing by 30° in longitude; allowing also for the latitude factor we may suppose that the τ_2 variation represents an aggregate of harmonics of degrees about 6. From Kaula's table, somewhat supplemented from Gaposchkin and Lambeck's (1970, 1971), I find the following sums of squares for degrees 4 to 8 in potential and gravity (all multiplied by 10^{-12}).

n	4	5	6	7	8	Total
U	2·25	0·63	0·81	0·19	0·23	4·11
g	20·3	10·1	20·2	6·8	11·2	68·6

It looks as if the decrease with n is rapid for the potential, hardly recognizable for gravity. (It must ultimately decrease, as otherwise g^2 would be infinite; but it is still slow up to $n = 8$.) The estimate of τ_2 would be about 8·3 mgal. I got 12·2 mgal. with a standard error of 2·3 mgal. The difference is not serious, but the revised value will be used.

The τ_1 variation would correspond to harmonics of degrees about 12, but the satellite data hardly reach this, and contributions up to about degree 20 would need to be considered. I think that my previous estimate of about 20 mgal. cannot be far wrong because it depends largely on comparison within the same surveys, and errors at base stations would mostly be eliminated.

The outstanding difference in the potential from the hydrostatic theory is in the main ellipticity term, for which J_2 is about $0\cdot001083$ and the hydrostatic value $0\cdot001072$, with a difference $0\cdot000011$. The coefficient in gravity would be this multiplied by mean g, say $0\cdot011$ gal. From Kaula's table we have the following results for A_{ns}, B_{ns}, the coefficients of normalized harmonics in gravity expressed as fractions of the mean.

$10^{-6}\times$	Compensated (km.)	Not compensated (km.)	Load from g (10^8 dynes/cm.^{2})	Load from surface form (10^8 dynes/cm.^{2})
A_{20}	$+5\cdot0$	$+4\cdot2$	$+0\cdot078$	$+0\cdot21$	$+0\cdot75$
A_{22}	$+2\cdot4$	$+2\cdot0$	$+0\cdot037$	$+0\cdot098$	$-0\cdot71$
B_{22}	$-1\cdot4$	$-1\cdot2$	$-0\cdot021$	$-0\cdot056$	$-0\cdot09$
A_{30}	$+1\cdot9$	$+0\cdot9$	$+0\cdot023$	$+0\cdot061$	$-0\cdot26$
A_{31}	$+1\cdot8$	$+0\cdot9$	$+0\cdot022$	$+0\cdot058$	$-0\cdot26$
B_{31}	$+0\cdot2$	$+0\cdot1$	$+0\cdot002$	$+0\cdot006$	$+0\cdot20$
A_{32}	$+1\cdot6$	$+0\cdot8$	$+0\cdot020$	$+0\cdot053$	$-0\cdot81$
B_{32}	$-1\cdot5$	$-0\cdot7$	$-0\cdot018$	$-0\cdot048$	$+0\cdot78$
A_{33}	$+1\cdot1$	$+0\cdot5$	$+0\cdot013$	$+0\cdot034$	$+0\cdot18$
B_{33}	$+2\cdot8$	$+1\cdot3$	$+0\cdot034$	$+0\cdot090$	$+0\cdot89$
A_{40}	$+1\cdot6$	$+0\cdot5$	$+0\cdot018$	$+0\cdot048$	$+0\cdot46$
A_{41}	$-1\cdot7$	$-0\cdot5$	$-0\cdot019$	$-0\cdot050$	$-0\cdot38$
B_{41}	$-1\cdot4$	$-0\cdot5$	$-0\cdot015$	$-0\cdot040$	$-0\cdot41$
A_{42}	$+0\cdot9$	$+0\cdot3$	$+0\cdot010$	$+0\cdot027$	$-0\cdot71$
B_{42}	$+1\cdot8$	$+0\cdot6$	$+0\cdot020$	$+0\cdot053$	$+0\cdot10$
A_{43}	$+2\cdot8$	$+0\cdot9$	$+0\cdot031$	$+0\cdot082$	$+0\cdot61$
B_{43}	$-0\cdot6$	$-0\cdot2$	$-0\cdot007$	$-0\cdot019$	$-0\cdot25$
A_{44}	$-0\cdot2$	$-0\cdot1$	$-0\cdot002$	$-0\cdot005$	$-0\cdot07$
B_{44}	$+1\cdot0$	$+0\cdot3$	$+0\cdot011$	$-0\cdot0029$	$+0\cdot77$
τ_2	±8	$\pm0\cdot9$	$\pm0\cdot08$	$\pm0\cdot23$	
τ_1	±20	$\pm0\cdot7$	$\pm0\cdot18$	$\pm0\cdot49$	

It is interesting to compare the coefficients in gravity with the surface load. The table gives the coefficients in the potential, the corresponding elevations of the surface on the hypothesis of complete compensation, on that of no compensation, the load on the latter hypothesis, and the load given by the surface form, all normalized.

In the fifth edition of this book the surface load was computed from Prey's table (1922). This was revised by Lee and Kaula (1967) and further by G. Balmino, Kaula and K. Lambeck (1973). Their results are given as equivalent heights for harmonics up to degree 8, in metres, assuming density of rocks $2\cdot65$ and of sea water $1\cdot03$ g./cm.^{3}. I convert their results above to equivalent surface load. The changes are on the whole not large, but appreciable.

Separate solutions are given, s for the rock surface and o for the oceans. The rock surface for this purpose includes the bottom of the oceans. The contribution from the land is therefore $s-o$. On inspection of Balmino, Kaula and Lambeck's list it is found that the signs of the terms in $s-o$ and o are the same in 21 cases out of 25, different in 4. This suggests that a lot of land structure is continued under the oceans. It may be connected with the opinion of Biswas (see p. 490) that oceans may have deepened and continents risen during geological time; it could be interpreted as a statement that some primitive differences of level have survived.

The signs in gravity and the surface load, impossible terms in gravity and the irrelevant A_{00} and A_{20} being omitted, are like in 10 cases, opposite in 8; there is no appreciable correlation between gravity and the surface load.

A by-product of the investigation is that the positions of the observing stations themselves have been found to need corrections; Kaula (1963a) finds shifts up to 280 metres.

The variation of height, if the inequalities were compensated, would be obvious on a map of the globe, especially if we remember that the estimates are root mean square and not extreme heights. Comparison with the harmonics given by Prey shows, as already mentioned, that the signs of the harmonics and the actual surface load have next to no correlation. The coefficients on the hypothesis of no compensation would not imply variations of height large enough to be obvious. The next column gives the corresponding coefficients in the load. The mean height of the land (with respect to the whole surface) is 0.23 km.; the mean for the sea is -2.68 km. The loads corresponding to the impossible terms in gravity are as follows:

$$A_{10} + 1.31 \qquad A_{21} + 0.61$$
$$A_{11} + 1.17 \qquad B_{21} + 0.75$$
$$B_{11} + 0.70$$

These differences in load must be balanced by differences of density below.

Two methods were used to estimate the strength required to support these loads. The core must be taken as devoid of strength in any case. From the absence of deep-focus earthquakes at depths over 0.1 times the radius it is worth while to consider also the hypothesis that there is no stress-difference below that depth. In the first place a simple elastic solution was found corresponding to the surface load on a uniform body, subject to the above restrictions. The maximum stress difference was located and estimated for each harmonic. This requires the solution of a cubic equation, which is sometimes easy; when it was not, the Mises function was calculated and the principle was used that this is nearly proportional to the square of the stress-difference.

In the second place additional functions were added to the stresses so as to satisfy the equations of equilibrium and to reduce the stress-difference at its maximum. As in the two-dimensional case it was found that little reduction at the maximum was possible before that somewhere else reached it.

A third method has been used by Kaula (1963b). In my first and second solutions no account was taken of disturbances of gravity other than by the surface load; and in particular there was none due to changes of density due to displacements of the level surfaces. This amounts to treating the shell as of uniform density. Kaula introduces alteration of density as an additional function to be estimated. The integral of the Mises function was then made a minimum for each harmonic subject to the two boundary conditions. In an incompressible body this would be an extension of the elastic solution. The solution is carried out on the supposition that the requisite strength is available throughout the shell.

In my solutions (apart from the elastic one) two cases were considered, one with strength down to the core, the other with a 10 % discontinuity in density at a depth of 0·1 of the radius and no strength below it. Continuity of pressure implies in the latter case that a larger or smaller discontinuity of density would be cancelled by smaller or larger displacements of the interface. In all cases, for low harmonics, the greatest stress-difference is at the core boundary or the interface. The ratio of the surface densities at the interface and the outer surface is usually about -1, but greater than $-(0\cdot9)^{-2}$, so that the total extra mass per unit area of the outer surface always has the same sign as that at the outer surface itself. The assumption of uniform mass per unit area does not correspond to least required strength. The ratios of the maximum stress-differences to $10^9 g_n \sqrt{S_n^2}$ and the directions of the places where they occur are as follows:

	Strength to core	Strength to depth $0\cdot1R$	Direction
p_2	9·6	18	(0, 0)
$p_2^2 \cos 2\lambda$	9·4	22	(90°) (0, 0)
p_3	7·2	14	(90°) (37°)
$p_3^1 \cos \lambda$	11·1	19	(60°, 0), (0, 0)
$p_3^2 \cos 2\lambda$	7·5	14	(90°) (0, 45°) (35°, 0)
$p_3^3 \cos 3\lambda$	6·8	15	(23°, 0) (0, 0)
$n > 3$	$\doteqdot 7$		

The directions are nearly the same in both cases.

For $n = 10$ the maximum stress-difference is close to the range of load; the ratio in the corresponding column is about 4.

Terms in $p_n^s(\cos s\lambda, \sin s\lambda)$ can be taken together. The results are as follows, multiplied by 10^7 dynes/cm.^{2}:

	Strength to core	Strength to depth $0\cdot1R$	Kaula
p_2	4·8	19	3·1
p_2^2	2·7	6	1·8
p_3	1·4	3	2·2
p_3^1	2·0	4	3·5
p_3^2	1·6	2	0·7
p_3^3	2·0	3	1·4

My values are based on Kaula's 1966 determinations of the gravitation potential, his on his 1963 ones, so they are not strictly comparable. His list goes to p_4^4 with values up to $1\cdot6 \times 10^7$ dynes/cm.^{2}. They contemplate strength available throughout the shell.

The elevations that correspond to the impossible harmonics in gravity are real and also imply stress-differences. With their compensation they must give no shift of the centre of mass and no contribution to the products of inertia about the mean axis of rotation, and of course to the external gravitational field. This can be achieved by compensation at any depth, and with a thin crust the mass of this will be nearly equal and opposite to that of the elevation; the stress- difference needed will be nearly equal to the maximum load, and the resultants will come by compounding by squares, say $1 \cdot 7 \times 10^8$ dynes/cm.² from the first harmonics, $0 \cdot 8 \times 10^8$ dynes/cm.² from the second harmonics. They need be available only through the crust and do not add greatly to the $1 \cdot 6 \times 10^9$ dynes/cm.² inferred in § 6·08.

The above takes no account of the harmonics in the topography. Those that correspond to terms in gravity show no correlation, and will also imply stress-differences as if they were compensated. This will imply an increase of, roughly, 3×10^8 dynes/cm.², but this is less than we have already inferred from the support of mountains.

The proper combination of the results for different harmonics is not very clear. The estimates are not alternatives; the largest is for p_2 (which did not arise in my 1943 treatment) and must be supported in any case. It gives maxima at the poles and equator. $p_2^2 \cos 2\lambda$ and $p_3^1 \cos \lambda$ also give maxima on the equator, so there will be places where they add up. Strength about 10^8 dynes/cm.² will be needed if available down to the core; about 3×10^8 dynes/cm.² for strength to depth $0 \cdot 1R$ and none below that. The fourth harmonics will increase these a little.

For the τ_1 and τ_2 variations we can take approximately

$$g\sigma = 2 \cdot 3 \times 10^9 g_n$$

and the maximum stress-difference is about equal to the maximum load. This would imply on the elastic theory mean square contributions of about $4 \cdot 6 \times 10^7$ and $1 \cdot 8 \times 10^7$ dynes/cm.² to the stress-differences. The maxima will be at least twice these, and again need to be added to those found from the low harmonics because there will be places where they add up. As these occur at small depths for these intermediate harmonics they will not affect the results greatly for the case of strength down to the core, where the greatest stress-differences are on the core boundary in any case. But it appears that for strength down to $0 \cdot 1R$ depth the factor to be applied is about 4, and that we need a strength of about $3 \cdot 7 \times 10^8$ dynes/cm.² at depths less than $0 \cdot 1R$. If strength is admitted at great depths we can manage with about $1 \cdot 7 \times 10^8$ at small depths and 10^8 down to the core. The whole calculation is rough but must be of the right order of magnitude.

The distribution of the contributions from the various harmonics is very different from that in my 1943 paper, but the final result is about the same. The prominence of the p_2 contribution, in comparison with all others, supports the suggestion of O'Keefe that the Earth may have adjusted itself approximately to a hydrostatic form when rotating more rapidly than now and not have readjusted itself.

The assumption that the discontinuity at depth $0 \cdot 1R$ is pushed down when the surface load is positive, just as that at 30 km. is, is somewhat doubtful. Recent work for both Europe and Japan indicates that the rapid change of properties begins about depth $0 \cdot 03R$. Also there is strong experimental evidence that it is not due to a change of material but to a change of crystalline form under pressure. In that case, we should expect it to be raised instead of lowered where the surface load is positive. Unless this is compensated by a reduction of density further down there will be strong correlation between surface elevation and gravity for the low harmonics, which does not exist. Perhaps Kaula's method could be modified to fit absence of strength at depths more than $0 \cdot 1R$ and to take account of a change of state at depth $0 \cdot 03R$.

The choice of $0 \cdot 1R$ as a standard depth such that there is no strength below it was indicated by two considerations. First, it is about the extreme depth of deep focus earthquakes, which suggests that the yield below that depth, if any, is by flow and not fracture. Secondly, it is about the depth where, if the Earth was once fluid, the cooling since solidification would become small. I still think these reasons valid. However, it is odd that the study of imperfections of elasticity strongly suggests that damping of vibrations becomes less at great depths, as if great depth favours approach to perfect elasticity. The data can probably be reconciled, but further knowledge of imperfections of elasticity at very high pressures would be desirable.

In any case, the additional terms in the gravitational potential must be due to inequalities in density. Cook (1963) points out that for a given extra mass per unit solid angle, at distance R_1 from the centre, the coefficient in the potential contains a factor $(R_1/a)^n$. If a change of density ρ' is spread through depth h the extra mass would be as $R_1^2 \rho' h$ and for given $\rho' h$ the term in the potential would be as R_1^{n+2}. To match the observed harmonics in the potential with variable densities near radius R_1, $\rho' h$ would have to vary like R_1^{-n-2}. If R_1 is the core radius we have roughly

n	2	4	8	10
$(R_1/a)^{n+2}$	0·09	0·027	0·0024	0·0007

By an overwhelming amount the additional terms in the gravitational field are most easily accounted for by putting the extra masses as near the surface as possible.

In addition gravity is nearly uniform though the shell, and the extra load, at whatever depth, must be borne by stress-differences roughly proportional to the load. This is proportional to $\rho' h$, so it is irrelevant how it is distributed between the two factors. Then the reciprocals of the factors in the above list would need to be applied to the stress-differences needed. These would therefore become overwhelming. They are kept as small as possible by having the inequalities of density as near the surface as possible, and to that extent the assumption that they are at depth $0 \cdot 1R$ makes them as small as possible. The variations, however, seem to have little relation to the effect of change of state.

Heiskanen and Meinesz (1958) state as a 'basic hypothesis of geodesy' that the product of the mean isostatic anomaly over an area into the area cannot exceed $30 \, \text{mgal.(Mm)}^2$. (Mm $= 10^6$ metres.) O'Keefe (1959, 1961) points out that this is completely disproved by the harmonics J_2, J_3, J_4; this product, judged from a figure, is about 20 times their extreme. He also applies their estimate of viscosity to the rate of subsidence of these harmonics and gets a time of order 1000 years for decrease in the ratio e to 1.

6·11. A note on the solutions for surface load in special cases.
For a surface load p over $x > 0$ and 0 for $x < 0$ we can use the stress function χ and polar co-ordinates. The stress components are (Love, 1906, p. 89)

$$p_{rr} = \frac{1}{r^2}\frac{\partial^2 \chi}{\partial \theta^2} + \frac{1}{r}\frac{\partial \chi}{\partial r}; \quad p_{\theta\theta} = \frac{\partial^2 \chi}{\partial r^2}; \quad p_{r\theta} = -\frac{\partial}{\partial r}\left(\frac{1}{r}\frac{\partial \chi}{\partial \theta}\right).$$

Since there are finite non-zero stresses near $r = 0$, we take

$$\chi = r^2 f(\theta),$$

whence $\quad p_{rr} = f''(\theta) + 2f(\theta); \; p_{\theta\theta} = 2f(\theta); \; p_{r\theta} = -f'(\theta),$

$$S^2 = (p_{rr} - p_{\theta\theta})^2 + 4p_{r\theta}{}^2 = \{f''(\theta)\}^2 + 4\{f'(\theta)\}^2.$$

On $\theta = 0$ and $\theta = \pi$, $p_{r\theta} = 0$, whence $f'(0) = f'(\pi) = 0$.
On $\theta = 0$, $p_{\theta\theta} = p$, whence $f(0) = \frac{1}{2}p$.
On $\theta = \pi$, $p_{\theta\theta} = 0$, whence $f(\pi) = 0$.

There is a solution $f(\theta)$ that makes S constant. We have

$$\frac{dS^2}{d\theta} = 2f''(\theta)\{f'''(\theta) + 4f'(\theta)\},$$

and if this is zero everywhere, f' must be constant or of the form $\lambda \cos 2\theta + \mu \sin 2\theta$. Neither of these separately satisfies the boundary conditions. Let us assume

$$2f' = \pm S \sin 2\theta, \quad f = A \mp \tfrac{1}{4}S \cos 2\theta \quad (0 \leqslant \theta \leqslant \alpha),$$
$$2f' = \pm S \sin 2\theta, \quad f = B \mp \tfrac{1}{4}S \cos 2\theta \quad (\pi - \beta \leqslant \theta \leqslant \pi),$$
$$f = C \pm \tfrac{1}{2}S\theta \quad (\alpha \leqslant \theta \leqslant \pi - \beta).$$

Continuity of f' gives $\alpha = \beta = \frac{1}{4}\pi$ and the signs must be the same in the first and third equations, opposite in the second. Then continuity of f gives

$$A = C \pm \tfrac{1}{8}\pi S,$$
$$B = C \pm \tfrac{3}{8}\pi S$$

(the upper or the lower sign being taken in both).

The conditions at $\theta = 0$, π give

$$A \mp \tfrac{1}{4}S = \tfrac{1}{2}p,$$
$$B \pm \tfrac{1}{4}S = 0,$$

whence $\qquad\qquad \tfrac{1}{2}p = \mp \tfrac{1}{4}S(1 + \tfrac{1}{2}\pi).$

$f''(\theta)$ and therefore p_{rr} are continuous; thus the solution gives no discontinuity of stress, only a change in its analytic form when θ crosses $\tfrac{1}{4}\pi$ or $\tfrac{3}{4}\pi$. On $z = 0$, however, $p_{rr} = p_{xx}$ is discontinuous with regard to x at $x = 0$; for x passing through 0 implies a change of θ from 0 to π. The stress is homogeneous in $0 \leqslant \theta \leqslant \tfrac{1}{4}\pi$ and in $\tfrac{3}{4}\pi \leqslant \theta \leqslant \pi$. In the intermediate sector $p_{rr} = p_{\theta\theta}$ and the lines of r constant and θ constant are lines of greatest shear stress.

With the convention that S and p have the same sign

$$f = \tfrac{1}{4}(1 + \pi)\,S + \tfrac{1}{4}S\cos 2\theta \qquad (0 \leqslant \theta \leqslant \tfrac{1}{4}\pi),$$
$$f = \tfrac{1}{4}(1 + \tfrac{3}{2}\pi)\,S - \tfrac{1}{2}S\theta \qquad\quad (\tfrac{1}{4}\pi \leqslant \theta \leqslant \tfrac{3}{4}\pi),$$
$$f = \tfrac{1}{4}S(1 - \cos 2\theta) \qquad\qquad (\tfrac{3}{4}\pi \leqslant \theta \leqslant \pi).$$

No other function $g(\theta)$ that satisfies the boundary conditions can give stress differences everywhere $< S$ (taken positive). If possible, suppose

$$G^2(\theta) = \{g''(\theta)\}^2 + 4\{g'(\theta)\}^2 \leqslant T^2 < S^2.$$

Take $T > 0$. Then at $\theta = 0$, $\theta = \pi$, we have $g = f$, $g' = f'$; $f''(0) = -S$, $f''(\pi) = S$; $-T \leqslant g''(0) \leqslant T$, $-T \leqslant g''(\pi) \leqslant T$. Hence $g''(0) - f''(0) > 0$, $g''(\pi) - f''(\pi) < 0$. Hence there is an interval $(0, \theta_1)$ where $g \geqslant f$ and (θ_2, π) where $g \leqslant f$; let θ_3 be the smallest value of θ where $g - f$ has a maximum and θ_4 the largest where $g - f$ has a minimum. In (θ_3, θ_4) take the longest interval (θ_5, θ_6) where $g' \leqslant f' \leqslant 0$. Then $g'(\theta_5) - f'(\theta_5) = 0$, $g'(\theta_6) - f'(\theta_6) = 0$; and there is a θ_7 in (θ_5, θ_6) where $g''(\theta) - f''(\theta) = 0$, $g'(\theta_7) - f'(\theta_7) < 0$; and $f'(\theta_7) < 0$. Hence

$$G^2(\theta_7) - S^2 = 4\{g'^2(\theta_7) - f'^2(\theta_7)\} > 0,$$

contrary to hypothesis.

For a load over $-l < x < l$, the two lines of singularity proceeding inwards meet at depth l, enclosing a triangle with homogeneous stress. The sectors are bounded by circular arcs starting at the vertex and meeting the other two lines of singularity; from the points where they meet two other lines proceed at $\tfrac{1}{4}\pi$ to the surface. In the region below this composite boundary the stress difference is less than S.

The theory of the extension of a crack is complicated. Burridge and Willis study an expanding elliptical crack. They infer (1969, p. 456) that a tension crack is likely to spread with the Rayleigh speed. For a shear crack they seem to get the S velocity and the Rayleigh one in different cases. See also Burridge (1969).

In elementary dynamics it is supposed that the coefficient of friction has the same value before and after slip starts. This looks experimentally false. If it was so the initial acceleration would tend to zero as the shear force tends to the critical value, but it appears that however carefully the force is varied the initial acceleration does not tend to zero. Burridge and Knopoff (1967) therefore distinguish between dynamic and static friction, and show that under a general stress only part of the stress may be relieved by a crack, while the rest is preserved and can contribute to later slips when the general stress is continued. They have a model that gives a convincing imitation of aftershocks.

CHAPTER VII

Nutations, the Bodily Tide, and Free Vibrations

O, that this too too solid flesh would melt,
Thaw and resolve itself into a dew!
SHAKESPEARE, *Hamlet*

7·01. Displacements of the Earth as a whole. It has been known since the time of Hipparchus that the longitudes of stars, measured from the intersection of the planes of the equator and ecliptic (known to astronomers as the first point of Aries), show a secular increase, attributable to a retrograde motion of the equator with respect to the ecliptic. This is called *precession*. The amount is about 5200″ per century, so that the 'first point of Aries' is now in Pisces! The first part of the explanation is that the attraction of the Sun on the Earth's equatorial bulge produces a couple tending to bring the pole into coincidence with the pole of the ecliptic. The second is that, on account of the gyroscopic effect of the Earth's rotation, this couple produces a motion of the Earth's axis about the pole of the ecliptic. The pole of the Moon's orbit also moves in a circle about the pole of the ecliptic in about 18·6 years, and the Moon's attraction consequently produces a similar effect on an average (about twice as great), but the fluctuations due to the variation of the inclination of the Moon's orbit to the equator give also an oscillation of the Earth's pole in 18·6 years, called the principal *nutation*. Further, since the Sun and Moon are sometimes north and sometimes south of the equator, the couples vary in amount and produce smaller nutations with periods of half a year and half a lunar (sidereal) month. The orbits also have eccentricities, and these introduce smaller nutations.

The rate of precession is of the form

$$p = \frac{C - A}{C} (\alpha + \beta M),$$

where C and A are the Earth's principal moments of inertia and M the mass of the Moon. The amplitude of the 18·6-year nutation in the obliquity of the axis is of the form

$$N = \frac{C - A}{C} \gamma M.$$

Here α, β and γ are known very accurately, and these provide a pair of equations for $(C - A)/C$ and M. Until about 1903 these were the only data for these two constants. Hinks determined M separately, and more accurately, from the lunar inequality. J. Jackson pointed out later (1930) that the nutation could be calculated from Hinks's M, and the observed value was definitely smaller.

The attraction of the Sun and Moon on the body of the Earth produces elastic deformations analogous to the tides in the ocean. In one respect they are more amenable to mathematical treatment because the longest free periods of elastic vibrations in the Earth are of the order of an hour, so that we can use an equilibrium theory. They produce a tilting of the surface, which can be measured in favourable conditions. These effects are known as the bodily tides.

Besides these forced motions there are free oscillations. A rigid body under no external forces can be in stable steady motion about an axis of either greatest or least moment of inertia. If it is disturbed from this state and released, the axis describes a cone about the axis of resultant angular momentum. The motion can be seen in a Rugby ball in flight or in a disk thrown across the room. It was studied by Euler and is hence known as the Eulerian or free nutation. The period is $A/(C-A)$ times the period of rotation as judged by an observer referring the displacements of the axis of figure to axes rotating with the mean angular velocity. For the Earth it would be about 305 days. An observer from outside, looking at the Earth, would make the period A/C sidereal days.

The displacements of the axis of figure relative to the stars give rise to variations of the altitudes of stars when they cross the meridian. The motion is known, rather unfortunately, as the variation of latitude, and astronomers usually mistakenly believe that their observations refer to the instantaneous axis. In geophysics we naturally regard the latitude as a fixed angle for a given place, not affected by motions of the Earth as a whole or by periodic deformations. The zenith at a place is the direction opposite to local gravity. The latitude is determined essentially from observations of the altitudes of stars; apart from complications it would be the mean of the altitudes of a circumpolar star when it crosses the meridian above and below the pole. But refraction is great at the lower passage and a substantial correction is needed for it. To estimate the variation of latitude it is best to use only observations near the zenith to reduce this effect.*

The International Latitude Service began in 1900. Originally there were six observatories with vertically mounted telescopes in latitude 39° 8' N., so that all could observe the same stars. Determinations of the apparent displacements of stars could be interpreted as due to displacements of the pole (in space) along the meridians of 0° and 90° E. In fact there have been several changes of programme, and only three stations have operated throughout the period. In addition to this work similar observations have been made at several other observatories, notably Greenwich, Washington and Pulkovo.

Before Chandler's work many astronomers had sought without success for a period of 305 days; but S. C. Chandler in 1891 found a small variation, apparently consisting of two parts, with periods of a year and 14 months

* A discussion is given in the English translation of Fedorov's memoir (1961).

respectively. Both had amplitudes of the order of 0·1″. It was then pointed out by Newcomb that elasticity would lengthen the period in comparison with that of a rigid body, and that the 14-monthly motion was really the Eulerian nutation with its period so modified. A full historical account was given by W. D. Lambert (1931).

The principle of Newcomb's argument can be seen as follows. The rotation of the Earth produces an elastic strain, which is always symmetrical about the instantaneous axis and does nothing to displace it. This is superposed on a permanent ellipticity, the two together making up the whole ellipticity. The permanent part affects the axis of rotation just as in a rigid body, and it is this part that determines the period. But both parts are attracted by the Sun and Moon, and contribute to the precession. Conversely, the lengthening of the period from 10 to 14 months gives a measure of the elastic strain.

For many of these phenomena it is enough to take the statical case, the limiting case when the disturbing forces are of very long period. This is because the longest periods of elastic vibrations are of the order of an hour; the shortest forced periods are about half a day, so that the elastic strains, for equal amplitudes of the disturbing force, would not be expected to differ by more than 1 %. This does not apply to the ocean, where the free periods are of the order of a day, and the complicated form of the boundaries leads to difficulties that have not yet been satisfactorily met. For the core, again, a constant displacement nearly parallel to the boundary would satisfy the conditions, and constitutes a new degree of freedom of very long period. This arises for all displacements that alter the direction of the axis of figure, and more detailed treatment is needed. The relevant tide components are those containing $\cos \lambda$ or $\sin \lambda$ as factors; those independent of the longitude (the fortnightly and semiannual tides) and those containing $\cos 2\lambda$ or $\sin 2\lambda$ (semidiurnal tides) are not seriously affected. There is an annual disturbance, probably due to meteorological phenomena.

All these phenomena are altered by elasticity; the (relative) amount depends on the period, and is totally negligible for precession but is very large for the free nutation. There is a long-standing anomaly for the 18·6-year nutation, the observed magnitude of which is definitely less than that calculated for a rigid Earth. All these effects have consequently needed correction for elasticity, and still more so for fluidity of the core and the ocean.

Gravitation due to the Sun and Moon produces, besides the well-known oceanic tides, distortions of the Earth itself. Associated quantities can be measured and compared with theoretical values based on suggested structures of the Earth.

Within the last few years a new class of data has become available. The great Chilean earthquake of 1960 May 22 and the Alaskan one of 1964 March 27 set up oscillations that continued for several weeks, with periods of the order 20 to 50 minutes. They were first noticed by Benioff (1954) for the Khamchatka earthquake of 1952 November 4. See also S. W. Smith

(1968). These are elastic vibrations of the Earth as a whole. Surface waves of long period have also been studied by harmonic analysis. Here again the observations can be compared with theory.

Since the greatest part of the elastic theory is common to all these phenomena it is convenient to treat them together.

One point where these phenomena give definite information is on the fluidity of the core. Although seismology and the figure of the Earth together give very detailed information about the elasticity within the Earth, they do not decide whether the failure to detect S through the core is due to true fluidity or to rapid absorption, under which S might be admitted but damped out before travelling far. The existence of the strong reflexion ScS is evidence for a discontinuity, but its existence alone does not decide whether any S is transmitted. The lengthening of the 14-monthly period is too great to permit this hypothesis; the reduction of amplitude of the 18·6-year nutation requires a rigidity less than 6×10^8 dynes/cm.².

7·02. The bodily tide; Love's numbers.

Any set of displacements u_i, supposed to have continuous derivatives, can be expressed in the form (Jeffreys, 1967b)

$$u_i = \sum_n F \frac{\partial K_n}{\partial x_i} + G x_i K_n + H \epsilon_{ikm} x_k \frac{\partial K_n}{\partial x_m}, \tag{1}$$

where K_n is a solid harmonic and F, G, H are functions of r, n and possibly t. The radial displacement can be written as

$$x_i u_i / r = q K_n / r, \tag{2}$$

where

$$q = nF + Gr^2. \tag{3}$$

The F, G notation was introduced by Love in his *Problems of Geodynamics*. The expression can for many purposes be written more conveniently

$$u_i = \sum_n \left\{ F \left(\frac{\partial K_n}{\partial x_i} - n \frac{x_i}{r^2} K_n \right) + q x_i K_n + H \epsilon_{ikm} x_k \frac{\partial K_n}{\partial x_m} \right\}.$$

In this form the q term gives a radial displacement, F and H transverse ones, F in the direction of the horizontal derivative of K_n, H perpendicular to both. H gives no dilatation. A pure rotation is represented by H constant and $n = 1$. If $K_1 = a_1 x_1 + a_2 x_2$, the displacements for $H = 1$ are found to be $(-a_2 x_3, a_1 x_3, a_2 x_1 - a_1 x_2)$, which are just those due to small rotations about the axes $01, 02$. Other forms of H give a shear stress over concentric spheres. In problems relating to approximate spheres this notation is much more convenient than the usual expression in spherical polar coordinates, on account of the facts that derivatives with regard to the coordinates commute and that there is no distinction between covariant and contravariant derivatives.

For the bodily tide and Love's numbers we consider a disturbing potential

$$U_2 = c K_2 = c r^2 S_2. \tag{4}$$

The conventional equilibrium tide is U_2/g at the surface. The actual vertical displacement at the surface is denoted by

$$hU_2/g = qK_2/r = hca^2S_2/g. \tag{5}$$

The deformation of the Earth produces a disturbance of the gravitational potential, whose value at the surface is kU_2. This has to be taken into account in evaluating F and G theoretically. But in any case the numbers h, k, introduced by Love, specify the surface elevation and the disturbance of potential due to any tidal potential of the second degree.

The horizontal displacements of a particle are

$$u_\phi = F(r)\frac{\partial}{r\,\partial\phi}(r^2S_2), \quad u_\lambda = F(r)\frac{\partial}{r\cos\phi\,\partial\lambda}(r^2S_2), \tag{6}$$

and their surface values are

$$u_\phi = F(a)\,a\frac{\partial S_2}{\partial\phi}, \quad u_\lambda = F(a)\,a\frac{\partial S_2}{\cos\phi\,\partial\lambda}, \tag{7}$$

which may be written

$$u_\phi = \frac{l}{g}\frac{\partial U_2}{\partial\phi}, \quad u_\lambda = \frac{l}{g}\frac{\partial U_2}{\cos\phi\,\partial\lambda}, \tag{8}$$

with

$$F(a) = lk_2a/g. \tag{9}$$

The notation l was introduced by Toshi Shida (Shida and Matsuyama, 1912), and its importance is implicit in papers by L. M. Hoskins (1920).

The total external gravitational potential is

$$U = \frac{ga^2}{r} + U_2 + \frac{ka^5}{r^5}U_2, \tag{10}$$

and the acceleration is $-g_i'$, where

$$g_i' = \frac{ga^2}{r^3}x_i - \left(1+\frac{ka^5}{r^5}\right)\frac{\partial U_2}{\partial x_i} + \frac{5ka^5}{r^7}x_iU_2. \tag{11}$$

At the displaced position of a particle originally at x_i this is to the first order

$$g_i'' = \frac{ga^2}{r^3}(x_i+u_i) - \frac{3ga^2}{r^5}x_ku_kx_i - \left(1+\frac{ka^5}{r^5}\right)\frac{\partial U_2}{\partial x_i} + \frac{5ka^5}{r^7}x_iU_2. \tag{12}$$

The resultant, to the first order, at $r = a$, is

$$g\left\{1 - \frac{2U_2}{ag}(1-\tfrac{3}{2}k+h)\right\}. \tag{13}$$

Thus gravity at a place moving with the Earth will show a tidal variation in intensity, which will be recorded by sufficiently sensitive instruments for measuring small differences in gravity.

There is also, for the harmonic p_2, an effect on the moment of inertia about the polar axis. For a pendulum kept swinging for a long time these effects will produce a periodic disturbance of time-keeping. These should be greatest for the semi-annual tide. I predicted this in 1928 (Jeffreys, 1928b) and thought it so far below the possibility of observation that I did not even ask for reprints of the paper! But modern gravimeters, which can be made very sensitive to small changes of gravity, show the semi-diurnal and diurnal tides clearly.

As for u_i, several terms in g_i'' are purely radial, but there are transverse components derived from the terms, at $r = a$,

$$\frac{g}{a} F(a) \frac{\partial}{\partial x_i} (r^2 S_2) - (1+k) \frac{\partial U_2}{\partial x_i} = -(1+k-l) \frac{\partial}{\partial x_i} (k_2 r^2 S_2), \qquad (14)$$

and these are

$$g_\phi = -(1+k-l) \frac{\partial U_2}{a \partial \phi}, \quad g_\lambda = -(1+k-l) \frac{\partial U_2}{a \cos \phi \partial \lambda}. \qquad (15)$$

A body of water with a natural period short compared with the tidal periods will set itself nearly so that its free surface is one of constant U. If ζ is the elevation of such a surface,

$$-g\zeta + (1+k) U_2 = \text{constant.} \qquad (16)$$

The differences of level cannot be directly measured, but the bottom of the container itself moves with the Earth, the elevation being hU_2/g. Hence if ζ' is the elevation of the free surface relative to the bottom

$$\zeta' - (1+k-h) U_2/g = \text{constant.} \qquad (17)$$

The constant is determined by the fact that the mass of water remains constant with respect to time.

Another method is to use a horizontal pendulum similar to many types of seismograph. Take axes x, y, z (z vertical) at the fixed end of the boom. The shaft and boom are in the xz plane. The shaft is inclined at i to the vertical. Then undisturbed gravity alone would give an acceleration $g \sin i$ along the boom. The shaft is rigidly attached to the ground. Then the tilt in the y direction (rising along y) is

$$\frac{h}{g} \frac{\partial U_2}{\partial y}$$

equivalent to an acceleration $-h \, \partial U_2/\partial y$ in y. The extra field gives an acceleration $(1+k) \, \partial U_2/\partial y$, and the total is $(1+k-h) \, \partial U_2/\partial y$. The deflexion is

$$\frac{1+k-h}{g \sin i} \frac{\partial U_2}{\partial y}.$$

It is theoretically possible to determine all of k, $1+k-h$, $1+k-l$, and $1 - \frac{3}{2}k + h$ from observation. k is related to the period of the free variation of latitude. $1+k-h$ and $1 - \frac{3}{2}k + h$ have been explained. $1+k-l$ arises as follows. In astronomical observations the local vertical is defined by the direction of gravity, and consequently shows disturbances depending on $1+k-l$. If the axis of greatest moment was fixed in space, the effect would be shown by variations of latitude and longitude in tidal periods. But the tidal potential also produces couples on the Earth as a whole, and some of these perturb the axis of greatest moment. In attempts to find displacements of the axis it may be necessary to apply corrections for local deflexions of the vertical. For the whole tidal potential the ratio U_2/ga, converted into an angle, reaches about $0.01''$ and is not negligible in work that depends on the analysis of a large number of observations.

7·03. Observations of $1+k-h$. In the determination of $1+k-h$ from (17) it is necessary for accurate results that the tide in the water shall satisfy an equilibrium theory, or at least that the dynamical correction shall be calculable, and the method cannot be applied to the ocean as a whole; in fact, we must seek for enclosed bodies of water with natural periods short compared with a day, and study variations of the differences of level between different parts of them. We can consider only differences of level between different parts, since the total mass of water remains constant.

The horizontal pendulum seems to have been used first by Zöllner and various workers in Italy and France, notably Rossi and d'Abbadie,* and by O. Hecker in Germany; a further attempt was made by G. H. and H. Darwin. There is great difficulty in applying this method. The instrument is under considerable stress, and drifting of the zero is difficult to avoid or to separate from a semi-diurnal or diurnal period. Also the instrument is a seismograph and is affected not only by earthquakes but by the microseismic disturbance. Schweydar and Hecker (1921), using horizontal pendulums at four German stations and one American one, got values of $1+k-h$ from 0·34 to 0·59 for N.-S. components, 0·62 to 0·73 on E.-W. ones. From various tide components at Freiburg Schweydar got from 0·61 to 0·97 for E.-W. components, 0·50 to 1·63 on N.-S. components. Michelson and Gale (1919) at Chicago used the oscillations of water in a long horizontal tube with vertical end-pieces, and got 0·69. Proudman (1925) suggested using the tide in Lake Baikal. S. F. Grace (1930, 1931) got $-0·42$ for the Red Sea, 0·54 for Lake Baikal, for the semi-diurnal tide. R. Sterneck (1928 *a, b*) had previously considered the diurnal tide in Lake Baikal and got 0·73. W. D. Lambert (1933, 1936, 1940, 1944) summarized other data in his 'Reports on Earth Tides'. Noteworthy results are shown on p. 293. The notation for tide components is given in Darwin's *Scientific Papers*, **1**, p. 5. If γ, σ, η, ϖ are the rates of the Earth's rotation, Moon's mean motion, Sun's mean motion, and motion of the lunar perigee, M_2 has speed $2(\gamma-\sigma)$; K_1, γ; O_1, $\gamma-2\sigma$; N_2, $2\gamma-3\sigma+\varpi$; S_2, $2(\gamma-\eta)$. In the notation of this book γ is replaced by ω, σ by n, and η by n'.

The results are fairly consistent except for the Red Sea. Even for enclosed masses of water, however, there is some doubt about the theory, because the ocean tide itself makes a contribution to the disturbance of gravitation potential, and its weight also produces deformations, and thus the method is still not free from the difficulty about the theory of the ocean tide, though this is appreciably reduced. The phase lags would be negligible for the bodily tide if the ocean tide was absent; the actual lags are often great. The Red

---

* For historical account see Darwin, *Scientific Papers*, **1**, Chapter 14.

Observer	Place	Instrument	Direction	Component	$1+k-h$
W. Schaffernicht	Marburg	H.P.	E.-W.	M_2	0·87
			N.-S.	M_2	0·65
J. Egedal and	Bergen	Water level		M_2	0·58
J. E. Fjelstad				K_1	0·60
				O_1	0·75
				N_2	0·66
				S_2	0·63
H. Lettau	Leipzig	Double H.P.			0·6
R. Tomaschek	Pillnitz	H.P.	N.-S.		0·64
			E.-W.		0·83
	Beuthen		N.-S.		0·73
			E.-W.		0·75
	Berchtesgaden		N.-S.		0·48
			E.-W.		0·72
R. Corkan	Liverpool	H.P.		M_2	0·77?
E. Nishimura (1950)	Barim, Manchuria	Tiltmeter	N.-S.	M_2	0·715
			E.-W.	M_2	0·672
			N.-S.	S_2	0·685
			E.-W.	S_2	0·595
			E.-W.	O_1	0·615

Sea is under special suspicion because it is connected to the Indian Ocean. The tides in bays may be specially magnified, and may produce a great disturbance on an enclosed or nearly enclosed body of water near them. The other places (except Liverpool) are all a long way from the nearest ocean. A reasonable value for $1+k-h$, omitting the Red Sea and Liverpool, is

$$1+k-h = 0·68 \pm 0·05.$$

7·031. Observations of $1+k-l$. For the four international variation of latitude stations, Carloforte, Ukiah, Tchardjui, and Cincinnati, Nishimura (1950) analysed the observations for the M_2 tide, and got respectively $1+k-l = 1·08 \pm 0·06$, $1·06 \pm 0·06$, $1·31 \pm 0·19$, $1·66 \pm 0·18$. The mean is $1·20 \pm 0·10$. There is some sign of a systematic difference between continental and coastal stations.

7·032. Observations of l. Sassa, Ozawa and Yoshikawa (1951) have measured l directly with a horizontal strain meter, and get $l = 0·040$.

7·04. The variation of latitude. Let us consider how a deformable rotating body would move if its rotation is slightly disturbed. The centre of mass is taken as origin, and axes of x_1, x_2, x_3 are taken through it, turning with an angular velocity whose components are θ_1, θ_2, θ_3. If the Earth was rigid and these axes were fixed in it, the velocity of a particle of it would be $-\epsilon_{ikm}x_k\theta_m$. If now m is the mass of a particle of the body and u_i its velocity we can choose θ_i so as to make the velocities of points rigidly attached to the axes as good an approximation as possible to the actual velocities; that is, we make

$$\Sigma m(u_i + \epsilon_{ikm}x_k\theta_m)(u_i + \epsilon_{ips}x_p\theta_s) \tag{1}$$

a minimum. The condition for this is

$$\Sigma m(x_m^2 \theta_i - x_i x_k \theta_k + \epsilon_{ikm} u_k x_m) = 0. \tag{2}$$

But $$\Sigma m \epsilon_{ikm} x_k u_m = h_i, \tag{3}$$

the angular momentum of the body, and

$$\Sigma m(x_m^2 \delta_{ik} - x_i x_k) = A_{ik}, \tag{4}$$

the inertia tensor of the body. Hence

$$A_{ik}\theta_k = h_i, \tag{5}$$

which is the same relation as for a rigid body. The θ terms here are identical with the H terms for $n = 1$. The equations of motion have the usual form

$$\frac{dh_i}{dt} - \epsilon_{ikm} h_k \theta_m = L_i, \tag{6}$$

where L_i is the applied couple, which may be treated separately. We are not considering it here. In the actual case the motion of the shell can be represented by a rotation together with a set of elastic displacements of the form of 7·02 (2); and the latter do not contribute to the angular momentum or to the mean rotational displacement of any spherical shell about the centre. In this way of looking at the matter L_i would include couples due to the reaction of the atmosphere and core on the shell. In our problem the motion is almost wholly about the axis x_3, and we can transform to Cartesian notation. The products of inertia F, G, H and the variable parts of the moments of inertia A, B, C may be treated as small, and so may the angular velocity components θ_1, θ_2. Then neglecting products of small quantities we have

$$\left.\begin{array}{l} A\dot{\theta}_1 - (B-C)\,\theta_2\theta_3 - \theta_3\dot{G} + F\theta_3^2 = 0, \\ B\dot{\theta}_2 - (C-A)\,\theta_1\theta_3 - \theta_3\dot{F} - G\theta_3^2 = 0, \\ \dfrac{d}{dt}(C\theta_3) \hspace{3.5cm} = 0. \end{array}\right\} \tag{7}$$

From the last, $\theta_3 = \omega = $ constant, with a first order correction. (8)

The direction cosines of the instantaneous axis are

$$l, m, 1 = \theta_1/\omega,\ \theta_2/\omega,\ 1, \tag{9}$$

whence $$\left.\begin{array}{l} A\dot{l} + (C-B)\,\omega m - \dot{G} + \omega F = 0, \\ B\dot{m} - (C-A)\,\omega l - \dot{F} - \omega G = 0. \end{array}\right\} \tag{10}$$

The displacements due to rotation are the same as those due to a potential $\frac{1}{2}\omega^2\varpi^2$, where ϖ is the distance from the instantaneous axis; and to the first order in l, m this is

$$\tfrac{1}{2}\omega^2\{(x^2+y^2+z^2) - (lx+my+nz)^2\} = \tfrac{1}{2}\omega^2(x^2+y^2) - \omega^2 z(lx+my). \tag{11}$$

The first term is independent of the time and gives only a constant contribution to the ellipticity. The second is a harmonic of degree 2, and produces an elastic deformation of the Earth such as to give an extra external gravitational potential

$$- k\omega^2 z(lx + my) \, a^5/r^5. \tag{12}$$

But the gravitational potential due to the deformed Earth is

$$f\left\{\frac{M}{r} + \frac{(A+B+C)\,r^2 - 3(Ax^2 + By^2 + Cz^2 - 2Fyz - 2Gzx - 2Hxy))}{2r^5}\right\} \tag{13}$$

by MacCullagh's formula; whence

$$3fF = -k\omega^2 ma^5, \quad 3fG = -k\omega^2 la^5, \tag{14}$$

and, with $A = B$,

$$\left.\begin{aligned}
\left(A + \frac{k\omega^2 a^5}{3f}\right) l + \left(C - A - \frac{k\omega^2 a^5}{3f}\right) \omega m &= 0, \\[2mm]
\left(A + \frac{k\omega^2 a^5}{3f}\right) \dot{m} - \left(C - A - \frac{k\omega^2 a^5}{3f}\right) \omega l &= 0.
\end{aligned}\right\} \tag{15}$$

The solution is
$$l = \alpha \cos \omega(t - t_0)/\tau, \quad m = \alpha \sin \omega(t - t_0)/\tau, \tag{16}$$

where
$$\tau = \frac{A + k\omega^2 a^5/3f}{C - A - k\omega^2 a^5/3f}, \tag{17}$$

and α, t_0 are arbitrary. The axis therefore moves in the body in a circular cone in a period of τ sidereal days. We have also, if μ is the m of Chapter IV, and we neglect quantities of order μ^2,

$$\mu = \frac{\omega^2 a}{g} = \frac{\omega^2 a^3}{fM}. \tag{18}$$

Then
$$\frac{k\omega^2 a^5}{3fA} = \frac{k}{3}\frac{\mu a^2 M}{A}, \tag{19}$$

which is small. Also
$$\frac{3}{2}\frac{C - A}{Ma^2} = e - \tfrac{1}{2}\mu, \tag{20}$$

and if
$$\frac{A}{C - A} = \tau_0 \tag{21}$$

$$k = \left(\frac{2e}{\mu} - 1\right)\left(1 - \frac{\tau_0}{\tau}\right), \tag{22}$$

a result given by Love (1909) and Larmor (1909). As it is based on the first-order theory of the figure of the Earth it is an approximation to the solution of (17), accurate to about 1 part in 300, which is amply sufficient. Thus the period of the free oscillation in latitude provides a direct determination of k.

The direction of the axis of angular momentum with respect to the axes used is $(Al, Am, C)/C$, so that the observed motion is a little smaller than is given by $(l, m, 1)$.

7·041. We can now proceed to consider the forced variation of latitude due to periodic variations of the products of inertia, that is, to parts of F and G not determined by (14) and therefore not due to elastic deformation by the shift of the axis. We denote these extra parts by F_1, G_1. It is convenient to introduce an 'axis of inertia', whose extremity is the 'pole of inertia', with direction cosines λ, μ, 1 defined by

$$\lambda = -\frac{G_1\tau}{A}, \quad \mu = -\frac{F_1\tau}{A}. \tag{23}$$

This axis is the axis of maximum moment of inertia of the body whose moments and products of inertia are

$$A, A, A\left(1+\frac{1}{\tau}\right), F_1, G_1, H. \tag{24}$$

Then (10) reduce to the simple form

$$\frac{\tau}{\omega}l+m = \mu, \quad \frac{\tau}{\omega}\dot{m}-l = -\lambda. \tag{25}$$

The terms arising from $\dot{F}$ and $\dot{G}$ are small compared with those in F, G when the period is large compared with a day.

Now l and m may be found from the polar motion, being, indeed, the component angular displacements of the pole. If then some explanation of the annual variation of latitude, or of any other forced vibration, is suggested, we can compare the two sides of (25) and obtain a quantitative test of the hypothesis.

Changes in F and G can arise from any process that causes matter to accumulate unsymmetrically in middle latitudes. The chief is the annual change of the distribution of air, which is shown by the great variation of pressure in Central Asia; the monsoons are a by-product. The pressure gives a means of calculating the variation of mass per unit area. It is sometimes thought that the annual variation of latitude is due to the variation of pressure; this is not so, both being due to the variation of mass per unit area. If the excess mass per unit area is σ, we have

$$F_1 = a^4\int_0^\pi\int_0^{2\pi} \sigma\sin^2\theta\cos\theta\sin\lambda\,d\theta\,d\lambda,$$

$$G_1 = a^4\int_0^\pi\int_0^{2\pi} \sigma\sin^2\theta\cos\theta\cos\lambda\,d\theta\,d\lambda, \tag{26}$$

and the calculation is straightforward if σ is known.

The variation of pressure disturbs the ocean surface. On an equilibrium theory the water would adjust itself so that the loss of mass of the water per unit area just balanced the gain by the air, so that σ over the oceans would be a function of time only and not of position. If this is right the allowance for the oceans is simple.

There are possible contributions from the accumulation of snow in winter

and an almost negligible one from annual changes of vegetation. It should
be noticed that mere annual changes of level would produce hardly any
effect; the effect of lifting a given piece of matter vertically through any
practicable distance is negligible in comparison with the effect of trans-
porting the same mass horizontally through 90° of longitude in latitude 45°.

7·042. Results of observation. The analysis of the observations of the
polar motion meets special difficulties. It was expected that when material
for several years had accumulated the annual and 14-monthly components
would be separable by a straightforward harmonic analysis. Actually there
is a considerable irregularity. This is not altogether surprising, because
the free motion should undergo damping for various reasons, and its
existence invites questions about its maintenance. Now an irregular
disturbance would tend to build up a free vibration. A simple model was
suggested by Yule (1927) in somewhat similar problems. Imagine a massive
pendulum, at which several boys discharge peashooters. The hits occur at
irregular intervals, but gradually build up a vibration in the natural period
of the pendulum. The amplitude attained is limited by the damping. But
the motion is due simply to the occurrence of an excess of impulses
producing motion near one phase rather than the opposite, and if the hits
are at random intervals there will come a time when the excess is the other
way and the motion may die down or be reversed. Harmonic analysis over
a long period may then fail to reveal the periodicity at all, though it may be
revealed by shorter intervals. Yule gave a method of analysis based on
correlations between displacements at different intervals of time for
estimating the free period, and it has been applied in several similar
phenomena.

Now the variation of latitude shows just the behaviour of the pendulum
in Yule's model. In the first place the irregular disturbance is there. Again,
the motion, even if smoothed, cannot be represented by a combination of
two periodic motions of constant amplitude; the amplitude of the free
motion shows great fluctuations. Consequently an analysis similar to
Yule's would seem appropriate. Unfortunately, the error of observation is
not negligible and the separation of the irregular disturbance from the
observational error is a matter of extreme difficulty.

L. W. Pollak (1927), Jeffreys (1940b), and Walker and Young (1955) gave
analyses for increasing intervals of time. The relevant quantities are the
covariances of the angular displacements l, m at different intervals of time.
Walker and Young (1957) found that the published data had been smoothed,
and smoothing alters the covariances and destroys the whole basis of the
method. Even then there are complications because the observing stations
were not always the same; the stars observed also changed, mostly on
account of precession, and the methods of reduction were not uniform.
They managed to recover the unsmoothed values, and I made a new determi-

nation (Jeffreys 1968*d*) using data from 1899 to 1967. The data are given in terms of *x*, *y*, where *x* is the apparent displacement of the pole toward Greenwich, *y* to 90° W. I prefer to use

$$l = x, \quad m = -y.$$

For each interval of 7 years from 1899 to 1961 I took means for corresponding months and analysed for terms of the form $c + a\cos L + b\sin L$, where L is the Sun's longitude from January 16. The apparent standard errors of *c*, *a*, *b* were about 0·010″, but the separate values fluctuated irregularly by far more than this would suggest. *c* for *l* ranged from $-0\cdot015''$ to $+0\cdot053''$, for *m* up to 1947 from $+0\cdot003''$ to $-0\cdot041''$. But from 1947 *c* for *m* increased to $-0\cdot164''$. This deserves special comment because several analyses have tried to fit a uniform rate of variation. The variation is unexplained, but the facts contradict a uniform rate of drift in either the pole or the observatories. The mean values for the annual term were

$$l = -0\cdot066''\cos L - 0\cdot062''\sin L$$

$$m = +0\cdot060''\cos L - 0\cdot045''\sin L.$$

Converting L to the Sun's longitude ⊙ (from the vernal equinox) gives

$$l = -0\cdot084''\cos\odot + 0\cdot033''\sin\odot$$

$$m = -0\cdot013''\cos\odot - 0\cdot073''\sin\odot.$$

Before attempting to analyse for the free period the drift (whatever it may be due to) and the annual part must be removed. The values of *c*, which are means for 7 years, were adjusted to values at the centres of the intervals and then interpolated by a spline formula (§ 2·3) to keep the derivative continuous, and then interpolated to 1-month intervals. The interpolated *c* and the means for the annual terms were then subtracted. The remainders were classified over 14-month intervals and analysed for terms of the form

$$c' + a'\cos\gamma_0 t + b'\sin\gamma_0 t,$$

where $\gamma_0 = 2\pi/14$ months. Data up to 1967 December were available and were treated similarly.

According to the Yule model we are considering disturbances in the 14-month intervals, each of which produces a damped harmonic variation approximately of period 14-months; this gives a relation between consecutive values of a' and b' with an error, and the equations of maximum likelihood give estimates of the damping and a correction to the period. Observational error had given much trouble in earlier treatments, but with 14-month intervals it turned out to have no effect on the period and only a small one on the damping. Various checks on the general behaviour indicated that the Yule model was substantially correct.

In my 1940 paper I used a method based on maximum likelihood, but also noticed that from 1908 to 1921 the free motion showed a steady decline in amplitude, as if there was little disturbance in that interval. In Yule's model this would correspond to the detection of an interval covering several periods when there were no hits. Then the analysis could be made by taking account only of observational error. This gave a time of relaxation of $15\cdot1 \pm 1\cdot8$ years. I later doubted whether the reduction of amplitude could be largely due to the disturbances having been in such directions as to reduce the amplitude. However B. Guinot (1972) appears to have used a similar method; he uses six observatories not in the I.L.S. programme and gets a decline of amplitude from 1911 to 1935, most of it being from 1911 to 1923. He gets a time of relaxation of $16 \pm 1\cdot0$ year (p.e.) after 1929. However I am doubtful for the same reason as for my 1940 method. It appears that disturbances at intervals less than a year are correlated, but that correlation is inappreciable at intervals of more than 14 months. My results are

Period = $433\cdot2 \pm 2\cdot2$ mean solar days = $434\cdot3 \pm 2\cdot2$ sidereal days.

Time of relaxation 23 years, with limits corresponding to the standard error of 14 to 73 years. Its probability is far from normally distributed; that of the reciprocal is nearly so, and would be $(0\cdot043 \pm 0\cdot029)$/year.

I remarked to Young that another 100 years' observations would be needed to make a considerable improvement in the uncertainty of the damping, and he thought 200 years.

L. Mansinha and D. E. Smylie (1967) estimate that earthquakes can account for the generation of the Chandler wobble and the pole shift. The annual variation of latitude shows surprisingly large variations over 7-year intervals, and so far no really satisfactory explanation is known.

The polar motion also affects time-keeping. Okazaki and Nasaka (1971) find solutions from observations with photographic zenith tubes and astrolabes from $1962\cdot0$ to $1968\cdot0$ and get a fair agreement with the results from latitude observations.

7·05. Interpretation of *h* and *k*. According to Kelvin's theory of the straining of a uniform sphere of rigidity μ by body force

$$h = \frac{5}{2}\frac{2g\rho a}{19\mu + 2g\rho a}, \quad k = \tfrac{3}{5}h, \quad l = \tfrac{1}{2}k = \tfrac{3}{10}h.$$

Kelvin argued from the height of the fortnightly tide and drew the conclusion that the Earth as a whole is more rigid than steel. As k is better determined than h the argument may be adapted to the above value of k; with $a = 6\cdot37 \times 10^8$ cm., $\rho = 5\cdot53$ g./cm.^{3}, $g = 980$ cm./sec.^{2},

$$\mu = (1\cdot53 \pm 0\cdot04) \times 10^{12} \text{ dynes/cm.}^2.$$

The rigidity of steel is about $0\cdot8$ or $0\cdot9 \times 10^{12}$ dynes/cm.^{2}. At the time when Kelvin wrote a rigidity of this order seemed remarkably large, but it can now be checked by the seismological evidence. The density just below the intermediate layer being taken as $3\cdot3$ g./cm.^{3}, and the velocity of S as

4·3 km./sec., the rigidity is $6·2 \times 10^{11}$ dynes/cm.². Within the rocky shell these values rise to about 5 g./cm.³ and 7·5 km./sec., giving a rigidity near the base of the shell of $2·8 \times 10^{12}$ dynes/cm.². Further, a rigidity of the latter order is correct for over 1000 km. of the shell. The average rigidity of the shell is therefore distinctly larger than that of a homogeneous Earth that would fit the variation of latitude. Further, the assumption of homogeneity tends to put the mass too far from the centre and therefore increases the effectiveness of tidal forces; so that effectively a smaller rigidity than Kelvin's is needed to fit the value of k for the actual Earth. But a larger one is indicated by the seismic evidence if we regard the core as having a rigidity comparable with that of the shell. $\lambda + 2\mu$ for the outer part of the core is about 6×10^{12} dynes/cm.². With λ, μ nearly equal, as for the shell, μ would be about 2×10^{12} dynes/cm.², increasing toward the centre.

Formal solutions of problems allowing for departure from homogeneity have been given by G. Herglotz, W. Schweydar, L. M. Hoskins, A. E. H. Love and L. Rosenhead. Herglotz (1905) found the deformation of an incompressible sphere of Wiechert's type, with uniform shell and core, for a disturbing potential of degree 2. Schweyder (1916) used parabolic density and rigidity distributions. Love allowed for compressibility in a homogeneous sphere. Hoskins (1920) used distributions expressed by power series and also included compressibility. Rosenhead's solution is an extension of Herglotz's when the disturbing potential is a solid harmonic of any degree. Herglotz's solution is the most suitable for comparison, as it takes account of the existence of a major discontinuity, and the effect of compressibility appears to be small. As a specimen of the results I may quote my own solution (1915a, p. 198) based on the original Wiechert distribution with densities 3·2 and 8·2; a pair of rigidities of the shell and core that would fit the variation of latitude are $3·3 \times 10^{11}$ and $16·5 \times 10^{11}$ dynes/cm.². The former is much less than the seismological value, and the latter is less than that just suggested for a solid core. Stoneley (1926) combining Wiechert's densities with Knott's velocities (based on the Zöppritz-Turner tables and extrapolated to the centre) got a calculated yielding about two-thirds of that observed. His adopted rigidities were too low in the outer half of the thickness of the shell, about right in the deeper parts. His method was numerical and could be applied to any distribution of density and rigidity, besides taking account of compressibility. All the methods agree in implying a much lower rigidity somewhere than seismology did.* A way out of the difficulty, however, is opened by the suggestion that the central core is fluid. With a distribution of density of Wiechert type, but with the actual radius of the core, and an average rigidity of the shell I made two solutions. In the first the rigidity of the core was taken equal to that of the shell and gave $k = 0·191$. In the other it was

---

* M. S. Molodensky (1953) states that similar results were found in 1910 by L. S. Leybenzon.

taken zero and gave 0·372. The former result is decidedly too low. The latter is too high, but it appears possible that it would be appreciably reduced if we allowed for heterogeneity within the layers, chiefly because the use of a uniform density in each still puts the mass too far out. A further study by L. Rosenhead (1929 a, b) took account also of oceans covering the Earth to different uniform depths, the core being supposed fluid.* The following values are representative:

	k	$1 + k - h$
Statical theory	0·270	0·720
Fortnightly tide:		
Depth 7,260 ft.	0·288	0·705
Depth 29,040 ft.	0·279	0·710
Semi-diurnal tide:		
Depth 7,260 ft.	0·119	0·468
Depth 29,040 ft.	0·799	0·093

The agreement between the fortnightly and statical values of k is close, and it may be inferred that the theoretical values for annual and 14-monthly disturbances will lie between them. It is then in good agreement with the observed value. But the values for semi-diurnal tides depend greatly on the adopted depth of the ocean, and it appears that until the theory of tides in the actual ocean is more complete it would be dangerous to draw any conclusions from observed values of $1 + k - h$ for semi-diurnal and diurnal tides. The effect of the actual ocean is probably much smaller because the tidal currents are much obstructed and deflected. The main conclusion is that the period of the free motion of the pole is in good agreement with the hypothesis that the core is fluid, and would not be in agreement with any rigidity of the core standing to its bulk-modulus in any ratio that occurs for normal solids.

7·051. A further complication in the allowance for a liquid core was noticed by Poincaré (1910). In all the solutions for elastic deformations it had been supposed that the inertial terms can be neglected in comparison with elastic terms, so that internal displacements will conform to a statical theory. My solution and Rosenhead's for a liquid core simply adapted these by putting the core rigidity zero. This, combined with the previous assumption, would imply that the core is not moving. Poincaré's solution, which had not been seen to be relevant (though part of it is reproduced in Lamb's *Hydrodynamics*), took account of possible motions of the core with a rigid shell, and found that the presence of a liquid core would *shorten* the free period (a result previously given by Kelvin and S. S. Hough), and also that it would reduce the amplitude of the 19-yearly lunar nutation (which is practically unaffected by allowance for elasticity alone). The amplitude of the latter, calculated for a rigid Earth, is $9·2272'' \pm 0·0012''$

* Larmor (1915) and I (1915b) found that the ocean has a considerable effect on the free period; for a rigid Earth the lengthening would be from 305 to 342 days. An equilibrium theory of the tides was used; Rosenhead used a full dynamical theory.

and the observed value about $9 \cdot 2109'' \pm 0 \cdot 0023''$. The difference seems genuine and had presented a serious problem, as J. Jackson emphasized (1930). Application of modern data (Jeffreys, 1948d) indicated that a fluid core would make a change in the right direction, but three or four times too large. These results made it necessary to investigate the whole matter afresh, taking account of both fluidity of the core and elasticity of the shell (Jeffreys, 1949a, 1950a). The result for the variation of latitude was that the previous theory was not far wrong; the rigidity of the shell appropriate to the observed period was changed only from $2 \cdot 0$ to $1 \cdot 9 \times 10^{12}$ dynes/cm.^{2}. The smallness of the change is due to the fact that the shell bends in such a way as to produce comparatively little motion in the core after all.

A great advance was made by H. Takeuchi (1950). He took two distributions of density and elastic properties near to those found by Bullen and solved the statical problem completely by numerical integration. His results are

$$k = 0 \cdot 290, \quad h = 0 \cdot 587, \quad l = 0 \cdot 068$$
$$k = 0 \cdot 281, \quad h = 0 \cdot 610, \quad l = 0 \cdot 082$$

according to the model used. He further considered the effect of assuming a rigidity in the core and found that there would be definite disagreement with observation if this exceeded 10^{10} dynes/cm.^{2}. As the bulk-modulus is of order 10^{13} dynes/cm.^{2} the ratio would be far removed from any for a normal solid in any case; but zero rigidity, as for a liquid, is consistent with the data.*

This, however, does not close the question, because the main result of the corrected theory of the fluid core is that a statical theory is incorrect for motions that change the direction of the axis of rotation and therefore for the variation of latitude and the nutations. The nutations are produced by the same forces as the diurnal tides. It is correct for other motions, including the semi-diurnal, fortnightly, and semi-annual tides. R. O. Vicente and I (Jeffreys and Vicente, 1957a, b), using Takeuchi's theory of the shell, have made two solutions. In both the core was replaced by a simplified model with the correct mass and moment of inertia. In one the core was taken as uniform and incompressible, with an extra particle at the centre. In the other a Roche model (quadratic law of density) was used, the variation of density being taken as wholly due to compression. The free period was about 392 days in both cases, but allowance for the ocean tide would increase this to about 430 days. The correcting factors to the 19-yearly nutation are $0 \cdot 9964$ and $0 \cdot 9989$, so that the compressibility of the core is

---

* M. S. Molodensky (1953) has found numerical solutions for sixteen different Earth models. The nearest to those used by Takeuchi are his models 6 and 12. These have slightly smaller densities, and therefore elastic moduli in the shell. Model 6 assumes homogeneity and incompressibility in the core. Model 12 has a heterogeneous core. The values of h, k, l are almost the same for both. He finds for all his models that k is nearly $\frac{1}{2}h$.

important. In the actual core about half the variation of density is due to compression and the rest (with some arbitrariness) to the extra density of the inner core. The correct results should therefore be about midway between those given by the two models, and would make the amplitude about 9·212″ or 9·209″ according as Spencer Jones's or Rabe's value of the solar parallax is adopted. The agreement with observation is reasonably satisfactory, though there are some further complications.

The calculated bodily tide numbers on the two models are as follows. n is the speed relative to the stars; that relative to the Earth is $-\omega + n$, where ω is the Earth's rate of rotation.

Tide component	n	h	k	l	$1-h+k$	$1+h-\frac{3}{2}k$
			Central particle model			
OO	$-\omega/13\cdot7$	0·590	0·244	0·082	0·654	1·224
	$-\omega/183$	0·523	0·218	0·084	0·695	1·196
K_1	0	0·492	0·206	0·086	0·714	1·183
P	$\omega/183$	0·555	0·231	0·082	0·676	1·209
O	$\omega/13\cdot7$	0·584	0·242	0·082	0·658	1·221
Semi-diurnal and long period		0·585	0·289	0·082	0·704	1·152
			Roche model			
OO	$-\omega/13\cdot7$	0·597	0·258	0·070	0·661	1·210
	$-\omega/183$	0·710	0·298	0·072	0·588	1·263
K_1	0	0·551	0·244	0·082	0·693	1·185
P	$\omega/183$	0·568	0·264	0·084	0·696	1·172
O	$\omega/13\cdot7$	0·603	0·261	0·078	0·658	1·211
Semi-diurnal and long period		0·598	0·273	0·082	0·675	1·188

The largest diurnal tides are K_1 and O; it appears that the most likely tide to show detectable differences from the values found for semi-diurnal and long-period tides is O.

Molodensky's values for his model 6 are $h = 0\cdot619$, $k = 0\cdot310$, $l = 0\cdot091$. These might be expected to agree with those for semi-diurnal tides and our central particle model, but are somewhat larger; this may be a consequence of differences of density assumed in the shell.

One warning is needed. The nutations are motions of the Earth's axis of figure with regard to an inertial frame, that is, practically one fixed with regard to the stars. Their speed is n. But the corresponding tides are referred to the rotating Earth and their speed is $-\omega + n$. For instance, K_1 as a tide has the precession as its astronomical counterpart; the tidal counterpart of the 19-yearly nutation is a pair of tides of speeds $-\omega \pm \omega/6800$, which have nearly the same bodily tide numbers as K_1. The tides O and OO correspond to the fortnightly nutation, P to the semi-annual nutation; whereas the fortnightly and semi-annual tides are long-period tides and give no nutation.

Extensive work on the bodily tide numbers has been done in connexion with the International Geophysical Year (actually lasting over several years). Dr P. Melchior of Uccle, Belgium, is organizer. Melchior's summary (1966) amended, is as follows:

		$1-h+k$			$1+h-\frac{3}{2}k$	
	Obs.	C.P. model	Roche model	Obs.	C.P. model	Roche model
K_1	0·747	0·714	0·693	1·143	1·183	1·185
P_1	0·721	0·676	0·696	1·148	1·209	1·172
O_1	0·676	0·658	0·658	1·160	1·221	1·211
Q_1	0·654	—	—	1·167	—	—

Melchior and Georis (1968, p. 283) consider the determinations of $1-h+k$ the better, partly because the instruments are better calibrated, partly because the contribution 1 is due to the direct attraction of the tide-raising body, and the remaining part is numerically less for $1+h-\frac{3}{2}k$.

Q_1 is a tide of speed $\gamma - 3\alpha + \tilde{\omega}$ and therefore contains the eccentricity of the Moon's orbit as a factor; so it is much smaller than the others.

Melchior and Venedikov (1968) also succeeded in finding the component M_3 of period $\frac{1}{3}$ lunar day arising from the p_3 term in the Moon's field. They find the following values for the M_3 tide (probable errors):

	Observed	Calculated (Longman, 1963)
$\gamma_3 = 1-h_3+k_3$	$0·8352 \pm 0·0156$	0·803
$\delta_3 = 1+\frac{3}{5}h_3-\frac{4}{5}k_3$	$1·0717 \pm 0·0140$	1·069

These are, of course, adapted to a third harmonic, for which h and k differ from their values for a second. In particular the gravity factor δ_3 replaces $\delta_2 = 1+h-\frac{3}{2}k$.

I mentioned earlier that in the theory of artificial satellites it has become necessary to allow for the perturbations by the Moon and Sun. R. R. Newton (1967) has gone a step further. The bodily tides raised in the Earth produce similar effects, and he derives consistent values from solar and lunar tides, and finishes with

$$k = 0·336 \pm 0·028.$$

The result differs from that based on the variation of latitude by about 1·5 times the standard error; I do not consider the discrepancy serious.

A difficulty in the theory of the bodily tide has been pointed out by I. M. Longman (1963) and solved by C. L. Pekeris and Y. Accad (1973). If Love numbers are calculated on a statical theory, that is, for a liquid core ($\mu = 0$) and for a disturbance with speed $\sigma = -\omega + n = 0$ relative to the Earth, it is found that the equations lead to an equation that breaks up into two factors; according to one the dilatation is zero, according to the other the Adams–Williamson relation holds. If the dilatation is zero it is found that only two adjustable constants are available to satisfy three boundary conditions. But the Adams–Williamson relation cannot be universal; any concentration of denser materials to the centre will upset it, and so will any departure of temperature from the adiabatic.

Pekeris and Accad assume σ small but not zero and study the behaviour of the solution. There is no difficulty if the Adams–Williamson relation is satisfied. If it is not, there is an additional term asymptotically of the form $\exp(Ar/\beta^{\frac{1}{2}}\sigma^{\frac{1}{2}})$, where β represents the amount of departure from the relation. In the 'unstable' case, where the density increases less rapidly inwards, this gives a boundary layer; in the 'stable' case there is a harmonic variation with r. In both cases this introduces an extra constant.

The phenomenon is similar to one in hydrodynamics. Classical hydrodynamics often gave wildly wrong answers, and this was attributed to neglect of viscosity. I showed (H. Jeffreys, 1930d) that, if viscosity is neglected but the correct condition that there is no slip over a solid boundary is retained, then no solid in contact with a liquid could rotate. This is a far more serious objection to classical hydrodynamics than the neglect of viscosity. In fact viscosity, by raising the order of the differential equations by 1, permits the satisfaction of the boundary conditions by introduction of a boundary layer, often leaving irrotational motion valid as an approximation in most of the fluid. The Longman difficulty is essentially that the result that the dilatation is zero in the case considered reduces the order by 1.

Bullard's theory of the Earth's magnetic field depends on convection in the core. If this is right, on account of the great scale of the movement, the distribution of temperature must be nearly adiabatic and the material must be thoroughly mixed. Then Longman's condition would be closely satisfied, but not for his reason.

7·06. Annual motion. Computations of the annual motion from meteorological data have been made by Spitaler (1897), myself (1916a) and Rosenhead (1929b). Rosenhead had access to the most recent data, but he seems to have gone astray in discussing the variation of level of the ocean. He quotes (p. 152) from Nomitsu and Okamoto (1927; see also 1932), a result that the amplitude of the surface level of the sea near Japan is about three times that given on the equilibrium theory by atmospheric pressure, and consequently subtracts for the ocean three times the effect computed from atmospheric pressure. It appears, however, that this extra variation is mainly due to changes of salinity, which affects the density of the water. The associated effect on the mass per unit area is not necessarily generally proportional to the change of atmospheric pressure, and even in Japan the change of level may be partly or wholly compensated by the change of density. It appears safest to omit this contribution. Otherwise, ⊙ being the Sun's longitude, Rosenhead gives the following contributions.

	λ	μ
Atmospheric motion	$+0{\cdot}0102''\sin\odot - 0{\cdot}0020''\cos\odot$	$+0{\cdot}0509''\sin\odot - 0{\cdot}0077''\cos\odot$
Snowfall	$+0{\cdot}0175''\sin\odot - 0{\cdot}0238''\cos\odot$	$+0{\cdot}0093''\sin\odot - 0{\cdot}0126''\cos\odot$
Vegetation	$-0{\cdot}0038''\sin\odot + 0{\cdot}0017''\cos\odot$	$-0{\cdot}0055''\sin\odot + 0{\cdot}0024''\cos\odot$
Total	$+0{\cdot}0239''\sin\odot - 0{\cdot}0241''\cos\odot$	$+0{\cdot}0547''\sin\odot - 0{\cdot}0179''\cos\odot$

These are calculated as for a rigid Earth. The load, however, deforms the Earth and the actual values will be somewhat less. Rosenhead works out the theory of this and finds that a factor 0·696 should be applied. Rosenhead used a Wiechert model adapted as in 4·04. A more accurate treatment by Longman (1963) based on Gutenberg's model gives a factor $1 + k_2' = 0.690$. The difference of the models makes surprisingly little difference. k_2' is defined by Munk and MacDonald (1960, pp. 29–30). Then in all the theoretical values are

$$\lambda = -0.017'' \cos \odot + 0.017'' \sin \odot,$$

$$\mu = -0.012'' \cos \odot + 0.038'' \sin \odot.$$

In terms of $\odot$ the observed values of l, m are as on p. 298. Then from 7·04 (25) with $\tau = 434$,

$$\lambda = +0.003'' \cos \odot + 0.018'' \sin \odot$$

$$\mu = +0.026'' \cos \odot + 0.027'' \sin \odot.$$

In several cases a coefficient in the free motion jumped in consecutive 14-month intervals by about 0·060", nearly as large as any coefficient in the mean annual motion. If we have regard to the uncertainties of the observed values alone, these coefficients have standard errors of about 0·011", and the agreement is about as good as can be expected. Accordingly, the main features may be considered explained. It remains possible, however, that variations of temperature and salinity in the ocean give appreciable annual redistribution of mass, and they merit further attention.

When the annual terms are determined from observations over separate intervals of 6 years, the matter does not look so satisfactory, because even the largest coefficients in x and y fluctuate by factors of over 2. One would think that corresponding variations in the greatest annual motion of the atmosphere could hardly have escaped mention. The irregular motion itself is quite unexplained. In one case (1927·5 to 1928·7) a coefficient in the free motion jumped within a period by 0·10", nearly the mean amplitude, and larger than any coefficient in the mean annual motion.

Munk and MacDonald (1960, p. 116), think that the effect of vegetation, small in any case, is overestimated. But (p. 131) the shift of ocean mass due to wind stress, and the seasonal variation of ground water may be appreciable.

They thought at first that the meteorological irregularities were enough to explain the maintenance of the free motion, but (p. 166) quote E. M. Hassan for the result that they fail to do more than 0·1 of what is needed.

Another computation, from meteorological data alone, was made by Schweydar (1919). The results are substantially similar and he concludes that the changes of the products of inertia due to the redistribution of air are adequate by themselves. It is necessary to apply corrections for the depression of the ocean surface when it is loaded, and also to allow for the deformation of the solid Earth by the loading. In my treatment and in

Rosenhead's the former correction is applied first; the result would be that, after the displacements of the solid surface are allowed for, the ocean surface would no longer be one of equilibrium under the surface pressure. Schweydar on the other hand applied the correction for bending only to the land surface, and the result is that the mass of the solid alters. This has been pointed out by A. Young (1951), who found the result by direct computation. Neither treatment is satisfactory and we need some way of applying the corrections simultaneously.

I do not understand how Schweydar's agreement came to be so good. Rosenhead's data for atmospheric pressure were taken from Sir Napier Shaw's *Manual of Meteorology*, itself based on the *Réseau Mondial* of the Meteorological Office, and are as good as any data could be. But by them-selves they would have given a serious difference of phase, as in Spitaler's work and mine. This was cancelled by the allowance for snow and ice, which Schweydar did not take into account. More information on snow and ice would be most valuable, as the data I had were very rough.

A. Young (1952) has applied Schweydar's method (with some corrections) to the data on air pressure from 1925 to 1930. He gets a moderate agreement with the observed variation of latitude. He makes an allowance for damping, which had not been done previously. However, the difference in phase persists.

7·07. Tidal variations of gravity. The disturbance of gravity is of the order of $10^{-7}g$, but can be measured by specially designed static gravimeters. R. Tomaschek and W. Schaffernicht in Marburg got $1 - \frac{3}{2}k + h$ from 0·24 to 0·95 (Lambert, 1933). O. H. Truman (1939) at Houston, Texas, got 1·13. Workers of the geophysical staff of the Gulf Research and Develop-ment Company, at 13 places from Utah to Venezuela, got values from 1·4 to 2·2 (Lambert, 1940). These determinations were all based on the semi-diurnal tides. Considerable phase lags are found as in the measurement of $1 + k - h$, and disturbance by oceanic tides is again probable. Further, there are discrepancies between quite close stations, for which no explana-tion is suggested. At five stations in France, R. Bollo and A. Gougenheim (1949 a, b) got values from 1·147 to 1·310, with no determinable shift of phase. They remark that one of their stations was near enough to the coast to be specially affected by ocean tides if such an effect exists.

B. Baars (1951) has summarized observations at 26 stations distributed over the world, taking all tide components together. He gives $1·24 \pm 0·18$, but the stated uncertainty seems to be the average uncertainty estimated for the separate stations. There appears to be no significant difference between stations, and if we regard them as giving independent estimates of a constant value the standard error derived from their scatter is $\pm 0·014$.

Takeuchi's (1950) theoretical values are 1·15 and 1·19.

Kuo and Jachens (1970) have measured the tidal terms in gravity for the tides M_2 and O_1 across the United States. Their data make it possible to eliminate most of the effects of the ocean tides. Kuo has since set up stations approximately on a line in Europe, from Cambridge to Torino, and plans another from Copenhagen to Madrid.

A curious result emerges if we consider a pendulum kept permanently swinging (Jeffreys, 1928 b). On account of the periodic changes in apparent gravity it will show a periodic change in its time-keeping, and this accumulates over a half-period of the disturbance. Hence the small long-period terms are magnified in importance. The most important would be the semi-annual one. This is associated with a periodic change in the moment of inertia and hence in the rotation. The effect, of order 6×10^{-3} sec., is on the verge of being observable with modern techniques, and may become important because long-period tides in the ocean are not difficult to calculate.

For a homogeneous Earth, if $k = 0.288$, we should have

$$1 + k - h = 0.808, \quad 1 - \tfrac{3}{2}k + h = 1.048, \quad 1 + k - l = 1.144.$$

But on account of heterogeneity, and for most tides on account of the effects of the ocean, it would be expected that actual values would differ widely from these and from one another.

7·08. Free vibrations. A great deal of work started with Benioff's recognition of free vibrations of the Earth as a whole. The problem overlaps that of the Love numbers. In either case a model of the Earth has to be set up, and four differential equations of the second order are formed. These break up into sets of three and one, the former for the variables F, q and K, the latter for H. The former type are called spheroidal, the latter toroidal or torsional. Conditions at the centre impose three conditions on F, q and K, one on H, and the remaining $3 + 1$ have to be fixed by conditions at the surface. In the case of Love numbers a datum is the field imposed from outside and its period is given. For free vibrations the period has to be found. A number of trial periods are adopted and the boundary conditions require that a determinant involving the period shall vanish. An interpolated period is chosen so that this will happen. The analysis follows similar lines in both cases. Previous solutions were by Love (1911) for a very simplified model, and by Stoneley (1926), who used a Wiechert distribution of density and Knott's velocities, based on the Zöppritz–Turner times and extrapolated to the centre. Stoneley used a polynomial approximation, equivalent to one in Legendre polynomials, and made the total energy a minimum. The result, though rough, was enough to indicate that the model gave too little tidal yielding. It was a little later that I showed that with Gutenberg's revised radius of the core and a fluid core, some approach to the actual value could be attained. The first numerical solution for an Earth model based on modern values of the P and S velocities and the density was by Takeuchi (1950). His work and Stoneley's were done on hand machines; much more rapid calculation can now be done on electronic machines, and so many solutions have been obtained for different

Earth models that it is impossible to tabulate all of them here. Representative ones are by Nowroozi (1965), L. E. Alsop and J. N. Brune (1965), Takeuchi and Dorman (1964), Pekeris, Alterman and Jarosch (1961), Landisman, Sato and Nafe (1965), Gilbert and Backus (1961, 1965, 1966), MacDonald and Ness (1961), Benioff, Press and Smith (1961), Alsop and Kuo (1964), Gilbert and MacDonald (1960), Alsop, Sutton and Ewing (1961a), Ness, Harrison and Slichter (1961), Alsop (1963),

In the results spheroidal modes are indicated by $_lS_n^s$, where n and s indicate the harmonic affected. In general a mode has nodal spheres, the number of which is indicated by l. In the straightforward treatment the periods are independent of s, but when the rotation of the Earth is taken into account s has an appreciable effect, analogous to the Zeeman effect in spectroscopy. This was noticed almost simultaneously by Pekeris, Alterman and Jarosch (1961), and by MacDonald and Ness (1961). If $\sigma_0(n)$ is the speed for $s = 0$, the relation is

$$\sigma_n^s = \sigma_0(n) + s\tau(n)\,\omega,$$

where for torsional oscillations $\tau(n) = 1/n(n+1)$; for the fundamental spheroidal ones

$$\tau(2) = 0\cdot395, \quad \tau(3) = 0\cdot183, \quad \tau(4) = 0\cdot099.$$

Important verifications of the theory are, first, that since the point of application of the disturbance is known the nodes of the separate oscillations can be found; these agree with observation. Secondly, the spheroidal modes should affect gravity and be recorded on gravimeters, the toroidal ones should not. This is verified. The rotational splitting is also verified.

Representative periods in minutes from Nowroozi (1965) are as follows. Calculations are for the model Bullen A for densities and Gutenberg's velocities. They are done also for three other models. Stars indicate amplitudes above the 95 % confidence level.

The general impression is of extremely good agreement, but doubts appear on inspection. Most authors give no statements of uncertainty. Landisman, Sato and Nafe (1965) derive them from comparison of results of different authors. All have applied Fourier analysis to a stretch of the record; some have found the discrete terms of a Fourier series fitting this stretch; others have assumed zero displacement outside it and worked out a Fourier integral; others have calculated Fourier components for the finite interval for periods that are not submultiples of the interval. All use only the resultant amplitude (or its square) as criterion of reality. Nowroozi's uncertainties are got, if T is the length of the interval used, by working out coefficients for periods T/k and taking $\frac{1}{4}(T_{k-1} - T_{k+1})$ as the uncertainty. A better way is to note that if there is a real period between T/k and $T/(k+1)$, the cosine and sine coefficients change sign for these values, and taking these together with those for $T/(k-1)$ and $T/(k+2)$ gives eight equations for three unknowns. It is quite possible, if the signs alternate in the pre-

Free vibrations

Spheroidal

Mode	Obs. (Alaskan)	Obs. (Chilean)	Calc.
2	54·14 ± 1·59	53·59	53·50
3	35·40 ± 0·68	35·62	35·32
4	—	25·82	25·53
5	—	19·86	19·65
6*	16·15 ± 0·14	16·07	15·92
7*	13·53 ± 0·10	13·52	13·43
8*	11·80 ± 0·07	11·76	11·73
9*	10·58 ± 0·06	10·57	10·53
10*	9·659 ± 0·051	9·629	9·650
11*	8·937 ± 0·043	8·917	8·950
12*	8·368 ± 0·038	8·362	8·383
13*	7·867 ± 0·033	7·875	7·887
14*	7·484 ± 0·030	7·465	7·475
15*	7·136 ± 0·027	7·102	7·108
16*	6·786 ± 0·025	6·778	6·782
17*	6·482 ± 0·023	6·492	6·492
18*	6·262 ± 0·021	6·231	6·230
19	6·178 ± 0·020	5·998	5·993
20*	5·901 ± 0·019	5·778	5·778
21*	5·613 ± 0·017	5·595	5·582
22	5·415 ± 0·016	5·410	5·402
23	5·230 ± 0·015	5·256	5·235
24	5·171 ± 0·014	5·105	5·080
25	4·949 ± 0·013	4·959	4·937
26	4·819 ± 0·012	4·827	4·802
27	4·720 ± 0·012	4·694	4·675
28	4·557 ± 0·011	4·588	4·557
29	4·490 ± 0·011	4·469	4·445

Toroidal

n	Alaskan	Chilean	Calc.
6	15·34 ± 0·13	15·36	15·36
7	13·73 ± 0·10	13·53	13·57
8	12·27 ± 0·08	12·25	12·22
9	10·96 ± 0·07	11·14	11·15
10	10·58 ± 0·06	10·32	10·28
11	9·689 ± 0·051	9·614	9·55
12	8·937 ± 0·043	8·975	8·93
13	8·368 ± 0·038	8·379	8·39
14	7·867 ± 0·033	7·942	7·93
15	—	7·527	7·51
16	7·081 ± 0·027	7·161	7·15
17	6·768 ± 0·025	6·819	6·81
18	6·528 ± 0·023	6·502	6·52
19	6·262 ± 0·021	6·246	6·24
20	6·016 ± 0·019	6·006	5·99
21	5·796 ± 0·018	5·769	5·77

dicted way, for the true uncertainty to be much less than that used by
Nowroozi. If they do not, it may be much more. (Jeffreys, 1964 *a*,
1967 *d*, Appendix C.)

Landisman, Sato and Nafe (1965) give extensive lists for various models,
for harmonics up to degree 25 and modes with up to 4 nodal spheres. The
following table for toroidal modes is rearranged from part of theirs: in $_lT_n$,
n refers to the degree of the harmonic, l is the number of nodal spheres. It
refers to a Jeffreys–Bullen A model.

n	l	Period (min.)	Obs.
1	1	13·42	—
	2	7·53	
	3	5·15	
2	0	43·51	44·12 ± 0·39
	1	12·56	
	2	7·38	
	3	5·10	
3	0	28·16	28·47 ± 0·06
	1	11·51	
	2	7·18	
	3	5·03	
4	0	21·56	21·76 ± 0·04
	1	10·44	
	2	6·92	
	3	4·94	
5	0	17·78	17·90 ± 0·02

The observed values have been computed straightforwardly from their Table 2a. They are slightly, but apparently genuinely, longer than the calculated ones.

For $n = 1$, $l = 0$ cannot arise because a motion of this type would give a change of angular momentum.

This work, including that on Love numbers, is a valuable check on the seismic velocities and the density distributions adopted. As just indicated, however, I think that it is possible that in some cases the observed values are not the best obtainable, and I distrust all the uncertainties. Also where distributions of velocity are attributed to me the velocities near the surface have been increased, and it is not clear that the consequences would agree with the travel times.

Landisman, Sato and Nafe make a substantial change in the distribution of density. In Bullen's model the density, except in the transition region beginning about depth $0·06R$, is based on the Adams–Williamson relation. It can however be taken as an unknown and adjusted to fit the periods. The result is that the density is nearly constant from depth 1600 km. down to the core. Pekeris (1965) has shown that for the higher modes the periods of toroidal modes depend almost entirely on the S velocities and hardly at all on the density. Any information on the density must come from the lower modes. Even for these the result is surprising and must be taken in relation to the unknown uncertainties of the observed periods. Bullen and Haddon (1967a) have introduced the radius of the core as an extra unknown and find that their conclusion can be avoided if this is increased by 15 km. The present apparent standard error of this is only 2·5 km., but this is based mainly on the residuals of ScS, and hence depends on the comparison of P with S, and then of S with ScS, so some additional uncertainty may enter. I have a greater doubt arising from the different results of calculation of the Love numbers; this is a simpler problem than that of the free periods. Results found by different authors differ by about 1 part in 30. Similar differences are found for the difference between statical and semi-diurnal tides, and for the effect of the outer layers. The ranges of uncertainty of the P and S times are of order 1 in 300. The ratio of the speeds of the slowest free vibration and the semi-diurnal tide is about 13, and we should expect the Love numbers for statical and semi-diurnal tides to differ in a ratio of the order of 1/200. For a simplified model Love (1911, p. 73) got about 1 in 500. Vicente and I found (1966), again, for a simplified model, that the effect of the surface layers should again be of this order.

In view of these unexpected results of calculation Vicente and I (1966) examined the differential equations in the forms given by various authors and found them equivalent. The usual form is by Alterman, Jarosch and Pekeris (1959 a, b). So I think the explanation must lie in the arithmetic. Perhaps it is as easy to misplace a decimal point on a machine as on a piece of paper. I think therefore that the next step should be that a number of users of high speed machines should use identical forms for α, β and ρ, as simple as possible, and compare results. Birch's model or even quadratic laws of density such as I used in the theory of the figure of the Earth (1963 b) would be the most suitable, as small corrections will need to be made later, and it is easier to introduce irregularities into a smooth function than to redistribute them in an already irregular one. At present the discrepancies in the Love numbers suggest errors of the same order in the calculated periods.

With a rigid shell and a liquid core there is an additional free period of speed near 1 day, referred to the Earth. For the central core model Vicente and I got the period $23^h 56^m 47^s$ sidereal time, a nearly diurnal nutation in the opposite sense to the Earth's rotation. The Roche model gave two such solutions, of periods $23^h 54^m 13^s$ and $24^h 9^m$, the latter actually longer than a solar day. The existence of these periods led to uncertainties in the calculation of the effects corresponding to the tide components O and OO. Pariiski (in R. O. Vicente, 1961) suggested a search for these motions, which was carried out by N. Popov (1963 a, b) and D. V. Thomas (1964). There is evidence for a period near $23^h 56^m 54^s$, but the uncertainty is considerable. See also Jeffreys and Vicente (1964). Slichter (1961) finds an additional period of 86^m, and suggests that it may represent an oscillation of a rigid inner core, with respect to the main core, but finds quantitative difficulties; the straightforward solution would need a density of 27 to 29 g./cm.[3] for the inner core.

For the modified Lomnitz law (p. 48) the extra term in the restoring coefficient gives a lengthening of the period, greater for the longer periods. The observed periods are about 1 per cent longer than the calculated ones, and this is of the order of magnitude of the effect to be expected (Jeffreys, 1967c). For the slowest free vibration of the Earth as a whole Q appears to be about 400. $\alpha = 0.2$ makes $\cot \frac{1}{2}\pi\alpha$ about 3; thus damping would lengthen the period by about 1 part in 130. The thickness of the shell is about 3000 km., so that the value inferred from free vibrations might be wrong by about 20 km.; which is fairly close to the change inferred by Bullen and Haddon.

In the circumstances it seems objectionable to use periods of free vibrations to correct the distributions of velocities of seismic waves. It may become possible when we know more about the radial distribution of the damping. Unfortunately the data are very incomplete, especially in the top 200 km. It seems worth while to call attention again to the method of making small corrections given above (pp. 75–6). Much more elaborate methods are in use but I see no reason to suppose that they are more accurate.

E. P. Fedorov has analysed about 135,000 observations of the International Latitude Service for the 18·6-yearly and fortnightly forced nutations. Popov has analysed for the semi-annual ones. In all previous discussions of observations the amplitudes in obliquity and longitude have been assumed to be in the ratio for a rigid body. This is not true for an Earth with a liquid core. Fedorov and Popov have estimated them separately. It is convenient to analyse the actual motions, which are elliptical, into two circular motions in opposite senses. Vicente and I (Jeffreys and Vicente, 1957 a, b) did the calculations for two models of the core. Both were chosen to make the mass and moment of inertia agree with a model by Bullard. In one, called the central particle model, the core was taken homogeneous and incompressible, with an extra particle at the centre. In the other, called the Roche model, a quadratic law of density was assumed and the variation was assumed to be entirely due to compression. It was thought that interpolation with weight 0·4 for the central particle and 0·6 for the Roche model would give a reasonable representation of the actual core. The results for the direction cosines of the axis of figure were (Jeffreys, 1959 b):

$$l = 9 \cdot 2118'' \cos \Omega + 0 \cdot 0971'' \cos 2\mathbb{D} + 0 \cdot 5535'' \cos 2\odot,$$

$$m = 6 \cdot 8399'' \sin \Omega + 0 \cdot 0897'' \sin 2\mathbb{D} + 0 \cdot 5023'' \sin 2\odot.$$

Fedorov's and Popov's values are (with some revision of the uncertainties)

$$l = (9 \cdot 198'' \pm 0 \cdot 004) \cos \Omega + (0 \cdot 0949'' \pm 0 \cdot 0027) \cos 2\mathbb{D}$$
$$+ (0 \cdot 578'' \pm 0 \cdot 004) \cos 2\odot,$$

$$m = (6 \cdot 853'' \pm 0 \cdot 004) \sin \Omega + (0 \cdot 0918'' \pm 0 \cdot 0027) \sin 2\mathbb{D}$$
$$+ (0 \cdot 533'' \pm 0 \cdot 004) \sin 2\odot.$$

The 2$\mathbb{D}$ terms are in good agreement. The calculated values for the 2$\odot$ terms are doubtful as there are resonances in the neighbourhood, which would be sensitive to the treatment of the inner core. The terms in Ω (the longitude of the moon's node) are curious; the agreement is much better than in any previous comparison, but the reductions of the coefficients from the rigid body value (that in l with the latest estimates about 9·225″) are in nearly the same ratio. With any liquid core model the reduction in the coefficient of $\cos \Omega$ would be more than in that of $\sin \Omega$. Actually one observed coefficient agrees well with the central particle, the other with the Roche model.

The main result of the comparison, however, is that the assumption of a liquid core, with allowance for the elasticity of the shell, removes most of the discrepancy.

As data for Love's numbers and free oscillations have been used to derive upper bounds for the rigidity of the core, it is worth while to point out that the reduction of the calculated 18·6-year nutation in comparison with that for a rigid Earth depends entirely on the time of transmission of an S wave through the core being long compared with a day. The velocity would need to be less than 0·08 km./sec. and the rigidity less than 6×10^8 dynes/cm.^{2}. This is far less than any bound obtained otherwise. A larger rigidity would again make the amplitude approximate to the value for a rigid Earth.

H. Kimura (1902a,b) noticed that the stations showed a variation of period 1 year and amplitude about 0·02″, the same in all longitudes. Explanations of this were attempted as due to some systematic error, such as refraction over the domes. However, a change in the method of observation led Wako (1970, 1972) to an explanation. The practice had been for a station to observe pairs of stars crossing the meridian near midnight. Later, pairs were also observed crossing it at 10 p.m. and 2 a.m. Thus data were available for the same date and stars differing by $\pm 30°$ in right ascension. Wako analysed the data for 1955–66 for a term of the form $a \sin(m\odot + n\alpha)$, and got $m = 2, n = -1$; $\odot$ is the Sun's longitude and α the right ascension of the star. By two methods of analysis he found

$$(0\cdot0137 \pm 0\cdot0019)'' \sin(2\odot - \alpha + 2°\cdot2 \pm 7°\cdot8)$$
$$(0\cdot0203 \pm 0\cdot0029)'' \sin(2\odot - \alpha + 4°\cdot3 \pm 8°\cdot1).$$

To understand these it is necessary to attend to the conditions of observation. Observations were made about midnight and refer to stars crossing the meridian then. They therefore refer, for each station, to constant values of $\odot - \alpha$ (but not to the same stars). Thus the results, if always made at exact midnight, would represent the $2\odot$ term as one in $\odot$, independent of longitude. For comparison, the values previously available (Jeffreys, 1959b) for the coefficient of the term in $\cos 2\odot$ are as follows:

Calculated	Rigid Earth	0·5528″
	Central particle core	0·5734″
	Roche model core	0·5403″
Observed	Fedorov–Popov	0·578″ ± 0·004″

The observations had been corrected as for a rigid Earth; an increase of 0·02″, as Wako gets, would bring the result into agreement with that for a core with a central particle and with Fedorov and Popov. It seems that the inner core, probably solid, is better represented by the central particle than by the too compressible one.

7·09. Periodic changes in the rotation of the Earth. There has been a great improvement recently in artificial standards of time-keeping; crystal clocks and other types can achieve a constancy to about 1 part in 10^9, and are actually more uniform than the Earth's rotation. Stoyko (1937) has found that the Earth's rotation has an annual variation. Van den Dungen, Cox and van Mieghem (1949) examined the effect of a periodic variation of the moment of inertia of the atmosphere about the polar axis, and therefore belonging to the same group of phenomena as the annual variation of latitude. This preliminary estimate was about a fifth of the observed effect. W. H. Munk and R. L. Miller (1950) investigated the matter in greater detail and found about 15 % of the observed effect. They examined also the change of angular momentum of the atmosphere due to seasonal variation of wind, and found that the reaction of this on the Earth would give an effect of the right order of magnitude. Y. Mintz and Munk, however, in a more detailed calculation, reduced the effect to about one-third. They have revised their work (1951, 1954); the astronomical determination of the variation has been reduced; and they conclude that with allowance for the effect of the bodily tide the results of observation are now in reasonable agreement with theory.

A. Young (1951) finds that the effect of displacement of the ocean reverses the sign of the change of moment of inertia found by van den Dungen, Cox and van Mieghem. The total effect is about a hundredth of that observed. He has also examined the effect of seasonal variations in the polar ice and finds a result of the same order of magnitude.

I found in the paper (1968b) on the tidal effect on gravity, that tides could also have an effect on the rotation of the Earth, the most important period being semi-annual. However this does not appear to have been detected.

7·10. Strain due to surface loading can be calculated by methods similar to those used for the bodily tide. This has been done in detail by I. M. Longman (1962, 1963, 1966). It is relevant to the Rosenhead correction in the annual variation of latitude. Longman used Gutenberg's model. A. Roy has recalculated the Love numbers and the effects of loading for the same model and gets close agreement with Longman. This is the first time two workers have got consistent results.

CHAPTER VIII

Imperfections of Elasticity: Tidal Friction

'Ye fyul', sez a chep, 'it's a bonny myun,
They've ketched and myed it a clock fyece.'
Tyneside song, 'The Fiery Clock Fyece'

8·01. Nature of tidal friction. This is a phenomenon associated with any departure of the Earth's material from perfect elasticity, or, in the case of the ocean and the core, from perfect fluidity. It takes place slowly, but is irreversible, and its effects accumulate over long periods and may have produced great changes in the rotation of the Earth and the orbit of the Moon since they first came into existence. It may be illustrated by the following two diagrams. Let m in Figs. 22 and 23 denote the Moon, while

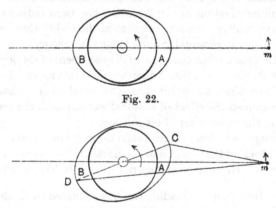

Fig. 22.

Fig. 23.

the circle enclosed by the thick line denotes the equatorial section of the solid Earth. The arrows indicate the directions of rotation and revolution. If the solid Earth was a perfect sphere and the ocean a perfect fluid, the tides raised in a deep ocean would be at A and B, vertically below the Moon and straight opposite to it, as shown by the continuous ellipse in Fig. 22. Moderate rotation would affect the height of the tides, but not their position with reference to the Moon. During the revolution of a particular point P on the Earth's surface, the water above it will still rise and fall when the rotation is taken into account, the maximum height being when P is just below or just opposite to the Moon, and the minima at the two points of quadrature. If, however, friction is present, it will delay the times of maximum and minimum elevation at P in accordance with the well-known effect of friction in making small oscillations lag in phase. Thus the highest tide at P will not occur till some time after it has passed A or B, and the form of the section of the water surface by the equator will resemble the ellipse in Fig. 23, the high tides being at C and D.

Now let us consider the attraction of the tides on the Moon. For simplicity the two high tides may be replaced by two heavy particles at C and D. The attraction of C on the Moon is along mC, and that of D is along mD. That of C is the greater, for it is nearer to m. Neither force acts accurately along the line joining the centre of the Earth to the Moon, and therefore both have components at right angles to it. That arising from C is evidently the greater, both because the resultant force is greater, and because the angle CmO is greater than DmO. Thus there will be on the whole a force on the Moon with a component in the direction of revolution.

Similarly, we may consider the attraction of the Moon on the two tidal protuberances. As the angular momentum of the Earth and Moon together must remain constant, this attraction tends to turn the Earth in a direction opposite to its rotation, which is therefore retarded.

Similar results will hold even if there is no ocean, but the interior of the Earth is imperfectly elastic. Bodily tides, like ocean tides, will tend to accelerate the Moon and retard the Earth.

8·02. Effects of tidal friction. The above simple discussion does not take into account the complications due to the irregular form of the oceans. But we shall see that so long as there is any dissipation the same secular effects follow from considerations of energy. The solar tides, again, behave similarly to those raised by the Moon, and therefore have to be taken into account in a quantitative discussion.

Let the masses of the Earth, Moon and Sun be M, m and m'. Let the mean angular velocities of the Moon and Sun about the Earth be n and n' and their distances c and c'. Let the couples acting on the Earth due to the lunar and solar tides be $-N$ and $-N'$. Let the Earth's rotation be ω and the principal moment of inertia about the polar axis C. Then

$$n^2 c^3 = f(M+m), \quad n'^2 c'^3 = f(M+m'), \tag{1}$$

f as usual being the constant of gravitation. Put

$$c = c_0 \xi^2, \quad n = n_0 \xi^{-3}, \quad c' = c'_0 \xi'^2, \quad n' = n'_0 \xi'^{-3}, \tag{2}$$

suffix 0 indicating the present value.

The angular momentum of the orbital motion of the Moon and Earth about the centre of mass of the two together is

$$\frac{M m c^2 n}{m+M} = \frac{M m c_0^2 n_0}{M+m} \xi. \tag{3}$$

That of the Earth's rotation about its axis is $C\omega$. To the couple $-N$ acting on the Earth's rotation must correspond one of $+N$ tending to increase the orbital angular momentum. The solar tides will have no secular effect on the Moon, for their period is different from that of the lunar tides, and in the course of a lunar synodic month they are presented to the Moon in all aspects. Thus they will accelerate the Moon in one half of the month as much as they retard it in the other half. Similarly, the lunar tides, in the long run, will not affect the Sun. Then altogether

$$\frac{Mmc_0^2 n_0}{M+m}\frac{d\xi}{dt} = N,$$ (4)

$$\frac{m'Mc_0'^2 n_0'}{m'+M}\frac{d\xi'}{dt} = N',$$ (5)

$$C\frac{d\omega}{dt} = -N-N'.$$ (6)

If E is the total mechanical energy in the system, it decreases at a rate equal to the sum of the rates of performance of work by the angular motions in overcoming the couples. This gives

$$-\frac{dE}{dt} = (N+N')\omega - Nn - N'n'.$$ (7)

Consider the effects of the lunar tides separately. These give

$$-\frac{dE}{dt} = N(\omega - n).$$ (8)

The left side is positive, since the couples arise from dissipation of energy. Thus N has the same sign as $\omega - n$. In the Earth-Moon system this is positive, and therefore N is positive. The argument is independent of any assumption about the form of the ocean or the nature of the imperfection of elasticity. Similarly, N' must be positive. It follows from (4) and (5) that the mean distances of the Sun and Moon must be increasing, and accordingly that their mean motions must be decreasing. From (6), the rate of rotation of the Earth must be decreasing.

Consider now the ratio of N to N'. If the x axis is towards the Moon, we have for the tide-raising potential due to the Moon (with the obvious changes from the argument of 4·08)

$$\frac{fm}{2c^3}(2x^2 - y^2 - z^2).$$ (9)

With spherical polar coordinates at the centre of the Earth, the longitude being measured from the direction of the Moon, this is

$$\frac{fmr^2}{2c^3}\{(\tfrac{1}{2} - \tfrac{3}{2}\cos^2\theta) + \tfrac{3}{2}\sin^2\theta\cos 2\lambda\}.$$ (10)

The first term, for a given point on the Earth, is independent of time, and therefore gives only a small permanent deformation. The second term produces the semi-diurnal tide, for it has maxima when the point agrees in longitude with the Moon or differs by 180°. Thus the tide-generating potential is

$$U = \frac{3}{4}\frac{fmr^2}{c^3}\sin^2\theta\cos 2\lambda. \tag{11}$$

The couple due to the attraction of the Moon on an element of the Earth is $\rho\frac{\partial U}{\partial\lambda}d\tau$, where $d\tau$ is an element of volume and ρ the local density. The total couple is therefore

$$-N = -\iiint\frac{3}{2}\frac{fmr^2\rho}{c^3}\sin^2\theta\sin 2\lambda\, r^2 dr\, d\omega, \tag{12}$$

taken through the Earth. Consider only the oceanic tide, treating the solid for a moment as rigid. The integral through the solid then vanishes. The height of the oceanic tide is proportional to U except for a lag due to tidal friction, so that it is of the form $H\sin^2\theta\cos 2(\lambda-\epsilon)$, where ϵ is the angle COA in Fig. 23, and the periodic part of $\int\rho\,dr$ is ρ times this, the relevant value of ρ being now the density of the ocean. Then if A is the Earth's radius

$$-N = -\iint\frac{3}{2}\frac{fmA^4}{c^3}\rho H\sin^4\theta\sin 2\lambda\cos 2(\lambda-\epsilon)\,d\omega$$

$$= -\tfrac{8}{5}\pi\frac{fmA^4}{c^3}\rho H\sin 2\epsilon. \tag{13}$$

A similar formula applies to the tide raised by the Sun. Now the periods of the two tides are not very different, and we may suppose that the respective values of H are nearly in the ratio of the coefficients in the total potentials and that the two lags are nearly equal. Then, nearly,

$$\frac{N}{N'} = \left(\frac{m/c^3}{m'/c'^3}\right)^2 = 4\cdot 9. \tag{14}$$

We readily see that the same arguments can be generalized so as to apply to the bodily tide, always supposing that the equations of motion are linear and that the system is far from resonance.

8·03. Non-linear friction. The argument needs modification if the equations of motion are non-linear, and it appears that this is the important case. In the open ocean there is little trouble, but most of the dissipation at present is in shallow marginal seas, where the currents are magnified by local conditions. The friction of the current over the bed is nearly proportional to the square of the velocity and directed against the velocity. This produces a reaction on the tide in the open ocean, and the tidal friction arises from the attraction of the Sun and Moon on the secondary tides produced by this reaction. We therefore need to know the parts of the frictional force that have periods equal respectively to the periods of the lunar and solar tides.

In a region where friction is considered, let u be the velocity of the lunar and v that of the solar current, and choose the origin of time so that we can take

$$u = a \sin pt, \quad v = b \sin rpt, \quad pt = x, \quad b/a = \nu. \tag{1}$$

b is less than $\tfrac{1}{2}a$, and their ratio is approximately the same all over the Earth. The frictional force can be written in a form that makes its direction explicit:

$$F = k \,|\, u + v \,|\, (u + v). \tag{2}$$

F can be expressed as the sum of a series of sines and cosines. We need only the terms whose periods are the same as those of the disturbing potentials, since others will give only periodic effects on the Sun and Moon. Denote them by $P \sin pt$ and $Q \sin rpt$. Then

$$\pi P = \lim_{n \to \infty} \frac{1}{n} \int_0^{2n\pi} F \sin pt \, d(pt), \tag{3}$$

$$\pi Q = \lim_{n \to \infty} \frac{1}{n} \int_0^{2n\pi} F \sin rpt \, d(pt), \tag{4}$$

with

$$x = pt, \tag{5}$$

$$F = ka^2 \,|\, \sin x + \nu \sin rx \,|\, (\sin x + \nu \sin rx). \tag{6}$$

ν is fairly small, and $\sin x + \nu \sin rx$ has the same sign as $\sin x$ except within parts of intervals $\sin^{-1} \nu$ on each side of a multiple of π. Then we can break up the range of integration into intervals from $m\pi + \sin^{-1}\nu$ to $(m+1)\pi - \sin^{-1}\nu$, and intervals from $m\pi - \sin^{-1}\nu$ to $m\pi + \sin^{-1}\nu$. Then

$$P = P_1 + P_2 \tag{7}$$

and

$$\tfrac{1}{2}\pi \frac{P_1}{ka^2} = \lim_{n \to \infty} \frac{1}{n} \sum_0^n \int_{m\pi + \sin^{-1}\nu}^{(m+1)\pi - \sin^{-1}\nu} (-1)^m (\sin x + \nu \sin rx)^2 \sin x \, dx, \tag{8}$$

$$\tfrac{1}{2}\pi \frac{P_2}{ka^2} = \lim_{n \to \infty} \frac{1}{n} \sum_0^n \int_{m\pi - \sin^{-1}\nu}^{m\pi + \sin^{-1}\nu} |\sin x + \nu \sin rx| (\sin x + \nu \sin rx) \sin x \, dx. \tag{9}$$

P_2 is clearly of order ν^4. Each integral in P_1 is the sum of a constant and a set of cosines, but such sums as

$$\frac{1}{n} \sum_{m=0}^n \cos rmx \tag{10}$$

tend to zero when r is not an integer. Hence several of the terms can be dropped at once, and we find

$$\tfrac{1}{2}\pi \frac{P}{ka^2} = \lim_{n \to \infty} \frac{1}{n} \sum_{m=0}^n (-1)^m \int_{m\pi + \sin^{-1}\nu}^{(m+1)\pi - \sin^{-1}\nu} \{(\tfrac{3}{4} + \tfrac{1}{2}\nu^2) \sin x - \tfrac{1}{4} \sin 3x\} \, dx$$

$$= \tfrac{4}{3} + \nu^2 + O(\nu^4). \tag{11}$$

Similarly, $$Q = Q_1 + Q_2, \qquad (12)$$

$$\tfrac{1}{2}\pi \frac{Q_1}{ka^2} = \lim \frac{1}{n}\Sigma(-1)^m \int_{m\pi+\sin^{-1}\nu}^{(m+1)\pi-\sin^{-1}\nu} \nu \sin x \, dx = 2\nu\sqrt{(1-\nu^2)}. \qquad (13)$$

For Q_2 we suppose $\sin^{-1}\nu$ small enough for its cube to be neglected. We have

$$\int_{m\pi-\sin^{-1}\nu}^{m\pi+\sin^{-1}\nu} |\sin x + \nu\sin rx| (\sin x + \nu\sin rx)\sin rx \, dx$$

$$\doteqdot \int_{-\sin^{-1}\nu}^{\sin^{-1}\nu} |(-1)^m\theta + \nu\sin rm\pi| \{(-1)^m\theta + \nu\sin rm\pi\}\sin rm\pi \, d\theta$$

the average of which is

$$\tfrac{5}{4}\nu^3 + O(\nu^4) + \text{periodic terms}. \qquad (14)$$

Hence to order ν^3 $$\frac{\pi P}{ka^2} = \tfrac{8}{3} + 2\nu^2, \quad \frac{\pi Q}{ka^2} = 4\nu + \tfrac{1}{2}\nu^3. \qquad (15)$$

The argument is independent of the value of r except that it is supposed small enough for x to be replaced in $\sin rx$ by its mean value within an interval of x of length $2\sin^{-1}\nu$. If $r\nu$ is large Q_2 will be less than is given by (14) (Q_1 is unaffected). For comparison of solar and lunar tides r is about $\tfrac{30}{29}$.

The rates of dissipation in the lunar and solar tides, per unit area, are then in the ratio

$$\tfrac{3}{2}\nu^2 \frac{(1+\tfrac{1}{8}\nu^2)}{1+\tfrac{3}{4}\nu^2}. \qquad (16)$$

By 8·02 (8), the ratio $$\frac{N'}{N} = \frac{3}{2}\frac{\nu^2(1+\tfrac{1}{8}\nu^2)}{1+\tfrac{3}{4}\nu^2} = \frac{1}{3\cdot7}, \qquad (17)$$

if we take the currents as in the ratio of the potentials.

The ratios are independent of the ratio of the periods, to this accuracy. I was somewhat doubtful about the analysis as a whole, especially the value of Q, and asked Dr J. Howlett to do a numerical check on the Atlas machine. The dissipation in the variation of latitude might also be important, and the check covered a range of ν from $1/2\cdot3$ to $1/20$, and of r from $13/14$ to $1/800$. The agreement was satisfactory. A verification has also been carried out by R. R. Newton, who finds however that the ratio is altered if the tidal motion is superposed on a steady current and will approximate to $1/4\cdot9$. I think the same would be true, if, as I believe, they are superposed on ordinary wave motion.

8·04. The secular accelerations of the Sun and Moon. Let us now consider the effect of the variations in the angular velocities on observations of astronomical phenomena. Suppose that ω, n, n' are given by observations near time zero, and that from the values found we infer that a known fixed star would cross the Greenwich meridian at time T if all these quantities were constant. Then the effect of the variation in the Earth's rotation is to put the Earth ahead in time T by $\frac{1}{2}T^2\frac{d\omega}{dt}$. Thus the meridian of Greenwich reaches a fixed star in time $T - \frac{1}{2}\frac{T^2}{\omega}\frac{d\omega}{dt}$ instead of in time T; in other words, the time of transit on a dynamical time-scale is hastened by $\frac{T^2}{2\omega}\frac{d\omega}{dt}$. The Moon moves among the stars with mean angular velocity n, and therefore the alteration of the time of the observation makes the Moon an angle $\frac{nT^2}{2\omega}\frac{d\omega}{dt}$ behind its calculated position when the star transits. On the other hand, the change in its own angular velocity puts it ahead by an angle $\frac{1}{2}T^2\frac{dn}{dt}$ in time T. Thus, in comparison with the positions inferred for time T, the Moon appears to have gained on the stars by $\frac{1}{2}T^2\left(\frac{dn}{dt} - \frac{n}{\omega}\frac{d\omega}{dt}\right)$; in other words it has an apparent secular acceleration $\frac{dn}{dt} - \frac{n}{\omega}\frac{d\omega}{dt}$. Let us denote this by ν. The Sun similarly will have a secular acceleration, which we shall denote by ν'. Then

$$\nu = \frac{dn}{dt} - \frac{n}{\omega}\frac{d\omega}{dt}, \tag{1}$$

$$\nu' = \frac{dn'}{dt} - \frac{n'}{\omega}\frac{d\omega}{dt}, \tag{2}$$

and any other astronomical body not affected directly by tidal friction, with mean motion n'' relative to the stars, will have a secular acceleration $-\frac{n''}{\omega}\frac{d\omega}{dt}$. The terms in $d\omega/dt$ represent an allowance for the fact that the rotation of the Earth provides the practical standard of time, which is different from dynamical time if the rotation is varying.

From 8·02 (2)

$$\frac{dn}{dt} = -3n_0\xi^{-4}\frac{d\xi}{dt}, \quad \frac{dn'}{dt} = -3n_0\xi'^{-4}\frac{d\xi'}{dt}, \tag{3}$$

whence

$$\nu = -3\frac{M+m}{Mm}\frac{N\xi^{-4}}{c_0^2} + \frac{N+N'}{C\omega}n_0\xi^{-3}, \tag{4}$$

$$\nu' = -3\frac{m'+M}{m'M}\frac{N'\xi'^{-4}}{c_0'^2} + \frac{N+N'}{C\omega}n_0'\xi'^{-3}. \tag{5}$$

So long as intervals of only a few thousand years are considered, ξ and ξ' may be taken equal to 1 without sensible error. Put

$$\frac{Mm}{M+m}\frac{c_0^2 n_0}{C\omega_0} = \kappa, \tag{6}$$

so that κ is the present ratio of the orbital angular momentum to the angular momentum of the Earth's rotation. With

$$C = 0{\cdot}33Ma^2, \qquad (7)$$

$$\kappa = 4{\cdot}89. \qquad (8)$$

Then
$$\nu = \frac{M+m}{Mmc^2}\{(\kappa-3)\,N + \kappa N'\}, \qquad (9)$$

where the zero suffixes have been dropped.

The ratio of the first term in ν' to the second is

$$\frac{3N'}{N+N'}\frac{c^2n}{c'^2n'}\frac{m}{M+m}\frac{1}{\kappa} \simeq 1 \times 10^{-4} \times 10^{-2} \times \tfrac{1}{5}, \qquad (10)$$

and is very small. Then
$$\nu' \doteqdot \kappa \frac{M+m}{Mm}\frac{n'}{n}\frac{N+N'}{c^2}. \qquad (11)$$

Now ν and ν' are capable of being found by comparing recent and ancient astronomical observations. In particular, it is evident that a determination of the time of occultation of a star by the Moon, or of the conjunction of the Moon with a star, will determine ν directly. Again, the Moon in time T will gain on the Sun in longitude by $\tfrac{1}{2}(\nu-\nu')\,T^2$ in comparison with the calculated motion, and therefore all lunar eclipses will occur earlier by $\dfrac{1}{2}\dfrac{\nu-\nu'}{n-n'}T^2$. Thus observations of the times of eclipses give $\nu-\nu'$.

The magnitude of an eclipse is determined by the distance of the Moon from the node at the time of conjunction or opposition to the Sun. The motion of the node has no part arising from tidal friction, and is given by the purely gravitational theory of the Moon's motion. The longitude of the Moon at the time of the eclipse is increased by $-\dfrac{1}{2}\dfrac{n(\nu-\nu')}{n-n'}T^2$ on account of the earlier time of the eclipse, but also by $\tfrac{1}{2}\nu T^2$ on account of its own secular acceleration. The total effect is therefore to increase the Moon's longitude, and therefore the longitude relative to the node, by $\dfrac{1}{2}\dfrac{n\nu'-n'\nu}{n-n'}T^2$. The magnitudes of the eclipses therefore give $n\nu'-n'\nu$. Since

$$n\nu' - n'\nu = n\frac{dn'}{dt} - n'\frac{dn}{dt}, \qquad (12)$$

the magnitudes of the eclipses do not depend on the rotation of the Earth. This is evident directly, since they involve only the positions of the Sun and Moon with respect to the centre of the Earth. Again, dn'/dt has been seen to be very small, so that the magnitudes of the eclipses give practically a direct determination of dn/dt. This applies to eclipses of the Moon, which are visible all over a hemisphere. Newton finds, however, that they give very little information. The magnitude of a solar eclipse depends on the position of the observer, and ω affects this; change of obliquity is also relevant.

Fotheringham (1920 a, b), in his table (1920 b, p. 124), uses the magnitudes of lunar eclipses to give an estimate of ν', not of ν, having apparently adopted a standard value of $\nu - \nu'$. He has a magnitude for a solar eclipse but does not give the separate equation based on it.

Observations of the time of passage of the Sun across the equator, when the precession of the equinoxes is known, give the secular acceleration of the Sun directly.

Many ancient observations made by Greek, Babylonian, Chinese and Egyptian astronomers have been discussed with a view to determining the secular accelerations. There is a complication here because purely gravitational causes make a substantial contribution to dn/dt; the secular change of the eccentricity of the Earth's orbit produced by the planets leads to a disturbance of the Moon's orbit, as was shown by Laplace. Laplace found only the first term of this theoretical secular acceleration, which appeared to be about equal to that indicated by the old observations. But J. C. Adams in a rediscussion found that the later terms made a serious correction, practically halving the calculated amount. There was an animated discussion on the matter, but there is no doubt that Adams was right. It is usual to tabulate the coefficient of $(T/\text{century})^2$ in the various quantities. J. K. Fotheringham (1920 and earlier papers) and K. Schoch (1926) have given careful discussions of the ancient data. Later discussions are by de Sitter (1927 a, b), Spencer Jones (1939), C. A. Murray (1957), Brouwer (1952 a, b) and van der Waerden (1961). From (9) and (11)

$$\frac{\nu}{\nu'} = \frac{\dfrac{\kappa - 3}{\kappa} N + N'}{N + N'} \frac{n}{n'}. \tag{13}$$

If N/N' was zero this would be $n/n' = 13\cdot3$. This is obvious without analysis, for it corresponds to the case where all the tidal friction is in the solar tides; and then dn/dt is zero and dn'/dt is insignificant. Thus the whole of the secular accelerations arise from variations of the rate of the Earth's rotation, and are therefore in the ratio of the angular motions.

If $N' = 0$, so that all the friction is in the lunar tides, the ratio is $5\cdot1$. With the theoretical ratios indicated in $8\cdot02$ (14) and $8\cdot03$ (17) we have

$$N/N' = 4\cdot9, \quad \nu/\nu' = 6\cdot8, \tag{14}$$

$$N/N' = 3\cdot7, \quad \nu/\nu' = 6\cdot9. \tag{15}$$

8·05. Observed secular accelerations. Fotheringham (1927) showed that there are fluctuations in the longitudes of the Sun, Moon and planets, which appear to be attributable to variations in the Earth's rate of rotation; that is, they are in about the ratios of the mean motions. The method used by de Sitter and Spencer Jones is to adopt a table of B, the discrepancy between the Moon's longitude and that given by Brown's tables (allowing for possible small corrections to the longitude and mean motion at a given instant) and to assume for any other body that the corresponding discrepancy is $Q(n''/n) B$, where n'' is the mean motion and Q is a constant to be determined. Brown's value from purely gravitational theory for the coefficient of T^2 is $6.01''/(1 \text{ century})^2$ relative to the stars, or $7.12''/(1 \text{ century})^2$ relative to the equinox. The difference is due to perturbations of the Earth's orbit by the planets, which give the equinox a secular acceleration. The chief data are the Sun's right ascension and declination, times of transits of Mercury and longitudes of Venus. De Sitter, combining all data, found that they were consistent with Q being the same for all and about 1.25 ± 0.03 (standard error). If the changes were entirely due to changes in the rotation of the Earth, Q would be 1. If they were due to variations in the intensity of tidal friction Q would be $13.3/6.3 = 2.1$ or $13.3/7.2 = 1.8$. De Sitter's value is intermediate and would suggest that tidal friction is fluctuating but that changes of the rotation, possibly due to changes of the moment of inertia, are more important. Spencer Jones gets $Q = 1.025 \pm 0.050$ or 1.062 ± 0.050 (standard error) according as he excludes or includes the Sun's right ascensions, which are suspected of systematic errors. Thus the value 1 is not excluded by the observations.

The facts are clear, as is shown in particular by the correspondence between the data shown in Spencer Jones's diagrams for 1680–1930. The explanation is another matter. It is not difficult to suggest explanations of a secular change in the intensity of tidal friction or in the Earth's moment of inertia. It is quite another matter to explain irregular changes of the amounts indicated. De Sitter points out that if the whole of the Central Asian mountains were displaced through their own height the change in C would give about one-quarter of the change of rotation that appears to have occurred in 1897, and to attribute some of the changes to changes in tidal friction would imply reversing its direction. The phenomenon invites comparison with the irregular part of the variation of latitude, which also appears attributable to changes in the moments and products of inertia, and of comparable amounts, of order 10^{-8} of the greatest moments of inertia. But the time-scales are quite different, being of the order of decades for the longitudes and months for the variation of latitude. Munk and Revelle (1952) have made a detailed analysis of the possibilities. They find that changes in winds, ocean currents and ice covering are quite inadequate. Crustal movements (assumed *ad hoc*) are not totally excluded, but would have to be much greater in Antarctica than anywhere else! Electromagnetic coupling of the shell to a core in convective movement could apparently do it, and would be consistent with Bullard's theory of the westerly drift of the Earth's magnetic field.

The uncertainty about the explanation of the modern observations is reflected in the determination of the present mean motions and hence in the amounts of the discrepancies from gravitational theory disclosed by the ancient observations. De Sitter's values are

$$\tfrac{1}{2}\nu = (5\cdot22'' \pm 0\cdot45'')/\text{century}^2,$$

$$\tfrac{1}{2}\nu' = (1\cdot80'' \pm 0\cdot24'')/\text{century}^2.$$

Spencer Jones infers that the secular accelerations at present are less than for the average of the last 2000 years, but adopts the observed ratio. On this basis he derives the present values

$$\tfrac{1}{2}\nu = (3\cdot11'' \pm 0\cdot85'')/\text{century}^2,$$

$$\tfrac{1}{2}\nu' = (1\cdot07'' \pm 0\cdot09'')/\text{century}^2.$$

The ratio ν/ν' is much less than either of those suggested by theory. De Sitter argues that the theory is incomplete because it treats the Sun and Moon as in the equatorial plane, and considers that the discrepancy might be removed by allowance for changes in the orbital elements other than the mean distances. It is true that such allowance is desirable, because important components of the tides are due to the inclinations, but it is difficult to see why it should greatly affect the ratio in question, because the periods of the principal tides raised by the Sun and Moon are not very different, and the ratio N/N' should be little affected. The ratio of the two terms in the angular momentum equations would hardly be affected by a factor different from 1 by more than the cosine of the obliquity of the ecliptic, which is 0·92. Accordingly, I think that the theoretical ratio, considered as such, cannot be far wrong.

C. A. Murray (1957) by comparison of observations between 1680 and now with ancient ones, estimates that the secular acceleration inferred from the ancient observations is about twice that from the modern ones, and suggests that the intensity of tidal friction may have changed through changes of shore lines and depths of the sea. I think it very unlikely that such changes can have been large enough to produce the effect.

B. L. van der Waerden (1961) notes that the longitudes of the Moon and Sun at time T should satisfy

$$\Delta L = x + yT + zT^2 + B^*,$$

$$\Delta L' = x' + y'T + z'T^2 + B^*/13\cdot37.$$

B^* is a consequence of variability of the Earth's rotation. This is not satisfactorily explained, but as a result of Spencer Jones's discussion of irregularities in the longitude of the Moon, planets and satellites, which are nearly in the ratios of their mean motions, its existence appears to be established. From modern observations it appears that values of B^* have a high correlation up to intervals of 20 years, but that changes in B^* in successive intervals of 24 years are substantially uncorrelated.

For modern data, we can eliminate B^*;

$$\Delta L - 13\cdot37\Delta L' = -(p + qT + rT^2)$$

which is observable; and $\qquad p = 13\cdot37x' - x$

with two similar relations. rT^2 is the true secular retardation of the Moon, the Earth's rotation having been eliminated. Van der Waerden gets from the modern data

$$p = 8\cdot7'', \quad q = 26\cdot7'', \quad r = 11\cdot2'' \pm 0\cdot9''/\text{century}^2.$$

He derives from Fotheringham and Schoch's values of z and z' the estimates $r = 15\cdot6''$ and $15\cdot1''/\text{century}^2$, and as their value of z' may well have a standard error of $0\cdot3$ that of the calculated r may well be consistent. Hence he infers that there is no contradiction between the ancient and modern values of r.

He considers that the best of the ancient data are for a lunar eclipse of -424 October 9, one of -382 December 23, and an occultation of Spica -282 November 8/9. Great importance had been attached to the solar eclipse of -128, called the eclipse of Hipparchus; but, following Neugebauer, he says that there is no record that Hipparchus actually observed this eclipse and he may have reported one of -309; so this eclipse must regretfully be discarded. The question is discussed at some length by Fotheringham (1920 *b*, pp. 11 and 12) who considers it most probable that the eclipse of Hipparchus was really that of -128.

In addition there are 25 eclipse observations between 829 and 1004, reported by Ibn Yunis. The data can be summarized as four values for -386, $+950$, $+1635$ and $+1962$, with fairly well estimated uncertainties. Then the modern values of the motions of the Sun are combined with these and the results are

$$x = -27\cdot4, \quad x' = -1\cdot40,$$

$$y = 24\cdot2, \quad y' = 3\cdot81,$$

$$z = 6\cdot28 \pm 0\cdot82, \quad z' = 1\cdot31 \pm 0\cdot10.$$

It is observed that there is a solar semidiurnal variation of pressure in the atmosphere, which is not of tidal origin, because it is far larger than the corresponding lunar tide, and must be thermal. The maxima are about 10 a.m. and 10 p.m. local time. Kelvin noticed that the attraction of the Sun on this wave would give an accelerating couple, and Holmberg (1952) estimates the consequent correction to the time as $2\cdot94T^2$ sec. The motions of the Moon and Sun during this time are $1\cdot61''T^2$ and $0\cdot12''T^2$, and adding these to z and z' above gives for the tidal effects on the Moon

$$(7\cdot89'' \pm 0\cdot84'')\, T^2$$

and for the Sun $\qquad (1\cdot43'' \pm 0\cdot11'')\, T^2.$

This overestimates the effect, since the body of the Earth yields elastically, and as in the theory of the annual variation of latitude a factor 0·69 should be applied: this factor was calculated by Rosenhead for a Wiechert model and confirmed by Longman for a much better model. The motions should then be $1·11''T^2$ and $0·08''T^2$. Adding these to z and z' above gives for the tidal effect on the Moon $(7·39 \pm 0·82)''\,T^2$ and for the Sun $(1·39 \pm 0·10)''\,T^2$.

The ratio is $5·3(1 \pm 0·13)$. The theoretical ratios for $N/N' = 4·9$ and $3·6$ are 6·8 and 7·0. The difference is about twice the apparent standard error. If we assume the ratio to be 6·8 the solution is $z = 8·3 \pm 0·5$, $z' = 1·3 \pm 0·08$, and restoring the thermal part we have

$$z = 7·2 \pm 0·5, \quad z' = 1·2 \pm 0·08.$$

Doubling these gives the apparent secular accelerations. The part of z due to tidal friction is nearly double previous estimates, and the couple and the rate of dissipation must also be doubled. The rate of the Earth's rotation was estimated to be reduced by the tides by 1 sec. in $1·2 \times 10^5$ years. This must now be halved. But if θ is the total rotation of the Earth, for the thermal effect alone,

$$\theta = 1296000''\,\frac{T}{1\,\text{day}} + 2·94 \times 15''\left(\frac{T}{\text{century}}\right)^2,$$

$$\dot{\theta} = 1296000''\,\frac{\dot{T}}{1\,\text{day}} + 2·94 \times 30''\,\frac{T\dot{T}}{(\text{century})^2},$$

which gives a shortening by 1 sec. in 5×10^5 years. The resultant effect is that the day lengthens by 1 sec. in about 70,000 years.

The thermal tide is explained by Pekeris (1937) as an effect of resonance, and if this is correct it would be unimportant during most of the Earth's history. The effect on the Earth's rotation was noted by Kelvin long ago but forgotten (see Kelvin and Tait, 1883, 1912, §8·30).

A puzzle has been that it is hard to see how a large magnification by resonance and a large shift of phase could both occur. The solution has been found by Chapman and Lindzen (S. Chapman and R. S. Lindzen, 1969). A brief comment by T. G. Cowling is 'The important thermal stimuli are those arising from heat absorbed by ozone, carbon dioxide, and water vapour well above the Earth's surface, and require no strong amplification by resonance to give the observed effects.'

For further discussion see §§8·071, 8·072.

8·06. Tidal friction in the ocean. We have now to consider whether there are sources of dissipation of energy that can account for the values of dE/dt indicated. Sir G. H. Darwin, whose development of the theory takes up most of the second volume of his collected papers, mentioned elastico-viscosity, but mostly treated the Earth as a viscous liquid. His methods can however be adapted to a wide range of other laws. In several passages he expresses doubt about whether a solid could show imperfection of elasticity under the small strains, of the order of 10^{-6}, actually involved in the bodily tides, and appeared to favour the hypothesis that the dissipation is really in the ocean. Actually neither hypothesis is acceptable. In a paper of 1915 I found the viscosity, on the elasticoviscous theory, needed to give the right dissipation; and then, applying this viscosity to the free variation of latitude, I found that the latter should be damped out in a few days (1915b). The order of magnitude of the dissipation in the ocean can be treated as follows. If $\bar\zeta$ is the height of the equilibrium tide, and there is not approximate resonance, the actual tide height will be comparable with $\bar\zeta$, and the velocity components will be of order $g\bar\zeta/\omega a$. The extreme value of $\bar\zeta$ is 26 cm., so that the velocities are of order 1 cm./sec. Now if the motion is turbulent the frictional force is $k\rho(u^2+v^2)$, directed against the resultant velocity, where ρ is the density of water and k is a numerical coefficient called the coefficient of skin friction and is about 0·002 (see, for example, Jeffreys, 1925). The dissipation is therefore at a rate $k\rho(u^2+v^2)^{\frac{3}{2}}$ per unit time per unit area, and this is about 0·004 erg/cm.² sec. The area of the whole ocean being about $3·7 \times 10^{18}$ cm.², the dissipation is of the order of 10^{16} ergs/ sec., which is a small fraction of that needed to account for the secular accelerations. Thus the chief part of the tidal dissipation cannot arise from skin friction in the open ocean.

The validity of this estimate depends on whether the Osborne Reynolds criterion for turbulence is satisfied. If the motion were purely viscous, the kinematic viscosity being ν, and if the influence of the boundary is considerable to a distance h from it, the criterion for purely viscous motion is that uh/ν shall be less than about 300 (Jeffreys, 1926d). For semi-diurnal motion in a viscous medium (Lamb, 1932, p. 590)

$$h^2 = \nu/\omega, \qquad uh/\nu = u/(\nu\omega)^{\frac{1}{2}} = 800.$$

Thus the criterion for viscous motion is not satisfied, and the motion is turbulent.

8·07. Tidal friction in shallow seas. The last argument assumes explicitly that the conditions are such that the tide height is not large compared with the equilibrium tide. If the form of the bottom is such that the tides are greatly magnified by resonance or by the shallow-water effect, the result will not hold. Such circumstances cannot hold in the ocean as a whole, but they do hold locally. Places where the tidal currents have greater velocities than the 1 cm./sec. inferred for the open ocean in the last section are well known to everybody. In such places the dissipation must greatly exceed that in the open ocean, since the rate of dissipation per unit area is proportional to the cube of the velocity. The question is whether the increase in the rate of dissipation per unit area is enough to compensate for the limited area of the regions concerned and make the total dissipation in them exceed that in the open ocean.

The question was answered in the affirmative by Sir Geoffrey Taylor (1919), who determined the rate of dissipation in the tides in the Irish Sea in two ways. One method is that just used for the open ocean, but using the observed velocities of the tidal currents to determine the dissipation per unit area. Integrating this over the whole sea gives the dissipation within it.*

The alternative method is to find the rate of work done on the sea by the ocean and the Moon together. This is found to be, on the average, positive. The energy in the sea is, however, not increasing, and therefore the energy supplied from outside is dissipated as fast as it enters. Thus the total rate of dissipation can be found. If we consider a line across the sea as the boundary, and take the axis of y along this line, the velocity across it is u (taken positive towards the sea). Let D be the depth of the water at any point. The pressure at any depth z below the mean position of the surface is $g\rho(z+\zeta)$, since the depth below the actual free surface is $z+\zeta$. Thus the entering water does work at a rate $g\rho(z+\zeta)u$ per unit area of the section, and the total rate of performance of work is

$$\int dy \int_{-z}^{D} g\rho(z+\zeta)\,u\,dz = \tfrac{1}{2}g\rho \int (D+\zeta)^2\,u\,dy. \tag{1}$$

In general ζ is small compared with D. In addition, the entering water brings in its own energy. Taking the mean sea-level as the zero of potential, we see that the mean potential within a column of height ζ and depth D is $-\tfrac{1}{2}g(D-\zeta)$, and the mass per unit area is $\rho(D+\zeta)$. The potential energy per unit area is therefore $-\tfrac{1}{2}g\rho(D^2-\zeta^2)$. Now the horizontal cross-section of the column that enters in unit time per unit length of the boundary is u. Hence the inflow of potential energy is

$$-\tfrac{1}{2}g\rho(D^2-\zeta^2)\,u \tag{2}$$

per unit length of the boundary.

The kinetic energy of the entering water evidently contributes energy at a rate
$$\tfrac{1}{2}\rho(D+\zeta)\,(u^2+v^2)\,u \tag{3}$$

per unit length. Combining these three sources of energy, we have for the whole supply of energy from the ocean

$$g\rho u(D\zeta+\zeta^2)+\tfrac{1}{2}\rho(D+\zeta)\,u(u^2+v^2) \tag{4}$$

per unit length of the boundary. Also u and v are of the order of $c\zeta/D$, where c is the velocity of a long wave in water of depth D, and

$$c^2 = gD. \tag{5}$$

---

* Tidal friction in shallow water due to resistance proportional to the square of the velocity was discussed provisionally by W. D. Macmillan (T. C. Chamberlin and others, 1909).

Then $D(u^2 + v^2)$ is of order $g\zeta^2$. Thus much the largest part of the entering energy is contributed by the term $g\rho u D\zeta$. Thus the whole rate of transfer across the boundary is very nearly

$$\int g\rho Du\zeta \, dy \qquad (6)$$

taken along the boundary. If this again is integrated with regard to the time over a whole period of the motion, we get the total inflow during a period; and if there is more than one side open to the sea the contributions from all must be added.

In addition the work done by the Moon is needed. The rate of performance of work is

$$\iint g\rho\bar{\zeta} \frac{\partial \zeta}{\partial t} \, dS \qquad (7)$$

taken over the whole area of the sea. This must be added to (6).

Taylor found the rate of dissipation in the Irish Sea by both these methods. He found that energy enters through St George's Channel and the North Channel at a mean rate of $6\cdot4 \times 10^{17}$ ergs/sec. The rate of performance of work on the sea by the Moon is, on an average, about $-4\cdot3 \times 10^{16}$ ergs/sec. The total, about 6×10^{17} ergs/sec., is absorbed by friction. The alternative method, in which the dissipation all over the sea is estimated from the velocities, gave $5\cdot2 \times 10^{17}$ ergs/sec., agreeing with the other within the limits of error of the determinations of velocity used.

This estimate refers to spring tides, when the tidal currents are at a maximum. To find the mean dissipation through the lunar month a correcting factor must be applied. Let θ be the phase of the lunar current and $(1-s)\theta$ that of the solar current; s is about $1/29$. Let the amplitudes be A and $A\nu$. Then the total current is

$$A\{\cos\theta + \nu\cos(1-s)\theta\} = A(1 + 2\nu\cos s\theta + \nu^2)^{\frac{1}{2}} \cos\left(\theta - \tan^{-1}\frac{\nu\sin s\theta}{1+\nu\cos s\theta}\right), \qquad (8)$$

which is now expressed as a simple harmonic motion with slowly varying amplitude and period. The amplitude at springs is $A(1+\nu)$. The dissipation is proportional to the cube of the current, and therefore the mean over the lunar day is proportional to the cube of the amplitude during the day. Thus the ratio of the mean dissipation to that at springs is the average of

$$(1 + 2\nu\cos s\theta + \nu^2)^{\frac{3}{2}}/(1+\nu)^3. \qquad (9)$$

If ν^6 is neglected, the numerator of this average is $1 + \frac{9}{4}\nu^2 + \frac{9}{64}\nu^4$. Assuming that the velocities vary in proportion to the vertical ranges, we have

$$\frac{1+\nu}{1-\nu} = 2\cdot3, \quad \nu = 0\cdot39,$$

$$\frac{1 + \frac{9}{4}\nu^2 + \frac{9}{64}\nu^4}{(1+\nu)^3} = 0\cdot51.$$

Thus the correcting factor is about 0·5, and the mean rate of dissipation in the Irish Sea is 3×10^{17} ergs/sec. This is about 30 times the amount estimated in 8·06 for the whole of the open ocean; and about 50 Irish Seas would be enough to account for the whole of the lunar secular acceleration.

Taylor's methods were extended by the present writer (1920*b*) to include most of the shallow seas of the globe. A discussion on similar data, and giving comparable results, was given by Heiskanen (1921). See also Lambert (1928).

For the Yellow Sea and Malacca Strait it was possible to apply both of Taylor's methods of calculating the dissipation, and the results were concordant. In most seas it was necessary to rely wholly on estimates based on the velocity. The Bering Sea appeared to be the most important region of all, but data were available for only part of it and this might not be representative. A revised solution is by G. R. Miller (1966). The unit is 10^{17} ergs/ sec.; contributions over 1 unit are listed.

European waters		Asiatic waters	
Bristol Channel	2·4	Bering Sea	2·4
English Channel	5	Japan and Islands	4·3
Irish Sea	3·2	Malacca Strait	7
North Sea	4·5	N. Bay of Bengal	6·0
Norway—Svaldbard	3·2	Sea of Okhotsk	21
		Oman (Persian Gulf)	1·6
		S. India	4
		Ryukyu Islands	6
		W. Australia	4·2
		S. Alaska	5

American waters		Pacific	
Bay of Fundy	2·3	Australia—Lesser Sunda Islands	15
Davis Strait	2	Barrier Reef	2·4
Gulf of California	4	Mindanau—New Guinea	0·2
Hudson Strait	12	Vancouver	1·6
N.E. Coast of S. America	5		
S.E. Coast of S. America	13		

The total is estimated as $1·7 \times 10^{19}$ ergs/sec.

There are some notable omissions from this list. The Mediterranean, Baltic and White Seas, and the Gulf of Mexico are so narrow at the entrances that little tide can enter. The South China Sea contributes little though there are strong tidal currents, but these are mainly diurnal. The diurnal tides arise from the inclination of the Moon's orbit to the equator, and though they affect the inclination they contribute little to the secular acceleration.

Some of the contributions are for open coasts, especially the southeast coast of South America. In my treatment two-thirds of the estimated dissipation was in the Bering Sea, and it was hard to see how the energy got in—with too heavy dissipation the currents themselves would be damped out. It now appears that the strong currents previously known there are very limited in area, as Munk and MacDonald (1960, p. 215) pointed out. The argument of §8·03 suggests an explanation. The ratio Q/P is nearly independent of the ratio of the periods. Ordinary waves breaking on the coast are almost wholly dissipated. Then the tidal components and those associated with the variation of latitude will be dissipated too, approximately in the ratio of their amplitudes. Hence open coasts can make a considerable contribution.

One consequence of the supposition of high dissipation along the coasts is a change in the adopted boundary condition. If h is the depth and v the velocity normal to the coast, the usual condition is that $hv \to 0$ at the coast. This is correct at the actual coast, but the usual equations of motion neglect damping and do not hold up to the coast. The corrected boundary condition in deep water is that there is no reflected wave. It becomes necessary, in the study of tides in the ocean, to modify the boundary condition accordingly. This is done by Proudman (1941) for a shelf of uniform depth h'; he gives

$$hu_n = h'u'_n = c'\zeta,$$

where h is the depth just outside the shelf, h' that on it, u_n and u'_n the corresponding normal velocities, and $c' = \sqrt{(gh')}$; ζ is the elevation of the surface. It explains one feature of tides that has been troublesome. The highest tides usually occur a little after new and full moon; that is, if κ is the lag of phase of the tide and γ its speed, $d\kappa/d\gamma$ is usually positive. For frictionless tides its average is zero in cases examined; but it is positive at most places with dissipation.

Friction in deep ocean is probably negligible; most of the dissipation is in shallow seas and probably along open shores. I had not considered the latter. It is a matter of observation, however, that waves coming up to the shore give hardly any reflected waves, except where the coast is an actual wall, as of a harbour. The energy is nearly all dissipated in breakers. Tidal motion superposed on ordinary waves is presumably dissipated in the same way. In the usual treatment the boundary condition is that at the coast $hu \to 0$, where h is the depth and u the normal velocity. I believe the correct condition to be that
$$u = -(g/h)^{\frac{1}{2}}\zeta$$

where h is the depth at a suitable distance from the coast, u the velocity *away* from the coast, and ζ the tide height (Jeffreys, 1968c).

Munk and MacDonald (p. 217) report on Groves and Munk (1958) who made a calculation of dissipation based on Dietrich's tidal charts. They get for the lunar tide

$$dE/dt = -3 \cdot 2 \times 10^{19} \, \text{ergs/sec.}$$

for the solar $-1 \cdot 0 \times 10^{19}$; and compare with astronomical values $-2 \cdot 7 \times 10^{19}$ and $-0 \cdot 6 \times 10^{19}$. They say that 'the agreement is better than we had any right to expect'.

8·071. Observed accelerations. Comparison with observation is difficult, and on the whole the difficulties increase as time goes on. The chief seems to be the variation of the rate of rotation of the Earth, which seems to be established by Spencer Jones's comparison of the anomalies in the longitudes of the Moon, Sun and planets. Various suggestions have been made to account for this, such as settling of iron from the base of the shell into the core. One that has almost certainly occurred is redistribution of water and ice due to climatic change. The retreat of Alpine glaciers in the last century or so is well known. Greenland was habitable by Europeans about A.D. 1000. Brooks (1926, 1949) comments on the total absence of mention of icebergs in the Icelandic sagas of the same period, indicating that the North Atlantic and probably the Arctic were free of ice at that time. This omission is perhaps not altogether convincing, since as far as I know there is also no mention of difficulties in language, though the Norsemen went to Britain, Ireland, Russia, Poland and Byzantium. Variation of n in historic times would be much more difficult to understand. It has often been suggested that depths of shallow seas may have altered, but it seems very unlikely that any important change has occurred in the last 2500 years.

The existence of some anomaly is most striking for what E. W. Brown called 'the great empirical term' in the Moon's motion. The diagram given by Munk and MacDonald (1960, p. 180) shows a deviation of about $+12''$ in 1680, rising to $+21''$ at 1770, decreasing to $-16''$ at 1909 with sharp fluctuations to 1950, reaching $-20''$. From Spencer Jones's comparison this must be attributed wholly to variations of $\dot{\omega}$, and has been eliminated in van der Waerden's r, which is $-\frac{1}{2}\dot{n}$.

The tides in the Earth, both oceanic and bodily, affect the external potential and hence the motion of an artificial satellite. The theory was given by Kozai (1965) and has been extended by Musen and Felsentreger (1973). The observations have been analysed by R. R. Newton (1968), who derives both the Love number k and the lags for the main lunar and solar tides. Applying these to the Moon and Sun gives the tidal couples, and Newton expresses the results as contributions to $\dot{\omega}/\omega$. However the lunar part leads to what is practically a direct determination of $\dot{n}$.

The theory of the ocean tides, with the actual form of the oceans, has been developed by Pekeris and Accad (1969) by numerical integration of the tidal equations. This is a tremendous piece of work. They assume a linear law of resistance, but this probably does not produce a large error, since the strong currents are in shallow water anyhow. More serious is that, as in classical tidal theory, elastic yielding of the solid Earth is neglected. This will apply a factor about $1+k-h$. According to the calculations of Rosenhead for a rough model and Longman for a much more detailed one this is about 0·69, and its square will appear in the dissipation. Later work by Hendershott has allowed for this in detail.

The problem started with the discussion of ancient observations. Their interpretation has been discussed in great detail by R. R. Newton (1970) and F. R. Stephenson (1972). The information is in many languages, Anglo-Saxon, Latin, Greek, Arabic, Chinese and Babylonian, but the authors either know these themselves or have access to people that do. Many of the statements are rejected, largely on account of 'assimilation'. An eclipse has often been interpreted as an omen and associated with a historical event, sometimes at a widely different date. An eclipse was associated by Snorre with the battle of Stiklestad, but it is clear that the date was wrong. The same applies to one associated with Caesar's crossing of the Rubicon. Others are as fictional as that in Rider Haggard's *King Solomon's Mines*. There are often doubts in the real eclipses about where the observer actually was at the time. There has been great difficulty over the 'eclipse of Hipparchus'. This was believed to be in 129 B.C., but Hipparchus did not say that he had observed it himself, and he may have used a report of one of three earlier ones. Newton gives weight 0·3 to the earlier ones and 0·1 to 129 B.C; Stephenson after much discussion decides that 129 B.C. is the only possible one. Fotheringham had adopted the last with some hesitation. The place of observation of a solar eclipse or an occultation is important, since the belt of totality is narrow. Change of the obliquity of the ecliptic needs to be allowed for.

Newton and Stephenson treat both $\dot{n}$ and $\dot{\omega}$ as unknowns to be estimated over different intervals of time. I think that at present it is worth while to treat $\dot{n}$ as constant but $\dot{\omega}$ as fluctuating. Brouwer suggested that variations of $\dot{\omega}$ were highly correlated at intervals of order 20 years but substantially independent outside such intervals, and this is adopted by van der Waerden. But the existence of the great empirical term suggests that 200 years would be a better choice. In fact van der Waerden reduces the observations to four groups at wide intervals, and his procedure would be consistent with this.

8·072. Theoretical discussion. In what follows I proceed on the hypothesis that $\dot{n}$ is constant, leading possibly to a mean value of $\dot{\omega}$ with variations that may be useful in other subjects, perhaps in study of climatic variation.

Values of the needed constants are as follows:

$$n = 2{\cdot}661\,699 \times 10^{-6}\,\text{radians/sec.}$$

$$\omega = 7{\cdot}2921 \times 10^{-5}\,\text{radians/sec.}$$

$$\text{Tropical year} = 3\,155\,692{\cdot}6\,\text{sec.}$$

$$a = 6{\cdot}378 \times 10^{8}\,\text{cm.}$$

$$c_0 = 3{\cdot}844 \times 10^{10}\,\text{cm.}$$

$$M = 5{\cdot}976\,06 \times 10^{27}\,\text{g.}$$

$$C = 0{\cdot}330 M a^2 = 80{\cdot}22 \times 10^{43}\,\text{g. cm.}^2$$

$$\frac{m}{M+m} = \frac{1}{82 \cdot 30}$$

$$\nu = \dot{n} - 0 \cdot 036\,501\dot{\omega}$$

$$\nu' = -\tfrac{1}{366}\dot{\omega} = -0 \cdot 002\,73\dot{\omega}$$

$$\kappa = \frac{m}{M+m}\frac{Mc_0^2 n_0}{C\omega_0} = 4 \cdot 881.$$

Constancy of angular momentum of the Earth–Moon system gives

$$\frac{Mmc^2 n}{M+m} + C\omega = \text{constant}$$

$$n = n_0 \xi^{-3}, \quad c = c_0 \xi^2$$

whence
$$\kappa \omega_0 \xi + \omega = \text{constant},$$

$$\kappa \dot{\xi} = -\dot{\omega}/\omega$$

$$\dot{n}/n = -3\dot{\xi}$$

whence
$$\dot{\omega} = \tfrac{1}{3}\kappa\frac{\omega}{n}\dot{n} = 44 \cdot 5\dot{n}.$$

The solar contribution to $\dot{\omega}$ is estimated as $1/4 \cdot 9$ of the lunar one, so that $44 \cdot 5$ should be replaced by $53 \cdot 5$. This gives

$$\nu = -0 \cdot 95\dot{n}, \quad \nu' = -0 \cdot 15\dot{n}, \quad \nu - \nu' = -0 \cdot 80\dot{n}.$$

Spencer Jones's estimate of the value of $\dot{n} = -2r$ for the last 250 years, used by van der Waerden, is equivalent to

$$\dot{n} = -(22 \cdot 4 \pm 1 \cdot 8)''/\text{century}^2$$
and would lead to

$$\dot{\omega} = -(1198 \pm 96)''/\text{cy}^2, \, \nu = (+21 \cdot 3 \pm 1 \cdot 7)''/\text{cy}^2.$$

$$\nu' = (+3 \cdot 4 \pm 0 \cdot 3)''/\text{cy}^2, \quad \nu - \nu' = (417 \cdot 9 \pm 1 \cdot 44)''/\text{cy}^2.$$

Van der Waerden's $2z$, as adjusted above, would give

$$\dot{n} = -(17 \cdot 9 \pm 1 \cdot 1)''/\text{cy}^2.$$

But since the estimate is on 2 degrees of freedom the probability of a departure of, say, over 3 times the nominal standard error is considerably greater than if the posterior probability followed the normal law.

R. R. Newton (1968) studied the perturbations of four artificial satellites to determine the disturbance of the potential due to the tides in the Earth, including the ocean. The effects of the lunar and solar tides M_2 and S_2 were found separately. He finds for the Love number k, $k_S = 0 \cdot 359 \pm 0 \cdot 042$,

$k_{\rm M} = 0.314 \pm 0.036$, and from the lags ϵ (corresponding to 2ϵ of § 8·02) $\kappa = k \sin \epsilon = 0.0117 \pm 0.0011$ for the Moon and 0.0051 ± 0.0017 for the Sun. These differ rather seriously, and Newton discussed the explanation to some extent. He estimates various contributions to $10^{11}\dot\omega/\omega$. With some adjustment he gives possible values -22.1 to -16.4 for the lunar tide. These on the hypotheses above would correspond to

$$\dot n = (-23.5'' \text{ to } -17.5'')/\text{century}^2.$$

This is consistent with Spencer Jones and van der Waerden. The immediate interest is that the perturbation of the external field leads practically to a direct determination of $\dot n$.

Newton and Stephenson in their discussions of ancient observations present the results in terms of $\dot n$ and $\dot\omega/\omega$; the last is, I think, unfortunate. The results are equivalent to the following; $\dot n$ and $\dot\omega$ are in seconds of arc/century^2:

	$\dot n$	$\dot\omega$
Newton	-42 ± 4	-1180 ± 140
Stephenson	-34.2 ± 1.9	-1213 ± 57

These would lead to $\nu = +1''/\text{cy}^2$ and $+10.1''/\text{cy}^2$ respectively. The uncertainties are not independent, but the maximum probability density for any linear function of the unknowns would occur at the right place. Newton's is inconsistent with all previous determinations of ν; Stephenson's is consistent. In fact Newton gives an equation for ν directly, from lunar conjunctions and occultations, $\nu = (8.7 \pm 1.9)''/\text{cy}^2$ and I think that if he has used it his uncertainties should have been greatly increased on account of a huge contribution to χ^2. Stephenson's $\dot n$ is substantially larger than what the two most direct methods appear to give.

Newton (1972) analyses observations listed by Al-Biruni from 829 to 1019 and gets an acceleration of the Earth's spin of -26.5 ± 5.8 parts in 10^9/century, a change of obliquity of $(-47.9 \pm 2.0)''$/century, and a lunar acceleration of $(-46.4 \pm 6.8)''$/century^2. He gets widely different values for different intervals, both for $\dot n$ and $\dot\omega$. It is very difficult to see how $\dot n$ can vary appreciably; there are possible explanations for $\dot\omega$.

In addition van Flandern (1970) has studied occultations of stars by the Moon in 1955–69, taking the time from an atomic clock. Thus the Earth's rotation does not arise. He had only a few years' observations and got

$$\dot n = (-52 \pm 16)''/\text{cy}^2.$$

The method is promising but more data are needed. L. V. Morrison (1972) considers that Spencer Jones's uncertainties should be multiplied by about 3.

There are considerable differences between estimates of the dissipation in the ocean tides. I take the unit as 10^{19} ergs/sec. We have the following, using § 8.04 where necessary.

Stephenson	5·1
Newton (from satellites)	2·88 to 2·16
Munk and MacDonald (1960)	$\begin{cases} 2\cdot7 \text{ from astronomy} \\ 3\cdot2 \text{ from tides} \end{cases}$
Van der Waerden	2·76 ± 0·22
Pekeris and Accad	6·3
Hendershott	3·0.

Since Pekeris and Accad assumed a rigid Earth I have suggested that elasticity would multiply all tidal phenomena by a factor about 0·69 and the dissipation by about the square of this. This agrees with Hendershott's determination, which allows for elasticity. Stephenson's value is remote from the others. It appears that the most reliable values all indicate dE/dt about $-2\cdot9 \times 10^{19}$ ergs/sec. This is about twice what I have found previously.

Several estimates of the effect of the thermal tide exist and are expressed in different ways. Holmberg gives the couple as $3\cdot7 \times 10^{22}$ dyne cm.; Munk and MacDonald give dE/dt for it, which leads to $3\cdot0 \times 10^{22}$ dyne cm. I know of no complete recalculation. The effect is larger than might be expected; resonance was suggested but found unsatisfactory, but Chapman and Lindzen have shown that it can be explained in terms of phenomena in the upper atmosphere.

In the above comparison, where $\dot\omega$ has been calculated from $\dot n$, the solar tide will increase it numerically by about 1 part in 4·9; the atmospheric tide will decrease it by about $54''/\text{cy}^2$. The results are reasonably consistent with this. See also Jeffreys (1975c).

8·08. Frictional reaction on the oceanic tides. If ζ is the elevation of the oceanic tide the potential energy per unit area is $\frac{1}{2}g\rho\zeta^2$. The average kinetic energy will be of the same order. If the actual tide is comparable with the equilibrium tide, the total energy is

$$\tfrac{1}{2}g\rho H^2 . 2\pi A^2 \int_0^\pi \sin^5\theta\, d\theta = \frac{16\pi}{15} g\rho H^2 A^2 = 4 \times 10^{23} \text{ergs}.$$

Comparing this with §8·07 we see that all the energy in the tide at any moment would be dissipated in the next 2×10^4 sec., say $\frac{1}{4}$ day. The dissipation is therefore so heavy that it must have a controlling effect on the oceanic tides comparable with that of inertia. The phase lag also must be a moderate angle. It actually becomes rather difficult to see how so much energy gets into the shallow seas to be dissipated.

It looks as if the ocean tide resembles a set of progressive waves converging on the regions of dissipation and being absorbed there, somewhat like ordinary sea waves approaching a beach. In the latter case nearly all the energy of the waves is absorbed on breaking, and there is hardly any reflected wave. See also p. 333.

The general conclusion is confirmed in the paper of Proudman (1941).

The work of Pekeris and Accad (1969), followed by Hendershott (1972, 1973), has shown that the actual tides are several times larger than the equilibrium tide, and that M_2 and presumably S_2 are near resonance. In that case the problem of how the energy gets into the shallow seas becomes much less serious. This may help to explain the discrepancy found by Newton between the lags in the tidal potentials due to the M_2 and S_2 tides, though he has examined the possibility without arriving at a definite solution.

8·09. Tidal friction in the past. If we adopt the results of 8·04, we have

$$\frac{d\omega}{dt} = -\frac{N+N'}{C} = -2\cdot5 \times 10^{-22}/\text{sec.}^2.$$

The present value of ω is $7\cdot3 \times 10^{-5}/1$ sec. Thus ω changes by 10^{-5} of its amount in 3×10^{12} sec., or 10^5 years. The day has probably lengthened by a second in the last 120,000 years. Thus tidal friction, historically speaking, is a slow process. But we have to consider intervals of the order of 10^9 years, and in such an interval the period of rotation would have increased by some hours.

If we attempt to take the extrapolation further back, we must have recourse to equations 8·02 (4) (5) (6). The couple N has been seen to be proportional to $(m/c^3) H \sin 2\epsilon$, and if we suppose ϵ to be constant and H to vary as m/c^3, as is reasonable, N is proportional to c^{-6} or ξ^{-12}. We denote the present value by N_0. The solar tidal friction is at present small compared with the lunar, and cannot have changed much, since the Sun's distance has hardly varied. Thus it must have been of smaller relative importance in the past. Hence we can drop N'. Then

$$\frac{Mm}{M+m} c_0^2 n_0 \frac{d\xi}{dt} = N_0 \xi^{-12}, \quad C\frac{d\omega}{dt} = -N_0 \xi^{-12}. \tag{1}$$

If we introduce κ as in 8·04 (6), (1) becomes

$$\kappa C\omega_0 \frac{d\xi}{dt} = N_0 \xi^{-12}, \tag{2}$$

$$\tfrac{1}{13}(1-\xi^{13}) = -\frac{N_0 t}{\kappa C\omega_0}. \tag{3}$$

Substituting our adopted values we find for the time when ξ was $0\cdot8$, and therefore the distance of the Moon 240,000 km.,

$$t = -4 \times 10^9 \text{ years.} \tag{4}$$

The time needed by the Moon to recede from its shortest distance to 240,000 km. would not exceed $\frac{1}{20}$ of this.

From 8·02 (4) and (6)

$$\frac{d}{dt}(\kappa\omega_0\xi+\omega) = 0, \tag{5}$$

whence

$$\kappa\xi+\omega/\omega_0 = \kappa+1 = 5\cdot82. \tag{6}$$

Now it is possible for n and ω to be equal. If this happens, $\omega = n_0\xi^{-3}$, and the condition to be satisfied is

$$\kappa\xi+\frac{n_0}{\omega_0}\,\xi^{-3} = 5\cdot82. \tag{7}$$

This equation has two real roots, $\xi = 1/5\cdot1$ and $\xi = 1\cdot20$. The former makes the common period of rotation and revolution 4·8 hr.; the latter makes it 47 days. Since $1\cdot2^{13}-1 = 10$ nearly, the time needed to reach the latter state would be of order 5×10^{10} years.

Urey (1951) had suggested that the inner parts of the shell contain free iron, and that this is still settling into the core. This would produce a secular decrease of the moment of inertia and accelerate the rotation. But the thermal tide is a known phenomenon and produces a similar effect.

An interesting observation recently made by J. W. Wells and C. T. Scrutton (in B. G. Marsden and A. G. W. Cameron, 1966, pp. 70–89) is that Devonian corals show strong bands corresponding to annual growth, but superposed on these are finer bands that may be monthly and daily. The first may be thermal, the others tidal. Wells (1963) reported 400 days in the Devonian year; Scrutton (1964) 30·6 days in the synodic month. Berry and Barker (1968) for the late Cretaceous give $29\cdot65 \pm 0\cdot18$ days in the month, $12\cdot49 \pm 0\cdot02$ months in the year, making 370·3 days in the year.

If we take $\dot\omega = -1400''/\text{cy}^2$ throughout, the change in ω in 370×10^6 years would be about 0·11 rev./day; in 60×10^6 years about 0·017 rev./day. The actual changes found are about 10 % and 1·5 %.

An attempt to check the method on modern corals is made by Lamar and Merrifield (1966), who found the bands running about 28 to the month. Perhaps the corals do not count so well after all! More detail is given by Runcorn (1964, 1970).

Similar results to those for corals have been found for bivalves (Runcorn, 1968).

8·10. The resonance theory of the origin of the Moon.

The period of a free oscillation of a fluid mass of the density of the Earth, the displacement of the surface at any instant being proportional to a surface harmonic of degree 2, is about 2 hr., depending somewhat on the distribution of density. Now if the Earth ever rotated in 4·8 hr. the period of the tide raised by the Sun would be 2·4 hr., very near the free period, and the agreement led Darwin to the suggestion that resonance actually occurred. According to this theory the Earth and Moon once formed a single fluid body, in which the originally small tide raised by the Sun increased gradually in amplitude until its height became comparable with the mean radius and the mass became unstable and broke into two. The present angular momentum of the Earth-Moon system, if the two were combined into one body, would make it rotate in about 4 hr.;.actually this would be reduced, but only by a small percentage, if past solar tidal friction is taken into account. *Prima facie* the agreement is evidence for the truth of the resonance theory. On fuller analysis, however, it breaks down for two reasons. In the first place, there would be a considerable discontinuity of velocity at the core boundary, and this would lead to considerable loss of energy by friction. As the rate of supply of energy from the tide-raising forces increases like the amplitude, and the dissipation by turbulence like the cube of the amplitude, there will be a critical height that would never be exceeded (Jeffreys, 1930a). The critical elevation would be only about $\frac{1}{17}$ of the radius, and no question of instability arises.

Another consideration, the importance of which was pointed out to me by E. W. Brown in a letter, is that the elementary theory of resonance supposes the amplitude not to become so large that higher powers of the displacements have to be retained. If it does, special treatment is necessary. Brown (1932) and Brown and Shook (1933) have given a partial solution, which I have extended (Jeffreys, 1959a). The problem, in the absence of damping, can be illustrated by means of the pendulum under periodic force. The equation of motion can be reduced to

$$\ddot{x} + p^2 \sin x = \epsilon \sin t$$

where ϵ is small. For p^2 not too near 1 there is an elementary solution of order ϵ. When $p^2 \gg 1$, if ϵ is neglected, it is possible to find a large amplitude that makes the speed 1. For small ϵ there are two solutions with amplitudes near this, one in phase with $\sin t$, the other opposite in phase, besides the elementary solution. The former is unstable, the latter stable. At a value of p^2 slightly greater than 1, the former coalesces with the elementary solution and both disappear for smaller p^2. The amplitudes at this stage, and at $p^2 = 1$, are of order $\epsilon^{\frac{1}{3}}$.

For $p^2 < 1$ there is only one solution; as p^2 passes through 1 and becomes large this solution passes continuously into the large stable solution for $p^2 \gg 1$. But it becomes large only when p^2 has become much greater than 1. This corresponds to the development of a large amplitude in a swing.

For $p^2 > 1$, and decreasing, when the elementary solution coalesces with the unstable solution, the motion becomes, effectively, an oscillation with speed not quite equal to 0 or p about the solution valid for $p^2 < 1$.

A small term in x^2 does not affect the main conclusions.

It appears that if the free period is originally shorter than the forced period the maximum amplitude would be of the order of $(A^2H)^{\frac{1}{3}}$, where H is the amplitude of the equilibrium tide and A the Earth's radius. The maximum amplitude would be nowhere near enough to produce instability. But if the free period is originally the longer and either period varies so that the free period becomes the shorter, a large amplitude will be worked up, but not near the stage where the periods agree. On the contrary, the large amplitude will occur only when the free period has become considerably the shorter. The latter is the case where the free period varies little and the rate of rotation is diminishing. But in the actual case the amplitude is still severely limited by dissipation.

An alternative theory of the origin of the Moon is that under sufficiently rapid rotation a fluid mass would become unstable even without external disturbance. Moulton, however, showed (T. C. Chamberlin and others, 1909, p. 152) that even for a homogeneous mass the available angular momentum in the Earth-Moon system would fall short of that needed for instability; and I found (1917, p. 122) that with allowance for central condensation still more angular momentum would be needed. Accordingly we cannot suppose that the Moon was ever part of the Earth. They may have been close together originally, and must have been closer together than they are now, but there is at present no way of estimating the original separation.

F. Nölke (in Gutenberg, 1931–6, **1**, p. 56) also pointed out that a fragment pulled out from the Earth would strike it again and be reabsorbed. The same difficulty has been found by R. A. Lyttleton (1938*a, b*, 1941) in studying disruption by rotation alone, and he concludes that the final result would not be a planet and satellite, but two independent planets.

An attractive addendum to the resonance theory was Osmond Fisher's theory that the separation took place when the Earth had already acquired a thin solid crust, and that the Pacific represents the scar left. But this theory falls with the resonance theory.

8·11. Tidal friction in the future. We have seen that under the action of the Moon's tides alone the Earth-Moon system is tending to such a state that each body keeps the same face permanently towards the other, the common period being about 47 days. The solar tides, however, will meanwhile have helped to delay the Earth's rotation, and the estimated common period and the time needed to reach the state in question will both be reduced. Further, the state is not final, because the solar tides will continue to lengthen the period of rotation, thus making it longer than the period of revolution of the Moon. A reference to 8·01 shows that when this happens a fixed point on the Earth's surface will move round the Earth more slowly than the point where the Moon is in the zenith, and the high tides, while still occurring after the Moon has passed the zenith or nadir, will be on the opposite sides of AB from C and D (see Fig. 23). Thus they will retard the Moon's revolution and make the Moon return to the Earth. This process will continue till the Moon is at last dragged down to such a distance that it will be broken up by the action of tides raised in it by the Earth. It will then ultimately form a set of separate bodies something like Saturn's ring, though it now appears that they will be fewer and more massive (Jeffreys, 1947c). The time needed for these changes is long, even by cosmogonical standards.

8·12. The history of the Moon's rotation. Let us now return to equation 8·02 (13). The height of the equilibrium tide is U/g, and its amplitude is therefore

$$\frac{3}{4}\frac{fmA^2}{c^3g} = \frac{3}{4}\frac{mA^4}{Mc^3}.$$

(1)

If then the tide has approximately its equilibrium height,

$$-\frac{d\omega}{dt} = \frac{N}{C} = 5\pi f\rho \sin 2\epsilon \left(\frac{m}{M}\right)^2 \left(\frac{A}{c}\right)^6.$$

(2)

We shall have recourse to this expression again in considering the rotations of the planets.

If we now use A, Ω for the radius and rate of rotation of the Earth, a, ω for the Moon, it appears from (2) that

$$\frac{d\omega/dt}{d\Omega/dt} = O\left\{\left(\frac{M}{m}\right)^4 \left(\frac{a}{A}\right)^6\right\},$$

(3)

provided elasticity is not so great as to affect the order of magnitude of the height of the tides, that the densities are of the same order, and the phase lags of the tides also. With actual values this ratio is about 17,000. The hypotheses assume that both bodies at the time considered were largely fluid. Thus if the Earth and Moon were originally fluid and rotated at about the same rate, the rotation of the Moon would approach the rate of revolution 17,000 times as fast as that of the Earth would. The Moon would be brought to present the same face always towards the Earth before the rotation of the Earth had been greatly affected.

The Moon would be unable to retain any atmosphere or water vapour, on account of its low gravity. The absence of the blanketing effect of the atmosphere would enable it to solidify somewhat earlier than the Earth. Henceforth any tidal friction in the Moon must have been in the bodily tide, for there can have been no shallow seas upon it. Also the elastic tide in the Moon must be about $\frac{1}{50}$ of the hydrostatic equilibrium tide. On both grounds the ratio 17,000 must have been much reduced after the Moon's solidification; it may, however, well be as great as 100.

We therefore have an explanation of how the Moon came to present the same face permanently to the Earth. As the rate of revolution decreased, tidal friction would always be able to step in and bring the rate of rotation into agreement with it.

We have, however, found no positive ground so far for supposing any appreciable amount of bodily tidal friction in the Earth, and accordingly there is no strong reason to assume it in the Moon. Let us consider, then, what would happen if the Moon was completely free from internal friction. Take an axis OX in the ecliptic, fixed in direction. Let the longest axis of the Moon make an angle θ with OX, and

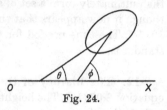

Fig. 24.

let the line joining the centres make an angle ϕ with OX. Neglect the inclinations of the Moon's orbit and equator to the ecliptic. Put

$$\phi = \theta + \psi. \tag{4}$$

The equation of rotation satisfied by θ is

$$C\ddot{\theta} = \frac{3fM}{c^3}(B-A)\cos\psi\sin\psi, \tag{5}$$

that is,

$$\ddot{\psi} + \frac{3fM}{c^3}\frac{(B-A)}{C}\cos\psi\sin\psi = \ddot{\phi}, \tag{6}$$

where A, B, C are the Moon's moments of inertia. But $\dot{\theta} = n$, and dn/dt is small. As $B > A$ there is a stable solution with ψ small. Also

$$\frac{fM}{c^3} = \frac{n^2 M}{M+m}, \tag{7}$$

and to sufficient accuracy

$$\ddot{\psi} + 3n^2\frac{B-A}{C}\psi = -\frac{dn}{dt}. \tag{8}$$

Since dn/dt varies very slowly and steadily over long intervals, a sufficiently accurate particular integral is

$$\psi = -\frac{C}{B-A}\frac{dn}{3n^2\,dt}. \tag{9}$$

Now at present

$$\frac{dn}{dt} = -3n_0\frac{d\xi}{dt}, \quad \kappa\Omega_0\frac{d\xi}{dt} + \frac{d\Omega}{dt} = 0, \tag{10}$$

whence

$$\frac{1}{3n^2}\frac{dn}{dt} = \frac{1}{\kappa n\Omega}\frac{d\Omega}{dt} = -4 \times 10^{-13}. \tag{11}$$

Also $(B-A)/C$ is about $0\cdot 0002$. Hence

$$\psi = 2 \times 10^{-9} = 4'' \times 10^{-4} \tag{12}$$

roughly. Thus even in the absence of any friction in the Moon, the recession of the Moon would produce a perfectly imperceptible deviation of the axis of least moment from the line of centres.

It may be pointed out that, small though this deviation is, it has an important dynamical effect. The Moon's longest axis pointing systematically to one side of the Earth causes the couple on the Moon produced by the Earth's attraction to be on an average negative. It is this couple that reduces the Moon's rate of rotation and keeps it equal to the period of revolution while the latter changes.

It may be thought that, even though the particular integral of (6) is small, the complementary functions may increase greatly; in other words, the amplitude of the free libration in longitude may become great. This would contradict the fact that this libration is imperceptible by observation. But we see that the period of the free libration is $\dfrac{2\pi}{n}\left\{\dfrac{C}{3(B-A)}\right\}^{\frac{1}{2}}$, or about 40 months, and remains for all time proportional to the period of revolution. Thus the change in the period of oscillation during a complete period is small compared with the period itself. In these circumstances it is known that an approximation valid for all time to the solution of

$$\frac{d^2y}{dt^2} + \chi(t)\,y = 0,$$

where $\chi(t)$ is always positive, is

$$y \sim \chi^{-\frac{1}{4}}\left\{A\cos\int_0^t \chi^{\frac{1}{2}}\,dt + B\sin\int_0^t \chi^{\frac{1}{2}}\,dt\right\}, \tag{13}$$

A, B being arbitrary.* Thus the amplitude is proportional to the square root of the current period. If it was small when the Moon revolved in one of our present days it would still only have been multiplied by about 5.

Hence to explain the facts that the Moon's periods of rotation and revolution are equal, and that its free libration in longitude is imperceptible, we need not assume that tidal friction is still operating in its interior. It is enough that tidal friction should have been enough to produce these conditions before, or soon after, solidification, which is highly probable. Once produced, they would be permanently maintained by the Earth's attraction on the Moon's equatorial protuberance.

8·13. Bodily imperfections of elasticity. We have so far found no evidence for these in the Earth for small stresses, but there are indications from the theory of tidal friction that they are, or have been, important in most satellites and in Mercury and probably Venus. Other indications may be sought in the Earth on the following lines. (1) Seismic waves should show damping as they travel; this applies especially to transverse waves, but even longitudinal ones are subject to some damping, and it is desirable also to see what limitation the transmission of *PKP* puts on the viscosity of the core. (2) The damping of the 14-monthly variation of latitude needs investigation. (3) Any long-continued departure from the hydrostatic state, such as we have discussed under stresses in the Earth, will lead to progressive deformation if elasticity is imperfect. The known inequalities imply stress-differences near or possibly somewhat over the laboratory

* The solution has been obtained by several authors in various ways. Cf. H. and B. S. Jeffreys (1972, §§17·12–17·13), H. Jeffreys (1968 *b*).

values of the crushing strengths, and it must not be supposed that any estimate of viscosity found from them is applicable to very small stresses; but as imperfection of elasticity certainly increases with the stress-difference it is safe to say that under small stress-differences the Earth will be more nearly perfectly elastic than under these large ones.

As the bodily tide implies strains of the order of 10^{-7}, as against 10^{-3} or so for rocks on the verge of fracture, it seems safe to suppose that the stress-strain relation is linear. The simplest linear relation that expresses imperfection of elasticity is the elasticoviscous law 1·04 (30); its extension, the law 1·04 (36), is the simplest law that expresses elastic afterworking, and reduces to the elasticoviscous law in the case $\tau' = \infty$. Accordingly, we shall suppose at present that the solution of a problem in imperfect elasticity can be derived from that of one in pure elasticity by replacing the rigidity by the operator, with $\tau' > \tau > 0$,

$$\mu \frac{1 + Q/\tau'}{1 + Q/\tau} \quad \left(\frac{\mu}{1 + Q/\tau}, \text{if } \tau' = \infty \right), \tag{1}$$

where Q denotes the operator of integration with regard to the time. In particular, for harmonic motions proportional to $e^{i\gamma t}$, μ is replaced by

$$\mu \frac{\tau}{\tau'} \frac{1 + i\gamma\tau'}{1 + i\gamma\tau} \quad \left(\frac{\mu i\gamma\tau}{1 + i\gamma\tau}, \text{if } \tau' = \infty \right). \tag{2}$$

For finite τ', this is effectively $\mu\tau/\tau'$ for long periods and μ for short ones, so that both for long and short periods elastic afterworking approximates to perfect elasticity. It reduces the effective rigidity for disturbances of long periods, but it is only for intermediate periods comparable with τ and τ' that appreciable departure from perfect elasticity is to be expected.

For infinite τ', the modified μ tends to μ for large γ, so that for short periods the material behaves as nearly perfectly elastic. For long periods it behaves like $\mu\tau i\gamma$, so that the material behaves like a liquid of viscosity $\mu\tau$.

As the disturbances to be examined include periods of a second or so (earthquake waves), half a day (tides), 14 months (free variation of latitude), and also geological effects that have persisted for 10^7 to 10^9 years, it appears that there is some hope of finding at any rate the orders of magnitude of τ and τ'. We begin with earthquake waves and the variation of latitude because they give positive evidence.*

Take first plane transverse waves. The periods are short, so we take $\gamma\tau$ large, with displacement proportional to $\exp i\gamma(t - x/\beta)$ for perfect elasticity. To allow for imperfect elasticity we must multiply β by

$$\left(\frac{\tau}{\tau'} \frac{1 + i\gamma\tau'}{1 + i\gamma\tau} \right)^{\frac{1}{2}} = 1 - \frac{1}{2i\gamma} \left(\frac{1}{\tau} - \frac{1}{\tau'} \right) + O\left(\frac{1}{\gamma^2} \right), \tag{3}$$

* The 19-yearly nutation, and the precession with a period of 25,000 years, are due to couples whose direction varies slowly in space, and therefore in about a day relative to the Earth. The corresponding deformations of the Earth belong to the group of diurnal tides.

and $i\gamma(t - x/\beta)$ is replaced by

$$i\gamma \left(t - \frac{x}{\beta} \right) - \frac{x}{2\beta} \left(\frac{1}{\tau} - \frac{1}{\tau'} \right). \tag{4}$$

The damping in a given distance is insensitive to the period. As a representative value we suppose that the second term reaches 1 when $x = 10^4$ km.; then with $\beta = 5$ km./sec.

$$\frac{1}{\tau} - \frac{1}{\tau'} = \frac{10^{-3}}{1^s}. \tag{5}$$

This may be too high, as some of the observed damping may be due to internal reflexion and scattering. The latter would explain the actual dependence of the damping on the period, for which this kind of imperfection of elasticity gives no help. (See p. 359.)

Longitudinal waves show damping of the same order of magnitude and will lead to a similar estimate. They give an interesting extreme value to the viscosity of the core. We introduce the bulk-modulus k, which is supposed to show no imperfection of elasticity. Then the equation of plane waves is

$$\frac{\partial^2 u}{\partial t^2} - \frac{k + \frac{4}{3}\mu}{\rho} \frac{\partial^2 u}{\partial x^2} = 0. \tag{6}$$

For viscosity we must replace μ by $\eta i\gamma$. Put $k/\rho = \alpha^2$, $\eta/\rho = \nu$. Then for a time factor $e^{i\gamma t}$ the distance factor is

$$\exp \left(-\frac{i\gamma x}{\alpha} - \frac{2}{3} \frac{\nu\gamma^2}{\alpha^3} x \right). \tag{7}$$

It is now known from observations with short-period vertical seismographs that PKP waves with periods of order 1^s pass through the core. We therefore take $x = 6 \times 10^8$ cm., $\alpha = 9 \times 10^5$ cm./sec., $\gamma = 6/1^s$, and suppose that for these values

$$\frac{2}{3} \frac{\nu\gamma^2}{\alpha^3} x = 1. \tag{8}$$

This gives $\qquad\qquad \nu = 5 \times 10^7$ cm.^{2}/sec. $\qquad\qquad$ (9)

This also may be regarded as an upper limit, especially since the damping of longitudinal waves, if any, would include the effect of the second coefficient of viscosity. With a density of 10 the viscosity η would be 5×10^8 c.g.s. Laboratory values for liquids are mostly of order $0 \cdot 01$ c.g.s.; 10^3 is given for treacle and 5×10^6 for shoemaker's wax. In the core pressure and temperature would affect the viscosity in opposite directions, and data for calculating the effects are incomplete, but it is improbable that the true value is anywhere near as large as 10^8 c.g.s. Nevertheless, the establishment of an extreme possible value from geophysical evidence is of interest. Bridgman (1931, pp. 330–356) finds that temperature reduces viscosity, while pressure increases it. The effect of pressure is very sensitive to the molecular complexity of the material; it is likely to be small for iron but large for silicates. Miki (1952) estimates values of order 10^{-2} or 10^{-3} cm.^{2}/sec. according as the core is liquid or gaseous. This is relevant to the possibility of convection currents in the core and their magnetic effects (cf. Bullard, 1950).

R. Gans (1972) in a critical discussion, partly theoretical and partly consisting of comparison of different materials, finds that the viscosity of the liquid core should be between $0 \cdot 037$ and $0 \cdot 185$ c.g.s. Limits for the kinematic viscosity would be about a tenth of these in cm.^{2}/sec.

If the core is convecting, it should have nearly an adiabatic distribution of temperature on account of its great size, and should also be thoroughly mixed. Then the Adams–Williamson distribution of density should be very accurate, and Longman's conclusion holds for a different reason.

8·14. The damping of the variation of latitude.

We have seen that the 14-monthly variation of latitude seems subject to such damping as would reduce its amplitude in a ratio e^{-1} in about 15 years. An explanation may be sought in tidal friction in seas, imperfections of elasticity in the shell, or viscosity in the core. We cannot reach a definite conclusion, but we can make enough progress to indicate what points need closer investigation. We consider first the energy available to be dissipated. Consider a state of rigid rotation; the kinetic energy is given by

$$2T = A(\omega_1^2 + \omega_2^2) + C\omega_3^2, \tag{1}$$

and the resultant angular momentum by

$$G^2 = A^2(\omega_1^2 + \omega_2^2) + C^2\omega_3^2. \tag{2}$$

The ultimate effect of dissipation would be to give a uniform rotation Ω about the C axis, without change of the resultant angular momentum, since the motion considered is a free one that could exist in the absence of any external disturbance. Then

$$G^2 = C^2\Omega^2, \tag{3}$$

and the energy will be $\frac{1}{2}C\Omega^2 = G^2/2C$. Hence the energy available to be dissipated is

$$T - G^2/2C = \frac{1}{2}\frac{A(C-A)}{C}(\omega_1^2 + \omega_2^2) = \frac{1}{2}\frac{A(C-A)}{C}\omega^2(l^2 + m^2) \tag{4}$$

$$\doteqdot 7 \times 10^{33}(l^2 + m^2)\,\text{ergs}. \tag{5}$$

Here since l and m are small $(l^2 + m^2)^{\frac{1}{2}}$ is the angle between the C axis (axis of figure) and the instantaneous axis of rotation.

The dissipation might be due to either (1) imperfect elasticity in the shell, (2) viscosity resisting slip of the shell over the core at the boundary, or (3) resistance in tidal currents in the ocean. It appears from the following arguments that (2) and (3) cannot account for more than a small fraction of it.

It is likely that the motion in the core is mainly according to the rules of classical hydrodynamics, but that the shearing of the shell over the core produces a boundary layer. The dissipation in this can be found by a method often used for viscous motion. If V is the difference of the tangential velocities of the core and shell when viscosity is neglected, the disturbance of velocity in the liquid due to the drag of the shell dies down in a depth of order $(\nu/\omega)^{\frac{1}{2}}$, where ν is the kinematic viscosity, and the rate of dissipation per unit area is of order $\rho V^2 \sqrt{(\omega\nu)}$. In the solution for a rigid shell and perfectly fluid core the ratio of the deviations of the axes of rotation of the core and shell from their mean position is small; this also holds for certain core models with an elastic shell (Jeffreys, 1949 a, 1950 a, 1956 b; Jeffreys and Vicente, 1957 a, b). The motion of the shell is given closely by uniform rotation about the instantaneous axis, which deviates from its mean position by about $\frac{1}{430}$ of the angle between it and the axis of figure. Then the rate of dissipation is of order

$$F = \kappa\rho(l^2 + m^2)\, a_1^4 \omega^2 \sqrt{(\omega\nu)}\,(1/430^2),$$

where a_1 is the radius of the core. We take $\rho = 9\,\mathrm{g./cm.^3}$, $a_1 = 3\cdot47 \times 10^8\,\mathrm{cm.}$, $\omega = 7\cdot29 \times 10^{-5}/\mathrm{sec}$. If the velocity of slip was the same everywhere the coefficient κ would be 4π. A more detailed treatment gives $\kappa = 8\pi/3$. ν appears to be of order $0\cdot01\,\mathrm{cm.^2/sec}$. (Gans, 1972). Then we find

$$F = 2\cdot7 \times 10^{19}(l^2 + m^2)\,\mathrm{ergs/sec}.$$

$$\frac{E}{F} = 2\cdot6 \times 10^{14}\,\mathrm{sec}.$$

$$= 0\cdot8 \times 10^7\,\mathrm{years}.$$

The time of relaxation for the amplitude is of order 2×10^7 years. A serious misprint in the fourth and fifth editions of this book (ω^3 for ω^2) was pointed out by Verhoogen (1974). He took ν as $0\cdot026\,\mathrm{cm.^2/sec}$. and κ as 4π. Bondi and Gold (1956) called attention to the factor $1/430^2$; it was omitted in previous editions and also by Verhoogen. With Gans's value of ν the thickness of the boundary layer is of the order 12 cm., so that the use of a boundary layer theory is justified. See also the note on p. 363.

In previous editions I have attempted to estimate the effect of dissipation by the 14-monthly tide in the ocean by comparison with the semi-diurnal tide. These arguments were not satisfactory. The theory of the semi-diurnal tide has since been improved by Pekeris and Accad (1969) and Hendershott (1972, 1973), but that of the 14-monthly tide is still very incomplete. A dynamical theory has not been worked out. The amplitude on an equilibrium theory is about $0\cdot5$ cm. Haubrich and Munk (1959) compare the theoretical equilibrium tide with the observed 14-monthly tide using 11 stations. They remark (p. 2383) that at three stations spectral analysis shows peaks at essentially the expected period, averaging about twice the amplitude; one is about the level of the corresponding noise while the remaining 7 are below the noise level. Later observations are discussed by Wunsch (1974). He finds in the North Sea and the Baltic about 4 times the equilibrium amplitude, but he is revising his work (Wunsch, 1975).

However, an upper bound for the dissipation can be found fairly easily by a method due to Bondi and Gold (1956). They take a simplified case of a rigid core and a rigid shell rotating relatively to each other, with friction at the boundary proportional to the rate of slip, and adjust the amount of friction to make the damping a maximum. A small frictional coefficient gives small damping, but so also does a large one, since it makes the core turn with the shell and reduce the slip. Thus for any moment of inertia of the core there is a maximum damping. This gives a time factor

$$\exp\left\{-\frac{\beta\alpha\omega t}{2(1+\alpha)}\right\}.$$

Here $\beta\omega$ is the speed of the free period considered and α is the ratio of the moments of inertia. For the core and shell, α is about 0·1 and the exponent is about $-t/(4 \text{ years})$, so that closer analysis makes the damping far less than the maximum. But for the oceans, compared with the rest of the Earth, α is about $0·4 \times 10^{-3}$ and the exponent is about $-0·4\pi \times 10^{-3}t/T$, where T is the free period $(2\pi/\beta\omega)$. The time of relaxation becomes about 1200 years, which is enough to say that damping in the ocean is negligible.

The argument is independent of the type of dissipation assumed; for any distribution giving a certain amount of damping (e.g. electromagnetic) there is one within Bondi and Gold's system that would give more. Bondi remarks that this argument assumes that the shell–core linkage involves no mechanism that amplifies the variable parts of the angular velocity of the shell. Rochester (1968) also finds electromagnetic damping in the core inadequate.

Alternatively, suppose that damping is due to elastic afterworking in the shell. If the Earth was perfectly rigid and the free period was $T_0 = 2\pi/\gamma_0$, the correction for elasticity could be expressed as a period $T = 2\pi/\gamma$, with

$$1-\gamma/\gamma_0 = k\lambda,$$

where λ is a known constant and k is nearly proportional to $1/\mu$. $1-\gamma/\gamma_0$ is about 0·3. Then we can express the effect of imperfection of elasticity by making a small change (compare §2·041):

$$\left(1-\frac{\gamma}{\gamma_0}\right) = k\lambda\{1-f(i\gamma)\}; \quad \Im\left(\frac{\gamma}{\gamma_0}\right) = k\lambda\,\Im f(i\gamma) \doteqdot \left(1-\frac{\gamma}{\gamma_0}\right)\Im f(i\gamma). \quad (15)$$

If $f(i\gamma)$ has a small positive imaginary part, $\Im(i\gamma)$ will make a contribution to the damping. I think that in some studies the factor $1-\gamma/\gamma_0$ has been omitted; at least it is not mentioned explicitly. (See note on p. 363.) Then if we modify μ according to 8·13(2)

$$1 - \frac{T_0}{T} = \frac{1}{3}\frac{\tau'}{\tau}\frac{1+i\gamma\tau}{1+i\gamma\tau'}, \tag{16}$$

and
$$\arg(1 - T_0/T) = -\tan^{-1}\gamma\tau' + \tan^{-1}\gamma\tau. \tag{17}$$

Damping in 15 years would correspond to

$$\frac{1}{T} = \frac{1}{440} + \frac{i}{15\times 365\times 2\pi}, \quad \arg\left(1 - \frac{T_0}{T}\right) = -\frac{1}{40}, \tag{18}$$

whence
$$\tan^{-1}\gamma\tau' - \tan^{-1}\gamma\tau = \tfrac{1}{40}, \tag{19}$$
if $2\pi/\gamma = 440$ days.

There are two classes of solution according as $\gamma\tau$ is large or small. For $\gamma\tau$ small,
$$\tau' - \tau = 1\cdot 8\,\text{days} = 1\cdot 5\times 10^5\,\text{sec}. \tag{20}$$

For $\gamma\tau$ large,
$$\frac{1}{\tau} - \frac{1}{\tau'} = 4\times 10^{-9}/\text{sec}. \tag{21}$$

The damping of seismic waves in the shell gave

$$\frac{1}{\tau} - \frac{1}{\tau'} = 10^{-3}/\text{sec}. \tag{22}$$

Combining this with (20) we find, nearly,

$$\tau = 10^3\,\text{sec.}, \quad \tau' = 1\cdot 5\times 10^5\,\text{sec}. \tag{23}$$

This looks absurd; it would make the final displacement under given stress about 150 times the initial displacement, whereas 1·6 is a more usual ratio even under large stresses. (21) contradicts (22) directly.

But if most of the damping in seismic waves in the shell is by scattering, we must reject (22) in any case. In the extreme case of elasticoviscosity ($\tau' = \infty$) (21) leads to $\tau = 2\cdot 5\times 10^8$ sec. $= 2700$ days, and $\mu\tau$, the effective viscosity, is of order 5×10^{20} g./cm. sec. If, as a pure guess, we assume instead $\tau' = 1\cdot 1\tau$, (20) leads to $\tau = 18$ days and (21) to $\tau = 270$ days.

A further test is provided by the amount of tidal friction. We at least know that the dissipation in the shell is a small fraction of the total amount. From 8·12(2), with $N/C = 2\times 10^{-22}/\text{sec.}^2$, we should have

$$2\epsilon = 0\cdot 016.$$

This is a considerable overestimate. Then if we take 2ϵ to be the true lag in the bodily semi-diurnal tide we have, for $2\pi/\gamma = 0\cdot 5$ day,

$$\tan^{-1}\frac{1}{\gamma\tau} - \tan^{-1}\frac{1}{\gamma\tau'} = 2\epsilon \ll 0\cdot 01.$$

With $\tau' = 1\cdot1\tau$, this leads to two solutions according as $\gamma\tau$ is large or small. One gives
$$\tau \gg 1 \text{ day,}$$
the other
$$\tau \ll 0\cdot01 \text{ day.}$$

The second is impossible. The first would be consistent with either $\tau = 18$ days or $\tau = 300$ days.

To summarize this rough discussion, the damping of the variation of latitude cannot be due to viscosity of the core or to turbulence in the overlying water. Other types of damping, notably electromagnetic, might just be sufficient in suitable circumstances. If not, we may be driven to suppose that the damping is due to imperfections of elasticity in the shell. This would necessitate that the observed damping of seismic waves is due to scattering and not to imperfections of elasticity. Elasticoviscosity, with an effective viscosity coefficient of order 10^{21}g./cm. sec., would fit the damping. Any finite viscosity would of course have to be considered in relation to the problems it raises, such as those treated in Chapter 6. With a simple type of elastic afterworking, in which the characteristic times τ, τ' are connected by $\tau' = 1\cdot1\tau$, there are two solutions, which make $\tau = 18$ and 300 days respectively. Either would make the loss of energy in the semi-diurnal tides in the shell small compared with the loss in shallow seas, and would therefore be consistent with the observed secular accelerations of the Sun and Moon. The ratio $1\cdot1$ is only a guess within plausible limits; all that is said for it is that it avoids some inconsistencies.

Since many planets and satellites that have neither shallow seas nor fluid cores do show features attributable to tidal friction, imperfection of elasticity in them seems to be the only explanation. A merit of the above discussion is that it suggests limits to the amount of such imperfection.

The modified Lomnitz law is considered in §8·16.

8·15. Tidal friction in other planets and satellites.

We have seen that tidal friction operates in four ways:

(1) Tides raised in the satellites by their primaries tend to make each keep the same face towards its primary.

(2) Tides raised in the primaries by the satellites alter the rates of rotation of the primaries.

(3) Tides raised in the primaries by the satellites also alter the distances of the satellites and hence their mean angular velocities of revolution.

(4) Solar tides affect all the rotations.

These effects will be considered separately.

In the first place we can exclude (2) for all satellites except the Moon. For tidal friction transfers angular momentum from the rotation to revolution, and if the rotation of a planet has been much affected in this way the angular momentum of the satellite's revolution would be comparable with that of the planet's rotation. This is true of no satellite except the Moon.

Secondly, the Moon and the greater satellites of Jupiter and Saturn keep constant faces towards their primaries (apart from a possible slow change for J III). These cannot for the most part be established by observation of markings, since the disks are too small; they are found from variations of brightness with position in the orbit. Iapetus in particular varies by about 2 magnitudes. Venus and Mercury have slow rotations. The satellites of Uranus have axes nearly perpendicular to the ecliptic and therefore nearly in their orbital planes.* The relation for the former type of satellite is explained qualitatively by tidal friction and in no other way. In Darwin's pioneer work most of the relevant formulae were found, but lack of knowledge of the age of the solar system and of quantitative data on imperfections of elasticity in solids prevented a quantitative test. We now know that the age is in the neighbourhood of 3×10^9 years $\doteqdot 10^{17}$ sec., and at last we have some positive information about imperfection of elasticity under small stress from the damping of the variation of latitude. The amount of the latter is doubtful by a factor of 2 or so, but can be used to give orders of magnitude.

For the terrestrial planets and their satellites the rigidity plays an important part, since it makes the actual bodily tide much less than the conventional equilibrium tide, especially for the smaller bodies. Examination shows (Jeffreys, 1957c) that the longer time-scale for elastic afterworking, and also the solution for elasticoviscosity, are quite unsatisfactory as explanations of the rotation of Mercury. If the original period was 6 days the periods could be brought into approximate coincidence in about 10^9 years; but if the period was of the order of 1 day this time would be multiplied by about 40. Further, when coincidence is approached the lag of the tide decreases again and the time needed is further prolonged. It would be unlikely that the actual close agreement could be produced in much less than 10^{11} years. There are similar difficulties for the Moon, but in this case they are possibly not insuperable. It is found, however, that the elasticoviscous solution would make the Moon's excess ellipticities subside with a relaxation time of order 400 years. These are theoretically directly connected with the inclination of the Moon's axis to its orbit, which has not changed noticeably in the last 200 years. The elasticoviscous solution must therefore be rejected in any case.

The short time-scale of elastic afterworking, in which the final yield under constant stress is approached in about 18 days, leads to no trouble. The periods of rotation and revolution would nearly coincide after about 10^9 years for Mercury and less for the Moon.

The satellites of the great planets have densities from about 0·8 to 3. Those with the smaller densities are probably mainly composed of ice. Results based on the Earth cannot therefore be applied to them. However, a rough guide is given by the fact that the most important factor in the angular

* I am indebted to Dr W. H. Steavenson for most of this information.

acceleration for similar compositions is $(n^2a)^2$, where n is the orbital angular velocity and a is the radius. I tabulate (diameter/period^2) ÷ (km./day^2).

Moon	5	Mimas	560	Ariel	80
		Enceladus	270	Umbriel	22
Phobos	160	Tethys	230	Titania	13
Deimos	6	Dione	110	Oberon	5
		Rhea	80		
J V	600	Titan	18	Triton	130
J I	1200	Hyperion	0·9		
J II	300	Iapetus	0·2		
J III	100	Phoebe	0·001		
J IV	18				
J VI	0·002				

Phobos and Deimos are probably stony, and thus analogous to the Moon. We infer that they should keep constant faces to Mars. For the rest the analogy fails; and the rotation of Iapetus seems to show clearly that the tidal lag is much greater than would be inferred from the lunar analogy. The tide in the Moon is only about 1/80 of the equilibrium value; if that in Iapetus approximated to the equilibrium value on account of low rigidity, and if the lag is also larger, the rotation could be explained. If so, the rotations of the satellites of Jupiter as far out as J IV and of Saturn as far as Iapetus present no difficulty. Triton gives none in any case.

The inclinations of orbital and rotation planes have been neglected in the simplified account given here, but were treated by Darwin, whose work was extended by Stratton (1906). For some physical properties the state where the axis of rotation of a satellite is in the orbital plane is actually stable. Thus the anomalous inclinations of the axes of the satellites of Uranus may be explicable.

We now come to effects on the orbits (Jeffreys, 1957 d). Secular accelerations of four satellites have been estimated from observation, as follows (Sharpless, 1945; van Woerkom, 1950; Kozai, 1956):

Phobos	$(+0·001882° \pm 0·000171°)$ $(t/\text{year})^2$
Deimos	$(-0·000266° \pm 0·000164°)$ $(t/\text{year})^2$
J V	$(+0·00034° \pm 0·00015°)$ $(t/\text{year})^2$
Mimas	$(+0·000278° \pm 0·000119°)$ $(t/\text{year})^2$

Uncertainties are probable errors. The corresponding values of $d\xi/dt$ are

Phobos	$(-1·1 \times 10^{-16})$ $(1 \pm 0·09)$/sec.
Deimos	$(+0·6 \times 10^{-16})$ $(1 \pm 0·6)$/sec.
J V	$(-0·3 \times 10^{-16})$ $(1 \pm 0·4)$/sec.
Mimas	$(-0·5 \times 10^{-16})$ $(1 \pm 0·4)$/sec.

On the face of it, if these are genuine, they imply considerable changes during the age of the system. But the theoretical values for Phobos and Deimos, the short time-scale of elastic afterworking being assumed for Mars, are -3×10^{-19}/sec. and $+1·2 \times 10^{-21}$/sec. (The negative value for Phobos arises from the fact that its period of revolution is shorter than that of the rotation of Mars.) For the long time-scale the rates are divided by 10. It appears therefore that tidal friction makes no appreciable contribution towards an explanation of the secular accelerations of Phobos and Deimos. Since the periods of revolution of J V and Mimas are longer than those of the primaries' rotation, their $d\xi/dt$ would be positive, so that the observed values have the wrong sign to be explained by tidal friction.

Actually these satellites are not those likely to be most affected by tidal friction. The maximum $d\xi/dt$, for any constitutions of the primaries, would be of the order of 10^{-13}/sec. for J I and Triton, 10^{-14}/sec. for J II and J V, Mimas, Enceladus, Tethys and Dione. (These take the lag of the semi-diurnal tide to be $\frac{1}{4}\pi$.) None has been claimed for J I, which has been observed longest, and it can hardly be more than 0.5×10^{-16}/sec. This would imply that it is possible (but not proved) that the orbit of J I may have been considerably enlarged by tidal friction, but the effect must be secondary for Triton, Mimas and Tethys and negligible for all others. In any case the tidal lags must be very small, as if Jupiter behaves nearly as a perfectly elastic solid or a perfect fluid.

Of the observed secular accelerations, three are not much more than their standard errors and hence their reality is doubtful. Kerr and Whipple (1954) have examined the possibility that they are due to accretion of small particles, but find this unsatisfactory. They appear, however, to consider no form of resistance except accretion, and it needs emphasis that resistance by passage through a gas does not involve accretion. When a rifle bullet has its velocity halved by air resistance, a mass of solid air equal to the original mass of the bullet is not found sticking to it. The effect of a gaseous resisting medium is very difficult to estimate; some of the problems were stated in the second edition of this book. The chief point is that each part of such a medium must be approximately in orbital motion about the Sun; we do not know how much this motion is perturbed near a planet. It is possible that the speed near a planet may be less than that of a satellite in a circular orbit at the same distance, and if so the gas could produce a resistance to the motion of a satellite and hence a secular acceleration. At any rate present knowledge does not exclude the possibility that the observed secular acceleration of Phobos is genuine and due to a gaseous resisting medium. If this interpretation is correct the secular acceleration found for Deimos must be spurious, but those of J V and Mimas may be real.

The rotation of Mercury has been considered above. The rate of change of rotation of Venus would be about seven times that of the Earth due to the Sun, and about twice that due to the Sun and Moon together, if the properties were similar. Urey remarks (1952, p. 113): 'Venus has no water in its atmosphere; but it is difficult to understand the chemical composition of its atmosphere without postulating a primitive atmosphere in which water was a prominent constituent.' The same may apply to Mars. But if shallow seas existed only in the early history of these planets it seems unlikely that the effect of tidal friction can have been anywhere nearly as great as for the Earth.

The general conclusion is that tidal friction has had a controlling influence on the rotations of all satellites except the outer (small) ones of Jupiter and Saturn. It has greatly altered the rotations of Mercury and the Earth, and possibly has had an appreciable influence on Venus and Mars. It seems unlikely to have affected the mean distances of satellites much, with the exception of the Moon and possibly J I.

Kaula (1969) considers the effect of the asymmetry of the Earth (especially the oceans) on the tidal couples, with special reference to the inclinations. The present data do not appear enough to give definite answers. See also Musen and Felsentreger (1972).

Darwin (1880) considered the effect of viscosity on the eccentricity and inclination of the Moon's orbit. When the friction is mostly turbulent it appears that his qualitative results are not greatly altered (Jeffreys 1961e). de/dt is in general positive except when the month is less than about $\frac{3}{2}$ days, and the inclination should be increasing. For small obliquities the obliquity increases if the month is more than twice the day, otherwise decreases.

8·16. The modified Lomnitz law. C. Lomnitz's experimental law makes the creep under constant stress increase like $q \log (at)$, with a constant, for t more than a small fraction of a second. My first paper on the subject (1958a) contains some mistakes and misprints; these are corrected in a joint paper (Jeffreys and Crampin, 1960). The damping of the 14-monthly nutation, with the modified law of index α, gives the equation

$$qa^{\alpha} = \frac{1}{60\alpha!} (1 \cdot 2 \times 10^{-5}/1 \text{ sec.})^{\alpha}.$$

My later determination would multiply this by about $0 \cdot 6 \pm 0 \cdot 3$. An S pulse arriving at $80°$ in about 20^m is estimated to reach half its final amplitude in about 2^s; two solutions are made according as the final amplitude is taken as 1 or $\frac{1}{2}$ times that for no damping. The results in the two cases are $\alpha = 0 \cdot 256$ and $0 \cdot 236$. If α was much less the beginning of an S pulse at $80°$ would be so spread out as to be unreadable.

The lags in a forced vibration, with the former values of α, are as follows:

Period (days)	Lag (radians)
0·25	0·0036
0·5	0·0044
1	0·0053
10	0·0095
100	0·0172
1000	0·0310
10000	0·0559
100000	0·1009
1000000	0·1818

The rate of recession of the Moon according to these values would be about 0·08 of what is needed to account for the secular acceleration.

The Moon's free librations would have relaxation times of $7·6 \times 10^5$, $2·5 \times 10^5$ and $4·1 \times 10^4$ years; thus the law explains why they are not detected. The Moon's rotation is satisfactorily explained.

If Mercury originally rotated in 1 day its rotation could be brought to its present rate in $\simeq 3 \times 10^9$ years. It could doubtfully be if the original rotation was much faster. It has been believed for a long time that Mercury, like the Moon, rotates in the orbital period. Recent results (Colombo, 1965; Colombo and Shapiro, 1966; Liu, 1966; Dyce, Pettengill and Shapiro, 1967) indicate $\frac{2}{3}$ of the orbital period, but that would not affect the conclusion.

The Moon's dynamical ellipticities would decrease by about 10% in 10^8 years and 17% in 3×10^9 years.

An uncompensated inequality of degree 2 on the Earth would be reduced to 50% in 10^8 years and 20% in 3×10^9 years.

For higher harmonics the flow is less, roughly as $1/n$. The indications are that no important part of isostatic adjustment can be attributed to linear creep, and fracture or flow near the elastic limit is presumably dominant.

In other cases the results do not differ much from those mentioned above for the short-scale exponential afterworking.

The two data used refer to the Earth's shell as a whole and combination of them should be legitimate. The results are essentially average properties. What is important is that a linear law, applied over an enormous range of time scale, explains many facts and leads to no contradiction.

Extrapolation to geological periods is not so drastic as might appear. The ratio of the period of the free nutation to the seismological periods is of order 2×10^7; extrapolation to a geological period requires only one such factor.

Speakers have suggested in discussions that even if the Lomnitz law or its modification agrees over the range of periods used there may be an additional creep proportional to the time but not detectable in this range. I should answer, first, that in scientific practice the onus of proof is always on the advocate of the more complicated hypothesis. When a law is found to fit over a range the best way of testing it further is to extrapolate over a greater range; and no serious error has been found.

What may necessitate further modification is the larger values of creep (smaller Q) estimated at depths of order $100 \, \text{km}$. There are signs however that these correspond to smaller values of α. If this is so the subsidence of gravity anomalies of short horizontal extent may be more rapid at first but slower later than on the hypothesis that q and α are independent of depth. This certainly needs investigation.

The quantities in the law do not explain the small amplitudes of S about $10°$. Taking the travel time as 200^s, corresponding to a distance of about $7°$, and the damping factor in the amplitude as e^{-k}, I find the following results (Jeffreys, 1965) for k:

Period (s)	$\alpha = 0.256$	$\alpha = 0.236$
1	0·176	0·251
2	0·105	0·148
5	0·053	0·073
10	0·032	0·043
20	0·019	0·025
50	0·0096	0·013
100	0·0057	0·0074
1000	0·0010	0·0013

For distance 20° these must be multiplied by about 3. It is clear that they do not account for any important part of the fluctuation of amplitude by factors of order 100. I should expect results for surface waves to be comparable. If absorption is the explanation of the small amplitudes it must apparently be in a very thin layer.

After my revision of the damping of the 14-monthly period (1968d), Crampin and I (Jeffreys and Crampin, 1970) recalculated the above results. α is $0·19 \pm 0·04$. There were no fundamental changes, but we gave a table showing the interval after the theoretical arrival of a unit S pulse to reach $\frac{1}{4}$, $\frac{1}{2}$ and $\frac{3}{4}$ of its limiting amplitude for $\alpha = 0·005$, $0·015$ and $0·025$. For $\alpha \to 0$ this time, for a travel time of 20 minutes, would be 50 to 100 seconds, and an S pulse would be unreadable.

Amplitudes of P and S at distances up to about 30° are given by L. Ruprechtova (1959). They diminish to about 12°, rise continuously to about 20°, and then decrease again. If the reduction up to 12° is attributed to the modified Lomnitz law it leads to $1/Q$ about 0·03 to a depth (below the Moho) of about 80 km. At greater distances the rays may penetrate a layer where the damping is much less severe and the distance travelled in the layer of strong damping diminishes (Jeffreys, 1973a).

But Bolt and Lomnitz (1967) from an explosion, observed in $3° < \Delta < 5·2°$, got for P Q about 320 ± 34; adapting to S by a factor $\frac{4}{9}$ still leaves the damping much less severe than in the above estimate.

Gibowicz (1972) uses data for P in New Zealand, and gets over similar ranges to Ruprechtova's values of Q from 70 to 180; forming $1/Q$ and applying a factor $\frac{9}{4}$ to adapt to S gives 0·32 to 0·12, overlapping my estimate.

Particularly important data would be periodic analysis for S up to 25° for stations near the same azimuth and possibly the higher modes of surface waves. There may be some complication for S at short distances on account of the variability with distance of the ratio of the incident and surface amplitudes.

Berzon, Passechnik and Polikarpov (1974) have studied amplitudes associated with ranges of frequency in P at distances from 10° to 100°. The results are very detailed. The average values of Q for P are as follows:

Depth (km.)	Q_P
100–760	710
760–2900	1200–1330

To adapt to S these must be divided by about $2 \cdot 2$.

Several estimates of damping have been made for seismic waves and free vibrations. D. L. Anderson and C. B. Archambeau (1964), Anderson 1967 *b* take their Q to be independent of frequency, thus corresponding to the original Lomnitz law, but think that it may have some dependence. Q^{-1} is the energy dissipated for advance of phase by a radian ÷ total energy carried in a wave-length. The damping factor in a wave-length is $e^{-2\pi/Q}$. Their estimates are as follows:

Depth (km.)	Q
0–400	100
400–800	200
800–1000	1000
1000–Core	2000

D. L. Anderson, A. Ben-Menahem and C. B. Archambeau (1965) give theoretical values for Rayleigh waves. From observations they get the following, and quote Fedotov (1963) for others.

Depth (km.)	A.–B.–A.	Fedotov
Upper layers	450	400 ± 200
70	80–100	80 ± 30
120	100–150	130 ± 60

Ben-Menahem (1965) for surface waves, both Love and Rayleigh, studies different periods and gives the results in the form

$$\gamma = \log_e \frac{A(\Delta)/A(\Delta')}{\Delta' - \Delta}, \quad Q = \frac{\omega}{2\gamma U},$$

where A is the amplitude, $2\pi/\omega$ the period, and U the group-velocity,

Period (s)	$\gamma \times 10^4$/km.
250–150	0·18–0·34 Rayleigh
200–150	0·27–0·37 Love
230–139	0·20–0·46 Rayleigh
277–139	0·17–0·47 Love
333–125	0·14–0·47 Rayleigh
125– 50	0·45–1·24 Rayleigh
125– 73	0·54–0·90 Love
227–125	0·21–0·55 Love
73– 50	0·88–1·33 Love
71– 55	0·92–1·27 Love

If Q was constant these should be nearly inversely as the period. In this range of frequencies this is not far wrong. I analysed Ben-Menahem's data in more detail in the hope of tracing a significant variation of Q with period. In uniform material $\gamma \times$ period should vary as (period)$^\alpha$. It turned out to be in the opposite direction to that given by the modified Lomnitz law. The

24-2

reason is that the longer periods penetrate to greater depths and imply that the damping decreases rapidly with depth, as is indicated also from the above studies of bodily waves at short distances. But we need extended studies of damping of waves of short periods.

Burton (1974) from the damping of Rayleigh waves gets values for Q with 95 % confidence limits (presumably about twice the standard error) as follows:

Period (sec.)	
20	285–700
30	175–770
40	191–∞

Burton and Kennett (1972) get values of $1/Q$ for Rayleigh waves of periods 15 to 65 seconds.

The formula for the damping of surface waves was due to Brune; a somewhat simplified proof is by Jeffreys and Ben-Menahem (1971).

Carpenter and Davies (1966), also Davies (1967), consider the effect of damping according to the logarithmic law on surface waves. The observations for these have been claimed to support Gutenberg's distribution of velocities against those of Jeffreys and Bullen; but they find that allowance for damping shifts Gutenberg's well beyond ours.

Marshall and Carpenter (1966) make estimates of Q for Rayleigh waves as follows:

Region	Period (sec.)	Q
E. Pacific	25	260
Northern hemisphere	20	405
(average)	30	285
	40	400

L. M. Dorman (1968) studied the damping of P in about the range $11° < \Delta < 16°$, by spectral analysis. The period with the largest amplitude was about 0.5^s. Q was estimated as about 475 in the upper 125 km. For S it would presumably need a factor $\frac{4}{9}$, giving Q about 210. If α is about 0·2, $\cot\frac{1}{2}\pi\alpha \doteq 3$, and the shift of phase with a travel time of 300^s would be 4^s to 5^s. This is of the order of the discrepancies found by Shimshoni and me in trying to fit the travel times.

Alsop, Sutton and Ewing (1961 b) give for the $_0S_2$ free vibration the formula

$$A(t) = A(t_0)\exp\left[-\pi(t-t_0)/QT\right],$$

where T is the period, 53 min., $Q = 370$. As this mode affects all the shell, it should be comparable with my average values with distance about 3000 km., and the expected damping exponent would be of order 5×10^{-3}. Thus their value is of the order that would be expected from this rough comparison.

There is thus strong independent evidence that damping really is much more severe in a thin layer, probably at depth about 70 km. It seems possible that with variable q with the Lomnitz law the behaviour of a composite body may imitate the modified law.

I should emphasize the importance of the damping of the free nutation, because it provides information for a far longer period than any of the seismic waves or free vibrations.

J. C. Savage (1965) tested various rocks and found that Q varies only by a factor of about 3 from a few to 10^6 cycles/sec. For igneous rocks it is about 100. He thinks it may be due to diffusion of heat at interfaces on compression.

Knopoff and MacDonald (1960) point out that Q is independent of period if the equation of motion is

$$\ddot{x} + \omega^2(a + b\,|x|\,\operatorname{sgn}\dot{x}) = 0.$$

In spite of the apparent simplicity of this, application to some of the problems discussed above appears to be difficult.

Some estimates of Q have been made from the width of the peak in amplitude; this may be dangerous as other effects may widen the peak. Others have been made by comparing amplitudes over intervals of time; this should be safe. Lomnitz (1962) has proposed a modification of his law; the logarithmic form is retained but a is no longer treated as large. He fits many data, including the damping of the free nutation, with $a = 0.055$/sec.; thus for periods under 20^s it approximates to elasticoviscosity, while q has a constant value of 0.018. With these values $1/Q$ varies with period from 0.0004 to 0.0250; he remarks that this is less than the range in silicate glasses between $0°$ and $600°$ C.

8·17. Relation to rotation of the core. There is clear sign in seismology that damping in a given distance is sensitive to period. The waves observed at short distances have periods of the order of 1^s or less. But P at large distances usually takes 2 or 3^s for the first swing (for recording on the same type of instrument); S perhaps 7^s. Apparently the short period components are damped out as they travel. This is particularly noticeable for the surface waves. The components with periods of a few seconds, seen on the Zürich and Strasbourg records of the Jersey earthquake (Plates I, II) have completely disappeared before $20°$.

Electromagnetic coupling between the core and the shell has been suggested as an explanation of several phenomena. These have been discussed by Rochester and Smylie (1965). They conclude that the maximum effect is $< \frac{1}{5}$ of that needed to explain changes in the length of the day; 10^{-5} to 10^{-4} of that needed to explain the damping of the free nutation; 10^{-3} of that needed to explain the observed amplitude. See also Rochester (1968, 1970).

It has always appeared that the core rotates with the shell. This is not quite obvious since the only way of transmitting to the core its share of the tidal friction couple is by friction at the boundary. Lyttleton and Bondi (1948) have examined the problem, which is somewhat complicated because a motion in meridian planes is set up. They find that the condition is satisfied, for purely viscous motion, if the kinematic viscosity is large compared with $1 \, \text{cm.}^2/\text{sec.}$ There is however reason to suppose it more like $0 \cdot 01 \, \text{cm.}^2/\text{sec.}$ Let the radius of the core be a; and the difference of angular velocities be $0 \cdot 1 \omega$. The average thickness of the boundary layer is of order $(\nu/\omega)^{\frac{1}{2}}$; then the Reynolds number $R = $ (range of velocity) (thickness of boundary layer)$/\nu = 0 \cdot 1 a(\omega/\nu)^{\frac{1}{2}}$, which will reach a usual condition for turbulence $R = 1000$ if $\nu < 2 \times 10^4 \, \text{cm.}^2/\text{sec.}$ roughly. Smaller ν will favour turbulence at much smaller velocities of slip.

If the motion is turbulent the frictional stress will be about

$$0 \cdot 002 \rho (0 \cdot 1 a \omega)^2$$

and give a couple of order a^3 times this. The moment of inertia is of order ρa^5; then we shall have

$$\frac{\dot{\omega}}{\omega} \simeq -2 \times 10^{-4} \omega = -2 \times 10^{-8}/\text{sec.}$$

Then the difference would be annihilated in something of the order of a year. It appears therefore that even with very small viscosity the core would keep closely in step with the shell.

Lyttleton and Bondi (1953) have also considered precession and find an odd behaviour, leaving it in some doubt whether the core can take part fully in the precession. However they treat the core boundary as spherical, and a small ellipticity of it has a great effect for couples tending to produce changes of long period in the direction of the Earth's axis. For the precession, even without viscosity, the shell grips the core in such a way as to compel it to move with the shell. The matter is treated in more detail by K. Stewartson and P. H. Roberts (1963, 1965).

8·18. Changes in the rotation of the Earth. These are related to the B term in the perturbations of the Sun's and Moon's longitudes, and have been attributed to convection in the core and coupling between the rotations of the core and shell. The matter has been investigated by J. B. Taylor (1963). He finds that for slow (long-period) convection there can be no such coupling. He tells me in a letter that the slower motions not covered in his theorems would have a period of order 50 years, and that such motions might produce a change in the length of the day of order $0 \cdot 01^s$. The period and magnitude are of the order of the variations detected.

The obliquity of the ecliptic (its inclination to the equator) has a secular change. Most of this is due to perturbations of the Earth's orbit by the planets, but there is an excess estimated by Duncombe (1958) as $-0 \cdot 287'' \pm 0 \cdot 029''$ per century. Aoki (1967) suggests that the equator also has a secular shift (besides the precession) and considers astronomical consequences related to the rotation of the galaxy. Sekiguchi (1967) considers possible explanations. Dissipation in the diurnal tides would produce a shift of the equator, but Darwin's work seems to imply one in the wrong direction. See also Kakuta and Aoki (1971). A rediscussion by Fricke (1971) suggests that the anomaly is due to incorrect values of precession and proper motions. Laubscher (1972) gives corrections $0 \cdot 036'' \pm 0 \cdot 006''$/cy to the obliquity and $- 0 \cdot 029'' \pm 0 \cdot 010''$/cy to the planetary precession.

A valuable discussion of the problems of this chapter is by W. Munk (1968).

Note to p. 349. Some errors and misprints were introduced in the 4th edition and carried over into the 5th, p. 325. There is a confusion of notation as (l_1, m_1) was used with two different meanings. In the discussion of dissipation in the core the notation was that of Jeffreys (1950a) and in (13) $(l_1^2 + m_1^2) \omega^2$ should be replaced by $\dot{l}_1^2 + \dot{m}_1^2$ and this is approximately equal to $(l^2 + m^2) \omega^2/430^2$ in the notation of that paper, where $(l^2 + m^2)^{\frac{1}{2}}$ is the angle between the axis of figure of the shell and its mean position; but for the purpose of this argument the difference between this and the $(l^2 + m^2)^{\frac{1}{2}}$ of p. 348, equation (5) of the present edition, may be neglected.

As this edition has been going through the press further discussion has appeared in *Nature*. See Rochester (1975), Yatskiv and Sasao (1975) and Verhoogen (1975).

In the argument about the ocean tide at the top of p. 325 of the 5th edition, (l_1, m_1) has a different meaning, but it should not have been introduced. The disturbing potential should be $\omega^2(lx + my)z$.

Note to p. 351. Physically, the greater part of the energy is kinetic energy of rotation as a rigid body. Only the elastic part is directly subject to damping. If the whole energy is used in calculating Q the damping of elastic waves for the actual period will be seriously underestimated.

CHAPTER IX

The Age of the Earth

Is there any thing whereof it may be said, See, this is new? it hath been already of old time, which was before us.

ECCLES. i. 10

9·01. Outline of methods. So far we have been considering chiefly the Earth's present state rather than its history. When we come to the latter the estimation of long intervals of time becomes necessary, and the problem is to find a suitable clock. What we need is to find a process that persists in the same direction and at a definite rate, and whose total amount in the interval considered can be found. Several processes have been suggested for the purpose. It will be necessary to examine how closely they conform to the requirements just stated.

1. We have seen that tidal friction suggests an age of the Moon of the order of 4×10^9 years.

2. The age of the ocean could be found if we knew the total amount of any dissolved constituent in it and the rate of transfer of that constituent to it.

3. The age of the ocean could also be found if we knew the total quantity of sedimentary rocks on the Earth and their rate of formation. Methods 2 and 3 will be called the denudational methods.

4. The Sun's radiation has apparently not varied greatly during geological time, for if it had there would have been considerable variations of the temperature of the Earth's surface. If we can find the total amount of the energy the Sun has radiated away, we can find an upper limit to the time it can have been radiating at its present rate, which gives an upper limit to the time since the Earth was associated with the Sun.

5. The time since the solidification of the Earth may be found if we know its law of cooling and certain facts about the initial and present distributions of temperature.

6. The radioactive elements are continually breaking up at known rates. If in a given specimen of rock we know the present amount of a radioactive element and the amount of its end-product, the age of the rock can be calculated. This method can give the age of an igneous rock of any age, and is not limited to estimating the whole age of the Earth.

7. Observations of distant nebulae show a systematic displacement of spectral lines to the red, and the simplest interpretation of this is that the nebulae are receding from us at a rate nearly proportional to the distance. Looking backwards, we should infer that about 4000 million years ago the nebulae were all very close together, and hence that the whole universe possibly came into existence about that time.

[364]

8. The planets are continually sweeping up meteors, and there is a disk about the Sun, which causes the faint Zodiacal Light, and appears to be dust and gas in comparable amounts. Collisions between dust particles would lead to a steady change in form.

9. On certain theories of the origin of the solar system a diffuse gas would have existed throughout the system at the outset. This would ultimately mostly be absorbed into the Sun, as a consequence of its own viscosity. In the meanwhile it would have reduced the eccentricities and inclinations of the planets' orbits. From the fact that these are now fairly small, but not extremely small, it may be inferred that the time needed for the gas to disappear and the time needed for it to reduce the eccentricities of the orbits to, say, half their original values were not very different. This provides an equation for the density, and hence the time may be calculated.

We may say at once that in certain circumstances Method 6 is quite satisfactory and 1 and 7 may be, though they cannot go into so much detail. None of the others is satisfactory.

9·02. Denudational methods. The amount of sodium in the ocean is fairly accurately known and so is, roughly, the total amount of sediments on the surface. We have a good deal of information about the rates of transport of sodium and sediments to the sea by rivers, and it appeared to former geologists that the estimation of the age of the ocean was a matter of simple division: that is, it was supposed that the rates have been constant. But the general rate of denudation depends on the slope of the land, the quantity, temperature and carbon dioxide content of the rain falling, and on the nature of the soil exposed. No quantitative relation is known between the rate of denudation and any one of these factors, nor do we know even approximately how any one of them has varied during geological time. We have much information about the type of rocks exposed in many places at various geological dates, but there is no geological date such that the nature of the solid surface then is known for all parts of the Earth. Information on the rainfall is still more vague. Consequently the most that could reasonably have been hoped for these methods is that they might suggest an order of magnitude.*

The amount of sodium carried to the sea annually is about $1·56 \times 10^{14}$ g., and the amount in the ocean is about $1·26 \times 10^{22}$ g. Hence at the present rate the ocean could have acquired its sodium content in 8×10^7 years. This is practically Joly's estimate of 1899. But much of the sodium carried to the sea is from the denudation of sedimentary rocks, and has therefore been in the sea before. Igneous rocks contain only about 2 % of the chlorine needed to combine with their sodium, and therefore it is probable that most of the chlorine in the ocean is of volcanic origin. If so, the amount of sodium corresponding to the chlorine found in rivers must be almost wholly derived from sedimentary

* Most of the following arguments are from A. Holmes (1913).

rocks and from airborne salt particles originating in sea spray. This amounts to about 60 % of the whole sodium carried by rivers. Hence the amount of new sodium is unlikely to exceed $6 \cdot 9 \times 10^{13}$ g. annually. If this value is adopted, the corresponding age of the ocean is $1 \cdot 8 \times 10^8$ years. A further increase is needed to allow for unchlorinated sodium in the sedimentary rocks.

Holmes (1926a) pointed out further that the current estimates of the sodium carried by rivers in solution alone are greater than the whole sodium content of the rocks being denuded. The explanation seems to be that the sodium is usually determined by difference and not directly, and is therefore affected by systematic errors of the estimates of the other metals. Consequently the method is vitiated from the start.

After an elaborate discussion of the method based on the accumulation of sediments, Holmes decided that the best way of using it is as follows. The igneous rocks at present exposed produce a cubic mile of sediments in about five years. The total volume of sediments is estimated at 7×10^7 cubic miles. Sediments derived from other sediments are not new and are excluded by this method. The age of the ocean is then estimated as $3 \cdot 5 \times 10^8$ years.

It might be supposed that these estimates would be too low, because there has been a great elevation of the land in many places not long ago geologically, and still more recently there has been a glaciation, which has converted many igneous rocks into easily denudable forms without their having been in the sea. If so, denudation is now faster than its average through geological time.

9·021. Thickness of sediments. Two considerations emerge from the composition of the ocean, whatever the time-scale may be. Assuming that average igneous rocks contain 2 % of sodium, we see that the mass of igneous rocks that would have to be denuded to account for the sodium in the ocean is about 6×10^{23} g. If we suppose that only half the sodium goes into solution, the rest being retained in the resulting sediments, this amount must be doubled. The surface of the Earth is about 5×10^{18} cm.², so that the average mass of igneous rocks converted into sediments is $2 \cdot 4 \times 10^5$ g./cm.². This would imply an average thickness of sediments of about 1 km. over the whole surface. The actual thickness varies greatly from place to place; if all the sediments were confined to the continents and the continental shelves, the mean thickness there would be about 3 km. The estimate suggests that there should be many places where primitive igneous rocks could be found at a not prohibitive depth. It should also be useful when the study of surface waves in seismology begins to take account of the sedimentary layer (Jeffreys, 1930b).

9·022. The sodium-chlorine balance. The other concerns the chlorine. It practically balances the sodium, but its origin is almost independent. It is generally supposed to be of volcanic origin, less from any direct measurements and analyses of volcanic emanations than because there is no other apparent source for most of it. Yet it must have kept pace with the sodium very closely during geological time, for a pronounced excess of either would have destroyed all life in the sea. How is it done? Little attention seems to have been paid to the problem beyond occasional remarks about the influence of calcium. It is true that a moderate excess of sodium would be dealt with by precipitation of calcium and magnesium carbonates, and a moderate excess of chlorine by redissolving them. Thus for sodium as hydroxide and calcium as bicarbonate

$$2\mathrm{NaOH} + \mathrm{CaH_2(CO_3)_2} = \mathrm{Na_2CO_3} + \mathrm{CaCO_3} + 2\mathrm{H_2O},$$

and for sodium as carbonate and magnesium as sulphate

$$\mathrm{Na_2CO_3} + \mathrm{Mg_2SO_4} = \mathrm{Na_2SO_4} + \mathrm{MgCO_3}.$$

But this only postpones the difficulty until all the calcium and magnesium have been precipitated or dissolved.* There is no reason in the apparent origins of sodium and chlorine why an excess of any amount in either direction might not accumulate, and the possibility of life becomes an accident.

9·03. The solar energy method. The radiation received from the Sun by a square centimetre of surface exposed normally to the rays at the Earth's distance from the Sun is 0·03 cal./sec. Taking the Earth's distance from the Sun as $1·5 \times 10^{13}$ cm., we see that the Sun must be losing energy at a rate of $3·3 \times 10^{33}$ ergs/sec. The source of this energy has been the subject of much inquiry. It was noticed by Helmholtz that any known source of chemical energy at this rate would exhaust itself noticeably even within historic time. Helmholtz produced a theory in which the energy is maintained by gravitation. Supposing the Sun to have been originally widely distended and to have collected to its present size under gravity, it is possible to calculate the loss of energy on reasonable hypotheses about the distribution of density. This energy would be converted into heat and radiated away. For a mass M of uniform density and radius a the energy thus derived from condensation is $\frac{3}{5}fM^2/a$, f being the constant of gravitation. If the Sun is denser towards the centre the numerical factor will be somewhat increased, but the order of magnitude remains unaffected. In the case of the Sun this energy amounts to $2·6 \times 10^{48}$ ergs, or $1·3 \times 10^{15}$ ergs for each gram of the Sun's mass. It would supply the actual radiation for about 8×10^{14} sec. or $2·5 \times 10^7$ years. It may be remarked that the energy of the most violent chemical reactions is of the order of 10^4 cal./g., or 10^{11} ergs/g., so that the gravitational energy is at any rate overwhelmingly important in comparison with any chemical source.

Up to about 1900 it appeared that this imposed a definite limit of about 20 million years to the age of the Sun, and therefore to the time the Earth can have been associated with it. The discovery of radioactivity showed,

* Just as the gold standard in economics prevents small financial crises but not great ones.

however, that other powerful sources of energy existed, and before long methods based on the study of radioactivity in rocks were indicating much longer intervals of time. Purely astronomical phenomena have also been found to point in the same direction. Individually none of them is quite decisive, but together they provide a means of coordinating so many facts that they must be taken seriously. The first was based on the period of the variable star δ Cephei. It has been seen that, given the mass and radius of a star, we can find its condensational energy. If, in addition, we know the rate of emission of energy we can find how fast the star should be contracting to provide this energy. Now δ Cephei executes a regular pulsation, whose period is theoretically proportional to the inverse square root of the density.* Thus if, in fact, the energy is derived from condensation the period should be altering at a calculable rate. Eddington found that the rate of decrease of the period should be 17^s annually. The observed rate of change is not more than $0\cdot1^s$ annually. Accordingly, the changes are proceeding in this star not more than $\frac{1}{170}$ as fast as the contraction theory indicates. Other Cepheid variables point to the same conclusion.

An extended time-scale also provides some hope of accounting for the correlation between the masses of stars and their velocities. This could be accounted for by frequent close approaches between stars; but the Helmholtz time-scale did not provide time for enough such approaches.

Radioactivity did not meet the difficulty directly, because a Sun of pure uranium would just about radiate at the actual rate, but it stimulated the search for other sources of subatomic energy. One of the first was the suggestion of Eddington (1926, pp. 292–317), that the direct coalescence of a proton (hydrogen ion) and an electron may result in their annihilation, nothing remaining but an electromagnetic wave travelling out with the velocity of light. (Jeans (1904) had made a similar suggestion to account for radioactivity, of course before the nature of α-particles was known.) As the masses arise from the electric charges, they disappear; and according to the theory of relativity a loss of energy W is associated with a loss of mass W/c^2, where c is the velocity of light. Hence the energy of radiation derived from the annihilation of mass in this way is calculable. On this basis Eddington was led to ages of the order of 10^{12} to 10^{13} years for the stars. This hypothesis has not stood further examination, but a more moderate one, also considered by Eddington, has. For this we must consider some questions of atomic theory, which we shall need again when we discuss radioactivity.

9·04. Atomic structure. Investigations of atomic structure have indicated that each atom in its normal state consists of an external structure

* The pulsation theory of Cepheid variation is due to Prof. H. Shapley (1914, 1918, 1919); theories of it are due to Eddington (1918, 1941 and 1926, p. 290). and E. A. Milne (1949). Recent work is by Christy (1962, 1964, 1966, 1968) and Stobie (1969).

of Z electrons each carrying a negative electric charge denoted by $-e$, associated with a nucleus carrying a positive charge $+Ze$, so that the atom as a whole is electrically neutral. Z is known as the atomic number and also specifies the position of the element in the chemical periodic table. The mass of the nucleus is nearly an integral multiple of a certain unit, which itself is near the mass m_p of a hydrogen nucleus (proton). The mass m_e of an electron is about $\frac{1}{1800}$ of that of a proton, so that the electrons contribute little to the total mass. The mass of the nucleus is more than $2Zm_p$ and near Am_p, where A is a positive integer usually greater than $2Z$, especially for the heavier atom. The nucleus may therefore be made up of A protons and $A - Z$ electrons. This straightforward interpretation has been somewhat modified by the discovery of the neutron, which is an uncharged particle with a mass near that of the proton; it is widely supposed now that the nucleus consists of Z protons and $A - Z$ neutrons. Since there is some evidence that a neutron is itself composed of a proton and an electron in a state of specially close association (much closer than in a hydrogen atom), the two interpretations are not necessarily inconsistent, and for our purposes it does not matter much which we adopt. A is near the chemical atomic weight on the scale $O = 16$; the atomic masses can be measured directly with the mass-spectrograph introduced by Aston, with greater accuracy than by chemical methods. The small discrepancies between the actual mass and $A(m_p + m_e)$ were attributed by Aston to changes of energy involved in building up the atoms, and by the principle of relativity they provide a way of calculating these changes.

A is not always the same for the same Z; when it is not, we have two or more types of atom with the same number of electrons outside the nucleus and the same chemical properties, but different masses. Such types are known as isotopes. The first such case found was lead. It is customary to distinguish isotopes by adding the value of A as an index; thus lead with $A = 208$ is denoted by ^{208}Pb, which also has an atomic weight close to 208. Aston found that many other elements, as they occur in nature, consist of mixtures of isotopes. It was already known that on the scale $O = 16$ most atomic weights were nearly integers, but there were many exceptions. Aston's work showed that these exceptional elements were all mixtures of isotopes, and removed the greatest obstacle to the understanding of the structure of the nucleus.

The atomic mass of hydrogen ($Z = 1$, $A = 1$) is 1·008. Now if other elements are formed from hydrogen, as these considerations suggest, 0·8 % of the mass disappears as electromagnetic energy, and this would be enough, if the Sun consists of hydrogen (90 % of it apparently does), to supply the Sun's radiation for $1·4 \times 10^{11}$ years, which is enough for any geophysical or astronomical requirement. The conversion of hydrogen into helium, also an important stellar constituent, has been examined in detail by Bethe (1939) and Gamow (1939), who find that it can take place in the conditions existing in stars, and it may be said at last that the generation of stellar energy is understood.

9·05. The method based on the cooling of the Earth's crust was applied by Kelvin. With modern data it suggests that the Earth was liquid about 20 million years ago, on the supposition that there is no supply of energy except solar radiation. If there is an internal source of energy it gives only a lower limit, and we shall see in the next chapter that it is best used now to provide information about the distribution of radioactivity.

9·06. Radioactivity. This consists of spontaneous changes in atomic nuclei, taking the form of emission of particles of high velocity or waves of high frequency. Study in the laboratory shows the particles to be of two types, α- and β-particles. Waves similar to light waves, but very short, are associated with them and are known as γ-rays. β-particles are found to be free electrons. α-particles carry a charge $+2e$, but when collected in appreciable quantities they are found to have formed helium, having in the meantime acquired electrons from their surroundings and restored electrical neutrality. They are in fact helium atoms (atomic weight 4, atomic number 2) with the two outer electrons missing. The naturally occurring radioactive elements are mostly of high atomic weight.

The loss of these charged particles from the nucleus entails changes in the latter. It is convenient to take the β emission first, though it is the less important. The loss of an electron increases the nuclear charge by e, without immediate change in the outer electrons. But the atom must restore its electrical neutrality. It cannot do so by acquiring a new nuclear electron, for the new electron would need a velocity as great as that of the β-particle to get in, and there are not many electrons with such velocities about. Consequently, it acquires a new outer electron, and shifts one place to the right in the periodic table. The most important instances are that isotopes of potassium and rubidium, by emission of β-particles, become converted into calcium and strontium respectively. There is no appreciable change of mass.

For α emission the nuclear charge is reduced by $2e$. The restoration of neutrality requires the loss of two outer electrons and the element shifts two places to the left in the periodic table. The expulsion of the particle is associated with loss of energy, so that the atomic weight decreases by the mass of a helium atom together with a calculable amount representing the loss of energy. The energy is large enough to be detected experimentally. The particle itself hits something in its surroundings and its energy is converted into heat, so that radioactivity is always associated with generation of heat.

It is possible for an atom to lose an α-particle and two β-particles. If this happens, whatever the order, the result is an atom with the same nuclear charge as the original one, but with atomic weight 4 units lower, and therefore is an isotope of the original one.

Radioactivity was discovered by Becquerel in 1896; he found that uranium salts gave out rays capable of fogging a photographic plate enclosed in opaque paper. Madame Curie carried out an elaborate investigation of the phenomenon and found that the original uranium ore was much more active, in proportion to the amount of uranium present, than a pure uranium compound, and accordingly inferred that the ore contained some other substance, still more active. She succeeded in 1898 in isolating this substance, which proved to be a new element, and was given the name of radium.

An astonishing fact was soon discovered about the occurrence of radium. It occurs in nature only in the presence of uranium, which itself never occurs without radium. The ratio of the masses of the two elements present in a sample of ore is always the same, except possibly in some of the most recent rocks, namely, $(3 \cdot 40 \pm 0 \cdot 03) \times 10^{-7}$ part of radium to one of uranium (Rutherford and Boltwood, 1906; S. C. Lind and L. D. Roberts, 1920). Such a constancy suggests a chemical combination, but the atomic weights of uranium and radium are about 238 and 226, and a chemical compound containing 1 atom of one element to about three million of another seemed preposterous.

A further discovery led to the explanation. A specimen of a radium compound enclosed in a sealed vessel was found to liberate a gas called 'radium emanation', the rate of formation being such that if initially 1 g. of radium was present, only $0 \cdot 5$ g. would be present 1500 years afterwards. The rest would be transformed into the emanation and into the various disintegration products of the emanation. All uraniferous ores are many thousands of years old, on any hypothesis, and therefore we have to explain why any radium exists at all; why it has not all broken up long ago. The explanation suggested by its invariable association with uranium and by the appreciable radioactivity of the latter even when all radium has been removed is that as fast as the radium breaks up new radium is generated by the break-up of the uranium itself. The suggestion was experimentally verified by Soddy, who prepared a specimen of uranium quite free from radium, kept it for some years, and was able to demonstrate the presence of radium at the end of the experiment.

Uranium, however, does not pass straight to radium, nor is the emanation the final product. The latter, in fact, survives only a few days. Suppose then that u atoms of uranium are present at time t, and that each uranium atom becomes in succession unit amounts of various recognizable stages $X_1, X_2, \ldots, X_n$. Suppose that the numbers of units of each of these stages present at time t are $x_1, x_2, \ldots, x_n$. Further, suppose that what has been proved for radium is true in general, namely, that the rate of break-up of any product is simply proportional to the quantity present, and accordingly that any product X_r generates per unit time $\kappa_r x_r$ units of the next product X_{r+1}. $1/\kappa_r$ may be called the *average life* of X_r. Then $u, x_1 \ldots x_n$ satisfy the following differential equations:

$$\left.\begin{aligned}
\frac{du}{dt} &= -\kappa u, \\[2mm]
\frac{dx_1}{dt} &= \kappa u - \kappa_1 x_1, \\[2mm]
\frac{dx_2}{dt} &= \kappa_1 x_1 - \kappa_2 x_2, \\[2mm]
&\cdots\cdots\cdots\cdots\cdots \\[2mm]
\frac{dx_n}{dt} &= \kappa_{n-1} x_{n-1}.
\end{aligned}\right\} \tag{1}$$

Suppose that initially there are no degradation products present, so that when t is zero $u = u_0, x_1 = x_2 = \ldots = x_n = 0$. The solutions of these equations are:

$$\left.\begin{aligned}
u &= u_0 e^{-\kappa t}, \\[2mm]
x_1 &= \frac{\kappa u_0}{\kappa_1 - \kappa} (e^{-\kappa t} - e^{-\kappa_1 t}) \\[2mm]
x_2 &= \frac{\kappa \kappa_1 u_0}{\kappa_1 - \kappa} \left\{ \frac{1}{\kappa_2 - \kappa} (e^{-\kappa t} - e^{-\kappa_2 t}) - \frac{1}{\kappa_2 - \kappa_1} (e^{-\kappa_1 t} - e^{-\kappa_2 t}) \right\},
\end{aligned}\right\} \tag{2}$$

the expressions becoming more complicated as later products are considered.* But a considerable simplification is possible if all degeneration products except the last are short-lived in comparison with uranium, so that $\kappa_1, \kappa_2, \ldots, \kappa_{n-1}$ are all large compared with κ, and t is so great that $1/t$ is less than the smallest of $\kappa_1, \ldots, \kappa_{n-1}$. The solution then approximates to

$$u = u_0 e^{-\kappa t}, \tag{3}$$

$$x_r = \frac{\kappa}{\kappa_r} u \quad (r = 1, \ldots, n-1), \tag{4}$$

$$x_n = \int_0^t \kappa_{n-1} x_{n-1} dt = u_0 (1 - e^{-\kappa t}). \tag{5}$$

Thus the amounts of all products present except the last remain in fixed ratios to one another and to the amount of uranium left, the ratios being such that the number of units of any product that break up in a given time is the same and equal to the number of atoms of uranium that break up in that time. Thus we have an explanation of the constancy of the

---

* The solution and approximations to it are most easily obtained by Heaviside's methods. Cf. Jeffreys (1931 *h*), H. and B. S. Jeffreys (1972, §8·11).

radium/uranium ratio. We know from experiments on radium that every year $\frac{1}{2280}$ of the radium present breaks up (V. F. Hess and R. W. Lawson, 1918). If r is the number of radium atoms in a specimen, and we allow for the difference in atomic weight between uranium and radium, we have

$$\frac{r}{u} = \frac{3 \cdot 4 \times 10^{-7} \div 226}{1 \div 238} = 3 \cdot 58 \times 10^{-7}$$

and

$$\frac{1}{\kappa} = \frac{1}{\kappa_r} \div \frac{r}{u} = 6 \cdot 37 \times 10^9 \text{ years.}$$

Knowing the rate of break-up of uranium, we shall now be able to find the time since the formation of any rock if we know the amounts of uranium and the end-product present. If indeed l denotes the number of units of the end-product,

$$t = \frac{1}{\kappa} \log \frac{u+l}{u}, \tag{6}$$

and if l/u is small an approximation will be

$$t = l/\kappa u. \tag{7}$$

Thus a chemical analysis of the rock should give the age when the end-product is identified.

The accumulation of experimental data on radioactivity has led to repeated revisions of the adopted constants. Probably the best values at present are as follows (A. O. Nier, 1939; E. C. Bullard, 1942, p. 45). In 1 g. of ^{238}U $1 \cdot 214 \times 10^4$ atoms disintegrate per second; in 1 g. of ^{226}Ra $3 \cdot 70 \times 10^{10}$ disintegrate per second. Using $6 \cdot 064 \times 10^{23}$ as the number of molecules per gram-molecule we find from these, for ^{238}U,

$$1/\kappa = 6 \cdot 65 \times 10^9 \text{ years,} \tag{8}$$

and for ^{226}Ra,

$$1/\kappa = 2297 \text{ years.} \tag{9}$$

The theoretical ratio of the masses present after a long time is

$$\frac{2297}{6 \cdot 65 \times 10^9} \times \frac{226}{238} = 3 \cdot 28 \times 10^{-7}. \tag{10}$$

It is customary to quote the 'half-period', such that

$$e^{-\kappa T} = \tfrac{1}{2}; \tag{11}$$

then

$$T = \log_e 2 \times \frac{1}{\kappa} = \frac{0 \cdot 693}{\kappa}. \tag{12}$$

The only redeeming feature of the extra factor, like that used in giving the 'probable error' of an estimate, is that it is fairly easy to take out again.

The argument given so far is independent of whether the various products considered are pure substances or not. All that has been assumed about them is that each is made up of units of similar composition, each unit having been derived from one atom of uranium. The units themselves may be composed of atoms, which need not be all alike. In fact the transitions consist in the loss of α- and β-particles, and at each step involving the loss of an α-particle the product consists of an atom with atomic weight reduced by 4 and an additional helium atom. The passage from uranium to radium involves the formation of three helium atoms. Radium in turn produces five successive disintegration products with loss of helium, and the final product from one uranium atom will be a substance of atomic weight 206 and 8 atoms of helium.

At the time when these discoveries were made no element with an atomic weight of 206 was known. It might in fact not be permanent, or the last stage might be the formation, for instance, of two atoms of atomic weight 103. The only metal other than the known disintegration products invariably found in uranium minerals was lead, the atomic weight of which, as ordinarily found, is 207·2. But direct determinations of the atomic weight of lead from uranium minerals were made in 1914 by Hönigschmid and St Horovitz, Richards and Lembert, and Maurice Curie. It was found to be 206·2, not far from the predicted atomic weight, and almost a whole unit lower than ordinary lead. It is not radioactive. Later determinations gave values near 206·05 (O. Hönigschmid and L. Birkenbach, 1923). Thus the end-product is identified; its unit consists of an atom of lead of atomic weight 206, with 8 atoms of helium. Thus the determination of the age of a uranium mineral requires the determination of the amount of uranium still present, and of the amount of helium or uranium lead present. When these are known the ratio l/u is determinable, and then (7) gives the age of the mineral.

The use of the lead/uranium ratio for finding the ages of minerals was first attempted by Boltwood (1907), who found that the lead/uranium ratio for minerals of the same geological age was approximately constant. The helium/uranium ratio was applied in 1908–10 by the fourth Lord Rayleigh (R. J. Strutt). Both methods have been extensively used. It will be seen that the applicability of either method to a given rock requires three conditions. First, the final product estimated must have been absent from the mineral when this was formed. There seems to be no reason to believe that original helium ever occurs in appreciable amounts in igneous rocks. Original lead is common, but it is often possible to attach a very high probability to its absence. Uranium in pitchblende is in the form of the oxide, $U(UO_4)_2$. No lead compound isomorphous with this occurs in these ores, and hence the lead and uranium must crystallize separately. When the crystals are too small for the analysis of a single crystal to be undertaken, as indeed is usually the case, it is more difficult to be sure that no crystals of a lead compound are intermingled with them. If, however, we confine our attention to ores containing a large percentage of uranium, we can be practically certain that the amount of original lead is small compared with the amount formed from uranium. Doubt can be dispelled either by an atomic weight determination or by a mass-spectrograph analysis, which finds the ratios of the amounts of different isotopes of lead present.

The second condition required is that radioactivity must be the only agency that has altered the composition of the mineral. All the lead or helium formed in the mineral must still be in it. Now helium is, after hydrogen, the most mobile substance in existence, and if the mineral has been exposed to the air loss by diffusion is certain. Even within the crust leakage into the surrounding rocks is probable. The sheer volume of the helium produces cracks in crystals, which encourage leakage. Holmes and V. S. Dubey (1929) have attempted to minimize this danger by using fine-grained igneous rocks, not necessarily containing much uranium. Diffusion then cannot do much harm because the concentration of the helium generated is nearly uniform from the start, though loss is possible at the margins. Further applications of the method have given inconsistent results, but it appears possible that it can, with due precautions, be successfully applied to magnetite (Hurley and Goodman, 1943). The lead method is impossible because original lead may be as abundant as uranium.

The helium method has also been used to find ages of iron meteorites. Arrol, Jacobi and Paneth (1942) gave values from 60 to 7000 million years for six. In later work a systematic error was detected (Paneth 1953) and it is probable that the greatest ages are about 4.5×10^9 years (C. Patterson, 1955, 1956; Russell and Allan, 1955).

Minerals altered by heat or water must be avoided in any case, since these encourage chemical separation of lead from uranium and diffusion of helium. Uranium and thorium themselves may be lost.

Thirdly, it must be possible to determine accurately the amount of lead or helium in the final product. The estimation of lead is easy, and presents no likelihood of serious error. In estimating helium, however, the specimen has first to be ground to a fine powder, which then has to be heated *in vacuo* to drive off the included helium. Leakage occurs to some extent during the powdering, and again the age found from the helium method is likely to be too low.

We thus see that while, with proper precautions in selecting the minerals to be examined, the lead/uranium ratio is likely to give correct determinations of the ages of rocks, the helium/uranium method is practically certain to give results too low.

The element thorium, of atomic weight 232, is also radioactive, one atom of it yielding in succession six α-particles. The final product is a lead of atomic weight 208. Direct determination of the atomic weight of lead from thorium ores gives values slightly below this (O. Hönigschmid, 1916), but most thorium ores also contain some uranium, which probably accounts for the difference.

Thorium was found by Rutherford and H. Geiger (1910) to break up at a rate of one part in 1.87×10^{10} per year. The number of thorium atoms disintegrating per gram per second is given by A. F. Kovarik and N. I. Adams (1938) as $4 \cdot 11 \times 10^3$ ($1 \pm 0 \cdot 02$). From this we derive

$$1/\kappa_{Th} = (20 \cdot 1 \times 10^9)\ (1 \pm 0 \cdot 02)\ \text{years.} \tag{13}$$

Consequently it might be thought that, like uranium, it could be used for the estimation of geological time. There are, however, rather serious difficulties.

As thorium is usually associated with uranium, it is best to have a formula for calculating the age of a mixed mineral from its lead content. Equation (7) implies, when we take account of the difference of the atomic weights of uranium and lead, that $7 \cdot 37 \times 10^9$ g. of uranium should produce 1 g. of lead per year. Similarly $20 \cdot 8 \times 10^9$ g. of thorium should be producing 1 g. of lead per year. Thus a mineral containing x g. of uranium and y g. of thorium is accumulating lead at a rate of

$$\frac{x}{7 \cdot 37 \times 10^9} + \frac{y}{20 \cdot 8 \times 10^9} = \frac{x + 0 \cdot 35 y}{7 \cdot 37 \times 10^9}\ \text{g./year.} \tag{14}$$

Hence if z is the number of grams of lead in the mineral the age should be

$$\frac{7 \cdot 37 \times 10^9 z}{x + 0 \cdot 35 y}\ \text{years.} \tag{15}$$

A small correction is needed for very old minerals. If

$$v = 1 \cdot 155 \frac{z}{x + 0 \cdot 35 y} \tag{16}$$

the estimate in (15) should be multiplied by $1 - \frac{1}{2}v + \frac{1}{3}v^2$ (Holmes and Lawson, 1927).

When this formula is applied to minerals containing little thorium, it is ordinarily found to give results consistent with geological methods; that is, of two minerals, the older by stratigraphical evidence also gives the greater age as calculated from the chemical analysis. Exceptions are few and usually easily explicable as due to original lead or metamorphism. Minerals rich in thorium however, give very irregular results, some giving ages as great as others found from uranium minerals of the same geological age, but many much lower (down to about a quarter) (Holmes, 1926 b). The variations are not systematic, and seem to exclude the possibility that they are due to any peculiarity in the behaviour of uranium or thorium, and to show that they are due to differences in the histories of individual specimens. Lead generated in a uranium mineral would come into existence surrounded by uranium and oxygen, and would form the highly insoluble lead uranate. On the other hand, thorates do not exist, and thorium lead would only form an oxide or silicate, except possibly when associated with uranium. Subsequent leaching by water would therefore tend to remove thorium lead but not uranium lead. Holmes, by a discussion of the atomic weights of specimens of lead from mixed minerals, was able to show that the ratio of the amounts of thorium and uranium lead is less than would be expected from the amounts of thorium and uranium, and to different extents strongly suggesting that thorium lead has been removed, especially from the minerals containing least uranium. Hence results based on minerals containing much thorium are to be treated with caution.

The following table relating the ages derived from radioactivity to the geological dates was given by Holmes (1947 b). It was derived by interpolating between the absolute dates given by radioactivity on the basis of uniform rates of deposition between them. Ages are in millions of years since the beginning of the formation.

Pleistocene	1	Triassic	182–196
Pliocene	12–15	Permian	203–220
Miocene	26–32	Carboniferous	255–275
Oligocene	38–47	Devonian	313–318
Eocene	58–68	Silurian	350
Cretaceous	127–140	Ordovician	430
Jurassic	152–167	Cambrian	510

It should be said that slight revisions of the estimated ages of minerals are continually being made, as experimental determinations of the radioactive constants improve and as more attention is given to the isotopic analysis to test for original lead and contamination more accurately.

Many pre-Cambrian rocks have been dated by this method, but as they are not fossiliferous the ordinary geological method of comparing ages in different regions fails. For pre-Cambrian geology the absolute age is the only method of comparison.

Potassium is feebly radioactive; an isotope of atomic weight 40 is partly transformed into ^{40}Ca and partly into ^{40}A. ^{40}Ca is ordinary calcium, which is always present, so this process is not useful for estimates of age. The rate of formation of argon has been a matter of much discussion, but appears now to be settled, and the A/K ratio has been used successfully. Rubidium also is radioactive, ^{87}Rb passing into ^{87}Sr, with a half period about 5×10^{10} years, and this process has been used successfully because rubidium and strontium are rare. $\kappa = 1 \cdot 4 \times 10^{-11}$/year, $1/\kappa = 7 \times 10^{10}$ year. As for uranium and potassium there are complications owing to the association of other isotopes of rubidium with the radioactive one.

According to L. H. Ahrens (1947), the greatest ages found from radioactivity are for pegmatites from northern Karelia (Finland) and south-east Manitoba (Canada). Several methods are available, and after a detailed discussion Ahrens decides that the most likely age for both is about 2100 million years. The rocks in question are intrusive into sediments, which contain pebbles of still more ancient granites, so that the true age of the crust is greater by at least a geological period. Holmes (1948) in a discussion of these two series, using data for the uranium series, concurs for Manitoba but gets about 1800 million years for Karelia. More recently rocks with ages approaching 3×10^9 years have been analysed, particularly in Australia and South Africa (Grant, 1953).

Sir George Darwin was interested in radioactivity from the start. He considered solar radioactivity in 1905 (G. H. Darwin, 1905). By taking tidal friction at its maximum possible rate he had estimated 5×10^7 years as the least possible age of the Moon, which was already greater than the Helmholtz and Kelvin estimates; but in his Presidential Address to the British Association in 1905 he said that this was likely to be a wild underestimate and that 10^9 years was quite likely.

9·07. Russell's method. Another method, based on one of H. N. Russell (1921), is as follows. If we assume that all the lead of average igneous rocks has been derived from uranium and thorium, since the formation of the Earth, we shall obtain an estimate of the age of the crust as a whole. But as there was probably some lead in the crust to start with the result will be in excess of the truth. I should prefer to state the argument as follows. The proportions of uranium, thorium and lead in average igneous rocks are given respectively as 6, 15 and 7·5 parts in a million. Aston has shown that ordinary lead consists mainly of three isotopes, of atomic weights 206, 207 and 208, in the proportions $4:3:7$. The first is identical with uranium lead, the last with thorium lead. The intermediate one is the end-product of the actinium series, which begins with the isotope of uranium of atomic weight 235. Ordinary lead comes from ores, chiefly galena, which occur in high concentration in very limited regions, and it is not definitely established that the widely dispersed lead (rock lead) found in small quantities in most rocks has the same isotopic composition. But if we assume that it has, rocks contain 2·2 and 3·8 parts per million of uranium and thorium lead. Applying the method to these separately we get for the age of the crust from uranium lead

$$t = 6 \cdot 37 \times 10^9 \log \frac{6/238 + 2 \cdot 2/206}{6/238} \text{ years}$$

$$= 2 \cdot 25 \times 10^9 \text{ years,}$$

and from thorium lead

$$t = 1 \cdot 87 \times 10^{10} \log \frac{15/232 + 3 \cdot 8/208}{15/232} \text{ years}$$

$$= 4 \cdot 6 \times 10^9 \text{ years.}$$

Either estimate separately is an upper limit, and we infer that the age of the crust is not more than $2 \cdot 2 \times 10^9$ years. About half the lead of atomic weight 208 would then be regarded as original lead not derived from thorium since the Earth's crust came into existence. The weak point of the argument is the assumption that the rock lead has the same isotopic constitution as ore lead. Holmes (1929), from the present abundance of uranium and thorium, inferred that the lead at present being produced by radioactivity has a mean atomic weight of 206·94 (which is another way of expressing the difference given above for the ages of uranium and thorium); that produced long ago, when uranium was relatively more abundant, might be as low as 206·8. But ore leads of all ages gave atomic weights of about 207·20. It is at any rate clear that ore lead is not wholly of radioactive origin, and some of it must have always been lead. Holmes, however, referred to a sublimate from Vesuvius, the atomic weight of lead in which was determined by Piutti and Migliacci as from 207·025 to 207·079, with a mean value of 207·05, and inferred that this lead before the eruption had been genuine rock lead, of which half was of radioactive origin and the rest ordinary.

The estimate of the abundance of rock lead used above was made by Clarke and Steiger (1914), who used a mixed sample made up from numerous specimens of rocks of different types. G. Hevesy and R. Hobbie (1931) get 30 parts in a million for a mixture of 58 granites and 5 parts in a million for a mixture of 67 gabbros. Taking a standard age of 1600 million years, Holmes (1931) finds that the radioactive lead would be $\frac{1}{8}$ and $\frac{1}{5}$ of the lead actually present. For the Whin Sill of Northumberland and Durham the ratio would be 1/60. E. B. Sandell and S. S. Goldich (1943, p. 182) give lead contents from 3 to 23 parts in a million for North American rocks with more than 63 % silica, from 4 to 10 per million for rocks with less. But the immediate consequence of Russell's argument as modified above is that general averages for all rocks are useless, and the minimum derived may well need to be multiplied by 3 or 4 (or divided by 2).

9·071. Rutherford's method. This is as follows. In general it is found that isotopes of odd atomic weight are less abundant than those of even atomic weight. Then ^{235}U should always have been less abundant than ^{238}U. At present the ratio is 1/139. The 'half-period' for it is $7·07 \times 10^8$ years, according to Segrè, corresponding to a decay constant of $9·8 \times 10^{-10}$/year. Prof. O. R. Frisch states privately that this is probably correct to a few per cent. According to these the time since ^{235}U was as abundant as ^{238}U would be 5000 million years. The age of the crust will not be more than this.

9·072. Holmes's method. This is based on the isotopic analyses of ore leads of various ages, mainly due to Nier and his collaborators. There is an isotope of lead of atomic weight 204, which is not of radioactive origin, and the abundances of the isotopes of atomic weights 206, 207 and 208 are expressed as multiples of that of atomic weight 204. These multiples are found to be correlated with the age of the ore, being on the whole less in the ores of the greatest ages. The suggestion is that the primitive lead contained all four isotopes, widely dispersed through the rocks, and that the quantities of the three higher ones have been increased progressively by radioactivity, samples of the mixture being extracted at intervals and concentrated to produce the ores. By an analysis that many readers have found obscure Holmes (1946, 1947*a*) arrived at the result that the age of the crust is about 3350 million years. The following is a revised statement of the problem.

If a, u_m are the original atomic abundances of ^{206}Pb, ^{238}U, expressed as multiples of that of ^{204}Pb, in a region of the crust, and t_0 is the age of the crust, then at time $-t_m$ the amount of ^{206}Pb present would be

$$x_m = a + u_m\{1 - \exp[-\lambda_1(t_0 - t_m)]\}. \tag{1}$$

If b, c, v_m, w_m refer similarly to original ^{207}Pb, ^{208}Pb, ^{235}U, Th, and y_m, z_m to values for ^{207}Pb, ^{208}Pb at time $-t_m$, we find two similar equations for y_m, z_m, where λ_2, λ_3, the decay constants of ^{235}U, Th, must replace λ_1. In the formation of the ore the lead was separated from the U and Th and its composition thenceforth underwent no further change. Thus each ore sample analysed gives three equations of the form (1).

If u_m, v_m, w_m were always the same, say u, v, w, we should have for n ores a set of $3n$ equations of condition to determine u, v, w, a, b, c, t_0. This simple presentation, however, breaks down, since it is easily seen to lead to no determinate answer. Comparison of ores with different t_m would determine $u \exp[-\lambda_1 t_0]$, but this does not determine t_0; for a change of t_0 would not alter this if compensated by a suitable change of u, and the resulting change of $a + u$ in turn would be compensated by a suitable change of a. But if u_m, v_m, w_m have been different in different regions solutions are possible.

Holmes assumes that elements originally had uniform isotopic composition. At present the ratio of atomic abundances of ^{238}U, ^{235}U is 139. Then, even if u_m, v_m are variable from place to place, we should still always have

$$\frac{u_m}{v_m} = 139\frac{\exp[\lambda_1 t_0]}{\exp[\lambda_2 t_0]}, \tag{2}$$

and by combining this with (1) and the corresponding equation for ^{207}Pb to eliminate u_m, v_m we have

$$\frac{x_m - a}{y_m - b} = 139\frac{\exp[\lambda_1 t_0] - \exp[\lambda_1 t_m]}{\exp[\lambda_2 t_0] - \exp[\lambda_2 t_m]}. \tag{3}$$

Then the estimates of x_m, y_m for different lead ores should give a set of equations for a, b, t_0. Holmes's solution and one by F. G. Houtermans (1947) rest on rather unsatisfactory graphical methods (Jeffreys, 1948*c*). A better way is due to E. C. Bullard and J. P. Stanley (1949). They rewrite (3) in the form

$$(x_m - a)(\exp[\lambda_2 t_0] - \exp[\lambda_2 t_m]) - 139(y_m - b)(\exp[\lambda_1 t_0] - \exp[\lambda_1 t_m]) = 0. \tag{4}$$

This is not satisfied exactly by the measured values, but inspection of the data for samples from neighbouring places leads to estimates of the uncertainties of x_m, y_m, and comparison of different age determinations leads to an uncertainty of t_m. Thus this family of equations have known uncertainties and can be solved by the method of least squares. There is, however, a difficulty, which arises in one form or another in all methods of solution. Data for ores from neighbouring places are averaged so that the equations will as far as possible have independent uncertainties, and thirteen equations result. If all are retained the result would be decidedly less than 2600 million years. If two are rejected the rest lead to

$$t_0 = 3290 \pm 200 \text{ million years,} \qquad (5)$$

and the residuals are consistent with the uncertainties estimated originally. Thus the result depends greatly on the treatment of two equations out of thirteen. It is unsatisfactory that this should be so, as the authors remark. Rejection of observations simply because they have unexpectedly large residuals is well known to be liable to lead to seriously incorrect results, combined with great underestimates of uncertainty. It is possible that the close agreement of those retained is accidental, and quite likely that the ores rejected are at least as typical as those retained.

These methods make no use of the ^{208}Pb data. A method that permits their use is as follows. Suppose that u_m, v_m, w_m are distributed about means with standard errors σ_1, σ_2, σ_3. Let the observations be grouped in ranges of t_m. Then the standard deviation of x_m from a world mean would be $\sigma_1\{1 - \exp[-\lambda_1(t_0 - t_m)]\}$. The estimated standard deviation from a set of n_m observations is s_{1m}, and according to a familiar approximate formula

$$\log \sigma_1 + \log\{1 - \exp[-\lambda_1(t_0 - t_m)]\} = \log s_{1m} \pm \frac{1}{\sqrt{\{2(n_m - 1)\}}}. \qquad (6)$$

Each range of t_m gives an equation of this type for each of ^{206}Pb, ^{207}Pb, ^{208}Pb, and hence a set of equations for σ_1, σ_2, σ_3, t_0. In addition, we should have

$$\sigma_1 \exp[(\lambda_2 - \lambda_1)t_0] = 139\sigma_2. \qquad (7)$$

I have attempted a solution on these lines (which would need a correction, made in the above account), but the number of data in the groups for large values of t_m is too small for it to be very satisfactory.

This method really supposes that the whole of the scatter of the measured x, y, z for ores of similar ages arises from the scatter of the amounts of ^{238}U, ^{235}U, Th for different parts of the primitive crust. If this and (2) were strictly true, ores of similar age should show an exact linear relation between x_m and y_m. This is not so; actually x_m is more closely correlated with z_m than with y_m within the age groups (Jeffreys, 1948c, 1949b). The variation of y_m is small in comparison with the others, and it appears that some other irregularity affects y_m, and would consequently increase the s_{2m} systematically. Since σ_2 is much less than σ_1, and the errors of observation are about the same, they alone may be sufficient explanation. This method, while possibly useful for ^{206}Pb and ^{208}Pb, must therefore be mistrusted for ^{207}Pb.

It may be noticed that combination of (6) and its companion with (7) would lead to an estimate of t_0 even if all the t_m were the same. R. D. Russell and D. W. Allan (1955) have applied this principle to specimens none of which was older than 320 million years, and get values of t_0 from 3·4 to 5·0 × 10⁹ years from different sets of data. They estimate $t_0 = (4\cdot3 \pm 0\cdot4) \times 10^9$ years. The meteoric leads, treated similarly, gave about the same value.

9·08. The expanding universe. The fundamental fact in the argument from the expansion of the universe is that many nebulae show a displacement of spectral lines to the red, to such an extent that if we had no specimens showing intermediate displacements the lines would hardly be identifiable with those found in nearer stars. This displacement is strongly correlated with the faintness of the nebula, which would be most naturally attributed mainly to distance. The actual calibration of the scale in terms of distance depends on some intermediate steps. For the nearer stars the distance is found from the parallax, which is the angle subtended at the star by the mean radius of the Earth's orbit about the Sun. The usual stellar unit of distance is the parsec, which is the distance such that the parallax is 1″, and is equal to 3·1 × 10¹³ km. Parallaxes less than about 0·01″ are too small to be measured, and other methods must be used to estimate greater distances. For double stars the period of revolution and the actual distance between the components, if known, would give the sum of the masses, and if we can estimate the masses otherwise (at worst simply assuming average stellar masses) we can invert the argument and infer the actual distance. But we can measure the angular distance, and the ratio gives the distance of the star from the Earth. Another method is based on the observed fact that the periods of Cepheid variables are closely correlated with the absolute luminosity (i.e. the brightness at a standard distance). Thus the absolute luminosity of a distant Cepheid can be found from the period, and comparison with the apparent luminosity gives the distance. Cepheid variables are found in many clusters and nebulae, and permit estimates of the distance. For nebulae at still greater distances these stars cannot be seen separately, and the only course possible is to assume that the whole luminosities of the nebulae are comparable. The luminosities of those at intermediate distances being known, the distances can be estimated from the apparent brightness. The uncertainty naturally accumulates at each step of the extrapolation. On the other hand, the differences in apparent brightness are far greater than the differences in intrinsic brightness found in nebulae at intermediate distances, and there is no reasonable doubt that they are mainly due to differences in distance. The shift of the spectral lines, which reaches about 0·3 of a wave-length for galaxies, and possibly 0·9 for quasars, is interpreted as a Doppler effect due to the velocity of recession. The velocity so found is proportional to the estimated distance within the limits indicated by known variations in intrinsic brightness. The velocity at a distance of 38×10^6 parsecs, or $1 \cdot 2 \times 10^{21}$ km., is estimated at 20,000 km./sec. If the nebulae have receded at uniform velocities, they would all have been close together about 6×10^{16} sec., or 2×10^9 years ago. The suggestion is that the universe came into existence by a vast explosion at this time.

Recent work has shown that there was a considerable systematic error owing to incorrect estimates of the parallaxes of the standard Cepheids, and all the distances should be multiplied by about $2 \cdot 5$. Thus the age of the universe is now believed to be about 5×10^9 years.

There has been much discussion of the interpretation of this result, especially in relation to various forms of the principle of relativity. There is no doubt that if the nebulae were too close together it would be impossible to neglect gravitational accelerations; and with the high velocities indicated it is necessary, in an accurate treatment, to distinguish between a nebula's present position and the place where it was when the light reaching us now started on its way. Further, differences of gravitational potential may not be negligible, and would themselves produce shifts of the spectral lines similar to those produced by a velocity in the line of sight. It is likely, however, that the simple estimate is at any rate of the right order of magnitude.

9·09. To sum up, we have the following indications relevant to the age of the Earth:

Method	Estimate (millions of years)	Initial condition
Tidal friction	2000–4000	Moon at its nearest distance
Sodium in ocean	> 180	Formation of ocean
Sedimentation	> 350	Formation of ocean
Sun's radiation	> 20	Formation of Sun
Temperatures in crust	> 20	Formation of Earth
Radioactivity of known rocks	> 3000	Crystallization of oldest rock
Lead/uranium ratio in crust	< 4000?	Formation of crust
Uranium 235	< 5000	Formation of crust
Lead ores	4500 ± 300	Formation of crust
Recession of nebulae	5000	Formation of Universe
Lead in meteorites	4200?	

There is general consistency with an age in the neighbourhood of 4000 million years, but we should pay attention to the probable intervals of time between the different events taken as starting-points. These cannot be altogether separated from questions relating to the origin of the solar system. Many theories of this have been proposed, and although considerable ingenuity is being devoted to the problem I do not think any theory existing at present quite satisfactory (Jeffreys, 1948*a*). The most important point for geophysical purposes is whether the Earth was formerly fluid. The most direct evidence comes from the differences of composition of the planets. If they had always been solid and had grown from small sizes by accretion of interstellar matter, they would apparently all have picked up very much the same mixture and their present compositions should be very similar. But if they were formerly heated the lighter constituents would be lost by the planets of smaller mass and retained by the heavier ones. Thus the Earth at present could barely retain free hydrogen, but retains water easily. The Moon and Mercury do not retain water now. Water would be lost by the Earth if the surface temperature was raised to about 5000°, and at a smaller temperature if the radius was also larger. Thus the fact that the Earth and possibly Venus have appreciable amounts of water, the smaller planets little or none, while the great planets appear to be very largely composed of materials of low molecular weight, is easily explained if all the planets were once at temperatures of the order of the boiling-point of silica, and there is no apparent alternative except independent origins.

The separation of the iron core would indicate former fluidity if it were not for W. H. Ramsey's theory of the core (see p. 420). Again, the inert gases are very rare in comparison with adjacent elements in the periodic table, and the natural explanation would be that they had no possible home in a heated Earth except in the atmosphere. Water could be retained in solution in magmas; nitrogen and carbon possibly offer greater difficulties, as has been argued by Urey (1951; see also Jeffreys, 1952 a).

When I wrote the first and second editions of this book, it seemed that Jeans's tidal theory of the origin of the Solar System with some modifications was so satisfactory that it could be made part of the basis of geophysics. I then noticed a fundamental difficulty, related to the circulation theorem of hydrodynamics. The forces considered were gravitational, and in such conditions the integral $\int u_i \, dx_i$ around a circuit of particles in the fluid cannot change. Consequently it would be expected that, when densities had become comparable, the rates of rotation of the bodies of the system would also be comparable. But the outer planets rotate fifty times as fast as the Sun. I suggested that the tidal encounter might be replaced by an actual collision. Then there would be rapid shear in a boundary layer, involving an intense rotation. The boundary layer would be stretched out as the disturber receded, and it appeared that its mass would be of the right order to account for the total mass of the planets and its angular momentum to account for their rotations. There was an obvious difficulty, that matter newly ejected from the Sun might be too hot to hold itself together, but it seemed that at least the larger planets could do so.

Further, if half the ejected matter failed to be condensed, it would be available for reducing the eccentricities of the orbits later; and this led to a reasonable time scale. Its ultimate fate would be reabsorption into the Sun, and would account, again in order of magnitude, for the Sun's rotation. No previous theory of the origin had made a correct quantitative prediction, even in order of magnitude.

However H. N. Russell immediately pointed out a difficulty, which would apply also to the original form of the tidal theory. In an elliptical orbit the semi-latus rectum cannot be more than twice the perihelion distance; that of the ejected matter could not be more than twice the Sun's radius. Later changes could only redistribute the angular momentum and therefore the semi-latera recta; hence there is no explanation of their present values.

Lyttleton met Russell's difficulty by supposing the Sun to have been a double star; he showed that in suitable conditions the disturbing body could collide with the companion and both it and the companion could leave the Sun and yet leave much of the ejected matter associated with the Sun. This still leaves the difficulty of how the rapidly moving and expanding matter could condense, as was forcibly stated by Spitzer (1939). I think myself that this would be reduced by liquefaction owing to adiabatic expansion, which would remove most of the internal pressure.

However, since about 1950, most cosmogonists have abandoned catastrophic theories and reverted to something like the planetesimal theory of Chamberlin and Moulton. The latter (really going back to Buffon) starts with a collision to create the matter revolving about the Sun, and this part of it is open to the same objection. But most of these hypotheses start with a disk, each part in orbital motion about the Sun. The planets and satellites are supposed to have been built up by accretion in this disk. The problems of stability in such a disk are extremely difficult, and none of them, so far as I am aware, has been properly solved.

In the first place, in a rotating medium, interchange of two regions specified by $r_1 < r < r_2$, $r_3 < r < r_4$, of equal mass, without change of total angular momentum, requires an increase of energy if the velocity decreases less rapidly than $1/r$. Thus a velocity proportional, or anywhere nearly proportional, to $r^{-\frac{1}{2}}$ makes a strong contribution to stability. It is analogous to an upward rise in temperature, as in problems of convection, and the close analogy was shown in a striking series of experiments by G. I. Taylor (1922). Local gravitation might make for instability, but from analogy with a result of Jeans the density would need to be of the same order as what we should have if the Sun was expanded into a sphere with the diameter of the system. This gives a lower bound to the angular momentum, which is about 2000 times the actual one. This assumes spherical distribution of density; the actual density in the disk might be more. But the thickness of the disk must at least correspond to the present inclinations of the orbits, and a factor not less than 1/10 may be applied. The conclusion is that gravitational condensation is impossible.

It is often argued that the Reynolds number of such a rotating mass would be many times that needed to produce instability. But this is totally irrelevant. The Reynolds criterion concerns a homogeneous fluid with no differences of temperature. With temperature increasing upward far greater velocities are needed for instability; in the present problem the departure from uniformity of the angular momentum per unit mass has a controlling effect.

On the other hand there are two other types of condensation. One is direct condensation, as in the formation of snowflakes. The other is accretion of freely moving solid particles.

The criterion for direct condensation for a gas on to the surface of a solid is that the density of the gas shall exceed the saturation vapour density. This was examined by A. L. Parson (1945), by a great extrapolation of Clapeyron's equation. It appeared that if all the mass of the planets was extended uniformly through a sphere the density would exceed this density at least 10^{30} times, for the most important materials. The factor is so large that it seems impossible for the material to remain gaseous except for the permanent gases; thus accretion would be the dominant process. Accretion is very complicated. It seems that at planetary velocities the dominant process is one given by Hoyle and Lyttleton (esp. Hoyle, 1946). If a gravitating body moves through a stationary medium, some of the particles are deflected so as to collide some distance behind the body and lose the transverse component of their velocities. The remaining velocity can be low enough to produce capture. A difficulty in the early form of the mechanism is that collision would not destroy the transverse velocities, but only convert them into random motion, that is, heat; and the average total velocity would still exceed that of escape. Hoyle and Lyttleton (1940 a, b), however, showed that if molecular hydrogen is present the radiation could be enough to dispose of the heat, and Bondi revised the theory on this basis. The process seems to be of great importance in astrophysics.

It seems doubtful, however, whether it can explain what is needed in the Solar System. In the first place, as I pointed out (1916c) and in the first two editions of this book, the chance of a small body hitting another small body is to that of its hitting a large one in about the ratio of the sum of the surfaces. If the total mass of the small bodies is comparable with that of a planet, their total surface must be much larger; hence long before a small body, in general, could hit a large one it would have time to hit many small ones, and at planetary velocities it would be volatilized. From Parson's result it now appears that the bodies would recondense; but in the process all departures from velocities in circular orbits would average out. There seems to be no hope that a planet, initially a small particle, could grow into a large one with appreciable orbital eccentricity and inclination.

It has been suggested (originally by Chamberlin) that the asymmetrical distribution of the continents could be produced by accretion. Now the volume of a continent 3000 km. in diameter with a root 30 km. thick would be about 20×10^7 km.^{3}, and the diameter of a spherical body of this volume would be about 400 km., that of a large asteroid. There is no evidence that large asteroids are of granitic composition—the evidence for the Earth and

Moon suggests that they are mainly of ultrabasic rocks, and so are stony meteorites. Interpolation suggests that the same is true for asteroids.

It would also be remarkable, if the Earth picked up five or six asteroids of this size, that there should be none at all left interior to the orbit of Mars.

If smaller bodies are considered, the rotation of the Earth would ensure that all sides would be equally well exposed to accretion, which would therefore be practically uniform. With planetesimals 1 km. in diameter, I estimated the probability of getting the Pacific Ocean in this way as of order 10^{-10^5}.

To sum up, I think that all suggested accounts of the origin of the Solar System are subject to serious objections. The conclusion in the present state of the subject would be that the system cannot exist. So far as it is relevant to geophysics the main question is whether the Earth was formerly fluid. On the catastrophic theories it was. On the accretion theories the answer could depend greatly on the rate of accumulation. Hoyle (1946) thinks it could be done. The existence of continents and the upward concentration of radioactivity seem to demand former fluidity, as will be explained later.

Hoyle (1944, 1945) makes the further modification that the solar companion might have been a supernova. In these conditions, he thinks that a larger fraction of the material might be captured by the Sun.

It appears possible that the Earth was originally solid and attained at least partial fluidity later. But many present concepts seem definitely wrong.

Matter coming in with a planetary velocity, say 40 km./sec., would generate a shock wave with a velocity of at least 6 km./sec., the immediate particle velocity being comparable with the wave velocity. The energy communicated would be 6/40 times that of the accreted matter, say 10^{12} ergs/g., or 2×10^4 cal./g. The wave would ultimately be dissipated as heat throughout the body. Thus if a planet grew by accretion it might well reach fusion temperatures. Even if the infall was at the velocity due to the Earth's attraction alone something like 1000° rise would occur. This would not be true for snowflake condensation; the only effect then would be adiabatic compression, and the rise of temperature would correspond to the adiabatic gradient, which is less than the gradient of melting-point (see p. 390).

A good summary including an account of the theory of shock waves is by N. M. Shott (1966). He notes, among other things, that the adiabatic velocity of a shock wave in extreme cases can reach about 40 km./sec.

A detailed summary of present theories of the origin of the solar system is by I. P. Williams and A. W. Crenin (1968).

9·10. Early history of the Earth. We take a temperature of 3000° as a starting-point. The rate of loss of heat by radiation from a body of radius a at absolute temperature ϑ is $4\pi\sigma a^2\vartheta^4$, where σ is Stefan's constant, 5×10^{-5} c.g.s. Centigrade units. The rate of loss per gram of the body is

$3\sigma\vartheta^4/\rho a$, ρ being the density. For the Earth this amounts to 4 ergs/sec. g., or 3 cal./g. year. The total energy due to the initial temperature could hardly exceed 6000 cal./g., while the energy of condensation to the present radius would be of order 10,000 cal./g. Thus if the Earth remained gaseous the whole of the initial thermal and condensational energy would have been radiated away in something like 5000 years. The method uses the present radius of the Earth, and in the gaseous state the radius would be greater and radiation more rapid. Release of energy by conversion of hydrogen into other elements does not become important except at the temperatures that occur in bodies of stellar dimensions. Hence we may infer that the Earth, initially gaseous, was liquid within about 5000 years of its formation.

With a surface temperature in the liquid state of 1500°, the rate of loss of heat per gram would be about 0·2 cal./year. The total loss of heat from liquefaction to complete solidification, including the latent heat, would hardly exceed 2000 cal./g. Provided, then, that there was free access of heat to the surface by convection, complete solidification would not take more than 10,000 years. If any part became so stiff as to stop convection, the surface would cool more rapidly, and the formation of a solid crust can be dated not later than 15,000 years after the formation of the Earth. The supply of heat to the surface thereafter would be by conduction, which is much slower, and the surface would cool rapidly to the temperature maintained by solar radiation. The formation of an ocean then became possible. This, however, neglects the complication due to the presence of water. In the solidification much of the water would be expelled as vapour, and there would be a dense atmosphere, including an appreciable fraction of the present ocean. This would be very opaque, and radiation into space would be determined mostly by the temperature at its outside. In the atmosphere itself a temperature gradient of the order of 5°/km. is to be expected, so that if this primitive atmosphere was, say, 200 km. deep, the outside could be at nearly the present temperature while the magma below it was still above melting point. In that case the time needed for solidification might have to be multiplied by some hundreds.

We shall discuss the method of solidification in the next chapter; it appears that it is connected with the transfer of radioactive elements towards the outside by repeated crystallization, and the time needed for solidification of the whole of the shell would certainly be much longer than that just derived. For both these reasons it is possible that the intervals, from general liquefaction to the formation of a crust, and from that to the general solidification of the shell, may have been millions rather than thousands of years. It remains possible, however, that both were short compared with the time since the existence of the Earth as a separate body.

At first sight it might appear that the Earth is obviously not older than the Universe, but this is not quite obvious. It would be conceivable that the Universe in its original state had contained the solar system for a long time, and that it then became unstable and proceeded to disperse. If so, the method based on the recession of the nebulae would give the time since the beginning of the dispersion, and the Earth might be considerably older.

On the hypothesis that the Moon was formerly part of the Earth, a minimum original distance can be calculated, and the estimate from tidal friction is a calculated time for the Moon to have receded to its present distance. The hypothesis now seems untenable, and the Moon's distance when it first came to have a definite orbit about the Earth may have been considerably larger. Accordingly, the age found from tidal friction should perhaps be regarded as an upper limit.

CHAPTER X

The Thermal History of the Earth

I know it's something humorous, but lingering.
W. S. GILBERT, *The Mikado*

10·01. Method of solidification. We have seen that apart from certain complications the radiation from the surface of the Earth when in a liquid state could have disposed of the internal heat, including the latent heat of solidification, in a few thousand years. The method of solidification is of some importance. So long as the whole was fluid the process was comparatively simple. The heavy material of the core quickly settled to the centre and stayed there. Liquids in general contract and become denser as they cool, and the matter cooled by radiation at the surface sank through the hotter liquid below, thus maintaining irregular convection currents and a continual supply of heat to the surface. It is probable, though not certain, that stirring by the currents would keep the whole of the rocky shell uniform in composition. The distribution of temperature can be estimated from thermodynamic considerations. If we consider unit mass of the substance, the inflow of heat dQ satisfies the conditions

$$dQ = dE + p\,dv = d(E + pv) - v\,dp = c_p\,d\vartheta + M\,dp, \qquad (1)$$

where E is the internal energy, v the volume, p the pressure, ϑ the absolute temperature, c_p the specific heat at constant pressure, and M is to be found. The condition that $d(E + pv)$ shall be a perfect differential gives

$$\frac{\partial}{\partial \vartheta}(M + v) = \frac{\partial c_p}{\partial p}, \qquad (2)$$

and the condition that dQ/ϑ shall be a perfect differential gives

$$\frac{\partial}{\partial p}\frac{c_p}{\vartheta} = \frac{\partial}{\partial \vartheta}\frac{M}{\vartheta}. \qquad (3)$$

Combining (2) and (3) we have

$$M = -\vartheta\frac{\partial v}{\partial \vartheta}, \qquad (4)$$

where v is regarded as a function of the pressure and temperature, so that $\partial v/\partial \vartheta$ is α/ρ, where α is the coefficient of thermal expansion by volume, and ρ is the density. Now if no heat is gained by conduction $dQ = 0$, and

$$\frac{d\vartheta}{dp} = -\frac{M}{c_p} = \frac{\alpha\vartheta}{\rho c_p}. \qquad (5)$$

This gives the rate of increase of temperature on adiabatic compression

(L. H. Adams*, 1924). In a fluid cooling by convection this condition is satisfied very accurately. If, further, x is the depth,

$$\frac{dp}{dx} = g\rho, \tag{6}$$

$$\frac{d\vartheta}{dx} = \frac{g\alpha\vartheta}{c_p}. \tag{7}$$

With $g = 981$ cm./sec.^{2}, $\alpha = 2 \times 10^{-5}$ per degree, $\vartheta = 1400°$, $c_p = 0.2$ cal./g. degree $= 8 \times 10^6$ ergs/g. degree, this gives

$$\frac{d\vartheta}{dx} = 3° \times 10^{-6}/\text{cm.} = 0.3°/\text{km.} \tag{8}$$

The estimate is rough, but will serve our purpose.

During the liquid state this temperature gradient would be maintained. If it became greater, convection currents would increase in vigour and redistribute the temperature adiabatically. If it became less, convection currents would be damped down by viscosity, cooling at the top would become more rapid, and the gradient would steepen. The whole of the liquid mass would therefore cool together.

The melting-point of the material would depend on the pressure and therefore on the depth. The effect of pressure on the melting-point is expressed by the equation

$$\frac{d\vartheta_0}{dp} = \frac{\vartheta_0}{L}\left(\frac{1}{\rho_1} - \frac{1}{\rho_2}\right), \tag{9}$$

where ϑ_0 is the melting-point, L the latent heat of fusion, and ρ_1, ρ_2 are the densities in the liquid and solid states. Thus in a liquid layer

$$\frac{d\vartheta_0}{dx} = \frac{g\vartheta_0}{L}\left(1 - \frac{\rho_1}{\rho_2}\right). \tag{10}$$

With the fairly typical values for silicate rocks

$$\vartheta_0 = 1300°, \quad L = 100 \text{ cal./g.} = 4 \times 10^9 \text{ ergs/g.}, \quad \rho_1/\rho_2 = 0.9,$$

this gives $3°/\text{km}$. Later work (Shimazu, 1954) suggests that both these values are too high, but confirms that the melting-point gradient is the greater, about $1°/\text{km}$.

The important point of (7) and (10) is that in a well-stirred magma, while the actual temperature and the melting-point both increase with depth, the melting-point increases the faster. As the fluid cools, therefore, the melting-point is first reached at the bottom, and solidification proceeds from the bottom upwards. Convective agitation in the fluid continues till solidification is complete; the rate of solidification is given by the condition that the latent heat of the solidifying liquid at the bottom in solidifying and the heat loss by the body of the liquid must supply the actual loss by radiation from the surface.

* The factor ρ is omitted from Adams's formulae.

The iron core hardly affects these considerations. There would always be continuity of temperature across its boundary, but the difference of density is too great for convective interchange between core and shell, though convection currents would exist in each and would maintain conditions of the type (7) or (10) in each. The shell and core would cool together till solidification began and stopped convection at the bottom of the shell. Thenceforward the core could cool only by conduction through the solid, and further loss of heat from it became unimportant.

The inference from laboratory data that solidification proceeded from the bottom upwards was made by L. H. Adams in the paper just cited. The contrary result had been given by Kelvin.

The discussion supposes that the material of the shell was originally thoroughly mixed and remained so, that the gradients given by (7) and (10) were constant, and that there was no internal generation of heat. The first postulate was probably true till a late state of solidification. The variability with depth of the quantities in (7) is probably not serious. But the factor $1 - \rho_1/\rho_2$ in (10) probably varies considerably. A pressure of $1\cdot2 \times 10^{11}$ dynes/cm.^{2}, which would be reached at a depth of 400 km., would raise the density of solid dunite by about 10 %. The compressibility in the liquid state is presumably greater, perhaps double, and it appears that the difference of density might disappear at a depth of a few hundred kilometres and the effect of pressure on the melting-point with it. In any case it is likely that $d\vartheta_0/dx$ is less at great depths than at the surface, and it looks possible that it may become less than the liquid adiabatic gradient. If so, the cooling liquid would reach the melting-point first at some intermediate depth and begin to solidify there. Above this level solidification would proceed as before. But below it the escape of heat could only take place by conduction through a gradually thickening solid layer, and a thick liquid layer might be preserved for a long time. The data on the bodily tide show that there is no such layer now.

A clue is given by some work of Bridgman (1914, 1915). In an experimental investigation of the behaviour of twenty-three substances at pressures up to 12×10^9 dynes/cm.^{2}, he confirms that the difference of density in the solid and liquid states decreases with pressure, but there is no sign that it ever disappears; the compressibilities tend to equality, while the latent heat sometimes increases, sometimes decreases, but in no case shows signs of vanishing at high pressures. The melting-point did not appear in any case to be tending to any finite maximum. (One substance, bismuth, contracts in melting and therefore has its melting-point lowered by pressure; it need not therefore be considered in this comparison.) Bridgman gives on p. 109 of the last paper a graph of $d\vartheta_0/dp$ against ϑ_0. The former is not usually reduced by more than 1 part in 3, while ϑ_0 itself is increased by about 200°. The melting-points at low pressures ranged from $-67°$ to $+114°$ (bismuth excluded). Thus the experimental range of variation of melting-point amounted roughly to doubling the absolute temperature. We may infer by analogy that in rocks a reduction of $d\vartheta_0/dp$ to two-thirds of its surface value will not take place till pressure has raised the melting-point by 1000° or 1500°. At the standard rate for surface conditions of 3°/km., this would imply a depth of 300–500 km. But the gradient will not fall below the adiabatic until this reduction has occurred several times over; assuming a geometrical law, we have

$$\frac{\log 10}{\log 1 \cdot 5} = 5 \cdot 7,$$

so that the depth required would be 2000 or 3000 km. But if, as seems more likely, $d\vartheta_0/dp$ tends to a steady value, the melting-point gradient will not vary so much, and will therefore exceed the adiabatic throughout the thickness of the shell. A smaller melting-point gradient at low pressures would affect the result in the same way. The fact that the shell is solid throughout therefore presents no difficulty of explanation; the natural extension of Bridgman's results indicates that the melting-point gradient everywhere exceeds the adiabatic, and therefore that the shell solidified from the bottom upwards in the Earth's infancy.

10·02. Separations of materials. The chief separations of materials during solidification would be the formation of the granitic and intermediate layers and of the ocean. It was formerly widely believed that in the liquid state even granite and basalt were immiscible and that the separation took place like that of oil from water. In work at the Geophysical Laboratory, Washington, it was found that, generally speaking, silicates in the liquid state mix perfectly freely, both at atmospheric pressure and under high pressure of water vapour (N. L. Bowen, 1915, p. 9). In 1927 J. W. Greig found that in certain cases magmas containing CaO, MgO or SrO and much silica, when fused at about 1700°, separate into two liquids. The more acidic of these is nearly pure silica. The other has 3–5 molecules of SiO_2 to two of the metallic oxide. It appears that these magmas are much too acidic to have any relation to the geophysical problem, as Greig points out (1927). Liquid immiscibility is therefore only an outside possibility.

The most probable method of separation is by crystallization. The usual (not universal) order of crystallization in a mixed magma appears to be that expressed by H. Rosenbusch,* namely, ores and oxides first, then the ferromagnesian minerals (olivine, augite, etc.), then the felspars, and lastly quartz. The existence of silica in the form of quartz indicates that it crystallized below 800° C.; in some rocks (pegmatites, not ordinary granites) the form of the quartz shows that it actually crystallized below 575° C. The maintenance of the quartz in the liquid state so far below its normal melting-point is to be attributed to the concentration of volatile constituents in it, mainly water. We notice that the outer layers shown by

* Quoted by Clarke (1924, p. 308).

Extrusion of water

seismology are so thin in comparison with the whole depth of the shell that if the whole was fused and stirred a refined chemical analysis would be needed to detect any difference. The first stage of the separation would therefore be the crystallization of olivine from a magma that was already nearly pure olivine, and this would settle to the bottom, being the densest constituent. The outer layers are then only the residuum, left when the residual magma had become too acidic to deposit more olivine. A verification that separation can occur in this way in natural conditions is due to L. R. Wager and W. A. Deer (1939). They studied a sill, mainly of olivine, in Greenland, where the forsterite (Mg_2SiO_4) predominated at lower levels, fayalite (Fe_2SiO_4) at higher, the denser material being at the top. Forsterite has the higher melting-point. It would separate out first and settle, leaving the magma above progressively enriched in iron. Thus in some circumstances melting-point can be more important than density.

We have seen that the primitive Earth was probably too hot to retain water vapour. The present water was therefore in solution inside the Earth. As cooling proceeded water would be extruded in the form of steam. At first this could not condense, since the critical temperature is 365° C., and much of it was probably lost. It has been found experimentally that at high temperatures and pressures water and silicates are miscible in all proportions, and if the present ocean was volatilized it would produce all the required conditions for perfect mixing, as Dr J. W. Evans pointed out (1919). From this argument alone we must suppose that most of the ocean was dissolved in the magma until a late stage. This would have some effect on the melting-point. J. H. L. Vogt (1926) gives the following temperatures when magmas of the respective types begin to deposit crystals: dunite, 1500–1600°; gabbro, 1250°; diorite, 1200°; syenite, 1100°; granite, 1000°. The effects of admixture of small quantities of other substances on the melting-points of several minerals are known.* The depression of the melting-point per gram-molecule of the solute per 100 g. of the mineral is about 400° for orthoclase and 640° for anorthite. The solute considered being water, of molecular weight 18, the great reductions of melting-point found for quartz in granites and other rocks must be taken to imply an amount of water of the order of 30 % of the quartz, or, say, 6–10 % of the granite as a whole. R. W. Goranson (1931, 1932) has shown that under a pressure of 4×10^9 dynes/cm.², corresponding to a depth of 15 km., a granite magma can hold 9 % of water in solution; at a quarter of this pressure it can hold 6 %, and its melting-point is reduced to 720°. Since on our hypothesis this is the material left over after the removal of the olivine, the concentration of water in the shell as a whole can hardly be expected to have exceeded about 1 %. A depression of the melting-point by 40° for the latter therefore seems to be the maximum allowable. Incidentally, this

---

* Boeke and Eitel (1923, p. 85). The theory is given in books on thermodynamics, for instance, that of Birtwistle (1927, pp. 115–116).

estimate of the original concentration of water in the upper layers is consistent with the hypothesis that most of the ocean was extruded from the upper layers during and since their solidification.

The effect of water on the solidification of the upper layers may be considerable, as G. F. S. Hills has pointed out to me. In 10·01 (9) $1/\rho_1$ and $1/\rho_2$ are the specific volumes in the two states. If the water content is considerable and the water remains in the liquid phase, $1/\rho_2$ should include the volume of the separated water, and if this is a fair percentage of the volume of the solid the value of $d\vartheta_0/dp$ will be considerably reduced. Then the melting-point might actually decrease with depth in the outer layers. Hence the outer layers may have solidified as a whole or even from the surface downwards. There is evidence to support this from the fact that high water content appears to favour solidification into the vitreous state, and there is as well the substantial, though hardly conclusive, evidence that the outer layers consist largely of material in this state.

Oxygen may have been in the primitive atmosphere as CO_2, nitrogen as NH_3. It is a remarkable fact, pointed out by Goldschmidt, that the oxygen is just about enough to oxidize all the combustible carbon in the crust, and there is a suggestion that all our oxygen has been formed from CO_2 by plants. The difficulty about this suggestion is that the lower plants, like animals, convert oxygen to CO_2, and it is not clear how the process could start. There appears to be no oxygen on Venus, so apparently the conditions for it to start did not arise there.

Recent biochemical work has given some hope of a solution. The atmospheres of the outer planets consist largely of CH_4 and NH_3. In presence of water and radiation these can synthesize, giving some of the simple organic compounds and even aminoacids, the bases of proteins. If, as is reasonable, the Earth once had such an atmosphere, a solution may be in sight (Bernal, 1967).

A curious thing is that this started with my inference in 1923 that the outer planets could not be red hot, as was generally believed, but must be at liquid air temperatures. My reason was that they would have had time to radiate away any conceivable store of original heat. Inspection of tables of physical constants suggested methane and ammonia as possible materials that could have densities of 0·6 to 1 at these temperatures; and they were soon identified in the spectra by R. Wildt. The history is an argument for the Unity of Science.

The general account given above is independent of the time-scale, so long as radiation can dispose of the heat transferred upward. The delay due to the formation of an atmosphere consisting largely of water vapour seems likely to be serious only in the later stages, but we are not in a position yet to estimate it.

We shall return later to the effect of internal heating during solidification.

10·03. Heating due to radioactivity. It has been explained already that the radioactive elements emit fast α- and β-particles and γ-rays, all of which are absorbed by the environment with production of heat. The heat can be directly measured; this was first done by H. H. Poole with pitchblende and afterwards by Rutherford and Robinson, and St Meyer and Hess, using radium and its degeneration products. The measurement is, however, one of extreme difficulty, since the temperature difference to be measured is very small. It is more accurate to calculate the heat generation from the energies of the particles and rays. It is found that all α-particles emitted by the same element have the same velocity; the masses are known; and the number emitted per second per gram of the element is a matter of direct counting. Hence the rate of production of energy by the α-particles can be calculated. The β-particles and γ-rays make smaller contributions, which can also be calculated. Where comparison has been made the agreement of the calculated and observed generations of heat is good (Lawson, 1927). The fullest recent discussion is by Bullard (1942). He arrives at the following values:

$7·0 \times 10^{-2}$ cal./sec. g. Ra ($0·72$ cal./g. U year, $2·20 \times 10^6$ cal./g. Ra year).

$6·3 \times 10^{-9}$ cal./sec. g. Th ($0·200$ cal./g. Th year).

$2·5 \times 10^{-13}$ cal./sec. g. K (8×10^{-6} cal./g. K year).

He remarks that the values for radium and thorium should be correct within 2 or 3 %, but that for potassium might be in error by 50 %. The estimates given for potassium have fluctuated greatly. Its radioactivity is due to the rare isotope ^{40}K. Different investigators have found very different values for the average energy of the β- and γ-particles emitted, and it seems probable that the best value for the heat emission is about (Jeffreys, 1950b)

$(7·2 \pm 1·1) \times 10^{-13}$ cal./g. K sec., or $(2·2 \pm 0·4) \times 10^{-5}$ cal./g. K year.

The difficulty about potassium is that it breaks up in two ways. One is by emission of a β-ray, giving ^{40}Ca. But also the nucleus can capture an electron from the innermost ring, giving ^{40}A. The frequency of the latter type of transition has been very difficult to determine. γ-rays are emitted and can be counted, but there has been a possibility that a further process is concerned. The most recent work discounts the latter possibility and seems to show that the only processes concerned are the β transition to ^{40}Ca, with a decay constant of $0·483 \times 10^{-9}$/year, and the γ transition to ^{40}A, with a decay constant of $0·065 \times 10^{-9}$/year. F. Birch (1951) in a critical discussion estimates the uncertainty as about 10 %, and the heat generation as about $2·7 \times 10^{-5}$ cal./g. K year. As the latter is only a shade above the sum of my estimate and its standard error I have not made a recalculation.

10·04. Radioactivity of rocks. The importance of this was first
pointed out by Rutherford and followed up by the fourth Lord Rayleigh
(then R. J. Strutt). Many later workers have added to knowledge of the
subject. The usual method is to compare the radioactivity of a specimen of
the rock with those of standard solutions containing radium and thorium,
so that the results are presented as grams of radium or thorium per gram of
rock. It is undesirable to reduce the radium series to its uranium equivalent, as this must adopt a standard value of the ratio of the amounts of
radium and uranium present, which is subject to revision.

The most striking feature of the results is that radium and thorium are
always present; the next is the variability of their amounts, even for rocks
of the same geological type. The latter needs emphasis because it is nearly
always ignored. Physicists might perhaps be excused for thinking that
granites are compounds of constant chemical composition, but when they
speak of *the* radioactivity of granite their remarks are liable to be copied
by geologists, who might be expected to know better. The variation is
considerable within the same region, and there are systematic differences
between regions. Systematic differences between experimenters have been
asserted, but while these may exist they are certainly not conspicuous in
comparison with the variations shown, except in a few special cases
(Jeffreys, 1936c; R. D. Evans and C. Goodman, 1941; Jeffreys, 1942b).
It is found that the radioactivities of igneous rocks of the same main geological type for the same region are distributed as if derived from a Pearson
law of Type III, according to which the chance that a variable x may fall
in a range dx is

$$\frac{1}{p!}\frac{(p+1)^{p+1}}{b^{p+1}}x^p \exp[-(p+1)x/b]dx. \tag{1}$$

When the law is taken in this form the expectation of x is b, and that of
$(x-b)^2$ is $b^2/(p+1)$. Given a set of n values of x, the best estimate of b is
the mean $\bar{x}$, and a first approximation to p is got by taking

$$\frac{\bar{x}^2}{p+1} = s^2 = \Sigma\frac{(x-\bar{x})^2}{n}. \tag{2}$$

This approximation is good when p is large. If it is small, as it often is,
a more accurate way of estimating p is available. For small p, s is comparable with $\bar{x}$; in a given region, for rocks of the same type, one specimen may
be 6–10 times as radioactive as another. The distribution in such a case
shows a concentration at values less than the mean, and a long tail extending
far above the mean.

In some recent work it is assumed that the chance of $\log x$ has a normal
distribution. The qualitative remarks about the Type III law apply also to
this. There has, so far as I know, been no comparison to see which fits better;
but in my study Type III was already quite satisfactory.

A straightforward classification of the data up to 1936 showed (Jeffreys, 1936c), what had been maintained for a long time by Holmes and other writers, that there was a general correspondence between mean radio-activity and acidity. But it also showed that the ratio $s/\bar{x}$ is correlated with rock type, being generally greater for granites than for basalts and gabbros, and greater for these than for dunites. The acid rocks are more variable than the basic ones, not only absolutely, but in comparison with the mean values. The following table gives the distribution of the ratios $s/\bar{x}$ for radium and thorium together, each region being taken as a unit. They are classified in ranges of 0·2, the centre of the range being given at the head of the column:

	0·1	0·3	0·5	0·7	0·9	> 1·0
Granites and granodiorites	1	6	8	4	2	0
Basalts	2	9	6	2	0	0
Plateau basalts	3	1	0	0	0	1
Eclogites	0	0	1	0	1	0
Peridotites	0	0	1	0	1	0
Dunites	2	0	0	0	0	0

The eclogites and peridotites each represent only two regions. The dunites analysed are so few that the data for the world have been taken together. The value over 1·0 for plateau basalts was suspected of systematic error on account of its isolation, and the author afterwards succeeded in tracing this to its source; this value therefore need not be considered further here. Plateau basalts are particularly extensive outpours that have occurred in a few regions; the data used are for those in the Deccan and Oregon.

As the data for several of the regions rest on only a few analyses, the results are subject to the usual uncertainty of sampling, and when allowance is made for this it is found that the data are consistent with the value of p always being the same for rocks of a given type, except that the basalts from oceanic islands, like the plateau basalts, stand out with specially small values of $s/\bar{x}$. On this assumption the estimates of p are as follows, with the corresponding values of $1/(p+1)$, to which $(s/\bar{x})^2$ would be expected to approximate in a long series of observations:

	Granite	Basalt (continental)	Plateau basalt	Oceanic basalt
p	$2·6 \pm 0·5$	$5·0 \pm 1·0$	27 ± 7	45 ± 15
$1/(p+1)$	$0·28 \pm 0·05$	$0·17 \pm 0·05$	$0·036 \pm 0·009$	$0·022 \pm 0·007$

Using these values it is possible to obtain estimates of the uncertainties of b as estimated from $\bar{x}$ corresponding to the standard error of the estimate when the normal law holds. In the following table regions with similar values have been taken together. The unit is 10^{-12} g./g. for Ra and 10^{-5} g./g. for Th.

Granites

North America, Greenland, Iceland, Scotland, Ireland, Japan

Ra $1·59 \pm 0·12$;   Th $0·81 \pm 0·08$

Finland   Ra $4·66 \pm 0·40$;   Th $2·80 \pm 0·24$

Alps   Ra $4·43 \pm 0·68$;   Th $3·30 \pm 0·50$

South Africa*   Ra $2·36 \pm 0·16$

Basalts

North America, Greenland, Iceland, Scotland, Ireland

Ra $0·96 \pm 0·06$;   Th $0·98 \pm 0·08$

England, Germany, France, Hungary

Ra $1·30 \pm 0·13$;   Th $0·88 \pm 0·10$

Plateau basalts

Ra $0·73 \pm 0·03$;   Th $0·52 \pm 0·02$

Oceanic island basalts

Ra $0·90 \pm 0·03$;   Th $0·46 \pm 0·03$

Dunites (world)

Ra $0·42 \pm 0·06$;   Th $0·33 \pm 0·03$

* From Immelmann (1934).

The work of Evans and Goodman showed much greater variability in a set of data mostly referring to North America, and must, I think, be interpreted as showing that North America is too large to be considered as a single region for this purpose. These summaries should therefore be applied to it with the greatest caution, if at all.

The later estimates are systematically less than the old ones, and it has been said that the earlier ones, mainly by Joly and Poole, were inflated by contamination. This seems impossible. If it was true the contamination could not be greater than the lowest values they found, and subtracting it would not remove the difference. It could be tested if modern methods were applied to the old specimens, which, I believe, still exist. In the meantime I think that the regional differences are real.

The differences in p may be of petrological importance. For want of anything better, rather than for any positive reason, it has been usual to think of the outer parts of the crust as made of a few uniform layers, from which surface igneous rocks are regarded as samples. If this was right, there is little reason why there should be much variation at all in the radio-activities of rocks of a given type, and it seems quite clear that either the primary magma is far from homogeneous or the surface rocks have been considerably altered on the way up. The rocks showing most resemblance to a primary magma would be the plateau and oceanic basalts. The dunites do in fact show a low variability, considering their scattered origins, but it is out of the question that surface dunites should be direct samples from the lower layer. They represent accumulations of olivine crystals from

Variation with region

basaltic magmas, and what they chiefly show is that when a basaltic magma deposits olivine crystals, the radioactive elements tend to remain in solution; a result of great general importance, for which there is much other evidence.

We can now estimate the rates of generation of heat by rocks in the groups considered. For potassium I take the content in granites as $3\cdot4\%$, for basalts $0\cdot76\%$, from Clarke's *Geochemistry*. The value for basalts is the mean for eight gabbros; values for effusive basalts may be much higher. Using Bullard's values given above for the rates of heat generation by uranium and thorium, and the revised value for potassium I find the following rates. The unit is 10^{-14} cal./g. sec. In the last column P cal./cm.^{3} sec. is the rate of generation of heat per unit volume.

Granites

Greenland, Iceland, Scotland, Ireland, Japan

Ra	Th	K	Total	$P \times 10^{14}$
$11\cdot1\pm0\cdot8$	$5\cdot1\pm0\cdot5$	$2\cdot4\pm0\cdot4$	$18\cdot6\pm1\cdot0$	50 ± 3

Finland

$32\cdot6\pm2\cdot8$	$17\cdot6\pm1\cdot5$	$2\cdot4\pm0\cdot4$	$52\cdot7\pm3\cdot1$	143 ± 8

Alps

$31\cdot0\pm4\cdot9$	$20\cdot8\pm3\cdot2$	$2\cdot4\pm0\cdot4$	$54\cdot2\pm5\cdot5$	147 ± 15

South Africa

$16\cdot5\pm1\cdot1$	?	$2\cdot4\pm0\cdot4$	27?	73?

Basalts

Greenland, Iceland, Scotland, Ireland

$6\cdot7\pm0\cdot4$	$6\cdot1\pm0\cdot4$	$0\cdot6$	$13\cdot5\pm0\cdot6$	38 ± 2

England, Germany, France, Hungary

$9\cdot1\pm0\cdot9$	$5\cdot5\pm0\cdot6$	$0\cdot6$	$15\cdot2\pm1\cdot1$	43 ± 3

Plateau basalts

$5\cdot1\pm0\cdot2$	$3\cdot3\pm0\cdot1$	$0\cdot6$	$9\cdot0\pm0\cdot2$	26 ± 1

Island basalts

$6\cdot3\pm0\cdot2$	$2\cdot9\pm0\cdot2$	$0\cdot6$	$9\cdot8\pm0\cdot3$	28 ± 1

Dunites

$2\cdot9\pm0\cdot4$	$2\cdot1\pm0\cdot2$	$0\cdot0$	$5\cdot0\pm0\cdot5$	17 ± 2

Many published determinations refer only to radium and leave thorium to be guessed from analogy. N. B. Keevil (1943) points out that a direct determination of the rate of α-ray emission from a rock, assuming a ratio $3\cdot5$ for Th/^{238}U, will give the rate of heat generation directly as accurately as we need, and only one measurement is necessary.

10·05. Rate of outflow of heat. It was noticed by Rayleigh (1906)
that the rate of supply of heat from the radium in average granite, if it was
uniform to a depth of some tens of kilometres, would supply the whole of
the heat being conducted out of the Earth, and the effect of radioactive
heating on the Earth's thermal state became a matter of the first importance.
Data for thorium and potassium were not available at the time. The most
unfortunate thing was the utter absence of good data on the rate of outflow
of heat. It is the product of the thermal conductivity and the vertical temperature gradient. Each of these showed great variation between different
determinations, and all that could be done was to take means and hope for
the best. As the amount of theoretical work became greater the position
became intolerable, and in 1935 the British Association appointed a committee to collect the data and obtain new ones. The first report, prepared
by D. W. Phillips (1936), revealed that though hundreds of measurements
of the conductivities of rocks and of the temperature gradient existed, there
was not a single place in the world where both had been measured, and
therefore no place where the rate of outflow of heat was known. Since then
much attention has been paid to the matter by the Department of Geodesy
and Geophysics, Cambridge, by E. M. Anderson in Scotland, by L. J.
Krige in South Africa, and by A. E. Benfield and F. Birch in the United
States.

The first result obtained was a verification that the equation of heat
conduction is true in natural conditions. The rate of transfer of heat across
unit area at depth x is $k\partial\vartheta/\partial x$, where k is the coefficient of thermal conductivity and ϑ is the temperature. At depths not affected by the daily
and annual variations of temperature this product should be independent
of the depth and of the time. In wells and borings that cut through several
different rocks, detailed measurements of temperature were made,* and
samples of rocks were taken away to have their conductivities measured.
It was found that the product $k\partial\vartheta/\partial x$ was constant in each hole used, though
the separate factors might vary each by a factor of 3. This was to be expected,
but the result indicated that determinations were possible. There are some
complications. In the British Isles it is necessary to apply a correction for
the residual effect of the Glacial Period (E. M. Anderson, 1934), which
produced an appreciable cooling to a considerable depth; the unevenness
of the land surface also produces some disturbance of underground temperature, for the temperature at the surface is atmospheric temperature,
which decreases with height more slowly than underground temperatures
do (Jeffreys, 1938b).

A. E. Benfield (1939) from five boreholes in Britain, one of which had
already been used by Anderson, found a mean heat outflow of (0.980 ± 0.116)

* The drilling itself disturbs the temperature considerably, and it is important
that the bore should have been left to itself long enough for steady conditions to
re-establish themselves. Cf. Bullard (1947).

$\times 10^{-6}$ cal./cm.^{2} sec. Correcting for the effect of the last glaciation, supposed to have ended 11,000 years ago, he found that the steady value would be $(1\cdot424 \pm 0\cdot091) \times 10^{-6}$ cal./cm.^{2} sec. Anderson (1940) rediscussed the data, with somewhat more detailed attention to probable changes of climate, and gave the following results for the individual bores and shafts, as corrected for climatic effects:

	Flow (10^{-6} cal./cm.^{2} sec.)
Balfour Bore (Fife)	1·20
Boreland Bore (Fife)	1·28
Blythwood Bore (near Glasgow)	1·75
South Balgray Bore (near Glasgow)	2·07
South Hetton Shaft and Bore (Durham)	1·82
Rose Bridge Colliery Shaft (Wigan)	1·34
Holford Bore (Cheshire)	1·43
Cambridge Bore	< 1?
Kentish Town Well and Bore (London)	< 1?
Hankham Bore (Sussex)	1·12
	Mean 1·500 ± 0·083

The Glasgow and Durham values are the highest, and Anderson suggests that this may be connected with the fact that they lie near Tertiary dikes belonging to the Mull swarm. L. J. Krige (1939) gave temperatures for six bores in South Africa, and Bullard (1939 a), after determining the conductivities, got $(1\cdot16 \pm 0\cdot09)$ cal./cm.^{2} sec. for the heat flow. H. P. Coster (1947), from eighteen bores in Persia, got a mean value of $(0\cdot87 \pm 0\cdot042) \times 10^{-6}$ cal./cm.^{2} sec. Assuming a rise of temperature of 5° 5000 years ago he got a corrected value of $(1\cdot18 \pm 0\cdot044) \times 10^{-6}$ cal./cm.^{2} sec., but suggested that the correction applied may be too large. Benfield (1947), using readings to a depth of 8680 ft. from a well in California, gets $(1\cdot29 \pm 0\cdot11) \times 10^{-6}$ cal./cm.^{2} sec. F. Birch (1947), using conductivities for analogous materials, gets from 1·0 to $1\cdot5 \times 10^{-6}$ cal./cm.^{2} sec. from five Colorado wells.

The differences between the regions are not great. This is noteworthy because the differences in the temperature gradients by themselves are famous. Values of about 30° C./km. used to be given as averages, mainly from European data, but the South African mines give about 10° C./km.; that is why it is possible for miners to work there at greater depths than elsewhere. The low temperature gradient is compensated by the high conductivity of most of the rocks, especially quartzite. The conductivities of quartzites range from about 0·011 to 0·018 c.g.s. as against 0·007 or so for granites. Sediments are very variable; values from 0·001 to 0·006 are common.

10·06. Upward concentration of radioactivity. According to the summary on p. 399, 1 cm.^{3} of average granite of the first group of regions supplies about $5\cdot0 \times 10^{-13}$ cal./sec. The total heat outflow in all the regions where it is determined is near 1×10^{-6} cal./cm.^{2} sec. Hence the whole of the heat outflow would be supplied by a layer of such granite about 20 km. thick, leaving nothing for conduction from the hot interior. The data for ordinary basalts lead to very similar results. Attempts have been made in many ways to reconcile the data. The simplest and the only satisfactory one is to suppose that radioactivity really is confined to an outer layer a few tens of kilometres thick, or at least that it falls off with depth so rapidly as to be effectively confined to such a depth. Several other lines of evidence support it. First, the geological and seismological evidence together indicate a passage from granite to dunite within such a depth, and from the known radioactivities of these rocks we should expect a reduction to under a third of the surface value. Secondly, we have seen that dunites are less radioactive than the basaltic magmas that they have presumably crystallized from, which suggests that the radioactivity of the greater part of the shell is very low. Thirdly, granite sometimes seems to have been fused several times at some depth, and to have risen towards the surface. In such cases we obtain samples of the top of the magma, at later and later stages; and it is found that the radioactive constituents are more abundant in the later specimens, indicating that the upper layers become enriched as time goes on. Thus Holmes (1926c) quotes the following contents by weight for Finland granites of decreasing age:

	Ra $(10^{-12} \times)$	Th $(10^{-5} \times)$	K $(10^{-2} \times)$
A	2·36	0·87	2·51
B and C	4·60	2·67	3·61
D	6·21	5·85	5·06

Fourthly, it can be shown easily that if the deep layers were as radioactive as the surface ones, the Earth could never have solidified; and the seismological evidence is perfectly clear that it is solid.

The explanation of the upward concentration of the radioactive elements is another matter; the immediate point is that its existence cannot be evaded without disastrous consequences. We shall see later that the considerations just offered are really an understatement of the case.

10·07. Radioactivity in early history. We can now return to the effect of radioactivity in the Earth during its early history. The total output of heat from the uranium and thorium series will not have varied greatly during the existence of the Earth; that from the ^{235}U series and potassium may have been somewhat greater in the past, because the decay constants of ^{235}U and ^{40}K are comparable with the age of the Earth, but their contributions are small now. Part of the heat being conducted out now is due to conduction from the heated interior, which would be going on if there was no radioactivity at all. The present supply of radioactive heat to the surface is not more than 2×10^{-6} cal./cm.^{2} sec. On the other hand, the Sun supplies 3×10^{-2} cal./cm.^{2} sec. to a surface normal to its rays, or an average of $0·75 \times 10^{-2}$ cal./cm.^{2} sec. over the surface of the Earth, and even at its present temperature the Earth radiates this away as fast as it receives it. In fact, this balance between radiation received from the Sun and lost into space is what determines the Earth's surface temperature. The contribution from heat conducted from the interior is only about 3×10^{-4} of the whole, and if it was cut off entirely the surface temperature would only drop by about $0 \cdot 03°$. This point must be emphasized because the belief that the Earth's surface temperature is now determined by internal heat, and may have been affected by changes of this within geological time, dies hard. In the early stages, when the surface temperature was high, the loss of heat was so rapid that radioactivity cannot be supposed to have delayed solidification (even if it was much greater than at present on account of larger contributions from ^{235}U and potassium). But as soon as viscosity became great enough anywhere to stop convection radioactivity would become a controlling influence. The equation of conduction of heat is

$$\rho c \frac{\partial \vartheta}{\partial t} = \frac{\partial}{\partial x}\left(k\frac{\partial \vartheta}{\partial x}\right) + \frac{\partial}{\partial y}\left(k\frac{\partial \vartheta}{\partial y}\right) + \frac{\partial}{\partial z}\left(k\frac{\partial \vartheta}{\partial z}\right) + P,$$

where ϑ is the temperature, ρ the density, c the specific heat, k the thermal conductivity, and P the rate of generation of heat per unit volume. If we neglect the curvature of the Earth and take the axis of x vertically downwards, we can omit derivatives with regard to y and z. Further, if the value of $d\vartheta_0/dx$ given by 10·01 (10) is maintained to great depths, and the conductivity is uniform, the condition just after solidification is

$$\rho c \frac{\partial \vartheta}{\partial t} = k\frac{\partial}{\partial x}\frac{\partial \vartheta_0}{\partial x} + P,$$

and the first term on the right vanishes. Thus the solid is heated up by radioactivity as if no conduction was disturbing the process. The terms neglected seem unlikely to be important. The inference is that if convection currents stopped radioactive heating would at once set in and raise the temperature till they started again; and when they started again the new heat would be carried up as fast as it was generated. It seems, in fact, that the effect of the heating in this case would be that the interior, instead of cooling till its viscosity was enough to prevent convection entirely, would cool very slightly less, till its viscosity was just enough to prevent all convection currents but the very feeble ones needed to carry off the new heat. The behaviour of the temperature up to the time when a solid surface was formed is hardly affected. The separation of the upper layers would carry a great deal of the radioactive matter to places near the surface, but so far we have given no reason why all of it should have been carried up. As soon as a solid surface formed it would cool rapidly and a truly solid crust would extend downwards till the temperature gradient in it was just enough to carry off the radioactive heat, and there need be no further change.

We now come to another complication arising from the actual method of solidification. We have supposed this to begin with the formation and settling of olivine crystals. Now it is well known that crystals when formed are usually pure substances. Pure ice crystallizes from salt water if it is cold enough. Isomorphism would affect this to the extent that the presence of both ferrous iron and magnesium in olivine is possible from the start, but this is a minor feature. In the state of slow convection that we have been considering, the fluid would consist of thin filaments entangled among a mass of crystals, and every time a given portion of fluid reached the top of the crystals it would produce more crystals, so that the materials still in solution would steadily disappear from the mass at the bottom. This process would be assisted by differences of pressure. In a mixture of solid and liquid at rest, such as a waterlogged sandstone, the stress in the water is hydrostatic; but the part of the weight of the solid not supported by the pressure of the water is supported by the grains themselves and transmitted between the grains by local stresses near their places of contact, which may be many times the average stress. Consequently, and especially in a substance near its melting-point, the grains would tend to be deformed in such a way as to extrude the liquid.* For both reasons it is probable that the whole of the liquid would be driven out, and that the material at the bottom would become pure solid olivine, gradually thickening upwards till the residual magma was sufficiently acidic to deposit crystals of other minerals. The radioactive substances would therefore become strongly concentrated upwards, and the process could stop only when solidification near the surface became complete and convection ceased altogether at the top on account of cooling to temperatures well below the melting-point.

Theoretical arguments bearing on this explanation have been given by V. M. Goldschmidt (1930). We have seen that the structure and composition of crystals is largely determined by the sheer size of the atoms concerned. Goldschmidt points out that the atoms of uranium and thorium are too large to fit into any silicate lattice, and consequently would remain in solution to the end. This provides a complete explanation of their association with the lighter rocks. It has also been confirmed by C. S. Pigott (1929), who has shown that in a granite at least half, possibly two-thirds, of the radium is on the surfaces of the mineral grains and in the interstices. Of the rest, it is strongly associated with the cleavage planes of the micas, the quartz and felspar crystals carrying very little. If the outer layers of the crust are largely glassy, as seems possible, the radioactive elements would probably be dissolved in the glass.

An explanation that has been suggested is that the high pressures concerned would by themselves inhibit radioactivity. Direct experimental test is of course impossible, but the geophysical evidence indicates a rapid decrease in the top 20 km. or so. Conditions at such depths can be imitated

---

* This process probably plays a large part in the formation of compact sediments. It is not necessary for this purpose that the depth should be anywhere near that where the mean pressure is equal to the crushing strength of the grains.

in the laboratory. Schuster (1907), F. D. Adams and A. S. Eve (1907) have shown that radioactivity is not affected by temperatures up to 2500° C. or pressures up to $2\cdot6 \times 10^{10}$ dynes/cm.². The latter corresponds to a depth near 80 km.

Holmes (1915a, p. 64) has suggested that volatile materials expelled from the lower layers carry the radioactive materials upwards by a sort of steam distillation. This process may play a part, but Goldschmidt's explanation seems more satisfactory.

We shall see in a moment that with given total radioactive heating in a solid shell the ultimate basal temperature is higher the more deeply the radioactive matter is buried, and it has often been inferred that in a well-stirred shell solidification could never have started. But the total radioactivity would in no case suggested be adequate to supply the radiation from a fluid surface, and it could not have more effect at the surface than to raise the temperature by an inappreciable amount. The considerations of 10·01 on the temperature gradients are independent of heat supply, and so long as general fluidity persisted the only effect would be a slight increase of the convection currents. Cooling would still take place and would lead to general solidification if it went on. The way out of the contradiction seems to be simply that in crystallization the fluid filaments left at any stage carried up the latent heat and the new heat by convection, and that as solidification proceeded the residual fluid was forced up by deformation of the crystals, so that solidification and the raising of the radioactive matter to the upper layers were part of the same process.

This process would apparently fail if solidification was to a glassy and not to a crystalline state, because this provides no apparent means of separation of materials. In that case the result would be that the cooling would proceed just far enough for the viscosity to permit sufficient convection to carry up the new heat. The fact that separation has happened may therefore be regarded as evidence that the shell is crystalline to the bottom.

10·08. Cooling of a solid Earth. We can now proceed to consider the cooling of a solid Earth. We take the time as zero when a solid crust formed. At that time the temperature everywhere was practically the melting-point. This increases with depth, but at present we shall take it constant and equal to the melting-point of the lower layer reduced to zero pressure. But when a solid crust formed the outer surface sank almost at once to the temperature maintained by solar radiation, which we may take as our zero of temperature. Anticipating the result that cooling has by this time become considerable down to a depth of order 300 km., we can suppose the depth affected great compared with the thickness of the radioactive layer and small compared with the radius of the Earth. The latter condition entitles us to treat the problem as one of flow of heat in one dimension, as for the cooling of a flat plate by radiation from the surface. The equation of heat conduction can therefore be taken as

$$\rho c \frac{\partial \vartheta}{\partial t} = \frac{\partial}{\partial x}\left(k\frac{\partial \vartheta}{\partial x}\right) + P. \tag{1}$$

We suppose that below a depth of order 20 km. radioactivity is negligible, so that P decreases rapidly with depth. k, c and ρ will be taken constant at great depths ($k = k_1$), but possibly having different values near the surface. The initial temperature is S. For positive t, and $x = 0$, $\vartheta = 0$. Then a solution of (1), independent of the time, is

$$k\frac{\partial \vartheta}{\partial x} = \int_x^\infty P\,dx, \tag{2}$$

$$\vartheta = \int_0^x \frac{1}{k(\xi)}\,d\xi \int_\xi^\infty P(\eta)\,d\eta = \vartheta_1, \tag{3}$$

say. ϑ_1 is zero at $x = 0$, but does not satisfy the initial conditions. If we write

$$\vartheta = \vartheta_1 + \vartheta_2, \tag{4}$$

$$\rho c \frac{\partial \vartheta_2}{\partial t} = \frac{\partial}{\partial x}\left(k\frac{\partial \vartheta_2}{\partial x}\right), \tag{5}$$

and at $t = 0$

$$\vartheta_2 = S - \vartheta_1. \tag{6}$$

Now ϑ_1 is constant below the radioactive layer. For $\displaystyle\int_\xi^\infty P(\eta)\,d\eta$ is zero when ξ is greater than the thickness of this layer; and then

$$\vartheta_1 = \int_0^\infty \frac{1}{k(\xi)}\,d\xi \int_\xi^\infty P(\eta)\,d\eta = S_0, \tag{7}$$

say. Thus initially

$$\vartheta_2 = S - S_0 + \alpha, \tag{8}$$

where α differs from zero only within the radioactive layer. When $x = 0$, $\vartheta_2 = \vartheta = 0$. Thus the conditions satisfied by (5), (8) specify a new problem, namely, that of the cooling of a mass with no internal generation of heat, and initial temperature $S - S_0 + \alpha$, after the temperature at $x = 0$ has been reduced suddenly to zero and maintained there. The effects of the term in α are negligible, because the heat that it represents is quickly conducted to the surface and lost. In general, in fact, we can neglect the thermal capacity of the upper layer. Thus in this layer we can take

$$k\frac{\partial \vartheta_2}{\partial x} = \text{constant} = k_1\beta, \tag{9}$$

say. Thus when x is a small fraction of 300 km.,

$$\vartheta_2 = \int_0^x \frac{k_1\beta}{k(\xi)}\,d\xi$$

$$= \beta x - \beta \int_0^x \left(1 - \frac{k_1}{k(\xi)}\right)d\xi. \tag{10}$$

Thus if $\partial \vartheta_2/\partial x$ is equal to β just below the upper layer the corresponding values of ϑ_2 at any time are such that linear extrapolation would make ϑ_2 equal to 0 at depth

$$x_0 = \int_0^\infty \left(1 - \frac{k_1}{k(\xi)}\right) d\xi. \tag{11}$$

We now introduce the thermometric conductivity

$$h^2 = k/c\rho, \tag{12}$$

and denote its value in the lower layer by h_1^2. The solution in the lower layer is

$$\vartheta_2 = (S - S_0) \operatorname{erf} \frac{x - x_0}{2h_1 t^{\frac{1}{2}}}, \tag{13}$$

where erf u is the error function, defined by

$$\operatorname{erf} u = \frac{2}{\sqrt{\pi}} \int_0^u e^{-v^2} dv. \tag{14}$$

It can be verified by direct differentiation that (13) satisfies (5). When u is large, $\operatorname{erf} u \to 1$. Thus at depths much greater than $2h_1 t^{\frac{1}{2}}$ we have nearly $\vartheta_2 = S - S_0$; and ϑ_2 for given x is equal to $S - S_0$ when t is small, as it should be. Also $\partial \vartheta_2/\partial x$ is such that, when $x - x_0$ is small, extrapolation to $x = x_0$ would make ϑ_2 zero. Hence, except in the radioactive layer,

$$\vartheta = \vartheta_1 + \vartheta_2 = S_0 + (S - S_0) \operatorname{erf} \frac{x - x_0}{2h_1 t^{\frac{1}{2}}}. \tag{15}$$

In the upper layers ϑ can be found from (3), (9) and (11). We need chiefly, however, only the rate of outflow at the surface. This is

$$\left(k \frac{\partial \vartheta}{\partial x}\right)_{x=0} = \int_0^\infty P(\xi) \, d\xi + k_1(S - S_0) \frac{1}{\sqrt{\pi}} \cdot \frac{1}{h_1 t^{\frac{1}{2}}}. \tag{16}$$

A useful interpretation of S_0 has been given by Bullard (1939*b*). If k is treated as constant, and we change the order of integration in (7), we have

$$kS_0 = \int_0^\infty d\eta \int_0^\eta P(\eta) \, d\xi$$

$$= \int_0^\infty \eta P(\eta) \, d\eta$$

$$= \int_0^\infty P(\eta) \, d\eta . \bar{\eta}.$$

Here $\int_0^\infty P(\eta) \, d\eta$ is the total rate of generation of heat contributing to the surface outflow, and $\bar{\eta}$ is the weighted mean depth of the generation. This shows directly the effect of burying the radioactive sources on the temperature, when the total rate of generation is kept the same.

Numerous exact solutions of problems of the cooling of the Earth's crust have been given,* but they all depend on the adoption of special forms for P. The present argument shows that within the accuracy needed the solution is independent of the form of P so long as the values of S_0 and $\int_0^\infty P(x)\, dx$ are maintained. This leads at once to two important results. First, if $S_0 < S$, the temperature will decrease with increasing time, tending ultimately to S_0. But if $S_0 > S$ the temperature at great depths will rise and fusion will recommence. In that case the conclusion contradicts the hypothesis because heat would then be carried up by convection. *Prima facie* the solidity of the shell shows that $S_0 < S$.

Secondly, granting that $S_0 < S$, the second term in (16) is positive, decreasing steadily as t increases. Then the observed value of the rate of outflow of heat gives an upper bound to $\int P\, dx$ and therefore to the total rate of production of heat.

Further, the latter statement is true even if $S_0 > S$. For if fusion occurs heat generated below the level of fusion is carried up by convection and reaches the surface more quickly. The liquid zone would extend upwards till the steepened temperature gradient in the outer rocks was again enough to carry off all the heat. There is a complication at this point because we should ordinarily have in this state a solid crust resting on a less dense liquid. As a pure problem of mechanics such a state would be stable if the crust was thick enough, and there is no reason why it should not be permanent. If it was too thin, or if it broke anywhere under some local disturbance, instability would arise and would lead to wholesale fractures of the outer crust. Solid blocks would be continually foundering and melting on the way down, while the fluid would actually come to the surface in places. The thermal balance at the surface would be as follows. The heat supply from the interior would be insufficient in any case to keep a large fraction of the surface fluid; at any moment most of it would be solid, the blocks being separated by veins of fluid. The diameters of these would be determined by the condition that the radiation from them and the conduction through the solid parts would together be able to dispose of the heat coming from within. This would be a tempting explanation of surface igneous activity, but unfortunately it requires a continuous connexion among the liquid parts at the surface, with the solid blocks separated. Actual igneous activity is always local and the crust has remained connected throughout geological time. Many hypotheses have been framed on the hypothesis that $S_0 > S$ but it seems inevitable that they must conflict with the transmission of transverse elastic waves throughout the shell and the known imperfections of isostasy. Further, it appears that the continual melting and resolidification would itself produce a new concentration

* Ingersoll and Zobel (1913); Holmes (1915*b*); L. H. Adams (1924); Jeffreys (1916*b*, 1921, 1927*a*); H. and B. S. Jeffreys (1972, § 20·06).

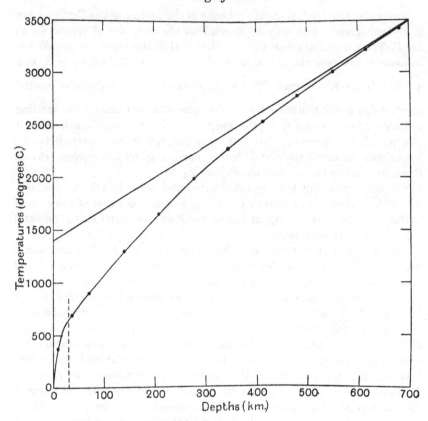

Fig. 25. Present distribution of temperature with depth. The
straight line represents the function $S + mx$.

(This diagram takes no account of cooling due to the m term, and in several
other respects is intended only to indicate the general features.)

of radioactivity upwards and bring S_0 down to less than S. There seems to
be no justification for the various theories of magmatic cycles that regard
the transmission of transverse waves as a temporary feature. Once a fluid
zone was established it would either be permanent or lead to such redistribution of radioactivity as would prevent the establishment of a second.

In the above account we have ignored the initial variation of temperature
with depth. If we continue to treat the problem as one of one-dimensional
flow, and the initial temperature was $S + mx$, with m constant, we have
already dealt with S, and need only calculate the effect of taking the initial
temperature as mx and add it on. But in a uniform solid mx is already
a solution of the equation of heat conduction, vanishing at $x = 0$, and
therefore, apart from a small correction near the surface, persists unchanged
for all time. However, this result is self-contradictory. If m is positive the

term implies a steady outflow of heat; therefore there must be cooling somewhere. The explanation is that for this term it is not accurate enough to treat the radius of curvature as infinite. If we do we automatically provide an infinite reservoir of heat and unlimited temperatures at great depths, and the heat coming out is supplied from this reservoir. The remedy is to allow for the Earth's radius. To match the initial value of $d\vartheta/dr$ at moderate depths and have a plausible form for the effect of pressure on the melting-point we should replace mx by

$$M = \frac{m}{2a}(a^2 - r^2),\tag{17}$$

and take the equation of conduction in the full form

$$\frac{\partial \vartheta}{\partial t} = h^2 \nabla^2 \vartheta.\tag{18}$$

From (17), this makes the initial rate of cooling

$$\frac{\partial \vartheta}{\partial t} = -\frac{3mh^2}{a},\tag{19}$$

which is the same at all depths; and

$$\vartheta_3 = \frac{m}{2a}(a^2 - r^2) - \frac{3mh^2t}{a}\tag{20}$$

is a solution of the equation of heat conduction. It does not, however, vanish for all time at $r = a$, and strictly needs further correction to allow for the boundary condition. An approximate solution under the condition that $ht^{\frac{1}{2}}$ is much less than a is*

$$\vartheta_3 = \frac{m}{2a}(a^2 - r^2) - \frac{3mh^2t}{a} + \frac{12mh^2t}{r}\Phi_2\left(\frac{a-r}{2ht^{\frac{1}{2}}}\right)\tag{21}$$

(which incidentally satisfies the boundary condition exactly). (See Appendix D.) At $r = a$

$$\frac{\partial \vartheta_3}{\partial r} \doteqdot -m + \frac{6mht^{\frac{1}{2}}}{\sqrt{\pi a}}.\tag{22}$$

At depths more than about $2ht^{\frac{1}{2}}$ the cooling is practically uniform as given by (19). The contribution to the rate of outflow of heat at the surface is still practically $k_1 m$, which should be added to (16).

In the one-dimensional theory the m term had appeared to give no cooling at all. We now see that it does lead to cooling, and as this cooling extends all the way to the centre, whereas that due to the $S - S_0$ term is effectively confined to depths of order $2ht^{\frac{1}{2}}$, the total losses of heat due to the two may well be comparable (Jeffreys, 1932a).

---

* $\Phi_2(u) = \displaystyle\int_u^\infty dv \int_v^\infty (1 - \mathrm{erf}\, w)\, dw$, also called iierf u by Hartree. The first integral is called $\Phi_1(v)$ or ierf v. The functions are closely related to the Hh functions (H. and B. S. Jeffreys, 1972, Chapter XXIII).

A possible lower estimate of m can be derived from Simon's interpretation of the inner core boundary as corresponding to the melting-point of iron. (See p. 421.) $m = 3°/\mathrm{km}$. in (17) would give a rise to the core boundary of about $7000°$; so if the actual temperature of the core is about $2400°$ higher than the melting-points of surface rocks m cannot be very different from $1°/\mathrm{km}$. This agrees with Shimazu's value.

In spite of the improved knowledge of the rate of outflow of heat we still cannot give a detailed quantitative statement of the distribution of temperature with depth, on account of the uncertainty of the vertical distribution of radioactivity. (16) gives an estimate of $\int P(\xi)\,d\xi$, on the supposition that the second term can be either neglected or approximately estimated. But the variation of temperature at a given depth with time depends on S_0, which depends on the vertical distribution of radioactivity as well as on the total amount. For instance, if $P = A$ down to depth H and $= 0$ for depths $> H$,

$$\int_0^\infty P(\xi)\,d\xi = AH, \qquad \int_0^\infty d\xi \int_\xi^\infty P(\eta)\,d\eta = \int_0^H A(H-\xi)\,d\xi = \tfrac{1}{2}AH^2.$$

But if at all depths $\qquad P(x) = A\,\mathrm{e}^{-x/H},$
we have

$$\int_0^\infty P(\xi)\,d\xi = AH, \quad \text{but} \quad \int_0^\infty d\xi \int_\xi^\infty P(\eta)\,d\eta = \int_0^\infty AH\,\mathrm{e}^{-\xi/H}\,d\xi = AH^2.$$

The two distributions give the same heat outflow, but the values of S_0, for uniform conductivity, differ by a factor of 2.

The result is an immediate consequence of Bullard's formulation (p. 408), since $\bar{x}$ is $\tfrac{1}{2}H$ in the first case and H in the second.

Another consequence is that for given total generation of heat and suitable distribution of heating, $\bar{x}$ can be large enough for S_0 to exceed the melting-point. We have applications of this (pp. 388, 425, 426, 469) in relation to the early history of the Earth, the occurrence of volcanoes in mountainous regions, and thermal blanketing by sediments and water.

Ingersoll and Zobel used an exponential distribution, and their solution was used by Holmes. He introduced the important principle (which he afterwards abandoned) of using the surface gradient and the surface radioactivity to provide data for the vertical distribution of radioactivity. I assumed in 1916 a uniform radioactivity in an upper layer, with none below it (the three-layer structure was not indicated by seismology till about ten years later). In the 1929 edition of this book I took a uniform upper layer, resting on an intermediate one with the radioactivity of average basalt and no radioactivity in the lower layer. But this is rather unsatisfactory. Where we find granite at or near the surface over extensive regions it seems fair to take surface granites as representative of the top of the upper layer. But it is dangerous to take recent granites or surface rocks of any other type as representative of rocks of similar composition at their normal depths, because they have almost certainly been enriched on the way up. There is no reason to suppose the lower layer to be quite free from radioactivity, though there is every reason to suppose that P in it continues to fall off with depth. Accordingly, I think that we cannot do better than to return to the exponential distribution.

The conductivity of granite is about 0·006 c.g.s., that of basalt about 0·004; we shall take the same value for dunite. With the latter datum, combined with a specific heat of 0·20 and a density of 3·3, we find $h_1^2 = 0·006$. We take $t = 2000$ million years $= 6 \times 10^{16}$ sec.; $m = 3 \times 10^{-5}$ °C./cm.; $S = 1400°$. With

$$P(x) = A \, \mathrm{e}^{-x/H}$$

(16), corrected by (22), is

$$\left(k \frac{\partial \vartheta}{\partial x}\right)_{x=0} = AH + \frac{k_1 S - AH^2}{\sqrt{(\pi h_1^2 t)}} + k_1 m.$$

With the British values

$$k(\partial \vartheta / \partial x) = 1·42 \times 10^{-6}, \quad A = 5·1 \times 10^{-13}, \quad \sqrt{(\pi h_1^2 t)} = 3·4 \times 10^7 \, \mathrm{cm}.$$

this gives $\quad H = 2·5 \times 10^6 \, \mathrm{cm}. = 25 \, \mathrm{km}., \quad S_0 = \dfrac{AH^2}{k_1} = 740°.$

With the South African values

$$k(\partial \vartheta / \partial x) = 1·16 \times 10^{-6}, \quad A = 7·3 \times 10^{-13},$$

$$H = 1·2 \times 10^6 \, \mathrm{cm}. = 12 \, \mathrm{km}.; \quad S_0 = 280°.$$

Even with the new information incorporated in this solution, the results are very uncertain. We have seen that if we took all the radioactivity to be in a uniform layer of depth H, we should get nearly the same value of H from the data, but S_0 would be halved. The granites analysed are in both cases from places a long way from the places where the temperatures have been measured, and for the South African ones the thorium content has not been measured and is taken from analogy. Again, as surface intrusions have come from some depth, it is hard to say whether they are more likely to be representative of the top or the bottom of the granitic layer. This question of sampling, so long as it remains undecided, is the greatest of the many sources of uncertainty in the quantitative treatment of the Earth's thermal history.

There seems to be a remarkable lack of determinations of conductivity for dunite, in spite of its importance. Birch, Schairer and Spicer (1942) give one, which is about the same as for granite. Birch (private communication) thinks that the differences between granite, basalt and dunite are small.

Schatz and Simmons (1972) measured the thermal conductivity of olivine in conditions corresponding to a depth of 400 km. in the shell, and got 0·020 cal./cm.^{2} sec. °C. They adopt 0·006 cal./cm.^{2} sec. °C. for the crust from G. J. F. MacDonald (1967).

It is particularly desirable that measures of heat outflow should be made in Finland for comparison, on account of the high mean radioactivity of the granites there.

As a representative value I take $S_0 = 500°$. Then for depths over about 30 km.

$$\vartheta_1 + \vartheta_2 = 500° + 900° \left(1 - \mathrm{erf}\, \frac{x - 15\,\mathrm{km.}}{380\,\mathrm{km.}} \right)$$

with sufficient accuracy. The main conclusion is that temperatures of several hundreds of degrees are to be expected within about 30 km. of the surface, but nevertheless there has been cooling through something like 600–1000°. The cooling drops off rapidly with depth and becomes inappreciable at a depth of about 700 km.

The cooling due to the m term is about $-3mh^2t/a$ for depths over 380 km.; with the data we have used it is about 50°. This is comparatively small, but it extends throughout the interior and its total effect may be of great importance.

The above account needs a further correction with modern information. The rate of generation of heat has been taken independent of time. This was justified when ^{238}U and Th were thought to be the only important sources and the age was taken to be about $1·6 \times 10^9$ years. But with an age of $4·5 \times 10^9$ years ^{235}U and ^{40}K would originally have been producing heat at rates comparable with ^{238}U, and the change of the latter itself is not negligible. A thorough study has been made by J. A. Jacobs and D. W. Allan (1956, and references there given) who have considered several different original distributions of radioactivity and temperature, and worked out the consequences numerically. From their results it would be possible to derive those for other distributions without difficulty—always supposing that the distribution of the radioactive elements in space has not changed with time.

The foregoing discussion concerns only continental conditions. We have seen that the study of gravity shows a definite difference of density between oceans and continents. This is checked by geology; granite is unknown in the islands of the Pacific proper, though it does occur in some of the marginal ones. Seismological evidence gives similar indications. Accordingly, it seems that of the materials investigated the oceanic island basalts are the most likely to be representative of the ocean floor. If then the radioactivity decreases with depth according to the exponential law, with a similar value of H to the continental ones, it may be expected that the rate of outflow of heat is of the order of half that in the continents and that S_0 also is halved. We should then expect the cooling below the oceans to be substantially greater than under the continents. On the other hand, it is conceivable that upward concentration of radioactivity stops when cooling reaches a definite amount, and in that case S_0 might have similar values in both cases. Then AH^2 has about the same value: A is less below the oceans; and accordingly AH is less and H greater. The temperature gradient in the ocean floor would still be less than in the continents and lower temperatures would still be expected through depths of the order of tens of kilometres, but no longer hundreds.

In specimen cores from the Pacific bottom, R. Revelle and A. E. Maxwell (1952) have found a heat flow about the same as on land. Bullard (1954) for a mean of five Atlantic stations gets about 20 % less than on land. The difference between oceanic and continental results is far less than anybody had expected, and no satisfactory explanation is known.

It might well be supposed that denudation has transferred a great deal of radioactivity to the ocean floor and insoluble compounds might have been formed. In that case, we should expect deep-ocean sediments to be highly radioactive. This has not, I think, been verified. But it has appeared (9·02) that the average thickness of granite denuded to form sediments would give only about 1 km. over the ocean floor, and this appears inadequate.

If this explanation is not available the similarity of average continental and oceanic values presents difficulties that are not, I think, sufficiently appreciated. The granitic layer is absent under the oceans; but this supplies most of the heat outflow in the continents. If we compare conditions below this layer the heat outflow will be of the order of three times greater below the oceans, and so will the temperature gradient. The lower thickness of the upper layers might be due to incomplete differentiation. This is hard to reconcile with the strength indicated by the ocean deeps and the fact that the velocity of P under the Pacfic is substantially the same as in Europe. As usual, convection is offered as an explanation, but this difficulty is quite independent of convection.

Such a difficulty forces us to consider the possibility of systematic error in the observations. For land determinations a correction for the after-effect of the glacial period has had to be made, and is a large fraction of the whole; and in any boring enough time has to be allowed for it to resume its temperatures after the disturbance produced by the boring itself. This may be weeks or months. The depth of a mine or boring may be up to 1 km. An ocean bed sample is extracted by a plunger and is usually not over 10 m. long; a much smaller quantity, with equally strong disturbance by friction, has to be tested sooner.

The effect of friction has been mostly eliminated. Temperatures are not measured in the plunger itself. A fin is attached to it, and a tiny element sensitive to temperature is attached to the fin on a slender strut. Continuous observations are made of the temperature, which appears to become steady in a few minutes and therefore to reach that of the surroundings. The theory of the method is due to Bullard (1954); see also Lister (1963).

If the oceanic values are correct and nevertheless we must accept the evidence that the rocks below the oceans are as strong as those below the continents, we may have to consider the possibility that the outflows of heat in the continents have been underestimated. We must recall that the estimates of the steady flow in the continents have already received a great increase from the directly observed values, on account of the allowance for the residual effect of the glacial period. But perhaps this allowance should be still greater!

Recent work indicates temperatures about the present in Britain for the last 10,000 years. Between about 10,000 and 50,000 years ago temperatures were mostly arctic. The rise may have been 15° to 20°. H. Godwin describes the data in *History of the British Flora*. There was a post-glacial warm period, possibly from 6000 to 1000 B.C.

J. Reitzel (1963) finds a remarkably uniform heat flow at 16 stations between Bermuda and Bahama Banks, given as $1·14 \times 10^{-6}(1 \pm 0·045)$ cal./cm.² sec. At the outer edge of this area they get 1·76; three, a few hundred km. to the northwest, close to the continent, give 1·17, 0·94 and 0·81.

A detailed map of heat flow over the Atlantic is given by Langseth, Le Pichon and Ewing (1966). It is fairly uniform within 1000 km. of continental margins; there is a much larger scatter in the Central Atlantic. It is thought that this may be due to variable gradient of temperature in the water and to irregularity of topography.

There is a mention of the time the probe was left in before extraction, presumably with reference to the need to establish equilibrium of temperature with the surroundings: but I think there may still be heating during extraction, which would be greater at the greater depths.

The authors remark that if there is drift there should be high heat flow in the wake, which is not observed.

With regard to temperatures below oceans, Meyerhoff, Meyerhoff and Briggs (1972, p. 668) quote a suggestion of MacDonald and Birch that the chemical constitution of the oceanic shell may be different from that below the continents. What concerns us here is the strength in any case.

Herzen and Uyeda (1963) used a long probe in front of the plunger. In two large regions the measures were about half the average oceanic value. In other places some reached $(4 \text{ to } 8) \times 10^{-6}$ cal./cm.² sec. Otherwise about a quarter were under $0·8 \times 10^{-6}$ cal./cm.² sec.

This violently unsymmetrical distribution implies that the normal law of error is far from being satisfied, and that the mean of the measures may deviate considerably from the true mean. Among other possibilities they mention that the rates of flow of heat were strongly correlated with elevation of the bottom above average. This suggests that the higher values are due to thin patches of strongly radioactive or even chemically active material, which would have little effect on temperatures at great depths.

I think that Herzen's suggestion of thermally active shallow patches is the least likely to conflict with other evidence.

It will be noticed that in any case the temperature approaches the melting-point most closely at the bottom of the intermediate layer. The importance of this was pointed out by E. M. Anderson (1934). It had been taken for granted that isostatic adjustment is by flow in the lower layer. Anderson remarked that the temperature is continuous, but the melting-point increases by about 200°. Therefore just above the interface the temperature is 200° nearer the melting-point than just below, and if it was high enough to weaken the lower layer seriously the intermediate layer just above would be thoroughly fluid. There are indications from the lunar craters that a liquid layer persisted for some time after the formation of a solid surface.

There remains a complication from the radioactivity of potassium. The average life of ^{40}K according to Birch's discussion would be $1·8 \times 10^9$ years. Shorter values found in previous discussions led to two difficulties. (1) They implied a generation of argon much larger than the amount in the atmosphere. (2) They made the heat generation by K in the Earth's early history much greater than that by U and Th together and it looked as if the Earth could not have solidified until most of the ^{40}K had disintegrated. These difficulties seem now to have been resolved. For the reasons indicated in 10·07, however, I think that this was never the correct conclusion even had the higher decay constants been correct. Even the higher estimates of the total radioactive heating are still small compared with the heating by solar radiation, and the argument of 10·07 stays unaltered. The conclusion would still be that crystallization proceeded until the concentration of radioactive elements in the upper layers became complete enough to permit general solidification, and therefore till $S_0 < S$; and if ideas of the distribution indicate that S_0 was greater than S in early geological time it is because they take too slow a decline of radioactivity with depth.

10·09. Reference should be made to a difficulty pointed out, but not published, by Holmes in 1917 and Lindemann (Lord Cherwell) in 1921. Average acid rock is producing helium at a rate of about 10^{-12} cm.^{3} per year per gram of rock. If the age of the Earth is 2×10^9 years and 10^5 cm. of rock have been denuded, the helium released would be of order 600 cm.^{3}/sq.cm. of surface, and would account for about 10^{-4} of the atmospheric pressure. The actual ratio is about 5×10^{-6}. It appeared that loss of helium from the atmosphere could not be appreciable (Jeans, 1921; Lennard-Jones, 1923) at the probable temperature of the upper atmosphere. It seems possible, however, that temperatures of about 1000° may occur at great heights, according to D. F. Martyn (1939), and at these helium could escape; and also that even a small amount of nitrogen and oxygen diffused through the solar system would suffice to knock enough helium atoms out to account for the loss (Lindemann, 1939).

These temperatures are now thoroughly confirmed by the distribution of density indicated by the resistance of the air to the motion of artificial satellites. King-Hele (1960a, Fig. VI, 4; 1960b) gives a provisional solution with temperatures rising from 1500° to 2000° K at heights of 400–600 km.

10·10. Other theories of the core and the method of solidification. A further group of arguments relating to the early history of the Earth, with possible consequences for its present internal constitution, may be considered here. W. Kuhn (1942) and Kuhn and Rittmann (1941) have held that the original matter, being a sample of solar matter, would be largely hydrogen, and argue from the law of diffusion that this could not be lost in the time available—or even in the whole age of the Earth. Consequently they held that the greater part of the Earth has nearly the composition of the Sun, and that there is no discontinuous change of properties. The fluidity of the core is attributed to the viscous term in the equation of elastico-viscosity. *PKP* at distances less than 140° is then attributed to waves of periods long compared with the period of relaxation. I find their arguments unconvincing. In a fluid cooling from the surface the convection currents set up effectively increase the coefficients of heat conduction and diffusion by enormous factors, and the mechanism for transfer of heat to the surface is equally available for the transfer of matter; consequently, the time needed for the loss of hydrogen would be comparable with that for solidification. Again, the density distribution must be explained somehow. Various theories (some confirmed by observation) indicate transitions of matter under high pressure to astonishingly high densities; but if high pressure was the cause of the elevation of the density of hydrogen to 9 g./cm.³ and more, the greater pressures in Jupiter and Saturn would make them denser than the Earth, and they are not. The introduction of elasticoviscosity does not help in explaining the seismic data. It would imply that transverse waves can enter the core freely and are absorbed in transit; this does not agree with the existence of the strong reflexion *ScS*. Also *PKP* at distances less than 143° shows short periods. Other arguments are given by R. Wildt (1947).

An extended discussion of the theory from the aspect of chemical thermodynamics is given by A. Eucken (1944a, b). He gives the argument for the delay of cooling by a cloudy atmosphere that I have used above. He also quotes Homer Lane for the result that in a contracting gaseous mass a large fraction of the potential energy becomes kinetic and the temperature therefore rises. This process ceases when the size of the particles prevents further compression; then the thermal energy is radiated and the mass cools. Clouds of condensed matter would evaporate in approaching the centre, but when cooling had proceeded far enough they would reach the centre, and the heavy materials would accumulate. Silicon has a saturation vapour pressure comparable with that of iron (those of other elements being mostly much higher), and is only saved from getting into the core by forming SiH_4. This would be decomposed by water at a later stage; then silica would be formed and the excess hydrogen lost. The critical temperature for iron would be about 9000°, and the actual temperature of the core may be in this neighbourhood. The discussion is very full and I can only attempt a brief summary. Eucken's general conclusion is that the identification of the core with iron is consistent with physical chemistry. Urey criticizes Eucken's mechanism but comes to the same conclusion.

The geophysical identification of the core with iron rests simply on the facts that iron gives a density in reasonable accordance with that indicated in the theory of the figure of the Earth, it is one of the commoner heavy elements in the stars, and that it would be liquid if in contact with olivine near the melting-point of the latter. Also a transition between rock and metal provides a natural explanation of the sharpness of the interface. But other identifications would be acceptable if they satisfy the geophysical conditions of density and compressibility. Kuhn and Rittmann find that the assumption of an iron core leads to a much higher iron content in the Earth, even relative to other metals of high melting-point, than in the Sun, and consider this an additional ground for rejecting this hypothesis.

In astrophysics the states of matter at high pressures and temperatures have become very important in the last forty years or so. At high temperatures the outer electrons are separated from their atoms, and the result is a gas consisting largely of free electrons, which make a substantial contribution to the gas pressure. The removal of the outer electrons from an atom also greatly reduces its effective size, so that the material can behave as a nearly perfect gas at densities such that liquefaction would take place at ordinary temperatures. At very high densities, however, even a very highly heated gas ceases to behave as a perfect gas, for reasons discussed in works on quantum theory under the name of pressure degeneracy. The temperatures and densities required, however, seem to be beyond what are likely to occur in the planets. The phenomena arise near the centres of normal stars and through most of the interiors of what are known as white dwarf stars, of which the companion of Sirius is the most famous example. But the temperature at the centre of the Earth is likely to be comparable with that of the Sun's atmosphere; that at the centre of Jupiter may be about that of the atmosphere of a star of Type B. In these conditions, though some of the outer electrons may be removed, there would be no drastic stripping of the atoms down to their innermost electrons such as has to be taken into account in the theory of stellar interiors. The further complication of pressure degeneracy of a gas consisting of stripped atoms does not arise; it happens at densities of order $10^4 \mathrm{g./cm.}^3$.

D. S. Kothari (1938), however, has pointed out that pressure degeneracy may occur at comparatively low temperatures. The essential point is simply that at high densities the atoms may be so closely pressed together that the outer electrons of different atoms interfere with one another in the way implied by the Pauli exclusion principle (which I shall not try to explain; but in atoms in their normal state it requires that there cannot be more than two electrons in a given orbit, and this type of restriction will become more serious when the packing implies that analogous restrictions will apply not only to a single atom but to the atom together with all its neighbours). Kothari has developed the consequences in some detail; one conclusion is that there should be a maximum possible radius for a cold body, however great its mass. The application to the planets is developed by J. G. Scholte (1947). He finds that for planets composed of hydrogen the mass-radius relation fits the outer planets well, and the inner planets are reasonably fitted by supposing them made of silicon. The relation, however, makes the pressure proportional to ρ^2, so that the bulk-modulus would be twice the pressure. The actual ratio in the core is about 5.

A milder form of pressure degeneracy has been suggested by W. H. Ramsey (1948). In a sense it is shown by all metals, even at ordinary temperatures and pressures. In a solid or liquid metal the atoms are too closely spaced for each atom to hold all the electrons associated with it in the gaseous state, and the outer electrons, instead of remaining each associated with one particular atom, roam about through the whole structure. Such a state explains the high electric conductivity of metals. Ramsey's suggestion is that at high pressures a material such as olivine, which is a bad conductor at ordinary pressures, may lose some of its electrons and effectively become a metal. He thinks that the central core, therefore, need not be a mixture of heavy metals, but olivine in this new state; and that the transition may be sharp enough to account for the actual discontinuity of properties at the core boundary. There is good reason to suppose that this can happen in hydrogen, and that it plays an important part in the physics of the great planets (R. Kronig, J. de Boer and J. Korringa, 1945). On this theory the inner core would be due to a second ionization. The final state would be a Fermi-Dirac gas as in white dwarf stars, but this would require higher pressures than occur in the planets.

Bullen (1946) makes an interesting suggestion. He notices that the bulk-moduli just outside and just inside the core are nearly equal. Further, though the values found for dk/dp are different, it appears that a little outside the core there is a substantial drop in dk/dp, corresponding to the straightening of the time-curves for P and S at a distance of about 95°. The value of dk/dp just above the level where this change takes place agrees well with that in the upper parts of the core. He is led to the suggestion that at high pressures the bulk-modulus may be the same for all substances at the same pressure. The anomaly in dk/dp just outside the core would then be regarded as an error of interpretation. The differential equations used in finding ρ and k assume uniformity of composition. If there is a zone where the density is increasing with depth on account of admixture of denser constituents as well as on account of compression, the wave-velocities would increase less rapidly than elsewhere even if the variation of k is smooth. Bullen therefore examines the hypothesis of continuity of dk/dp in detail. The drop in its value outside the core would be attributed to the admixture of denser materials. One explanation would be that the sulphides inferred by Goldschmidt are present, but are in solid solution in the silicates instead of forming a separate layer. Bernal tells me, however, that 'sulphur ions have a radius of 1·8A. as against 1·3A. for oxygen ions, and, though at high pressures the former may be reduced to 1·0A. and the latter to 0·7A. they would still be too different to form mixed crystals easily'. An increase of iron at the expense of magnesium in the silicates would have a similar effect, but it appears that in olivine magmas the crystals first formed are the poorest in iron; the lower melting-point is more important than the higher density (Wager and Deer, 1939).

Bullen's hypothesis may also help to explain the drop of the velocity of P just outside the inner core (if it exists), by new admixture of dense materials. With the known velocity and an extrapolated k this suggests a value for the density of the inner core, which could only be guessed in previous treatments. There is in any case a substantial jump in the velocity of P at the boundary of the inner core. This could be explained if there is rigidity, α^2 being then $(\lambda + 2\mu)/\rho$ instead of $(\lambda + \frac{2}{3}\mu)/\rho$. Bullen and Haddon (1967b) estimate that β may be 3·8 km./sec. at the top of the inner core and 2·9 km./sec. at the centre.

J. A. Jacobs (1954) suggested that a formerly fluid Earth, on account of differences between the adiabatic and melting-point gradients, might have solidified first at the centre and the outside, leaving a fluid layer trapped between. Lubimova (1956) inferred from thermodynamical considerations that the inner core is below melting point.

Birch (1961b, 1964) inferred from experiments on shock waves that the Earth's central density is unlikely to be much greater than 13 g./cm.³. He also (1968, p. 143) says that rocks at core pressures in shock waves show no sign of transformation to core densities. Bullen, using the Adams–Williamson relation, found (1965a) that he could manage with a density of 12·3. Bolt (1972) inferred from other seismological evidence that it is < 14 g./cm.³.

The solidity of the inner core would have no direct seismological consequences, because even if S waves could be generated in it they would have to be converted into P to get out. There might be an indirect consequence, because P waves reaching the inner core would be partly converted into S and give a new set of derived waves. These might be observable, but it seems likely that they would be small (Bullen, 1950). Freeman Gilbert finds that the periods of some free vibrations are sensitive to rigidity in the inner core. J. A. Jacobs (1953) suggests that a solid inner core may be a consequence of the normal process of solidification. F. E. Simon (1953) gives about 3200 ± 100° K. for the melting-point of iron at the pressure at the core boundary, $3900 \pm 250°$ K. at the boundary of the inner core and $4000 \pm 300°$ K. at the centre. These values are plausible geophysically. If the core is iron they set a lower bound to the temperature, and if the inner core is solid iron the temperature $3900°$ K. can be taken as an estimate of it.

It is not necessary to the argument that the hypothesis should be true for all substances. It does seem to be nearly true for the materials of the shell and the outer part of the core, and this would be enough to justify the explanation of the straightening of the time-curves about $95°$ as due to mixture.

Mixture is also considered by Ramsey, who shows that it permits us to adopt a density change of 0.3 instead of 0.6 at the $20°$ discontinuity, thus bringing it into agreement with the value suggested by crystallography.

Ramsey follows Kuhn and Rittmann in arguing that the hypothesis of an iron core leads to too high an iron content in the Earth as a whole. This difficulty would be removed by Lyttleton's theory that the planets were formed in two stages, the second disruption occurring after the heavier matter had settled to the centre. Eucken's appeal to Homer Lane's argument has an application to Hoyle's theory of the growth of planets by accretion.

A striking result found by Ramsey is that with certain properties of a material, a mass within a certain range can rest in three different states, two with cores and one without a core. Suppose that for pressure $p < p_c$, the density is ρ_0; if $p > p_c$ the density is $\rho_1 = \lambda\rho_0$. The radii of the core and the outside are R_c, R; the mass is M. For a sufficiently large mass of density ρ_0 the pressure p_c will just be reached at the centre. The radius and mass in this state are R_0, M_0. Then we find in the shell

$$p = \tfrac{4}{3}\pi f \rho_0^2 \left\{ \tfrac{1}{2}(R^2 - r^2) + (\lambda - 1)R_c^3 \left(\frac{1}{r} - \frac{1}{R} \right) \right\}, \tag{1}$$

and if $p = p_c$ when $r = R_c$, we have

$$p_c = \tfrac{4}{3}\pi f \rho_0^2 \left\{ \tfrac{1}{2}(R^2 - R_c^2) + (\lambda - 1)R_c^3 \left(\frac{1}{R_c} - \frac{1}{R} \right) \right\}. \tag{2}$$

If $R_c = 0$ and $p = p_c$ at the centre, $R = R_0$; whence

$$R^2 - R_c^2 - R_0^2 + 2(\lambda - 1)\left(R_c^2 - \frac{R_c^3}{R} \right) = 0. \tag{3}$$

The mass is
$$M = \tfrac{4}{3}\pi\rho_0 \{ R^3 + (\lambda - 1)R_c^3 \} \tag{4}$$

and
$$M_0 = \tfrac{4}{3}\pi\rho_0 R_0^3. \tag{5}$$

(3), (4) are a pair of equations to determine R, R_0, given M. For given R_c the left side of (3) is an increasing function of R, negative for $R = R_c$ and positive for R large. Hence each value of R_c gives one possible value of R and hence determines M by (4). We also find

$$\left(1 + (\lambda - 1)\frac{R_c^3}{R^3}\right)\frac{dR}{dR_c} = 3(\lambda - 1)\frac{R_c^2}{R^2} - (2\lambda - 3)\frac{R_c}{R}, \tag{6}$$

and $$\frac{R_0^3}{M_0}\frac{dM}{dR_c} = \frac{-3(2\lambda - 3)RR_c + 12(\lambda - 1)R_c^2 + 3(\lambda - 1)^2 R_c^5/R^3}{1 + (\lambda - 1)R_c^3/R^3}. \tag{7}$$

For R_c small $M \to M_0$, $R \to R_0$. Then for R_c small, if $\lambda > \frac{3}{2}$ this expression is negative and therefore $M < M_0$. But from (3) we find for R_c large

$$R \doteqdot R_c + R_0^2/2\lambda R_c, \tag{8}$$

and the numerator of (7) is approximately $3\lambda^2 R_c^2$. Further, the numerator of (7) can vanish for only one positive value of R/R_c, by the rule of signs. Also $$\frac{d}{dR_c}\frac{R}{R_c} = \frac{-R/R_c + 2(\lambda - 1)R_c^2/R^2 - (2\lambda - 3)R_c/R}{R_c\{1 + (\lambda - 1)R_c^3/R^3\}}.$$

The second term in the numerator is $< 2(\lambda - 1)R_c/R$; hence

$$\frac{d}{dR_c}\frac{R}{R_c} < \frac{-R/R_c + R_c/R}{R_c\{1 + (\lambda - 1)R_c^3/R^3\}} < 0.$$

Hence for planets with cores there is a minimum possible mass; but this minimum, say M', is less than M_0. Let the corresponding R_c be R'. For $M < M'$ there is no solution with a core. For $M' < M < M_0$ there are two suitable values of $R_c > 0$, and equilibrium is also possible for $R_c = 0$. Of these three configurations the one with no core and the one with the larger core are stable, the other unstable (Ramsey and Lighthill, 1950). Lighthill shows that the condition of incompressibility does not affect the results. For $M > M_0$ there is only one solution, which is continuous with the stable solution for $M' < M < M_0$. Hence if $\lambda > \frac{3}{2}$ no mass can be stable with a positive core radius less than R'. The relations are shown diagrammatically in Fig. 26, the stable states being shown by full lines. For the transition at the Earth's core boundary λ is about $\frac{5}{3}$. Ramsey finds that with the properties indicated by seismology M' is $0.793M_E$, $M_0 = 0.795M_E$, $R' = 1000$ km. The critical mass is below the mass of Venus, but the variation of mean density with mass predicted by the theory at masses between M_E and $0.8M_E$ is so rapid as to provide a convincing explanation of the fact that Venus is definitely less massive than the Earth in spite of having only a very slightly smaller radius.

The theory has been developed further in several papers by Ramsey and Bullen and applied to other planets. The doubtful point, in my opinion, is that the theory has so far given only an order of magnitude for the pressure at the transition. If this should turn out to be greater than the pressure at the centre of the Earth, which is about 2.6 times that at the core boundary, the theory would lose all basis. On the other hand if more

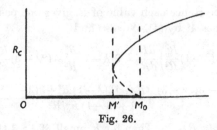

Fig. 26.

detailed treatment should predict the pressure at the core boundary within, say, 10 %, and also indicate the right value for the density jump, the theory would be irresistible.

Ramsey gives arguments that the core cannot be iron. These are discussed by Birch (1952) who finds them unconvincing. There is some sign that if the properties of the core are extrapolated to zero pressure the density would be about $6\cdot4$ g./cm.^{3}, whereas that of liquid iron at its melting-point is about $7\cdot0$ g./cm.^{3}. This could be explained, however, either by some admixture of lighter materials or by a temperature about 1000° above the melting-point of iron, which is quite possible.

Ramsey in a further paper (1951) has applied the theory for hydrogen to the outer planets, with the striking result that Jupiter and Saturn must be about 80 % hydrogen. Uranus and Neptune contain less but nevertheless have a mean molecular weight of about 4. This application depends only on the equation of state for hydrogen, which has been fairly accurately calculated.

B. Miles and Ramsey (1952) have considered several models, with different distributions of the heavier elements. The perturbations of the inner satellites enable us to decide fairly closely between these models. I found (1954 b) that for Saturn the best fit was given by their model S_2, with about $0\cdot18$ of the mass concentrated near the centre. Message (1955) found for Jupiter that the best was their J_3, with $0\cdot05$ of the mass near the centre.

Ramsey's theory has been revived by Lyttleton (1963, 1965). He notes first that the relation between bulk modulus and pressure agrees closely with the linear form
$$k = a + bp,$$
where $b = 3\cdot5$. Above depth $0\cdot06R$, $a = 1\cdot1673 \times 10^{12}$; from $0\cdot06R$ to the core boundary $a = 2\cdot1505 \times 10^{12}$; in the core $a = 1\cdot3416 \times 10^{12}$ dynes/cm.^{2}. That is, the core is more compressible than the greater part of the shell, for given pressure. With this relation the Adams–Williamson relation can be reduced to a case of Emden's equation. The relation between pressure and density takes the form
$$p = \frac{a}{b}\left\{\left(\frac{\rho}{\rho_u}\right)^b - 1\right\},$$
where ρ_u is the uncompressed density of the material. He says that over a range of temperature of some thousands of degrees thermal expansion would affect the density by as much as a difference of pressure of order 10^9 dynes/cm.^{2}. I think 10^{10} would be nearer, but in any case temperature is much less

important than pressure. The whole of his theory assumes that the core boundary is a change of state, and therefore that at the same pressure the liquid form is much denser than the solid. The same difficulties therefore arise as with Ramsey's theory. However, the consequences seem worth working out. A distribution of density is calculated. If the values of a and b for the shell are adopted throughout, the radius (for the same mass) comes out about 270 km. more than the actual one. The suggestion is that the Earth was originally cold, and that small radioactivity at great depths produced liquefaction with reduction of volume. This could give much more crustal shortening than thermal contraction could. But it is fundamental to the theory that the material is denser in the liquid state. This is true of water. But in Bridgman's experiments the only such material was bismuth; for all others, though the difference in density diminished with increasing pressure, there was no sign that it would ever disappear.

Bullen (1973, 1975, pp. 378–85) discusses a further theory of the core. O. G. Soroktin (1971) has shown that Fe_2O is unstable at ordinary pressures, but stable at high ones. He supposed that in the outer core $2FeO$ would change to $Fe_2O + O$. Bullen supposes that in a planet large enough for the critical pressure to be reached at the centre, there would be a zone where Fe_2O is stable, but at lower pressures it would break up into $FeO + Fe$; FeO could be absorbed into the shell and Fe would sink to the inner core. Bullen applies this to the Earth and other planets and the results seem satisfactory. See MacDonald (1962), Kovach and Anderson (1965). Lyttleton (1970) gives an account of his views, with a comparison with the other terrestrial planets.

10·11. Blanketing by sediments. Where deep sediments have been formed and afterwards uplifted, it is always found that some alteration has taken place in them. This may be simply consolidation by compression and by solution and redeposition. But there may be great chemical changes of types associated with high temperature, and it is possible to infer a good deal about the temperature reached from the minerals actually present. Some of these changes occur at surprisingly low temperatures. The degree of metamorphism in the deepest widely extended deposits of the Scottish Highlands corresponds to the formation of an iron garnet. C. E. Tilley (1926, p. 50) says: 'The absence of almandine in the normal contact aureole, and the visible evidence of its replacement when introduced into the sphere of igneous intrusions, conspire to demonstrate that the iron garnet is unstable under the conditions prevailing in normal contact metamorphism.' The temperature of formation of this mineral must naturally be lower than that of its decomposition.* Tilley puts the temperature of formation at about 250° C. It seems that temperature alone will not produce this mineral, shearing stress being also needed, but without heating no stress will produce it. It seems rather remarkable that shearing stress should aid the formation

* Cf. also G. L. Elles and C. E. Tilley (1930).

of crystals that are about as nearly spherical as crystals can be. I have heard it suggested that they act as ball bearings.

Though the average thickness of sediments over the continents is likely to be about 2 km., there are places where the thickness is much greater, and the thickness of known deposits in places is certainly in the neighbourhood of 10 km. Two explanations of the rise of temperature in such regions are often mentioned. One is that it is due to direct compression, the heat being supposed not to have had time to be conducted away while deposition was taking place. This is easily disposed of; the rise of temperature on adiabatic compression by the weight of 10 km. of granite would be about 1°. The other is that each isothermal surface has a 'normal level', and that if rocks are displaced vertically they take up the normal temperature appropriate to their new level. But the notion of a normal level for an isothermal surface is equivalent to a statement that heat conduction across vertical planes is much more important than across horizontal ones; in other words, that the horizontal extent of the inequalities is much less than their vertical extent, which might be true for dikes but not for regional metamorphism.

There are actually three reasons why deep sedimentation should tend to raise the temperature (Jeffreys, 1931c). Two concern the quantity S_0 considered in 10·08.

(1) Radioactivity of the sediments themselves makes a contribution.

(2) The burial of the underlying radioactive rocks to a greater depth increases S_0, since this is proportional not only to the total amount of radioactivity but to its mean depth.

(3) The heat present in the crust before sedimentation started will continue to be conducted out, and even without any change in S_0 would tend to establish the normal temperature gradient in the sediments; and as the base of the sediments was at the surface and is now deeply buried it naturally becomes hotter.

None of these effects is immediate, and detailed calculation is needed to find their relative importance. The first is not great on account of the comparatively small depth and the low radioactivity of most sediments. It is not likely to exceed 60°. For the second, with a fairly representative set of values of 10 km. for the sediments, 15 km. for the radioactive layer below, supposed uniform and sufficiently radioactive to make the ultimate gradient 28°/km., the rise of temperature at the base of the sediments is found to be as follows:

Time (million years)	Temperature rise (° C.)
1·3	23
2·0	35
3·6	58
8·1	102
32	175
130	225
∞	280

The time needed to give a basal temperature approaching the ultimate one is of the order of a geological period.

The third part begins by raising the temperature at the base rapidly, but the contribution rises to a maximum of about 90° in 13 million years and then tends to zero. Its value after 130 million years is about 25°. Thus the most important part is the second; this represents heat generated by radio-activity in the original crust after sedimentation started and then conducted into the sediments. Altogether the rise of temperature at the base of the sediments might reach 300° in a geological period but is unlikely to be much over 250°. The data are of course subject to the usual variability, but they were not chosen to give agreement with the petrological evidence. It may be concluded that the evidence of thermal metamorphism is consistent with the hypothesis that the rise of temperature is due to thermal blanketing by the sediments themselves. The effect is not immediate, but takes something like a geological period to approach its full amount.

Associated with the rise in S_0 will be a rise in the temperatures of the matter at all levels, which would presumably weaken the intermediate and lower layers greatly below the sediments.

10·12. Radioactivity in the Earth's early history. If the radio-active elements were originally uniformly distributed, and the Earth cold, could they have generated enough heat to melt it? The generation per unit volume would be

$$\frac{4\pi AHa^2}{\frac{4}{3}\pi a^3} = \frac{3AH}{a};$$

with $AH = 10^{-6}$ cal./cm.^{2} sec. this gives about 50×10^{-16} cal./cm.^{3} sec., or 150 cal./cm.^{3} in 1000 million years. The heat capacity per cm.^{3} would be about 1. Thus, in 2000 million years the present radioactivity would not be enough to melt originally cold rocks if it was uniformly distributed.

Birch gives an average of $0·33 \times 10^{-6}$ cal./g. year for stony meteorites. If this extended to depth 500 km. the generation of heat would be $1·6 \times 10^{-6}$ cal./cm.^{2}; this would account for the heat outflow. The total generation in 2000 million years would be 660 cal./g. This is of the order of magnitude of that needed to raise the rocks to the melting point and partly to fuse them. An initially cold state so far would therefore not be inconsistent with a former fluid state, followed by magmatic differentiation and formation of continents by convection.

The age of the Earth, based on the oldest known rocks, appeared to be about 2×10^9 years when this book was originally written; and this was short enough in comparison with the average lives of U and Th for the growth of lead with time to be treated as nearly linear. However the later work, especially that on lead isotopes, indicates that $4·5 \times 10^9$ years is nearer. Also ^{235}U and ^{40}K have average lives much shorter than those of ^{238}U and Th. This suggests that in the early history of the Earth the generation of heat may have been much more than now. Data for ^{238}U, Th and ^{40}K have been given. There seem to be no direct measures of heat generation by ^{235}U, but Prof. O. R. Frisch has kindly made an estimate for me, based on the difference of masses between ^{235}U and $^{207}Pb + 7He$. He estimates that the total emission of energy per U atom in the ^{235}U series is 0.94 ± 0.06 of that in the ^{238}U series.

If the age T is 4×10^9 years, we have

	^{238}U	^{235}U	Th	^{40}K
κT	0.60	3.92	0.20	2.22
$e^{\kappa T}$	1.8	50.4	1.2	9.2

The present ratio of $^{235}U/^{238}U$ is $\frac{1}{139}$. I use the values of heat generation for Greenland, Iceland, Scotland, Ireland and Japan. The value for uranium is $(11\cdot2 \pm 0\cdot9) \times 10^{-14}$ cal./g. sec. But this must be interpreted as $11\cdot1 \times 10^{-14}$ from ^{238}U and $0\cdot077 \times 10^{-14}$ from ^{235}U, and the corresponding values 4×10^9 years ago would be $20\cdot4 \times 10^{-14}$ and $3\cdot9 \times 10^{-14}$. Then the totals would be

U^{238}	U^{235}	Th	K^{40}	Total
$10^{-14} \times 20\cdot4$	3·9	6·2	22·1	52·2

as against 18·9 found previously. This would multiply S_0 of 10·08 by about 3, the greater part of the difference coming from ^{40}K. An immediate consequence is that the base of the upper layers could not have solidified. Further, the generation of heat would be about $1\cdot9 \times 10^{-5}$ cal./g. year, or, with a specific heat of 0·3, about $0\cdot6 \times 10^{-4}$ deg./year, and a rise of 1000 deg., in the absence of conduction, would take about $1\cdot7 \times 10^7$ years.

This result suggests that it would be interesting to consider what would happen if the present radioactive elements, with their quantities adapted to 4×10^9 years ago, were uniformly distributed through the shell. The estimate given above would need to be multiplied by 3. A rise of 1000° would take about 2×10^9 years. This estimate, though extremely rough, is highly relevant to the hypotheses that suppose the Earth to have been formed by accretion of cold matter. If this was of uniform composition it is quite possible that radioactive heating could produce fusion in a fraction of the age of the Earth. Then the process of fractional crystallization, combined with cooling from the surface, could produce the upward concentration of radioactivity and convective formation of the continents.

A very instructive article on the thermal history is by F. Birch (1965).

CHAPTER XI

The Origin of the Earth's Surface Features

'You couldn't deny that, even if you tried with both hands.'
'I don't deny things with my *hands*', Alice objected.
'Nobody said you did', said the Red Queen. 'I said you couldn't if you tried.'
'She's in that state of mind', said the White Queen, 'that she wants to deny *something*—only she doesn't know what to deny!'
'A nasty, vicious temper', the Red Queen remarked.

LEWIS CARROLL, Through the Looking-Glass

11·01. The principal problems. The conspicuous superficial phenomena that require physical explanation are the difference between continents and ocean basins, the formation of mountains, and the various types of igneous activity. Mountain formation being considered first, the folding and thrusting shown in mountain ranges imply a shortening of their transverse sections, and therefore a reduction of total area and a reduction of the radius.

Of the many sources of contraction that have been suggested, only two have been shown adequate to account for any appreciable fraction of the crumpling that has occurred; these are thermal contraction and changes in the rotation of the Earth.

There are two alternative forms of the contraction theory, a fracture theory and an elastic instability theory. It will be necessary to discuss later what is meant by elastic instability, but it must be remarked now that many criticisms of the contraction theory implicitly identify it with the elastic instability theory, which, as far as I know, has never had any explicit supporters, and is in fact completely unsatisfactory.

The differences between continents and ocean basins express even more widespread inequalities, with the additional complication that they appear to be associated with differences of chemical composition and not only with mechanical history.

Igneous activity is going on now in many places, and has been prominent in many others in the past.

11·021. Thermal contraction. It was seen in the last chapter that different parts of the interior have cooled since solidification by different amounts. In cooling they must have contracted in volume in different ratios, and in this way have set up a state of stress. The mathematical discussion of the character of the stress distribution was due originally to Dr C. Davison (1887) and Sir G. H. Darwin (1887).* Consider the Earth at

---

* The contraction theory qualitatively goes back to Newton; cf. Brewster, *Memoirs of Sir Isaac Newton*, II, Appendix 4.

some instant during its cooling, and consider the effect of the cooling that goes on during some further interval. We must consider both the term in $S - S_0$ in 10·08 and the term in m; the former has not led to important cooling yet at depths more than about 500 km., whereas the latter, though it does not reach such large amounts, takes comparable values all the way to the centre. The temperature at the outer surface, however, is practically fixed. Consequently in any case the surface must be shortened if it is to continue to fit the interior. At the depth of most rapid cooling, on the other hand, cooling alone would make a sphere too small to fit the interior, and such a sphere must be stretched. There will be an intermediate level such that cooling produces just enough contraction to maintain a fit. This level was called by Davison the *level of no strain*.

The behaviour under the stresses produced will depend on the distribution of strength. If this is low there will be practically continuous flow and conditions will remain nearly hydrostatic. If strength is considerable stresses will build up until they reach the strength, and then there will be fracture, which will be recognized by an earthquake. This will relieve the stresses locally, but many earthquakes will be needed to relieve those over a complete sphere. In either case the final result will be very much the same; hydrostatic stress will be maintained everywhere within the limits indicated by the strength, and in particular there must be fractures at the surface of sufficient extent to enable the surface to fit the contracted interior.

It should be noticed at once that this argument is confirmed by the indications of departures from hydrostatic stress in the interior. The range of depth covered by deep-focus earthquakes, and the range through which gravity anomalies indicate stress-differences comparable with, though less than, the strengths of surface rocks, are about 600 km., which is about the range through which cooling through 100° or more since solidification is indicated. Consequently there is evidence of the existence of such stresses over just the range of depth where the strength is likely to be able to support them and where contraction is likely to produce them.

11·022. The amount of compression available. Let us now apply the above considerations to a quantitative discussion of the amount of mechanical adjustment that must have occurred in the Earth. Consider a shell of internal radius r and thickness dr. Let its coefficient of linear expansion be n, where n may be variable, and let its initial density be ρ. Let the rise of temperature be v; this will of course be negative. Then v is a function of r. The density of the shell, if we ignore the small change due to compressibility, will become $\rho(1 - 3nv)$. The inner radius becomes $r(1 + \alpha)$ and the outer $r(1 + \alpha) + dr\left\{1 + \dfrac{\partial}{\partial r}(r\alpha)\right\}$. Thus the mass after the changes is

$$4\pi r^2(1 + \alpha)^2 \rho(1 - 3nv)\left\{1 + \frac{\partial}{\partial r}(r\alpha)\right\} dr = 4\pi\rho r^2 dr\left\{1 + 2\alpha + \frac{d}{dr}(r\alpha) - 3nv\right\}, \quad (1)$$

neglecting squares and products of α and v. But the mass is unaltered. Hence we have the equation of continuity

$$2\alpha + \frac{d}{dr}(r\alpha) - 3nv = 0. \tag{2}$$

Given the cooling, this is a differential equation to determine α.

Now if a shell simply expanded without stretching, its radius would increase by rnv instead of $r\alpha$, so that the stretching required to make it continue to fit the interior is $r(\alpha - nv) = rk$, say. It is k that concerns us more directly. Then from (2) we have

$$\frac{d}{dr}(kr^3) = -r^3 \frac{d}{dr}(nv). \tag{3}$$

At the centre kr^3 will vanish; hence

$$k = -\frac{1}{r^3} \int_0^r r^3 \frac{d}{dr}(nv)\,dr \tag{4}$$

$$= -nv + \frac{1}{r^3} \int_0^r 3r^2 nv\,dr. \tag{5}$$

At great depths the cooling arises mainly from the m term, and nv will be nearly constant, so that k will be small. Hence the cooling is associated with little deformation of the deep interior. But at the outer surface v is zero, and values of v at all depths make contributions to k.

Unfortunately, the experimental determination of the coefficients of expansion of rocks is difficult. They usually disintegrate when heated to some temperature near 800°, and show an expansion, not completely reversed on cooling, which is partly due to the formation of cavities. In former works of mine Fizeau's determinations were used; according to these

$$n = \epsilon + \epsilon' V,$$

where V is in °C., and

$$\epsilon = 7 \times 10^{-6}/1°\,\mathrm{C.}, \quad \epsilon' = 2{\cdot}4 \times 10^{-8}/(1°\,\mathrm{C.})^2.$$

These give a linear expansion of $1{\cdot}9\%$ between 0° and 1000° C. N. E. Wheeler (1910) gives $2{\cdot}73\%$ for granite and $1{\cdot}54\%$ for olivine dolerite. Day, Sosman and Hostetter (1914), in a critical investigation, find a complicated behaviour at high temperatures, with volume expansions of the order of 10% between 900° and 1250° C.

Birch, Schairer and Spicer give the following data. Coefficients of linear expansion between 20° and 100°, granites $(8 \pm 3) \times 10^{-6}/1°$, basalts $(5{\cdot}4 \pm 1) \times 10^{-6}/1°$. Glasses (volumetric) at 20°, about 20×10^{-6}, 1200°, about 90×10^{-6}. Total expansions by volume between 800° and 1000°, orthoclase $0{\cdot}47\%$, augite $0{\cdot}69\%$, diopside $0{\cdot}69\%$, olivine $0{\cdot}78\%$. The change of density on melting for diabase ($=$ dolerite) is 8–9 %.

It is probable that conductivity increases with depth and that the coefficient of expansion decreases. In particular there is reason to suppose that at high pressures and temperatures heat transfer by radiation exceeds that by ordinary conduction (Lubimova, 1958). It does not appear however that these effects are very important. The effect of increase of conductivity would be more rapid tendency to uniformity of temperature, but since some parts gain as much heat as others lose, the effect on the total contraction would be small. Change of the coefficient of expansion at great depths would not make much change in volume, since the change of temperature is small (except for the m term). So it seems that conditions in the top 600 km. or so must play a dominant part in any case.

But the results can give only the order of magnitude of the actual contraction within the Earth, partly because there is considerable uncertainty about the actual amount of cooling, partly because not much is known about the effect of pressure on the coefficients of expansion. An estimate of order of magnitude, however, is useful. The results of 10·08 suggested that the $S - S_0$ term gives an average cooling of order 500° through depths down to about 400 km. This, with a volume contraction of 5 %, would imply a reduction in radius of about 20 km., and therefore about 130 km. in circumference. To this should be added allowances for contraction in crystallization and for loss of volume due to extrusion of water. The cooling below the oceans may well be twice that below the continents. We can therefore say that the surface compression over a great circle is of the order of 200 km., but it may easily be half or twice this. The depth of the level of no strain is about 100 km. For given amount of cooling due to the $S - S_0$ term the depth increases like $t^{\frac{1}{2}}$, and therefore the total compression since consolidation also increases like $t^{\frac{1}{2}}$.

The contribution from the m term is still more difficult to estimate on account of uncertainty of the coefficient of expansion. This term gives a cooling of the order of 45° except close to the surface. If we allow a 1 % decrease in radius due to it, it implies a crustal shortening of 400 km. This is not likely to be too high, because the material everywhere is only a little below the melting-point, a condition that usually makes for a large coefficient of expansion. This contribution increases uniformly with time, and therefore is specially important because it meets a difficulty repeatedly raised by Holmes. The importance of the m term was not noticed until 1932, and if crustal shortening is entirely due to the $S - S_0$ term it would suggest that the rate of mountain formation should have greatly decreased during geological time. This does not appear to agree with the facts. Holmes maintained that the rate has increased. No theory that has been suggested gives a hint of an explanation of an increase, and I think that any increase indicated in Holmes's data is not significant, and in any case could well be attributed to more detailed knowledge of the later mountain systems. Nevertheless, the existence of a large term that gives a uniform rate of contraction removes a major difficulty.

The mechanism described above is essentially that of Davison, whose paper was published forty years before the discovery of deep-focus earthquakes and was written when the Earth, at depths exceeding 50 km. or so, was widely believed to be fluid. Consequently it implicitly assumes plastic adjustment at great depths, which is not essential to the principle. The zone below the level of no strain is in fact subject to horizontal tension* and can adjust itself by sliding fracture; and the non-elastic extensions at the level of most rapid cooling would be about 10 times the general shortening of the surface. Thus the mechanism in a solid Earth is much more complicated than is supposed above. The first step would presumably be a fracture near the level of most rapid cooling; this could be relieved by normal faults reaching right up to the surface. But the outer surface was already too small to fit the interior, and this adjustment is an effective increase in the size of the interior except in the immediate neighbourhood of the mighty faults. Thus the compression available for mountain building may well be in the neighbourhood of 11 times that contemplated above.

The possible importance of fracture near the level of most rapid cooling was tentatively suggested as an explanation of rift valleys in the third edition of this book. It has been developed by A. L. Hales (1953*a, b*). He actually gets a crustal shortening about the same as I did; this is because he uses a much smaller coefficient of thermal expansion. He regards geosynclines as the remains of the depressions made when the original faults reached the surface, and gets a satisfactory estimate of their cross-section.

It should be noted that modern theories suggest a smaller coefficient of expansion and a smaller melting-point gradient and both probably decrease with depth. If so, the contribution from the m term also will increase less rapidly than in proportion to t.

Recent determinations of thermal expansion (Birch *et al.*) seem to agree better with the earlier ones than with those found by Day *et al.* The latter may be due to formation of cracks, which would not occur under high pressure. On the other hand the olivine-spinel transition may be relevant. If a liquid at high temperature cooled first to the rhombic form, and passed to the cubic form as it cooled, a contraction of about 10 % would be available. The equations may be non-linear, as in the case of solidification from the surface with allowance for latent heat; but in the latter case the solution is surprisingly simple (Jeffreys, 1927*a*, esp. pp. 11–16).

Fractures near the layer of most rapid cooling would be expected to be at about 45° to the vertical. This seems to agree with the fact, noted by Benioff and Gutenberg, that deep earthquakes seem to lie on steeply inclined planes, reaching the surface near mountain ranges. See also § 11·101.

A detailed analysis of the thermodynamical aspects has been carried out

* Since all the stresses are in fact pressures, this must always be understood as meaning that the pressure across a vertical surface is less than that across a horizontal one.

by F. Birch (1952). There is evidence from analogy concerning the coefficient of thermal expansion, which should be calculable within a factor 2, and it seems probable that the one that I adopted is too high; that used by Hales may be too low.

For isothermal compression of most solids, even for quite large compressions, it is found that the pressure satisfies closely

$$P = 3K_0 f(1+2f)^{\frac{5}{2}},$$

where

$$\rho/\rho_0 = (1+2f)^{\frac{3}{2}}.$$

Here ρ_0 is the density at zero pressure and K_0, f are constants for the material (at given temperature).

Other information can be derived from a result of Grüneisen, that the parameter

$$\gamma = \frac{k\alpha}{\rho c_v},$$

where k is the isothermal bulk-modulus, α the coefficient of thermal expansion, ρ the density, and c_v the specific heat, always lies between 0·5 and 2. An equation of state can be derived, and many interesting quantitative consequences have been found by use of modern theories of the solid state (Birch, 1952; Shimazu, 1954; Miki, 1952, 1954; Verhoogen, 1953).

Darwin examined the heat supply due to the damping of rotation in bodily tides, and decided that it was not enough to affect the heat outflow considerably.

On any contraction theory the reduction in radius implies a loss of potential energy, which will reappear as heat. The thermodynamics of this has been examined by Lapwood (1952) who finds that its effect on the amount of contraction is small.

If oceanic water has come from the shell it would account for a reduction of about 5 km. in the radius. If it varies greatly with position it might account for considerable variation of vertical displacement. This is emphasized in a paper by L. S. Dillon (1974).

11·03. Compression needed. Direct measurement of the folding and thrusting in many of the great mountain systems has been carried out. Up to 1920 or so estimates quoted were 40–50 miles in the Appalachians, 25 miles in the Rocky Mountains of British Columbia, 10 miles in the Coast Range of California, and 74 miles in the Alps (Pirsson and Schuchert, 1915, p. 361; repeated, except for the Alps, in the 1920 ed.). The tendency of later work was to discover new thrust planes and enormous overfolds, and to increase all these estimates. Thus Heim (1921, **2**, 50) gave 200–300 km. for the compression involved in the Tertiary folding of the Alps; A. Keith (1923) gives 320 km. for the Appalachians, and G. P. Mansfield

(1923) mentioned one thrust with a displacement of 35 miles (56 km.) in the Rockies.

There is, however, another type of evidence that indicates that much of the observed folding and thrusting must be due to something other than crustal shortening. The Alps, Rockies and Appalachians are all systems where isostasy gives a fair approximation to the facts. Let us suppose that these regions were also in an approximately isostatic state before the mountains were formed. Take the undisturbed structure to consist of 1 km. of sediments of density 2·4, 15 km. of granite of density 2·6, and 18 km. of intermediate rock of density 2·9, the whole resting on dunite of density 3·3. Suppose that by crustal shortening a region with this structure was halved in area and the outer layers therefore doubled in thickness. The extra thickness added would be 34 km., but isostasy would be restored if only 28·4 km. of the lower layer was displaced. Thus the residual height after adjustment would be 5·6 km. This is more than the maximum height of the Alps and far more than their mean height. Yet Heim's estimate involves a shortening in a ratio of 3 or 4 to 1, and even higher values are given by some other writers.

Such a result makes it necessary to re-examine our premises. The present relief in mountainous regions is not due directly to the processes involved in their formation; it is due to the carving out of deep valleys, at the expense, in the first instance, of the sedimentary layer. The mean thickness has therefore been reduced, and balance is restored by the inflow of new heavy matter. The loss of a kilometre from the top implies the addition of 0·73 km. below, so that the mean level is lowered by 0·27 km. But the denudation is mainly at the expense of the slopes and valleys; if the mountain tops had been denuded by the average amount for the region they would never have become mountain tops. The inflow by itself would raise the mountain tops by 0·73 of the average depth denuded, and if the summits are denuded by less than this the effect of denudation and compensation together will be to make the summits higher. Presumably the ratio is much less than 0·73, and we may reasonably infer that the height of the mountain tops above sea-level is greater than when the mountains were formed, by an amount of the order of half the mean depth removed, or, say, a quarter of the elevation of the mountain tops above the neighbouring valleys (Jeffreys, 1927a; Nansen, 1927). An important confirmation of this argument was given by L. R. Wager (1937). The River Arun rises north of the Himalayas, cuts through a gorge between Everest and Kunchinjunga, and runs south to join the Ganges. The height of its source is 26,000 ft., and it falls to a level of about 13,000 ft. in Tibet, but to form its gorge it must have cut through rocks at a height of 18,000 ft. The gorge itself is 6000 ft. deep. This would be possible only if the gorge has risen since the river cut it. That is, the river cut its gorge when the pass between Everest and Kunchinjunga was not more than 13,000 ft. high. The general rise is attributed to the process just described. Wager considered the alternatives that the gorge was cut by a glacier and that the river has cut back through the mountains and beheaded streams on the north side, but found them unsatisfactory.

In the Alps and Rockies the height of most of the highest peaks is about 4 km., that of the valleys about 1 km. Hence we may suppose that the primitive height was about 3·2 km., and the crustal shortening would be in a ratio of 1·66 to 1. The present widths at the widest points may be taken as 100 km., so that the crustal shortening indicated is in the neighbourhood of 60–70 km.

No help is to be obtained from the hypothesis that the upper layers are much thicker than I have assumed above. If the upper layer was granite and 60 km. thick, doubling its thickness would give an elevation of no less than 12 km., and the crustal shortening needed to form the Alps would be only 0·27 of the present width. Replacing the lowest 20 km. of the granite by basalt reduces the discrepancy only a little.

It must be recalled that the ranges in question are just those where isostasy is most nearly true, and the failure of the geological estimates of compression to agree with those indicated by isostasy can have only one explanation: that the measured compression includes something that is not crustal shortening, and is therefore unsuited for comparison with the shortening indicated by theory.

The Himalayas and the ranges to the north present an apparent difficulty. The elevation is greater, and so is the width. If we apply the same argument to the Himalayas alone the shortening needed comes out at about 160 km., to which a comparable amount should be added for the Karakorum. The Altai and Tian Shan are of Hercynian (Permo-Carboniferous) age, and do not enter into the comparison. But it looks as if the shortening in Tertiary times across the two ranges flanking Tibet was about five times that shown anywhere else in the Tertiary systems, and this seems geometrically impossible. There are, however, two possible escapes. The Alps and Rockies have been taken as typical simply because they are regions where isostasy is well verified and the general structure is known from seismology. We have gravity data from only the south side of the Himalayas, and none at all for the Tibet plateau and the Karakorum. We have no determinations of thickness of the upper layers from any part of the region. Accordingly, it is permissible to suppose, as far as we know at present, either that the thickness was specially great in these regions before the Himalayan uplift, or that at some stage in the process (possibly even the present) isostasy has been seriously wrong. An expedition to measure gravity on the plateau is desirable.

If we take the estimates given by isostasy, then as there have been three major epochs of mountain formation since the Cambrian and probably about five earlier ones, we are led to a total crustal shortening in the neighbourhood of 500 km. This is very near the amount indicated by the thermal contraction theory. The agreement is of course accidental; on account of the various uncertainties in the quantitative data used in the theory the theoretical value may be wrong by a factor of 3, and on account of those of the crustal structure the estimate of the actual contraction may be wrong by a factor of 2, though it is unlikely to be. But what we can say from the comparison is that there is no evidence that thermal contraction is insufficient to account for all the mountains that now exist or have existed, and in any case it is a major contributor to their explanation; and that until some other agency is shown to be of comparable importance the presumption is that contraction accounts for practically the whole. The argument is precisely the same as that for tidal friction in shallow seas as an explanation of the secular acceleration of the Moon, and in the latter application it has never been questioned.

The linear compression of granite under crushing stress nearly sufficient to break it is about 10^{-3}. Hence on any contraction theory the Earth's circumference would be shortened at any general fracture by something like 40 km., which is again of the order of magnitude of the shortening indicated from isostasy.

G. M. Lees (1953) has maintained that the geological estimates do represent crustal shortening and that a radial shortening of the order of 1000 km. is needed. Other geological writers have denied the possibility of horizontal outflow of the type considered here. However, there is a simple answer to such arguments. It is maintained here that the volume extruded at the surface was comparatively small, say corresponding to a cross-section of order 50×60 km.[2]. The alternative to having the movement shallow would be that each chain implies shortening of the order of 300 km. going down through most of the shell. If such a disturbance took place the extruded matter would be of the order of 1000 times the volume of a mountain chain, and no explanation has been offered of where such matter may have gone.

11·04. Distribution of mountains in time and place. We have considered cooling and readjustment to take place together. Actually in a solid with a finite strength differences of cooling would produce stress-differences, which would grow until they reached the strength of the material, and then be mainly relieved by fracture. The processes would therefore really alternate; there would be long intervals of quiescence, separated by short intervals when general fracture is taking place so as to restore approximate hydrostatic stress. Thus any theory that takes account of the properties of solids automatically explains the observed intermittence of mountain formation.

In a body under general shear stress, failure of elasticity begins at one place, and the immediate effect is to increase the stress-difference in the neighbourhood several times. Consequently, if a whole region has been near breaking-point, one fracture will make the local stress-differences far exceed the strength, and therefore the fractures will extend rapidly through the region until all the stress-differences have been brought down to below the strength. The number of fractures would be small, because it is easier to extend an old fracture than to start a new one. Now in our problem we are considering a spherical shell too large to fit its interior. It is then a geometrical problem to find whether a readjustment is possible that will reduce all great circles of the sphere by about the same amount. This can be done. For simplicity, suppose that there is crumpling across the meridian of Greenwich by the same amount everywhere, and also across the equator from 90° E., through 180° E., to 90° W. Every great circle of the sphere would intersect both of these 180° arcs, and the shortenings of different great circles would be comparable. Such a distribution is remarkably like what the Tertiary mountains show. We have the great Pacific system running the whole length of the west coast of America, and roughly at right angles to it the system including the Pyrenees, Alps, the Balkan mountains, and passing through the Himalayas to Indo-China. Special local complications (especially weakening of the crust by heating in regions of heavy sedimentation, as we saw in 10·11) might well produce exceptional curvatures, but the main features are just what we might expect on a contraction theory.

Under the oceans there is presumably a range of depth where the cooling has been notably more than under the continents. We cannot say yet whether this range is to be measured in tens or hundreds of kilometres, but the essential point is its existence. In addition, it appears that basic rocks are on the whole stronger than acidic ones. On account both of the greater intrinsic strength and the probable lower temperature, the rocks below the oceans must be stronger than those below the continents. Now where the compressed ocean floor abuts on a compressed continent the weaker will be the first to give way; the continent margin will be driven inwards, and its rocks piled up over those further inland, forming ranges of mountains parallel to the coast. These correspond closely to the mountains of the Pacific coast of America. This remarkable system is therefore explicable as a natural consequence of the contraction theory.

To sum up, thermal contraction predicts the correct order of magnitude of the total crustal shortening indicated by mountain systems (200 km. as against 500 km., but either estimate can easily be wrong by a factor 2); it also explains the intermittence of mountain formation in time, and the amount of shortening in any one period of mountain formation; and it accounts for the general features of the distribution of the Tertiary system of mountains, which is the system whose distribution is most fully known.

11·05. Changes in the Earth's rotation. We have seen that a simple calculation based on the present rate of the secular acceleration of the Moon's motion implies that 1600 million years ago the Earth rotated in about 0·84 of our present day; if the Moon was ever part of the Earth the period just after separation was about 5 hr. The ellipticity of figure is proportional to the square of the angular velocity, so that on the above estimates the ellipticity 1600 million years ago was about $\frac{1}{210}$, and the original ellipticity might be as low as $\frac{1}{13}$. Now the radius of the Earth in colatitude θ is $R\{1 + e(\frac{1}{3} - \cos^2\theta)\}$. We are not at the moment considering variations of the mean radius, but it appears that the equatorial radius contains a variable part $\frac{1}{3}eR$. This would imply a contraction of the equatorial circumference by 18 km. in the last 1600 million years, and by 1000 km. since the Moon was formed (cf. Chamberlin, 1909, p. 48). The effect during known geological time is not very important, but if the Moon was ever much closer to the Earth the effect in the early stages may have been considerable. On the other hand, the effect in the polar regions is a stretching. The rate of tidal friction containing the inverse sixth power of the Moon's distance as a factor, its effects must in any case have been concentrated in the early history of the Earth, and mountains as we know them cannot be attributed to it.

In addition to the change of ellipticity there is a general symmetrical contraction, noticed by Dr J. W. Evans and evaluated by Stoneley (1924a). This would have shortened every great circle by about 70 km. since the earliest times. But this effect is inseparably connected with the much larger effect of change of ellipticity, and the total effect, with its stretching of the crust over the polar regions, is qualitatively different from the one that arises in our immediate problem. It appears that change of rotation, whatever its influence may have been, is unlikely to have determined directly the formation of existing mountains.

11·06. Elastic instability. This is a phenomenon of the first importance in engineering,* and may have some geophysical importance, though not in our present problem. A familiar example is in the compression of a postcard by thrust from the ends. So long as the thrust is not too great the card remains plane. But when it exceeds a certain amount the card bends. The plane form is still one of equilibrium, but it is no longer stable. The energy needed to bend the card is supplied by the inward motion of the two ends; for in bending the arc remains nearly unaltered and the chord is shortened. The phenomenon is not fracture or even flow; if the card is released it returns to its original plane form.

In building construction the phenomenon arises in the design of columns. At first sight it appears that all that is needed in the design of a pillar is that the load per unit area of the section should be less than the crushing strength. But this condition may be satisfied, and yet the application of the load may lead to considerable bending without immediate fracture.

* It is usually called elastic stability by engineers. Nobody knows the reason for this.

The redistribution of stress in bending may or may not lead to such additional stresses that fracture or flow is produced as a secondary feature (especially in such materials as brass), but in any case the consequences are disastrous. Consequently, in the design of any member of a building or other framework to support thrust, it is necessary to attend to the possibility of this kind of instability. For the sake of economy pillars and struts must be made as thin as is consistent with safety. The resistance to bending depends on the moment of inertia of the section, not on its area, and consequently, with a given area of section, it is best to design it so as to put the material as far as possible from the centre. Consequently, especially in metal frameworks, tubes are used in preference to rods, or the section is made in the form of the letter H or I.

Elastic instability is not confined to rods. It can occur under thrust in plates, cylindrical tubes and even conical shells. It is very difficult to produce in a body approximating to a sphere. Low rigidity and small thickness favour elastic instability. If the thickness is comparable with the length, it can be shown that elastic instability will not occur unless the external stress is comparable with the rigidity. In materials that can be stretched or compressed by half their length without fracture, elastic instability can occur even in this case; that is why it is difficult to carve a piece of meat that contains no bone. But with the ratios of rigidity to crushing strength characteristic of building materials it will be impossible to produce elastic instability at all unless the thickness is small compared with the length; if the thickness and the length are comparable, fracture will take place before instability.

For spherical shells there is an additional difficulty. Any ruled surface can be deformed by bending in such a way that the straight lines on it remain straight, and if a plate or shell of such form is bent the energy comes wholly from the bending, the surface half-way through remaining unstretched. But a spherical shell cannot be deformed without stretching the sphere half-way through, and the stretching of this makes an important contribution to the energy needed. The consequence is that elastic instability cannot occur in a spherical shell except under disturbances of such small extent that the curvature of the sphere can be neglected, and then the stress needed is increased on account of the reduced superficial extent.

Thus elastic instability requires rather special conditions; either the strength must be comparable with the rigidity, or the thickness must be small compared with the length. Even in the latter case it will be difficult to produce unless the body approximates to a ruled surface, which usually means a rod, plate or cylindrical shell. As it happens, however, these conditions are easily produced in the laboratory, and numerous experimental results have been given with the intention of illustrating the behaviour of the Earth's crust under the kind of stress that would result from internal contraction. The results from such models could, however, be applied to the Earth only if it was shown either that they are not due to elastic instability or that the terrestrial phenomena considered are so due. Otherwise the model is irrelevant to the problem. We must therefore consider whether elastic instability can arise in the Earth's crust. We already know that there will be special difficulties arising from the fact that the boundary is nearly a sphere, but we give the theory its best chance by neglecting the curvature. We must, however, take into account the fact that any vertical movement of the Earth's surface requires additional energy to overcome the gravitational field. (See also Jeffreys, 1932*f*; Southwell, 1936.)

We are reduced, therefore, to considering a thin plate under two-dimensional distortion in a vertical plane. The horizontal distance from the end is x, the vertical displacement y, assumed small compared with the length. The length is

$$\int \left\{ 1 + \left(\frac{dy}{dx} \right)^2 \right\}^{\frac{1}{2}} dx, \tag{1}$$

taken between the ends. The length does not alter when the plate is bent, because the longitudinal stress does not alter. Hence the ends approach by a distance $\int \frac{1}{2} (dy/dx)^2\, dx$, and if Q is the thrust per unit breadth it does work

$$\frac{1}{2} Q \int \left(\frac{dy}{dx} \right)^2 dx \tag{2}$$

per unit breadth. The plate is supposed broad, and the elastic energy involved in bending, per unit area, is

$$\frac{E}{24(1 - \sigma^2)} d^3 \left(\frac{d^2 y}{dx^2} \right)^2, \tag{3}$$

where E is Young's modulus, σ Poisson's ratio, and d is the thickness. For a narrow beam, bending of a longitudinal section is associated with a bending of transverse sections in the opposite sense, and the effect of this on the energy is to remove the factor $1 - \sigma^2$, but this is near 1 anyhow. When the crust is elevated a distance y a force $g\rho y$ per unit area is needed to support the weight of the extra column of underlying material of density ρ. The gravitational energy is therefore $\frac{1}{2} g\rho y^2$ per unit area. The condition that the work done by the end-thrust shall supply at least the potential energy needed is then

$$\int \left[\frac{1}{2} Q \left(\frac{dy}{dx} \right)^2 - \frac{E d^3}{24(1 - \sigma^2)} \left(\frac{d^2 y}{dx^2} \right)^2 - \frac{1}{2} g\rho y^2 \right] dx \geqslant 0. \tag{4}$$

For instability there must be a form of y such that this is positive. The easiest way of finding a condition for this is to suppose y analysed into a series of sines or cosines of multiples of x; integrals of their products do not increase indefinitely with the length of the plate, while those of their squares do, and hence disturbances of different wave-lengths can be treated separately. If then

$$y = a \cos \kappa x, \tag{5}$$

the integral is positive if

$$Q\kappa^2 > \frac{Ed^3}{12(1 - \sigma^2)} \kappa^4 + g\rho, \tag{6}$$

and this will be satisfied for some real value of κ provided that

$$Q^2 > \frac{1}{3} \frac{Ed^3 g\rho}{1 - \sigma^2}. \tag{7}$$

Now Q cannot exceed the thrust involved if a stress equal to the crushing strength of granite, say 10^9 dynes/cm.^{2}, exists through depth d; that is, $Q \leqslant 10^9 d$. Also $E/(1 - \sigma^2)$ for granite is about 7×10^{11} dynes/cm.^{2}, and ρ may be taken as $3 \cdot 3$ g./cm.^{3}. With these values we find that we cannot have elastic instability before fracture unless

$$d < 1 \cdot 2 \times 10^3 \text{ cm.} \tag{8}$$

With this value of d, κ is given by the condition that the two terms on the right of (6) are equal; then the wave-length of the disturbance is

$$2\pi/\kappa = 8 \times 10^4 \text{ cm.} \tag{9}$$

It follows that the combined effect of elasticity and gravity is such as to prevent elastic instability unless the thickness of the crust is less than something of the order of 12 m. It is immaterial for this purpose whether by the crust we mean the granitic layer, or the upper layers together, or the material above the level of no strain. If there is a compacted layer anywhere more than 12 m. thick it will suffice, and fracture will occur first.

At every point we have taken the conditions most favourable to elastic instability; besides neglecting the curvature of the Earth (which may actually be legitimate since the critical wave-length is so short) we have totally neglected the rigidity of the lower layer. We conclude that as an explanation of geological processes elastic instability is out of the question.

It is serious, however, in attempts to imitate geological conditions by models. 12 m. is a short distance in the Earth, and so is the critical wave-length of $0 \cdot 8$ km. They are very big distances in a laboratory. Consequently, it is very hard to illustrate the geological conditions by direct experiment, and results derived from models should always be mistrusted unless it is specifically shown that the conditions exclude elastic instability. The solution was given by M. Smoluchowski (1909); it was rediscovered by S. Goldstein (1926). Both give other applications.

The most familiar model is that of the shrivelled apple, and it is from it that some writers have inferred that the effect of contraction would be a general minute puckering of the crust rather than a few large mountain systems. But this is a case of elastic instability; the skin is not fractured.

An interesting one was given by A. J. Bull (1932), who studied a sheet of collodion on a stretched sheet of rubber. When the rubber was allowed to contract the collodion became contorted into a regularly repeated pattern, and Bull drew the usual inference. But the collodion was not broken. His model is in fact an excellent illustration of elastic instability, and has nothing to do with our problem. What is really astonishing is that this argument has been reproduced, with Bull's photograph, but with no mention of my reply (1932*f*), which appeared shortly afterwards in the same journal.

11·07. Mountain structure. So far we have only indicated that the extensive overfolding and thrusting found by geologists in the great mountain systems, and also in some old ones such as the Scottish Highlands, which were once as great, includes something that is not crustal shortening. The geological evidence on the matter is quite convincing, and we must seek another interpretation of it. Now there is a well-known geological phenomenon on a smaller scale that shows considerable folding of surface rocks without there having been any shortening of the underlying rocks at all. This is hill-creep (see Plate VIIc). Soft rocks on a hillside are usually found to have slipped downwards, with much internal contortion. Even hard rocks may be affected; strata dipping into a hill are often found to have been dragged over where they emerge, and in consequence geological students are warned not to judge the dip of rocks from that in small exposures a few feet deep on a hillside. On a larger scale similar structures are found in boulder clay (Plate IX).* These have been produced by the shearing action of ice flowing over the clay. In these phenomena hard rocks are not affected for more than a few feet below the surface. Now except for scale the great overfolds in mountains are very similar, and the comparison suggests that they also may be due directly to slipping down a slope under gravity, the slope itself having been produced in the original elevation.

We clearly need more analysis of what is likely to happen when a fracture is formed at the surface. In the first place, since the stress-differences arise through the outside being too large to fit the interior, they can be relieved only by a displacement that reduces the outer surface; a vertical fracture or a horizontal splitting will not do. There must be either an oblique fracture with sliding along the thrust plane, or a general thickening by continuous flow. Relief by thrusting would usually be at an angle of somewhat less than 45° to the horizontal, and to achieve a movement of the order of 60 km. in this way would imply that the whole of the crust to a depth of 40 or 50 km. was overthrust, leaving the upper edge overhanging by a similar height. What would be happening meanwhile in the deep parts would depend on whether they broke or flowed; but some type of adjustment is necessary.

* For other formations due to gravitational action on slopes, see J. V. Harrison and N. L. Falcon (1934, 1936).

The suggested distortion of the outer surface is our immediate concern. Clearly it could never be completed, because the stresses introduced by the height of the overhanging part would far exceed the strength. In fact, it may appear that the thrusting has made matters worse. But this is not so; it has relieved the stress-differences around most of the surface, which are not primarily due to gravity, and on a body where gravity was smaller it would provide the complete solution. What it has done is to replace them by an intense local system of stress-differences, which are a secondary feature arising from gravity, and can be relieved by local adjustment.

It is likely that this would usually be by flow. Cold rocks at atmospheric pressure usually give way by oblique fracture, but hot ones do so by flow, even when they have only been buried by 10 km. or so, and we must expect this in regions of deep sediments and also in the deeper parts of the granitic layer and the intermediate one. The lower layer may escape being thrust over the intermediate one, and the intermediate over the upper, by yielding within themselves so as to give a local thickening instead. It is, however, very unlikely that the properties of the various layers would be in just such an adjustment that the yield would proceed so as to give directly a uniform thickening of the upper and intermediate layers, with just enough outflow in the lower layer to give isostasy. It is more likely that the failure in the upper layers would result in their being crushed before the lower layer has had time to adjust itself to the local load by outflow. Then there might be a local thickening of the upper layers by 20 km. or so on a belt about 60 km. wide, leaving a plateau nearly 20 km. in height, a remarkable picture.

In fact, whether the first stage took the form of continuous flow or a huge thrust, it could not persist. If it was ever achieved it would still leave stresses several times the crushing strengths. The solution seems clear when the problem is stated. The first stage would be interrupted several times by further fractures or flow within the elevated material. If it was by fracture the result would be a complicated system of thrusts and block faults, possibly with occasional complete overturning. If it was by flow, huge sheets of material would spread out over the surrounding country, just like the nappes found in the Alps and Highlands (Plate VIII). Normally, this would take place in several stages, and successive nappes could be formed and spread one over another. In either case the field evidence measures the folding and thrusting of the surface rocks, which is far larger than the original crustal shortening.

We do not yet understand why flow of a solid under shear stress ordinarily leads to violent contortion rather than to roughly linear flow. The final appearance often suggests turbulent fluid motion, but it seems impossible to maintain that the velocities were high enough to produce turbulence according to the usual Reynolds criterion for a fluid. Some sort of 'plastic instability' must be involved. This certainly occurs in the simple case of the stretching of a uniform metal rod. If one part becomes slightly thinner than the rest, extra tension is thrown on it per unit cross-section, and it extends more rapidly. Hence the rod ordinarily parts by flowing near one section, until that section becomes so thin that it breaks. The same sort of thing may happen under shear stress, especially in stratified material, but no theory has been worked out (Plates VII, VIII, IX).

Such a theory also explains some minor features. Such a material as putty, when stretched, may undergo a large continuous yield; but it may break if this goes on long enough. Here we are regarding the nappes as originally behaving like putty, though probably they would be much stiffer. Then it is possible that the end of a nappe may break off and slide down the slope by itself, forming a detached piece. Such detached pieces are known and called klippen. On the usual theory that a nappe has been driven forward entirely by horizontal thrust through it, such pieces would be impossible, and to explain them it becomes necessary to imagine the original nappe to have been long enough to make a continuous connexion to the present position of the klippe and to suppose that the intermediate part has been removed—for which there is no evidence.

Again, it is sometimes found that a nappe lies on the opposite side of the main core of the range from the remainder. This is called back-folding, and is also very difficult to understand when the nappes are supposed to be due to horizontal thrusting—it appears to imply that the material extended at both ends under forces tending to shorten it. But we should expect back-folding if the flow is due to sliding from an elevated region; such flow may well take place in both directions from the main elevation. How much of it there would be would depend on the degree of asymmetry of the elevation. If the region rose symmetrically nappes might be formed equally on both sides. If it approached the asymmetry of an oblique fracture the flow would be practically one-sided. As in fact back-folding is exceptional, we may infer that the original rise was usually very asymmetrical.*

The flow of the rocks would presumably be slow, for, though weak, they would have a high viscosity and would be moving in comparatively thin sheets. Isostatic adjustment might well keep pace approximately with the spreading. The outflow from the central region would thin the sediments there and possibly expose the granitic layer, and a central massif of granite, showing signs of considerable metamorphism, would remain. The existence of the gneissic core of Mont Blanc is again in harmony with the theory.

Geological evidence for gravitational sliding in the northern Apennines is given by B. M. Page (1963).

King Hubbert and Rubey (1959; 1961 *a, b*) have suggested that flow at low slopes may be due to lubrication by included water. This can certainly

---

* A somewhat fuller account is in Jeffreys (1931*f*). Actual conditions would probably be much more variable than the diagrams show. In particular the diagrams show an abnormal amount of back-folding.

happen in recent sediments. Details of their argument have been criticized
(Laubscher, 1960; Birch 1961*a*). I doubt myself whether this can be applicable to mountain nappes. They are usually metamorphosed. Did this happen
before or after extrusion of the water? If before, we have to suppose that
rocks buried by 10 km. or so had not become compacted. If after, we have
to suppose that the metamorphism occurred in thin layers near the surface.
Miss G. L. Elles once remarked to me, with regard to some Highland nappes,
that the Ballachulish slate (really a shale) acts as a lubricant. I think myself
that the presence of internal weak layers is an important part of the process.

M. R. Mudge (1968) notices that laccoliths in the U.S.A. are usually
overlain by mudstones. He infers that cracks in the mudstones at depths 1
to 2 km. are rapidly healed, and that rising magma has to spread along the
bedding planes.

Mountain formation may occur several times on about the same site, as
for the Appalachians and the Altai. It is possible that the explanation of
this may be related to that of the lowering of mountains. Below an old range
blanketing has delayed cooling, and under general cooling this is where the
fracture by tension below would start. Hence an old range is a place where
a geosyncline may develop and the range goes down, possibly below sea
level, and more sediments can be deposited. Then in readjustment a new
range may form alongside and overflow the geosyncline; at any rate it seems
possible that the two problems are really one problem.

A. E. J. Engel, C. G. Engel and R. G. Havens (1965) give evidence that
sub-oceanic material has been differentiated in the formation of island
basalts.

Henshaw and Zen (1965) argue that when sediments consolidate the
pressure in the interstitial water may reach that due to the mean density of
sediments + water. If there is a layer semipermeable to salts there will be an
osmotic pressure, which may make the difference and make the whole float.
They estimate that pressures of order 300×10^6 dynes/cm.^{2} are not out of
the question.

11·08. Non-isostatic vertical movements. Before the idea of isostasy was introduced into geophysics it was natural to suppose thick sediments to have been formed in deep troughs in the ocean bed. If later folding
lifted them up as mountain ranges the later history would be fairly simple;
the ranges would be denuded till they became hills and finally plains.
The great difficulty about this process is that sedimentation often went on
with only slight interruptions till depths of thousands of feet were reached,
without, apparently, any great change in the depth of the water. For instance, the Carboniferous limestones of the Howgill area (near Sedbergh,
Pennines), if we can judge from the ecology of present-day marine organisms, must have been deposited in clear water not more than 600 ft.
deep; and this depth must have been maintained while 3000 ft. of limestone

were formed. (I am indebted to Professor W. B. R. King for the data.) In the
Coal Measures of South Wales, about 8000 ft. of sandstones and shales were
formed, with occasional coal, indicating that during this period the surface
was sometimes above and sometimes below sea-level, but never more than
about 60 ft. from it in either direction. O. T. Jones (1938) estimated that in
the formation of the lower Palaeozoic in Wales the pre-Cambrian floor was
depressed to 35,000 ft. below sea-level; but there are no genuinely deep
ocean deposits. Some of the shales may have been deposited in about 1000 ft.
of water, but the depth fluctuated about a much smaller value. Such a
sequence of events cannot be represented by the filling up of a trough with its
bottom at a fixed level. The notion of isostasy suggested an explanation, for
it appeared that the weight of sediments themselves would depress the floor
and that the top could remain at a nearly constant level. But A. Morley
Davies (1918) pointed out that this explanation is inadequate. A correction
on one point was given by E. M. Anderson (1918).

If sediments of density 2·3 are deposited on land, the effect of outflow
of matter of density 3·3 below so as to restore isostasy would be to lower the
surface by 0·70 km. for each kilometre of material added, leaving the outer
surface raised by 0·30 km. Thus 3·3 km. of sediments could be deposited in
a broad valley 1 km. deep before it was filled up.

If sediments of density ρ_1 are deposited in water of density ρ_2, and k is
the depth of the sediments, x the depression of the original surface, and ρ_0
the density in the lower layer, the mass per unit area is increased by

$$\rho_1 k - (k-x)\rho_2 - \rho_0 x = (\rho_0 - \rho_2)(k-x) - (\rho_0 - \rho_1)k,$$

and the condition for compensation is that this shall vanish. Hence

$$k = \frac{\rho_0 - \rho_2}{\rho_0 - \rho_1}(k - x).$$

But $k - x$ is the reduction in the depth of the water, and therefore this
equation gives the depth of sediments capable of being deposited in water
of given depth. With our previous values, and $\rho_2 = 1$, the coefficient is 2·3;
thus 2·3 km. of sediments could be deposited in water originally of depth
1 km. This estimate would be too high if the deposition is over a region less
than 50 km. or so in width. The ratio is utterly unrepresentative of what
went on during the formation of the Carboniferous system of Britain, and
of many other systems. We need a ratio more like 10 or 30 to 1.

The ratios would be increased appreciably if we took for ρ_2 the density of
the intermediate layer, say 2·9, but still nowhere near enough. The geological evidence is clear that the conditions considered must have shown
considerable departures from isostasy; the bottom must have gone down
by almost the thickness of the sediments, and therefore the region was
lightened by something like the weight of a kilometre of rock for 3 km. of

sediments. Pushing down from the top will not explain it; it must also have been pulled down from within.

The greater part of the sediments of Britain were formed by streams flowing to the south-east and therefore from a continent situated in the present North Atlantic. The present depth in this region is about 2 km. Denudation can lower a region, but it cannot unaided bring it below sea-level.* This continent must also have been pulled down by some force from the interior.

While isostasy does not remove the difficulty about deep sediments, it creates one about the later history of mountains. We have already seen that about 3 km. of sediments have been removed from Alpine valleys by denudation since the original compression, which took place about 30 million years ago. If denudation had proceeded at this rate since the Carboniferous period, say 250 million years ago, 25 km. of material would have been removed. This is greater than the whole probable thickness of the granitic layer. The actual conditions in old mountain ranges are quite different. The mountains of Wales and Scotland, the Appalachians and the Urals may be taken as typical. Old rocks, including gneisses that may have been derived from the granitic layer, are indeed usually exposed in the centres of such ranges; but the extraordinary fact is that old sediments survive, even though they must have borne the full effect of the denudation. This seems to show that the average rate of denudation cannot have been more than a small fraction of that in Tertiary and post-Tertiary times, even when the regions compared are mountainous in both cases.

This argument says more than the one we have used in discussing the denudational methods of estimating the age of the crust, because in the latter we appealed to the existence of the Tertiary mountains to explain why the present rate of denudation may be several times the average of the past. We are now considering ranges that have been mountains much longer.

It seems unlikely that the climatic conditions can have varied greatly. The intensity of the atmospheric circulation is mostly determined by solar radiation, and if a range of given height is in the way of a current of moist air of given intensity the rainfall will probably be about the same. The inference to be drawn is that the real height of the old ranges, during most of the time since their formation, has been much less than that of the Alps, and that denudation has been small because the relief has been low. The original height, if we can judge by the degree of folding, was comparable with that of the Alps, but it seems likely that they were lowered to something approaching their present height in something under 100 million years from their formation.

A further argument, not depending on assumptions about the rate of denudation, is provided by the fact that some mountain ranges showing

* More accurately, below the lower limit of wave erosion, perhaps 20–40 m. below low tide.

great folding have been denuded so far as to be almost indistinguishable from plains. How can a mountain range be so much lowered? We have seen that if compensation takes place by adjustment in the lower layer, the removal of 3 km. of sediments only lowers the outer surface by about 0·8 km. Thus if a mountain range was originally 3 km. in average height, and the sediments composing it were 3 km. thick, it could not be lowered to the level of the surrounding country until first the 3 km. of sediments, and then 11 km. of granite, had been removed. Here again we have a result inconsistent with the presence of old sediments in the ranges in question.

The only way of reconciling these facts seems to be that denudation, accompanied by adjustment in the lower layer, is not the only cause of the lowering of mountains. If the granitic or the intermediate layer can move outwards under the excess of pressure due to the weight of the mountains, the outer surface can be lowered without heavy denudation. If such outflow reduced the thickness of the granitic layer by 10 km., the outer surface would be lowered by 2 km. If the intermediate layer became 20 km. thinner, a similar result would follow.

The normal temperature being nearer the melting-point in the intermediate layer than in any part of the granitic layer, we may naturally suppose that most of the outflow takes place in the intermediate layer. This would in any case have been weakened by thermal blanketing due to the sediments themselves, which would therefore favour such a process. But in view of the evidence that there were serious departures from uniformity of mass distribution before the formation of the mountains, it seems undesirable to insist too much on its maintenance afterwards.

The above considerations depend on the supposition that isostatic adjustment was able to keep pace with the surface changes, so that the departure from uniformity of mass distribution at any instant was inappreciable. There would actually be a time lag, as in all dissipative processes, but this, if anything, makes matters worse. If a mountain system was once compensated, denudation would lighten it, and if adjustment failed to keep pace the result would be negative gravity anomalies. Unfortunately, the isostatic and free-air anomalies in Wales and the Highlands are prevailingly positive.

11·081. Loading when a layer is subject to phase change under pressure. Let the density of the material at low pressures be ρ_1, and suppose that at a standard pressure it changes to a state of density ρ_2. Initially the outer surface and the interfaces are supposed flat. A thickness k of material of density ρ_0 is added at the surface. Horizontal movement is neglected. Let the depression of the original surface be x and let the interface between the two states be y below where it was. Then the conservation of mass gives

$$\rho_1(y-x)-\rho_2 y = 0,$$

and the condition of unchanged pressure at the interface gives

$$gk\rho_0 + g(y-x)\rho_1 = 0.$$

The solution is $\quad x = (\rho_2 - \rho_1)\rho_0 k/\rho_1\rho_2; \quad y = -\rho_0 k/\rho_2.$

The additional pressure at a given place in the lower region is simply $g\rho_0 k$.

The conditions suppose that the phase change under pressure is faster than the adjustments due to imperfections of elasticity. The outer surface is not much depressed; the interface *rises*. My attention was called to this point in a lecture by Sir Edward Bullard. It would have relevance to the effects of loading in displacing either the 20° discontinuity or the Mohorovičić discontinuity if the latter represents a phase change from some form of basalt to eclogite. The matter is developed further by J. F. Lovering (1958).

11·09. The hypothesis of finite viscosity. In the foregoing account I have proceeded on the hypothesis that the materials of the shell have a finite strength, and that they follow the laws of elasticity so long as the stress-difference does not reach that strength. The estimated stress-differences given in Chapter VI will then be regarded as least possible values of the strength. But so far as they may also be regarded as the stress-differences surviving after larger ones have been relieved, they may be close approximations to the actual strength. For in relief of stress, once the stress-differences are brought below the strength, there is no need for further adjustment.

There is however, a group of geophysical hypotheses that regards the materials as capable of flow under any stress-difference, however small; on these the stress-differences found are incidental to viscous flow. The elasticoviscous law will suffice to formulate these hypotheses. Further, so long as we are dealing with long-continued stresses, the elastic part of the strains will be small compared with the viscous part, and it will suffice to replace the elasticoviscous law by that of viscosity.

Given the viscous law, any excess load on the surface will produce outflow below, which will tend to remove the excess load. Two cases concern us. If the lower layer, which is deep, is viscous, such flow will account for the main features of isostasy. But we have also seen that some features of the later history of mountain ranges suggest that flow in the intermediate layer is important. It is therefore necessary to investigate also the effects of a finite viscosity in a layer whose thickness may be small compared with the horizontal scale of the loads.

11·091. Surface loading of a deep layer. For harmonic loading we can derive the solution from that of 6·041 by replacing μ by the viscosity η and re-interpreting w as the vertical velocity. If the surface elevation at any instant is $a(t)\cos\kappa x$, we have $\sigma = -\rho a \cos\kappa x$, and the vertical velocity is

$$\cos\kappa x \frac{da}{dt} = -\frac{g\rho a}{2\eta\kappa}\cos\kappa x. \tag{1}$$

Hence a varies with the time like $\exp(-gt/2\nu\kappa)$, where ν is the kinematic viscosity. We notice that the inequalities of greatest wave-length subside fastest.

Now consider an elevated strip such that at $t = 0$ the elevation ζ is 1 from $x = -l$ to $x = +l$ and zero elsewhere. Then initially

$$\zeta = \frac{1}{\pi}\int_0^\infty \{\sin\kappa(x+l) - \sin\kappa(x-l)\}\frac{d\kappa}{\kappa}, \tag{2}$$

and at any later time

$$\zeta = \frac{1}{\pi}\int_0^\infty \{\sin\kappa(x+l) - \sin\kappa(x-l)\}\exp\left(-\frac{gt}{2\nu\kappa}\right)\frac{d\kappa}{\kappa}. \tag{3}$$

This is expressible in terms of Bessel functions. We have (H. and B. S. Jeffreys, 1972, § 21·022)

$$\mathrm{Kh}_n(x) = \frac{1}{\pi}\int_0^\infty \exp\left\{-\tfrac{1}{2}x\left(\lambda + \frac{1}{\lambda}\right)\right\}\lambda^n\frac{d\lambda}{\lambda}, \tag{4}$$

whence with $a > 0$, b real,

$$\int_0^\infty e^{-i\kappa b}\, e^{-a/\kappa}\frac{d\kappa}{\kappa} = \int_0^\infty \exp\left\{-\tfrac{1}{2}x\left(\lambda + \frac{1}{\lambda}\right)\right\}\frac{d\lambda}{\lambda}$$

$$= \pi\mathrm{Kh}_0\{2\sqrt{(iab)}\}, \tag{5}$$

where $x = 2\sqrt{(iab)}$, $\lambda = \sqrt{(ib/a)}\,\kappa$. Hence (H. and B. S. Jeffreys, 1972, § 21·09)

$$\frac{1}{\pi}\int_0^\infty \sin\kappa b\, e^{-a/\kappa}\frac{d\kappa}{\kappa} = -\Im\mathrm{Kh}_0\{2\sqrt{(iab)}\}$$

$$= -\mathrm{khei}_0\, 2\sqrt{(ab)}. \tag{6}$$

Then
$$\zeta = \mathrm{khei}_0\sqrt{\frac{2gt(x-l)}{\nu}} - \mathrm{khei}_0\sqrt{\frac{2gt(x+l)}{\nu}}. \tag{7}$$

When $u \to 0$ through positive values, $\mathrm{khei}_0(u) \to -\tfrac{1}{2}$. It tends to zero rapidly as u increases, vanishing first at $u = 3\cdot9$, and never afterwards exceeds $0\cdot007$. For negative u we have

$$\mathrm{khei}_0(u) = -\mathrm{khei}_0(-u).$$

Hence as t increases, $x - l$ remaining small, the first term in (7) tends to $-\tfrac{1}{2}$ for x slightly more than l and to $+\tfrac{1}{2}$ for x slightly less than l. The discontinuity of unity assumed at $t = 0$ persists for all time. The successive forms of the surface for increasing t will be of the types shown in Fig. 27 (which is not to scale).

A possible criticism of this solution is that it assumes no shear stress over the upper surface. With a thin strong upper layer this condition should be replaced by that of no slip at the interface. But we saw in 6·041 that the solution actually gives no horizontal displacement at the interface, so that the two different boundary conditions actually lead to the same result.

Applications of these principles have been made by N. A. Haskell (1935,

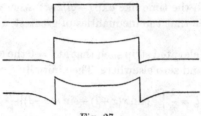

Fig. 27.

1936, 1937), E. Niskanen (1939) and Meinesz. The loading of the glaciated regions of Scandinavia and North America would be expected to have led to outflow; thus the removal of the ice would leave them underloaded, and inflow and consequent rise of the land surface would follow. Now these regions are actually rising, and detailed work has been done on the rate of rise. Haskell used data from Nansen, and assumed an initial rate of uplift proportional to exp $(-b^2r^2)$, r being the distance from the centre of the glaciated area. He inferred that ν is about 3×10^{21} cm.²/sec. Niskanen, from more numerous data, got $1 \cdot 1 \times 10^{22}$ cm.²/sec.

11·092. Outflow of an intermediate layer. Consider the case of a strong upper layer resting on a viscous intermediate layer, and this on a lower layer with a sufficiently low viscosity for us to treat it as hydrostatic. The reason for taking these conditions is that we wish to know the consequences of the suggestion that the lowering of mountains is due to outflow in the intermediate layer, the lower layer flowing freely enough to keep pace with all changes above it. Then if u is the horizontal velocity, $u = 0$ at the top of the intermediate layer and $\partial u/\partial z = 0$ at the bottom. The equation of motion is

$$\nu \frac{\partial^2 u}{\partial z^2} = \frac{1}{\rho} \frac{\partial p}{\partial x}. \tag{8}$$

The elevation of the surface is ζ, and the disturbance of pressure due to it is $g\rho\zeta$. The boundaries are then $z = -\zeta$ and $z = -\zeta + H$. Then

$$\frac{\partial^2 u}{\partial z^2} = \frac{g}{\nu} \frac{\partial \zeta}{\partial x}, \tag{9}$$

which is independent of z. Hence to the first order

$$u = -\tfrac{1}{2}z(2H - z)\frac{g}{\nu}\frac{\partial \zeta}{\partial x}, \tag{10}$$

and the rate of outflow per unit width in the y direction is, to the first order,

$$\int_0^H u\,dz = -\frac{1}{3}\frac{gH^3}{\nu}\frac{\partial \zeta}{\partial x}. \tag{11}$$

But

$$\frac{\partial \zeta}{\partial t} = -\frac{\partial}{\partial x}\int_0^H u\,dz = \frac{1}{3}\frac{gH^3}{\nu}\frac{\partial^2 \zeta}{\partial x^2}. \tag{12}$$

PLATE VII

(a) Contortion with fracture, near Leek, Staffordshire.

(b) Continuous distortion and fracture combined. Den's Door, Broadhaven, Pembrokeshire. This is probably a case where the stress was enough to produce fracture after a time but not at once, flow occurring in the intermediate interval.
(*Crown copyright reserved. Geological Survey photograph reproduced by permission of the Controller, H.M. Stationery Office*)

(c) Folding in shale due to hill-creep. Matlock, Derbyshire.

PLATE VIII

Kinlochewe thrust plane→
Cambrian quartzite→

Torridonian→

←Lewisian
←Thrust plane

(a)

Cambrian infold   Torridonian infold
↑   ↑

Thrust plane→

←Thrust plane

(b)

Panoramic view, Craig Roy, S.E. end of Loch Maree. The
normal order of succession, proceeding upward, is Lewisian
gneiss, Torridonian sandstone, and Cambrian quartzite.
Here a displacement along the Kinlochewe thrust plane
has brought Lewisian rocks, with infolded Cambrian and
Torridonian, above the Torridonian and Cambrian. The
rocks above the thrust plane constitute the nappe.

(By permission of H.M. Geological Survey)

Thus any original inequality spreads according to the equation of heat conduction.

In particular, if, at $t = 0$, $\zeta = a \cos \kappa x$,

$$\zeta = a \exp\left(-\frac{1}{3}\frac{gH^3\kappa^2 t}{\nu}\right) \cos \kappa x, \qquad (13)$$

and decay is most rapid for the shorter wave-lengths.

If, at $t = 0$, $\zeta = 1$ when $-l < x < l$, and otherwise is 0, then at later times

$$\zeta = \mathrm{erf}\left\{\left(\frac{3\nu}{4gH^3t}\right)^{\frac{1}{2}}(x+l)\right\} - \mathrm{erf}\left\{\left(\frac{3\nu}{4gH^3t}\right)^{\frac{1}{2}}(x-l)\right\}. \qquad (14)$$

Successive stages in the decay are illustrated as in Fig. 28.

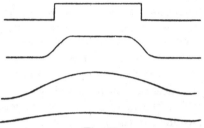

Fig. 28.

The actual geophysical problem contains several further complications, which have been neglected in order to display the contrast between the two methods of decay in the most striking way. Terms of the second order have been neglected in both solutions, and so long as this is justifiable they are independent of the vertical scale. In the first problem it is not correct in the boundary conditions close to the discontinuities, which will presumably be smoothed off over distances comparable with the original height. But so long as the elevated region is broad we see that for flow in a deep layer the strata will sag down in the middle; in a thin intermediate layer their highest point will be in the middle.

To get an order of magnitude for the viscosity of the intermediate layer we take $2l$ for the width of a mountain range, say 10^7 cm., and t between 10^7 and 3×10^8 years, that is, 3×10^{14} sec. $< t < 10^{16}$ sec. Then according to (14) considerable change would occur in time t at the centre of the range if

$$1{\cdot}3 \times 10^{23} < \nu < 4 \times 10^{24}\ \mathrm{cm.}^2/\mathrm{sec.}$$

On the theory of the lowering of mountains by outflow in the intermediate layer, $t = 30$ million years would be reasonable, and $\nu = 4 \times 10^{23}$ cm.²/sec. With this viscosity in the lower layer an uncompensated range would become approximately compensated in about 6×10^{14} sec. $= 20$ million years; but with the lower viscosity indicated in the discussion of the Fenno-Scandian uplift the flow in the lower layer would be able to keep pace with the changes above it, and so far the assumption that the lower layer

can be treated as hydrostatic in treating the viscosity of the intermediate
one is consistent.

11·093. Convection theories. A necessary condition for equilibrium
in a viscous fluid is that the density shall be uniform over the level surfaces.
There is strong reason to suppose that the temperature is not uniform over
such surfaces, and the differences of temperature will give rise to convection
currents. These in turn may be of two types. If the system as a whole is
stable, small variations of supply of heat will give rise to slow rising currents
where the temperature is high and sinking ones where it is low, the distribution being determined by that of the supply of heat. This condition could
arise, for instance, if the material is not uniform in composition, so that the
undisturbed density increases with depth. But a fluid cooling from the top
is usually unstable. In the elementary account usually given it is argued
that if a layer of denser fluid rests on one of lighter fluid, the potential
energy would be decreased if the two were interchanged, and therefore the
system is unstable. The argument is not quite complete, however, because
a proof of instability requires that there should be a kinematically possible
motion that would effect the interchange. The layers cannot be interchanged by simply pushing one through the other—this would require an
amount of energy to effect the compression that would far exceed the
energy available from the interchange. The only possible motions are such
that the lower fluid rises in some places and the upper sinks in others. But
then the relative motion is resisted by viscosity, and conduction between the
cool sinking currents and the warm rising ones tends to reduce the differences
of temperature that maintain them; thus the currents might be damped
down and the distribution kept stable.

The main fact that such distributions are usually unstable is familiar
enough. Explicit recognition that they are not always so is due to H. Bénard
(1901), and other examples are given by A. R. Low (1929) and by D. Brunt
(1925). In a shallow layer of fluid heated below, currents do not arise until the
vertical temperature gradient reaches a critical value. When this is passed,
a sort of honeycomb structure is developed. The surface is marked out into
hexagons or irregular pentagons, the fluid rising at the edges and sinking
in the centres, while slow currents pass from the edges to the centres at the
top. This condition in turn breaks down if the heating at the bottom is too
great; the pattern then disappears and is replaced by an irregular system of
eddy currents as in the boiling of a kettle. It is this last stage that is represented in the most familiar cases of fluids heated at the bottom or cooling
from the top. The theory of the initiation of the cellular convection has been
developed by various authors (Rayleigh, 1916; Jeffreys, 1926*d*, 1928*d*;
A. R. Low, 1929; D. G. Christopherson, 1940; A. Pellew and R. V. Southwell, 1940; H. and B. S. Jeffreys, 1972, §14·062; S. Chandrasekhar, 1952;
Jeffreys, 1956*a*). Workers in hydrodynamics usually refer to the prob-

lem as the 'porridge problem'. Porridge cooked in a single saucepan, if not properly stirred, can char at the bottom when below boiling-point at the top; viscosity can keep it stable with a temperature difference of some hundreds of degrees within a few centimetres.

If the disturbance of temperature contains a factor $\cos(lx + my)$, the result is that for any value of $l^2 + m^2$ there is a critical value of the excess β of the lapse rate of temperature (the vertical temperature gradient with the sign reversed) over the adiabatic, such that disturbances with this value will grow. There is one value that makes this a minimum. For $m = 0$ the distance between the rising and sinking currents is nearly equal to the depth. The criterion is that

$$\frac{g\alpha\beta H^4}{\kappa\nu} > \lambda,$$

where g is gravity, H the depth, κ the coefficient of thermometric conductivity, and ν the kinematic viscosity. λ is a constant depending on the boundary conditions. In the problems that have been solved it is of order 1000. If β exceeds the value given by this equation there will be a range of values of $l^2 + m^2$ such that disturbances can grow. The surprisingly large value of λ is connected with the fact that the problem depends on a differential equation of the sixth order. $6! = 720$.

If the supply of heat is such that the lapse rate for pure conduction would exceed the critical value, the vertical currents will carry heat up at a rate $\rho c w \, \partial\vartheta/\partial z$ per unit area, where β is the density, c the specific heat, w the vertical velocity, and ϑ the disturbance of temperature. The theory in its original form is a first-order one, but this second-order effect can be calculated from the first-order theory. The principle was introduced by Hales (1935), and is a contribution to pure hydrodynamics.

If there is a general current in one direction, produced, for instance, by tilting the vessel, the hexagons are immediately replaced by strips parallel to the current. This is due to the fact that for other modes the current systematically brings cold fluid below warm, and hence reduces the differences of the mean temperatures in vertical columns.

The matter is taken much further by P. H. Roberts and K. Stewartson in Donnelly, Herman and Pricogine (1965).

For a horizontal layer of fluid, according to the first-order theory, all disturbances containing a factor $\cos lx \cos my$ can arise first for the same value of β so long as $l^2 + m^2$ is the same. Thus a disturbance with $l = m = p$ could arise equally readily with $l = p\sqrt{2}$, $m = 0$. If there is a superposed general current the latter is the type that actually appears. For originally stationary liquid the disturbance that actually arises is in hexagonal cells except close to the boundary. Pellew and Southwell point out that this can be represented as a superposition of three ridge patterns inclined at 120°, of the form

$$\cos\kappa x + \cos\tfrac{1}{2}\kappa(x + \sqrt{3}y) + \cos\tfrac{1}{2}\kappa(x - \sqrt{3}y).$$

The preference for the hexagonal pattern is presumably to be attributed to the neglected terms of the second order. These have not been worked out, but the presumption is that they make for the greatest symmetry in a space-filling polygonal form. There is reason to suppose that the roll pattern ($m = 0$) is the more stable at small heat flows and changes to hexagons at somewhat larger ones.

In a circular vessel none of these forms will satisfy the conditions at the edge. However the argument can be stated in terms of the condition that the supply of energy through the rising of the lighter and sinking of the heavier material shall balance the dissipation by viscosity. (This is obscured by the use, as is now customary, of the vertical velocity instead of the disturbance of temperature as the fundamental variable. The change of variable transfers the viscosity into the term depending on the generation of energy. See Jeffreys (1956 a).) The contributions to these are proportional to the area, edge effects to the perimeter, and hence, when the diameter is large compared with the depth, the only effect of the edge will be a distortion of the pattern near it. This is observed.

In problems relating to a sphere we similarly find that in the first-order theory all the $2n + 1$ harmonics of degree n can appear at once. If we suppose that the actual one will have the greatest possible symmetry, all of degree 1 have the same sign all over a hemisphere. For higher n it is easiest to think in terms of the corresponding polynomials. For $n = 2$, the zx, zy ones will not arise as they give only a change of the the the polar axis, $x^2 - y^2$ and xy keep the same sign for up to $180°$, and addition of a term in $2z^2 - x^2 - y^2$ will increase the difference between poles and some points on the equator. Hence $\sin^2\theta(\cos 2\lambda, \sin 2\lambda)$ gives as much symmetry as is possible. For $n = 3, 4$ and 5 the best solutions will have the symmetries of the regular tetrahedron, cube or octohedron, and dodecahedron or icosahedron. The easiest way of getting the corresponding polynomials is to consider equal particles at the vertices and work out the potential.

For $n = 3$ the vertices can be at

$$(0, 0, a), \quad \left(0, \frac{2\sqrt{2}}{3}a, -\tfrac{1}{3}a\right), \quad \left(\sqrt{\tfrac{2}{3}}a, -\frac{\sqrt{2}}{3}a, -\tfrac{1}{3}a\right), \quad \left(-\sqrt{\tfrac{2}{3}}a, -\frac{\sqrt{2}}{3}a, -\tfrac{1}{3}a\right).$$

The potential contains the third harmonic

$$\tfrac{4}{9}(2z^2 - 3x^2 - 3y^2)z + \sqrt{2}(y^3 - 3x^2y)$$
$$= \tfrac{8}{9}r^3\left\{(\tfrac{5}{2}\cos^3\theta - \tfrac{3}{2}\cos\theta) - \frac{1}{\sqrt{2}}\sin^3\theta\sin 3\lambda\right\}.$$

For $n = 4$ cubic symmetry is given by the fourth harmonic

$$x^4 + y^4 + z^4 - \tfrac{3}{5}(x^2 + y^2 + z^2)^2$$
$$= \tfrac{1}{20}r^4(35\cos^4\theta - 30\cos^2\theta + 3) + \tfrac{1}{4}r^4\sin^4\theta\cos 4\lambda.$$

For $n = 5$ there will be harmonics of degree 5.

All these harmonics retain their symmetry under rotation, so that they generate a double infinity for $n = 1$ and a triple one for the others.

There can be no such symmetry for $n > 5$, for it would imply impossible regular solids. But there are semiregular solids with regular faces of two forms, one of them having 12 pentagons and 80 triangles. Two of them exist in mirror-image forms. For a larger number of faces the best we can do seems to be related to a problem studied by Coxeter (1962). How many non-overlapping circles of given radius can be put on a sphere? Their centres will specify a pattern, but joining them up will yield faces of different forms.

Allowance for rotation is made by Roberts (1968). It is found that except for the slowest rotations the easiest mode to excite is asymmetrical about the axis of rotation. In rapid rotation the most easily excited mode is small everywhere except near a cylinder, of radius about half that of the sphere, and coaxial with the axis of rotation.

Roberts's work may help to complete the theory of Bullard and Elsasser on the maintenance of the Earth's magnetic field. It is shown that the dipole field can be produced as in a dynamo by motions in the core; and Bullard has shown that the radioactivity needed to produce such motions need not be enough to upset the general theory of the thermal state of the shell. To complete the theory it is necessary to show first that suitable currents can be produced; and also that, given a small initial field, it can grow. Either part is difficult, but a complete theory would, apparently, require them both to be treated together.

The acceptance of a finite viscosity in the lower layer leads to the various convection theories. Of these I shall discuss those of C. L. Pekeris (1935) and A. L. Hales (1935) as representative of the main types.

In the theory of Pekeris the currents are supposed to arise from deep-seated differences of temperature due to inequalities of distribution of radioactivity of continental extent. Any such variation of temperature over level surfaces would produce currents in a medium of finite viscosity. It is thus a forced motion in a stable system. The phenomena are very similar to those of meteorology and of ocean currents, except that on account of the long time-scale and the high viscosity it is possible to neglect the inertia terms. There would be rising currents in the hotter regions and sinking ones in the cooler ones, with horizontal ones to maintain continuity. Eight cases are worked out in detail. With plausible values for the temperature differences and a kinematic viscosity of 3×10^{21} cm.^{2}/sec., velocities of the order of 1 cm./year would arise, and there would be shearing stresses of the order of 10^7 dynes/cm.^{2} on the under side of the crust. A particle would complete its cycle in something of the order of 10^9 or 10^{10} years. The normal stresses would be of order 10^8 dynes/cm.^{2}, and radial displacements of the order of 2 km. would arise at the outer surface,* but would be nearly com-

---

* Note that in Tables III and IV of the paper u is the radial displacement as defined in (40), not the velocity as in (1).

pensated, the corresponding gravity anomalies being of order 20 mgal. Elevation is associated with positive gravity anomalies. The stresses and elevations are nearly independent of the adopted viscosity, while the velocities are inversely proportional to it. The stresses and elevations for given inequalities of temperature are less than for an elastic solid with the same temperature distribution. This was to be expected because the elevation of the surface due directly to expansion in the elastic theory is partly removed by outflow on a viscous theory. What was not expected was that the reduction would be only by a factor of 10 or so.

An important feature is that the permanent difference of temperature between the poles and the equator would penetrate to great depths while the Earth was fluid and produce a general circulation, the surface currents being away from the equator. If the difference of temperature was about 60°, as at present, the tangential stress would reach about 10^8 dynes/cm.², irrespective of the adopted viscosity. The actual value would probably be a good deal less, since the absolute temperature would be about five times what it is now and therefore, by Stefan's law, the outgoing radiation would be some hundreds of times as much. To simplify matters very much, let the present mean temperature be 300° A. and the rate of radiation per unit surface σT^4. Suppose that the polar temperature was once 1500° A.; then the rate of radiation at the poles would be $\sigma(1500^4)$. At the equator the temperature T_e would be sufficiently higher to dispose of the solar radiation as well, so that, if we neglect all horizontal transfer of heat,

$$\sigma T_e^4 = \sigma(1500^4) + 2\sigma(300)^4,$$

whence $T_e - 1500°$ would be of order 1°. Thus it is likely that the stress arising as a residual effect of solar radiation in the early stages would be of order 10^6 dynes/cm.². The factor 2 is needed because the average receipt of radiation per unit area at the equator is about twice that for the whole surface.

In Hales's theory heating is supposed uniform but enough in amount to make the spherically symmetrical state unstable. It is found that the conditions are satisfied if ν is about 10^{22} or 10^{23} dynes/cm.². Surface elevations are not worked out. A particle would complete its revolution in about 10^{10} years.

The two theories rest on completely different principles and data, and it is remarkable that the results are so similar.

11·094. Criticism of the finite viscosity theories. The nearest approach to positive evidence for a finite viscosity in the lower layer is that provided by the Fenno-Scandian uplift. But this interpretation, if it is to be generalized to the Earth as a whole, invites further check; otherwise, it may be only a phenomenon of the particular region, even though this is a large one. If it is true in general, it will follow that every region of positive gravity anomalies is sinking, and every region of negative ones rising. In any mountain system undergoing denudation, either compensation should keep pace with denudation and there would be no systematic isostatic gravity anomalies, or it would fail to keep pace and the gravity anomalies would be negative. Where loads have been added recently, there should similarly be either no gravity anomalies or systematically positive ones. At every point these consequences are contrary to the facts. Reference to the map of gravity anomalies shows that, unless there is some serious systematic error in the gravity observations, Fenno-Scandia itself is interior to a region of prevailing positive anomalies and therefore ought to be sinking.

Even within Fenno-Scandia the rate of rise is far from being closely correlated with the gravity anomalies (Jeffreys, 1940c, 1975b). There is a similar rise (about 2·6 ft./century) in the neighbourhood of Lake Superior, for which a similar explanation is offered, but an awkward circumstance given by W. A. Johnston (1939) is that carvings made from 1741 to 1753 near the north of Hudson Bay, well inside where the ice-sheet was, show no definite sign of a rise. If there is any, it was less than 0·4 ft. between 1893 and 1934, and certainly much less than 5 ft. in 200 years. The rise is certainly less than in the Lake Superior region, which was near the edge of the ice-sheet. Judged by the values of the anomalies of gravity of large horizontal extent over the globe, neither those of Fenno-Scandia nor of Canada are exceptional; but larger anomalies, such as those in India, have not been found associated with systematic vertical movements.

The outstanding objection is based on the use of a finite viscosity in the lower layer to account for the compensation of mountain systems. Assuming that a gravity anomaly 100 km. in horizontal extent has had time to be reduced to half its original amount, which many would regard as a conservative estimate, it would follow from 11·091 (1) that one of 3000 km. extent would have been reduced in the same time to $2^{-30} = 10^{-9}$ of its original amount. Since such anomalies exist we should have to regard them as either very recently produced or of enormous original amount. Conversely, if we admit that flow has produced only a moderate reduction in the widespread inequalities, it ceases to be available as an explanation of the compensation of mountain systems, which is where, if anywhere, there is positive evidence for weakness in the lower layer.

Cyprus is a region of strong positive gravity anomalies (Bullard and Mace, 1939) but is not sinking, and has risen since the Pleistocene. Other parts of the Mediterranean region show similar anomalies, notably Sicily; some have risen in historic times, others sunk, while some have done first one and then the other. A famous example, illustrated in several geological textbooks, is the temple of Serapis at Pozzuoli, Italy. Marks of marine organisms are visible on the columns several metres above the ground, showing that in historic times the land has been depressed below the sea by several metres and elevated again.

The test from regions of heavy deposition was suggested by Barrell, who expected that large positive anomalies would exist over the Nile and Niger deltas. Detailed work has not been done on these; but Burrard (1918) and Bowie (1924) found slightly negative mean isostatic anomalies over the Ganges and Mississippi deltas.

As to regions where denudation has been long-continued and great, there are systematically positive anomalies over the Welsh mountains and the Highlands of Scotland. These objections would apply equally to the hypothesis that compensation proceeds by viscous flow in the intermediate layer. This hypothesis also does not agree well with the fact that the largest systematic anomalies of gravity anywhere are those of the Meinesz strips, since it gives most rapid decay for anomalies of short wave-length. In fact, with a viscosity small enough to permit appreciable flow in either the intermediate or the lower layer we should expect either the very short or the very long wave-lengths to have been removed; whereas observation shows that the respective inequalities are of comparable magnitudes, and so are those of a whole range of intermediate wave-lengths.

We have already discussed the peculiar phenomena of long-continued deposition without important change in the depth of the water; in these the bottom has sunk by more than the surface loading and the appropriate flow will explain.

The suggestion of systematic convection currents complicates the question; and it is noteworthy that the shear stresses on the base of the crust, acting in the same direction over long distances, would suffice to break the crust, and this is proposed by Holmes as an explanation of mountain formation. But, given the heat supply and the viscosity assumed, the currents could never stop. Thus the convection theories show no sign of an explanation of the intermittence of mountain formation; while the difficulties about long-continued deposition, positive gravity anomalies in regions of long denudation, and negative ones in regions of heavy deposition, and vertical movements in the directions opposite to what the loading would suggest, remain as great as ever.

I am not asserting that rocks behave as perfectly elastic even under small stresses; it appears that under any stress some elastic afterworking and hysteresis occur. What I say is that (1) at numerous points the facts are contrary to what we should expect if viscous flow was of dominating importance; (2) they are at no point contrary to what we should expect if the rocks at great depths have a non-zero strength and flow is negligible unless the stress-differences exceed the strength; and (3) the latter hypothesis leads directly to explanations of many of the outstanding facts.

The forms of the convection theory developed by Meinesz (1933, 1947, 1948) and Griggs (1939) verbally accept the notion of a non-zero strength but use the mathematics of finite viscosity. Meinesz considered his form consistent with thermal contraction. They are therefore intermediate in character between the theory developed here and the convection theories just described. I think, however, that the introduction of a non-zero strength will completely alter the solution. For instance, let us take the problem discussed in 6·042 and consider what will happen if the viscosity becomes finite only when the stress-difference reaches a certain value. In the corresponding elastic problem the greatest stress-differences are over a half-cylinder. The displacements will be purely elastic till the stress-differences reach the critical value, and flow will then begin. It will evidently lead to the squeezing out of the plastic zone, so that this will be pushed out on to the surface near the margin of the loaded region. There will be no plastic flow at all at great distances.

It should be noticed that if any part of the vertical variation of density is due to differences of composition, convection will be totally stopped unless the temperature gradient is increased sufficiently to cancel this part also. If viscosity, instead of following the elasticoviscous law, follows Lomnitz's law or my modification, thermal instability cannot occur (Jeffreys, 1958 d). Any disturbance of the hydrostatic state would be damped out. With these laws the stresses are still mainly determined by the variation in level of the isotherms, and if the law gives a rate of flow under constant stress tending to zero with time, the supply of heat by inflow to the ascending currents will tend to zero, and will not be able to balance the loss by conduction. Thus if convection ever started it would die out. The evidence for such laws is now overwhelming.

Advocates of convection seem never to mention the criterion on p. 455 I can suppose only that they include any thermally produced movements as convection; in particular fracture could be included. They never state their laws explicitly. But, as shown in the mention of Meinesz's and Griggs's forms, there is every reason to suppose that the results for other forms would be totally different from what they claim to explain.

When all the evidence is taken together it shows that the hypothesis of viscous flow, always tending to produce exact isostasy, is false and contradicts both the geological and the geophysical evidence. Uplift and depression of the solid surface take place, but not in accordance with this hypothesis. Elastic deformation by surface loading is not an effective substitute, because it would always be less than the isostatic deformation, whereas the actual deformation is often greater. In fact, the idea of the lower layer as yielding passively to stresses imposed from outside must be abandoned. Some of the phenomena at least must be attributed to changes taking place spontaneously in the lower layer, the outer layers adapting themselves to the lower instead of conversely. This gives an immediate explanation of the formation of deep sediments. If the ocean bottom near the coast sank slowly of its own accord the resulting depression would be progressively filled up by sediments, so that new sediments would always be deposited in shallow water. The sediments would cancel most of the negative gravity anomaly produced by the depression of the bottom, and isostasy would remain a very rough first approximation to the truth. Bowie maintained something similar. O. T. Jones has described such regions as having a 'sinking feeling'. The existence of the great ocean deeps, which we saw to be beyond the range of consistency with detailed isostasy, is in agreement with a theory of spontaneous depression of the ocean floor. The difference between an ocean deep and a geosyncline is that the latter is formed near enough to land to become filled with sediments, the former is not.

For these reasons it is urgent that the common use of 'equilibrium' to mean that some isostatic hypothesis is in approximate agreement with the gravity anomalies in a region should be abandoned. 'Equilibrium' has a definite meaning in mechanics, and every problem considered in Chapter VI is one of equilibrium in the proper sense of the word. The usages are habitually mixed so as to infer that regions of negative gravity anomalies are always rising and regions of positive anomalies sinking, and this is simply false. The use of the term 'isostatic movement' as if there were no vertical movements except those that tend to establish isostasy is equally objectionable and has been out of date for about 50 years. It distracts attention from the important matter of understanding subcrustal spontaneous stresses. Non-isostatic vertical movements are not fully understood, but it seems probable that many of them are due to differences of cooling.

11·10. Unsymmetrical contraction. The theory of the cooling of the Earth was naturally worked out first on the hypothesis of symmetry, but on account of the probable differences of radioactivity between different regions this hypothesis can only be regarded as provisional. The actual amounts of the differences are still unknown, but it is not unlikely that the basal temperature S_0 below the oceans is not more than half what it is below the continents. Such asymmetrical cooling would lead to a corresponding stress-system and the Earth would be pulled out of shape. The formal solution is of the same type as that of straining under bodily force (Jeffreys, 1932c). When the differences of cooling are confined to a depth small compared with the radius the result is practically that the whole of the volume contraction is pushed into the radius, so that the differences of elevation produced are nearly three times what would be expected from the linear contraction. On account of changes of density, on the other hand, the disturbances of the gravitational field would not be more than a tenth of those due to the surface inequality alone. Consequently unsymmetrical cooling would tend to give spontaneous changes of level very similar to those inferred in the last section. The data do not permit a calculation of the total amount, but it is possible that differences of level of 10 km. might arise in this way.

The solution treats the problem as one of pure elasticity, and the strains indicated are far more than any known rock could stand. Thus, as in the case of symmetrical cooling, there would be stages when the deep rocks yielded by fracture or flow and the stresses were mainly relieved. Thus there would probably be major revolutions when the surface inequalities were subsiding, separated by intervals when the inequalities were gradually growing, but punctuated by minor revolutions involving probably deep-focus earthquakes without much vertical displacement.

The possibility of asymmetry in the m term in the original temperature leads to little additional difficulty in the formal solution, but the data for a quantitative comparison are completely lacking.

There is strong evidence (Biswas 1969) that the oceans have become deeper during geological times. If the cooling has been greater below them this would be expected.

Whatever the explanation may be, it must be consistent with the fact that inequalities of height and gravity anomalies under the oceans are of the same order as in the continents. The inference from these that the strengths are also comparable is a direct consequence of the equations of equilibrium of a solid, and is the most directly relevant information available.

11·101. Island arcs and the growth of continents. Off the coast of Alaska and Asia there are several chains of islands; the Aleutians, Kuriles, Japan, and the Lu-Chu Islands. P. Lake (1931 *a, b*) noticed that these lie approximately on circles, and suggested that they were formed by fracture on planes inclined to the surface. Gutenberg and Richter, however, have found that shallow and deep earthquakes taken together appear to lie on steeply inclined surfaces. Coulomb (1945), Byerly and Benioff find that on the whole the first motion in an earthquake on such a surface is in opposite directions according as the earthquake is deep or shallow. J. Tuzo Wilson (1950 *a, b*, 1951) and Scheidegger and Wilson (1950) associate this with the fact that the contraction theory leads to excess horizontal tension below the level of no strain and to compression above it. They therefore replace Lake's plane fractures by cones. They trace arcuate formations also in the great mountain systems. Their work goes into much detail and cannot be adequately described here. A full account is in Section 4 of Kuiper's *The Solar System*, vol. 2 (1954). One important feature is that they (with P. M. Hurley) trace a progressive diminution of age of the rocks of North America as we recede from the oldest rocks of the Lake Superior region; it looks as if North America grew outwards by successive elevations, and they regard this as mainly due to the formation of successive systems of arcs around the margins.

Wilson (1957) points out that the present rate of production of lava would in the existence of the Earth have been enough to produce the whole of the matter above the Mohorovičić discontinuity. Incidentally the resulting compression of the outer surface would give a substantial increase of that available for mountain formation.

Muehlberger, Denison and Lidiak (1967), Goldich, Muehlberger, Lidiak and Hedge (1966) give an extensive discussion of basement rocks in North America. The ages have mostly been determined by the K/A and Rb/Sr ratios. It appears that at least half the area was already a continent 2500 million years ago, and that there were extensive superposed igneous out-flows, mainly granitic, at ages down to about 1100 million years. On the whole the ages decrease from north to south. This may be regarded as con-firmation that the differentiation that formed the upper layer was still going on 1000 million years ago, and that the cooling of the Earth began about then. See also P. M. Hurley (1968).

11·11. Source of earthquake energy. Gutenberg and Richter (1949, Table 8, p. 23) estimate the total release of energy in earthquakes as about 10^{26} ergs/year. We compare this with the energies likely to be available from accumulated strain and from gravity. The crustal shortening in a period of mountain formation is of order 50 km. on 40,000 km., and Young's modulus is about $1·7 \times 10^{12}$ dynes/cm.^{2}. This makes the energy per unit volume about 10^6 ergs/cm.^{3}. If this is uniform over the Earth and to a depth of 300 km. the total energy is about 10^{32} ergs. The last great period of mountain forma-tion was about 3×10^7 years ago; if we take the average interval as 10^8 years the energy available is of order 10^{24} ergs/year and far less than Gutenberg and Richter's estimate. It has been claimed on, I think, grounds similar to this that a contraction theory is inadequate.

However, I think a serious modification is needed. In the contraction theory the crustal shortening does not give the energy available to produce mountains; it absorbs energy and does not produce it. The relative exten-sions at great depths are likely to be 10 times the shortening that remains, and a factor 100 might be applied. Actually this is too large because an extension of 10^{-2} could never be built up, and in the time needed to produce it a hundredth of the energy would be released ten times. The process would begin with deep shocks followed by a succession approaching the surface, as in the theory of Wilson and Scheidegger.

The average interval between great periods of mountain formation may well be 3×10^8 years instead of 10^8 as used above. But the time since the last period is more like 3×10^7 years, and the present may be unusually active. We have already had a warning in relation to estimates of the age of the Earth from denudation! At any rate the discrepancy may be by a factor of 3 instead of 100 and is much less serious than it appeared.

Another mechanism requires examination. I have suggested (1947b) that many normal earthquakes (all of those used in constructing the 1954 travel times for European earthquakes) are in the sedimentary layer and are due to stresses produced in the carving out of deep valleys. Suppose that the range of height is 4 km., over a region 100×1000 km. Then the gravitational energy is of order

$$\tfrac{1}{2} g \rho h^2 (\text{area}) = 24 \times 10^{28} \text{ ergs}.$$

This would supply 8×10^{21} ergs/year for 3×10^7 years. It seems unlikely therefore that fracture due to gravity can account for much of the total output; but it may be reasonable for earthquakes in mountainous regions.

11·12. The origin of continents and ocean basins. From the petrological difference between the oceans and the continents, which is well confirmed by the distribution of gravity, and to some extent by the study of surface waves, we must regard the formation of the continents as a petrological problem. We have to explain why the granite became concentrated in about a third of the surface. It clearly could not occur while the Earth was thoroughly fluid, or since the greater part of it became solid. Thus we are limited to events during the process of solidification. We seem to be restricted to (1) a failure or partial failure of magmatic differentiation in the oceanic regions, so that the granite has crystallized out in the continents but remained dissolved under the oceans; or (2) an actual separation during solidification, so that the granite tended to collect together in patches.

The first suggestion does not obviously explain the difference of density; dissolved quartz and orthoclase, whatever form they took in the solution, would reduce the density of the magma, and the general effect on the level surfaces would probably be almost the same as if these constituents separated out on the top. The radioactive elements would be spread through a greater depth and the basal temperature would be higher below the oceans. This does not accord with the fact that the deepest earthquake foci are below the oceans. On the whole, the suggestion seems unlikely.

The second suggestion requires horizontal transfer during solidification. It was made by G. F. S. Hills (1934, 1937, 1947; cf. Jeffreys, 1934). It depends essentially on cellular convection adapted to a sphere. Hills made a comparison with the formation of scum in the boiling of jam, which is swept to the sides by the convection current spreading from the centre. He suggested that in the solidification of the Earth the granitic particles were collected into separate masses in this way. The crucial question is whether such a mechanism could lead to the kind of asymmetry that is shown by the existence of the land and water hemispheres, no previous explanation of this having been found satisfactory. A case was examined by Miss M. E. M. Bland and me (Jeffreys and Bland, 1951). We adapted the theory for a horizontal layer of fluid to a sphere. We took the viscosity uniform and the original differences of density small. We found that for given viscosity, with the temperature gradient gradually increasing, the first mode to be excited would be such that the disturbance of temperature over a sphere is proportional to a spherical harmonic of degree 1. In this state the currents would sweep through the centre and return over the outside, which is just what is needed. Conversely, this mode of disturbance would be the last to persist in a body whose viscosity was gradually increasing. The distribution of currents indicated in this solution suggested that in the actual Earth the presence of a dense core would seriously obstruct the currents and fundamentally change the type of disturbance most easily excited. I therefore examined the same problem as modified by the introduction of a dense fluid core (Jeffreys, 1952 b). As expected, the harmonics of degree 1 are no longer the easiest to excite; the easiest are of degree 3 or 4. This, however, is not a fatal objection. So long as several types of disturbance can be excited, it seems that the most widespread will become dominant given sufficient time. In the plane case a general current obliterates all modes such that the disturbance varies along it. In the formation of vortices between two streams of liquid, for instance, small ones are formed first but gradually coalesce to form big ones. In the formation of waves on water by wind, the longest waves capable of being excited become dominant as the waves travel. It seems that the crucial question about the formation of the land and water hemispheres is not whether the disturbances of degree 1 are the easiest to excite, but whether they can be excited at all; and the answer to this is in the affirmative.

The objection to this version of the theory is that continuously increasing viscosity suggests passage to a glassy state without separation of materials. I have therefore also examined a form in which a porous mass of crystals has been formed and residual magma percolates through the interstices. The equations for this problem had been given by Lapwood (1948) and solved by him in the case of a horizontal layer. In the spherical case the results were much as before; the easiest disturbances to excite are not those corresponding to first harmonics, but first harmonics can be excited, and that seems enough for the purpose.

There would be still a possibility that the instability could arise in the stage when olivine crystals were forming and sinking freely through the magma. This variant would be much more difficult and has not been treated.

It seems that some form of Hills's theory is likely to give a satisfactory explanation of the land and water hemispheres. At a late stage in the solidification the harmonics of degree 1 would be damped out and those of degrees 2 and 3 could survive. At this stage, if not before, the main continental mass might be disrupted into separate continents. At the last stage these harmonics also would be damped out and the continental masses would be permanent.

The objections to convection made earlier do not apply here, since the elasticoviscous law or that of pure viscosity might well once have been true if the Earth passed through a fluid or semi-fluid stage.

P. H. Roberts (1965) has studied convection in a sphere with reference to the differences between three principal moments of inertia and between the corresponding radii, when the disturbance corresponds to a harmonic of degree 2. If R and I are the mean radius and moment of inertia he gets

$$\frac{A-I}{a-R} = \frac{B-I}{b-R} = \frac{C-I}{c-R} = -\lambda\frac{I}{R},$$

where $\lambda = 0.2006$. For the Moon he quotes Baldwin that the difference of the equatorial and polar radii is about 2 km., giving an ellipticity ϵ of about 10^{-3}; H, the corresponding ratio for the moments of inertia is 6×10^{-4}. Then $H/\epsilon \doteqdot 0.6$. This is too great for the theory.

The argument assumes elasticoviscosity, which for other reasons is quite unsatisfactory. Yet it has been claimed that the Moon's excess ellipticities are maintained by convection, which would be impossible with a law that is better supported. What Roberts has shown is that even if conditions that would permit convection are assumed, it still fails to account for the facts.

It has been pointed out (especially Lingenfelter and Schubert, 1973) that the Earth, Moon, Mars and Venus all show displacements of the centre of mass from the centre of figure. The explanation offered above may be applicable to all.

11·13. Terrestrial magnetism. This phenomenon has until recently been closely associated with atmospheric electricity. The first work that brought it into relation with the mechanical aspects of Earth structure was the discovery that periodic changes in the magnetic field, attributable to induced currents in the shell, indicated a great increase of electrical conductivity at about the depth of the 20° discontinuity, and this in turn led to suggestions about the vertical distribution of temperature. (Lahiri and Price, 1939).

The main problems of terrestrial magnetism, apart from currents induced from outside, are first, the existence of the main doublet field, and then the secular change. Since the magnetic axis is not far from the geographical axis, many attempts have been made to explain the main field as a consequence of rotation, but with little success until recently. Elsasser and Bullard have, however, made notable progress. In suitable conditions the Earth's core could act as a dynamo: that is, a small magnetic field acting on a moving body can induce such electric currents as will augment the field. A source of power is needed to drive the dynamo, and this is attributed to motions due to thermal instability. Bullard (1950) finds that traces of radioactivity in the core could be sufficient to produce such motions without disturbing the thermal state of the shell to any important extent. The problem is exceedingly complicated; in the cases previously discussed maintenance of stability of a fluid heated below is due to viscosity and heat conduction, but here rotation and the electromagnetic forces themselves have to be allowed for. However, solutions have been obtained that do lead to a general magnetic field. Further, the core eddies have been found to have a general drift to the west, and this is in accordance with the greater part of the secular variation (Bullard, 1949; Bullard and Gellman, 1954; Cowling, 1957, Chapter 5). The magnetic axis is actually inclined at about 11° to the geographic axis, and this is a difficulty in any theory that derives the field essentially from the rotation. The magnetic pole has apparently described about $\frac{3}{4}$ of a circle in the last 300 years, but the centre is still well away from the pole of rotation.

The greater part of the shell is probably too hot to have permanent magnetism, and the greater part of the field must in any case arise from the core; the doublet field inside the core must be of the order of 10 gauss and the eddy fields of the order of 500 gauss.

It has often been suggested that magnetic interaction between the core and the shell may have appreciable effects on the period and damping of the variation of latitude. I have worked out a very simplified case, but find that at the most the effects would be of the order of a millionth of those needed to matter. See also § 8·17.

11·14. Forces on a floating continent. Imagine a block of depth h and density ρ floating in a liquid of density ρ_0. The base of the block is at a depth $\rho h/\rho_0$ below the surface of the fluid; the height of the centre of mass above that of the displaced fluid is $\frac{1}{2}(1-\rho/\rho_0)h$. The potential energy due to the presence of the block instead of an equal mass of liquid is therefore $\frac{1}{2}(1-\rho/\rho_0)mgh$, where m is the mass. If then g is not the same all over, the potential energy is least where gravity is least, and therefore there is a force tending to displace a floating body towards the regions where gravity is least. In particular, a floating body will tend to move towards the equator. This was first noticed by Eötvös (1912; cf. Lambert, 1921). With the formula for gravity
$$g = g_0(1 + 0·00529 \sin^2 \phi),$$
the equator-ward force per unit area is
$$\left(1 - \frac{\rho}{\rho_0}\right)\rho \frac{h^2}{R} g_0 \times 0·00529 \cos \phi \,|\sin \phi\,|.$$
With
$$\rho = 2·7, \quad \rho_0 = 3·3, \quad h = 15 \text{ km.,}$$
$$R = 6·4 \times 10^8 \text{ cm.,} \quad g_0 = 980 \text{ cm./sec.}^2,$$
the maximum value of this is about 4400 dynes/cm.².

There is also a force between continents due to the disturbance of gravity at the centre of each produced by the other. If a continent is truly floating the compensation tends to reduce gravity around it, but the effect falls off rather rapidly with distance. The force between two continents will be an attraction, but much less than the Eötvös force. For two continents of radius 2000 km., with their centres 90° apart, the attraction is 10^{-2} dynes/cm.² (Jeffreys, 1926c).

We have already commented on the Pekeris stress due to the original difference of temperature between the poles and the equator. This would be of order 10^8 dynes/cm.² at the stage we are considering. It is much larger than the Eötvös stress.

It therefore appears likely that if continents were formed in the way suggested by Hills they would drift towards the poles. So long as it appeared that the Eötvös force was far the most important acting on a floating body

PLATE IX

(*By permission of H.M. Geological Survey*)

Section of contorted drift, with Cromer Till at base. Beeston Regis, Norfolk.

PLATE X

Craters produced by exploding a mixture of KClO₃ and magnesium powder in a sheet of wet plaster of Paris. Note the flat depressed floor of the largest crater and the roughly hemispherical form of some of the smaller ones. (Photograph by S. Mohorovičić.)

it was difficult to see how the continents, if formed during solidification, could have failed to collect around the equator and stay there. It can hardly be said that complete concentration about the poles is altogether satisfactory, but it is at any rate less remote from the facts than an equatorial belt.

A question that has not yet been examined is whether the convective process could thicken the continents to their actual thickness in spite of gravity. To investigate this the theory would have to be taken to the second order, and the very doubtful rate of cooling would be an important factor.

Hills maintains that the continents actually drifted to the poles under the Pekeris force and then back to the equator under the Eötvös force, subsequently breaking up and drifting apart. I do not follow the later parts of his argument.

11·15. Vulcanism and intrusion. The formation of volcanoes, dikes, and sills, and other elevations of rock magma, in a liquid or partly liquid state, to the surface or near it requires explanation. On the theory of Chapter X the present temperature at the base of the intermediate layer may be about 600° C., and that at the base of the granitic layer about 300° C. Dry basalt and granite do not soften at temperatures below 1000° C. or more, and this value may have to be increased at these depths to allow for the effect of pressure on the melting-point. The generation of liquid granite and basalt is therefore a definite problem; so definite, indeed, that it formed Holmes's principal reason for abandoning the theory of the Earth's thermal history that he had played a leading part in creating. I do not think, however, that the difficulty is so serious. In the first place, we cannot admit that fusion in the upper layers is a normal condition. Vulcanism is local and occasional, not perpetual and world-wide, and the behaviour of the Earth in supporting mountains, transmitting transverse waves, and permitting oceanic tides shows quite definitely that the upper layers are solid at present. Secondly, Jaggar's observations (1917) on the crater of Kilauea showed that the basaltic magma there was truly fluid at temperatures of 750–850°, the reduced melting-point being due to volatile constituents. Day (1925) estimates that the Kilauea lava preserves some mobility down to about 600° C., though when it has solidified it does not flow under its own weight till it is reheated to about 1300° C. G. A. MacDonald (1963) gives a summary for Hawaiian lavas. He quotes from Jaggar 1000° just below the surface, 860° 1 m. down, with a rise to 1170° at the bottom, with depth 13 m. 1350° is found in flaming grottoes. Viscosities of order 10^4 are quoted. Elevation of magma to the surface being the exceptional condition, we must suppose the normal temperature in the intermediate layer to be below that of newly risen magmas. From this point of view the temperatures inferred for the intermediate layer seem highly satisfactory. Indeed it seems to me that the real difficulty is in the opposite direction to that felt by Holmes. Direct measurements of the viscosity of rocks at high temperatures have not been carried out, but Trouton and Andrews (1904) found that soda glass at 710° had a viscosity of 4×10^{10}, and at 575° one of $1 \cdot 1 \times 10^{13}$. Extrapolation to 500° suggests a viscosity of 2×10^{14}. We have seen that the kinematic viscosity of the intermediate layer is over 10^{22}, and the ordinary viscosity over 3×10^{22}. The legitimacy of the comparison needs experimental check, but on the face of it the intermediate layer is stiffer than we might expect from the theory of Chapter X, not softer.

Our task is to explain how the intermediate layer, which we shall assume for the purpose of discussion to be tachylyte, comes to be heated up by the requisite 100° at least to make it capable of rising through the granitic and intermediate layers, and how, having been so heated, it manages to solidify again. There is not the slightest difficulty about accounting for the heating. Formation of deep sediments, and still more the formation of mountains, would lead to thermal blanketing, and the basal temperatures would rise. Doubling the thicknesses would multiply S_0 by 4. Consequently, we should expect fusion to occur in the intermediate layer, and possibly even in the granitic layer, below a mountain chain. The squeezing out of the intermediate layer already mentioned, and the occurrence of large intrusions of granite and andesite in mountainous regions, are therefore to be expected.

The difficulty, if we appeal to radioactivity as a source of extra heat, is to see how the phenomenon could stop. Such heating is steady, and steady heating in a liquid or solid normally leads to a distribution of temperature tending to a steady state such that conduction and convection remove the new heat as fast as it is generated. Fusion once produced would therefore be permanent. But the areas of the surface covered by extinct volcanoes, bathyliths and plateau basalts must be hundreds of times that now occupied by active volcanoes. The extinct volcano presents a much more difficult problem than the active one. If general fusion of the intermediate layer occurred over a wide area, the outpouring would submerge the granitic layer and increase S_0, and the process would apparently continue till the whole of the intermediate layer was squeezed up to the surface. Meanwhile the granitic layer in turn would be heated up, and apparently nothing could stop the process except renewed concentration of the radioactive elements to the surface. The high radioactivities of Alpine granites might be held to confirm this; but the Finnish ones are as high and the region is not mountainous. In any case even the greatest outbreaks of igneous activity do not appear to have been on anything like the scale to be expected if they had been due to excess of radioactive heating.

It appears that the sources of heat required must be local and temporary. That they are local is confirmed by the great difference of level of the lavas in two neighbouring craters in Hawaii (Daly, 1926). Chemical activity is capable of producing such differences, for Jaggar shows in the paper just quoted that the lava in the fountains of the Kilauea crater is at about 1100°, though the temperature is variable; the normal temperature some distance down was 750–850°, and Jaggar attributes the difference to chemical action with the atmosphere. A. L. Day (1925) has gone a long way towards a constructive theory of vulcanism, in which chemical action between the gases expelled from the crust plays a leading part. The gases actually emitted at the surface are such as would interact if they were mixed, and it is certain that they must meet within the crust. Granting the facts, the explanation of local fusion when gases from different sources in the lower layer meet within the intermediate layer seems adequate. It has the great recommendation that when a given source of gas is exhausted or blocked up the fusion will cease, and the igneous activity will come to an end.

CHAPTER XII

Special Problems

'The art of narrative consists in concealing from your audience everything
it wants to know until after you expose your favourite opinions on topics
foreign to the subject. I will begin, if you please, with a horoscope located
in the Cherokee Nation; and end with a moral tune on the phonograph.'

O. HENRY, *Cabbages and Kings*, p. 81

12·01. Foreshocks and aftershocks. A large earthquake is usually
followed by other smaller ones from the same epicentre, called aftershocks.
Sometimes these run to thousands, as for the great Tokyo earthquake of
1923 September 1. They tend to decrease in frequency as time goes on and
come to an end in a few months. Even years afterwards, however, other
shocks may happen in the same place. Such late repetitions are not generally
classified as aftershocks.

Foreshocks are small shocks, which normally increase in intensity and
frequency until a major earthquake occurs. Usually they are not so numerous as aftershocks and are spread over a shorter interval of time.

The elastic rebound theory, in its simplest form, does not explain foreshocks and aftershocks. On it a gradually increasing stress would produce
no fracture at all until the stress-difference somewhere reached the strength.
There would then be fracture, which would redistribute the stresses and in
general increase them. The time taken for the redistribution would be of the
order of that taken by a P or an S wave to traverse the region. The extent
of the region for this purpose may be taken as not more than the uncertainty
of an epicentre, say $0 \cdot 1°$, since neither field methods nor instrumental
determinations indicate any change in epicentre. Thus there would be
further fractures within a few seconds, and the whole region would be
shattered until all the stress-differences in it were brought below the
strength. The whole process would be over within a minute at the outside.

With an extension suggested by the work of Phillips and Griggs, however,
the theory provides an explanation. We have to distinguish now between
the stress-difference needed to produce immediate fracture and that needed
to give fracture if it is maintained for a long time (usually hours or days).
With a stress-system increasing continuously the latter type is the important
one. Deformation would be substantially elastic (that is, apart possibly
from elastic afterworking and hysteresis) until the fundamental strength is
reached. Then a certain amount of plastic adjustment will follow and lead
after some time to a fracture. The resulting readjustment of stresses would
suffice, however, only to bring them below the amounts needed for immediate fracture, and if the general stress in the region was not relieved further
flow would follow and possibly lead to further fractures. In this way the

[472]

series of shocks, of diminishing intensity and frequency, might well come to be spread over months.

Foreshocks would arise in places where the fundamental strength was lower than the average. Fracture at them would tend to increase the stress-difference in the rest of the region and thus hasten the general fracture, which is the main earthquake.

Data for the aftershocks of the Tango (Japan) earthquake of 1927 March 7 were given by N. Nasu (1929). There were 1071 up to June 8. In an analysis (1938 c) I found that up to March 31 the chance of an earthquake occurring at the place in an interval of time dt was of the form $dt/(t-\beta)$, where β was 2^{d} $1^{\mathrm{h}} \pm 10^{\mathrm{h}}$ after the principal shock.* On March 31 (April 1 in Japanese time) a large aftershock occurred and was followed by a new series superposed on the first. For this new series a similar law was followed with β 3^{d} $4^{\mathrm{h}} \pm 1^{\mathrm{d}}$ 5^{h} before the principal shock. The anomalously late β in the first series is presumably due to a slight departure in the first few days from the simple law adopted.† This form being adopted as a standard of comparison, and the frequencies grouped over convenient ranges of time, there was no sign of anything but random departures from the law. That is, there was no sign that the chance of an earthquake in a given interval of time depended on anything but the time since the main shock, and there was no determinable periodic variation superposed on the law. With the exception of the large repetition on March 31 there was no sign that any aftershock tended to encourage or discourage another after it. It appeared that the whole series must be attributed to readjustments directly connected with the stress distributions left by two main shocks.

Similar results for other series have been found by E. Wanner (1937, 1941) who, however, in later work found that the rule of independence is not general. In the latter series there was no dominating shock but shocks appeared to occur in clusters.

It has been noticed by Imamura in Japan that progressive tilting of the ground occurs before a great earthquake. This would agree with the above interpretation of the behaviour of rocks on the verge of fracture. Such tilting has been used for prediction of earthquakes, and so has the occurrence of foreshocks. No such method, however, can predict with reasonable accuracy, say within half a day, when a large earthquake is to be expected, and many people hold that, until this can be done, prediction would be worse than useless. They say that a warning that there is a possibility of a large earthquake in a particular week, if it was acted on, would dislocate all activity in the region for a week, and the consequences might be as serious as those of the earthquake itself. Leith and Sharpe (1936) point out that aftershocks of deep-focus earthquakes are very rare.

---

* This form is due originally to F. Omori (1894), whose data were mainly from the Mino-Owari earthquake of 1891 October 28.

† A law of the form $dt/(t-\beta)^{1-k}$ was tried, but k was too small to be satisfactorily determined.

12·02. Periodicity. Numerous series of data on the occurrence of earthquakes have been analysed statistically with the object of detecting periodicities. Turner alone had eight papers on this subject in the first two volumes of the *Geophysical Supplement*. C. G. Knott made several analyses in his *Physics of Earthquake Phenomena*. When harmonic analysis is applied to a series of frequencies, the coefficients found are not in general zero even if the events are genuinely random, and it is necessary to have a proper statistical criterion to test whether any coefficient is larger than would be likely to occur by chance. The criterion usually adopted, when any has been, is that given by Schuster; this is equivalent to Pearson's χ^2 test for two degrees of freedom (Jeffreys, 1967d, p. 103). It is usually found in analysis of earthquake frequencies that the Schuster critical amplitude is exceeded, and many periodicities have been claimed to exist in consequence. But an amplitude larger than would be expected by chance may occur without there being any genuine periodicity; any sort of departure from independence of the events might produce such an effect. It is certain that if allowance is made for aftershocks, the criterion has to be modified in such a way that the margin of significance is greatly reduced. Further, many of the series used are incomplete, and there is a possibility that periods may be introduced by human selection. At present there seems to be no convincing evidence for any periodicity.

There is no question of a *dominating* periodicity, that is, of all earthquakes being concentrated in one half of a period, leaving none in the other half. The question is simply whether there is a significant excess for one half over the other.

It might easily be expected that dominating periodicities would exist. There are periodic changes of stress due to solar and lunar tides, and possibly diurnal and annual ones of meteorological origin. These correspond to strains of the order of 10^{-7}. If we take the lunar semi-diurnal tide as a model, and consider a secularly increasing strain, we should expect that if the increase in half a day is less than 10^{-7} the strain would always first reach a critical value in the half day when the lunar part of the stress is in the same direction as the general stress. Hence the lunar semi-diurnal periodicity would dominate the occurrence of earthquakes if strains increase by less than 3×10^{-4} per year; which is more like the actual variation in a geological period. The fact is, however, that it does not.

Knott treated the matter by selecting as periods several different 'months': the sidereal, synodic, anomalistic months and the period of the Moon from node to node. Some of these are theoretically associated with tidal deformations of the Earth, others are not. He found amplitudes above the Schuster limit in some of the latter type and inferred that, whatever the amplitudes might be due to, they were not due to dynamical effects of the Moon.

Oldham (1918, 1921, 1922) suggested that the general stress, when it reaches a certain amount, does not produce instant fracture, but a much more rapid change of form; and the tidal stresses superposed on this will not necessarily exceed the variation in a period. This suggestion is in complete accordance with the explanations of foreshocks and aftershocks suggested above and with Imamura's observations of tilting.

Solar effects, if gravitational, would be expected to be less than the corresponding lunar ones. Thermal effects are larger, but they are inappreciable except within a few metres of the surface. In loose sediments they would lead only to grains rolling over one another, and in compact rocks at most to superficial cracking. The latter occurs but does not give rise to seismic waves.

A method of analysis for periodicities is given in H. and B. S. Jeffreys (1972) and in more detail in Jeffreys (1964a, 1967d, Appendix C).

12·03. Earthquakes and faults. The formation of mountain systems must have involved earthquakes of a frequency and extent beyond anything we now know. Present earthquakes are usually in regions of steep slopes and may be attributed to the stresses near such slopes produced by gravity.

The commonest type of dislocation seen at the surface is the normal fault, in which the rocks on one side appear to have slipped down over those on the other, the inclination being such that the horizontal extent is increased. The relative vertical displacement may be anything from a few millimetres to a kilometre. In some faults, known as tear faults, the relative displacement is nearly horizontal and the plane of the fault nearly vertical. A thrust or reversed fault is one where the slip shortens the horizontal extent. O. T. Jones (1933) pointed out that the differences between the three types of fault could be explained in terms of the relative magnitudes of the vertical stress and the two horizontal ones. Close to the surface a true tension is possible, but not at depths over a few kilometres, and less for most sediments.

Earthquakes are often associated with known faults, and the presumption is that every fault has been the seat of an earthquake, some of many. The great Californian earthquake of 1906 was on a tear fault; there were slips along a great crack in the ground extending over 430 km., the relative displacements of the two sides being practically horizontal and reaching 7 m. Such relative displacements (in different directions) have been recorded in many earthquakes; but no known earthquake has given a displacement comparable with those in the larger faults. These must be attributed either to successions of earthquakes producing displacements along the same fault or to single earthquakes more intense than any we know. Even when the ground has been broken it is not always true that the whole displacement has been in a single movement, because there is usually a series of aftershocks between the original earthquake and a survey, and each of them may have contributed.

Conditions in faulting are extremely variable; faults occur even in loose sand. But it is at least possible to give some indications that may help to specify the conditions.

First take the purely elastic case. If the stress difference at breaking is S, the shear stress across the plane of fracture is $\frac{1}{2}S$, and is suddenly reduced to 0. We take the axis of x perpendicular to this plane. The fracture produces a pair of S waves travelling away from it. The solution is that the displacement on the side of positive x is

$$v = \frac{1}{2}\frac{S\beta}{\mu}\left(t - \frac{x}{\beta}\right) \quad (x < \beta t); \qquad v = 0 \quad (x > \beta t). \tag{1}$$

Thus the faces slip in opposite directions with velocity $\frac{1}{2}S\beta/\mu$, and the relative velocity is twice this. We consider compact rocks, so that representative values will be $S = 10^9$ dynes/cm.², $\mu = 3 \times 10^{11}$ dynes/cm.², $\beta = 3$ km./sec. Then the relative slip velocity is about 1000 cm./sec., and an apparent throw of 100 m. will occur in 10 sec. Friction and reduction of the stress by spreading in three dimensions might increase this estimate.

Now friction generates heat, and if this slip takes place under considerable normal stress much heat will be developed. Most faults give little or no indication of this, but some contain a fine-grained material known as flinty crush-rock or pseudotachylyte, which is held to have been made amorphous by the heat (T. J. Jehu and R. M. Craig, 1923, p. 430). It is rather difficult to see why this is not commoner; apparently the displacements are usually slow enough for the heat to be conducted away. If the frictional stress over the fault is F and the velocity of slip w, the rate of supply of heat is Fw mechanical units per unit area, or Fw/J heat units, where J is the mechanical equivalent of heat. If this supply begins at time 0 the rise of temperature will be (Jeffreys, 1942d)

$$\vartheta = \frac{hFw}{kJ}\left(\frac{t}{\pi}\right)^{\frac{1}{2}}, \tag{2}$$

where k is the conductivity and h^2 the thermometric conductivity. If H is the total slip and T the time taken, $w = H/T$, and

$$\vartheta = \frac{hFH}{kJ(\pi T)^{\frac{1}{2}}}. \tag{3}$$

For a mean depth of 1 km. and coefficient of friction $\frac{1}{2}$, F will be about 10^8 dynes/cm.²; take $k = 0.006$, $h = 0.08$, $H = 10^4$ cm., $J = 4 \times 10^7$, $\vartheta < 1000°$. We find that, if melting does not occur,

$$T > 2.4 \times 10^4 \text{ sec.} = 7 \text{ hr.}$$

The time suggested is very sensitive to the data used, since they all enter through their squares; if we used a mean depth of 200 m. the time would be divided by 25. It appears, however, that the time of formation of a fault of throw 100 m., indicated by the absence of pseudotachylyte, is much more than the time indicated by the probable rate of slip after fracture. It appears to follow that faults of this magnitude are not usually formed in a single stage.

If pseudotachylyte is formed, we can estimate its probable thickness. The heat energy supplied in a displacement of 100 m., with the values used above, is 2×10^4 cal./cm.^{2}. About 500 cal./g. would suffice to melt most rocks, so that the mass of the fused layer would be about 40 g./cm.^{2}, corresponding to a thickness of about 16 cm. This is a little too large because some of the heat would be taken up by rock that never reached the melting-point. It is, however, of the order of magnitude of the thicknesses of pseudotachylyte that are found, and the hypothesis that the heat generated by friction was used in metamorphosing the rocks is tenable provided the fault was produced in a single stage.

If pseudotachylyte is not formed, we may estimate a maximum value for H. We can put
$$H/T = 1000 \text{ cm./sec.,}$$

combine with (3), and solve for H and T. The result is $H = 4$ cm., $T = 0.004$ sec. Thus a fault with throw 100 m. containing no pseudotachylyte would have to be formed by about 2000 separate movements. This may be too high, because the shear stress needed to produce new slip on an imperfectly welded crack will be less than that needed to start a new crack; the slip velocity will then be less and heat conduction will be better able to remove the heat in the time available for any given displacement. In any case 4 cm. is nearer to the order of magnitude of the displacements in most single earthquakes, and it is satisfactory that the commoner kind of earthquake should theoretically be associated with the commoner kind of fault.

This estimate of displacement can be used to suggest a size for an earthquake focus. This would have to be interpreted as the extent of a fault face where slipping takes place in a single stage. If the linear dimensions are l cm. and the displacement 4 cm., the elastic energy liberated is of order $4Sl^2$. The energy in a large earthquake is of order 10^{22} ergs. With our previous value of S this gives l cm. $= 16$ km. This is reasonable. A focus of this size would emit a pulse such that the time of the first swing and recovery would be about 5 sec., which is of the order of the time as usually recorded by a seismograph.

On the whole, in spite of the great variability of the quantities that enter into the calculation, the results are reasonable. (See also E. M. Anderson, 1938, 1942.)

In elementary mechanics the coefficient of friction in sliding is assumed to be the upper bound of the ratio of the tangential to normal stress. This is not strictly true. Usually when slipping starts there is a non-zero acceleration; the friction during sliding is a little less than just before sliding starts. This suggests that the change of stress in faulting may be much less than the strength. This is related to a theory of foreshocks and aftershocks due to Burridge and Knopoff (1967) with experimental illustration.

12·04. The permanence of the continents. The geological record shows abundant evidence that the land surface has risen or sunk in places relative to the sea, and the outlines of the continents have certainly changed considerably. It seems, however, that most of these changes have been near the margins. Genuinely deep-ocean deposits do not occur within the continents. To some extent this is satisfactory, because the most likely explanation of changes of level, as far as we can see, is differential cooling, which is unlikely to be reversed. It would be very awkward for the theory developed in the last chapter if it could be shown that some extensive region was once land, has become deep ocean, and become land again. But no such case is known. The ancient shields of Siberia, Canada and Africa appear to have been land since pre-Cambrian time and the Pacific Ocean to have been ocean. Minor oscillations of the order of hundreds of feet are shown in unconformities. In part these are probably incidental to mountain formation, but a general mechanical explanation is lacking.

12·05. Shifts of the Earth's axis. We have to distinguish between changes of the direction of the axis of rotation in space and changes relative to the Earth; for the latter our natural standard of reference would be the ancient shields, but it will be necessary also to consider whether these are likely to have moved with respect to one another.

The axis of rotation is certainly moving in space; its most prominent motion is the precession of the equinoxes, and various small oscillations known as nutations are superposed on this. None of these is of geological importance. The principal part of the precession is due to the attraction of the Sun and Moon on the Earth's equatorial protuberance, and is a motion of the pole parallel to the plane of the Earth's orbit. The latter plane itself is moving slowly owing to the perturbations due to the planets. According to Leverrier's solution the inclination of the Earth's orbit to the plane of the resultant angular momentum of the system can approach but not exceed 5°. E. W. Brown casts some doubt on the applicability of the theory over intervals of more than about 10^8 years. But as the precessional motion is parallel to the actual orbital plane it appears that slow disturbances of the orbital plane will be mostly transmitted directly to the axis of rotation without altering the mean inclination over long intervals. There are also disturbances due to tidal friction. The theory of Chapter VII is a simplified one, since we have neglected the inclinations and eccentricities of the orbits. These, especially the inclinations, introduce additional components into the tides, and all of these are affected by friction. The investigation of their effects formed the subject-matter of most of the second volume of Darwin's *Scientific Papers*, and he was able to show that they could have brought the Earth-Moon system to its present state from an initial state with the Earth and Moon close together, the Moon's orbit nearly circular, and in the plane of the equator, and the equator at any rate much less inclined to the ecliptic than it is now. At least the signs of his effects seem correct with modern results on dissipation.

Slow deformations of the Earth itself present a complicated problem, chiefly on account of the difficulty of specifying a set of axes in the body with sufficient precision. If we adopt the method of 7·04, the axes taken in the body are such that there is no resultant angular momentum due to the deformation. The problem has been treated in detail by Darwin (3, 1–46) (see also Tisserand, 2, Ch. xxx). The main conclusion is that the deflexion of the axis of greatest moment due to deformation is always small. Darwin says (§ 3): 'If the elevation of a large continent proceeds at the rate of 2 feet in a century,... the greatest angle made by the instantaneous axis with the axis of figure is comparable with $\frac{1}{376}''$, a quantity beyond the power of observation. On the score of these terms, the instantaneous axis will therefore remain sensibly coincident with the axis of figure.'

The question then reduces to: Has the axis of figure moved geographically with regard to the surface? It is supposed that enough of the Earth retains its form sufficiently closely to be taken as a standard of reference for the motion of the axis of figure, and Darwin considers the effect on the axis of figure (the axis of greatest moment of inertia) produced by changes of level of the remainder. He shows that with the most favourable distribution, if half the surface was raised by 10,000 ft., the axis might be displaced by 8°. For an area of the size of Africa or South America it might be 1° or 2°. There is nothing in these results to suggest that large displacements of the axis of figure or the axis of rotation within the Earth can have occurred.*

The Eötvös force provides a mechanism not considered by Darwin. Isostatically compensated elevations tend to move toward the equator, and if the crust as a whole remains unbroken it will tend to set itself so that the elevations, as far as possible, are collected about the equator. The effect is slow, but could become important given sufficient time. From the variation of latitude data we have seen that, if the damping is due to elastico-

---

* Lambert (1931, p. 267) points out that there is a mistake in Darwin's integration of the differential equations in § 5, which takes account of a possible viscosity. In the unnumbered equations at the foot of p. 13 and in the middle of p. 14,

$$\text{for} \quad \nu(\nu^2 - \mu^2) \qquad \text{read} \quad \frac{\nu(\nu^2 - \mu^2)}{\nu^2 + \mu^2},$$

$$\text{for} \quad -2\mu\nu^2 \qquad \text{read} \quad -\frac{2\mu\nu^2}{\nu^2 + \mu^2},$$

$$\text{for} \quad \frac{u\nu(\nu^2 - \mu^2)}{\nu^2 + \mu^2} \qquad \text{read} \quad \frac{u\nu(\nu^2 - \mu^2)}{(\nu^2 + \mu^2)^2}$$

$$-\frac{2\mu u\nu^2}{\nu^2 + \mu^2} \qquad \qquad -\frac{2\mu u\nu^2}{(\nu^2 + \mu^2)^2},$$

$$\text{for} \quad -u\nu^2 \qquad \text{read} \quad -u\nu^2/(\nu^2 + \mu^2),$$

$$\text{for} \quad u\nu^2/(\nu^2 + \mu^2)^{\frac{1}{2}} \quad \text{read} \quad u\nu^2/(\nu^2 + \mu^2)^{\frac{3}{2}}.$$

Darwin's conclusion that large displacements of the instantaneous axis from the axis of figure could arise in a viscous Earth is reversed by this correction.

viscosity, the viscosity is probably of the order of 5×10^{20} c.g.s. With this value, a shear stress of 4000 dynes/cm.^{2} would produce distortion at a rate of about $0 \cdot 8 \times 10^{-17}$/sec. The crust would be displaced through a radian in something like 3000 million years. If we took the viscosities inferred (wrongly, in my opinion) from the Fenno-Scandian uplift this time must be multiplied by about 100. Walker and Young's first result (1955) would divide it by 5.

In fact, consistently with any current interpretation of isostasy, the greater part of the rate of shear due to the Eötvös force cannot be within the range of depth where the viscosity is 10^{22} or more. It must be at depths over 600 km., where the lower value may apply. Then the appropriate Earth model approximates to that considered in 6·10, that of a thick crust of finite strength, of thickness of order 600 km., overlying an elasticoviscous interior. The amount of the gravity anomalies corresponding to harmonics of degrees 2 and 3 provide our only means of estimating what the forces tending to produce crustal displacement are likely to be.

If there is a strong layer of thickness H and the stresses are so distributed as to require the minimum strength, we know from 6·10 (6) that the disturbance of gravity for height h' contains factors $h'H$. The Eötvös forces on the elevation and the compensation will also differ in absolute magnitude by a similar factor. Hence the Eötvös force for given disturbance of gravity is nearly independent of the thickness of the strong layer.

For the resultant over the whole crust to tend to turn it round, the corresponding gravity anomalies would have to correspond to the surface harmonics $\sin\phi \cos\phi$ $(\cos\lambda, \sin\lambda)$. My analysis showed that these are not significant at present; but any internal process that produces the other low harmonics might well produce these too and set up a displacement of the pole tending to annul them; its ultimate effect would be to bring the crust so that its maximum moment of inertia is about the axis of angular momentum. 20 mgal. for the amplitude of the gravity disturbance is not unreasonable; and if somewhat larger values were occasionally reached rates of displacement large enough to be of interest to palaeoclimatology and palaeomagnetism might follow.

T. Gold (1955b) points out that in the conditions supposed the relevant axis of maximum moment of inertia is that of the crust alone; any part of the differences of the moment of inertia that would be annulled in the conditions of long-continued flow is not relevant. Thus (merely for illustration) in 7·04 (9) the free period is determined by the part of $C - A$ that is not due to the stresses; and in the present type of displacement the moments of inertia of the viscous interior do not affect the problem. It appears in fact that the products of inertia produced by geological changes within the crust might be comparable with the relevant part of $C - A$, and if so there is no apparent limit to the amount of crustal displacement that might occur, given sufficient time, since F in 7·04 might be comparable with this part. Gold, I think, weakens his case by producing arguments in favour of complete isostasy; if the whole crust was weak any excess masses would immediately redistribute themselves by flow within the crust and the mechanism of the process breaks down. It is desirable that a more detailed quantitative analysis of the time-scale of the process should be made. (See also Munk, 1956.)

It has been seen in 8·13 that there are great difficulties in attributing the damping of the variation of latitude to elasticoviscosity. If it is rejected the whole explanation of polar wandering breaks down.

12·06. Continental drift. We next come to the question of whether large displacements of parts of the surface relative to one another have occurred. Displacements of the order of 100 km. such as are involved in mountain formation are not in doubt. But F. B. Taylor and A. Wegener (1924) and their numerous followers have maintained that there have been displacements of the order of thousands of kilometres; in particular, that America was originally joined to Europe and Africa and has drifted to its present position since the Cretaceous period.

The evidence for such a theory is mainly palaeontological. The chief method of comparing the ages of strata in different parts of the world is the comparison of fossils, and there is no doubt that in the main biological evolution has followed much the same course everywhere. There are notable exceptions, for instance the abundance of marsupials in Australia and their rarity everywhere else. But on the whole the appearance of a type in one region is followed within much less than a geological period by its appearance in widely separated regions. If the type was continental in habitat geologists were led to postulate land connexions, which often required long 'land bridges' across present oceans: if it was marine sea connexions had to be assumed. The results were not always satisfactory. The number of connexions, each devised for a special purpose, became uncomfortably large (cf. J. W. Gregory, 1929, 1930); and in some cases a land connexion across an ocean destroyed a necessary sea connexion between different parts of the ocean. Explanations of the disappearance of the land bridges were also lacking.

Wegener's principal contribution was a definite declaration that widespread changes of level are impossible and that the whole theory of land bridges is untenable. He maintained that the land connexions existed because what are now separate continents were once in juxtaposition, and that they have disappeared because the continents have drifted apart. He proposed a mechanical explanation of the drift, which we shall proceed to examine.

The argument against changes of level was based on isostasy. Continents were supposed to consist of a material similar to granite, the *sal* of Suess (renamed *sial* by Wegener), floating in a denser medium, Suess's *sima*, which also formed the ocean floor. Given the densities of the two materials, the elevation of the land surface would be determined by the thickness of the *sial* layer, and it would be impossible to alter it, except by long-continued denudation, and even this could not bring the surface below sea-level. On this analysis a land region could never be converted into deep ocean, as the land bridge theory requires.

The argument evidently assumes perfect isostasy; without this there is no reason to suppose the elevation of the surface determined by the local thickness of the *sial*. Wegener's book was published in 1915, just after Barrell's series of papers, which showed decisively that there are widespread imperfections of isostasy and that these demand stress-differences in the lower layer of the order of a seventh of the strength of granite to explain them. The more modern data discussed in Chapter VI indicate that this fraction must be increased to about a third. It also assumes that the respective densities are constant and makes no allowance for variations due to differences of temperature and composition. We have seen in Chapter XI that plausible variations of temperature could lead to changes of relative level of the order of 10 km. Accordingly the argument is unacceptable.

The explanation proposed for the drift was that the Eötvös force tending to make a floating body move towards the equator, would, on a rotating planet, produce a steady drift to the west. This looks surprising at first sight, less so at second sight, but then serious difficulties appear. It is a fact that winds are mainly nearly at right angles to the pressure gradient, that is, along the isobars, and at a height of 1 km. or so the agreement is usually very close; and Wegener was primarily a meteorologist. The explanation is simple for slow motion of a fluid of small viscosity. If u, v are the horizontal components of velocity and X, Y the components of force per unit mass, the equations of motion reduce in these conditions to

$$\frac{\partial u}{\partial t} - 2\omega\sin\phi\, v = X, \quad \frac{\partial v}{\partial t} + 2\omega\sin\phi\, u = Y.$$

If the motion is steady or of long period, so that rates of change of velocity relative to the Earth's surface are insufficient to produce considerable changes in a day, the $\partial/\partial t$ terms are small compared with the ω terms, and (u, v) is nearly perpendicular to (X, Y).

However, this argument neglects viscosity, and takes no account of the fact that the Eötvös force is a shear applied to the surface, and breaks down even with the data used by Wegener. At the time when he wrote, the secular acceleration of the Moon was still usually attributed to bodily tidal friction, and the elasticoviscous law was used to express it. On this hypothesis the viscosity was shown to be of order 10^{16} c.g.s. However, in 1915 I showed that with such a viscosity the free variation of latitude would be damped out in a few days, and that the correct value could not be less than 10^{20} c.g.s. The present situation is that there is no positive evidence that appreciable tidal friction occurs anywhere except in shallow seas. The theory of steady motions affected by both rotation and viscosity was given by V. W. Ekman (1905) and somewhat extended by myself (1920a), in application to ocean currents produced by wind. In these conditions for small viscosity the horizontal motion due to a shear stress over the surface decreases rapidly with depth, changing in direction as it does so. At the surface, with uniform viscosity, it is at 45° to the shear stress, and this is approximately verified for ocean currents. It becomes inappreciable at depths more than about $(\nu/\omega)^{\frac{1}{2}}$, which with Wegener's viscosity would be about 10^{10} cm.—far more than the radius of the Earth. Viscosity is therefore much more important than rotation, and the conditions approximate to those obtained by neglecting the rotation terms. In that case the velocity is in the direction of the applied force, that is, towards the equator. From the variation of latitude data alone we infer that the viscosity is more than 5×10^{20} c.g.s., so that a shear stress of 4000 dynes/cm.^{2} would produce distortion at a rate under 0.8×10^{-17}/sec. The crust would be displaced through a radian in something like 3000 million years. If we take the viscosities inferred from the Fenno-Scandian uplift this time must be multiplied by about 100. Accordingly displacements of geological import- ance due to the Eötvös force are out of the question.

If isostatic compensation was complete the pole would tend to settle so as to make the continents lie as symmetrically as possible about it. Then the present distribution would imply a pole near Hawaii, as Munk and MacDonald point out. The motion inferred from palaeomagnetism has been *away* from Hawaii.

It does not appear that recent accounts of continental drift place much emphasis on this force, but its influence survives in the claims that recent changes of longitude have been determined and support continental drift. These never come from long-established observatories, which would be in the best position to find secular changes of longitude if they existed. In most cases the earlier determinations were made by pioneer expeditions, and the difficulties of accurate determinations of longitude in such condi- tions are serious. But apart from the wrong assumption that the Eötvös force can produce drift of the amount required, and the dynamical mistake of supposing that if it could the drift would be wholly to the west, there is no reason why deformations of the crust should show themselves wholly in longitude. Latitude is not particularly difficult to measure in a pioneer expedition, and there is a curious lack of claims of displacements in latitude. As a matter of fact there is one. Lambert (1922a, b) found from 1900 to 1917 a motion of the North Pole with an average of 0·0062″ annually along the meridian of 81° W. Wanach (1927) found somewhat smaller rates but also towards America. Lambert says (1928, p. 268): 'A displacement of this sort, amounting to perhaps 1 minute of arc in 12,000 years, would be of little use in explaining the last advance of the ice in Pleistocene time, for this occurred not very many times 12,000 years ago, and a displacement of the pole by only a few minutes of arc would have only a negligible effect on the climate. Moreover, we have no assurance that even this slow quasi-secular displacement existed in the past nor that it will continue in the future.' We may add that the phenomenon was noticed first at Ukiah, a little north of San Francisco, and close to the fault that produced the great earth-quake of 1906. The displacement due to this would subtend about 0·2″ at the Earth's centre. The difficulty of principle alluded to by Lambert is serious; because, before we could assert that any progressive change, noticed since there have been observatories with the necessary accuracy, can be extrapolated over 10^5 or 10^8 years, we should need to know that it does not represent some known phenomenon that there is every reason to regard as intermittent and unlikely ever to exceed, say, the displacement on a known fault.

Much more observational material is now available. Munk and MacDonald (p. 177) remark, following Melchior, that the changes in the direction of motion of the pole occurred just when the star catalogue was changed. These changes are unavoidable because, on account of precession and proper motions, stars cease to pass near enough to the zenith to be observed in the telescopes used. But they are a major source of trouble, as Fedorov also found in his analysis for the nutations. Markowitz however thinks there are real displacements of the pole. Surveys later have not confirmed progressive displacement at Ukiah; the most likely station to have a local movement has been said to be Mizusawa. But Markowitz does not think that a displace-ment of any single station will explain the results. Yumi thinks that both Ukiah and Mizusawa have moved. But in my recent analysis of the varia-tion of latitude I found (1968d) irregular movements in l of order 0·05″ over the whole interval; those in m were similar up to about 1947, since when they have passed 0·2″. The facts are contrary to any hypothesis of a uniform rate of drift.

It must be noticed that the above argument supposes the motion of a continent resisted only by the viscosity of the lower material. No account at all is taken of the resistance of the ocean floor to the blocks being pushed through it. It appears, in fact, that Wegener, having assumed a finite viscosity for the *sima* below the continents, where it would be at a high temperature, light-heartedly did the same for it on the ocean floor, where the temperature is about 0° C. The ocean floor shows irregularities of level of the same order of magnitude as those within the continents, and must have a comparable strength to prevent them from flattening out. Now a hori-zontal stress of 4000 dynes/cm.² over a continent of radius 2000 km. would give a total force of $4000\pi(2 \times 10^8)^2$ dynes. This is resisted by the strength of the lithosphere, say 40 km. thick. An average stress of 10^5 dynes/cm.²

around a cylinder of radius 2000 km. through the lithosphere would suffice to prevent all relative movement. The strength at the surface is of order 10^9 dynes/cm.^{2}. Even that inferred for the lower layer from gravity anomalies would suffice with a considerable margin. Hence the Eötvös force, and *a fortiori* any smaller force, even on the hypothesis of a finite viscosity at a depth of 40 km., would not make continents move relatively to one another. The effect would be a rotation of the whole crust over the interior tending to bring the greatest continents as nearly as possible symmetrically distributed about the equator, but the rate of rotation would be slower than that given above (p. 483), because now the motion would be resisted by the viscosity of the whole interior and not only by that of the parts below the continents. In fact we can apply to the theory as proposed by Wegener the words used by Dutton about the thermal contraction theory 'It is quantitatively insufficient and qualitatively inapplicable. It is an explanation which explains nothing which we wish to explain.' But whereas the objections to the thermal contraction theory mentioned by Dutton have been answered since he wrote, those to continental drift have not been answered.

The evidence that it is impossible has not till recently been based on an alternative theory; it was that acceptance of the one condition that might make it possible leads to a host of contradictions.

The Pacific mountains are often mentioned in support of continental drift. This is really astonishing. If a continental block is being pushed through an ocean floor with no strength, why should the latter not simply get out of the way? But if it is being pushed against an ocean floor stronger than the continents, for which we saw reason in 11·04, the Pacific mountains are at once explicable.

The idea of a weak ocean floor right up to the surface is absolutely fundamental in Wegener's presentation; and though later advocates of some form of continental drift, where they attempt any kind of explanation, depart to some extent from Wegener, it seems to me that this idea is essential to the theory. It is inconsistent with the existence of ocean deeps, and it is hard to see how any mountains at all could be formed without something strong pushing the continents from all sides.

The convection theories are more satisfactory than Wegener's because they provide for larger stresses. Presumably because they appeal to a finite viscosity, they are widely supposed to provide an explanation of continental drift. But the effect of a system of convection currents would be to sweep floating material towards the places where the currents sink. It seems conceivable that the stresses in suitable conditions might be enough to break the crust, in which case one block might be pushed over the edge of another, with either folding or overthrusting. It does not seem likely, however, that intermittence of mountain formation or continental drift can be explained in this way. The piling up would stop when the gravitational forces tending to make the thickened regions spread out balanced the shearing forces due to the currents, and in a medium of fixed composition that would be the end of the matter. Local thickening in the actual crust might lead to heating in a geological period, and perhaps to refusion, but apparently this would only lead again to an upward movement of the radioactive materials and to a new permanent state.

It is argued that if evidence from palaeontology and meteorology proves that continental drift has taken place, evidence from geophysics that it is impossible is beside the point.* If I admitted the premises I might accept the conclusion, while still maintaining that it is remarkable that the advocates of continental drift have not produced in fifty years an explanation that will bear inspection. But I must reject the whole attitude that maintains that any type of scientific evidence can by itself be so completely demonstrative as to require the rejection of any evidence that appears to conflict with it. If evidence is conflicting, the scientific attitude is to look for a new idea that may reconcile it. The question here is whether the new idea is to be found in palaeontology, meteorology, or in the mechanics of the Earth itself. I think myself that it is least likely to be in the last, because it is the easiest subject of the three. The mere fact that it uses mathematical technique is trivial; mathematics is a tool for testing whether a given set of hypotheses accounts for a given set of facts. I am encouraged in this attitude by my experience in several public discussions on the question. It has always happened that after several distinguished palaeontologists have presented evidence favouring continental drift, some other equally distinguished ones have proceeded to point out other facts that are made more difficult to explain. I rather tentatively suggested in the second edition of this book that possibly migration of species often took place without any land connexion at all, but by spores, seeds, or eggs floating across on vegetable refuse in oceanic currents. Some reviewers declared this impossible (thus neatly reversing the argument mentioned in the first sentence of this paragraph), but I have been encouraged by the thorough examination of the evidence for former land connexions between the present continents given by G. G. Simpson (1943) and by the arguments given by D. Lack (1947). The hypothesis that land forms can migrate only by land connexion appears to lead to contradictions at numerous points. Migration across the sea would of course be rare, but as it would explain the phenomenon without needing to happen more than a few times its rarity is no objection. In fact, it is argued by Lack that migration by land connexions would imply greater similarity of fauna than actually occurs. His problem is to explain why, of all the groups of birds of the American continent, only some closely related finches are represented in Galapagos Islands.

* As there is still a tendency to represent the opposition as one between geology and geophysics, it is worth while to mention that the fundamental arguments used here are due to the geologists Barrell and Morley Davies.

Charles Darwin, in the *Origin of Species*, reported several striking results on the germination of seeds after long immersion in sea water, and concluded that 'the seeds of about 10/100 plants of a flora, after being dried, could be floated across a space of sea 900 miles in width, and would then germinate'. He also reported that a hibernating *Helix pomatia* was put in sea water for 20 days, and perfectly recovered. If 1 in 1000 were able to survive for 2700 miles it would probably explain as much as needs to be explained.

When a pit is dug for industrial reasons and left alone it fills with water. Before long it usually contains fish and water snails. These are believed to have been carried as eggs on the feet of birds. There were no birds before the Jurassic, but there were large flying insects in the Carboniferous. In the Carboniferous there were also plenty of large cryptogamic trees, logs from which could well have served to transport animals. The occurrence of a small reptile, *Mesosaurus*, in South Africa and South America, has been claimed as conclusive proof of drift. Meyerhoff and Meyerhoff (1972 *a*, p. 290) point out that it was not a freshwater but a marine animal, and give evidence that smaller reptiles *have* crossed the Atlantic.

Of the recent Antarctic reptilian finds, Meyerhoff tells me, only three Triassic taxa out of about 48 found are the same as three Triassic African forms; there are more than 150 Triassic taxa. One of the taxa common to Africa and Antarctica is aquatic. The almost total lack of similarity among the remaining forms is certainly not evidence for continental drift.

In a discussion O. T. Jones (1960) quoted arguments that the tillite (ancient boulder clay) of South Africa is really a desert deposit.

Axelrod (1963) writes: 'No paleobotanical evidence (exists) that unequivocally supports palaeomagnetic data, which ostensibly show that the continents have drifted so much as 60° to 70° of latitude since the Carboniferous. Several lines of evidence in climatic symmetry, local sequences of floras in time, distribution of forests with latitude, evolutionary (adaptive) relations, all seemingly unite to oppose the palaeomagnetic evidence for major movement.'

Climate is clearly not a function of latitude only. I take it that Axelrod's argument against large variations in the distribution of climate with latitude does not forbid similar changes of mean temperature over a parallel in all latitudes. There would be no definite astronomical evidence against this. Changes in the distribution of mountains would certainly affect climate. The dry belts east of the Rockies and the Siwaliks (Punjab) are due to moisture having fallen out of westerly winds in crossing the mountains. The formation of the Hercynian system (Permian) is probably responsible for the salt beds of Cheshire and South Germany and the Yellow (desert) Sands of Northumberland and Durham.

Barber *et al.* (1959) report on logs transported up to 10,000 miles, and say: 'It does not seem impossible for seeds to be overgrown in the wood, or, as

Darwin observed, trapped and sealed into the interstices of the roots and then transported from one continent to another. This remote chance, which has presumably been available for the 100 million years since the origin of the angiosperms, may be part of the explanation of the puzzling floral similarities of the southern tips of the continents.'

King-Hele (1963) quotes from Erasmus Darwin evidence for transport of live seeds from N. America to Norway; they germinated.

The difficulty about long migrations across the ocean can be mitigated in many cases by supposing that they took place along chains of islands.

The meteorological evidence comes partly from former glaciations, of which the Permo-Carboniferous is the most famous, and partly overlaps with palaeontology, since types of plants are found as fossils in latitudes far from those that would appear appropriate to them. The ready acceptance of this evidence springs mainly, I think, from the popular belief that meteorology is an easy subject; everybody knows when it is raining. But also everybody (in this country at any rate) finds fault with the weather forecasts if he reads them. The latter fact is the true indication of the difficulty of meteorology—even experts in it may go badly wrong in trying to predict for 12 hr. or less. Turbulence in fluid motion, even in its simpler cases, apart from a few empirical rules, is comparable in difficulty with nuclear physics. In the atmosphere we have all the problems of turbulence complicated by the importance of rotation and the proximity of a nearly spherical boundary. Though progress is being made, the facts remain that we have only the roughest possible explanation of the general circulation of the atmosphere, and that the problem of predicting the path of a cyclone of temperate latitudes is a modification, with additional complexities, of that of predicting the motion of an individual eddy in a turbulent wind channel. In the second edition of this book, Appendix C, I gave an account of what can be inferred about explanations of climatic variation by the methods yet available; I see little to modify in this, but feel less hopeful than I did then about the adequacy of the methods, and am inclined to think that ancient climates are more likely to tell us something about meteorology than vice versa.

Perhaps the best-known argument for continental drift is the alleged fit of South America into the angle of Africa. On a moment's examination of a globe this is seen to be really a misfit by about 12°. The coasts along the arms could not be brought within hundreds of kilometres of each other without distortion. The widths of the shallow margins of the oceans lend no support to the idea that the forms have been greatly altered by denudation and deposition; and if the forms had been altered by folding there would be great mountain ranges pointing towards the angles and increasing in height or width, or both, away from the angles, which is not the case. Similar misfits are found in comparing North America with Europe (Lake, 1922). The submersion of the North Atlantic continent is avoided by having

North America and Europe in contact. The resulting picture of both continents developing from each other's sediments is admirable. A petrological comparison of the regions alleged to have been in contact leads to a further set of inconsistencies (H. S. Washington, 1923). It should be said that most present supporters of continental drift admit that Wegener's reconstruction is more or less unsatisfactory; but they continue to republish his maps with little or no indication of how they would alter them.

Bullard, Everett and Smith (1965) agree that the misfit mentioned above is genuine, though even since their paper maps have been published that suggest that the fit is good. However they call attention to the shallow region off the coast of South America, which widens greatly to the south. In consequence the 500-fathom lines could be made to fit much better. However this reconstruction involves a new problem. Why should the South American continental shelf have gone down? Vertical movement is needed on a larger scale than in the instances we have already mentioned. In suitable conditions either steep or nearly horizontal displacements can be explained, but not both at once.

Lyustikh (1965, 1967) shows (Fig. 2) a series of many very similar outlines of different coasts. With many to choose from it would not be surprising if two selected ones should face each other across an ocean.

I acknowledge that if Greenland is taken as part of North America the coasts of America and of Europe and Africa are roughly parallel; but if this is taken to imply former contact it may well have been in the convective stages of continent formation. If an original continent was formed by a first harmonic it could conceivably have been torn apart later by the third and fourth. Too much must not be expected from such a hypothesis, since there have certainly been later changes of form due to mountain formation, vertical movement, and probably continental growth.

Meyerhoff and Meyerhoff (1972, pp. 316–318) and Meyerhoff, Meyerhoff and Briggs (1972, p. 682) quote many discoveries and determinations of ages of suboceanic rocks. Perhaps the most striking was of trilobites of Cambrian age described by Furon in 1949 from two separate localities. Many ages going back $(1699 \pm 55) \times 10^6$ years have been determined by the potassium–argon method. The evidence should dispose finally of any idea that the Atlantic did not exist before the Cretaceous.

Meyerhoff and Meyerhoff (1972*a*, pp. 316–18) and Meyerhoff, Meyerhoff deposits) and that of coals, which require a warm and damp climate. He concludes that the distribution, apart from variations due to the elevation of mountains, the zonation in latitude has moved north and south from time to time, but about equally in all longitudes. This implies that the pole of rotation has not moved greatly since Cambrian time. It appears that climatic variation is a consequence of variation of solar radiation. Brooks (1926, 1949) long ago showed that quite a small variation in this can produce a considerable change in temperature, mainly because if ice forms much of the radiation is reflected instead of being absorbed.

Biswas (1969) is particularly concerned with evidence that the continents have risen and the ocean floors have sunk during geological time.

Evidence from rock magnetism interpreted as in favour of polar wandering or continental drift has accumulated in the last 20 years. It is found that the direction of magnetization of rock specimens in many places is different from the present direction of the Earth's magnetic field there, and it is inferred that the direction of the Earth's field relative to the surface has changed. For some time the results for rocks of the same geological age were consistent with a shift of the magnetic axis, and on the assumption that the magnetic axis was always near to the axis of rotation they were held to show that the axis of rotation has shifted in the Earth. Examination of suggested causes of displacement had led to nothing satisfactory, but the mechanical arguments against polar wandering are nothing like so strong as those against continental drift.

However, work by Blackett (1956) has shown rock magnetization in India and Australia in directions far from the direction of the field indicated by measures in the northern continents, and it has been asserted that they prove continental drift. But the evidence for irregularity of the ocean floor is now stronger than ever; and I shall continue to believe that a viscous material would flatten out under gravity however many physicists treat it as of no importance.

When I last did a magnetic experiment (about 1909) we were warned against careless handling of permanent magnets, and the magnetism was liable to change without much carelessness. In studying the magnetism of rocks the specimen has to be broken off with a geological hammer and then carried to the laboratory. It is supposed that in the process its magnetism does not change to any important extent. I have been told that the magnetic minerals, magnetite and haematite, stand ill-treatment better than steel does. But I remain doubtful. In fact the modern study started with the announcement that many old rocks showed magnetism in opposite directions to the present field, but later work appeared to show that the magnetism at neighbouring geological dates appeared to concentrate about a direction and its opposite; since then reversals have been ignored, it being usually supposed that the Earth's general field is liable to sudden reversal but not to intermediate shifts. The reason for the reversal, it being supposed genuine, remains unknown.

J. W. Graham (1956) makes an interesting suggestion. It is known that under non-hydrostatic stress solids can become magnetized; the phenomenon is known as magnetostriction. In general the effect is not permanent, passing off after the stress is removed. Graham suggests that this may not be so when the stress-difference persists for a long time. If so, it would explain why the directions of magnetization of rocks over a wide area are nearly coincident but depart from the general field, without there needing to be either polar wandering or continental drift. He gives examples that appear difficult to explain on either of the latter hypotheses. Meyerhoff points out that the scatter of positions found for the magnetic pole in the same geological period exceeds the whole width of the Atlantic.

No help would be derived by rejecting the conclusions from the damping of the variation of latitude. It would be quite out of the question to take this as more rapid. If it were less so, all the difficulties pointed out here would become more serious than ever.

The success of the Lomnitz law of imperfect elasticity, or the modified form of it, would make discussion of hypotheses based on elasticoviscosity pointless if adequate attention was paid to it. Convection continues to be treated as a cure for nearly all ills. It has not yet been claimed to provide an explanation of the data used in the new laws. The alleged displacement of America since the Cretaceous would correspond to a velocity of order 10^{-6} cm./sec., always in one direction. This offers no hope of appreciably affecting motions periodic in from a second to a year.

A review of a collective work on continental drift by S. K. Runcorn and others is by G. J. F. MacDonald (1963). An extensive discussion opened by Blackett appears also in *Phil. Trans.* A **258**, 1965.

Egyed maintains that the Earth has expanded during geological time. C. H. Barnett (1962) notes that to make the continents cover the whole surface, the initial surface would have to be about $\frac{1}{4}$ of the present one. He tests this by outlining the continents on a $4\frac{1}{2}$ in. sphere on to rubber sheets and transferring to a 3 in. sphere. With suitable sliding a fair fit is obtained. The following points however should be noted. A portion of a spherical surface cannot be transferred to one of different radius without distortion, and the quality of the fit must depend on the distribution of the stretching. If a fit is obtained in this way, it is clear evidence that the continents have been distorted, and the case for Wegener's fit (bad enough already) disappears.

Again, the interpretation of the palaeomagnetic and fossil evidence requires the separation of America from Europe and Africa to have occurred since the Permian and possibly since the Cretaceous. If this was due to a general expansion we may well ask why it was so long before this happened.

The increase in volume would imply that the previous density was about 40. The only hope of an explanation of the change is a change of the constant of gravitation, by a larger factor still. There is however no adequate confirmation.

Magnetic anomalies on the Atlantic ridge and off the coast of California have been interpreted as due to displacements of the order of 100 km. Whether this is correct or not, such displacements are of the order of those found in mountains, for which explanations exist; there is no reason to suppose it legitimate to extrapolate to displacements of whole continents for thousands of miles.

A volcano off Iceland has been claimed as support for drift, but there are volcanoes on the land, and whatever the explanation of volcanoes may be there is nothing surprising about the occurrence of a new one near by. These are all examples of the reckless claims that get into the newspapers.

12·061. Sea floor spreading and plate tectonics. For a recent bibliography, see McKenzie and Weiss (1975). To sum up some results in previous chapters, I recall that:

(1) Thermal contraction in a cooling Earth can produce stress-differences capable of producing fracture, and this would normally be at about 45° to the vertical. Heated rocks elevated to the surface can flow approximately horizontally under gravity. The amount and distribution of mountains and the formation of nappes are explained by the theory. The vertical distribution of deep earthquakes is also explained.

(2) No other mechanism has been shown capable of producing stress-differences of the right magnitude. It should be noticed that any sort of imperfection of elasticity at stress-differences below the strength tends to reduce the stress-differences and hence the likelihood of fracture. Fracture takes place in spite of it, not because of it. Further, to explain the variety of geological phenomena, we need as much variety as we can get; and fracture provides it.

(3) The only suggested explanation of the fact that nearly all the land is in one hemisphere is convection while the Earth was in at least a partly fluid state. Differentiation during this stage explains also the upward concentration of radioactivity. It implies also that the ocean basins existed from a very early stage, possibly 10^9 years from the Earth's formation. The force between floating continents is an attraction, not a repulsion; absurdly small, but in the opposite direction to the supposed continental drift.

(4) Many old mountain systems appear, from the amount of folding in them, to have been originally of height comparable to the Alps, but have been denuded to heights of order 1 km. If this was done and isostasy was maintained, the denudation should have removed all sediments and cut deeply into the granitic layer. This has not happened. The only explanation is that they have been pulled down from below.

(5) Geophysicists that support continental drift, and even those geologists that follow them, seem completely uninterested in vertical movement, which has taken place on a tremendous scale. Geodesists still assume that the Earth is always tending to a state of perfect isostasy, in spite of the evidence that great changes in the opposite direction have continued over intervals of order 10^7 years.

(6) Radioactivity may well have regional variations. Stresses due to differences of cooling would account for fractures at depths of some hundreds of kilometres, leading to the surface and accounting for changes of level of the observed order.

Advocates of convection and related hypotheses make no reference to the cases where the theory of thermal instability is properly worked out, nor to its experimental verification. Some actually admit that the proper equations for their hypotheses are unknown, but consider that does not matter. In this respect Wegener was better than his modern admirers; he did suggest a mechanism, even though it is hopelessly inadequate in amount and would give movements in the wrong direction.

The alleged evidence now available is largely from slow movements over a few years detected in geodesy. Extrapolations of these are freely made over intervals of order 10^8 years. Now we know of plenty of earthquakes with displacements of a few metres; mountains have required shortenings of the crust of order 50 km., possibly more in the Himalayan region. But these movements have happened and then stopped. There is no geophysical evidence whatever that they can continue at a uniform rate over thousands of kilometres. The longest series of observations that might be relevant are those of the International Latitude Service from 1899 to 1967. Averages over intervals of 7 years in the displacements of the pole (with reference to a standard origin) towards Greenwich fluctuated irregularly between $-0.015''$ and $+0.053''$; to N. America from 1899 to 1947 between $-0.039''$ and $+0.003''$. But up to 1961 the latter decreased to $-0.164''$ and to $-0.225''$ in 1967. The uncertainties are about $0.005''$. This behaviour is completely inconsistent with a uniform rate of movement in either the pole or the observatories. ($1''$ of arc would correspond to about 31 metres at the surface.) The evidence for spreading of the sea floor is magnetic (Dietz, 1961; Vine and Matthews, 1963; Vine, 1966). Along lines nearly perpendicular to the Mid-Atlantic Ridge, the magnetic anomalies show reversals in sign, which are identified with reversals of the general magnetic field. It is supposed that the matter flowed out horizontally from the ridge, and that the reversals correspond to changes of the field where it started. I know of no evidence that there were not successive outflows at or near the planes where the anomalies are now. It is claimed that the anomalies are in straight lines, but detailed maps show that they are very irregular.

The supposed extension of the sea floor would imply an increase of the area of the Earth's surface and hence of its radius. It has been suggested that the expansion could be due to a reduction of the constant of gravitation and consequent relief of pressure. This has been connected with cosmological theories. But the geological evidence appealed to requires that the spreading began in the Cretaceous, and no cosmological theory suggests that the gravitational 'constant' remained approximately the same for most of the Earth's age and halved during the last $1.5\,\%$. The explanation by convection originally assumed the convection to extend through the whole of the shell. This seems, however, to have been abandoned in recent writings, which take the depth of the convective layer to be of order 400 km. However, in thermal convection the motion is in cells, whose radii are of the same order as the depth. There is not the faintest sign of a motion in the same direction for thousands of kilometres. If the meaning of 'convection' is extended to include non-linear behaviour, this must be expansion or contraction of a solid with a stress difference near the strength, and the result would approximate to 45° fractures. Again there is no sign of horizontal displacements in one direction of the magnitude assumed. This is of course the Lake–Wilson theory of island arcs, which is part of mine.

All these considerations support the hypothesis that the main influence in tectonics is not any effect of forces at small depths but of changes at much greater depths, probably of order 400–700 km.

Several authors give a map showing the reconstruction of Gondwanaland according to supposed continental drift; this is reproduced by Meyerhoff and Teichert (1971, p. 313). South Africa, South America, Antarctica, Australia and India are supposed to have been in contact. All had glaciers in the Permo-Carboniferous. But most of these glaciers would have been far inside the supposed continent and the moisture needed to form them could not have reached them. Brooks (1926) long ago studied the present distribution of ice in detail and showed the importance of distance from the sea in controlling it.

A. A. and H. A. Meyerhoff (1972a) infer that the 'so-called linear magnetic anomalies are features which first formed in Archaean times and have been modified since'. One of their arguments is that some of them extend into the continents, where the rocks are not later than Devonian. They say that the anomalies are approximately concentric with Archaean shields.

They also refer to measures of Grønlie and Ramberg (1970) showing basins west of Norway with 7000–10000 m. of sedimentary rocks, with marine strata certainly of Jurassic and probably Permian age. These basins extend to the Mid-Atlantic Ridge. The sediments in them are flat. Therefore sea-floor spreading during the Mesozoic and Cainozoic is not possible in the North Atlantic.

An extensive criticism is by V. V. Beloussov (1970). I cannot go into detail but mention a few points. It is found that the present normal magnetic field has lasted 0.7×10^5 years. Before that the field was reversed for 1.7×10^6 years; before that there was a normal field lasting 1.1×10^6 years. Before that the geomagnetic time scale is undetermined. If spreading had taken place at a uniform rate the corresponding distances should be in the ratios 1.0, 2.4, 1.6. Of four Atlantic regions three agree very roughly with these ratios, but one is completely discordant. The northern Pacific is discordant. Six different sections in the South Pacific are all discordant.

In any case the rates of spreading required differ greatly in different regions.

Other oddities are found near Iceland and the Aleutian arc. In the latter the strips of magnetic anomalies show a bend at right angles. Near Iceland they continue on land and connect up with Pleistocene volcanoes.

Beloussov quotes Byrd (1967) and Van Andel (1968), who consider that the mid-ocean ridge is formed of successive basalt layers, the younger being the narrower.

A. A. Meyerhoff and Teichert (1971, p. 286) say: 'For each Palaeozoic and Mesozoic faunal similarity "explained" by joining the continents, several dissimilarities are left unexplained ... The floral similarities are explained readily by wind and insect dispersal, as are today's remarkable similarities in many parts of the Southern Hemisphere ... Most Permo–Carboniferous glacial ice caps were close to present coastlines, or within reach as today of moisture-laden winds from the Atlantic, Indian, and Pacific Oceans.'

Wesson (1970, 1972) gives very detailed accounts of several sides of the problem. I might call special attention to his reference to Horai (1969), who suggests that the observed equality of oceanic and continental heat flow may be an effect of the last ice age.

The Meyerhoffs (1972a, p. 319) give a comparison of A. H. Voisey (1958) of the eastern coast of Australia (rotated) and that of North America. The fit is better than any of Lyustikh's. Also (pp. 337–59) the magnetic anomalies are approximately concentric with Archaean shields.

Biswas (1969) finds no evidence for extensive glaciation within India in Permian times. The rocks interpreted as tillites appear to be mainly due to wind erosion. He calls attention to places within 100 miles in the same latitude, which would have to be interpreted as coral reefs and glaciers at the same time.

He is largely concerned with oceanic influence on climate. One paragraph (1969, p. 4) seems confused. It reads 'Heat conductivity of water is considerably less than that of rocks. The higher the areal extent of oceans and seas, the more uniform should be the distribution of temperature due to solar radiation.' Ordinary conductivity in the ocean is much less than that due to turbulence, and deep currents play a large part in heat transfer. The second sentence is correct for this reason.

He is impressed by the extensive areas that were once sea and are now land, and attributes the change to deepening of the oceans and rise of the continents throughout geological time. This may be related to Meyerhoff's emphasis on variation of the widths of climatic zones. Incidentally, if thermal contraction is greater under the oceans Biswas's conclusion would be consistent.

As the papers of Meyerhoff and his collaborators amount to about 250 pages, including about 40 pages of references (both for and against his views) I have been able to give only a few abstracts.

95 % of evaporites (anhydrite, gypsum, rock salt, potash salts) from late pre-Cambrian to Miocene are in belts that now receive < 100 cm. of annual rainfall; the positions of the desert belts have remained nearly constant for 8×10^8 years; the evaporite zones since pre-Cambrian and the coal since Devonian have been symmetrical with respect to the present axis of rotation. The mean latitudes and widths of the zones have not altered greatly.

Notably the thermal equator has been systematically north of the equator all the time. Deviations from symmetry have time and space relations with the distribution of warm ocean currents and the tectonic history.

It is pointed out by Rezanov (1968) that the palaeomagnetic reconstruction puts Japan and the Kuznetsk basin of Central Asia in the same place in middle Cretaceous time! The Meyerhoffs say (1972*a*, p. 319) that there are many such inconsistencies. I may perhaps say that I had long expected that they would exist, but actual cases seem to have been long in turning up.

Creer and Ispir (1970) have used several series of observations of rocks formed during reversals of the magnetic field. It has been supposed that the field just clicks over, but they find that the magnetic pole usually wanders continuously from north to south or back. In one case it nearly reached the equator and then returned. These changes took short times geologically; but the rates of movement far exceed those inferred by believers in continental drift. What is clear is that palaeomagnetism is a far more complicated matter than has yet been assumed.

There have always been many critical geologists, for instance P. Lake, H. Stille, O. T. Jones and W. B. R. King. But few of them stated their objections in print or in full. The American Association of Petroleum Geologists has published Memoir 12: *North Atlantic Geology and Continental Drift* and Memoir 23: *Plate Tectonics – Assessments and Reassessments*.

Plate tectonics is a further development. The essential feature is that the land is supposed to be composed of strong plates (usually called *rigid*, apparently on the principle that a word of two syllables is to be preferred to one of one syllable, even if it has already a different meaning). These are supposed to be capable of displacement over the weaker substratum. The problem is then, if the sea floor is spreading, where does it go? The answer proposed is that it is dragged down under the plates by the convection current.

The mechanics of this process is difficult. In the first place, there are what seem overwhelming objections to thermal instability involving stress differences below the ordinary strength of the material, which we know, from the support of mountains and ocean deeps and from the analysis of gravity, must be about 2×10^9 dynes/cm.^{2} down to 50 km. and 3×10^8 dynes/cm.^{2} down to, probably, 700 km. or so, and possibly 10^8 dynes/cm.^{2} down to the core. Secondly, even a material originally granular, at depths of order 2 km., would have its grains deformed under the pressures, and would become thoroughly welded. This motion of the material in contact with continental matter would be resisted by the full shearing strength, which would be equal to that in the plate. It is suggested that basaltic material from the sea floor would change to eclogite and that the increase of density would help; but the density of eclogite is nearly the same as that of olivine and the suggestion gives little help. The advocates of this hypothesis may be thinking of sticking a knife into butter. But the inserted matter will have at least the stress-differences of its surroundings, and if the materials are of comparable strength it will either shatter or crumple at once.

When these considerations are taken into account it is clear, in the first place, that stress-differences at least of the ordinary strength are needed, and such stress-differences can be achieved by thermal contraction or possibly expansion. Then the theory differs from the one I gave in 1932 only in having a plate travelling downwards instead of a fault extending upwards.

Stacey (1969, p. 210) estimates from the San Francisco earthquake of 1906 that the strength near the surface was about 9×10^7 dynes/cm.^{2}, and supposes that it must decrease with depth. This is in flat contradiction to the fact that the support of mountains needs at least 2×10^9 dynes/cm.^{2} to a depth of order 50 km.

By a thermodynamic argument he infers that to get a velocity of 4 cm./year in convection an upper bound to the strength of the lower mantle is $1 \cdot 4 \times 10^7$ dynes/cm.^{2}, and says that the departure of the ellipticity from the hydrostatic value implies one of 10^8 dynes/cm.^{2}. The latter is about my value (p. 280). Anyhow on this ground he infers that convection cannot be in the lower mantle and must be in the crust. Since my argument is that the actual law of creep makes a steady velocity impossible I can accept the result.

P. J. Wyllie (1971) assumes a lithosphere 160 km. thick dipping at 45°. If such a thickness is intruded it must raise the surface by 200 km. or so, and the excess would have to be denuded away. The difficulty that I mention about the lowering of mountains is multiplied by 40 or so. I should say that Wyllie admits that there may be something wrong somewhere. Incidentally, to push the plate under such a load would require an enormous force; in plate tectonics this is overlooked, though the authors make much of the difference in density between basalt and eclogite.

The methods used in argument on this subject need comment. The inferences made by those advocating the theory (perhaps 'tectonicians' would be an appropriate term) are habitually stated as facts, with no reference to any objections that have been made. It has been stated repeatedly, not only in scientific journals but in newspapers and textbooks that their beliefs are 'universally accepted'. The reader of this section will find that this is not so. For example, although Stacey (1969) gives references to me they are not to any constructive result that I have given. Attacks on Meyerhoff's work did give rise to an article in *Nature Physical Science* (**239** (1972), 137–8) in which the writer defended the right of 'heretics' such as Beloussov, Meyerhoff and myself to have our views heard and seriously considered.

12·07. Growth of water. An ocean presumably formed at a very early stage in the Earth's history. But volcanoes are continually ejecting steam, which must ultimately be added to the ocean. The estimation of its amount would be difficult, but it is quite possible that a considerable fraction of the ocean has been added during geological time.

An interesting set of formations that may be relevant to the problem are discussed by H. H. Hess (1946). In the North Pacific, mostly between 10° and 25° N., 145° and 180° E., there are numerous elevations, which Hess calls *guyots*. Their shapes suggest volcanic cones with the tops smoothed off at levels from 1 to 2 km. below the ocean surface and therefore several kilometres above the normal ocean floor. The interpretation suggested is that they are genuine volcanic cones that have been denuded to their present form; but, if so, the ocean surface must have been something like 1·5 km. below its present position. They are certainly older than the atolls, and Hess thought that they are pre-Cambrian. If the quantity of water in the ocean has increased we have an immediate explanation.

Hess gives a different explanation, based on work of Ph. H. Kuenen (1937, 1941). Kuenen estimates that the rate of sedimentation in the deep ocean is about 1 cm. in 10,000 years for the red clay and 1 cm. in 5000 years for the Globigerina ooze, which would imply an accumulation on the sea bottom of the order of 1 km. since the pre-Cambrian. Hess supposes that the ocean surface has risen because the bottom was raised by these sediments. The estimate is of the same order as we got from the amount of sodium in the sea, and it seems possible that both processes have played an important part.*

12·08. The surface features of the Moon. † The most conspicuous features on the Moon's surface are the craters and the maria. The latter are extensive plains, looking darker than the rest; most of the rest is very irregular and thickly covered with ring-like formations mostly known as craters, though some of the largest are also known as walled plains. A few craters are also to be seen in the maria, but they are very much more sparse than outside them. From some of the larger craters long bright streaks extend. These lie approximately along great circles. No folded mountain ranges have been recognized, though there are long ridges sufficiently like mountain ranges to have received the names of the Alps and the Apennines.

The craters in fact bear little resemblance to most present terrestrial volcanoes. The characteristic form of the latter is conical, the slope being determined by similar considerations to scree slopes. It should be said that the latter are usually said to represent the 'angle of repose', which is not strictly true. A mass of fragmentary matter can certainly be built up to such a slope that friction between particles ceases to be able to preserve equilibrium. But the possible slope is capable of wide variation and the

* I understand that some later work reports cretaceous fossils on the guyots. If so, both these explanations would be inadequate.

† The treatment follows that of previous editions, without taking account of lunar exploration since 1969.

uniformity of slope of screes is not explained. But the formation of screes is essentially a dynamical matter. A stone falling at the top and bouncing or rolling down is gaining energy from gravity, but losing it by impact. If the former is the greater the stone will go to the foot; if the latter, it will be stopped after a few bounces. Thus a steep slope will be made less steep by building up from the bottom; a gentle one will be made more steep by accumulation at the top. There will be a critical slope such that all parts rise uniformly. It is easy to see that this slope should be nearly uniform. The loss of energy at an impact is the product of the kinetic energy and a numerical factor depending on the coefficients of restitution and friction and on the shapes of the stones. They will fall in varying orientations, but on an average this factor will have a definite value for each material. Then the condition that the loss of kinetic energy on impact must on the whole balance the gain by gravity between impacts determines the slope. But both contain gravity as a factor, and hence the steady slope should be uniform and independent of gravity. It will probably be a little less than the statical repose angle. For volcanoes solid fragments make up far more of the ejected matter than lava flows, and similar considerations apply. On the Moon gravity is about a sixth of what it is on the Earth, but this will not affect the angle of a scree. A lunar volcano, if produced in a similar way to a terrestrial one, would be conical with a small crater on the top, and the slope of the sides would be very similar. But the actual lunar craters show little sign of scree slope, and the crater itself makes up the greater part of the area. Some craters have a central peak. In some early accounts this was regarded as the essential part of the volcano, and the rim was supposed to consist of ejected matter. This fails to account for the flat floor. On the other hand W. W. Campbell detected depressions at the top of some central peaks, which may therefore be true volcanoes, formed after the main crater.

A discussion of geological possibilities was given by Daly (1946), who finally abandoned them all and adopted the hypothesis that the craters are due to the infall of small independent bodies. Daly, however, did not mention what seem to me to be the nearest terrestrial analogues, the ring dikes and cone sheets of Ben Nevis, Glencoe, Arran and Ardnamurchan (discovered about 1909; for references and mechanical discussion, see E. M. Anderson, 1936, 1942). In these a fracture appears to have started internally and extended to the surface in a cone; magma is ejected along the margin. The diameters are of the order of 4–10 miles. Kilauea, in Hawaii, is a large lava lake of similar dimensions, not at a high elevation. See Plate II (5) in vol. 14 of Chambers's Encyclopaedia. The diameters of lunar craters reach something like 100 miles, but the difference of scale does not appear to be a serious objection. The angle of the cone is determined by conditions of fracture, and the diameters would depend mainly on the depth where the fracture started. This is a point where the difference of gravity on the Earth and Moon might well have a determining influence.

There are signs that the activity on the Moon took place in three main stages. In the first the irregular elevated parts were produced, possibly by a convective process similar to that considered by Hills for the explanation of the continents; in the second great outpours formed the maria, which appear in places to have encroached on the rims of earlier craters; and in the third the craters with bright rays were formed. The last are certainly later than the maria because the rays cross the maria. A few craters were also formed within the maria.

The bright streaks by themselves seem to show that the associated craters were formed by explosion, the streaks representing showers of ejected dust, and the flat floors that the sites were partly filled up by inflowing magma. Now in the early history of the Moon, as in that of the Earth, it may be supposed that the expulsion of water dissolved under pressure played a large part. If rising steam reached the top the release of pressure would accelerate the volatilization, and the result would be an explosion; and if the underlying material was still semi-fluid there seems to be no reason to look further for an explanation. See Plate I (6) in vol. 14 of Chambers's Encyclopaedia for an illustration of this phenomenon on a small scale in Yellowstone Park.

A rough test of whether internal explosions could have produced the bright streaks is as follows. On a horizontal plane the maximum range of a projectile with initial velocity V is V^2/g. Some of the bright streaks may be 1000 km. long, and lunar gravity is about 180 cm./sec.². This makes V about 1·3 km./sec., of the order of the velocity of a shell. Allowance for curvature of the surface and variation of intensity of gravity would decrease this. The Moon's radius is about 1500 km., and the velocity in a circular orbit above the surface would be about 1·6 km./sec.; that is, this velocity would make the possible range over the surface infinite.

The velocity of efflux of a gas into a vacuum is somewhat indefinite, since the molecules come off with different velocities. The mean velocity of a water molecule at 0°C is about 0·6 km./sec., and at 1000° would be about 2·2 times this or 1·3 km./sec. About 5 % of the molecules might come off with twice this velocity, and in any case an appreciable quantity of dust could be carried off at similar velocities. At any rate it appears that explosion with expulsion of water vapour at magmatic temperatures could give velocities of the right order.

Krakatoa sent material into the stratosphere. The height of the stratosphere near the equator is about 30 km. This would imply a vertical velocity of at least 0·8 km./sec. The actual initial velocity would be greater since energy would be lost through air resistance.

An alternative, due originally to G. K. Gilbert (1892) strongly supported by L. J. Spencer (1933, 1937), and accepted by Daly (1946), Urey (1952) and others, is that the craters are due to impact of small bodies moving in orbits about the Earth or the Sun. This is not so different from the last hypothesis as might appear. The relative velocity of a meteor or asteroid in a highly eccentric orbit about the Sun and striking the Moon would be of the order of 40 km./sec. That of a terrestrial satellite might be anything from zero to 1 or 2 km./sec. Even 1 km./sec. would correspond to a kinetic energy of 5×10^9 ergs/g. and with a specific heat of $0 \cdot 3$ dissipation of this energy would lead to a rise in temperature of about 400°. The dissipation of planetary velocities by impact would certainly lead to complete volatilization. There would be some penetration before volatilization was complete, and the effect of an impact at a planetary velocity would be indistinguishable from that of an explosion below the surface, because it would *be* one. Accordingly it seems that it would be impossible to decide between these two theories from any evidence about the present appearance of the craters, and other types of evidence are needed. That explosions could give something very like the craters has been shown by S. Mohorovičić (1928), who exploded pieces of a mixture of magnesium powder and potassium chlorate under a plastic surface (Plate X).

The actual method of conversion of energy into heat on impact does not appear to have been explained, but seems to be as follows. We are concerned with velocities above that of a longitudinal wave. Then in each body there will be a cone of discontinuity of velocity and stress. In a solid this implies stress-differences. Direct heating by adiabatic compression is probably fairly small; the main effects will be flow and fracture. In the latter case the material will be pulverized and solid friction will produce liquefaction. In either case further dissipation in the region of flow will give volatilization. When the shock waves have become attenuated in these various ways they can spread further into the remaining solids as ordinary elastic waves.

To explain the flat floors we must suppose that the craters were formed while there was still general fluidity a short distance below the surface, and this in itself suggests that the larger craters were formed at an early stage in the Moon's existence. We cannot expect to find corresponding features on the Earth, because they would have been covered or denuded away. In itself the indications that the largest craters are also among the latest would agree with the idea of a progressive thickening of the crust—the deeper the explosion the more rarely would stresses arise to break the crust, and the larger would the craters be when explosion did occur. But even the largest could probably have been formed by explosion at a depth of 50 km. or so, and the time needed for solidification to that depth would probably not be more than 10^7 years, perhaps much less.

This argument would be consistent with the theory that the explosions were due to the liberation of internal gases; the loss of the gases itself would play an important part in promoting solidification. The two necessary conditions, a thickening crust and decreasing activity of the principal stimulus, would naturally go together on this hypothesis. But it makes it very difficult to accept the impact theory, since there is no apparent reason

33

why the wandering bodies should have become exhausted at the right time. On this theory there appears to be no reason why the maria should not be thickly peppered with craters.

The impact theory would require the infall of a few large bodies and many small ones in the first stage; then a few enormous ones to make the maria; a few large ones again to make the bright-ray craters; and then almost complete cessation. It seems very difficult to get such behaviour out of an essentially random process, particularly since collisions among the bodies themselves would tend to make them smaller and probably more numerous as time went on.

The absence of survivors suggests so perfect a vacuum pump that not a single molecule remained. A body 12 km. in diameter at the Moon's distance would appear to us as of stellar magnitude 0, i.e. somewhat brighter than Vega, one of diameter 1 km. as of magnitude 5. They would be easily visible if they existed. In some variants of the meteoritic theory the bodies are supposed to have been satellites of the Earth, so that any survivors would be visible all the time.

A possible loophole might be that the maria might be recent. But flows on such a scale would suggest a continuous fluid layer, and with such a layer near the surface the excess ellipticity would have subsided. There are again possible escapes from the difficulty, but I feel that the meteoritic theory needs too many narrow escapes.

Large meteors are known to have occasionally produced considerable holes in the Earth, the Meteor Crater of Arizona and the Siberian Meteor of 1908 June 30 being famous examples (Russell, Dugan and Stewart, 1945, pp. 455–457). The holes are of the form of half an ellipsoid, not flat-bottomed. There may well be occasional falls such as these on the Moon, which might be visible to us if they were on the unilluminated portion, but there appear to be no records.

Baldwin (1949) makes a very careful discussion of the impact theory, with particular attention to meteoritic craters on the Earth. He estimates (pp. 91–92) that on the Earth one crater or group of pits is formed per 5000 years. Allowance for the difference in size would give one per 60,000 years on the Moon, or something of the order of 30,000 in 2×10^9 years. He considers that this is of the order of the number of very small craters on the Moon; but these are younger than the main features.

Baldwin's account of the physics of impact and his comparison with explosions is most valuable, especially since there *are* meteor craters on the Earth. However, I think that (1) he underestimates the difficulty of the time sequence on the Moon, (2) he overestimates the difference between the striking craters and cauldron subsidences (ring dikes and cauldron subsidences are almost synonyms), (3) while he argues against explosion due to pressure generated internally, he does not mention Krakatoa, which at least showed that such things can happen. Some supporters of the impact theory maintain that the greater part of the masses of the Earth and Moon was acquired by accretion of large bodies, forming the large craters and the continents at a late stage. This must be totally rejected, for reasons given in 9·09.

Estimates of the present rate of meteoric infall are conflicting. From earlier data I estimated (1933) that it would cover the Earth to a depth of 2×10^{-5} cm. in 100 million years. This is so small that it appeared to provide a satisfactory explanation of why the bright streaks on the Moon, which are too shallow to cast a shadow, have failed to become buried and rendered invisible.

Watson (1945, p. 115) gives numbers per magnitude. With a correction to his table, in which a factor of 10 has been dropped, the mass per stellar magnitude falling per day is 110 kg., for magnitude -3 and fainter, down to magnitude 10. A meteor of magnitude 30 would be expelled from the solar system by radiation pressure; so the rate certainly falls off at that magnitude. If we extend the data over the range from magnitude -10 to $+30$ we get a total of 4400 kg./day. Watson suggests about 550 kg./day for those brighter than magnitude 10 (including the meteorites proper); this makes about 5000 kg./day in all. The covering would be about 1 cm. in 10^{10} years. This is substantially more than my previous estimate but still very small.

Petterson and Rotschi (1952) estimate from the nickel content of ocean sediments that the present rate is 1 cm./10^7 years; but though Watson's extrapolation is certainly rough it is hard to believe that it can be wrong by a factor of 1000; and estimates of the covering on the Moon as large as 1 km. have been given.

Even the lower estimate leads to an optical difficulty. It is natural to think of the deposited material as dust, which it would be on the Earth. If so we should expect the whole surface to reflect light like dust and there-fore equally well everywhere. This is plainly false. The maria are as dark as a blackboard. Most silicates when pulverized appear light grey. Urey (1952) suggests that the material covering the maria may be largely ferrous sulphide. Gold (1955a) favours the formation of dust in a quite different way but suggests darkening by prolonged ultraviolet radiation. Another possibility is that we may be quite wrong in thinking of the material as dust. However small the particles hitting the Moon at a planetary velocity may be, they would presumably penetrate and leave a fresh surface. Oberbeck and Quaide (1967, 1968) say that 86 % of the area is covered with fragments 5 to 15 metres thick, with an infilling of particles under 1 mm. diameter, slightly cohesive. I am indebted to Dr G. Fielder for the references.

The variations of temperature of portions of the Moon's surface during an eclipse have occasionally been measured and give a clue to the thermal conductivity. J. J. Gilvarry (1958) finds that a mountainous region near the south pole has a conductivity in the neighbourhood of 3×10^{-6} c.g.s. This would agree with dust and is far too low for a solid rock or even pumice. A region that included the Mare Vaporum gave a similar value. Urey (1956) has criticized Gold's mechanism for the formation of dust. It does appear, however, that there is evidence for a covering of some sort of dust; but unless this dust has stayed close to where it was originally formed it seems difficult to understand the great differences of reflecting power between different parts of the surface. J. Green (1971) examines the possibility that the crater Copernicus is a caldera, and finds it satisfactory.

I do not condemn the impact theory outright, though I do condemn some forms of it that are vigorously advocated. However, most presentations of it imply that the only igneous phenomena on the Earth are volcanoes of Vesuvius type, and this is grossly misleading, and so far as I know none of them pay attention to the statistical aspects of the theory.

The changes of temperature between day and night on the Moon far exceed those on the Earth, and it has been argued that thermal splintering would have produced scree slopes, which are not observed, unless the whole of the surface features were formed very recently. But direct observation in arid climates and laboratory experiment show that thermal splintering does not occur even at improbably large temperature ranges, unless assisted by water (E. Blackwelder, 1933).

Three explanations of the lack of folded mountains on the Moon appear to be available. (1) The upward concentration of radioactivity may be less complete. (2) The m term in the initial temperature contains gravity as a factor and would therefore be smaller than in the Earth. (3) On account of the lack of denudation by water, mountains would not acquire the deep carving that they do in the Earth. Thus the elevated strips on the Moon may resemble what our mountains were when first formed. I do not suggest that all these explanations should be accepted at once.

A very detailed discussion of lunar features, on geological lines, is given by J. E. Spurr (1945, 1948).

As this work has been condemned for no apparent reason except that it conflicts with the meteoritic theory, I thought it desirable to get an opinion from a petrologist, and consulted Prof. C. E. Tilley. He says 'I have no fixed ideas on the subject but I am glad to say that in my opinion a case can be made out for a volcanic origin of these craters (cf. Kilauea and Mauna Loa) as is indeed made by Spurr'. There has been much later work; the internal explosion theory appears to be gaining ground. See Fielder (1961, 1965, 1966), Marcus (1966 a, b), Fielder and Marcus (1967) and Fielder *et al.* (1972).

12·09. Geysers. In the problems (described in 11·09) of the stability of a layer of fluid heated below the horizontal extent of the system was taken to be large; an individual cell has dimensions comparable with the depth, so that the region was supposed broad enough to accommodate many cells.

But a similar problem of stability arises for a vertical tube of liquid heated below. Hales (1937) has studied this and gets a criterion of the same form, the radius of the tube replacing H. The application suggested is to the theory of intermittent boiling. With a sufficiently narrow tube it would be possible for water to be stable while near air temperature at the top and boiling at the bottom. With a slight increase of temperature, boiling would start at the bottom. This would expel some of the water and reduce the pressure, thus permitting more to boil, so that the result would be an explosive outburst. When this is over, new water percolates in and the cycle recommences. On the other hand, with a wider tube the heating would produce currents all the time and boiling would be continuous. Thus this type of instability suggests an explanation of the difference between steady and intermittent geysers. T. Thorkelsson (1928 a, b, 1940) shows that the actual outbreak does not correspond to boiling at the bottom, but to the release of dissolved gases; however, the principle is the same. There used to be a toy based on this principle, consisting of two glass bulbs, connected by a narrow tube, the lower being partly filled with a volatile liquid. Heating the lower bulb with the hand produced a cyclic process of the same type.

12·10. Rift valleys. These are long depressed strips, bounded by faults. The most famous is that of East Africa, which extends, with some branching and intermission, from Lake Nyasa to the Jordan Valley. The Rhine and the Leine (on which Göttingen stands) are other examples. The geological question is whether the bounding faults are normal or reversed. They were originally supposed to be normal and due to the strip having fallen in owing to tension from the sides. General tension near the surface is difficult to explain, and the existence of normal faults in any case calls for more theoretical attention than it has received. Bullard's study of gravity in East Africa (1938) was largely devoted to the Rift, and he found systematically negative isostatic anomalies over it. Hence the floor is lighter than normal, and should be tending to rise. It is difficult to see how any reactions across normal faults could resist this; the slightest rise of the floor would open the fault, and slipping up would then be unresisted. Accordingly Bullard considered the alternative hypothesis due to Wayland, that the faults are reversed as in the second figure, so that the Rift could be a feature of general compression. Here the force from the sides could have pushed the floor down and could continue to hold it down. Bullard considered the matter quantitatively and found the agreement satisfactory. The projecting edges would naturally break off and bury the faults under scree, so that it would be hard to test the question by direct geological observations.

Some geologists have, however, objected that major faults in general branch to a considerable extent, and therefore that if the main faults are reversed there should be subsidiary faults showing thrusting towards the valley; and these are not observed. This objection seems serious.

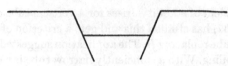

Fig. 29. Double normal fault.

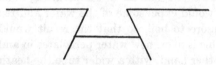

Fig. 30. Double reversed fault.

In view, however, of the evidence that many crustal movements are consequences of changes at great depths, it seems worth while to suggest a third alternative. Owing to the fact that Davison's examination of the thermal contraction theory came forty years before the discovery of deep-focus earthquakes, the theory has usually assumed plastic adjustment to stress-differences at great depths, which is no longer essential to it. The zone below the level of no strain is in fact subject to tension and can adjust itself by fracture. If it does, the outer parts will be *pulled down* and may in certain conditions develop what look like normal faults, but are not due to horizontal tension.

Without wishing to insist on this interpretation, I think that something on these lines would help to explain ocean deeps and possibly the contortion of the ancient shields.

APPENDIX A

Critical Stress-Difference

Take the directions of the principal stresses as axes and let $p_{11} > p_{22} > p_{33}$. Let the direction cosines of the normal to a given plane be n_1, n_2, n_3. Let l_i, m_i denote the direction cosines of two perpendicular lines in the plane. Then the normal stress is p_{nn}, with

$$p_{nn} = p_{11} n_1^2 + p_{22} n_2^2 + p_{33} n_3^2. \tag{1}$$

The shear-stress component across the plane and in the direction l_i is

$$p_{nl} = p_{11} n_1 l_1 + p_{22} n_2 l_2 + p_{33} n_3 l_3, \tag{2}$$

with a similar expression for p_{nm}. Then the shear stress in a direction making angles ϕ, $\phi - \frac{1}{2}\pi$ with l_i, m_i is $p_{nl} \cos \phi + p_{nm} \sin \phi$, the maximum of which is $(p_{nl}^2 + p_{nm}^2)^{\frac{1}{2}}$. Denote this by P. Then

$$P^2 = p_{11}^2 n_1^2 (l_1^2 + m_1^2) + 2 p_{11} p_{22} n_1 n_2 (l_1 l_2 + m_1 m_2) + \dots. \tag{3}$$

But

$$l_1^2 + m_1^2 + n_1^2 = 1, \quad l_1 l_2 + m_1 m_2 + n_1 n_2 = 0; \tag{4}$$

hence

$$P^2 = p_{11}^2 n_1^2 + p_{22}^2 n_2^2 + p_{33}^2 n_3^2 - (p_{11} n_1^2 + p_{22} n_2^2 + p_{33} n_3^2)^2. \tag{5}$$

We want P^2 to be a maximum subject to $n_1^2 + n_2^2 + n_3^2 = 1$; hence for all permissible variations

$$p_{11}^2 n_1 \delta n_1 + p_{22}^2 n_2 \delta n_2 + p_{33}^2 n_3 \delta n_3 - 2 p_{nn} (p_{11} n_1 \delta n_1 + p_{22} n_2 \delta n_2 + p_{33} n_3 \delta n_3) = 0, \tag{6}$$

$$n_1 \delta n_1 + n_2 \delta n_2 + n_3 \delta n_3 = 0. \tag{7}$$

Hence there is a λ such that

$$n_1 (p_{11}^2 - 2 p_{nn} p_{11} - \lambda) = 0, \tag{8}$$

with two similar equations.

First suppose that none of n_1, n_2, n_3 is zero. Then the equation

$$x^2 - 2 p_{nn} x - \lambda = 0 \tag{9}$$

has three different roots p_{11}, p_{22}, p_{33}, which is impossible. Hence the plane is parallel to at least one principal axis.

If $n_1 = 0$, we have

$$\lambda = p_{22}^2 - 2 p_{nn} p_{22} = p_{33}^2 - 2 p_{nn} p_{33}.$$

Hence $p_{22} = p_{33}$, which is excluded, or

$$p_{22} + p_{33} - 2 p_{nn} = 0, \tag{10}$$

that is,

$$p_{22}(1 - 2 n_2^2) + p_{33}(1 - 2 n_3^2) = 0.$$

But

$$(1 - 2 n_2^2) + (1 - 2 n_3^2) = 2 - 2(1 - n_1^2) = 0.$$

Hence, again since $p_{22} \neq p_{33}$,
$$n_2^2 = n_3^2 = \tfrac{1}{2}. \tag{11}$$

Then
$$P^2 = \tfrac{1}{2}(p_{22}^2 + p_{33}^2) - \tfrac{1}{4}(p_{22} + p_{33})^2 = \tfrac{1}{4}(p_{22} - p_{33})^2. \tag{12}$$

If n_2 or $n_3 = 0$ we proceed similarly; then the greatest value of P^2 is given by taking
$$n_2 = 0, \quad n_1^2 = n_3^2 = \tfrac{1}{2}, \quad P^2 = \tfrac{1}{4}(p_{11} - p_{33})^2. \tag{13}$$

Take
$$n_1 = \frac{1}{\sqrt{2}}, \quad n_2 = 0, \quad n_3 = \frac{1}{\sqrt{2}}; \quad l_1 = -\frac{1}{\sqrt{2}}, \quad l_2 = 0, \quad l_3 = \frac{1}{\sqrt{2}};$$
$$m_1 = 0, \quad m_2 = 1, \quad m_3 = 0. \tag{14}$$

Then
$$p_{nl} = \tfrac{1}{2}(p_{33} - p_{11}), \quad p_{nm} = 0, \tag{15}$$

and therefore the greatest shear stress is across a plane through the axis of intermediate principal stress and equally inclined to the two extreme principal stresses; and its magnitude is half the difference of the extreme principal stresses. The direction of the greatest shear stress is along the normal to the intermediate principal stress and in the plane of shear.

In Anderson's modification it is supposed that slip begins when the greatest shear stress exceeds the sum of a constant of the material and a frictional resistance, i.e. when
$$|P| \geqslant S + \mu p_{nn}. \tag{16}$$

Then (6) is replaced by
$$\frac{1}{P}\{p_{11}^2 n_1 \delta n_1 + p_{22}^2 n_2 \delta n_2 + p_{33}^2 n_3 \delta n_3$$
$$- 2p_{nn}(p_{11} n_1 \delta n_1 + p_{22} n_2 \delta n_2 + p_{33} n_3 \delta n_3)\}$$
$$+ 2\mu(p_{11} n_1 \delta n_1 + p_{22} n_2 \delta n_2 + p_{33} n_3 \delta n_3) = 0. \tag{17}$$

Subject to (7) it follows that
$$n_1\left(\frac{p_{11}^2 - 2p_{11}p_{nn}}{P} + 2\mu p_{11} - \lambda\right) = 0, \tag{18}$$

with similar equations; and as before one of $n_1, n_2, n_3 = 0$. Take $n_2 = 0$. Then
$$p_{11} + p_{33} - 2p_{nn} + 2\mu P = 0. \tag{19}$$

Also
$$P^2 = p_{11}^2 n_1^2 + p_{33}^2 n_3^2 - (p_{11} n_1^2 + p_{33} n_3^2)^2 = n_1^2 n_3^2 (p_{11} - p_{33})^2. \tag{20}$$

Hence
$$p_{11}(1 - 2n_1^2) + p_{33}(1 - 2n_3^2) \pm 2\mu n_1 n_3 (p_{11} - p_{33}) = 0.$$

Put $n_1 = \sin\theta$; then
$$(p_{11} - p_{33})(\cos 2\theta \pm \mu \sin 2\theta) = 0,$$

whence
$$\tan 2\theta = \pm 1/\mu. \tag{21}$$

When $\mu = 0$, $\theta = \pm \tfrac{1}{4}\pi$, agreeing with the Coulomb-Hopkins theory. When $\mu = 1$, θ may be $\tfrac{1}{8}\pi$ or $\tfrac{3}{8}\pi$. It is clear that the former arises when the normal stress is a thrust and the latter when it is a tension.

Subject to (16), $|P|$ must be greater than μp_{nn}. But the mean stress cannot be much different from $g\rho h$, where h is the depth; say 3×10^8 dynes/ cm.^{2} per km. Thus, at a depth of 3 to 6 km., the mean stress is greater than what appears to be the strength of the material. If μ is of order 1, as is often assumed, fracture would be impossible at greater depths. Since all evidence indicates that deep earthquakes are fractures, it appears that μ in (16) must decrease greatly with depth. But in that case (21) shows that θ must tend to $\pm \frac{1}{4}\pi$, and the fractures will be at about 45° to the extreme principal stress. For any hypothesis that involves changes of volume at intermediate depths an extreme stress will be nearly horizontal and the fractures will be at about 45° to the vertical. This is in agreement with the account on p. 433.

APPENDIX B

The Straining of an Elastic Sphere

We begin by considering a uniform sphere. The results are of historical importance and remain interesting in suggesting orders of magnitude.

1. Surface loading. Let the surface stress be radial and equal to WK_n, where K_n is a solid harmonic of degree n and W is constant. The stress-components p_{ri} across the surface are $WK_n x_i/r$. The elastic equations are

$$(\lambda + \mu)\, \partial \Delta / \partial x_i + \mu \nabla^2 u_i = 0. \tag{1}$$

Put
$$q = x_i u_i. \tag{2}$$

Then q/r is the radial displacement. The stress-components are

$$p_{ik} = \lambda \Delta \delta_{ik} + \mu \left(\frac{\partial u_k}{\partial x_i} + \frac{\partial u_i}{\partial x_k} \right), \tag{3}$$

and the components across a sphere are

$$p_{ri} = \frac{x_k}{r} p_{ik} = \lambda \Delta \frac{x_i}{r} + \frac{\mu}{r} \left(x_k \frac{\partial u_i}{\partial x_k} + x_k \frac{\partial u_k}{\partial x_i} \right)$$

$$= \lambda \Delta \frac{x_i}{r} + \frac{\mu}{r} \left(\frac{\partial q}{\partial x_i} - u_i + r \frac{\partial u_i}{\partial r} \right). \tag{4}$$

There are three types of solution related to a solid harmonic K_n.

ϕ *type:*
$$u_i = A \frac{\partial K_n}{\partial x_i}. \tag{5}$$

A is constant. Then $\Delta = 0$ and (1) are satisfied;

$$q = A x_i \frac{\partial K_n}{\partial x_i} = n A K_n, \tag{6}$$

$$p_{ri} = \frac{\mu A}{r} \left\{ n \frac{\partial K_n}{\partial x_i} - \frac{\partial K_n}{\partial x_i} + (n-1) \frac{\partial K_n}{\partial x_i} \right\}$$

$$= 2(n-1) \frac{\mu}{r} A \frac{\partial K_n}{\partial x_i}. \tag{7}$$

ω *type:*
$$u_i = B r^2 \frac{\partial K_n}{\partial x_i} + C x_i K_n. \tag{8}$$

B, C are constant. Then

$$\nabla^2 u_i = 2\{(2n+1)B + C\} \frac{\partial K_n}{\partial x_i}, \tag{9}$$

$$\Delta = \{2nB + (n+3)C\} K_n, \tag{10}$$

$$q = (nB + C) r^2 K_n. \tag{11}$$

[510]

From (1) $2\{n\lambda + (3n+1)\mu\}B + \{(n+3)\lambda + (n+5)\mu\}C = 0,$ (12)

$$p_{ri} = \frac{\lambda x_i}{r}\{2nB + (n+3)C\}K_n + \frac{\mu}{r}\bigg[(2nB+C)r^2\frac{\partial K_n}{\partial x_i}$$

$$+ \{2nB + (n+2)C\}x_i K_n\bigg].$$ (13)

χ *type:* $\qquad u_i = \epsilon_{ikm}x_k\dfrac{\partial K_n}{\partial x_m},$ (14)

$$\Delta = 0,$$ (15)

$$\nabla^2 u_i = 0, \quad q = 0.$$ (16)

The equations of equilibrium hold automatically. Also

$$p_{ri} = \frac{\mu}{r}\epsilon_{ikm}(n-1)x_k\frac{\partial K_n}{\partial x_m}.$$ (17)

This stress is perpendicular to x_i, so that the stress is everywhere tangential to a sphere with centre O.

If the stress over $r = a$ is normal and proportional to K_n, we can satisfy the boundary conditions by combining (7) and (13) so as to make the coefficient of $\partial K_n/\partial x_i$ zero. Thus the χ type of solution does not arise. It does arise, however, when there is tangential stress at $r = a$.

Any solution of (1) can be expressed in terms of the ϕ, ω, and χ types. We have, from (1), $\nabla^2\Delta = 0, \quad \nabla^2\operatorname{curl}\mathbf{u} = 0.$ (18)

Thus Δ and $(\operatorname{curl}\mathbf{u})_i$ can be expressed as sums of solid harmonics. If $\Delta = DK_n$, (1) are satisfied by an ω solution, say $\mathbf{v}_1$; but the ω solutions are restricted by (12). Put

$$\mathbf{u} = \mathbf{v}_1 + \mathbf{v},$$ (19)

$$\operatorname{div}\mathbf{v}_1 = \Delta.$$ (20)

The ω solutions lead to

$$(\operatorname{curl}\mathbf{v}_1)_i = \epsilon_{ikm}(2B-C)x_k\frac{\partial K_n}{\partial x_m}$$ (21)

which gives $\nabla^2\operatorname{curl}\mathbf{v}_1 = 0$. Then

$$\operatorname{div}\mathbf{v} = 0, \quad \nabla^2\operatorname{curl}\mathbf{v} = 0$$ (22)

hence $\mathbf{v}$ is a curl; and from (1)

$$\nabla^2\mathbf{v} = 0.$$ (23)

Then a particular solution is

$$\mathbf{v} = \mathbf{v}_2 = \operatorname{curl}\mathbf{F},$$ (24)

where $\mathbf{F}$ has to be chosen so that $\nabla^2\operatorname{curl}\mathbf{F} = 0$. This is satisfied if

$$F_i = \frac{r^2}{2(2n+1)}\frac{\partial K_n}{\partial x_i}.$$ (25)

Then
$$(\mathbf{v}_2)_i = \epsilon_{ikm}\frac{\partial}{\partial x_k}\frac{r^2}{2(2n+1)}\frac{\partial K_n}{\partial x_m} = \epsilon_{ikm}\frac{x_k}{2n+1}\frac{\partial K_n}{\partial x_m}, \tag{26}$$

which is of χ type. Put
$$\mathbf{u} = \mathbf{v}_1 + \mathbf{v}_2 + \mathbf{v}_3. \tag{27}$$

Then
$$\nabla^2\mathbf{v}_3 = 0, \quad \operatorname{curl}\mathbf{v}_3 = 0. \tag{28}$$

Hence $\mathbf{v}_3$ is a solution of the ϕ type.

It is not necessary that $\mathbf{v}_1$, $\mathbf{v}_2$, $\mathbf{v}_3$ should be derived from the same K_n. In general, a given set of boundary conditions will require a sum of different harmonics.

We shall speak of two vector functions u_i, v_i of position on a sphere as orthogonal if
$$\iint u_i v_i dS = 0. \tag{29}$$

We show that pairs of solutions of types (ϕ, ϕ), (ϕ, ω), (ω, ω), (χ, χ) are orthogonal if derived from orthogonal K_n, K_p; and pairs of the types (ϕ, χ), (ω, χ) are orthogonal without exception. As an ω solution on $r = a$ reduces to the form $B\partial K_n/\partial x_i + Cx_i K_n$, it is enough to replace an ω solution by $x_i K_n$.

(ϕ, ϕ): note first that for integration through a sphere

$$\iiint\frac{\partial K_n}{\partial x_i}\frac{\partial K_p}{\partial x_i}d\tau = \iiint\frac{\partial}{\partial x_i}\left(K_n\frac{\partial K_p}{\partial x_i}\right)d\tau = \iint\frac{x_i}{r}K_n\frac{\partial K_p}{\partial x_i}dS = \frac{p}{a}\iint K_n K_p dS. \tag{30}$$

Also
$$\iiint\frac{\partial K_n}{\partial x_i}\frac{\partial K_p}{\partial x_i}d\tau = \int dr\iint\left(\frac{r}{a}\right)^{n+p-2}\left(\frac{\partial K_n}{\partial x_i}\frac{\partial K_p}{\partial x_i}\right)_{r=a}\frac{r^2}{a^2}dS$$
$$= \frac{a}{n+p+1}\iint\frac{\partial K_n}{\partial x_i}\frac{\partial K_p}{\partial x_i}dS. \tag{31}$$

Hence ϕ solutions derived from orthogonal K_n, K_p are orthogonal. If $K_p = K_n$,
$$\iint\left(\frac{\partial K_n}{\partial x_i}\right)^2 dS = \frac{n(2n+1)}{a^2}\iint K_n^2 dS. \tag{32}$$

(ϕ, ω):
$$\iint\frac{\partial K_n}{\partial x_i}x_i K_p dS = \iint nK_n K_p dS = \left(0, n\iint K_n^2 dS\right), \tag{33}$$

according as K_n, K_p are orthogonal or identical.

(ω, ω):
$$\iint x_i K_n \cdot x_i K_p dS = \iint r^2 K_n K_p dS = \left(0, a^2\iint K_n^2 dS\right). \tag{34}$$

(ϕ, χ):
$$\iint\frac{\partial K_n}{\partial x_i}\epsilon_{ikm}x_k\frac{\partial K_p}{\partial x_m}dS = a\iiint\frac{\partial}{\partial x_k}\left(\epsilon_{ikm}\frac{\partial K_n}{\partial x_i}\frac{\partial K_p}{\partial x_m}\right)d\tau$$
$$= a\iiint\epsilon_{ikm}\left(\frac{\partial^2 K_n}{\partial x_i\partial x_k}\frac{\partial K_p}{\partial x_m} + \frac{\partial K_n}{\partial x_i}\frac{\partial^2 K_p}{\partial x_k\partial x_m}\right)d\tau$$
$$= 0. \tag{35}$$

(ω, χ): $$\iint x_i K_n \epsilon_{ikm} x_k \frac{\partial K_p}{\partial x_m} dS = 0. \tag{36}$$

(χ, χ): $$\iint \epsilon_{ikm} x_k \frac{\partial K_n}{\partial x_n} \epsilon_{ips} x_p \frac{\partial K_q}{\partial x_s} dS$$

$$= \iint (\delta_{kp}\delta_{ms} - \delta_{ks}\delta_{mp}) x_k x_p \frac{\partial K_n}{\partial x_m} \frac{\partial K_q}{\partial x_s} dS$$

$$= \iint \left(r^2 \frac{\partial K_n}{\partial x_m} \frac{\partial K_q}{\partial x_m} - x_k x_m \frac{\partial K_n}{\partial x_m} \frac{\partial K_q}{\partial x_k} \right) dS$$

$$= \iint \left(a^2 \frac{\partial K_n}{\partial x_m} \frac{\partial K_q}{\partial x_m} - nq K_n K_q \right) dS$$

$$= (0, n(2n+1) - n^2) \iint K_n^2 dS$$

$$= (0, n(n+1)) \iint K_n^2 dS, \tag{37}$$

according as K_n, K_q are orthogonal or identical.

We have seen that any solution of (1) is expressible in terms of a sum of ϕ, ω, χ solutions; and for each the surface values of u_i, p_{ri} reduce to the forms indicated. The problem is, if u_i or p_{ri} are given over the boundary, to express them in the proper forms. The only difference from most expansions in orthogonal functions arises from the incomplete orthogonality of the ϕ, ω types. But if

$$(u_i)_{r=a} = \sum_n \left(A'_n \frac{\partial K_n}{\partial x_i} + C_n x_i K_n + D_n \epsilon_{ikm} x_k \frac{\partial K_n}{\partial x_m} \right), \tag{38}$$

$$\iint u_i \frac{\partial K_n}{\partial x_i} dS = \left(\frac{n(2n+1)}{a^2} A'_n + n C_n \right) \iint K_n^2 dS, \tag{39}$$

$$\iint u_i x_i K_n dS = (n A'_n + a^2 C_n) \iint K_n^2 dS. \tag{40}$$

Thus A'_n, C_n are found by solving a pair of simultaneous equations. For an ω solution B_n/C_n is specified, whether u_i is taken to be a displacement or replaced by a stress-component p_{ri}; thus when C_n is found it gives the part of A'_n arising from the ω solution and the rest gives the ϕ solution.

The only part of the displacement that gives any angular momentum for a sphere is the χ solution for $n = 1$ or -2. For the ϕ type

$$\iiint \epsilon_{ikm} x_k u_m d\tau = \iiint \epsilon_{ikm} x_k \frac{\partial K_n}{\partial x_m} d\tau. \tag{41}$$

The integrand is a sum of solid harmonics, not of degree 0, and the integral over a sphere vanishes. For the ω type we have immediately

$$\iiint \epsilon_{ikm} x_k x_m K_n d\tau = 0. \tag{42}$$

For the χ type, each component is a harmonic of degree n; hence when it is multiplied by x_k and integrated over a sphere it gives 0 except for $n = 1$ or -2. $n = -2$ is excluded by continuity at the centre for a complete sphere. For $n = 1$ the χ motion is a rigid rotation.

This result justifies the assumption implicit in 7·04, that the elastic strains do not contribute to the angular momenta to the order of accuracy attempted. For more detailed discussion see Jeffreys (1949a).

2. Surface loading of a sphere. We now return to

$$(p_{ri}) = WK_{na}\frac{x_i}{r}, \tag{1}$$

where

$$2(n-1)\frac{\mu}{a}A + \frac{\mu}{a}(2nB+C)a^2 = 0, \tag{2}$$

$$2n(\lambda+\mu)B + \{(n+3)\lambda + (n+2)\mu\}C = W. \tag{3}$$

Also

$$2\{n\lambda + (3n+1)\mu\}B + \{(n+3)\lambda + (n+5)\mu\}C = 0. \tag{4}$$

If

$$D = 2(2n^2+4n+3)\lambda\mu + 2(2n^2+2n+2)\mu^2, \tag{5}$$

we find

$$DB = -\{(n+3)\lambda + (n+5)\mu\}W, \tag{6}$$

$$DC = 2\{n\lambda + (3n+1)\mu\}W, \tag{7}$$

$$(n-1)DA = Wa^2\{(n^2+2n)\lambda + (n^2+2n-1)\mu\}, \tag{8}$$

$$Dq = WK_n\left[a^2\frac{n}{n-1}\{(n^2+2n)\lambda + (n^2+2n-1)\mu\} - r^2(n+1)\{n\lambda + (n-2)\mu\}\right]. \tag{9}$$

If we neglect $1/n$ we have $\quad q = \tfrac{1}{4}WK_n(a^2-r^2) \tag{10}$

for all values of λ, μ. The terms of order $1/n$ do *not* vanish at $r = a$. K_n itself becomes small at a depth of order a/n.

For $n = 2$,

$$q = \frac{WK_2}{\mu}\frac{(8\lambda+7\mu)a^2 - 3\lambda r^2}{19\lambda + 14\mu}. \tag{11}$$

As $K_2 \propto r^2$, the ellipticities of surfaces originally spherical decrease outwards, contrary to what happens in the hydrostatic theory of the figure of the Earth (where, however, variation of the ellipticity is wholly due to variation of density, which we are not considering here). For $r = 0$ and $r = a$ we have for $\mu q/a^2 WK_n$

λ/μ	$r = 0$	$r = a$
∞	$\frac{8}{19}$	$\frac{5}{19}$
1	$\frac{5}{11}$	$\frac{4}{11}$

The ratios of these according as λ/μ is taken as 1 or ∞ are $\frac{88}{85}$ and $\frac{55}{76}$ respectively. Accordingly compressibility may make differences of the order of 10–20 % in the results.

3. Deformation by bodily force. Let the bodily force be derived from a gravitational potential $k_n K_n$. The equations of equilibrium are

$$(\lambda + \mu)\frac{\partial \Delta}{\partial x_i} + \mu \nabla^2 u_i = -\rho k_n \frac{\partial K_n}{\partial x_i}. \tag{1}$$

The ϕ, ω, and χ types are complementary solutions of these equations. In addition a particular integral is got by taking

$$u_i = \frac{\partial \phi}{\partial x_i}, \tag{2}$$

whence

$$(\lambda + 2\mu)\frac{\partial}{\partial x_i}\nabla^2 \phi = -\rho k_n \frac{\partial K_n}{\partial x_i}. \tag{3}$$

Now

$$\nabla^2(r^m K_n) = m(2n + m + 1)r^{m-2}K_n. \tag{4}$$

Hence it is sufficient to take the contributions from the particular integral to be derived from

$$\phi = -\frac{\rho k_n}{\lambda + 2\mu}\frac{r^2 K_n}{2(2n+3)}, \tag{5}$$

which gives

$$\Delta = \nabla^2 \phi = -\frac{\rho k_n K_n}{\lambda + 2\mu}, \tag{6}$$

$$q = -\frac{\rho k_n}{\lambda + 2\mu}\frac{n+2}{2(2n+3)}r^2 K_n, \tag{7}$$

$$p_{ri} = -\frac{\rho k_n}{(\lambda + 2\mu)r}\left[\frac{(n+1)\mu}{2n+3}r^2\frac{\partial K_n}{\partial x_i} + \left(\lambda + \frac{2(n+1)\mu}{2n+3}\right)x_i K_n\right]. \tag{8}$$

p_{ri} is to vanish for $r = a$. Taking the complementary functions as before we have

$$D = 2(2n^2 + 4n + 3)\lambda\mu + 2(2n^2 + 2n + 2)\mu^2, \tag{9}$$

$$DB = -\frac{\rho k_n}{\lambda + 2\mu}\left(\lambda + \frac{2(n+1)\mu}{2n+3}\right)\{(n+3)\lambda + (n+5)\mu\}, \tag{10}$$

$$DC = \frac{2\rho k_n}{\lambda + 2\mu}\left(\lambda + \frac{2(n+1)\mu}{2n+3}\right)\{n\lambda + (3n+1)\mu\}, \tag{11}$$

$$(n-1)DA = n\rho k_n a^2\{(n+2)\lambda + (n+1)\mu\}, \tag{12}$$

$$\frac{q}{\rho k_n K_n} = \frac{n^2 a^2\{(n+2)\lambda + (n+1)\mu\} - n(n-1)r^2\{(n+1)\lambda + n\mu\}}{2(n-1)\mu\{(2n^2 + 4n + 3)\lambda + (2n^2 + 2n + 2)\mu\}}. \tag{13}$$

For n large this expression reduces to $(a^2 - r^2)/4\mu$, irrespective of λ/μ. For $n = 2$ it becomes

$$\frac{2a^2(4\lambda + 3\mu) - r^2(3\lambda + 2\mu)}{\mu(19\lambda + 14\mu)}, \tag{14}$$

reducing to $\dfrac{5\lambda+4\mu}{\mu(19\lambda+14\mu)}$ or $\dfrac{8\lambda+6\mu}{\mu(19\lambda+14\mu)}$ for $r=a, r=0$.

When $\lambda/\mu=1$ and ∞ these expressions become

λ/μ	$r=a$	$r=0$
1	$3/11\mu$	$14/33\mu$
∞	$5/19\mu$	$8/19\mu$
Ratio	$\frac{55}{57}$	$\frac{132}{133}$

The effects of compressibility in this problem are therefore only 1–4 %.

4. Incompressible gravitating sphere. The smallness of the effect of compressibility in the last problem encourages us to assume incompressibility when we wish to take account of the mutual attraction of the deformations. This is a great simplification because it removes disturbances of the field due to internal changes of density. We can also take the density uniform; to do so while admitting compressibility would be to assume that the naturally lighter materials are at the greater depths. (This circumstance makes the allowance for compressibility in parts of Love's *Geodynamics* unsatisfactory.)

We take the surface form as given by

$$r=a(1+\epsilon S_n), \quad S_n=K_n/r^n. \tag{1}$$

There is a gravitation potential $U_n=k_n K_n$ arising from other bodies. Let U_0 be the undisturbed gravitation potential and p_0 the corresponding pressure. These may not be small; but

$$\frac{\partial p_0}{\partial x_i}=\rho\frac{\partial U_0}{\partial x_i}. \tag{2}$$

The stress on an element in the actual state is taken to be $-p_0\delta_{ik}+p_{ik}$, and p_{ik} is related to the displacement according to the usual rules. $\Delta=0$, but $\lambda\Delta$ may not tend to 0 as $\lambda/\mu\to\infty$. Let $\lambda\Delta=-p$. Then

$$p_{ik}=-p\,\delta_{ik}+2\mu e_{ik}. \tag{3}$$

At $r=a$ the normal stress is a tension

$$-g\rho a\epsilon S_n. \tag{4}$$

In addition the surface inequality produces a potential

$$\frac{4\pi f\rho}{2n+1}\frac{r^n\epsilon S_n}{a^{n-2}}=\frac{3g}{2n+1}\frac{\epsilon}{a^{n-1}}K_n, \tag{5}$$

which must be added to U_n. Then, keeping the highest power of λ, we have

$$D = 2(2n^2 + 4n + 3)\lambda\mu, \tag{6}$$

$$DB = -\rho\left(k_n + \frac{3g}{2n+1}\frac{\epsilon}{a^{n-1}}\right)(n+3)\lambda + (n+3)\lambda\frac{g\rho\epsilon}{a^{n-1}}, \tag{7}$$

$$DC = 2\rho\left(k_n + \frac{3g}{2n+1}\frac{\epsilon}{a^{n-1}}\right)n\lambda - 2n\lambda\frac{g\rho\epsilon}{a^{n-1}}, \tag{8}$$

$$(n-1)DA = n(n+2)\rho a^2\left(k_n + \frac{3g}{2n+1}\frac{\epsilon}{a^{n-1}}\right) - n(n+2)\lambda\frac{g\rho\epsilon}{a^{n-3}}, \tag{9}$$

$$q = \frac{n^2 a^2(n+2) - n(n^2-1)r^2}{2(n-1)\mu(2n^2+4n+3)}\left\{\rho\left(k_n + \frac{3g}{2n+1}\frac{\epsilon}{a^{n-1}}\right) - \frac{g\rho\epsilon}{a^{n-1}}\right\}K_n. \tag{10}$$

We notice that A, C, B, q all contain a factor

$$k_n' = k_n - \frac{2(n-1)}{2n+1}\frac{g\epsilon}{a^{n-1}}. \tag{11}$$

If $k_n' = 0$ the deformation due to the surface load cancels that due to the external field.

The factor $n-1$ in the denominator in all these problems means that for $n = 1$ the force applied is in the same direction at all points, and no statical solution is possible.

At the surface $q = a^2\epsilon S_n$; then (10) for $r = a$ gives an equation for ϵ:

$$\left(1 + \frac{ng\rho a}{(2n^2+4n+3)\mu}\right)\epsilon = \frac{n(2n+1)}{2(n-1)(2n^2+4n+3)\mu}\rho a^{n-2}k_n. \tag{12}$$

For the Earth we may take as an average value

$$\frac{\mu}{\rho} = (6 \times 10^5\,\text{cm./sec.})^2,$$

and for $n = 2$

$$\frac{19\mu}{2g\rho a} \doteq \frac{19 \times 36 \times 10^{10}}{2 \times 1000 \times 6 \times 10^8} \doteq 6.$$

Thus even for $n = 2$ rigidity plays a much larger part than gravity in determining the amount of deformation.

If there is no external force but the surface elevation above the sphere is given,

$$\left.\begin{aligned}
A &= -\frac{n(n+2)}{(2n+1)(2n^2+4n+3)}\frac{g\rho\epsilon}{a^{n-3}\mu}, \\[2mm]
B &= \frac{(n+3)(n-1)}{(2n+1)(2n^2+4n+3)}\frac{g\rho\epsilon}{a^{n-1}\mu}, \\[2mm]
C &= -\frac{2n(n-1)}{(2n+1)(2n^2+4n+3)}\frac{g\rho\epsilon}{a^{n-1}\mu}.
\end{aligned}\right\} \tag{13}$$

5. Internal stress. In (13) we are considering a free surface

$$r = a(1 + \epsilon S_n),$$

and supposing the stresses related to a consistent set of internal displacements by the elastic equations. These equations are therefore appropriate to the estimation of the internal stress, when the external form is measured, and the internal displacements are supposed to be elastic displacements produced by the load.

If

$$u_i = (A + Br^2)\frac{\partial K_n}{\partial x_i} + Cx_i K_n, \tag{1}$$

we find

$$p_{ik} = \lambda \Delta \delta_{ik} + \mu \left\{ 2(A + Br^2)\frac{\partial^2 K_n}{\partial x_i \partial x_k} + (2B + C)\left(x_k \frac{\partial K_n}{\partial x_i} + x_i \frac{\partial K_n}{\partial x_k}\right) + 2C\delta_{ik}K_n \right\}. \tag{2}$$

Terms in δ_{ik} do not affect the stress-differences.

Take

$$K_2 = 2x_3^2 - x_1^2 - x_2^2. \tag{3}$$

By symmetry we need consider only the plane $x_2 = 0$; then the relevant parts of the stresses are given by

$$\frac{p_{11}}{\mu} = -4(A + Br^2) - 4(2B + C)x_1^2, \tag{4}$$

$$\frac{p_{22}}{\mu} = -4(A + Br^2), \tag{5}$$

$$\frac{p_{33}}{\mu} = 8(A + Br^2) + 8(2B + C)x_3^2, \tag{6}$$

$$p_{12} = p_{23} = 0, \tag{7}$$

$$\frac{p_{13}}{\mu} = 2(2B + C)x_1 x_3, \tag{8}$$

while

$$2B + C = -\tfrac{3}{2}C, \quad A = 2Ca^2, \quad A + Ba^2 = \tfrac{3}{4}Ca^2. \tag{9}$$

p_{22} is a principal stress; the others are μ times the roots of

$$\begin{vmatrix} -4(A + Br^2) + 6Cx_1^2 - \varpi & -3Cx_1 x_3 \\ -3Cx_1 x_3 & 8(A + Br^2) - 12Cx_3^2 - \varpi \end{vmatrix} = 0. \tag{10}$$

At $r = a$, $x_1 = 0$, $\varpi = -4(A + Ba^2)$ or $8(A + Ba^2) - 12Ca^2$,

$$\text{stress-difference} = \mu \,|\, 12(A + Ba^2 - Ca^2)\,| = 3\mu \,|\, C \,|\, a^2. \tag{11}$$

At $r = a$, $x_3 = 0$, $\varpi = 8(A + Ba^2) = 6Ca^2$, or $\varpi = -4(A + Ba^2) + 6Ca^2 = 3Ca^2$, while $p_{22} = -3\mu Ca^2$; hence

$$\text{stress-difference} = 9\mu \,|\, C \,|\, a^2. \tag{12}$$

At $r = 0$, $\varpi = -4A$ or $8A$,

$$\text{stress-difference} = \mu \,|\, 12A \,| = 24\mu \,|\, C \,|\, a^2. \tag{13}$$

With more detailed examination it is found that the stress-difference is less than $24\mu \,|\, C \,|\, a^2$ except at the centre.

For a surface inequality $a\epsilon S_2$, where

$$S_2 = \epsilon(\tfrac{1}{3} - \cos^2\theta), \tag{14}$$

we have

$$C = -\frac{4}{3.5.19}\frac{g\rho\epsilon}{a\mu}, \tag{15}$$

and the stress-difference at the centre is

$$\tfrac{32}{95}g\rho a\epsilon. \tag{16}$$

For the Moon

$$g\rho a \doteqdot 1000 \times 5\cdot 5 \times 6 \times 10^8 \left(\frac{3\cdot 3}{5\cdot 5} \times \frac{3}{11}\right)^2 = 0\cdot 9 \times 10^{11} \tag{17}$$

and

$$\epsilon \doteqdot \tfrac{1}{1500}. \tag{18}$$

Hence the stress-difference at the centre is about 2×10^7 dynes/cm.².

6. For a non-uniform sphere with a spherical distribution of properties the solutions above need modification. Also the specification of displacement has been Eulerian, that is, u_i refers to the particle at x_i, displaced from $x_i - u_i$. It is more convenient now to take the particle at x_i to be displaced to $x_i + u_i$. This is a Lagrangian specification. The differential equations are not altered to the first order in u_i in the absence of initial stress, but when there is large initial stress there are differences of order u_i in the stresses for given x_i. In particular the term (4) does not arise because x_i with this specification is already at the free surface. The great advantage of the method is that we have to evaluate several integrals, and with the Lagrangian specification these are all through the same (standard) position of the body. We take the F, q part and the H part separately (Jeffreys and Vicente, 1966, 1967); the energies contain no terms in their products. For the F, q part the kinetic energy is given by

$$2T = \iiint \rho \left\{ \dot{q}\frac{x_i K_n}{r^2} + \dot{F}\left(\frac{\partial K_n}{\partial x_i} - \frac{x_i}{r^2}K_n\right)\right\}^2 d\tau$$

$$= \iiint \rho \left\{\frac{\dot{q}^2}{r^2} + n(n+1)\frac{\dot{F}^2}{r^2}\right\} K_n^2 \, d\tau$$

$$= M\int_0^a \rho r^{2n}\{\dot{q}^2 + n(n+1)\,\dot{F}^2\}\, d\tau, \tag{1}$$

where $d\omega$ is an element of solid angle and $\dot{M} = \iint S_n^2 \, d\omega$.

34-2

The work function is given by (Jeffreys and Vicente, 1957 a, b, (12), p. 144):

$$2W = \iiint \rho \left(u_i u_k \frac{\partial^2 U_0}{\partial x_i \partial x_k} + u_i \frac{\partial u_k}{\partial x_i} \frac{\partial U_0}{\partial x_k} - u_i \frac{\partial U_0}{\partial x_i} \frac{\partial u_k}{\partial x_k} + u_i \frac{\partial U_1}{\partial x_i} + \tfrac{1}{2} u_i \frac{\partial U_2}{\partial x_i} \right) d\tau$$

$$- \iiint (\lambda \Delta K_n \, \delta_{ik} + 2\mu e_{ik}) \frac{\partial u_i}{\partial x_k} d\tau. \quad (2)$$

Here U_0 is the undisturbed gravitational potential, U_1 that due to disturbing bodies, and U_2 the additional potential due to the deformation of the Earth itself. The factor $\tfrac{1}{2}$ in $\tfrac{1}{2} u_i \partial U_2 / \partial x_i$ arises because each element of displacement contributes to U_2, so that this term is quadratic in the displacements. Transforming to the F, q notation we find

$$2W = -M \int_0^a r^{2n} \lambda \left\{ q' + (n+1) \frac{q - nF}{r} \right\}^2 dr$$

$$- M \int_0^a r^{2n} \mu \left\{ n(n+1) F'^2 + 2n(n-2)(n+1) \frac{FF'}{r} \right.$$

$$+ n(n+1)(3n^2 - 2n - 2) \frac{F^2}{r^2} + 2q'^2 + 4(n-1) \frac{qq'}{r} + (3n^2 - 3n + 6) \frac{q^2}{r^2}$$

$$\left. + \frac{2n(n+1)}{r} F'q + 2(n+1) n(n-4) \frac{Fq}{r^2} \right\} dr$$

$$+ M \int_0^a \rho r^{2n} \left\{ q^2 \left(r \frac{d}{dr} \frac{U_0'}{r} - \frac{U_0}{r} + \frac{2n(n+1)}{r} qF U_0' \right) \right\} dr$$

$$+ M \int_0^a [\rho r^{2n} c \{ nq + n(n+1) F \} + \tfrac{1}{2} \rho r^{2n} \{ rqK' + nq(K - c) \}$$

$$+ F(K - c) n(n+1)] dr. \quad (3)$$

Here we have taken

$$U_1 = cK_n, \quad U_2 = (K - c) K_n. \quad (4)$$

For the H terms we have

$$2T = n(n+1) M \int_0^a \rho \dot{H}^2 r^{2n+2} dr, \quad (5)$$

$$2W = -n(n+1) M \int_0^a \mu r^{2n} \{ r^2 H'^2 + 2(n-1) rHH' + (n-1)(2n+1) H^2 \} dr. \quad (6)$$

The only part of the displacement that gives any angular momentum for a sphere is the H solution for $n = 1$ or -2. For the q type the displacement is radial. For the F type

$$\iint \epsilon_{ikm} x_k \left(\frac{\partial K_n}{\partial x_m} - \frac{n x_m}{r^2} K_n \right) dS = \iint \epsilon_{ikm} x_k \frac{\partial K_n}{\partial x_m} dS. \quad (7)$$

The integrand is a sum of solid harmonics and its integral over a sphere vanishes. For the H type each component is a harmonic of degree n; hence when it is multiplied by x_k and integrated over a sphere it gives 0 except for $n = 1$ or -2. For $n = 1$, with constant H, it is a rigid rotation (and the corresponding part of W vanishes). $n = -2$ for a complete sphere would be excluded by continuity at the centre, but it could arise for a uniform shell.

Most writers, following Alterman, Jarosch & Pekeris (1959 a, b), use spherical polar coordinates and use variables differing from the present ones by powers of r. The variables q, F, H, K satisfy differential equations of the second order.

The differential equations become

$$\rho \ddot{F} = \lambda\Delta + \mu \left\{(n+2)\frac{F'}{r} - (n^2 + 2n + 2)\frac{F}{r^2} + (n+4)\frac{q}{r^2}\right\}$$

$$+ \frac{d}{dr}\left(\frac{X_4}{r}\right) + \rho Kc + \rho\frac{q}{r}U'_0, \quad (8)$$

$$\rho \ddot{q} = (n-1)\lambda\Delta + \mu \left\{(2n+2)\frac{q'}{r} + (n^2 - n - 6)\frac{q}{r^2} - n(n+1)\frac{F'}{r}\right.$$

$$\left.- n(n^2 - 3n - 4)\frac{F}{r^2}\right\} + \frac{d}{dr}\frac{X_2}{r} - \rho\left(\frac{4U'_0}{r} + 4\pi f\rho\right)q + \rho n(n+1)U'_0\frac{F}{r}$$

$$+ n\rho Kc + \rho r Kc, \quad (9)$$

$$\rho \ddot{H} = \mu\left(\frac{n+3}{r}H' + \frac{n-1}{r^2}H\right) + \frac{d}{dr}\frac{X_8}{r}, \quad (10)$$

$$K'' + \frac{2n+2}{r}K' = 4\pi f\rho\frac{(n+2)q - n(n+1)F}{r^2} + \frac{d}{dr}\left(4\pi f\rho\frac{q}{r}\right). \quad (11)$$

Where
$$X_2 = r^2\lambda\Delta + \mu\{2q'r + (2n-2)q\},$$

$$X_4 = \mu\{rF' + (n-2)F + q\},$$

$$X_6 = rK' - 4\pi f\rho q,$$

$$X_8 = \mu\{rH' + (n-1)H\}.$$

By a device given by AJP, Vicente and I transform (1966) these to eight equations of the first order that do not contain derivatives of the mechanical properties. X_2, X_4, X_8 are related to the stress components. These variables arise more directly when the equations of motion are derived by Hamilton's principle from the energies. The differential equations become, with $\partial^2/\partial t^2 = -\sigma^2$,

$$q' = \frac{1}{(\lambda + 2\mu)\,r}\left[X_2 - \{(n+1)\,\lambda + 2(n-1)\,\mu\}\,q + n(n+1)\,\lambda F\right],$$

$$-\sigma^2\rho q = -\left\{\frac{4\mu(3\lambda + 2\mu)}{\lambda + 2\mu} + 4\rho U_0'\,r\right\}\frac{q}{r^2} + \frac{(n-2)\,\lambda + 2n\mu}{(\lambda + 2\mu)\,r^2}\,X_2 + \frac{X_2'}{r}$$

$$+ n(n+1)\left\{\frac{2\mu(3\lambda + 2\mu)}{\lambda + 2\mu} + \rho U_0'\,r\right\}\frac{F}{r^2} - n(n+1)\frac{X_4}{r^2}$$

$$+ n\rho K + \rho X_6,$$

$$F' = \frac{1}{r}\left\{\frac{X_4}{\mu} - (n-2)\,F - q\right\},$$

$$-\sigma^2\rho F = \left\{\frac{2\mu(3\lambda + 2\mu)}{\lambda + 2\mu} + \rho r U_0'\right\}\frac{q}{r^2} + \frac{\lambda X_2}{(\lambda + 2\mu)}\,r^2$$

$$- \frac{2\mu}{(\lambda + 2\mu)}\,r^2\{(2n^2 + 2n - 1)\,\lambda + 2(n^2 + n - 1)\,\mu\}\,F + \frac{X_4'}{r}$$

$$+ \frac{n+1}{r^2}\,X_4 + \rho K,$$

$$K' = \frac{1}{r}(4\pi f\rho q + X_6),$$

$$X_6' = -\frac{1}{r}\{4\pi f\rho q + 4\pi f\rho n(n+1)\,F + (2n+1)\,X_6\},$$

$$X_8 = \mu\{rH' + (n-1)\,H\},$$

$$-\rho\sigma^2 H = \frac{X_8'}{r} + \frac{n+2}{r^2}\,X_8 - (n-1)\,(n+2)\frac{\mu H}{r^2}.$$

U_0 is the undisturbed gravitational potential; at the surface $U_0' = -g$.

The differential equations have a singularity at $r = 0$. Four constants remain arbitrary there. Using a zero suffix for $r = 0$ we find:

$$q_0 = nF_0, \quad X_{20} = 2n(n-1)\,\mu F_0, \quad X_{40} = 2(n-1)\,\mu F_0, \quad X_6 = -4\pi f\rho nF_0.$$

K_0 is arbitrary; and the other constant is D, giving additional parts

$$q_1 = (n+1)\,\{n\lambda + (n-2)\,\mu\}\,Dr^2,$$

$$F_1 = \{(n+3)\,\lambda + (n+5)\,\mu\}\,Dr^2,$$

$$\Delta_1 = -2(n+1)\,(2n+3)\,\mu D,$$

$$X_{21} = r^2\lambda\Delta_1 + 2(n+1)\,\mu q_1$$

$$= \mu r^2(n+1)\,D\{(2n^2 - 2n - 6)\,\lambda + 2(n+1)\,(n-2)\,\mu\},$$

$$X_{41} = \mu(nF_1 + q_1)$$

$$= \{n(2n+4)\,\lambda + (2n^2 + 4n - 2)\,\mu\}\,\mu Dr^2,$$

$$X_{61} = -4\pi f\rho n(n+1)\,(\lambda + \mu)\,Dr^2,$$

$$K_1 = -4\pi f\rho(n+1)\,\mu Dr^2,$$

$$X_{80} = (n-1)\,\mu H_0.$$

The D solution is not the only one that contains r^2; the F_0 and K solutions will also introduce r^2 in the next approximation. Most solutions start from $r = 0$ and develop power series to start the integration. This is not strictly necessary; we can take trial values of F_0, K_0, D and H_0 as first approximations, and carry out the first two steps of the integration from the formulae. With a central difference method these will be corrected as we proceed. An example (for a second-order equation) is given by Jeffreys and Jeffreys (1972, foot of p. 295).

All the variables are continuous at interfaces. X_2, X_4, X_8 vanish at the free surface. The boundary condition on K is

$$X_6 + (2n+1)(K-c) = 0.$$

H and X_8 are independent of the other variables, and vanish if there is no shear stress over the surface. Displacements depending on them are often called toroidal, others spheroidal.

In the estimation of the period of a free vibration the method (for the F, q type) would be to find the F_0, D and K solutions for several trial values of σ. These will not in general satisfy the boundary conditions, but their consistency gives a determinantal equation; interpolation to make the determinant zero will give a better value of σ. Given an approximate form of F and q, Rayleigh's principle will give a much better value of σ^2 when the energies are given in an explicit form.

Rayleigh's principle can also be used to estimate small corrections to the model. It would probably be convenient to consider them in the order β, α, ρ, since this is the order of sensitivity of the period to changes of the model (but α does not arise for toroidal modes). There is however some trouble for changes of ρ, since any change of ρ alters U_0 through the whole sphere, and this must be allowed for.

J. S. Derr (1969a) combines many determinations of free periods and gives estimates and standard errors up to degree 99. He also (1969b) works out theoretical values for a model and adapts it to the observed values. Wiggins (1968) has carried out a similar analysis, and further treatments are by D. L. Anderson, Dziewonski and Gilbert.

APPENDIX C

Castigliano's Principle

For a solid strained from a state of zero stress the strain energy per unit volume is $\frac{1}{2}p_{ik}e_{ik}$. Using the relations

$$2\mu e_{ik} = p_{ik} - \frac{\lambda}{3k}p_{mm}\delta_{ik}, \tag{1}$$

$$p_{ik} = \frac{1}{3}p_{mm}\delta_{ik} + p'_{ik}, \tag{2}$$

we can write this as
$$\frac{1}{2}p_{ik}e_{ik} = \frac{1}{18k}p_{mm}^2 + \frac{p'_{ik}p'_{ik}}{4\mu}. \tag{3}$$

This gives the strain energy in terms of the stresses. If a body is self-strained (i.e. if the stress does not disappear even when all external forces are removed), as in an elastic band gripping a rod, the p_{ik} satisfy the equations of equilibrium but could not be made to vanish everywhere by any set of small displacements. The strain energy could, however, be made to disappear by breaking up the body into pieces. Then (3) remains a correct expression for the energy that would be lost in this process.

There are six independent p_{ik}, and the equations of equilibrium are three relations between them. Hence there is a three-fold infinity of solutions of the equations of equilibrium if we admit self-strained states.

Let a body be strained from a state of zero stress by displacements u_i, under forces ρX_i per unit volume. Part of the boundary is under given stress, part is undisplaced. Then

$$\frac{\partial p_{ik}}{\partial x_k} + \rho X_i = 0, \quad p_{ik} = \lambda \Delta \delta_{ik} + 2\mu e_{ik}. \tag{4}$$

Let a body exist in the same configuration under the same external forces but with a stress system $p_{ik} + q_{ik}$. Then

$$\frac{\partial q_{ik}}{\partial x_k} = 0. \tag{5}$$

The strain energy is
$$V = \iiint \left\{ \frac{1}{18k}(p_{mm} + q_{mm})^2 + \frac{1}{4\mu}(p'_{ik} + q'_{ik})^2 \right\} d\tau$$
$$= V_0 + V_1 + V_2, \tag{6}$$

say, where
$$V_1 = \iiint \left\{ \frac{1}{9k}p_{mm}q_{nn} + \frac{1}{2\mu}p'_{ik}q'_{ik} \right\} d\tau$$

$$= \iiint (\tfrac{1}{3}\Delta q_{mm} + q'_{ik}e'_{ik}) \, d\tau$$

$$= \iiint \left\{ (q_{ik} - \tfrac{1}{3}q_{mm}\delta_{ik}) e'_{ik} + \tfrac{1}{3}\Delta q_{mm} \right\} d\tau$$

[524]

$$= \iiint \{q_{ik}(e_{ik} - \tfrac{1}{3}e_{mm}\delta_{ik}) + \tfrac{1}{3}\Delta q_{mm}\}\, d\tau$$

$$= \iiint q_{ik} e_{ik}\, d\tau = \iiint q_{ik}\frac{\partial u_i}{\partial x_k}\, d\tau$$

$$= \iint l_k q_{ik} u_i\, dS - \iiint u_i \frac{\partial q_{ik}}{\partial x_k}\, d\tau. \tag{7}$$

The volume integral vanishes, by (5). For parts of the boundary where the stress across the boundary is prescribed, $l_k q_{ik} = 0$. For parts where the displacement is zero, $u_i = 0$. Hence

$$V_1 = 0. \tag{8}$$

This holds also for any part where u_i is perpendicular to q_{ni}, as for free slipping against a smooth fixed surface. Also

$$V_2 = \iiint \left(\frac{1}{16k}q_{mm}^2 + \frac{1}{4\mu}q'_{ik}q'_{ik}\right) d\tau \geqslant 0. \tag{9}$$

Hence $$V \geqslant V_0. \tag{10}$$

Hence, of all states consistent with a given configuration and given external forces, the one with the least strain energy is the one derived by elastic strain from an unstressed state.

APPENDIX D

Cooling of a Sphere with initial temperature $m(a^2-r^2)/2a$,
surface temperature zero

The equation of heat conduction is

$$\frac{\partial \vartheta}{\partial t} = h^2 \nabla^2 \vartheta. \tag{1}$$

We use operational methods and put $p = h^2 q^2$; the subsidiary equation is

$$\frac{d^2}{dr^2}(r\vartheta) - q^2 r\vartheta = -q^2 r \frac{m}{2a}(a^2 - r^2), \tag{2}$$

and the solution vanishing at $r = a$ is

$$r\vartheta = \frac{3m}{q^2}\frac{\sinh qr}{\sinh qa} + \frac{mr}{2a}(a^2 - r^2) - \frac{3mr}{aq^2}. \tag{3}$$

The interpretation convenient if $2ht^{\frac{1}{2}} \ll a$ is

$$r\vartheta = \frac{3m}{q^2}e^{-q(a-r)}(1 - e^{-2qr})(1 + e^{-2qa} + e^{-4qa} + \ldots) + \frac{mr}{2a}(a^2 - r^2) - \frac{3mr}{aq^2}. \tag{4}$$

Now*

$$\frac{1}{q^2}e^{-qx} = 4h^2 t \Phi_2\left(\frac{x}{2ht^{\frac{1}{2}}}\right) = 4h^2 t\,\mathrm{iierf}\left(\frac{x}{2ht^{\frac{1}{2}}}\right). \tag{5}$$

Φ_2 is very small for large arguments; hence if $2ht^{\frac{1}{2}} \ll a$

$$\vartheta \doteq \frac{12mh^2 t}{r}\Phi_2\left(\frac{a-r}{2ht^{\frac{1}{2}}}\right) + \frac{m}{2a}(a^2 - r^2) - \frac{3mh^2 t}{a}, \tag{6}$$

and at $r = a$

$$\frac{\partial \vartheta}{\partial r} \doteq \frac{6mht^{\frac{1}{2}}}{a}\Phi_1(0) - m$$

$$= -m + \frac{6mht^{\frac{1}{2}}}{\sqrt{\pi}.a}. \tag{7}$$

* H. and B. S. Jeffreys (1972), p. 569; D. R. Hartree (1936).

APPENDIX E

Long-period Tides

A statement often found in works on tidal theory is that the equilibrium theory is liable to be invalidated by the possibility of free steady motions, but again becomes a good approximation if there is friction. This statement, at first sight, seems surprising because any oscillatory motion, of whatever period, is treated in the theory independently. I have not found any published argument for it, but I think that something like the following is intended.

Take a gyroscopic system with two degrees of freedom and some damping, so that the equations of motion are

$$\ddot{x} + 2k\dot{x} - 2\omega\dot{y} + m^2x = X,$$

$$\ddot{y} + 2k\dot{y} + 2\omega\dot{x} + n^2y = Y.$$

If there is no friction and a free steady motion is possible, the reduced equations

$$-2\omega\dot{y} + m^2x = 0,$$

$$2\omega\dot{x} + n^2y = 0,$$

have a solution with $\dot{x}$, $\dot{y}$ constant and not both zero; whence m or $n = 0$. Take $n = 0$, and also take $Y = 0$, so that there is no force acting directly on y. Then if $X \propto \exp(i\gamma t)$, we find

$$\frac{x}{-\gamma^2 + 2ki\gamma} = \frac{X}{(\gamma^2 - 2ki\gamma - m^2)(\gamma^2 - 2ki\gamma) - 4\omega^2\gamma^2}$$

that is,

$$x = \frac{X}{m^2 + 2ki\gamma - \gamma^2 + \dfrac{4\omega^2\gamma^2}{\gamma^2 - 2ki\gamma}}.$$

When $\gamma \to 0$, $\dfrac{x}{X} \to \dfrac{1}{m^2}$ in consequence of the fact that when $k \neq 0$ the last term in the denominator is of order γ.

The matter has been considered in more detail by Proudman (1960) who finds that the annual tide and the 14-month one associated with the free nutation should follow the equilibrium rule fairly well, but the fortnightly one should not. The semi-annual one is doubtful.

APPENDIX F

Some Needs of Geophysics

Much further work is needed before we can understand several important facts, some of which have been known for fifty years or more.

First, we need a general survey of gravity, such that every point of the surface is within 5° of a place where gravity has been measured. Artificial satellites have given good determinations of the gravitational potential at least to all harmonics of degree 4 (probably 8 and possibly 16), and far better than gravity surveys alone can give. But gravity can still give valuable information for the contributions from the higher harmonics to the distribution of potential over limited regions, notably for the correction of estimates of the mean radius of the Earth for deflexion of the vertical. The function that arises in Stokes's theorem keeps the same sign over long intervals of distance, so that unsurveyed regions are liable to make considerable contributions over regions where gravity is well observed. But if Stokes's function is modified by omission of harmonics up to degree 4 it keeps the same sign over much shorter distances and the contributions from the higher harmonics would be much better determined.

Second, on any theory of mountain formation it is hard to understand why so many of the major mountain chains are in western China or near its borders. A factor of 2 in the apparent crumpling on different great circles would be easy to understand, but the ratio is more like 5 or possibly 10. It might be due to specially great thickness of the lighter upper layers there; this could be tested by a gravity survey.

Third, since the present standard times of travel of seismic waves were produced, the number of observations of P (longitudinal waves) has greatly increased, chiefly through the use of new stations with instruments of high magnification and short period (1^s or so). This has been mainly for the rapid and accurate determination of epicentres. But the accuracy for S (transverse waves) has diminished. The reason is that the prevailing periods in S are of 4^s to 7^s and the new instruments are insensitive to them. In many respects S is more interesting than P, and the new stations should be equipped with instruments capable of recording these periods. This applies specially to the phase SKS, which provides the most direct evidence on velocities at small depths in the core. There is practically no direct information on times of S at short distances under the oceans. An expedition with temporary instruments to a seismic region in the Pacific could probably produce useful results in a year.

Provisionally I think that the cost of this work would be between 1 and

10 % of that of sending up one artificial satellite and 0·01 % of that of a landing on the Moon.

Fourth, with any law of imperfection of elasticity, if vertical movements are due to differences of surface load, they should be fairly directly related to gravity anomalies. So far as I know no comparison has ever been made except for Fenno–Scandia, and there is no evidence that the conclusion drawn from it are general; there is a good deal of evidence against it. Geodesists have a habit of speaking of any vertical movement as isostatic, which begs the question of spontaneous internal change. We need a detailed comparison. (See Jeffreys, 1975b.)

Fifth, we know a great deal about elasticity when the stresses are well below the strength, and it is not difficult to adapt the results to any linear law of imperfect elasticity. But near the elastic limit the stress–strain relations are far from linear, even before fracture. The great distortions found in geology are understood only to the extent that something like them can be obtained in the laboratory and can be seen in the convolutions of boulder clay, where we have more direct knowledge of the forces acting. A quantitative theory of plastic instability is needed. A simple case is that of a plastic rod under tension ('Plasticine' will do). If it becomes slightly thinner in one place, the tensile force per unit area is increased there, and flow becomes faster, until the specimen is thin enough there to break.

Sixth, the clearest evidence about imperfection of elasticity at small stress-differences (corresponding to elastic strains of order 10^{-4} or less) is from the damping of elastic waves and of the free (14-monthly) nutation. This corresponds to a law according to which the creep under constant stress applied at time 0 increases like about $t^{0\cdot2}$. Any power of t less than the first would imply that rate of yield under constant shear stress would decrease with time, and conversely that constant rate of yield would need increasing shear stress. In particular the law forbids thermal instability (convection).

Perhaps the most difficult problem in geophysics is the maintenance of the sodium-chlorine balance in the sea. We also need to know much more about how volcanoes became extinct.

I think that the strength of the ocean floor, indicated most strongly by the deeps, shows that there may be some systematic error in the measures of outflow of heat, and I suggest a possible one. Another might be that if no granite was formed in the original differentiation it would provide no refuge for the radioactive elements, and that in the upper layers they are more concentrated towards the surface than in the continents. It seems doubtful whether any of the borings below the oceans have reached the relevant depth.

APPENDIX G

Statistical Methods

Considerable difficulty is introduced into all branches of physics by inadequate numerical methods. Not more than one estimate in five is accompanied by a proper estimate of uncertainty, and the result is that without an analysis of the original observations it is often impossible to judge whether two estimates of a quantity agree or give evidence of a systematic difference. Even the main result often differs from the least squares solution by more than the standard error of the latter, usually owing to graphical methods. It is astonishing that experimenters will spend months in making a series of observations and grudge the day or so needed to present the results in an intelligible form.

The so-called 'probable error' should be abandoned; the factor 0·6745 applied in computing it has to be removed before any modern significance test is applied. It is also important that computations should be made to an accuracy of two figures in the standard error. The number of degrees of freedom (number of observations less number of parameters estimated) should be given explicitly, at any rate if it is less than 20, because the posterior probability distribution of an estimate in such conditions differs appreciably from the normal law, but according to a known rule (the t rule).

The simplest test of consistency among several estimates is provided by Pearson's χ^2. If there are n observed values,

$$\chi^2 = \sum_{r=1}^{n} (O_r - C_r)^2 / \sigma_r^2,$$

where C_r is a calculated value, $O_r \pm \sigma_r$ (standard error) the observed value. If the errors of the O_r are independent and derived from the normal law, the expectation of χ^2 is $n - m = \nu$, where m is the number of parameters that have been estimated by least squares. ν is called the number of degrees of freedom. If the hypotheses are satisfied there is about a 2/3 chance that χ^2 will lie between $\nu \pm \sqrt{(2\nu)}$, and about a 19/20 chance that it will lie between $\nu \pm 2\sqrt{(2\nu)}$. Most departures from the conditions assumed tend to increase χ^2, and if a value in these intervals is found the hypothesis under consideration may be provisionally accepted. The most important departures are presence of an unconsidered systematic variation and correlation between the errors. Tests for these and further details are given in my *Theory of Probability*.

It is possibly still necessary to answer the statement that a standard error indicates an excessive claim of precision. It is necessarily based on the particular assumptions used in analysing the data—such as, for instance, that the observed values of arrival times are derived from a linear function of the distance together with a random error. If these are wrong the probability of a large error is correspondingly increased. In that sense the standard error is a minimum estimate of uncertainty. But without it there is no way of comparing different sets of data for consistency at all, because a large discrepancy could always be attributed to errors of observation if we have no separate estimate of the scatter of the estimates that such errors would be likely to produce.

Since the above was written more estimates are accompanied by standard errors, but these are still often ignored. It has been said, for instance, that my 1942 estimate of the equatorial radius is disproved by the recent determinations, but the difference from mine is less than the standard error of mine alone. The visual and dynamical parallaxes of the Moon were said to be in disagreement; the difference was 1·4 times the standard error, twice the standard error being about the amount usually treated as worth considering. See also Jeffreys (1957a).

The usual practice among statisticians is to speak of 5 % and 1 % limits; this means the following. Let O be an observed and C a value calculated according to a certain hypothesis; then if the chance, on the hypothesis, that the deviation would exceed $O - C$ is less than 5 %, the hypothesis is dubious, and, if the chance would be less than 1 %, the hypothesis is almost certainly wrong. I have pointed out (Jeffreys, 1967d, p. 385) that $O - C$ *is* what is observed and it is therefore not exceeded. What the argument implies is that a hypothesis that may be true is rejected because it has *not* predicted something that has *not* happened; really a remarkable procedure. But in fact the method works reasonably well. My method, based on the simplicity postulate and the principle of inverse probability, does however give fairly similar answers. In practice all methods give the approximate result that the hypothesis is doubtful if a deviation from expectation exceeds twice the standard error and almost certainly wrong if it exceeds three times the standard error. My method explains why, and goes into more detail, but the refinements are seldom needed.

The method of uniform reduction. Seismological observations usually show a frequency distribution with a central group corresponding to a normal law with a standard error of 1 to 3 seconds, but the numbers at deviations over twice or three times the standard error are more than the normal law predicts. There are various reasons for this. One is that a new phase is usually superposed on a background that is already disturbed, and the observer has to decide which new onsets are distinct phases and which are merely parts of the background. If a phase has a weak beginning but oscillates with increasing amplitude a later oscillation may be read as the beginning; but a careful observer may allow for this and read, for instance, a sharp microseism for P. One method of treatment is to say that a deviation more than, say, $2 \cdot 5\sigma$ is highly improbable, to reject all observations that deviate from the mean by more than this, and compute a mean and standard error from those that are left. An objection to such treatment is that the decision about a doubtful observation may shift the mean by a large fraction of its apparent standard error. More serious is the fact that the recalculation of the standard error after the rejection makes a considerable reduction in its estimate, and may lead to doubtful assertions of differences between means in comparison of different sets. In seismology there is usually a long trail of large residuals and a rule of rejection makes large and dubious alterations. In such a case we may suppose that the law has the form

$$P(dx|\alpha, h, H) = \left\{ \frac{(1-m)h}{\sqrt{\pi}} \exp\left\{-h^2(x-\alpha)^2\right\} + mg(x-\beta) \right\} dx$$

where mg is always small,

$$\int_{-\infty}^{\infty} g(x-\beta)\,dx = 1$$

and g varies little within ranges of order $1/h$. (It is convenient at this stage to use the precision constant h rather than the standard error σ to reduce the number of divisions needed.)

A first approximation can be got by examining the central groups. If m is zero we can estimate α roughly and consider the numbers of observations n_1, n_2 in ranges $-k_1 < x-\alpha < k_1$, $-k_2 < x-\alpha < k_2$, and form

$$\frac{\operatorname{erf} hk_1}{\operatorname{erf} hk_2} = \frac{n_1}{n_2}.$$

This is an equation for h, where

$$\operatorname{erf} x = \frac{2}{\sqrt{\pi}} \int_0^x e^{-u^2}\,du$$

and is tabulated by Milne-Thomson and Comrie (1931).

The essential point is that m and g' are both small, so that derivatives of mg are small of the second order. We have

$$\log L = S \log\left[\frac{(1-m)h}{\sqrt{\pi}} \exp\left\{-h^2(x-\alpha)^2\right\} + mg(x-\beta) \right]$$

$$\frac{1}{L}\frac{\partial L}{\partial \alpha} \doteq S\, 2(1-m)\frac{h^3}{\sqrt{\pi}}(x-\alpha)\exp\left\{-h^2(x-a)^2\right\}$$

and if we write $\quad w^{-1} = 1 + \dfrac{m}{1-m}\dfrac{\sqrt{\pi}}{h} g(x-\beta)\exp h^2(x-\alpha)^2$

we have nearly $\qquad \dfrac{1}{L}\dfrac{\partial L}{\partial \alpha} = S\, 2h^2 w(x-\alpha)$

$$\frac{h}{L}\frac{\partial L}{\partial h} = Sw(1-2h^2)(x-\alpha)^2.$$

Thus α and h can be estimated in the usual way, with weights w; and since w is near 1 when $(x-\alpha)/\sigma$ is moderate and very small when it is large, the second derivatives are affected little by variation of w.

We note also that the ratio of the two terms in w^{-1} is nearly the ratio of the densities of observations at large and moderate values of $|x-\alpha|$. This can be found by inspection and denoted by μ, and then

$$w^{-1} = 1 + \mu \exp h^2 (x-\alpha)^2.$$

This is nearly equivalent to reducing the numbers of observations in all ranges, including the central ones, by the mean over a convenient outlying range, and for that reason the method is called *uniform reduction* (Jeffreys, 1936a, 1967d, pp. 215–16).

The method was applied originally to seismic travel times. Over a range from 20° to 105° the data were grouped in 5° intervals; the whole of the departures from the 5° means were treated together and weights were derived. These were then applied to the separate 5° intervals.

Smoothing. The seismological tables give values of a function t over a long range of Δ; there is no theoretical form for the function, but we may at least expect that it varies smoothly over long ranges of Δ. Some writers have divided the range up and fitted straight lines over the sections and claimed a satisfactory fit with observation. But if so there is a discontinuity in $dt/d\Delta$ at every junction. If this was so the amplitude would be infinite at every junction and zero everywhere else. All that this means is just that the curvature makes a non-significant difference in a short interval.

Various formulae for smoothing exist, such as adding $\frac{1}{4}$ of the second difference or subtracting $\frac{1}{12}$ of the fourth. These give an appearance of smoothness, which, however, is not due to increased accuracy but to correlation between consecutive errors. We need a method that gives smoothness and also increases accuracy.

The relevant property of the means over intervals of, usually, 5° is that the second differences are mostly not significant; but if those at intervals of 15° are taken they are clearly genuine. In any interval it is possible to choose two values of Δ, say Δ_1 and Δ_2, such that if we fit a linear function of Δ by least squares the uncertainties of t at Δ_1 and Δ_2 are independent of each other and that the values of t there would not be affected by inclusion of a square term. These values of t are called *summary values*, and are more accurate than the means over intervals of 1°. Wider intervals can be used when convenient. The results can then be interpolated by divided differences. The method might smooth out real irregularities, but this can always be checked by forming χ^2 for the original data against the interpolated values. Details of the method are in *Theory of Probability*, pp. 223–7.

APPENDIX H

Nomenclature and Units

1. Notation. The Royal Society committee on notation recommends capital letters K, E, G for K, E, μ to give uniformity. Most symbols have to be used in more than one sense in physics, and all these are used in more than one sense in this book. They give no symbol for λ, and if formulae containing it are transformed to contain k instead they become cluttered up with extra factors $\frac{2}{3}$ and $\frac{4}{3}$. They use G also for the gravitational constant; this would make it impossible to write a paper on the bodily tide. I use an italic f; a roman f had long been used in dynamical astronomy and led to no confusion. It is suggested that f might be mistaken for a function, but (1) a function is associated with an argument in brackets, and (2) in over fifty years I have not met a case where some other letter is not already suggested for any function associated with f.

2. S.I. units. These are based on the metre, kilogram and second instead of the centimetre, gram and second. It is influentially proposed that these should be adopted for general use. There has been insufficient consultation of scientists in general for their opinions. When subsidiary units are wanted it is proposed that new names should be adopted for powers of 10^3 up to $10^{\pm 9}$.

If the new standards are adopted the whole of the standard tables of physical constants will need to be rewritten. The density of water will be 1000 kilograms/metre³. It is not recorded whether the compilers and publishers have been consulted.

In geophysics we are concerned with lengths from about 10^{-8} cm. to 6×10^8 cm., just within the range 10^{-9} to 10^9; in metric units the Ångström becomes 10^{-10} metres. But nuclear physics is concerned with smaller lengths than the Ångström, and within the solar system we are concerned with distances of up to 10^{15} cm.

The units in this edition, as in previous ones, are those of the c.g.s. system. Conversion factors may be found in *Changing to the Metric System* (1969, HMSO) and there is a table of selected conversion factors in the *Journal of the Geological Society* (1972, **128**, after p. 310).

BIBLIOGRAPHY AND AUTHOR INDEX

(Numbers in brackets refer to pages in the text where the works are mentioned.
References are given similarly to other mentions of persons.)

The following abbreviations are used for some titles frequently referred to.

AJ	*Astronomical Journal.*
BAGMT	British Association Seismological Committee, Gray-Milne Trust.
BGSA	*Bulletin of the Geological Society of America.*
BSSA	*Bulletin of the Seismological Society of America.*
CP	Jeffreys, *Collected Papers.* Gordon and Breach, London.
GBG	*Gerlands Beiträge zur Geophysik.*
GJRAS	*Geophysical Journal of the Royal Astronomical Society.*
JGR	*Journal of Geophysical Research.*
MNGS	*Monthly Notices of the Royal Astronomical Society, Geophysical Supplement.*
MNRAS	*Monthly Notices of the Royal Astronomical Society.*
PRSA	*Proceedings of the Royal Society, Series A.*
QJRAS	*Quarterly Journal of the Royal Astronomical Society.*
TAGU	*Transactions of the American Geophysical Union.*
U.S.C. and G.S.	*United States Coast and Geodetic Survey.*

Adams, F. D. (19)
ADAMS, F. D. and EVE, A. S. 1907 *Nature, Lond.* **76**, 269. (406)
Adams, J. C. (324)
ADAMS, L. H. 1924 *J. Wash. Acad. Sci.* **14**, 459–472. (391, 392, 409)
    1931 *GBG*, **31**, 315–321. (103)
ADAMS, L. H. and GIBSON, R. E. 1926 *Proc. Nat. Acad. Sci., Wash.* **12**, 275–283. (103)
    1929 *Proc. Nat. Acad. Sci., Wash.* **15**, 713–724. (103)
ADAMS, L. H. and WILLIAMSON, E. D. 1923a *J. Franklin Inst.* **195**, 475–529. (101)
    1923b *J. Wash. Acad. Sci.* **13**, 418–428. (205)
ADAMS, R. D. 1968 *BSSA*, **58**, 1933–1947. (119)
    1969 *BSSA*, **59**, 1419–1420. (119)
    1972 *BSSA*, **62**, 1063–1071. (109)
ADAMS, R. D. and RANDALL, M. J. 1964 *BSSA*, **54**, 1299–1313. (158, 214)
ADAMS, W. S. 1941 *Astrophys. J.* **93**, 11–23. (254)
AHRENS, L. H. 1947 *Nature, Lond.* **160**, 874–875. (377)
AIG 1967 *Bull. géod. Publ. Spec., Geodetic Reference System 1967.* Bureau central de l'Association Internationale de Géodésie, 19 rue Auber, Paris (9e). (261)
AIRY, G. B. 1855 *Phil. Trans.* **145**, 101–104. (226)
AKI, K. and KAMINUMA, K. 1963 *Bull. Earthq. Res. Inst., Tokyo,* **41**, 243–259. (100)
AKIM, E. L. 1966 *Dokl. Akad. Sci., U.S.S.R.* **170**, no. 4, 799–802. (220)
ALDER, B. J. and WAINWRIGHT, T. E. 1957 *J. Chem. Phys.* **27**, 1208–1209. (20)
ALEXANDER, S. S. 1967 *I. S. A. Abstract* no. 11, Zürich meeting. (160)
ALSOP, L. E. 1963 *BSSA*, **53**, 483–501. (309)

[535]

35-2

ALSOP, L. E. and BRUNE, J. N. 1965 *JGR*, **70**, 6165–6173. (309)

ALSOP, L. E. and KUO, J. 1964 *Ann. di Geof.* **20**, 286–300. (309)

ALSOP, L. E., SUTTON, G. H. and EWING, M. 1961*a* *JGR*, **66**, 631–641. (309)

1961*b* *JGR*, **66**, 2911–2915. (360)

ALTERMAN, Z. 1965 *GJRAS*, **9**, 121–152. (134)

1966 *GJRAS*, **11**, 189–224. (134)

ALTERMAN, Z., JAROSCH, H. and PEKERIS, C. L. 1959*a* *PRSA*, **252**, 80–95. (312, 521)

1959*b* *PRSA*, **252**, 219–241. (312, 521)

ANDERSON, D. L. 1967*a* *GJRAS*, **13**, 9–30. (215)

1967*b* *GJRAS*, **14**, 135–164. (359)

ANDERSON, D. L. and ARCHAMBEAU, C. B. 1964 *JGR*, **69**, 2071–2089. (359)

ANDERSON, D. L., BEN-MENAHEM, A. and ARCHAMBEAU, C. B. 1965 *JGR*, **70**, 1441–1448. (359)

ANDERSON, D. L. and HARKRIDER, D. G. 1968 *BSSA*, **58**, 1407–1500. (79)

ANDERSON, D. L. and KOVACH, R. L. 1964 *Proc. Nat. Acad. Sci., Wash.* **51**, 168–172. (160)

ANDERSON, E. M. 1905 *Trans. Geol. Soc. Edinb.* **8**, 387–402. (14)

1918 *Geol. Mag.* p. 192. (447)

1934 *GBG*, **42**, 133–159. (401, 417)

1936 *Proc. Roy. Soc. Edinb.* **56**, 128–157. (17, 499)

1938 *Proc. Roy. Soc. Edinb.* **58**, 242–251. (477)

1940 *Proc. Roy. Soc. Edinb.* **60**, 194–209. (402)

1942 *The Dynamics of Faulting.* Oliver and Boyd, Edinburgh. (477, 499, 506)

ANDERSON, E. M. and RADLEY, E. G. 1915 *Quart. J. Geol. Soc.* **71**, 205–217. (104)

ANDERSON, O. L. 1973 *JGR*, **78**, 4901–4914. (215)

ANDERSON, O. L. and NAFE, J. E. 1965 *JGR*, **70**, 3951–3963. (214)

ANDERSON, O. L. and SOGA, N. 1967 *JGR*, **72**, 5754–5757. (215)

AOKI, S. 1967 *Publ. Astr. Soc., Japan*, **19**, 585–595. (362)

ARNOLD, E. P. 1966 Ph.D. thesis, Cambridge. (63, 75, 125, 126, 127, 128, 145, 146, 163, 164) *Director, I.S.C.* (84)

ARNOLD, E. P., JEFFREYS, H. and SHIMSHONI, M. 1963 *GJRAS*, **8**, 12–16. *CP*, **2**, 660. (124)

ARROL, W. A., JACOBI, R. B. and PANETH, F. A. 1942 *Nature, Lond.* **149**, 235–238. (375)

Aston, F. (369, 378)

Atkinson, R. d'E. (210)

AXELROD, D. J. 1963 *JGR*, **68**, 3257–3263. (487)

BAARS, B. 1951 *N. V. de Bataaf. Petr. Maat.*, Geol. Dept., The Hague. (307)

Balavadze, B. K. (99)

BALCHAN, A. S. and COWAN, G. R. 1966 *JGR*, **71**, 3577–3588. (210)

BALDWIN, R. B. 1949 *The Face of the Moon.* Chicago University Press. (467, 502)

BALMINO, G., KAULA, W. M. and LAMBECK, K. 1973 *JGR*, **78**, 478–481. (278)

BARBER, H. N., DADSWELL, H. E. and INGLE, H. D. 1959 *Nature, Lond.* **184**, 203–204. (487)

BARBER, N. F. and URSELL, F. 1948 *Phil. Trans.* A, **240**, 527–560. (168)

BARNETT, C. H. 1962 *Nature, Lond.* **195**, 447–448. (491)

Barrell, J. (272, 459, 482, 486)

BARRELL, J. 1914 *J. Geol.* **22**, 28–48, 145–163, 209–236, 289–314, 441–468, 537–555, 655–683, 729–741. (232)

1915 *J. Geol.* **23**, 27–44, 425–443, 499–515. (232)

Bastings, L. (88)

BASTINGS, L. 1935 *PRSA*, **149**, 88–103. (106)

BATCHELOR, G. K. and DAVIES, R. M. (eds.) 1956 *Surveys in Mechanics* (Cambridge Monographs in Mechanics and Applied Mathematics), pp. 250–351. (3)

BATEMAN, H. 1910 *Phil. Mag.* (6), **19**, 576–587. (51)

Båth, M. (140)

BÅTH, M. 1967 *Handbook on Earthquake Magnitude Determinations.* Seism. Inst Uppsala. (60)

BÅTH, M. and STEFANNSON, R. 1966 *Ann. di Geof.* **19**, 119–130. (147)

BEAUFILS, Y., MECHLER, P. and ROCARD, Y. 1970 *C.R. Acad. Sci. Paris*, **270**B, 926–928. (122)

BEILBY, G. T. 1921 *Aggregation and Flow of Solids.* Macmillan. (20)

Bellamy, E. F. (83, 111)

BELOUSSOV, V. V. 1970 *Tectonophysics*, **9**, 489–511. (494)

BÉNARD, H. 1901 *Ann. Chim. Phys.* (7), **23**, 62–144. (454)

Benfield, A. E. (214)

BENFIELD, A. E. 1939 *PRSA*, **173**, 428–450. (401)

1947 *Amer. J. Sci.* **245**, 1–18. (402)

Benioff, H. (308, 433, 463)

BENIOFF, H. 1954 *TAGU*, **35**, 984–985. (288)

BENIOFF, H., PRESS, F. and SMITH, S. W. 1961 *JGR*, **66**, 605–619. (309)

BEN-MENAHEM, A. 1965 *JGR*, **70**, 4641–4651. (359)

BEN-MENAHEM, A. and VERED, M. 1973 *BSSA*, **63**, 1611–1636. (67)

Bernal, J. D. (421)

BERNAL, J. D. 1936 *Observatory*, **59**, 268. (206)

BERNAL, J. D. 1967 *The Origin of Life.* Weidenfeld and Nicolson, London. (395)

Bernard, P. (168)

Berroth, A. (239)

BERRY, W. B. N. and BARKER, R. M. 1968 *Nature, Lond.* **217**, 938. (340)

BERZON, I. S., KOGAN, S. D. and PASSECHNIK, I. P. 1972 *Earth Plan. Sci. Letters*, **16**, 166–170. (157)

BERZON, I. S., PASSECHNIK, I. P. and POLIKARPOV, A. M. 1974 *GJRAS*, **39**, 603–611. (358)

BETHE, H. A. 1939 *Phys. Rev.* **55**, 434–456. (369)

BILBY, B. A., COTTRELL, A. H. and SWINDEN, K. H. 1963 *PRSA*, **272**, 304–314. (92)

BINGHAM, E. C. 1922 *Fluidity and Plasticity.* McGraw-Hill. (18)

Birch, F. (216, 312, 413, 427)

BIRCH, F. 1938 *BSSA*, **28**, 49–56. (99)

1943 *Bull. Geol. Soc. Amer.* **54**, 263–285. (104)

1947 *Amer. J. Sci.* **245**, 733–753. (402)

1951 *JGR*, **56**, 107–126. (396, 417)

1952 *JGR*, **57**, 227–286. (103, 130, 210, 424, 434)

1961*a* *Bull. Geol. Soc. Amer.* **72**, 1411–1443. (446)

1961*b* *GJRAS*, **4**, 295–311. (215, 421)

1961*c* *JGR*, **66**, 2199–2239. (104)

1963 *Solids under Pressure*, ed. W. Paul and D. M. Warschauer, pp. 137–162. McGraw-Hill, New York. (210)

538   Bibliography and Author Index

BIRCH, F. 1964 *JGR*, **69**, 4377–4388. (214, 421)
    1965 *BGSA*, **76**, 133–153. (416, 428)
    1968 *Phys. Earth Plan. Int.* **1**, 141–147. (210, 421)
    1972 *GJRAS*, **29**, 373–385. (161, 210)
    1973 *Observatory*, **93**, 219. (210)
BIRCH, F. and BANCROFT, D. 1938 *J. Geol.* **46**, 59–87, 113–141. (104)
BIRCH, F., SCHAIRER, J. F. and SPICER, H. C. 1942 *Handbook of Physical Constants*, Geol. Soc. Amer. Special Papers, no. 36. (21, 26, 130, 413, 431)
    See also Clark, S. P. 1966
BIRTWISTLE, G. 1927 *The Principles of Thermodynamics*. Cambridge University Press. (394)
BISWAS, B. 1969 *Rising Continents, Deepening Ocean Basins and their Changing Configuration*. B. Biswas, Calcutta. (278, 463, 490, 495)
BLACKETT, P. M. S. 1956 *Nature, Lond.* **178**, 1085–1086. (490)
BLACKETT, P. M. S. *et al.* 1965 *Phil. Trans.* A, **258**, 1–215. (491)
BLACKWELDER, E. 1933 *Amer. J. Sci.* **26**, 97–113. (504)
BOEKE, H. E. and EITEL, W. 1923 *Grundlagen d. phys.-chem. Petrographie*. Borntraeger. (21, 394)
BOLLO, R. and GOUGENHEIM, A. 1949a *Ann. Geophys.* **5**, 176–180. (307)
                            1949b *C.R. Acad. Sci., Paris*, **229**, 983–984. (307)
BOLT, B. A. 1959 *GJRAS*, **2**, 190–198. (110, 158)
    1964 *BSSA*, **54**, 191–208. (158, 214)
    1965 *Nature, Lond.* **107**, 967–969. (158)
    1968 *BSSA*, **58**, 1305–1324. (158)
    1970 *GJRAS*, **20**, 367–382. (108, 119, 159)
    1972 *Phys. Earth Plan. Int.* **5**, 301–311. (108, 161, 421)
BOLT, B. A., DOYLE, H. A. and SUTTON, D. J. 1958 *GJRAS*, **1**, 135–145. (101)
BOLT, B. A. and LOMNITZ, C. 1967 *BSSA*, **57**, 1093–1114. (358)
BOLT, B. A., NIAZI, M. and SOMERVILLE, M. R. 1970 *GJRAS*, **19**, 299–305. (127)
BOLT, B. A. and NUTTLI, O. 1966 *JGR*, **71**, 5977–5985. (123)
BOLT, B. A. and QAMAR, A. 1972 *Phys. Earth Plan. Int.* **5**, 400–402. (161)
BOLTWOOD, B. B. 1907 *Amer. J. Sci.* (4), **23**, 77–88. (374)
BONDI, H. and GOLD, T. 1956 *MNRAS*, **115**, 41–46. (349, 350)
BORN, M. and GREEN, H. S. 1946 *PRSA*, **188**, 10–18. (20)
                       1947a *Nature, Lond.* **159**, 251–254. (20)
                       1947b *PRSA*, **189**, 103–117. (20)
                       1947c *PRSA*, **190**, 455–474. (20)
BORNE, G. V. D. 1904 *Nachr. Ges. Wiss. Göttingen*, pp. 1–25. (28, 138)
BOSCOVICH, R. J. 1755 *De litteraria expeditione...ad dimitiendos duos meridiani gradus et corrigendam mappam geographiam*, Rome. (226)
BOUGUER, P. 1749 *La Figure de la Terre*. Paris. (224, 225)
BOWEN, N. L. 1915 *J. Geol.* **23**, Suppl. 1–91. (393)
BOWER, D. R. 1966 *JGR*, **71**, 487–493. (190)
Bowie, W. (462)
BOWIE, W. 1917 *U.S.C. and G.S. Spec. Publ.* no. 40. (229)
           1924 *U.S.C. and G.S. Spec. Publ.* no. 99. (460)
BRAGG, W. L. 1937 *The Atomic Structure of Minerals*. Cornell University Press. (24)
Bridgman, P. W. (5, 19, 425)
BRIDGMAN, P. W. 1914 *Phys. Rev.* **3**, 126–141, 153–203. (392, 425)

BRIDGMAN, P. W. 1915 *Phys. Rev.* **6**, 1–33, 94–112. (392)

1925 *Amer. J. Sci.* (5), **10**, 359–367. (102)

1931 *The Physics of High Pressure.* Bell. (347)

1935 *Phys. Rev.* **48**, 825–847. (275)

1951 *Bull. Geol. Soc. Amer.* **62**, 533–534. (208)

BROOKS, C. E. P. 1926, 1949 *Climate through the Ages.* Benn, London. (334, 489)

BROUGHTON EDGE, A. B. and LABY, T. H. 1931 *The Principles and Practice of Geophysical Surveying.* Cambridge University Press. (165)

Brouwer, D. (173, 201, 253, 335)

BROUWER, D. 1950 *Les constantes fondamentales d'Astronomie*, CNRS Paris, 5–18. (254)

1952a *Proc. Nat. Acad. Sci., Wash.* **38**, 1–12. (324)

1952b *AJ*, **57**, 125–146. (324)

1959 *AJ*, **64**, 378–397. (259)

Brown, E. W. (172, 173, 220, 325, 334, 478)

BROWN, E. W. 1904 *MNRAS*, **64**, 524–534. (220)

1932 *Rice Inst. Pamphl.* no. 1. (341)

Other references are given in the 4th edition of this book.

BROWN, E. W. and SHOOK, C. A. 1933 *Planetary Theory.* Cambridge University Press. (341)

BROWNE, B. C. 1937 *MNGS*, **4**, 271–279. (190)

BROWNE, B. C. and COOPER, R. I. B. 1950 *Phil. Trans.* A, **242**, 243–310. (190)

BRUNE, J. N., BENIOFF, H. and EWING, M. 1961 *JGR*, **66**, 2895–2910. (78)

BRUNT, D. 1925 *Nature, Lond.* **115**, 299–301. (454)

BRUTON, R. H., CRAIG, K. J. and YAPLEE, B. S. 1959 *AJ*, **64**, 325. (255)

BUCHBINDER, G. G. R. 1965 *BSSA*, **55**, 441–462. (159)

1968 *JGR*, **73**, 5901–5923. (157)

1971 *BSSA*, **61**, 429–456. (161)

BUCHBINDER, G. G. R., WRIGHT, C. and POUPINET, G. 1973 *BSSA*, **63**, 1699–1708. (160)

BULL, A. J. 1932 *Geol. Mag.* **69**, 73–75. (443)

Bullard, E. C. (103, 305, 313, 325, 400, 412, 450, 457, 467, 505)

BULLARD, E. C. 1938 *Phil. Trans.* A, **237**, 237–271. (190, 505)

1939a *PRSA*, **173**, 474–502. (402)

1939b *MNGS*, **4**, 534–536. (408)

1942 *MNGS*, **5**, 41–47. (373, 396)

1947 *MNGS*, **5**, 127–130. (401)

1948 *MNGS*, **5**, 186–192. (202, 212)

1949 *PRSA*, **197**, 433–453. (467)

1950 *MNGS*, **6**, 36–41. (348, 467)

1954 *PRSA*, **222**, 408–429. (415)

1957 *Verh. K. Ned. Geol. Mijnb Gen.* **18**, 23–41. (214)

BULLARD, E. C. and COOPER, R. I. B. 1948 *PRSA*, **194**, 322–347. (234)

BULLARD, E. C., EVERETT, J. E. and SMITH, A. G. 1965 *Phil. Trans.* A, **258**, 41–51. (489)

BULLARD, E. C. and GASKELL, T. F. 1941 *PRSA*, **177**, 476–499. (166)

BULLARD, E. C., GASKELL, T. F., HARLAND, W. B. and KERR GRANT, C. 1940 *Phil. Trans.* A, **239**, 29–94. (166)

BULLARD, E. C. and GELLMAN, H. 1954 *Phil. Trans.* A. **247**, 213–278. (467)

BULLARD, E. C. and MACE, C. 1939 *MNGS*, **4**, 473–480. (459)

540   *Bibliography and Author Index*

BULLARD, E. C. and STANLEY, J. P. 1949 *Ver. finn. geod. Inst.* no. 36, pp. 33–40. (380)

Bullen, K. E. (71, 116, 118, 130, 143, 205, 210, 216, 302, 311, 360, 421, 423)

BULLEN, K. E. 1933 *Constants of Seismological Observatories.* BAGMT. (72)

    1934 *MNGS*, 3, 190–201. (72)

    1936a *N.Z. J. Sci. Tech.* 18, 493–507. (98)

    1936b *MNGS*, 3, 395–401. (206)

    1937a *MNGS*, 4, 143–157. (69, 115)

    1937b *MNGS*, 4, 158–164. (69, 71)

    1937c *Trans. Roy. Soc. N.Z.* 67, 122–124. (212)

    1938a *MNGS*, 4, 317–331. (115)

    1938b *MNGS*, 4, 332–335. (115)

    1938c *MNGS*, 4, 469–471. (71, 115)

    1938d *Table for Converting Geographic to Geocentric Distances.* BAGMT. (71)

    1938e *N.Z. J. Sci. Tech.* B, 19, 497–519. (98)

    1938f *N.Z. J. Sci. Tech.* B, 19, 519–522. (98)

    1938g *N.Z. J. Sci. Tech.* B, 20, 31–43. (98)

    1938h *N.Z. J. Sci. Tech.* B, 20, 61–66. (98)

    1939a *Proc. 6th Pacific Sci. Congr.* pp. 103–110. (98, 212)

    1939b *MNGS*, 4, 578–582. (143)

    1939c *MNGS*, 4, 583–593. (120)

    1939d *Trans. Roy. Soc. N.Z.* 69, 188–190. (212)

    1940a *Trans. Roy. Soc. N.Z.* 70, 137–139. (212)

    1940b *BSSA*, 30, 235–250. (212)

    1941 *Trans. Roy. Soc. N.Z.* 71, 164–166. (212)

    1942 *BSSA*, 32, 19–29. (210, 212)

    1945 *MNGS*, 5, 91–98. (56, 64)

    1946 *Nature, Lond.* 157, 405. (420)

    1947 *Introduction to the Theory of Seismology.* Cambridge University Press. (136)

    1950 *MNGS*, 6, 125–128. (421)

    1956 *MNGS*, 7, 214–217. (213)

    1960a *GJRAS*, 3, 258–269. (53)

    1960b *GJRAS*, 3, 354–359. (60)

    1963a *Introduction to the Theory of Seismology*, 3rd edn, Cambridge University Press. §8·7 (60), p. 231. (212)

    1963b *GJRAS*, 7, 584–592. (208)

    1965a *GJRAS*, 9, 233–252. (214, 421)

    1965b *GJRAS*, 9, 265–274. (65)

    1973 *Nature, Lond.* 243, 68–70. (425)

    1975 *The Earth's Density.* London, Chapman & Hall. (210, 212, 425)

BULLEN, K. E. and HADDON, R. A. W. 1967a *Nature, Lond.* 213, 574–576. (159, 311, 313)

    1967b *Phys. Earth Plan. Int.* 1, 1–13. (212, 421)

    1967c *Proc. Nat. Acad. Sci., Wash.* 58, 846–852. (212)

    1969 *Phys. Earth Plan. Int.* 2, 35–49. (159)

    1970 *Phys. Earth Plan. Int.* 2, 342–349. (159)

Bibliography and Author Index 541

BULLEN, K. E. and HADDON, R. A. W. 1973a *GJRAS*, **35**, 31–38. (161)
    1973b *Phys. Earth Plan. Int.* **7**, 199–202. (71)
BURRARD, S. 1918. *Prof. Papers Surv. India*, no. 17. (460)
BURRIDGE, R. 1962 *PRSA*, **270**, 144–154. (134)
    1963 *PRSA*, **276**, 367–400. (134)
    1969 *Phil. Trans.* A, **265**, 353–381. (92, 284)
BURRIDGE, R. and KNOPOFF, K. 1967 *BSSA*, **57**, 341–372. (285, 477)
BURRIDGE, R. and WILLIS, J. R. 1969 *PCPS*, **66**, 443–468. (284)
BURTON, P. W. 1974 *GJRAS*, **36**, 167–189. (360)
BURTON, P. W. and KENNETT, B. L. N. 1972 *Nature Physical Science, Lond.* **238**, 87–90. (360)
BUTCHER, J. G. 1876 *Proc. Lond. Math. Soc.* (1), **8**, 103–135. (7)
BYERLEE, J. D. 1967 *JGR*, **72**, 3639–3648. (17)
Byerly, P. (111, 118, 147, 463)
BYERLY, P. 1926 *BSSA*, **16**, 209–265. (113).
    1942 *Seismology*, Prentice-Hall. (83, 136)
BYERLY, P. and WILSON, JAMES T. 1935 *BSSA*, **25**, 223–246. (58, 98)
BYRD, E. E. 1967 Thesis, M.I.T. (494)
CALLANDREAU, O. 1889 *Ann. Obs. Paris*, pp. 1–84. (201)
    1897 *Bull. Astron.* **14**, 214–217. (201)
Caloi, P. (88)
Campbell, W. W. (499)
CAPUTO, M. 1965 *Bull. Géod.*, p. 197. (262)
    1967 *The Gravity Field of the Earth*. Academic Press, New York and London. (262)
CAPUTO, M., HARRISON, J. C., VON HUENE, R. and HELFER, M. D. 1963 *JGR* **68**, 3273–3282. (190)
CAPUTO, M. and PIERI, 1968 *Ann. di Geof.* **21**, 123–149, (262)
CARDER, D. S. 1964 *BSSA*, **54**, 2270–2294. (125)
CARDER, D. S. *et al.* 1962 *BSSA*, **52**, 977–1077. (60, 101)
CARDER, D. S., TOCHER, D., BUFE, C., STEWART, S. W., EISLER, J. and BERG, E. 1967 *BSSA*, **57**, 573–590. (163)
CARPENTER, E. W. and DAVIES, D. 1966 *Nature, Lond.* **212**, 134–135. (360)
CASSINIS, G. 1930 *Bull. géod.* pp. 40–49. (181, 187)
CAVENDISH, H. 1772–4 Ms. note; Papers, **2**, 1921, 404. (226)
CHAMBERLIN, T. C. *et al.* 1909 *The Tidal and other Problems*. Carnegie Inst. Publ. no. 107. (330, 342, 439)
Chandler, S. C. (287)
CHANDRASEKHAR, S. 1952 *Phil. Mag.* (7), **43**, 1317–1329. (454)
Chapman, S. (119)
CHAPMAN, S. and LINDZEN, R. S. 1969 *Space Science Rev.* **10**, 1–188. (328)
CHOWDHURY, D. K. and FRAZIER, C. W. 1972 *TAGU*, **53**, 1046. (158)
    1973 *JGR*, **78**, 6021–6027. (158)
CHRISTOPHERSON, D. G. 1940 *Quart. J. Math.* **11**, 63–65. (454)
CHRISTY, R. F. 1962 *Astrophys. J.* **136**, 887–902. (368)
    1964 *Rev. Mod. Phys.* **36**, 555–571. (368)
    1966 *Astrophys. J.* **144**, 108–179. (368)
    1968 *QJRAS*, **9**, 13–39. (368)
Clairaut, A. C. (184, 200)

542   Bibliography and Author Index

CLAIRAUT, A. C. 1743 *Théorie de la figure de la terre*. Paris. (196)

CLARK, S. P. 1966 *Handbook of Physical Constants* (revised edition). *Geol. Soc. Amer. Mem.*, 97. (21)

Clarke, A. R. (201)

CLARKE, F. W. 1924 *Data of Geochemistry*. U.S. Geol. Survey. (21, 393, 400)

CLARKE, F. W. and STEIGER, G. 1914 *J. Wash. Acad. Sci.* **4**, 58–62. (379)

CLEARY, J. 1969 *BSSA*, **59**, 1399–1405. (127)

CLEARY, J. and HADDON, R. A. W. 1972 *Nature, Lond.* **240**, 549–551. (161)

CLEARY, J. and HALES, A. L. 1966 *BSSA*, **56**, 467–489. (162)

Coker, E. G. (19)

COLOMBO, P. 1965 *Nature, Lond.* **208**, 575. (210. 357)

COLOMBO, P. and SHAPIRO, I. I. 1966 *Astrophys. J.* **145**, 296–307. (210, 357)

Comrie, L. J. (68, 71, 72, 73)

COMRIE, L. J. 1938 *Geoc. Direction Cosines of Seismological Observatories*. BAGMT. (72)

1941 *MNGS*, **5**, 27–29. (75)

CONRAD, V. 1925 *Mitt. ErdbKomm. Wien*, no. 59. (88)

1928 *GBG*, **20**, 240–277. (88)

COOK, A. H. 1959a *GJRAS*, **2**, 199–214. (184)

1959b *GJRAS*, **2**, 222–238. (262)

1963 *Nature, Lond.* **198**, 1186. (282)

1965 *Metrologia*, **1**, 84–114. (255)

1967 *Phil. Trans.* A, **261**, 211–252. (255)

1970 *MNRAS*, **150**, 187–194. (220)

COSTER, H. P. 1947 *MNGS*, **5**, 131–145. (402)

1948 *MNGS*, **5**, 193–199. (120)

COTTRELL, A. H. 1962 'Fracture' in *Properties of Reactor Materials*. Butterworth, London. (92)

1963 *PRSA*, **276**, 1–18. (92)

COULOMB, C. A. DE 1776 *Mem. math. phys. de l'acad. roy.* **7**, 343–382. (14, 18)

COULOMB, J. 1945 *Ann. Géophys.* **1**, 244–255. (463)

COWLING, T. G. 1957 *Magnetohydrodynamics*. Interscience Tracts. (467)

COXETER, H. S. M. 1962 *Trans. New York Acad. Sci.* **11**, 24, 320–331. (457)

CREER, K. M. and ISPIR, Y. 1970 *Phys. Earth Plan. Int.* **2**, 283–293. (496)

Curie, M. S. (371)

CURTIS, A. R. and SHIMSHONI, M. 1970 *BSSA*, **60**, 1077–1087. (79)

CWILONG, B. M. 1947 *J. Glaciology*, **1**, 53. (20)

DALY, R. A. 1926 *Our Mobile Earth*. Ch. Scribner's Sons. (470)

1946 *Proc. Amer. Phil. Soc.* **90**, 104–119. (499, 500)

DARBYSHIRE, J. and M. 1957 *MNGS*, **7**, 301–307. (169)

Darwin, C. (487)

Darwin, G. H. (264, 292, 328, 340, 353, 354, 363, 429, 434)

DARWIN, G. H. 1880 *Phil. Trans.* **171**, 713–891. (356)

1887 *Phil. Trans.* A. **178**, 242–249. (429)

1900 *MNRAS*, **60**, 82–124. (201)

1905 British Association Presidential Address, *Scientific Papers*, vol. 4, p. 547. Cambridge University Press. (378)

1907–16 *Scientific Papers* (5 vols.), Cambridge University Press. (292, 478, 479)

Darwin, H. (292)

Davies, A. Morley (486)
DAVIES, A. MORLEY 1918 *Geol. Mag.* pp. 125 and 233. (447)
DAVIES, D. 1967 *GJRAS*, **13**, 421–424. (360)
DAVISON, C. 1887 *Phil. Trans.* A, **178**, 231–242. (429, 432, 506)
DAY, A. L. 1925 *J. Franklin Inst.* **200**, 161–182. (469, 471)
DAY, A. L., SOSMAN, R. B. and HOSTETTER, J. C. 1914 *Amer. J. Sci.* (4), **37**, 1–39. (431)
Deacon, G. E. R. (168)
Debye, P. (43)
DERR, J. S. 1969a *BSSA*, **59**, 2079–2100. (523)
    1969b *JGR*, **74**, 5202–5220. (523)
DIETZ, R. S. 1961 *Nature, Lond.* **190**, 854–857. (493)
DILLON, L. S. 1974 *Amer. Assn Petroleum Geol.* Mem. 23, ed. C. F. Kahle, 167–239. (434)
DONNELLY, R. J., HERMAN, R. and PRICOGINE, I. 1965 *Non-equilibrium Thermodynamics, Variational Techniques, and Stability*, pp. 125–164. Chicago University Press. (455)
DORMAN, J., EWING, M. and OLIVER, J. 1960 *BSSA*, **50**, 87–115. (130, 146)
DORMAN, L. M. 1968 *JGR*, **73**, 3877–3883. (360)
DORMAN, L. M. and LEWIS, B. T. R. 1970 *JGR*, **75**, 3357–3365, 3367–3386. (235)
    1972 *JGR*, **77**, 3068–3077. (235)
DOYLE, H. A. and HALES, A. L. 1967 *BSSA*, **57**, 761–772. (162)
DUNCOMBE, R. L. 1958 *Astr. papers of the American Ephemeris*, 16, part 1. (362)
VAN DEN DUNGEN, F. H., COX, F. J. and VAN MIEGHEM, J. 1949 *Bull. Acad. Belg. Cl. Sci.* **35**, 642–655. (315)
DUTTON, C. E. 1889 *Bull. Wash. Phil. Soc.* **11**, 51–64. (227, 485)
    1925 *J. Wash. Acad. Sci.* **15**, 359–369 (reprint of above). (227, 485)
DYCE, R. B., PETTENGILL, G. H. and SHAPIRO, I. I. 1967 *AJ*, **72**, 351–359. (210, 357)
DZIEWONSKI, A. M. and GILBERT, F. 1971 *Nature, Lond.* **234**, 465–466. (216)
Eckert, W. J. (173)
ECKERT, W. J. 1965 *AJ*, **70**, 787–792. (220)
EDDINGTON, A. S. 1918 *MNRAS*, **79**, 2–22. (368)
    1926 *The Internal Constitution of the Stars*. Cambridge University Press. (368)
    1941 *MNRAS*, **101**, 182–194. (368)
EKMAN, V. W. 1905 *Ark. Math. Astr. Phys.* **2**, 11. (483)
ELLES, G. L. and TILLEY, C. E. 1930 *Trans. Roy. Soc. Edinb.* **56**, 621–646. (425)
Elsasser, W. M. (457, 467)
ENGDAHL, E. R. and FLINN, E. A. 1969 *Science*, **163**, 177–179. (119)
ENGEL, A. E. J., ENGEL, C. G. and HAVENS, R. G. 1965 *BGSA*, **76**, 719–739. (446)
EÖTVÖS, R. VON 1912 *Verh. d. 17. Konf. d. Int. Erdmessung*, p. 111. (468)
ERGIN, K. 1967 *JGR*, **72**, 3669–3688. (160)
ESPINOSA, A. F. 1966 *Ann. di Geofisica.* **19**, 415–425. (160)
EUCKEN, A. 1944a *Nachr. Ges. Wiss. Göttingen*, pp. 1–25. (418)
    1944b *Naturwissenschaften*, pp. 112–121. (418)
Euler, L. (287)
Evans, J. W. (439)
EVANS, J. W. 1919 *Observatory*, pp. 165–167. (394).
EVANS, R. D. and GOODMAN, C. 1941 *Bull. Geol. Soc. Amer.* **52**, 459–490. (397)
Everest, G. (225)

EWING, W. M., CRARY, A. P. and RUTHERFORD, H. M. 1937 *Bull. Geol. Soc. Amer.* **48**, 753–802. (166)

EWING, W. M., PRESS, F. and JARDETSKY, W. S. 1957 *Elastic Waves in Layered Media.* McGraw-Hill. (132)

EWING, W. M., WORZEL, J. L., HERSEY, J. B., PRESS, F. and HAMILTON, G. R. 1950 *BSSA*, **40**, 233–242. (166)

FAYE, H. 1880 *C.R. Acad. Sci., Paris*, **90**, 1443–1446. (226)

FEDOROV, E. P. 1958 *Nutatsiya i Vynuzhdennoye Dvizheniye Polyusov Zemli.* Acad. Sci. Ukraine *SSR.* (313, 484)
 1961 *Nutation and Forced Motion of the Earth's Pole.* (Translation of above, by B. Jeffreys.) Pergamon Press. (287, 313, 484)

FEDOTOV, S. A. 1963 *Izv. Akad. Nauk U.S.S.R., Ser. Geofiz.* 509–520. (359)

FERMOR, L. L. 1913 *Rec. Geol. Surv. India*, **43**, 41–47. (23)

FIELDER, G. 1961 *MNRAS*, **123**, 15–26. (504)
 1965 *MNRAS*, **129**, 351–361. (504)
 1966 *MNRAS*, **132**, 413–422. (504)

FIELDER, G., FRYER, R. J., TITULAER, C., HERRING, A. K. and WISE, B. 1972 *Phil. Trans.* A, **271**, 361–409. (504)

FIELDER, G. and MARCUS, A. 1967 *MNRAS*, **136**, 1–10. (504)

FISCHER, I. 1959 *JGR*, **64**, 73–84. (255)
 1960 *JGR*, **65**, 2067–2076. (255)

Fisher, O. (342)

Fisher, R. A. (131)

Fotheringham, J. K. (327, 335)

FOTHERINGHAM, J. K. 1920*a* *MNRAS*, **80**, 578–581. (324)
 1920*b* *MNRAS*, **81**, 104–126. (324)
 1927 *MNRAS*, **87**, 142–167. (324)

Freedman, H. W. (163)

FRICKE, W. 1971 *Astron. and Astrophys.* **13**, 298–308. (363)

Friedlander, F. G. (133)

FRIEDLANDER, F. G. 1954 *Comm. Pure and Appl. Math.* 7, 705–732. (54, 109)

Frisch, O. R. (379, 428)

FUCHS, K., MAYER-ROSA, D. and LIEBAU, F. 1971 *Zs. f. Geophys.* **37**, 937–942. (208)

FURON, R. 1949 *C.R. Acad. Sci. Paris*, **228**, 1509–1510. (489)

GAMOW, G. 1939 *Nature, Lond.* **144**, 575–577. 620–622. (369)

GANS, R. F. 1972 *JGR*, **77**, 360–366. (348, 349)

GAPOSCHKIN, E. M., ed. 1974 *Smiths. Astr. Obs. Spec. Rep.* 353. (247)

GAPOSCHKIN, E. M. and LAMBECK, K. 1970 *Smiths. Astr. Obs. Spec. Rep.* 315. (259, 277)
 1971 *JGR*, **76**, 4855–4881. (259, 277)

GASKELL, T. F. and SWALLOW, J. C. 1951 *Nature, Lond.* **167**, 723. (166)

GIBOWICZ, S. J. 1972 *N.Z. J. Geol. Geophys.* **15**, 336–359. (358)

Gilbert, F. (140, 421)

GILBERT, F. and BACKUS, G. 1961 *Proc. Nat. Acad. Sci.* 47, 362–371. (309)
 1965 *Rev. of Geophys.* 3, 1–9. (309)
 1966 *Geophysics*, **31**, 326–332. (309)

GILBERT, F. and MACDONALD, G. J. F. 1960 *JGR*, **65**, 675–693. (309)

GILBERT, G. K. 1892 *Bull. Wash. Phil. Soc.* **12**, 241–292. (500)

GILVARRY, J. J. 1958 *Astrophys. J.* **127**, 751–762. (503)

Glennie, E. A. (234)

GODWIN, H. 1956, 1975 *History of the British Flora*. Cambridge University Press. (416)

Gogna, M. L. (63, 126, 159, 163)

GOGNA, M. L. 1967 *GJRAS*, **13**, 503–527. (125, 126, 162)

    1968 *GJRAS*, **16**, 489–514. (160)

    1973 *GJRAS*, **33**, 103–126. (125, 126)

GOGUEL, J. 1951 *Ann. Géophys.* **7**, 1–6. (250)

GOLD, T. 1955a *MNRAS*, **115**, 585–604. (503)

    1955b *Nature, Lond.* **175**, 526–529. (480)

GOLDICH, S. S., MUEHLBERGER, W. R., LIDIAK, E. G. and HEDGE, C. E. 1966 *JGR*, **71**, 5375–5438. (463)

Goldschmidt, V. M. (212, 395, 406, 421)

GOLDSCHMIDT, V. M. 1930 *Naturwissenschaften*, **18**, 999–1013. (405)

    1931 *Nachr. Ges. Wiss. Göttingen*, pp. 184–190. (207)

GOLDSTEIN, S. 1926 *Proc. Camb. Phil. Soc.* **23**, 120–129. (442)

GORANSON, R. W. 1931 *Amer. J. Sci.* (5), **22**, 481–502. (394)

    1932 *Amer. J. Sci.* (5), **23**, 227–236. (394)

    1940 *Bull. Geol. Soc. Amer.* **51**, 1023–1034. (19)

GOUDAS, C. C. 1967 *AJ*, **72**, 955–956. (220)

GRACE, S. F. 1930 *MNGS*, **2**, 273–296. (292)

    1931 *MNGS*, **2**, 301–309. (292)

Gräfe, H. (88)

GRAHAM, J. W. 1956 *JGR*, **61**, 735–739. (490)

GRANT, F. S. 1953 *Observatory*, **73**, 227–228. (378)

GREEN, J. 1971 *JGR*, **76**, 5719–5731. (504)

GREGORY, J. W. 1929 *Quart. J. Geol. Soc.* **85**, lxviii–cxxii. (481)

    1930 *Quart. J. Geol. Soc.* **86**, lxxii–cxxxvi. (481)

GREIG, J. W. 1927 *Amer. J. Sci.* **13**, 1–44, 133–154. (393)

GRIFFITH, A. A. 1920 *Phil. Trans.* A, **221**, 163–198. (20)

Griggs, D. T. (472)

GRIGGS, D. T. 1936 *J. Geol.* **44**, 541–577. (6)

    1939 *Amer. J. Sci.* **237**, 611–650. (460)

    1940 *Bull. Geol. Soc. Amer.* **51**, 1001–1022. (19)

GRØNLIE, G. and RAMBERG, J. B. 1970 *Norsk. Geol. Tidskr.* **50**, 375–391. (494)

GROVES, G. and MUNK, W. 1958 *J. Marine Res.* **17**, 199–214. (333)

Grüneisen, E. (434)

GUINOT, B. 1972 *Astron. and Astrophys.* **19**, 207–214. (299)

Gutenberg, B. (69, 87, 88, 99, 105, 106, 110, 117, 127, 130, 153, 158, 162, 205, 306, 308, 309, 360, 433, 463)

GUTENBERG, B. 1915 *Veröff. ZentBur. int. seism. Ass.* (59, 87)

    1926 *Z. Geophys.* **2**, 24–29. (121)

    1931–6 *Handbuch d. Geophysik* Borntraeger. (83, 342)

    1932 *GBG*, **35**, 6–50. (97)

    1944 *BSSA*, **34**, 85–102. (36)

    1945 *Amer. J. Sci.* **243A**, 285–313. (121)

    1948 *BSSA*, **38**, 121–128. (121, 145)

    1953 *BSSA*, **43**, 223–232. (145)

    1954 *BGSA*, **65**, 337–348. (146)*

    1955 *GSA* Special Paper 62, 19–34. (146)

* The Journal is wrongly given in some of Gutenberg's references.

GUTENBERG, B. 1959 *Ann. di Geof.* **12,** 439–460. (146)

GUTENBERG, B. and RICHTER, C. F. 1933 *GBG,* **40,** 380–389. (68)

1934 *GBG,* **43,** 56–133. (121)

1935 *GBG,* **45,** 280–360. (121)

1936*a* *GBG,* **47,** 73–131. (121)

1936*b* *BSSA,* **26,** 341–390. (121)

1937 *BSSA,* **27,** 157–183. (121)

1938 *MNGS,* **4,** 363–372. (109)

1939*a* *GBG,* **54,** 94–136. (109, 117)

1939*b* *BGSA,* **50,** 1511–1528. (112)

1939*c* *BSSA,* **29,** 531–537. (121)

1941 *GSA Spec. Pap.* no. 34, pp. 1–131. (113)

1942 *BSSA,* **32,** 163–191. (136)

1943 *BSSA,* **33,** 269–279. (92)

1945*a* *BSSA,* **35,** 3–12. (136)

1945*b* *BSSA,* **35,** 57–69. (136)

1945*c* *BSSA,* **35,** 117–130. (136)

1945*d* *BGSA,* **56,** 603–668. (113)

1946 *TAGU,* **27,** 776. (137)

1949 *The Seismicity of the Earth.* Princeton University Press. (113, 120, 464)

1956*a* *BSSA,* **46,** 105–146. (60)

1956*b* *Ann. di Geof.* **9,** 1–15. (60)

GUTENBERG, B., WOOD, H. O. and BUWALDA, J. P. 1932 *BSSA,* **22,** 185–246. (104, 167)

HALES, A. L. 1935 *MNGS,* **3,** 372–379. (455, 457)

1937 *MNGS,* **4,** 122–131. (505)

1953*a* *MNGS,* **6,** 460–466. (433)

1953*b* *MNGS,* **6,** 488–493. (433)

HALES, A. L. and SACKS, I. S. 1959 *GJRAS,* **2,** 15–33. (101)

HARKRIDER, D. G. and ANDERSON, D. L. 1966 *JGR,* **71,** 2967–2980. (141)

HARRISON, J. V. and FALCON, N. L. 1934 *Geol. Mag.* **71,** 529–539. (443)

1936 *Quart. J. Geol. Soc.* **92,** 91–102. (443)

HARTREE, D. R. 1931 *PRSA,* **131,** 428–450. (38)

1936 *Mem. Proc. Manchester Lit. Phil. Soc.* **80,** 85–102. (526)

HASKELL, N. A. 1935 *Physics,* **6,** 265–269. (451)

1936 *Physics,* **7,** 56–61. (452)

1937 *Amer. J. Sci.* (5), **33,** 22–28. (452)

Hassan, E. M. (306)

HAUBRICH, R. and MUNK, W. H. 1959 *JGR,* **64,** 2373–2388. (349)

HAWKES, L. 1929 *Nature, Lond.* **123,** 244; **124,** 225. (20)

HAYES, R. C. 1935 *N.Z. J. Sci. Tech.* **17,** 1–10. (98)

1936 *Bull. Dom. Obs. N.Z.* no. 101. (98)

Hayford, J. F. (Ch. V *passim*)

HAYFORD, J. F. 1909 *The Figure of the Earth and Isostasy.* U.S.C. and G.S. (186, 225)

1910 *Supplementary Investigation in* 1909. U.S.C. and G.S. (201, 225, 250)

HAYN, F. 1907 *Abh. Sächs. Ges. Wiss.* **30,** 1–103. (218)

Heaviside, O. (372)

Hecker, O. (149, 292)

HEIM, A. 1921 *Geologie d. Schweiz* (2 vols.). Tauchnitz, Leipzig. (434)

Heiskanen, W. (186, 190, 228, 230, 233, 241)

HEISKANEN, W. 1921 *Ann. Acad. Sci. Fenn.* 18A, 1–84. (332)

1924 *Veröff. Finn. Geodät. Ass.* pp. 1–96. (227)

1938 *Ann. Acad. Sci. Fenn.* 51 = *Publ. Isost. Inst. Int. Ass. Geod.* no. 1. (240)

1957 *TAGU*, 38, 841–848. (246)

HEISKANEN, W. and MEINESZ, V. 1958 *The Earth and its Gravity Field.* McGraw-Hill. (283)

Helmert, F. R. (177, 201, 232, 239, 249)

HELMERT, F. R. 1911 *S.B. preuss. Akad. Wiss.* pp. 10–19. (186, 190)

Helmholtz, H. L. F. von. (367)

Hendershott, M. C. (334)

HENDERSHOTT, M. C. 1972 *GJRAS*, 29, 389–403. (339, 349)

1973 *TAGU*, 54, 76–86. (339, 349)

HENRIKSEN, S. W. 1960 *Ann. Int. Geophys. Year*, 12, 197–198. (203)

HENSHAW, B. B. and ZEN, E-AN. 1965 *BGSA*, 76, 1379–1386. (446)

HERGLOTZ, G. 1905 *Z. Math. Phys.* 52, 275–299. (300)

1907 *Phys. Z.* 8, 145–147. (51)

HERRIN, E. *et al.* 1968 *BSSA*, 58, 1193–1351. (125, 163)

HERRIN, E. and TAGGART, J. N. 1962 *BSSA*, 52, 1037–1046. (162)

1968 *BSSA*, 58, 1325–1337. (123, 164)

HERSCHEL, J. 1837 *Proc. Geol. Soc.* 2, 597. (226)

HERZEN, R. VON and UYEDA, S. 1963 *JGR*, 68, 4219–4250. (416)

HESS, H. H. 1946 *Amer. J. Sci.* 244, 772–791. (498)

See also McKenzie and Weiss (1975), bibliography. (492)

HESS, V. F. and LAWSON, R. W. 1918 *S.B. Akad. Wiss. Wien.* 127, 1–55. (373)

HEVESY, G. and HOBBIE, R. 1931 *Nature, Lond.* 128, 1038–1039. (379)

HEY, J. S. and HUGHES, V. A. 1959 *Paris Symposium on Radioastronomy*, pp. 13–18. (255)

Hill, G. W. (239)

HILL, M. N. 1957 *Physics and Chemistry of the Earth*, vol. 2, pp. 129–163. Pergamon Press. (167)

HILL, M. N. and SWALLOW, J. C. 1950 *Nature, Lond.* 165, 193. (166)

HILL, M. N. and WILLMORE, P. L. 1947 *Nature, Lond.* 159, 207. (165)

HILL, R. 1950 *The Mathematical Theory of Plasticity.* Oxford University Press. (19, 269)

1954 *J. Mech. Phys. Solids*, 2, 278–285. (269)

Hiller, W. (88)

Hills, G. F. S. (395, 466, 468)

HILLS, G. F. S. 1934 *Geol. Mag.* 71, 275–276. (465)

1937 *Pan-Amer. Geol.* 67, 161–168. (465)

1947 *The Formation of Continents by Convection.* Arnold. (465)

Hinks, A. R. (201, 286)

Hipparchus. (286, 327, 335)

HIRVONEN, R. A. 1956 *TAGU*, 37, 1–8. (246)

HODGSON, E. A. 1932 *BSSA*, 22, 38–49, 270–287. (98)

Hodgson, J. H. (60)

HOLMBERG, E. R. R. 1952 *MNGS*, 6, 325–330. (327)

Holmes, A. (87, 417, 432, 460, 469)

HOLMES, A. 1913 *The Age of the Earth*. Harpers. (365)
    1915*a* *Geol. Mag.* pp. 60–71. (406)
    1915*b* *Geol. Mag.* pp. 102–112. (409)
    1926*a* *Nature, Lond.* **118**, 586. (103, 366)
    1926*b* *Phil. Mag.* (7), **1**, 1055–1074. (376)
    1926*c* *Geol. Mag.* **63**, 317–318. (403)
    1929 *Nature, Lond.* **124**, 477–478. (378)
    1931 *Nature, Lond.* **128**, 1039–1040. (379)
    1946 *Nature, Lond.* **157**, 680–684. (380)
    1947*a* *Nature, Lond.* **159**, 127–128. (380)
    1947*b* *Trans. Geol. Soc. Glasg.* **21**, 117–152. (377)
    1948 *Trans. Edinb. Geol. Soc.* **14**, 176–194. (377)

HOLMES, A. and DUBEY, V. S. 1929 *Nature, Lond.* **123**, 794–795. (375)

HOLMES, A. and LAWSON, R. W. 1927 *Amer. J. Sci.* (5), **13**, 327–344. (376)

HÖNIGSCHMID, O. 1916 *Z. Elektrochem.* **22**, 18–23. (375)

HÖNIGSCHMID, O. and BIRKENBACH, L. 1923 *Ber. dtsch. chem. Ges.* **56**, 1837–1839. (374)

HOPKINS, W. 1849 *Trans. Camb. Phil. Soc.* **8**, 456–466. (14)

HORAI, K. I. 1969 *Earth and Plan. Sci. Letters*, **6**, 39–42. (495)

HORI, G. 1960 *AJ*, **65**, 291–300. (259)

HOSKINS, L. M. 1920 *Trans. Amer. Math. Soc.* **21**, 1–45. (290, 300)

Hough, S. S. (301)

HOUTERMANS, F. G. 1947 *Z. Naturforsch.* **2***a*, 322–328. (380)

Howlett, J. (321)

HOYLE, F. 1944 *Proc. Camb. Phil. Soc.* **40**, 256–258. (387)
    1945 *MNRAS*, **105**, 175–178. (387)
    1946 *MNRAS*, **106**, 406–422. (386, 387)

HOYLE, F. and LYTTLETON, R. A. 1940*a* *Proc. Camb. Phil. Soc.* **36**, 325–330. (386)
    1940*b* *Proc. Camb. Phil. Soc.* **36**, 424–437. (386)

HUGHES, D. S. and CROSS, J. H. 1951 *Geophysics*, **16**, 577–593. (130)

HUGHES, D. S. and MAURETTA, C. 1957 *Geophysics*, **22**, 23–31. (130)

Hughes, H. (119, 120)

Hughes, J. S. (83, 132)

HUGHES, J. S. 1936 *B. A. Seism. Comm. Rep.* (116)

HUNTER, J. DE GRAAFF 1932 *MNGS*, **3**, 42–51. (232, 248)
    1935 *Phil. Trans.* A, **234**, 377–431. (251)
    1951 *PRSA*, **206**, 1–16. (175)

HURLEY, P. M. 1968 *Geochem. Cosmochem. Acta*, **32**, 1025–1030. (464)

HURLEY, P. M. and GOODMAN, G. 1943 *Bull. Geol. Soc. Amer.* **54**, 305–324. (375)

IDELSON, N. and MALKIN, N. 1931 *GBG*, **29**, 156–160. (191)

IMMELMANN, M. N. S. 1934 *Phil. Mag.* (7), **17**, 1038–1047. (399)

INGERSOLL, L. R. and ZOBEL, O. J. 1913 *Mathematical Theory of Heat Conduction*. Ginn. (409)

IZSAK, J. G. 1961 *AJ*, **66**, 226–229. (259)

JACKSON, J. 1930 *MNRAS*, **90**, 733–742. (286, 302)

JACOBS, J. A. 1953 *Nature, Lond.* **172**, 297–298. (421)
    1954 *Nature, Lond.* **173**, 258. (421)

JACOBS, J. A. and ALLAN, D. W. 1956 *Nature, Lond.* **177**, 155–157. (414)

JAGGAR, T. A. 1917 *Amer. J. Sci.* (4), **44**, 208–220. (469)

JEANS, J. H. 1904 *Nature, Lond.* **70**, 101. (368)
1921 *Dynamical Theory of Gases.* ch. 15. Cambridge University Press. (417)
JEFFREYS, B. 1965 *GJRAS*, **10**, 141–146. (180, 257)
JEFFREYS, H. *CP* denotes *Collected Papers*: vol. 1, 1971; vol. 2, 1973; vol. 3, 1974; vol. 4, 1975; vols. 5 and 6 in press. Gordon and Breach, London.
1915*a* *Mem. R. Astr. Soc.* **60**, 187–217. *CP*, **3**, 71. (222, 300)
1915*b* *MNRAS*, **75**, 648–658. *CP*, **4**, 3. (301, 329)
1916*a* *MNRAS*, **76**, 499–525. *CP*, **3**, 463. (305)
1916*b* *Phil. Mag.* (6), **32**, 575–591. *CP*, **4**, 359. (409)
1916*c* *MNRAS*, **77**, 84–112. *CP*, **3**, 105. (386)
1917 *MNRAS*, **78**, 116–131. *CP*, **4**, 29. (342)
1920*a* *Phil. Mag.* (6), **39**, 578–586. *CP*, **5**, (483)
1920*b* *Phil. Trans.* A, **221**, 239–264. *CP*, **4**, 57. (332)
1921 *PRSA*, **100**, 122–149. *CP*, **4**, 377. (409)
1924 *MNGS*, **1**, 121–124. *CP*, **3**, 3. (212)
1925 *Phil. Mag.* (6), **49**, 793–807. *CP*, **5**. (329)
1926*a* *MNGS*, **1**, 321–334. *CP*, **1**, 95. (36, 37)
1926*b* *MNGS*, **1**, 385–402. *CP*, **1**, 285. (59)
1926*c* *MNGS*, **1**, 412–424. *CP*, **4**, 85. (468)
1926*d* *Phil. Mag.* (7), **2**, 833–844. *CP*, **4**, 455. (329, 454)
1926*e* *Proc. Camb. Phil. Soc.* **23**, 472–481. *CP*, **1**, 125. (67)
1926*f* *B. A. Seism. Comm. Rep.* (107)
1927*a* *GBG*, **18**, 1–29. *CP*, **4**, 275. (409, 433, 435)
1927*b* *MNGS*, **1**, 483–494. *CP*, **1**, 303. (59, 88)
1928*a* *MNGS*, **1**, 500–521. *CP*, **1**, 319. (107, 113)
1928*b* *MNGS*, **2**, 56–58. (290, 308, 315)
1928*c* *MNGS*, **2**, 101–111. *CP*, **1**, 137. (45)
1928*d* *PRSA*, **118**, 195–208. *CP*, **4**, 469. (454)
1928*e* *Proc. Camb. Phil. Soc.* **24**, 19–31. *CP*, **1**, 15. (20)
1928*f* *Geol. Mag.* **65**, 280. (227)
1930*a* *MNRAS*, **91**, 169–173. *CP*, **4**, 99. (341)
1930*b* *GBG*, **26**, 58–60. *CP*, **4**, 405. (366)
1930*c* *Proc. Camb. Phil. Soc.* **26**, 101–106. *CP*, **1**, 30. (105)
1930*d* *PRSA*, **128**, 376–393. *CP*, **5**. (305)
1931*a* *GBG*, **30**, 336–350. *CP*, **1**, 157. (67)
1931*b* *MNGS*, **2**, 318–323. *CP*, **1**, 149. (47)
1931*c* *MNGS*, **2**, 323–329. *CP*, **4**, 313. (426)
1931*d* *MNGS*, **2**, 329–348. *CP*, **1**, 343. (113)
1931*e* *MNGS*, **2**, 407–416. *CP*, **1**, 377. (66, 132)
1931*f* *Geol. Mag.* **68**, 435–442. *CP*, **4**, 409. (445)
1931*g* *Cartesian Tensors.* Cambridge University Press. (105)
1931*h* *Operational Methods in Mathematical Physics.* Cambridge University Press. (372)
1931*i* *GBG*, **31**, 378–386. *CP*, **3**, 9. (175)
1932*a* *MNGS*, **3**, 6–9. *CP*, **4**, 321. (411)
1932*b* *MNGS*, **3**, 30–41. *CP*, **3**, 415. (265)
1932*c* *MNGS*, **3**, 53–59. *CP*, **4**, 419. (462)
1932*d* *MNGS*, **3**, 60–69. *CP*, **3**, 429. (265, 269, 274)
1932*e* *PRSA*, **138**, 283–297. *CP*, **1**, 37. (10, 11)

JEFFREYS, H. 1932f Geol. Mag. **69**, 321–324. CP, **4**, 427. (443)

1933 Nature, Lond. **132**, 934. CP, **5**. (503)

1934 Geol. Mag. **71**, 276–280. CP, **4**, 431. (465)

1935a MNGS, 3, 253–261. CP, **1**, 177. (143)

1935b MNGS, 3, 271–274. CP, **1**, 187. (69)

1935c MNGS, 3, 310–343. CP, **2**, 1. (116)

1935d Proc. Phys. Soc. **47**, 455–458. CP, **1**, 173. (165)

1936a Bur. Centr. Séism. Trav. Sci. no. 14, 3–86. CP, **2**, 36. (115, 533)

1936b MNGS, 3, 401–422. CP, **2**, 121. (98, 119)

1936c GBG, **47**, 149–171. CP, **4**, 199. (397)

1936d Proc. Roy. Soc. Edinb. **56**, 158–163. CP, **4**, 438. (17)

1937a MNGS, **4**, 1–13. CP, **3**, 189. (209, 220, 222)

1937b MNGS, **4**, 50–61. CP, **2**, 197. (206, 296)

1937c MNGS, **4**, 62–71. CP, **3**, 203. (209, 210)

1937d MNGS, **4**, 165–184. CP, **2**, 210. (116)

1937e MNGS, **4**, 196–225. CP, **2**, 243. (89, 92, 95, 97, 98, 148)

1937f MNGS, **4**, 13–39. CP, **2**, 169. (132)

1937g GBG, **49**, 393–401. CP, **2**, 233. (132)

1938a MNGS, **4**, 281–308. CP, **2**, 331. (108, 117)

1938b MNGS, **4**, 309–312. CP, **4**, 330. (401)

1938c GBG, **53**, 111–139. CP, **2**, 301. (473)

1939a MNGS, **4**, 424–460. CP, **2**, 359. (117, 150)

1939b MNGS, **4**, 498–533. CP, **2**, 409. (117, 120, 159)

1939c MNGS, **4**, 537–547. CP, **2**, 447. (117)

1939d MNGS, **4**, 548–561. CP, **2**, 459. (109)

1939e MNGS, **4**, 594–615. CP, **2**, 525. (52, 214)

1940a BSSA, **30**, 225–234. CP, **2**, 548. (123)

1940b MNRAS, **100**, 139–155. CP, **3**, 503. (297, 298)

1940c Geogr. J. **95**, 384–385. (459)

1941a MNGS, **5**, 1–22. CP, **3**, 215. (238)

1941b Phil. Mag. (7), **32**, 365–368. CP, **1**, 67. (270)

1942a MNGS, **5**, 33–36. CP, **2**, 561. (110, 120, 153)

1942b MNGS, **5**, 37–40. CP, **4**, 225. (397)

1942c MNRAS, **102**, 194–204. CP, **3**, 241. (201)

1942d Geol. Mag. **79**, 291–295. CP, **4**, 445. (476)

1942e Proc. Camb. Phil. Soc. **39**, 48–51. CP, **1**, 191. (133)

1943a MNGS, **5**, 55–66. CP, **3**, 253. (241)

1943b MNGS, **5**, 71–89. CP, **3**, 441. (265, 274)

1947a MNGS, **5**, 99–104. CP, **2**, 565. (92)

1947b MNGS, **5**, 105–119. CP, **2**, 571. (91, 464)

1947c MNRAS, **107**, 260–262. CP, **3**, 267. (343)

1948a MNRAS, **108**, 94–103. CP, **5**, (383)

1948b MNGS, 5, 219–247. CP, **3**, 277. (248, 249, 252)

1948c Nature, Lond. **162**, 822–823. CP, **4**, 231. (380, 381)

1948d MNRAS, **108**, 206–209. CP, **3**, 523. (302)

1949a MNRAS, **109**, 670–687. CP, **3**, 527. (302, 349, 514)

1949b Nature, Lond. **164**, 1046. (381)

1949c MNGS, 5, 398–408. CP, **3**, 307. (255)

1950a MNRAS, **110**, 460–466. CP, **3**, 547. (302, 349, 363)

JEFFREYS, H. 1950*b* *Ann. Géophys.* **6**, 10–17. *CP*, **4**, 239. (396)

1951 *MNRAS*, **111**, 410–412. *CP*, **3**, 21. (212)

1952*a* *Nature, Lond.* **169**, 260–262. (384)

1952*b* *MNGS*, **6**, 272–277. *CP*, **4**, 507. (466)

1952*c* *MNGS*, **6**, 316–318. *CP*, **3**, 25. (185)

1952*d* *MNGS*, **6**, 348–364. *CP*, **2**, 587. (121)

1953 *Bull. géod.* **30**, 331–338. *CP*, **3**, 29. (181, 193)

1954*a* *MNGS*, **6**, 557–565. *CP*, **2**, 605. (123, 164)

1954*b* *MNRAS*, **114**, 433–436. *CP*, **3**, 345. (202, 424)

1955 *Quart. J. Mech. Appl. Math.* **8**, 448–451. *CP*, **3**, 37. (181, 193)

1956*a* *Quart. J. Mech. Appl. Math.* **9**, 1–5. *CP*, **4**, 513. (454, 456)

1956*b* *MNRAS*, **116**, 362–364. (349)

1957*a* *MNRAS*, **117**, 347–355. *CP*, **6**, (531)

1957*b* *MNRAS*, **117**, 475–477. *CP*, **3**, 355. (218, 219)

1957*c* *MNRAS*, **117**, 506–515. *CP*, **4**, 105. (353)

1957*d* *MNRAS*, **117**, 585–589. *CP*, **4**, 117. (354)

1957*e* *MNGS*, **7**, 332–337. *CP*, **1**, 220. (38, 105)

1958*a* *MNRAS*, **118**, 14–17. *CP*, **4**, 123. (356)

1958*b* *GJRAS*, **1**, 92–95. *CP*, **4**, 129. (12)

1958*c* *GJRAS*, **1**, 154–161. *CP*, **2**, 614. (124)

1958*d* *GJRAS*, **1**, 162–163. *CP*, **4**, 135. (461)

1958*e* *GJRAS*, **1**, 191–197. *CP*, **2**, 623. (124)

1959*a* *Quart. J. Mech. Appl. Math.* **12**, 124–128. *CP*, **6**. (341)

1959*b* *MNRAS*, **119**, 75–80. *CP*, **3**, 591. (313, 314)

1959*c* *GJRAS*, **2**, 42–44. *CP*, **3**, 43. (246)

1959*d* *PRSA*, **252**, 431–435. *CP*, **1**, 71. (17)

1961*a* *GJRAS*, **6**, 115–117. *CP*, **1**, 227. (78)

1961*b* *MNRAS*, **122**, 421–432. *CP*, **3**, 371. (218, 220)

1961*c* *GJRAS*, **6**, 493–508. *CP*, **2**, 631. (124)

1961*d* *Smithsonian Astr. Obs. Special Report*, **79**. *CP*, **3**, 359. (246)

1961*e* *MNRAS*, **122**, 339–343. *CP*, **4**, 157. (356)

1962*a* *GJRAS*, **6**, 550–552. *CP*, **2**, 648. (145)

1962*b* *GJRAS*, **7**, 212–219. *CP*, **2**, 652. (124)

1962*c* *GJRAS*, **7**, 412–414. *CP*, **1**, 230. (67, 122)

1963*a* *Smithsonian Contrib. to Astrophysics*, **6**, 205–212. *CP*, **3**, 359. (246)

1963*b* *GJRAS*, **8**, 196–202. *CP*, **3**, 47. (202, 312)

1964*a* *BSSA*, **54**, 1441–1444. *CP*, **6**, (139, 310, 475)

1964*b* *GJRAS*, **8**, 541. *CP*, **3**, 55. (185, 262)

1965 *Nature, Lond.* **208**, 675. *CP*, **4**, 183. (358)

1966 *GJRAS*, **12**, 111. *CP*, **2**, 685. (76)

1967*a* *MNRAS*, **136**, 311–312. *CP*, **3**, 401. (220)

1967*b* *GJRAS*, **12**, 465–468. *CP*, **1**, 77. (289)

1967*c* *Nature, Lond.* **215**, 1365–1366. *CP*, **4**, 187. (312)

1967*d* *Theory of Probability*. Oxford, Clarendon Press. (74, 75, 115, 131, 139, 163, 241, 245, 310, 475, 531, 533)

1968*a* *GJRAS*, **15**, 249–251. *CP*, **2**, 687. (163)

1968*b* *Asymptotic Approximations*. Oxford, Clarendon Press. (43, 54, 345)

1968*c* *GJRAS*, **16**, 253–258. *CP*, **5**. (333)

JEFFREYS, H. 1968*d* *MNRAS*, **141**, 255–268. *CP*, **3**, 641. (79, 298, 358, 484)

1971 *MNRAS*, **153**, 73–81. *CP*, **3**, 403. (218, 219)

1972 *Icarus*, **17**, 404–405. (Review of Stacey, *Physics of the Earth*.) (497)

1973*a* *GJRAS*, **33**, 237–238. *CP*, **6**. (358)

1973*b* *Scientific Inference*, Cambridge University Press. (131)

1974 *Amer. Assn Petroleum Geol.* Mem. 23, ed. C. F. Kahle, 395–405. *CP*, **6**. (496)

1975*a* *GJRAS*, **40**, 23–27. *CP*, **6**. (48)

1975*b* *Journ. Geol. Soc.* **131**, 323–325. *CP*, **6**. (459, 529)

1975*c* *QJRAS*, **16**, 145–151. *CP*, **6**. (338)

JEFFREYS, H. and BEN-MENAHEM, A. 1971. *GJRAS*, **24**, 1–2. *CP*, **6**. (360)

JEFFREYS, H. and BLAND, M. E. M. 1951 *MNGS*, **6**, 148–158. *CP*, **4**, 495. (465)

JEFFREYS, H. and BULLEN, K. E. 1935 *Bur. Centr. Séism. Trav. Sci.* no. 11. *CP*, **1**, 439. (113, 114)

1940 *Seismological Tables*. BAGMT. (102, 118, 121, 153)

JEFFREYS, H. and CRAMPIN, S. 1960 *MNRAS*, **121**, 571–577. *CP*, **4**, 139. (12, 356)

1970 *MNRAS*, **147**, 295–301. *CP*, **4**, 189. (358)

JEFFREYS, H. and HUDSON, J. A. 1965 *GJRAS*, **10**, 175–179. *CP*, **1**, 245. (46)

JEFFREYS, H. and JEFFREYS, B. S. 1946, 1950, 1956, 1962, 1966, 1972 *Methods of Mathematical Physics*. Cambridge University Press. (1, 43, 45, 52, 77, 79, 133, 178, 181, 345, 372, 409, 411, 451, 454, 475, 523, 526)

JEFFREYS, H. and LAPWOOD, E. R. 1957 *PRSA*, **241**, 455–479. *CP*, **1**, 195. (133–4)

JEFFREYS, H. and SHIMSHONI, M. 1966 *GJRAS*, **10**, 515–524. *CP*, **2**, 665. (135)

JEFFREYS, H. and SINGH, K. 1973 *GJRAS*, **32**, 423–438. *CP*, **6**. (164)

JEFFREYS, H. and VICENTE, R. O. 1957*a* *MNRAS*, **117**, 142–161. *CP*, **3**, 555. (254, 302, 313, 349, 520)

1957*b* *MNRAS*, **117**, 162–173. *CP*, **3**, 577. (254, 302, 313, 349, 520)

1964 *Nature, Lond.* **204**, 120–121. *CP*, **3**, 597. (312)

1966 *Mém. Acad. Roy. de Belgique*, **37**, 3–30. *CP*, **3**, 603. (312, 519, 521)

1967 *Acad. Roy. Belg. Bull. Cl. Sci.* (5), **53**, 926–933. *CP*, **3**, 633. (79, 519)

JEHU, T. J. and CRAIG, R. M. 1923 *Trans. Roy. Soc. Edinb.* **53**, 419–441. (476)

JOHNSTON, W. A. 1939 *Amer. J. Sci.* **237**, 94–98. (459)

Joly, J. (365, 399)

Jones, H. Spencer. (173, 210, 303, 325, 326, 334)

JONES, H. SPENCER 1928 *Ann. Cape Obs.* **10**, pt. 8. (254)

1939 *MNRAS*, **99**, 541–558. (324)

1941 *MNRAS*, **101**, 356–366. (201)

1942 *Mem. R. Astr. Soc.* **66**, 11–66. (201)

Jones, O. T. (462, 487, 496)

JONES, O. T. 1933 *Observatory*, **56**, 82–85. (475)

1938 *Quart. J. Geol. Soc.* **94**, lx–cx. (447)

1960 *GJRAS*, **3**, 128–129. (487)

JULIAN, B. R., DAVIES, D. and SHEPPARD, R. M. 1972 *Nature, Lond.* **235**, 317–318. (216)

KAKUTA, C. and AOKI, S. 1971 *I.A.U. Symposium*, **48**, 192–194. (363)
KAMINUMA, K. 1966 *Bull. Earthq. Res. Inst. Tokyo*, **44**, 495–510. (100)
Kaula, W. M. (244, 246, 255)
KAULA, W. M. 1963*a JGR*, **68**, 4967–4978. (279)
    1963*b JGR*, **68**, 5183–5190. (279, 282)
    1966 *JGR*, **71**, 5303–5313. (193, 244, 259, 277, 278, 280)
    1969 *AJ*, **74**, 1108–1114. (356)
KEEVIL, N. B. 1943 *J. Geol.* **51**, 287–300. (400)
KEILIS-BOROK, V. I. and YANOVSKAYA, T. D. 1967 *GJRAS*, **13**, 223–234. (160)
KEITH, A. 1923 *Bull. Geol. Soc. Amer.* **34**, 335. (434)
Kelvin, Lord (11, 43, 265, 272, 299, 301, 327, 370, 392)
KELVIN, LORD 1878 *Trans. Geol. Soc. of Glasgow*, **6**, 38–49. (272)
KELVIN, LORD and TAIT, P. G. 1883, 1912 *Treatise on Natural Philosophy*. Cambridge
    University Press. (204, 328)
KERR, F. J. and WHIPPLE, F. L. 1954 *AJ*, **59**, 124–127. (355)
KIMURA, H. 1902*a AJ*, **22**, 107–108. (314)
    1902*b Astr. Nachr.* **158**, 234–239. (314)
KING, D. W., HADDON, R. A. W. and CLEARY, J. W. 1974 *GJRAS*, **37**, 157–174.
    (161)
King, W. B. R. (447, 496)
King-Hele, D. G. (256)
KING-HELE, D. G. 1958 *PRSA*, **247**, 49–72. (173)
    1960*a Satellites and Scientific Research*. Routledge and Kegan
    Paul. (418)
    1960*b Nature, Lond.* **187**, 490–491. (418)
    1963 *Erasmus Darwin*, p. 90. Macmillan. (488)
KING-HELE, D. G. and COOK, G. E. 1965 *GJRAS*, **10**, 17–27. (257)
KING-HELE, D. G., COOK, G. E. and SCOTT, D. W. 1967 *Planetary and Space
    Science*, **15**, 741–769. (258)
KING HUBBERT, M. and RUBEY, W. W. 1959 *BGSA*, **70**, 115–206. (445)
    1961*a BGSA*, **72**, 1441–1452. (445)
    1961*b BGSA*, **72**, 1581–1594. (445)
KNOPOFF, L. and GILBERT, F. 1961 *BSSA*, **51**, 35–49. (54, 109)
KNOPOFF, L. and MACDONALD, G. J. F. 1960 *JGR*, **65**, 2191–2197. (361)
Knott, C. G. (300, 308, 474)
KNOTT, C. G. 1899 *Phil. Mag.* (5), **48**, 64–97. (36)
    1908 *Physics of Earthquake Phenomena*. Oxford University Press.
    (41, 474)
    1919 *Proc. Roy. Soc. Edinb.* **39**, 158–208. (52)
KOGAN, S. D. 1960 *Akad. Nauk. USSR, Inst. of Physics of the Earth*, no. 3, 371–380.
    (159)
Kosminska, J. P. (100)
Kossmat, F. (231)
KOTHARI, D. S. 1938 *PRSA*, **165**, 486–500. (419)
KOVACH, R. L. and ANDERSON, D. L. 1964 *BSSA*, **54**, 161–182. (146)
    1965 *JGR*, **70**, 2873–2882. (425)
KOVALEVSKY, J. (ed.) 1965 *Bull. Astr.* **25**, 1–330. (259)
KOVARIK, A. F. and ADAMS, N. I. 1938 *Phys. Rev.* **54**, 413–421. (376)
KOZAI, Y. 1956 *Publ. Astr. Soc. Japan*, **8**, 91–93. (354)
    1965 *Publ. Astr. Soc. Japan*, **17**, 395–402. (334)

KOZIEL, K. 1948 *Acta Astron. Crakow* (a), **4**, 61–193. (219)

    1966 *PRSA*, **296**, 248–253. (218)

    1967 *Icarus*, **7**, 1–28. (218)

KRIGE, L. J. 1939 *PRSA*, **173**, 450–474. (401, 402)

KRONIG, R., DE BOER, J. and KORRINGA, J. 1945 *Physica*, **12**, 245–256. (420)

KUENEN, PH. H. 1937 *Amer. J. Sci.* (5), **43**, 457–468. (498)

    1941 *Amer. J. Sci.* **239**, 161–190. (498)

KUHN, W. 1942 *Naturwissenschaften*, **30**, 689–696. (418)

KUHN, W. and RITTMANN, A. 1941 *Geol. Rdsch.* **32**, 215–256. (418, 422)

KÜHNEN, F. and FÜRTWÄNGLER, P. 1906 *Veröff. Preuss. Geod. Inst.* **27**. (189)

KUIPER, G. P. 1954 *The Solar System* (2). *The Earth as a Planet.* Chicago University Press. (463)

Kulikov, K. A. (254)

KUO, J. T. and JACHENS, R. C. 1970 *Science*, **168**, 968–971. (307–8)

LACK, D. L. 1947 *Darwin's Finches.* Cambridge University Press. (486)

LAHIRI, B. N. and PRICE, A. T. 1939 *Phil. Trans.* A, **237**, 509–540. (119, 467)

Lake, P. (496)

LAKE, P. 1922 *Geol. Mag.* **59**, 338–346. (488)

    1931a *Geogr. J.* **78**, 149–160. (463)

    1931b *Geol. Mag.* **68**, 34–39. (463)

LAMAR, D. L. and MERRIFIELD, P. M. 1966 *JGR*, **71**, 4429–4430. (340)

Lamb, H. (111, 132)

LAMB, H. 1904 *Phil. Trans.* A, **203**, 1–42. (67)

    1932 *Hydrodynamics*, 6th edn. Cambridge University Press. (263, 301, 329)

LAMBERT, W. D. 1921 *Amer. J. Sci.* (5), **2**, 129–158. (468)

    1922a *AJ*, **34**, 103–110. (483)

    1922b *U.S.C. and G.S. Spec. Publ.* no. 80. (483)

    1928 *Bull.* **11**, Sect. Ocean. Int. Res. Council. (332, 483)

    1931 *Bull. U.S. Nat. Res. Coun.* no. 78, ch. 16. (192, 288, 479)

    1933 *Trav. Ass. Géod. Int.* Report on Earth Tides. (292, 307)

    1936 *Trav. Ass. Géod. Int.* Report on Earth Tides. (292)

    1940 *U.S.C. and G.S. Spec. Publ.* no. 223. (292, 307)

    1944 *Trav. Ass. Géod. Int.* Report on Earth Tides. (292)

    1945 *Amer. J. Sci.* **243**A, 360–392. (181)

    1960 *GJRAS*, **3**, 360–366. (184)

LAMBERT, W. D. and DARLING, F. W. 1931 *Bull. géod.* 334–340. (187)

    1936 *U.S.C. and G.S. Spec. Publ.* no. 199. (193)

Landisman, M. (140, 153)

LANDISMAN, M. and MUELLER, S. 1966 *GJRAS*, **10**, 525–538. (123)

LANDISMAN, M., SATO, Y. and NAFE, J. 1965 *GJRAS*, **9**, 439–502. (309, 310, 311)

LANDOLT-BÖRNSTEIN, *Physikalische Tabellen.* Springer, Berlin. (21)

LANGSETH, M. G., LE PICHON, X. and EWING, M. 1966 *JGR*, **71**, 5321–5355. (416)

Laplace, P. S. (196, 324)

Lapwood, E. R. (147, 149)

LAPWOOD, E. R. 1948 *Proc. Camb. Phil. Soc.* **44**, 508–521. (466)

    1949 *Phil. Trans.* A, **242**, 63–100. (67, 133)

    1952 *MNGS*, **6**, 401–407. (434)

    1955 *MNGS*, **7**, 135–146. (93)

LAPWOOD, E. R. and GOGNA, M. L. 1970 *GJRAS*, **20**, 383–390. (125)

Larmor, J. (10)

LARMOR, J. 1909 *PRSA*, **83**, 89–96. (295)

        1915 *Proc. Lond. Math. Soc.* (2), **14**, 440–449. (301)

LAUBSCHER, H. P. 1960 *BGSA*, **71**, 611–628. (446)

LAUBSCHER, R. E. 1972 *Astron. and Astrophys.* **20**, 407–414. (363)

LAWSON, R. W. 1927 *Nature, Lond.* **119**, 277–278 and 703–704. (396)

Lee, A. W. (45, 90)

LEE, A. W. 1932*a* *MNGS*, **3**, 85–105. (168)

        1932*b* *MNGS*, **3**, 105–116. (168)

        1934 *MNGS*, **3**, 238–252. (145, 168)

        1935 *PRSA*, **149**, 183–199. (168)

LEE, W. H. K. and KAULA, W. M. 1967 *JGR*, **72**, 753–758. (278)

Lees, G. M. (103)

LEES, G. M. 1953 *Quart. J. Geol. Soc.* **109**, 217–257. (437)

LEET, L. D. 1938 *BSSA*, **28**, 45–48. (99)

LEET, L. D. and EWING, W. M. 1932*a* *Phys. Rev.* (2), **39**, 868. (104)

        1932*b* *Physics*, **2**, 160–173. (104)

Legendre, A. M. (196)

Lehmann, I. (51, 124, 125, 130)

LEHMANN, I. 1930 *GBG*, **26**, 402–412. (108)

        1934 *Medd. Geod. Inst. København*, **5**. (56, 118)

        1936 *Bur. Centr. Séism. Trav. Sci.* **14**, 3–31. (109)

        1953 *BSSA*, **43**, 291–306. (108, 127)

        1955 *Ann. di Geof.* **8**, 351–370. (124)

        1962 *BSSA*, **52**, 519–526. (146)

        1964 *BSSA*, **54**, 123–129. (101)

        1970 *GJRAS*, **21**, 359–372. (161)

LEHMANN, I. and PLETT, G. 1932 *GBG*, **31**, 38–77. (107, 114, 127, 131)

LEITH, A. and SHARPE, J. A. 1936 *J. Geol.* **44**, 877–917. (112, 473)

LENNARD-JONES, J. E. 1923 *Trans. Camb. Phil. Soc.* **22**, 535–556. (417)

Leverrier, U. J. J. (478)

Leybenzon, L. S. (300)

LIEBAU, F. 1971 *Bull. Soc. fr. Minéral. Cristallogr.* **94**, 239–249. (208)

LIEBAU, F. and HESSE, K. F. 1971 *Zs. f. Kristallog.* **133**, 213–224. (208)

LIEBERMANN, L. N. 1949 *Phys. Rev.* **75**, 1415–1422. (3)

LIEBERMANN, R. C. and RINGWOOD, A. E. 1973 *JGR*, **78**, 6926–6932. (215)

LIND, S. C. and ROBERTS, L. D. 1920 *J. Amer. Chem. Soc.* **42**, 1170. (371)

LINDEMANN, F. A. (LORD CHERWELL) 1939 *Quart. J. R. Met. Soc.* **65**, 330–336. (417)

LINGENFELTER, R. E. and SCHUBERT, G. 1973 *The Moon*, **7**, 172–180. (467)

LISTER, C. R. B. 1963 *GJRAS*, **7**, 571–583. (415)

LIU, H. S. 1966 *JGR*, **71**, 3099–3100. (210, 357)

Lomnitz, C. (356, 461)

LOMNITZ, C. 1956 *J. Geol.* **64**, 473–479. (11)

        1957 *J. Appl. Phys.* **28**, 201–205. (11)

        1962 *JGR*, **67**, 365–368. (361)

LONCAREVIČ, B. D. 1963 *Nature, Lond.* **198**, 23–24. (190)

Longman, I. M. (305, 328, 334, 348)

LONGMAN, I. M. 1962 *JGR*, **67**, 845–850. (315)
     1963 *JGR*, **68**, 485–496. (304, 306, 315)
     1966 *GJRAS*, **11**, 133–137. (315)
LONGUET-HIGGINS, H. C. and WIDOM, B. 1964 *Molecular Physics*, **8**, 549–556. (20)
LONGUET-HIGGINS, M. S. 1950 *Phil. Trans.* A, **243**, 1–35. (169)
LOVE, A. E. H. (87, 289, 290, 300, 308, 314)
LOVE, A. E. H. 1906 *Mathematical Theory of Elasticity.* (283)
     1909 *PRSA*, **82**, 73–88. (295)
     1911 *Some Problems of Geodynamics.* Cambridge University Press.
     (42, 289, 308, 312, 516)
LOVERING, J. F. 1958 *TAGU*, **39**, 947–955. (450)
LOW, A. R. 1929 *PRSA*, **125**, 180–195. (454)
LUBIMOVA, E. A. 1956 *Dokl. Akad. Nauk SSSR*, **107**, 55–58. (421)
     1958. *GJRAS*, **1**, 115–134. (432)
Lyttleton, R. A. (210, 385)
LYTTLETON, R. A. 1938*a* *MNRAS*, **98**, 633–645. (342)
     1938*b* *MNRAS*, **98**, 646–650. (342)
     1941 *MNRAS*, **101**, 349–351. (342)
     1963 *PRSA*, **275**, 1–22. (424)
     1965 *MNRAS*, **129**, 21–39. (424)
     1970 *Advances in Astronomy and Astrophysics*, **7**, 83–145. (425)
LYTTLETON, R. A. and BONDI, H. 1948 *Proc. Camb. Phil. Soc.* **44**, 345–359. (361)
     1953 *Proc. Camb. Phil. Soc.* **49**, 498–515, (362)
LYUSTIKH, E. N. 1965 *Bulletin of the Moscow Society for investigating Nature (Geology)*, **40** (1), 5–27. (489)
     1967 *GJRAS*, **14**, 347–352. (489)
MACDONALD, G. A. 1963 *BGSA*, **74**, 1071–1078. (469)
MacDonald, G. J. F. (416, 483)
MACDONALD, G. J. F. 1959 *JGR*, **64**, 1967. (414)
     1962 *JGR*, **67**, 2945–2974. (425)
     1963 *TAGU*, **44**, 602–603. (491)
MACDONALD, G. J. F. and NESS, N. F. 1961 *JGR*, **66**, 1865–1911. (309)
Macelwane, J. B. (62)
MACELWANE, J. B. 1923 *BSSA*, **13**, 13–69. (113)
     1930 *GBG*, **28**, 165–227. (106, 108)
     1933 *Bull. Nat. Res. Coun.* no. 90, ch. 13. (142)
MACELWANE, J. B. and SOHON, F. W. 1932, 1936 *Theoretical Seismology.* John Wiley and Sons. (83, 88)
McEVILLY, T. V. 1964 *BSSA*, **54**, 1997–2015. (100)
McKENZIE, D. and WEISS, N. 1975 *GJRAS*, **42**, 131–174. (492)
MACMILLAN, W. D. 1909 *Publ. Carneg. Instn*, no. 107, pp. 69–75. (330)
MANSFIELD, A. P. 1923 *Bull. Geol. Soc. Amer.* **34**, 268. (434)
MANSINHA, L. 1964 *BSSA*, **54**, 369–376. (92)
MANSINHA, L. and SMYLIE, D. E. 1967 *JGR*, **72**, 4731–4744. (299)
MARCUS, A. 1966*a* *Icarus*, **5**, 190–200. (504)
     1966*b* *MNRAS*, **134**, 269. (504)
MARSDEN, B. G. and CAMERON, A. G. W. (eds.) 1966 *The Earth–Moon System.* Plenum Press. (340)
MARSHALL, P. D. and CARPENTER, E. W. 1966 *GJRAS*, **10**, 549–550. (360)
MARTYN, D. F. 1939 *Quart. J. R. Met. Soc.* **65**, 328–330. (417)

Maskelyne, N. (224)

MATUZAWA, T. 1928 *Bull. Earthq. Res. Inst. Tokyo*, **5**, 1–28. (98)

1929 *Bull. Earthq. Res. Inst. Tokyo*, **6**, 117–229. (98)

MATUZAWA, T., YAMADA, K. and SUSUKI, T. 1929 *Bull. Earthq. Res. Inst. Tokyo*, **7**, 241–260. (98)

Maxwell, J. C. (7)

MAXWELL, J. C. *Papers*, **2**, 26. (10)

MECHLER, P. 1969 *Phys. Earth Plan. Int.*, **2**, 93–104. (122)

MECHLER, P. and ROCARD, Y. 1964 *C.R. Acad. Sci. Paris*, **259**, 2269–2271. (122)

1970 *C.R. Acad. Sci. Paris*, **270**B, 298–300. (123)

1971 *C.R. Acad. Sci. Paris*, **272**D, 3120–3121. (123)

See also Beaufils, Mechler and Rocard (1970)

Meinesz, F. A. Vening (189, 234, 236, 241, 452)

MEINESZ, F. A. VENING 1932 *Gravity Expeditions at Sea, 1923–1930.* (233)

1933 *Proc. K. Akad. Wet. Amst.* **36**, 372–377. (460)

1947 *Proc. K. Akad. Wet. Amst.* **50**, 237–245. (460)

1948 *Quart. J. Geol. Soc.* **103**, 191–207. (460)

MEINESZ, F. A. VENING, UMBGROVE, J. H. F. and KUENEN, PH. H. 1934 *Gravity Expeditions at Sea, 1923–1932.* (233)

MEINESZ, F. A. VENING and WRIGHT, F. E. 1930 *Publ. U.S. Naval Observatory*, (2), **13**, Appendix 1. (233)

MEISSNER, E. 1926 *Proc. 2nd Congr. Appl. Math. Zürich*, pp. 3–11. (77)

Melchior, P. (484)

MELCHIOR, P. 1966 and later papers *Ass. Int. Géodesie, Commission des marées terrestres, Bulletin des informations*, 46. (304)

MELCHIOR, P. and GEORIS, B. 1968 *Physics of Earth and Planetary Interiors*, **1**, 267–287. (304)

MELCHIOR, P. and VENEDIKOV, A. 1968 *Physics of Earth and Planetary Interiors*, **1**, 363–372. (304)

MESSAGE, P. J. 1955 *MNRAS*, **115**, 550–557. (202, 424)

MEYERHOFF, A. A. 1970 *J. Geol.* **78**, 1–51, 406–444. (489)

MEYERHOFF, A. A. and MEYERHOFF, H. A. 1972*a* *Amer. Assn of Petroleum Geologists Bull.* **56**, 269–336, 337–359. (487, 489, 494–496)

1972*b* *J. Geol.* **80**, 34–60. (495)

MEYERHOFF, A. A., MEYERHOFF, H. A. and BRIGGS, R. S. 1972 *J. Geol.* **80**, 663–692. (489)

MEYERHOFF, A. A. and TEICHERT, C. 1971 *J. Geol.* **79**, 285–321. (494, 495)

MICHAEL, W. H. 1970 *The Moon*, **1**, 484–485. (220)

MICHE, R. 1944 *Ann. Ponts et Chaussées*, pp. 25–78. (168)

MICHELSON, A. A. and GALE, H. G. 1919 *J. Geol.* **27**, 585–601. (292)

MIKI, H. 1952 *J. Phys. Earth*, **1**, 67–74. (348, 434)

1954 *J. Phys. Earth*, **2**, 1–3. (434)

MILES, B. and RAMSEY, W. H. 1952 *MNRAS*, **111**, 427–447. (424)

MILLER, G. R. 1966 *JGR*, **71**, 2485–2489. (332)

Miller, J. C. P. (45)

MILNE, E. A. 1949 *MNRAS*, **109**, 517–527. (368)

Milne, J. (82, 83)

MILNE-THOMSON, L. M. and COMRIE, L. J. 1931 *Four-figure Mathematical Tables.* Macmillan, London. (532)

MINTZ, Y. and MUNK, W. H. 1951 *Tellus*, **3**, 117–121. (315)

                    1954 *MNGS*, **6**, 566–578. (315)

MISES, R. 1928 *Z. angew. Math. Mech.* **8**, 161–185. (15)

MOHOROVIČIĆ, A. 1909 *Jb. Met. Obs. Agram.* **9**, 1–63. (86)

                1922 *Rad. Jugoslav. Akad.* **226**, 94–190. (113)

Mohorovičić, S. (88, 128)

MOHOROVIČIĆ, S. 1914 *GBG*, **13**, 217–240. (52, 87)

                1916 *GBG*, **14**, 187–198. (52, 87)

                1928 *Arh. Hemiju Farmaciju, Zagreb*, **2**, 66–76. (501)

MOLODENSKY, M. S. 1953 *Trud. Geof. Inst., Sbornik Staty, Moscow*, **19**, 3–52. (300, 302, 303)

MORRISON, L. V. 1971 *Nautical Almanac Office, Tech. Notes*, 22, 23.

                1972 *IAU Symposium* 47, *The Moon*, 395–401. (337)

MORRISON, L. V. and SADLER, F. McB. 1969 *MNRAS*, **144**, 129–141 (220)

MUDGE, M. R. 1968 *BGSA*, **79**, 327. (446)

MUEHLBERGER, W. R., DENISON, R. E. and LIDIAK, E. G. 1967 *Am. Assn Petroleum Geologists*, **51**, 2351–2380. (463)

MÜLLER, G. 1973 *JGR*, **78**, 3469–3490. (161)

Munk, W. H. (483)

MUNK, W. H. 1956 *Nature, Lond.* **177**, 551–554. (481)

                1968 *QJRAS*, **9**, 352–375. (363)

MUNK, W. H. and MACDONALD, G. J. F. 1960, 1975 *The Rotation of the Earth: a Geophysical Discussion*. Cambridge University Press. (306, 333, 334, 484)

MUNK, W. H. and MILLER, R. L. 1950 *Tellus*, **2**, 93–101. (315)

MUNK, W. H. and REVELLE, R. 1952 *MNGS*, **6**, 331–347. (325)

Murnaghan, F. D. (6)

MURRAY, C. A. 1957 *MNRAS*, **117**, 478–482. (324, 326)

MUSEN, P. and FELSENTREGER, T. 1972 *Goddard Space Flight Center*, Greenbelt, Md. (356)

                1973 *Celestial Mechanics*, **7**, 256–279. (334)

MUSKAT, M. 1933 *Physics*, **4**, 14–28. (67)

Nakano, H. (133)

NAKANO, H. 1925 *Jap. J. Astr. Geophys.* **2**, 1–94. (67)

Nansen, F. (452)

NANSEN, F. 1927 *Avh. norske VidenskAkad.* **92**, no. 12. (435)

NASU, N. 1929 *Bull. Earthq. Res. Inst. Tokyo*, **6**, 245–331. (473)

NESS, N. F., HARRISON, J. C. and SLICHTER, L. B. 1961 *JGR*, **66**, 621–629. (309)

Neugebauer, O. (327)

NEUMANN, F. 1933 *TAGU*, **14**, 329–335. (118)

Newcomb, S. (262, 288)

NEWLANDS, M. 1950 *MNGS*, **6**, 109–124. (45)

                1953 *Phil. Trans.* A, **245**, 213–308. (68)

Newton, R. R. (321, 323)

NEWTON, R. R. 1967 *Johns Hopkins Univ. Tech. Mem. TG* 905. (304)

                1968 *GJRAS*, **14**, 505–539. (334, 336)

                1970 *Ancient Astronomical Observations*, Johns Hopkins Press. (335)

                1972 *Mem. RAS*, **76**, 99–128. (337)

Nguyen Hai. (134)

NGUYEN HAI 1961 *Ann. Geophys.* **17**, 60–66. (110, 158, 214)

                1963 *Ann. Geophys.* **19**, 285–346. (158, 214)

Bibliography and Author Index 559

NIAZI, M. 1973 BSSA, 63, 2035-2046. (164)
Nier, A. O. (379)
NIER, A. O. 1939 Phys. Rev. 55, 150-153. (373)
NISHIMURA, E. 1950 TAGU, 31, 357-376. (293)
NISHIMURA, E., HISHIMOTO, T. and KAMITSUKI, A. 1956 Tellus, 8, 329-334. (101)
    1958 Tellus, 10, 137-144. (101)
NISKANEN, E. 1939 Publ. Isost. Inst. Helsinki, 6. (452)
    1942 Publ. Isost. Inst. Helsinki, 12. (265)
NOMITSU, T. and OKAMOTO, M. 1927 Sci. Mem. Kyoto, A, 10, 125-161. (305)
    1932 Sci. Mem. Kyoto, A, 15, 123-129. (305)
NOWROOZI, ALI A. 1965 JGR, 70, 5145-5156. (309, 310)
OBERBECK, V. R. and QUAIDE, W. L. 1967 JGR, 72, 4697-4704. (503)
    1968 JGR, 73, 5247-5270. (503)
OKAZAKI, S. and NASAKA, M. 1971 Ann. Tokyo Astr. Obs., 12, 175-187. (299)
O'Keefe, J. A. (256, 281)
O'KEEFE, J. A. 1959 JGR, 64, 2389-2394. (203, 283)
    1961 JGR, 66, 1992-1993. (283)
O'KEEFE, J. A., SQUIRES, R. K. and ECKELS, A. 1959a Science, 129, 565. (173)
    1959b AJ, 64, 245-253. (173)
Oldham, R. D. (88, 106, 205)
OLDHAM, R. D. 1899 Mem. Geol. Surv. India, 29. (86)
    1900 Phil. Trans. A, 194, 135-174. (86)
    1906 Quart. J. Geol. Soc. 62, 456-475. (86)
    1918 Quart. J. Geol. Soc. 74, 99-104. (474)
    1921 Quart. J. Geol. Soc. 77, 1-3. (474)
    1922 Quart. J. Geol. Soc. 78, lv-lxii. (474)
OLDROYD, J. G. 1946 Phil. Mag. (7), 37, 648-651. (270)
    1947 Proc. Camb. Phil. Soc. 43, 100-105, 383-405, 521-532. (18)
OLIVER, J. and EWING, M. 1957 BSSA, 47, 187-204. (146)
OMORI, F. 1894 J. Coll. Sci. Imp. Univ. Tokyo, 7, 111-200. (473)
ORLIN, H. et al. 1966 Gravity Anomalies, Unsurveyed Areas. Nat. Acad. Sci. Publ. 1357. (190)
Orowan, E. (12)
PAGE, B. M. 1963 BGSA, 74, 655-672. (445)
PANDIT, B. J. and SAVAGE, J. C. 1973 JGR, 78, 6097-6099. (11)
PANETH, F. A. 1953 Observatory, 73, 225-227. (375)
PARSON, A. L. 1945 MNRAS, 105, 244-245. (386)
PATTERSON, C. 1955 Geochim. Cosmochim. Acta, 7, 151-153. (375)
    1956 Geochim. Cosmochim. Acta, 10, 230-237. (375)
Pearson, K. (131, 474, 530)
PEKERIS, C. L. 1935 MNGS, 3, 343-367. (457)
    1937 PRSA, 158, 650-670. (328)
    1955 Proc. Nat. Acad. Sci. 41, 469-480, 629-639. (67)
    1956 Proc. Nat. Acad. Sci. 42, 439-443. (67)
    1965 Proc. Nat. Acad. Sci. 53, 1254-1261. (311)
    1967 PRSA, 297, 449-458. (133)
PEKERIS, C. L. and ACCAD, Y. 1969 Phil. Trans. A, 265, 413-436. (334, 339, 349)
    1973 Phil. Trans. A, 273, 237-260. (304)
PEKERIS, C. L., ALTERMAN, Z. and JAROSCH, H. 1961 Phys. Rev. 122, 1692-1700. (309)

560   *Bibliography and Author Index*

PELLEW, A. and SOUTHWELL, R. V. 1940 *PRSA*, **176**, 312–343. (454)

PENTILLÄ, E. 1972 *Ann. Acad. Sc. Fenn.* A III, *Geol. Geof.* **110**, 5–38. (122)

PETIT, F. 1849 *C.R. Acad. Sci., Paris*, **29**, 729–734. (224)

PETTERSON, H. and ROTSCHI, H. 1952 *Geochim. Cosmochim. Acta*, **2**, 81–90. (503)

Phillips, D. W. (472)

PHILLIPS, D. W. 1931 *Trans. Instn Min. Engrs*, **80**, 212–242. (6)

    1932 *Trans. Instn Min. Engrs*, **82**, 434–450. (6)

    1934 Ph.D. thesis, Cambridge (unpublished). (6)

    1936 *B.A. Rep., Thermal Conductivities of Rocks Comm.*, pp. 258–291. (401)

PHILLIPS, P. 1905 *Phil. Mag.* (6), **9**, 513–531. (12)

PIGOTT, C. S. 1929 *Amer. J. Sci.* (5), **17**, 13–34. (405)

PILANI, W. L. and KNOPOFF, L. 1964 *BSSA*, **54**, 19–40. (116)

PIRSSON, L. V. and SCHUCHERT, C. 1915 *Textbook of Geology*. John Wiley and Sons. (434)

PIZZETTI, P. 1913 *Prin. d. Teoria Mecc. d. Figure d'Equilibrio d. Planetti*. Pisa. (181, 184, 187)

Plaskett, H. H. (83)

POINCARÉ, H. 1910 *Bull. Astron.* **27**, 321–356. (301)

POISSON, S. D. 1829 *Mem. Acad. Sci., Paris*, **8**, 623–627. (28, 86)

    1831 *Mem. Acad. Sci., Paris*, **10**, 578–603. (28)

POLLAK, L. W. 1927 *GBG*, **16**, 108–194. (297)

Poole, H. H. (396)

POPOV, N. 1963*a Nature, Lond.* **196**, 1153. (312)

    1963*b Soviet Astronomy*, **7**, 422. (312)

PRANDTL, L. 1920 *Nachr. Ges. Wiss. Göttingen*, pp. 74–85. (269)

PRATT, J. H. 1855 *Phil. Trans.* **145**, 53–100. (225)

    1859 *Phil. Trans.* **149**, 745–778. (227)

PREY, A. 1922 *Abh. Ges. Wiss. Göttingen*, **11**, no. 1, 3–29. (245, 278)

PROUDMAN, J. 1925 *Observatory*, pp. 386–388. (292)

    1941 *MNGS*, **5**, 23–26. (333, 338)

    1960 *GJRAS*, **3**, 244–249. (527)

QAMAR, A. 1973 *BSSA*, **63**, 1073–1105. (161)

Rabe, E. (173, 303)

RABE, E. 1950 *AJ*, **55**, 112–126. (209, 254)

RABE, E. and FRANCIS, M. P. 1967 *AJ*, **72**, 852–864. (210, 256)

Radau, R. (201, 212)

RADAU, R. 1885 *C.R. Acad. Sci., Paris*, **100**, 972–974. (199)

RAITT, R. W. 1956 *Bull. Geol. Soc. Amer.* **67**, 1623–1639. (166)

    1957 *U.S. Geol. Survey Prof. Paper*, 260–268. (167)

RAMAN, C. V. and VENKATESWARAN, C. S. 1939 *Nature, Lond.* **143**, 798. (10)

Ramsey, W. H. (120, 384, 422, 423)

RAMSEY, W. H. 1948 *MNRAS*, **108**, 406–423. (420)

    1951 *MNRAS*, **111**, 427–447. (424)

RAMSEY, W. H. and LIGHTHILL, M. J. 1950 *MNRAS*, **110**, 325–342. (423)

RANDALL, M. J. 1970 *GJRAS*, **21**, 441–445. (125, 157)

    1971 *GJRAS*, **22**, 229–234. (125, 126)

RAPP, R. H. 1970 *Ohio State Univ., Dept of Geodetic Studies, Rept*, **134**, 4. (190)

Rasch, G. (51)

RAYLEIGH (3RD BARON = STRUTT, J. W.) 1887 *Proc. Lond. Math. Soc.* (1), **17**, 4–11. (41, 86)

1894 (2 vols.) *Theory of Sound.* Macmillan. (38, 270)

1916 *Phil. Mag.* (6), **32**, 529–546. (454)

Rayleigh (R. J. Strutt) (374, 397)

RAYLEIGH (4TH BARON = STRUTT, R. J.) 1906 *PRSA*, **77**, 472–485. (401)

1940 *Nature, Lond.* **145**, 29. (20)

Read, H. H. (103)

REID, H. FIELDING 1910 *Univ. Calif. Publ. Bull. Dep. Geol. Sci.* **6**, 413–433. (27)

Reiner, M. (10)

REINER, M. 1943 *Ten Lectures on Theoretical Rheology.* Jerusalem: Rubin Mass. (18)

REITZEL, J. 1963 *JGR*, **68**, 5191–5196. (416)

REVELLE, R. and MAXWELL, A. E. 1952 *Nature, Lond.* **170**, 199–200. (415)

Reynolds, O. (329)

REZANOV, G. A. 1968 *Soviet Geol.* **3**, 35–48. (English Translation *Int. Geol. Rev.* **10**, 765–776.) (496)

Reznichenko, Y. V. (100)

RICHTER, C. F. 1935 *BSSA*, **25**, 1–32. (60, 136)

Rideal, E. K. (137)

RIECKER, R. E. and SEIFERT, K. E. 1964 *BGSA*, **75**, 571–574. (18)

RINGWOOD, A. E. 1956 *Nature, Lond.* **178**, 1303–1304. (207)

1958*a Geochim. Cosmochim. Acta.* **13**, 303–321. (207)

1958*b Geochim. Cosmochim. Acta.* **15**, 18–29. (207)

1958*c Geochim. Cosmochim. Acta.* **15**, 195–212. (207)

1958*d Bull. Geol. Soc. Amer.* **69**, 129–130. (207)

1959 *Geochim. Cosmochim. Acta.* **16**, 192–193. (207)

1962*a JGR*, **67**, 4005–4010. (207)

1962*b Nature, Lond.* **196**, 883–884. (207)

1966*a Advances in Earth Science*, 287–399. (207)

1966*b Geochemica Acta.* 41–104. (207)

1969 *Earth and Plan. Sci. Letters*, **5**, 401–412. (207)

1970 *Phys. Earth Plan. Int.* **3**, 109–135. (207)

RINGWOOD, A. E. and MAJOR, A. 1966 *Earth and Plan. Sci. Letters*, **1**, 351–357. (208)

1970 *Phys. Earth Plan. Int.* **3**, 89–108. (207)

ROBERSON, R. E. 1957 *J. Frank. Inst.* **264**, 180–201, 269–285. (173)

ROBERTS, P. H. 1965 *Mathematika*, **12**, 128–137. (466)

1968 *Phil. Trans. A*, **263**, 93–117. (457)

Roche, E. (196)

ROCHESTER, M. G. 1968 *J. Geomagnetism and Geoelectricity*, **20**, 387–402. (350, 361)

1970 *Earthquake Displacement Fields and the Rotation of the Earth*, eds. L. Mansinha, D. E. Smylie and A. E. Beck, pp. 3–13, 136–148. Reidel, Dordrecht. (361)

1975 *Nature, Lond.* **255**, 655, erratum and note by B. Jeffreys, 1975, *Nature, Lond.* **257**, 828. (363)

ROCHESTER, M. G. and SMYLIE, D. E. 1965 *GJRAS*, **10**, 289–316. (361)

Rohrbach, W. (151)

ROHRBACH, W. 1932 *Z. Geophys.* **8**, 113–129. (142)

ROSENHEAD, L. 1929*a MNGS*, **2**, 140–170. (301)

1929*b MNGS*, **2**, 171–196. (301, 305, 328, 334)

ROSENHEAD, L. *et al.* 1954 *PRSA*, **226**, 1–69. (3)

Routh, E. J. (222)

RUNCORN, S. K. 1964 *Nature, Lond.* **204**, 823–825. (340)
    1968 *Nature, Lond.* **218**, 459. (340)
    1970 *Palaeogeophysics*, Ch. 3, Academic Press. (340)

RUPRECHTOVA, L. 1959 *Trav. Inst. Geophys. de l'acad. Tcheck des Sci.*, no. 106. (135, 358)

Russell, H. N. (385)

RUSSELL, H. N. 1921 *PRSA*, **99**, 84–86. (378)

RUSSELL, H. N., DUGAN, R. S. and STEWART, J. Q. 1945 *Astronomy*, **1**. (502)

RUSSELL, R. D. and ALLAN, D. W. 1955 *MNGS*, **7**, 80–101. (375, 382)

Rutherford, Lord (379, 396, 397)

RUTHERFORD, LORD and BOLTWOOD, B. B. 1906 *Amer. J. Sci.* (4), **22**, 1–3. (371)

RUTHERFORD, LORD and GEIGER, H. 1910 *Phil. Mag.* (6), **20**, 691–704. (375)

Sadler, D. H. (220)

SADLER, F. McB. and MORRISON, L. V. 1969 *MNRAS*, **144**, 129–141. (220)

Sagan, C. (210)

SANDELL, E. B. and GOLDICH, S. S. 1943 *J. Geol.* **51**, 99–115, 167–189. (379)

SASSA, K., OZAWA, I. and YOSHIKAWA, S. 1951 *Int. Ass. Geodesy, Brussels Assembly.* (293)

SAVAGE, J. C. 1965 *JGR*, **70**, 3935–3949. (361)

SCHATZ, J. F. and SIMMONS, G. 1972 *JGR*, **77**, 6966–6983. (413)

SCHEIDEGGER, A. E. and WILSON, J. TUZO 1950 *Proc. Geol. Ass. Canada*, **3**, 167–190. (463)

SCHMERWITZ, G. 1938 *Z. Geophys.* **14**, 351–390. (89)

SCHOCH, K. 1926 *D. seculäre Accel. d. Mondes u. d. Sonne.* Berlin: Privately printed. (324)

SCHOLTE, J. G. 1947 *MNRAS*, **107**, 237–241. (420)
    1956 *Koninkl. Ned. Meteor. Inst.* **65**, 1–55. (109)

Schuster, A. (474)

SCHUSTER, A. 1907. *Nature, Lond.* **76**, 269. (406)

SCHWEYDAR, W. 1916 *Veröff. preuss. Geodät. Inst.* **66**, 1–51. (300)
    1919 *Preuss. Akad. Wiss. Ber.* **20**, 357–366. (306)

SCHWEYDAR, W. and HECKER, O. 1921 *Veröff. ZentBur. int. Erdmess.*, N.F., **38**. (292)

SCRASE, F. J. 1931 *PRSA*, **132**, 213–235. (111)
    1933 *Phil. Trans.* A, **231**, 207–234. (108, 116)

SCRUTTON, C. T. 1964 *Palaeontology*, **7**, 552–558. (340)

SEKIGUCHI, N. 1957 *Publ. Astr. Soc. Japan*, **19**, 596–605. (362)

SEZAWA, K. 1935 *Bull. Earthq. Res. Inst. Tokyo*, **13**, 245–250. (144)

SHAPLEY, H. 1914 *Astrophys. J.* **40**, 448–465. (368)
    1918 *Astrophys. J.* **48**, 279–294. (368)
    1919 *Astrophys. J.* **49**, 24–41. (368)

SHARPLESS, B. P. 1945 *AJ*, **51**, 185–186. (354)

Shaw, J. J. (82)

SHAW, W. NAPIER 1926–31, 1932, 1936 *Manual of Meteorology.* Cambridge University Press. (307)

SHIDA, TOSHI and MATSUYAMA, M. 1912 *Mem. Coll. Sci. Engng, Tokyo*, **4**, 277–284. (290)

SHIMAZU, Y. 1954 *J. Earth Sci.* **2**, 15–172. (120, 391, 412, 434)

Shimshoni, M. (129, 164)

SHIMSHONI, M. 1967 *GJRAS*, **13**, 471–481. (63)

   1971 *GJRAS*, **23**, 139–160. (162)

SHOR, G. G. 1964*a* *Nature, Lond.* **201**, 1207–1208. (167)

   1964*b* *JGR*, **69**, 1627–1637. (167)

SHOTT, N. M. 1966 *J. Geol. Education*, **14**, 149–166. (387)

SIMON, F. E. 1953 *Nature, Lond.* **172**, 746–748. (412, 421)

SIMPSON, G. G. 1943 *Amer. J. Sci.* **251**, 1–31, 413–429. (486)

Sitter, W. de (173, 185, 202, 325, 326)

SITTER, W. DE 1924*a* *Bull. Astr. Inst. Netherlds*, **2**, 97–108. (184, 201)

   1924*b* *Proc. K. Akad. Wet. Amst.* **27**, 1–15. (201)

   1927*a* *Bull. Astr. Inst. Netherlds*, **4**, 21–38. (324)

   1927*b* *Bull. Astr. Inst. Netherlds*, **4**, 57–61. (201, 219, 324)

SITTER, W. DE and BROUWER, DIRK 1938 *Bull. Astr. Inst. Netherlds*, **8**, 21–23.
  (201)

SLICHTER, L. B. 1961 *Proc. Nat. Acad. Sci.* **47**, 186–190. (312)

SMITH, STUART 1968 *BSSA*, **58**, 1701–1702. (288)

SMITH, W. B., SHAPIRO, I. I. and ASH, M. E. 1966 *AJ*, **71**, 871–872. (210)

SMOLUCHOWSKI, M. V. 1909 *Bull. Int. Acad. Sci. Krakau*, Part 2, 3–20. (442)

Soddy, F. (371)

SOLLINS, A. D. 1947 *Bull. géod.* pp. 279–300. (193)

SOMIGLIANA, C. 1933*a* *Atti Acad. Sci. Torino*, **69**, 345–357. (181)

   1933*b* *Bull. géod.* pp. 178–187. (181)

SOROKTIN, O. G. 1971 *Dokl. Akad. Nauk SSSR*, **198**, no. 6, 1327–1330. (425)

SOUTHWELL, R. V. 1936 *Theory of Elasticity*, Ch. 13. Oxford University Press.
  (441)

SPARKS, NEIL R. 1936 *BSSA*, **26**, 13–27. (99)

SPENCER, L. J. 1933 *Geogr. J.* **81**, 227–248. (500)

   1937 *Nature, Lond.* **139**, 655–657. (500)

SPITALER, R. 1897 *Denkschr. Akad. Wiss. Wien*, **64**, 633–642. (305)

SPITZER, L. 1939 *Astrophys. J.* **90**, 675–688. (385)

SPURR, J. E. 1945, 1948 *Lunar Catastrophic History*, 3 vols. Lancaster, Pa; Science
  Press. Concord, N.H.; Rumford Press. (504)

STACEY, F. D. 1969 *Physics of the Earth*. Wiley, New York. (497)

STAMOU, P. and BÅTH, M. 1974 *Phys. Earth Plan. Int.* **8**, 317–331. (162)

STAUDER, W. and BOLLINGER, G. 1963 *BSSA*, **53**, 661–680. (101)

Steavenson, W. H. (353)

STECHSCHULTE, V. C. 1932 *BSSA*, **22**, 81–137. (108)

STEPHENSON, F. R. 1972 Ph.D. thesis, Newcastle upon Tyne. (335)

STERNECK, R. 1928*a* *Ann. Hydrogr.* **56**, 221–225. (292)

   1928*b* *Verh. Magn. Met. Obs. Irkutsk*, **2–3**, 147–152. (292)

STEWART, R. M. 1973 *JGR*, **78**, 2588–2597. (210)

STEWARTSON, K. and ROBERTS, P. H. 1963 *J. Fluid Mech.* **17**, 1–20. (362)

   1965 *Proc. Camb. Phil. Soc.* **61**, 279–288.
   (362)

STOBIE, R. S. 1969 *MNRAS*, **144**, 461–535. (368)

Stokes, G. G. (43)

STOKES, G. G. 1849 *Trans. Camb. Phil. Soc.* **8**, 672–695; *Coll. Papers*, **2**, 131–171.
  (177, 190)

Stoneley, R. (83, 103, 143, 151)

STONELEY, R. 1924*a* MNGS, **1**, 149–155. (439)
    1924*b* PRSA, **106**, 416–428. (43)
    1926 MNGS, **1**, 356–359. (300, 308)
    1928 MNGS, **1**, 527–532. (150)
    1931*a* MNGS, **2**, 349–362. (99)
    1931*b* GBG, **29**, 417–435. (111)
    1935*a* MNGS, **3**, 262–271. (142)
    1935*b* Proc. Camb. Phil. Soc. **31**, 360–367. (142)
    1937*a* MNGS, **4**, 43–50. (142)
    1937*b* B. A. Rep. Seism. Comm. (116)
    1938 BSSA, **29**, 191–195. (99)
    1939*a* MNGS, **4**, 461–469. (116)
    1939*b* MNGS, **4**, 562–569. (138)
    1948 BSSA, **38**, 263–274. (150)
    1951 Int. Séism. Assn Comptes rendus, **10**, 1–12. (85)
    1955 MNGS, **7**, 71–75. (151)
STONELEY, R. and HOCHSTRASSER, U. 1956 MNGS, **7**, 279–288. (143)
    1961 GJRAS, **4**, 197–201. (143)
STONELEY, R. and TILLOTSON, E. 1928 MNGS, **1**, 521–527. (150)
STOYKO, N. 1937 C.R. Acad. Sci., Paris, **205**, 79–81. (315)
STRATTON, F. J. M. 1906 MNRAS, **66**, 374–402. (354)
SUBIZA, G. P. and BÅTH, M. 1964 GJRAS, **8**, 496–513. (158)
Suess, E. (23, 87, 481)
TAKEUCHI, H. 1950 TAGU, **31**, 651–689. (302, 307, 308)
TAKEUCHI, H. and DORMAN, J. 1964 JGR, **89**, 3429–3441. (309)
TAKEUCHI, H., SAITO, M. and KOBAYASHI, N. 1962 JGR, **67**, 2831–3839. (78)
Tatel, H. E. (147, 149)
TATEL, H. E., ADAMS, L. H. and TUVE, M. A. 1953 Proc. Amer. Phil. Soc. **97**, 658–669. (124)
Taylor, F. B. (481)
TAYLOR, G. I. 1919 Phil. Trans. A, **220**, 1–33. (330)
    1922 Phil. Trans. A, **223**, 289–343. (385)
    1934 PRSA, **145**, 1–18. (16)
TAYLOR, G. I. and QUINNEY, H. 1931 Phil. Trans. A, **230**, 323–362. (16)
TAYLOR, J. B. 1963 PRSA, **274**, 274–283. (362)
THOMAS, D. V. 1964 Nature, Lond. **201**, 487. (312)
THORKELSSON, T. 1928*a* Phil. Mag. (7), **5**, 441–444. (505)
    1928*b* On thermal activity at Rejkjanes, Iceland, Rit. Visindafélag Íslendinga, (Societas Scientiarum Islandica), **3**, 1–42. (505)
    1940 On thermal activity in Iceland and geyser action. Visindafélag Íslendinga, (Societas Scientiarum Islandica), **25**, 1–139. (505)
THULIN, A. 1960 Ann. Géophys. **16**, 105–127. (255)
Tilley, C. E. (103, 504)
TILLEY, C. E. 1926 Miner. Mag. **21**, 34–50. (425)
TILLOTSON, E. 1938 GBG, **52**, 377–407. (116)
    1939 MNGS, **3**, 537–547. (117)
TISSERAND, F. 1889–1896 Mécanique Celeste, 4 vols. Paris: Gauthier-Villars. (199, 479)
TODHUNTER, I. 1873 Mathematical Theories of Attraction. Macmillan. (226)
Tomaschek, R. (307)
Treskov, A. A. (147, 149)

TROUTON, F. T. and ANDREWS, E. S. 1904 *Phil. Mag.* (6), **7**, 347–355. (470)

TRUMAN, O. H. 1939 *Astrophys. J.* **89**, 445–562. (307)

TSUBOI, C. 1937 *Bull. Earthq. Res. Inst. Tokyo*, **15**, 650–653. (234)

TSUBOI, C. and FUCHIDA, T. 1937 *Bull. Earthq. Res. Inst. Tokyo*, **15**, 636–649. (234)

Turner, H. H. (83, 86, 113, 474)

TURNER, H. H. 1915*a* *MNRAS*, **75**, 530–541. (72)

            1915*b* *B. A. Rep. Seism. Comm.* (107)

            1922 *MNGS*, **1**, 1–13. (111)

            1926 *MNGS*, **1**, 425–446. (113)

Tuve, M. A. (147, 149)

Tvaltvadze, G. K. (99)

Tyrrell, G. W. (21, 22)

Urey, H. C. (419)

UREY, H. C. 1951 *Geochim. Cosmochim. Acta.* **1**, 209–277. (340, 384)

            1952 *The Planets*. Yale University Press. (355, 500, 503)

            1956 *Observatory*, **76**, 232. (504)

VAN ANDEL, T. H. 1968 *J. Marine Res.* **26**, 144–161. (494)

VAN FLANDERN, T. C. 1970 *Astron. J.* **75**, 657. (337)

VANEK, J. and STELZNER, J. 1959 *GBG*, **68**, 75–89. (135)

            1960*a* *Ann. di. Geof.* **13**, 393–407. (135)

            1960*b* *Nature, Lond.* **187**, 491–492. (135)

VERHOOGEN, J. 1953 *J. Geophys. Res.* **58**, 337–346. (434)

            1974 *Nature, Lond.* **249**, 334–335. (349)

            1975 *Nature, Lond.* **255**, 655–656. (363)

Veystsman, P. S. (100)

VICENTE, R. O. 1961 *Physics and Chemistry of the Earth*, **4**, 251–280. (312)

VINE, F. J. 1966 *Science*, **154**, 1405–1415. (493)

VINE, F. J. and MATTHEWS, D. H. 1963 *Nature, Lond.* **199**, 947–949. (493)

VOGT, J. H. L. 1926 *Econ. Geol.* **21**, 207–233. (394)

VOISEY, A. H. 1958 *Continental Drift, a Symposium*, University of Tasmania, pp. 162–171. (495)

Volet, C. (255)

WADATI, K. 1928 *Geophys. Mag., Tokyo*, **1**, 162–202. (111)

WADATI, K. and MASUDA, K. 1933 *Geophys. Mag., Tokyo*, **7**, 270–305. (61)

            1934 *Geophys. Mag., Tokyo*, **8**, 187–194. (61)

WAERDEN, B. L. VAN DER 1961 *AJ*, **66**, 138–147. (324, 326, 327, 334, 335)

WAGER, L. R. 1937 *Geogr. J.* **89**, 239–250. (435)

WAGER, L. R. and DEER, W. A. 1939 *Medd. om Grønland*, **105**, no. 4. (394, 421)

WAKO, Y. 1970 *Publ. Astr. Soc. Japan*, **22**, 525–544. (314)

            1972 *IAU Symposium*, **48**, 189–191. (314)

WALKER, A. M. and YOUNG, A. 1955 *MNRAS*, **115**, 443–459. (297)

            1957 *MNRAS*, **117**, 119–141. (297)

Walker, G. W. (142)

WALKER, G. W. 1921 *Phil. Trans.* A, **222**, 45–56, (112)

WALL, R. E., TALWANI, M. and WORZEL, J. 1966 *JGR*, **71**, 487–493. (190)

WANACH, B. 1927 *Z. Geophys.* **3**, 102–105. (483)

Wanner, E. (88)

WANNER, E. 1937 *GBG*, **50**, 85–99, 223–228. (473)

            1941 *Verh. Schweiz. Naturf. Ges.* pp. 102–103. (473)

WASHINGTON, H. S. 1923 *J. Wash. Acad. Sci.* **13**, 339–347. (489)

WATSON, F. G. 1945 *Between the Planets*. Harvard books on Astronomy. (503)

Wegener, A. (23, 481, 482, 485, 491)
WEGENER, A. 1924 *The Origin of Continents and Oceans*, trans. J. G. A. Skerl. Methuen. (481)
WELLS, J. W. 1963 *Nature, Lond.* **197**, 948. (340)
WESSON, P. S. 1970 *QJRAS*, **11**, 312–340. (495)
    1972 *J. Geol.* **80**, 185–197. (495)
    1974 *Amer. Assn Petroleum Geol.* Mem. 23, ed. C. F. Kahle, 146–154, 448–462. (496)
WHEELER, N. E. 1910 *Trans. Roy. Soc. Canada*, **4**, 19–44. (431)
Whipple, F. J. W. (28, 61, 108)
WHIPPLE, F. J. W. and LEE, A. W. 1935 *MNGS*, **3**, 287–297. (168)
WHITE, R. E. 1971 *GJRAS*, **24**, 109–118. (101)
Wiechert, E. (28, 87, 196, 204, 212, 300, 308)
WIECHERT, E. 1897 *Nachr. Ges. Wiss. Göttingen*, pp. 221–243. (204)
    1907 *Nachr. Ges. Wiss. Göttingen*, pp. 415–529. (36, 64)
WIECHERT, E. and GEIGER, L. 1910 *Phys. Z.* **11**, 294–312. (51)
WIGGINS, R. A. 1968 *Phys. Earth Plan. Int.* **1**, 201–266. (523)
Wildt, R. (395)
WILDT, R. 1947 *MNRAS*, **107**, 84–102. (418)
WILKINS, G. A. 1965 *QJRAS*, **6**, 70–73. (259)
WILLIAMS, I. P. and CRENIN, A. W. 1968 *QJRAS*, **9**, 40–62. (387)
Willmore, P. L. (84, 96, 99, 123, 148, 149)
WILLMORE, P. L. 1949 *Phil. Trans.* A, **242**, 123–151. (94, 137)
WILSON, JAMES T. and BAYKAL, O. 1948 *BSSA*, **38**, 41–53. (143, 144)
WILSON, J. TUZO 1950a *Trans. Amer. Geophys. Un.* **31**, 101–114. (463)
    1950b *Proc. Geol. Ass. Canada*, **3**, 141–166. (463)
    1951 *Proc. Roy. Soc. Tasmania*, **1950**, 85–111. (463)
    1954 See also Kuiper, Section 4. (463)
    1957 *Nature, Lond.* **179**, 228–230. (463)
    See also McKenzie and Weiss (1975), bibliography. (492)
WOERKOM, A. J. J. VAN. 1950 *Astron. Papers of American Ephemeris*, **13**, Part 1. (354)
Woollard, G. J. (189)
WORZEL, J. L. 1965 *Pendulum Gravity Measurements at Sea*, 1936–59. John Wiley and Sons. (190)
WRIGHT, J. U., CARPENTER, E. W. and SAVILL, R. A. 1962 *JGR*, **67**, 1157–1160. (60)
WRINCH, D. and JEFFREYS, H. 1923 *MNGS*, **1**, 15–22. *CP*, **1**, 252. (87)
WUNSCH, C. 1974 *GJRAS*, **39**, 539–550. Errata: 1975 *GJRAS*, **40**, 311. (350)
WYLLIE, P. J. 1971 *The Dynamic Earth*, p. 363. Wiley, New York. (497)
YAKOVKIN, A. A. 1952 *Trans. I.A.U.* **8**, 231. (219)
YATSKIV, YA. S. and SASAO, T. 1975 *Nature, Lond.* **255**, 655. (363)
YOUNG, A. 1951 *Bull. Acad. Roy. Belg. Cl. Sci.* 37, 728–751. (307, 315)
    1952 *Bull. Acad. Roy. Belg. Cl. Sci.* (5), **38**, 824–837. (307)
YULE, G. UDNY 1927 *Phil. Trans.* A, **226**, 267–298. (297)
ZHONGOLOVITCH, I. D. 1952 *Publ. Res. Inst. Theor. Astron.* (*Acad. Sci. USSR*), **3**, 1–126. (245)
    1957 *Bull. Inst. Theor. Astron. Moscow*, **6**, 505–523. (245)
Zöppritz, K. (87)
ZÖPPRITZ, K. 1907 *Nachr. Ges. Wiss. Göttingen*, pp. 529–549. (41, 86)

SUBJECT INDEX

Accretion, 355, 384, 386, 502
Adiabatic variation of temperature, 390
Adiabatic wave-velocities, 105
Adopted values of constants, 259
Aftershocks, 285, 472, 474, 475, 477
Age of crust, 380
Amplitudes, see Earthquakes
Anaseism, 28
Anisotropy, 131
Anticentre, 62
Approximations, symbols for, x
Asthenosphere, 232
Astronomical constants
   1948 adjustment, 252
   1965 revision, 259
Atomic structure, 368
Azimuth, 73
   variation of travel times with, 122, 129

Basalt, 22, 398, 413, 431
Bikini explosion, 125, 137, 167
Blanketing
   by sediments, 425
   by water, 388
Bodily tide, 289, 299, 346, 510
   heating by, 434.
   Longman's anomaly, 304, 348
Bodily waves in a sphere, 49
Bouguer anomaly, 229, 231–233
Bouguer gravity formula, 225, 251
Bromwich integral, 133
Bulk modulus, 4, 101, 105, 420
Burton explosion, 92, 96, 137, 165

Carbon, 384
Castigliano's theorem, 269, 524
Caustics, 54, 106, 108, 110
Cepheid variables, 368, 382
Channel waves, 105
Clairaut's equation, 196
Climate, 487
   variation of, 448, 487, 489, 494
Coda, 132, 136, 140
Co-geoid, 175, 190, 193, 248
Compensation (isostatic), 227
   distribution, 238
   of oceans, 235
   see also Isostasy; Surface loading
Compound phases, 60, 87, 120, 133
   see also Notation in seismology

Compressibility, see Bulk modulus
Compression
   available, 430
   needed, 434
Continental drift, 481–492
Continents
   forces on floating, 468
   growth of, 463
   origin of, 465
   permanence of, 478
Contraction
   thermal, 429
   through extrusion of water, 434
   unsymmetrical, 462
Convection, 454, 465, 491, 493, 496
Cooling of sphere (m term), 411, 431,
   432, 526
Core, 54, 86, 87, 106, 117, 120, 158, 205,
   418
   Bullen's theory, 420–424
   central particle model, 302, 313
   diffraction by, 108, 109
   discontinuities in, 160
   fluidity, 106, 300; allowance for in
      theory of nutations, Chapter VII
   inner, 110, 158, 212; solidity, 421
   Kuhn–Rittmann theory, 418
   radius, 159, 311
   Ramsey's theory, 120, 384, 420–424
   Roche model, 302, 313
   rotation, 361
   viscosity, 347–349
Crack, 16
   extension of, 284
Creep, 7
   Lomnitz's law, 11, 12, 148, 356
   Modified Lomnitz law, 12, 48, 148, 356
   see also Elastic afterworking
Crust, 270
Crystal structure, 24, 405
Curtsey, 147
Cusp, 60, 95, 124

Damping, 47, 147, 346, 347, 358
   of variation of latitude, 348–352, 482
   see also Earthquakes, damping of
      waves
Deep foci, 71, 111, 137, 433
Deformation by bodily force, 515
Delay-depth coefficient, 90, 165